Applied Statistics II

Third Edition

To my students: Past, present, and future.

Sara Miller McCune founded SAGE Publishing in 1965 to support the dissemination of usable knowledge and educate a global community. SAGE publishes more than 1000 journals and over 800 new books each year, spanning a wide range of subject areas. Our growing selection of library products includes archives, data, case studies and video. SAGE remains majority owned by our founder and after her lifetime will become owned by a charitable trust that secures the company's continued independence.

Los Angeles | London | New Delhi | Singapore | Washington DC | Melbourne

Applied Statistics II

Multivariable and Multivariate Techniques

Third Edition

Rebecca M. Warner

Professor Emerita, University of New Hampshire

Los Angeles | London | New Delhi
Singapore | Washington DC | Melbourne

FOR INFORMATION:

SAGE Publications, Inc.
2455 Teller Road
Thousand Oaks, California 91320
E-mail: order@sagepub.com

SAGE Publications Ltd.
1 Oliver's Yard
55 City Road
London, EC1Y 1SP
United Kingdom

SAGE Publications India Pvt. Ltd.
B 1/I 1 Mohan Cooperative Industrial Area
Mathura Road, New Delhi 110 044
India

SAGE Publications Asia-Pacific Pte. Ltd.
18 Cross Street #10-10/11/12
China Square Central
Singapore 048423

Acquisitions Editor: Helen Salmon
Editorial Assistant: Megan O'Heffernan
Content Development Editor: Chelsea Neve
Production Editor: Laureen Gleason
Copy Editor: Jim Kelly
Typesetter: Hurix Digital
Proofreader: Jennifer Grubba
Indexer: Michael Ferreira
Cover Designer: Gail Buschman
Marketing Manager: Shari Countryman

Printed in the United States of America

ISBN 978-1-5443-9872-3

This book is printed on acid-free paper.

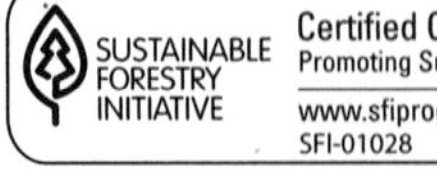

20 21 22 23 24 10 9 8 7 6 5 4 3 2 1

BRIEF CONTENTS

Preface xix

Acknowledgments xxi

About the Author xxiii

CHAPTER 1 • The New Statistics 1

CHAPTER 2 • Advanced Data Screening: Outliers and Missing Values 29

CHAPTER 3 • Statistical Control: What Can Happen When You Add a Third Variable? 64

CHAPTER 4 • Regression Analysis and Statistical Control 99

CHAPTER 5 • Multiple Regression With Multiple Predictors 133

CHAPTER 6 • Dummy Predictor Variables in Multiple Regression 187

CHAPTER 7 • Moderation: Interaction in Multiple Regression 215

CHAPTER 8 • Analysis of Covariance 254

CHAPTER 9 • Mediation 289

CHAPTER 10 • Discriminant Analysis 309

CHAPTER 11 • Multivariate Analysis of Variance 353

CHAPTER 12 • Exploratory Factor Analysis 398

CHAPTER 13 • Reliability, Validity, and Multiple-Item Scales 464

CHAPTER 14 • More About Repeated Measures 509

CHAPTER 15 • Structural Equation Modeling With Amos: A Brief Introduction 539

CHAPTER 16 • Binary Logistic Regression 583

CHAPTER 17 • Additional Statistical Techniques 625

Glossary 644

References 657

Index 666

DETAILED CONTENTS

Preface xix

Acknowledgments xxi

About the Author xxiii

CHAPTER 1 • The New Statistics 1

1.1 Required Background 1
1.2 What Is the "New Statistics"? 2
1.3 Common Misinterpretations of *p* Values 2
1.4 Problems With NHST Logic 4
1.5 Common Misuses of NHST 5
1.5.1 Violations of Assumptions 6
1.5.2 Violations of Rules for Use of NHST 6
1.5.3 Ignoring Artifacts and Data Problems That Bias *p* Values 7
1.5.4 Summary 7
1.6 The Replication Crisis 8
1.7 Some Proposed Remedies for Problems With NHST 9
1.7.1 Bayesian Statistics 9
1.7.2 Replace $\alpha = .05$ with $\alpha = .005$ 10
1.7.3 Less Emphasis on NHST 10
1.8 Review of Confidence Intervals 10
1.8.1 Review: Setting Up CIs 10
1.8.2 Interpretation of CIs 11
1.8.3 Graphing CIs 12
1.8.4 Understanding Error Bar Graphs 13
1.8.5 Why Report CIs Instead of, or in Addition to, Significance Tests? 13
1.9 Effect Size 14
1.9.1 Generalizations About Effect Sizes 14
1.9.2 Test Statistics Depend on Effect Size Combined With Sample Size 16
1.9.3 Using Effect Size to Evaluate Theoretical Significance 17
1.9.4 Use of Effect Size to Evaluate Practical or Clinical Importance (or Significance) 18
1.9.5 Uses for Effect Sizes 19
1.10 Brief Introduction to Meta-Analysis 20
1.10.1 Information Needed for Meta-Analysis 20
1.10.2 Goals of Meta-Analysis 20
1.10.3 Graphic Summaries of Meta-Analysis 21
1.11 Recommendations for Better Research and Analysis 22
1.11.1 Recommendations for Research Design and Data Analysis 23
1.11.2 Recommendations for Authors 23
1.11.3 Recommendations for Journal Editors and Reviewers 24

1.11.4 Recommendations for Teachers of Research Methods and Statistics 24
1.11.5 Recommendations About Institutional Incentives and Norms 25
1.12 Summary **25**

CHAPTER 2 • Advanced Data Screening: Outliers and Missing Values **29**
2.1 Introduction **29**
2.2 Variable Names and File Management **29**
2.2.1 Case Identification Numbers 29
2.2.2 Codes for Missing Values 29
2.2.3 Keeping Track of Files 30
2.2.4 Use Different Variable Names to Keep Track of Modifications 31
2.2.5 Save SPSS Syntax 31
2.3 Sources of Bias **31**
2.4 Screening Sample Data **34**
2.4.1 Data Screening Need in All Situations 34
2.4.2 Data Screening for Comparison of Group Means 34
2.4.3 Data Screening for Correlation and Regression 35
2.5 Possible Remedy for Skewness: Nonlinear Data Transformations **36**
2.6 Identification of Outliers **37**
2.6.1 Univariate Outliers 37
2.6.2 Bivariate and Multivariate Outliers 38
2.7 Handling Outliers **40**
2.7.1 Use Different Analyses: Nonparametric or Robust Methods 40
2.7.2 Handling Univariate Outliers 41
2.7.3 Handling Bivariate and Multivariate Outliers 41
2.8 Testing Linearity Assumption **41**
2.9 Evaluation of Other Assumptions Specific to Analyses **44**
2.10 Describing Amount of Missing Data **44**
2.10.1 Why Missing Values Create Problems 44
2.10.2 Assessing Amount of Missingness Using SPSS Base 45
2.10.3 Decisions Based on Amount of Missing Data 46
2.10.4 Assessment of Amount of Missingness Using SPSS Missing Values Add-On 48
2.11 How Missing Data Arise **50**
2.12 Patterns in Missing Data **51**
2.12.1 Type A and Type B Missingness 51
2.12.2 MCAR, MAR, and MNAR Missingness 51
2.12.3 Detection of Type A Missingness 52
2.12.4 Detection of Type B Missingness 52
2.13 Empirical Example: Detecting Type A Missingness **53**
2.14 Possible Remedies for Missing Data **56**
2.15 Empirical Example: Multiple Imputation to Replace Missing Values **57**
2.16 Data Screening Checklist **59**
2.17 Reporting Guidelines **60**
2.18 Summary **61**
Appendix 2A: Brief Note About Zero-Inflated Binomial or Poisson Regression **61**

CHAPTER 3 • Statistical Control: What Can Happen When You Add a Third Variable? 64

3.1 What Is Statistical Control? 64

3.2 First Research Example: Controlling for a Categorical X_2 Variable 66

3.3 Assumptions for Partial Correlation Between X_1 and Y, Controlling for X_2 71

3.4 Notation for Partial Correlation 72

3.5 Understanding Partial Correlation: Use of Bivariate Regressions to Remove Variance Predictable by X_2 From Both X_1 and Y 74

3.6 Partial Correlation Makes No Sense if There Is an $X_1 \times X_2$ Interaction 77

3.7 Computation of Partial *r* From Bivariate Pearson Correlations 79

3.8 Significance Tests, Confidence Intervals, and Statistical Power for Partial Correlations 81

- 3.8.1 Statistical Significance of Partial *r* 81
- 3.8.2 Confidence Intervals for Partial *r* 81
- 3.8.3 Effect Size, Statistical Power, and Sample Size Guidelines for Partial *r* 81

3.9 Comparing Outcomes for $r_{Y1.2}$ and r_{Y1} 82

3.10 Introduction to Path Models 82

3.11 Possible Paths Among X_1, Y, and X_2 83

3.12 One Possible Model: X_1 and Y Are Not Related Whether You Control for X_2 or Not 86

3.13 Possible Model: Correlation Between X_1 and Y Is the Same Whether X_2 Is Statistically Controlled or Not (X_2 Is Irrelevant to the X_1, Y Relationship) 87

3.14 When You Control for X_2, Correlation Between X_1 and Y Drops to Zero 87

- 3.14.1 X_1 and X_2 Are Completely Redundant Predictors of Y 87
- 3.14.2 X_1, Y Correlation Is Spurious 88
- 3.14.3 X_1, Y Association Is Completely Mediated by X_2 89
- 3.14.4 True Nature of the X_1, Y Association (Their Lack of Association) Is "Suppressed" by X_2 90
- 3.14.5 Empirical Results Cannot Determine Choice Among These Explanations 90

3.15 When You Control for X_2, the Correlation Between X_1 and Y Becomes Smaller (But Does Not Drop to Zero or Change Sign) 90

3.16 Some Forms of Suppression: When You Control for X_2, $r_{1Y.2}$ Becomes Larger Than r_{1Y} or Opposite in Sign to r_{1Y} 91

- 3.16.1 Classical Suppression: Error Variance in Predictor Variable X_1 Is "Removed" by Control Variable X_2 91
- 3.16.2 X_1 and X_2 Both Become Stronger Predictors of Y When Both Are Included in Analysis 94
- 3.16.3 Sign of X_1 as a Predictor of Y Reverses When Controlling for X_2 94

3.17 "None of the Above" 95

3.18 Results Section 95

3.19 Summary 96

CHAPTER 4 • Regression Analysis and Statistical Control 99

4.1 Introduction 99

4.2 Hypothetical Research Example 101

4.3 Graphic Representation of Regression Plane 102

4.4 Semipartial (or "Part") Correlation 103
4.5 Partition of Variance in Y in Regression With Two Predictors 104
4.6 Assumptions for Regression With Two Predictors 107
4.7 Formulas for Regression With Two Predictors 111
4.7.1 Computation of Standard-Score Beta Coefficients 111
4.7.2 Formulas for Raw-Score (*b*) Coefficients 112
4.7.3 Formulas for Multiple R and Multiple R^2 112
4.7.4 Test of Significance for Overall Regression: *F* Test for H_0: R = 0 113
4.7.5 Test of Significance for Each Individual Predictor: *t* Test for H_0: $b_i = 0$ 113
4.7.6 Confidence Interval for Each *b* Slope Coefficient 114
4.8 SPSS Regression 114
4.9 Conceptual Basis: Factors That Affect the Magnitude and Sign of β and *b* Coefficients in Multiple Regression With Two Predictors 117
4.10 Tracing Rules for Path Models 118
4.11 Comparison of Equations for β, *b*, *pr*, and *sr* 120
4.12 Nature of Predictive Relationships 121
4.13 Effect Size Information in Regression With Two Predictors 122
4.3.1 Effect Size for Overall Model 122
4.3.2 Effect Size for Individual Predictor Variables 122
4.14 Statistical Power 123
4.15 Issues in Planning a Study 124
4.15.1 Sample Size 124
4.15.2 Selection of Predictor and/or Control Variables 124
4.15.3 Collinearity (Correlation) Between Predictors 124
4.15.4 Ranges of Scores 125
4.16 Results 126
4.17 Summary 129

CHAPTER 5 • Multiple Regression With Multiple Predictors 133
5.1 Research Questions 133
5.2 Empirical Example 136
5.3 Screening for Violations of Assumptions 136
5.4 Issues in Planning a Study 136
5.5 Computation of Regression Coefficients With *k* Predictor Variables 138
5.6 Methods of Entry for Predictor Variables 140
5.6.1 Standard or Simultaneous Method of Entry 141
5.6.2 Sequential or Hierarchical (User-Determined) Method of Entry 141
5.6.3 Statistical (Data-Driven) Order of Entry 141
5.7 Variance Partitioning in Standard Regression Versus Hierarchical and Statistical Regression 142
5.8 Significance Test for an Overall Regression Model 144
5.9 Significance Tests for Individual Predictors in Multiple Regression 145
5.10 Effect Size 148
5.10.1 Effect Size for Overall Regression (Multiple R) 148
5.10.2 Effect Size for Individual Predictor Variables (sr^2) 148
5.11 Changes in F and R as Additional Predictors Are Added to a Model in Sequential or Statistical Regression 149

5.12 Statistical Power 150
5.13 Nature of the Relationship Between Each X Predictor and Y (Controlling for Other Predictors) 150
5.14 Assessment of Multivariate Outliers in Regression 151
5.15 SPSS Examples 152
5.15.1 SPSS Menu Selections, Output, and Results for Standard Regression 152
5.15.2 SPSS Menu Selections, Output, and Results for Sequential Regression 156
5.15.3 SPSS Menu Selections, Output, and Results for Statistical Regression 163
5.16 Summary 167
Appendix 5A: Use of Matrix Algebra to Estimate Regression Coefficients for Multiple Predictors 168
5.A.1 Matrix Addition and Subtraction 170
5.A.2 Matrix Multiplication 170
5.A.3 Matrix Inverse 172
5.A.4 Matrix Transpose 173
5.A.5 Determinant 174
5.A.6 Using the Raw-Score Data Matrices for X and Y to Calculate *b* Coefficients 177
Appendix 5B: Tables for Wilkinson and Dallal (1981) Test of Significance of Multiple R^2 in Forward Statistical Regression 180
Appendix 5C: Confidence Interval for R^2 183

CHAPTER 6 • Dummy Predictor Variables in Multiple Regression 187
6.1 What Dummy Variables Are and When They Are Used 187
6.2 Empirical Example 189
6.3 Screening for Violations of Assumptions 190
6.4 Issues in Planning a Study 193
6.5 Parameter Estimates and Significance Tests for Regressions With Dummy Predictor Variables 194
6.6 Group Mean Comparisons Using One-Way Between-S ANOVA 194
6.6.1 Sex Differences in Mean Salary 194
6.6.2 College Differences in Mean Salary 195
6.7 Three Methods of Coding for Dummy Variables 197
6.7.1 Regression With Dummy-Coded Dummy Predictor Variables 197
6.7.2 Regression With Effect-Coded Dummy Predictor Variables 200
6.7.3 Orthogonal Coding of Dummy Predictor Variables 205
6.8 Regression Models That Include Both Dummy and Quantitative Predictor Variables 208
6.9 Effect Size and Statistical Power 210
6.10 Nature of the Relationship and/or Follow-Up Tests 210
6.11 Results 210
6.12 Summary 211

CHAPTER 7 • Moderation: Interaction in Multiple Regression **215**
7.1 Terminology 215
7.2 Interaction Between Two Categorical Predictors: Factorial ANOVA 217
7.3 Interaction Between One Categorical and One Quantitative Predictor 219
7.4 Preliminary Data Screening: One Categorical and One Quantitative Predictor 220
7.5 Scatterplot for Preliminary Assessment of Possible Interaction Between Categorical and Quantitative Predictor 221
7.6 Regression to Assess Statistical Significance of Interaction Between One Categorical and One Quantitative Predictor 224
7.7 Interaction Analysis With More Than Three Categories 226
7.8 Example With Different Data: Significant Sex-by-Years Interaction 228
7.9 Follow-Up: Analysis of Simple Main Effects 230
7.10 Interaction Between Two Quantitative Predictors 232
7.11 SPSS Example of Interaction Between Two Quantitative Predictors 234
7.12 Results for Interaction of Age and Habits as Predictors of Symptoms 234
7.13 Graphing Interaction for Two Quantitative Predictors 236
7.14 Results Section for Interaction of Two Quantitative Predictors 242
7.15 Additional Issues and Summary 243
Appendix 7A: Graphing Interactions Between Quantitative Variables "by Hand" 244

CHAPTER 8 • Analysis of Covariance **254**
8.1 Research Situations for Analysis of Covariance 254
8.2 Empirical Example 257
8.3 Screening for Violations of Assumptions 259
8.4 Variance Partitioning in ANCOVA 263
8.5 Issues in Planning a Study 264
8.6 Formulas for ANCOVA 265
8.7 Computation of Adjusted Effects and Adjusted Y* Means 266
8.8 Conceptual Basis: Factors That Affect the Magnitude of SS_{Aadj} and $SS_{residual}$ and the Pattern of Adjusted Group Means 266
8.9 Effect Size 267
8.10 Statistical Power 268
8.11 Nature of the Relationship and Follow-Up Tests: Information to Include in the "Results" Section 268
8.12 SPSS Analysis and Model Results 269
    8.12.1 Preliminary Data Screening 269
    8.12.2 Assessment of Assumption of No Treatment-by-Covariate Interaction 270
    8.12.3 Conduct Final ANCOVA Without Interaction Term Between Treatment and Covariate 273
8.13 Additional Discussion of ANCOVA Results 278
8.14 Summary 279
Appendix 8A: Alternative Methods for the Analysis of Pretest–Posttest Data 281
    8.A.1 Potential Problems With Gain or Change Scores 282

CHAPTER 9 • Mediation 289

9.1 Definition of Mediation 289

9.1.1 Path Model Notation 289

9.1.2 Circumstances in Which Mediation May Be a Reasonable Hypothesis 290

9.2 Hypothetical Research Example 290

9.3 Limitations of "Causal" Models 290

9.3.1 Reasons Why Some Path Coefficients May Be Not Statistically Significant 291

9.3.2 Possible Interpretations for Statistically Significant Paths 291

9.4 Questions in a Mediation Analysis 292

9.5 Issues in Designing a Mediation Analysis Study 292

9.5.1 Types of Variables in Mediation Analysis 292

9.5.2 Temporal Precedence or Sequence of Variables in Mediation Studies 293

9.5.3 Time Lags Between Variables 293

9.6 Assumptions in Mediation Analysis and Preliminary Data Screening 293

9.7 Path Coefficient Estimation 294

9.8 Conceptual Issues: Assessment of Direct Versus Indirect Paths 295

9.8.1 The Mediated or Indirect Path: *ab* 296

9.8.2 Mediated and Direct Path as Partition of Total Effect 297

9.8.3 Magnitude of Mediated Effect 298

9.9 Evaluating Statistical Significance 298

9.9.1 Causal-Steps Approach 298

9.9.2 Joint Significance Test 299

9.9.3 Sobel Test of H_0: $ab = 0$ 299

9.9.4 Bootstrapped Confidence Interval for *ab* 300

9.10 Effect Size Information 301

9.11 Sample Size and Statistical Power 302

9.12 Additional Examples of Mediation Models 302

9.12.1 Multiple Mediating Variables 302

9.12.2 Multiple-Step Mediated Paths 303

9.12.3 Mediated Moderation and Moderated Mediation 303

9.13 Note About Use of Structural Equation Modeling Programs to Test Mediation Models 305

9.14 Results Section 305

9.15 Summary 306

CHAPTER 10 • Discriminant Analysis 309

10.1 Research Situations and Research Questions 309

10.2 Introduction to Empirical Example 320

10.3 Screening for Violations of Assumptions 320

10.4 Issues in Planning a Study 322

10.5 Equations for Discriminant Analysis 323

10.6 Conceptual Basis: Factors That Affect the Magnitude of Wilks' Λ 326

10.7 Effect Size 327

10.8 Statistical Power and Sample Size Recommendations 327

10.9 Follow-Up Tests to Assess What Pattern of Scores Best Differentiates Groups 328

10.10 Results 329
10.11 One-Way ANOVA on Scores on Discriminant Functions 345
10.12 Summary 348
Appendix 10A: The Eigenvalue/Eigenvector Problem 349
Appendix 10B: Additional Equations for Discriminant Analysis 351

CHAPTER 11 • Multivariate Analysis of Variance 353

11.1 Research Situations and Research Questions 353
11.2 First Research Example: One-Way MANOVA 354
11.3 Why Include Multiple Outcome Measures? 355
11.4 Equivalence of MANOVA and DA 356
11.5 The General Linear Model 357
11.6 Assumptions and Data Screening 358
11.7 Issues in Planning a Study 360
11.8 Conceptual Basis of MANOVA 361
11.9 Multivariate Test Statistics 364
11.10 Factors That Influence the Magnitude of Wilks' Λ 367
11.11 Effect Size for MANOVA 367
11.12 Statistical Power and Sample Size Decisions 367
11.13 One-Way MANOVA: Career Group Data 368
11.14 2 × 3 Factorial MANOVA: Career Group Data 375
11.14.1 Follow-Up Tests for Significant Main Effects 378
11.14.2 Follow-Up Tests for Nature of Interaction 385
11.14.3 Further Discussion of Problems With This 2 × 3 Factorial Example 389
11.15 Significant Interaction in a 3 × 6 MANOVA 389
11.16 Comparison of Univariate and Multivariate Follow-Up Analyses 393
11.17 Summary 394

CHAPTER 12 • Exploratory Factor Analysis 398

12.1 Research Situations 398
12.2 Path Model for Factor Analysis 400
12.3 Factor Analysis as a Method of Data Reduction 401
12.4 Introduction of Empirical Example 404
12.5 Screening for Violations of Assumptions 405
12.6 Issues in Planning a Factor-Analytic Study 407
12.7 Computation of Factor Loadings 408
12.8 Steps in the Computation of PC and Factor Analysis 410
12.8.1 Computation of the Correlation Matrix **R** 411
12.8.2 Computation of the Initial Factor Loading Matrix **A** 411
12.8.3 Limiting the Number of Components or Factors 412
12.8.4 Rotation of Factors 413
12.8.5 Naming or Labeling Components or Factors 414
12.9 Analysis 1: PC Analysis of Three Items Retaining All Three Components 414
12.9.1 Finding the Communality for Each Item on the Basis of All Three Components 416
12.9.2 Variance Reproduced by Each of the Three Components 419

12.9.3 Reproduction of Correlations From Loadings on All Three Components 419

12.10 Analysis 2: PC Analysis of Three Items Retaining Only the First Component 420

12.10.1 Communality for Each Item on the Basis of One Component 420

12.10.2 Variance Reproduced by the First Component 423

12.10.3 Cannot Reproduce Correlations Perfectly From Loadings on Only One Component 423

12.11 PC Versus PAF 424

12.12 Analysis 3: PAF of Nine Items, Two Factors Retained, No Rotation 425

12.12.1 Communality for Each Item on the Basis of Two Retained Factors 428

12.12.2 Variance Reproduced by Two Retained Factors 432

12.12.3 Partial Reproduction of Correlations From Loadings on Only Two Factors 433

12.13 Geometric Representation of Factor Rotation 434

12.14 Factor Analysis as Two Sets of Multiple Regressions 438

12.14.1 Construction of Factor Scores for Each Individual (F_1, F_2, etc.) From Individual Item z Scores 438

12.14.2 Prediction of z Scores for Individual Participant (z_{Xi}) From Participant Scores on Factors (F_1, F_2, etc.) 439

12.15 Analysis 4: PAF With Varimax Rotation 440

12.15.1 Variance Reproduced by Each Factor at Three Stages in the Analysis 442

12.15.2 Rotated Factor Loadings 442

12.15.3 Example of a Reverse-Scored Item 444

12.16 Questions to Address in the Interpretation of Factor Analysis 445

12.17 Results Section for Analysis 4: PAF With Varimax Rotation 447

12.18 Factor Scores Versus Unit-Weighted Composites 449

12.19 Summary of Issues in Factor Analysis 451

Appendix 12A: The Matrix Algebra of Factor Analysis 454

Appendix 12B: A Brief Introduction to Latent Variables in SEM 457

CHAPTER 13 • Reliability, Validity, and Multiple-Item Scales 464

13.1 Assessment of Measurement Quality 464

13.1.1 Reliability 464

13.1.2 Validity 464

13.1.3 Sensitivity 465

13.1.4 Bias 466

13.2 Cost and Invasiveness of Measurements 466

13.2.1 Cost 466

13.2.2 Invasiveness 466

13.2.3 Reactivity of Measurement 467

13.3 Empirical Examples of Reliability Assessment 467

13.3.1 Definition of Reliability 467

13.3.2 Test-Retest Reliability Assessment for a Quantitative Variable 468

13.3.3 Interobserver Reliability Assessment for Scores on a Categorical Variable 470

13.4 Concepts From Classical Measurement Theory 471

13.4.1 Reliability as Partition of Variance 473

13.4.2 Attenuation of Correlations Due to Unreliability of Measurement 474

13.5 Use of Multiple-Item Measures to Improve Measurement Reliability 476
13.6 Computation of Summated Scales 478
13.6.1 Assumption: All Items Measure the Same Construct and Are Scored in the Same Direction 478
13.6.2 Initial (Raw) Scores Assigned to Individual Responses 478
13.6.3 Variable Naming, Particularly for Reverse-Worded Questions 479
13.6.4 Factor Analysis to Assess Dimensionality of a Set of Items 480
13.6.5 Recoding Scores for Reverse-Worded Items 482
13.6.6 Summing Scores Across Items to Compute a Total Score: Handling Missing Data 482
13.6.7 Comparison of Unit-Weighted Summed Scores Versus Saved Factor Scores 485
13.7 Assessment of Internal Homogeneity for Multiple-Item Measures: Cronbach's Alpha Reliability Coefficient 489
13.7.1 Conceptual Basis of Cronbach's Alpha 489
13.7.2 Empirical Example: Cronbach's Alpha for Five Selected CES-D Items 490
13.7.3 Improving Cronbach's Alpha by Dropping a "Poor" Item 494
13.7.4 Improving Cronbach's Alpha by Increasing the Number of Items 494
13.7.5 Other Methods of Reliability Assessment for Multiple-Item Measures 496
13.8 Validity Assessment 496
13.8.1 Content and Face Validity 497
13.8.2 Criterion-Oriented Validity 498
13.8.3 Construct Validity: Summary 499
13.9 Typical Scale Development Process 500
13.9.1 Generating and Modifying the Pool of Items or Measures 500
13.9.2 Administer the Survey to Participants 501
13.9.3 Factor-Analyze Items to Assess the Number and Nature of Latent Variables or Constructs 501
13.9.4 Development of Summated Scales 502
13.9.5 Assess Scale Reliability 502
13.9.6 Assess Scale Validity 502
13.9.7 Iterative Process 503
13.9.8 Create the Final Scale 503
13.10 A Brief Note About Modern Measurement Theories 503
13.11 Reporting Reliability 504
13.12 Summary 504
Appendix 13A: The CES-D 505
Appendix 13B: Web Resources on Psychological Measurement 506
CHAPTER 14 • More About Repeated Measures 509
14.1 Introduction 509
14.2 Review of Assumptions for Repeated-Measures ANOVA 509
14.3 First Example: Heart Rate and Social Stress 510
14.4 Test for Participant-by-Time or Participant-by-Treatment Interaction 510
14.5 One-Way Repeated-Measures Results for Heart Rate and Social Stress Data 513
14.6 Testing the Sphericity Assumption 516
14.7 MANOVA for Repeated Measures 517

14.8 Results for Heart Rate and Social Stress Analysis Using MANOVA 518
14.9 Doubly Multivariate Repeated Measures 518
14.10 Mixed-Model ANOVA: Between-S and Within-S Factors 522
14.10.1 Mixed-Model ANOVA for Heart Rate and Stress Study 522
14.10.2 Interaction of Intervention Type and Times of Assessment in Hypothetical Experiment With Follow-Up 524
14.10.3 First Follow-Up: Simple Main Effect (Across Time) for Each Intervention 526
14.10.4 Second Follow-Up: Comparisons of Intervention Groups at the Same Points in Time 528
14.10.5 Comparison of Repeated-Measures ANOVA With Difference-Score and ANCOVA Approaches 529
14.11 Order and Sequence Effects 530
14.12 First Example: Order Effect as a Nuisance 532
14.13 Second Example: Order Effect Is of Interest 534
14.14 Summary and Other Complex Designs 536

CHAPTER 15 • Structural Equation Modeling With Amos: A Brief Introduction **539**

15.1 What Is Structural Equation Modeling? 539
15.2 Review of Path Models 540
15.3 More Complex Path Models 542
15.4 First Example: Mediation Structural Model 544
15.5 Introduction to Amos 545
15.6 Screening and Preparing Data for SEM 546
15.6.1 SEM Requires Large Sample Sizes 546
15.6.2 Evaluating Assumptions for SEM 547
15.7 Specifying the Structural Equation Model (Variable Names and Paths) 547
15.7.1 Drawing the Model Diagram 548
15.7.2 Open SPSS Data File and Assign Names to Measured Variables 550
15.8 Specify the Analysis Properties 552
15.9 Running the Analysis and Examining Results 554
15.10 Locating Bootstrapped CI Information 557
15.11 Sample Results Section for Mediation Example 561
15.12 Selected Structural Equation Model Terminology 562
15.13 SEM Goodness-of-Fit Indexes 564
15.14 Second Example: Confirmatory Factor Analysis 565
15.14.1 General Characteristics of CFA 565
15.15 Third Example: Model With Both Measurement and Structural Components 569
15.16 Comparing Structural Equation Models 574
15.16.1 Comparison of Nested Models 574
15.16.2 Comparison of Non-Nested Models 574
15.16.3 Comparisons of Same Model Across Different Groups 575
15.16.4 Other Uses of SEM 576
15.17 Reporting SEM 576
15.18 Summary 577

CHAPTER 16 • Binary Logistic Regression 583

16.1 Research Situations 583

16.1.1 Types of Variables 583

16.1.2 Research Questions 583

16.1.3 Assumptions Required for Linear Regression Versus Binary Logistic Regression 584

16.2 First Example: Dog Ownership and Odds of Death 584

16.3 Conceptual Basis for Binary Logistic Regression Analysis 585

16.3.1 Why Ordinary Linear Regression Is Inadequate When Outcome Is Categorical 586

16.3.2 Modifying the Method of Analysis to Handle a Binary Categorical Outcome 588

16.4 Definition and Interpretation of Odds 588

16.5 A New Type of Dependent Variable: The Logit 590

16.6 Terms Involved in Binary Logistic Regression Analysis 591

16.6.1 Estimation of Coefficients for a Binary Logistic Regression Model 592

16.6.2 Assessment of Overall Goodness of Fit for a Binary Logistic Regression Model 592

16.6.3 Alternative Assessments of Overall Goodness of Fit 594

16.6.4 Information About Predictive Usefulness of Individual Predictor Variables 594

16.6.5 Evaluating Accuracy of Group Classification 595

16.7 Logistic Regression for First Example: Prediction of Death From Dog Ownership 596

16.7.1 SPSS Menu Selections and Dialog Boxes 596

16.7.2 SPSS Output 600

16.7.3 Results for the Study of Dog Ownership and Death 606

16.8 Issues in Planning and Conducting a Study 607

16.8.1 Preliminary Data Screening 607

16.8.2 Design Decisions 608

16.8.3 Coding Scores on Binary Variables 608

16.9 More Complex Models 610

16.10 Binary Logistic Regression for Second Example: Drug Dose and Sex as Predictors of Odds of Death 611

16.11 Comparison of Discriminant Analysis With Binary Logistic Regression 620

16.12 Summary 620

CHAPTER 17 • Additional Statistical Techniques 625

17.1 Introduction 625

17.2 A Brief History of Developments in Statistics 625

17.3 Survival Analysis 627

17.4 Cluster Analyses 628

17.5 Time-Series Analyses 630

17.5.1 Describing a Single Time Series 630

17.5.2 Interrupted Time Series: Evaluating Intervention Impact 632

17.5.3 Cycles in Time Series 634

17.5.4 Coordination or Interdependence Between Time Series 636

17.6 Poisson and Binomial Regression for Zero-Inflated Count Data 638

17.7 Bayes' Theorem 639
17.8 Multilevel Modeling 640
17.9 Some Final Words 641

Glossary 644

References 657

Index 666

PREFACE

The second edition contained a review of basic issues, bivariate statistics, and multivariable and multivariate methods. The material from the second edition has been divided into two volumes, and substantial new material has been added.

Volume I of the third edition (*Applied Statistics I: Basic Bivariate Techniques* [Warner, 2020]) includes expanded coverage of basic concepts, all the bivariate techniques covered in the second edition, and factorial analysis of variance (ANOVA). To streamline chapters, technical information that had been in the body of chapters in the previous edition was moved to end-of-chapter appendices.

Volume II of the third edition (this book) begins with "adding a third variable" as an introduction to statistical control and a gentle introduction to path models. This material bridges a gap between introductory books (which often include little mention of statistical control) and intermediate or advanced books (which often assume an understanding of statistical control). Even if your primary focus is multivariable analyses, you may find Volume I useful. It includes numerous examples of common non-normal distribution shapes and guidance on how to evaluate when non-normal distribution shapes may be problematic. The use of rating scale data (such as Likert scales) in parametric analyses such as t tests and correlation is discussed. Interpretation and graphing of confidence intervals are explained. End-of-chapter appendices provide technical information not available in most introductory textbooks.

Volume II includes all the multivariable and multivariate techniques from the second edition. The moderation chapter has been revised, and five new chapters have been added. These include chapters about the New Statistics; evaluating outliers and missing values, including discussion of multiple imputations; further applications of repeated-measures ANOVA; an introduction to structural equation modeling and bootstrapping; and a brief overview of additional techniques including survival analysis, cluster analysis, zero-inflated binomial and Poisson regression, time-series analysis, and Bayes' theorem.

DIGITAL RESOURCES

Instructor and student support materials are available for download from **edge.sagepub.com/warner3e**. **SAGE edge** offers a robust online environment featuring an impressive array of free tools and resources for review, study, and further explorations, enhancing use of the textbook by students and teachers.

SAGE edge for students provides a personalized approach to help you accomplish your coursework goals in an easy-to-use learning environment. Resources include the following:

- Mobile-friendly **eFlashcards** to strengthen your understanding of key terms
- **Data sets** for completing in-chapter exercises
- **Video resources** to support learning of important concepts, with commentary from experts in the field

SAGE edge for instructors supports your teaching by providing resources that are easy to integrate into your curriculum. SAGE edge includes the following:

- Editable, chapter-specific **PowerPoint® slides** covering key information that offer you flexibility in creating multimedia presentations
- **Test banks** for each chapter with a diverse range of prewritten questions, which can be loaded into your LMS to help you assess students' progress and understanding
- **Tables and figures** pulled from the book that you can download to add to handouts and assignments
- **Answers to in-text comprehension questions**, perfect for assessing in-class work or take-home assignments

Finally, in response to feedback from instructors for *R* content to mirror the SPSS coverage in this book, SAGE has commissioned *An R Companion for Applied Statistics II* by Danney Rasco. This short supplement can be bundled with this main textbook.

The author welcomes communication from teachers, students, and readers; please e-mail her at rmw@unh.edu with comments, corrections, or suggestions.

ACKNOWLEDGMENTS

Writers depend on many people for intellectual preparation and moral support. My understanding of statistics was shaped by exceptional teachers, including the late Morris de Groot at Carnegie Mellon University, and my dissertation advisers at Harvard, Robert Rosenthal and David Kenny. Several people who have most strongly influenced my thinking are writers I know only through their books and journal articles. I want to thank all the authors whose work is cited in the reference list. Authors whose work has particularly influenced my understanding include Jacob and Patricia Cohen, Barbara Tabachnick, Linda Fidell, James Jaccard, Richard Harris, Geoffrey Keppel, and James Stevens.

Special thanks are due to reviewers who provided exemplary feedback on first drafts of the chapters:

For the first edition:

David J. Armor, *George Mason University*

Michael D. Biderman, *University of Tennessee at Chattanooga*

Susan Cashin, *University of Wisconsin–Milwaukee*

Ruth Childs, *University of Toronto*

Young-Hee Cho, *California State University, Long Beach*

Jennifer Dunn, *Center for Assessment*

William A. Fredrickson, *University of Missouri–Kansas City*

Robert Hanneman, *University of California, Riverside*

Andrew Hayes, *The Ohio State University*

Lawrence G. Herringer, *California State University, Chico*

Jason King, *Baylor College of Medicine*

Patrick Leung, *University of Houston*

Scott E. Maxwell, *University of Notre Dame*

W. James Potter, *University of California, Santa Barbara*

Kyle L. Saunders, *Colorado State University*

Joseph Stevens, *University of Oregon*

James A. Swartz, *University of Illinois at Chicago*

Keith Thiede, *University of Illinois at Chicago*

For the second edition:

Diane Bagwell, *University of West Florida*

Gerald R. Busheé, *George Mason University*

Evita G. Bynum, *University of Maryland Eastern Shore*

Ralph Carlson, *The University of Texas Pan American*

John J. Convey, *The Catholic University of America*

Kimberly A. Kick, *Dominican University*

Tracey D. Matthews, *Springfield College*

Hideki Morooka, *Fayetteville State University*

Daniel J. Mundfrom, *New Mexico State University*

Shanta Pandey, *Washington University*

Beverly L. Roberts, *University of Florida*

Jim Schwab, *University of Texas at Austin*

Michael T. Scoles, *University of Central Arkansas*

Carla J. Thompson, *University of West Florida*

Michael D. Toland, *University of Kentucky*

Paige L. Tompkins, *Mercer University*

For the third edition:

Linda M. Bajdo, *Wayne State University*

Timothy Ford, *University of Oklahoma*

Beverley Hale, *University of Chichester*

Dan Ispas, *Illinois State University*

Jill A. Jacobson, *Queen's University*

Seung-Lark Lim, *University of Missouri, Kansas City*

Karla Hamlen Mansour, *Cleveland State University*

Paul F. Tremblay, *University of Western Ontario*

Barry Trunk, *Capella University*

I also thank the editorial and publishing team at SAGE, including Helen Salmon, Chelsea Neve, Megan O'Heffernan, and Laureen Gleason, who provided extremely helpful advice, support, and encouragement. Special thanks to copy editor Jim Kelly for his attention to detail.

Many people provided moral support, particularly my late parents, David and Helen Warner; and friends and colleagues at UNH, including Ellen Cohn, Ken Fuld, Jack Mayer, and Anita Remig. I hope this book is worthy of the support they have given me. Of course, I am responsible for any errors and omissions that remain.

Last but not least, I want to thank all my students, who have also been my teachers. Their questions continually prompt me to search for better explanations—and I am still learning.

Dr. Rebecca M. Warner
Professor Emerita
Department of Psychology
University of New Hampshire

ABOUT THE AUTHOR

Rebecca M. Warner is Professor Emerita at the University of New Hampshire. She has taught statistics in the UNH Department of Psychology and elsewhere for 40 years. Her courses have included Introductory and Intermediate Statistics as well as seminars in Multivariate Statistics, Structural Equation Modeling, and Time-Series Analysis. She received a UNH Liberal Arts Excellence in Teaching Award, is a Fellow of both the Association for Psychological Science and the Society of Experimental Social Psychology, and is a member of the American Psychological Association, the International Association for Statistical Education, and the Society for Personality and Social Psychology. She has consulted on statistics and data management for the World Health Organization in Geneva, Project Orbis, and other organizations; and served as a visiting faculty member at Shandong Medical University in China. Her previous book, *The Spectral Analysis of Time-Series Data*, was published in 1998. She has published articles on statistics, health psychology, and social psychology in numerous journals, including the *Journal of Personality and Social Psychology*. She has served as a reviewer for many journals, including *Psychological Bulletin*, *Psychological Methods*, *Personal Relationships*, and *Psychometrika*. She received a BA from Carnegie Mellon University in social relations in 1973 and a PhD in social psychology from Harvard in 1978. She writes historical fiction and is a hospice volunteer along with her Pet Partner certified Italian greyhound Benny, who is also the world's greatest writing buddy.

CHAPTER 1

THE NEW STATISTICS

1.1 REQUIRED BACKGROUND

This book begins with analyses that involve three variables, for example, an independent variable, a dependent variable, and a variable that is statistically controlled when examining the association between these, often called a covariate. Later chapters describe situations that involve multiple predictors, multiple outcomes, and/or multiple covariates. The bivariate analyses covered in introductory statistics books are the building blocks for these analyses. Therefore, you need a thorough understanding of bivariate analyses (i.e., analyses for one independent and one dependent variable) to understand the analyses introduced in this book.

The following topics are covered in Volume I (*Applied Statistics I: Basic Bivariate Techniques* [Warner, 2020]) and most other introductory statistics books. If you are unfamiliar with any of these topics, you should review them before you move forward.

- The use of frequency tables, histograms, boxplots, and other graphs of sample data to describe approximate distribution shape and extreme outliers. This is important for data screening.
- Understanding that some frequently used statistics, such as the sample mean, are not robust against the impact of outliers and violations of other assumptions.
- Computing and interpreting sample variance and standard deviation and the concept of degrees of freedom (*df*).
- Interpretation of standard scores (*z* scores) as unit-free information about the location of a single value relative to a distribution.
- The concept of sampling error, indexes of sampling error such as SE_M, and the way sampling error is used in setting up confidence intervals (CIs) and statistical significance tests.
- Choice of appropriate bivariate statistics on the basis of types of variables involved (categorical vs. quantitative and between-groups designs vs. repeated measures or paired or correlated samples).
- The most commonly used statistics, including independent-samples *t*, between-*S* analysis of variance (ANOVA), correlation, and bivariate regression. Ideally, you should also be familiar with paired-samples *t* and repeated-measures or paired-samples ANOVA. The multivariate and multivariate analyses covered in this book are built on these basic analyses.
- The logic of statistical significance tests (null-hypothesis statistical testing [NHST]), interpretation of *p* values, and limitations and problems with NHST and *p* values.

- Distributions used in familiar significance tests (normal, t, F, and χ^2) and the use of tail areas to describe outcomes as unusual or extreme.
- The concept of variance partitioning. In correlation and regression, r^2 is the proportion of variance in Y that can be predicted from X, and $(1 - r^2)$ is the proportion of variance in Y that cannot be predicted from X. In ANOVA, $SS_{between}$ provides information about proportion of variance in Y that is predictable from group membership, and SS_{within} provides information about variance in Y that is not predictable from group membership.
- Effect size.
- The difference between statistical significance and practical or clinical importance.
- Factors that influence statistical power, particularly effect size and sample size.

1.2 WHAT IS THE "NEW STATISTICS"?

In the past, many data analysts relied heavily on statistical significance tests to evaluate results and did not always report effect size. Even when used correctly, significance tests do not tell us everything we want to know; misuse and misinterpretation are common (Greenland et al., 2016). Misuse of significance tests has led to selective publication of only results with $p < .05$; publication of these selected results has sometimes led to widespread reports of "findings" that are not reproduced when replication studies are performed. The focus on "new" and "statistically significant" outcomes means that we sometimes don't discard incorrect results. Progress in science requires that we weed out mistakes, as well as make new discoveries.

Proponents of the "New Statistics" (such as Cumming, 2014) do not claim that their recommendations are really new. Many statisticians have called for changes in the way results are evaluated and reported, at least since the 1960s (including but not limited to Cohen, 1988, 1992, 1994; Daniel, 1998; Morrison & Henkel,1970; and Rozeboom, 1960). However, practitioners of statistics are often slow to respond to calls for change, or to adopt new methods (Sharpe, 2013).

The main changes called for by New Statistics advocates include:

1. Understanding the limitations of significance tests.
2. The need to report effect sizes and CIs.
3. Greater use of meta-analysis to summarize effect size information across studies.

All introductory statistics books I know of cover statistical significance tests and CIs, and most discuss effect size. Adopting the New Statistics perspective does not require you to learn anything new. New Statistics advocates only ask you to think about topics such as statistical significance tests from a more critical perspective. Even though you have probably studied CIs and effect size before, review can be enlightening. This chapter also includes a brief introduction to meta-analysis.

1.3 COMMON MISINTERPRETATIONS OF *p* VALUES

Advocates of the New Statistics have pointed out that misunderstandings about interpretation of p values are widespread. In a survey of researchers that asked which statements about p values they believed to be correct, large numbers of them endorsed incorrect interpretations (Mittag &

Thompson, 2000). Statistics education needs to be improved so that people who use NHST understand its limitations.

There are numerous problems with p values that lead to misunderstandings.

1. A p value cannot tell us what we want to know. We would like to know, on the basis of our data, something about the likelihood that a research hypothesis (usually an alternative hypothesis) is true. Instead, a p value tells us, often very inaccurately, about the probability of obtaining the values of M and t we found using our sample data, given that the null hypothesis is correct (Cohen, 1994).

2. Common practices, such as running multiple tests and selecting only a few to report on the basis of small p values, make p values very inaccurate information about risk for Type I decision error.

3. Even if we follow the rules and do everything "right," there will always be risk for decision error. Ideal descriptions of NHST require us to obtain a random sample from the population of interest, satisfy all the assumptions for the test statistic, have no problems with missing values or outliers, do one significance test, and then stop. Even if we could do this (and usually we can't), there would still be nonzero risks for both Type I and Type II decision errors. Because of sampling error, there is an intrinsic uncertainty that we cannot get rid of.

4. There is a fairly common misunderstanding that p values tell us something about the size, strength, or importance of an effect. Published papers sometimes include statements like "with $p < .001$, the effect was highly significant." In everyday language, *significant* means important, large, or worthy of notice. However, small p values can be obtained even for trivial effects if sample N is large enough. We need to distinguish between p values and effect size. Chapter 9 in Volume I (Warner, 2020) discusses this further.

From Volume I (Warner, 2020), here are examples of some things you should not say about p values. A more complete list of misconceptions to avoid is provided by Greenland et al. (2016).

Never make any of the following statements:

- $p = .000$ (the risk for Type I error can become very small, but in theory, it is never 0).

- p was "highly" significant. This leads readers to think that your effect was "significant" in the way we define *significant* in everyday language: large, important, or worthy of notice. Other kinds of effect size information (not p values) are required to evaluate the practical or clinical significance of the outcome of a study.

- p was "almost" significant (or synonymous terms such as *close to* or *marginally* significant). This language will make people who use NHST in traditional ways, and New Statistics advocates, cringe.

- For "small" p values, such as $p = .04$, we cannot say:

  Results were not due to chance or could not be explained by chance. (We cannot know that!)

  Results are likely to replicate in future studies.

  The null hypothesis (H_0) is false.

  We accept (or have proved) the alternative hypothesis.

We also cannot use $(1 - p)$, for example $(1 - .04) = .96$, to make probability statements such as:

> There is a 96% chance that results will replicate.
>
> There is a 96% chance that the null hypothesis is false.

- For p values larger than .05, we cannot say, "Accept the null hypothesis."

The language we use to report results should not overstate the strength of the evidence, imply large effect sizes in the absence of careful evaluation of effect size, overgeneralize the findings, or imply causality when rival explanations cannot be ruled out. We should never say, "This study proves that. . . ." Any one study has limitations. As suggested in Volume I (Warner, 2020): It is better to think about research in terms of degrees of belief. As we obtain additional high-quality evidence, we may become more confident of a belief. If high-quality inconsistent evidence arises, that should make us rethink our beliefs.

We can say things such as:

- The evidence in this study is consistent with the hypothesis that . . .
- The evidence in this study was not consistent with the hypothesis that . . .

Hypothesis can be replaced by similar terms, such as *prediction*.

Misunderstandings of p values, and what they can and cannot tell us, have been one of several contributing factors in a "replication crisis."

1.4 PROBLEMS WITH NHST LOGIC

The version of NHST presented in statistics textbooks and used by many researchers in social and behavioral science is an amalgamation of ideas developed by Fisher, Neyman, and Pearson (Lenhard, 2006). Neyman and Pearson strongly disagreed with important aspects of Fisher's thinking, and probably none of them would endorse current NHST logic and practices. Here are some commonly identified concerns about NHST logic.

1. **NHST turns an uncertainty continuum into a true/false decision.** Cohen (1994) and Rosnow and Rosenthal (1989) argued that we should think in terms of a continuum of likelihood:

   A successful piece of research doesn't conclusively settle an issue, it just makes some theoretical proposition to some degree more likely. . . . How much more likely this single research makes the proposition depends on many things, but not on whether p is equal to or greater than .05: .05 is not a cliff but a convenient reference point along the possibility-probability continuum. (Cohen, 1994)

   Surely, God loves the .06 nearly as much as the .05. (Rosnow & Rosenthal, 1989)

   One way to avoid treating .05 as a cliff is to report "exact" p values, as recommended by the American Psychological Association (APA) Task Force on Statistical Inference (Wilkinson & Task Force on Statistical Inference, APA Board of Scientific Affairs, 1999). The APA recommended that authors report "exact" values, such as $p = .032$, instead of a yes/no judgment of whether a result is significant or nonsignificant on the basis of $p < .05$ or $p > .05$. The possibly annoying quotation marks for "exact" are meant as a reminder that in practice, obtained p values often seriously underestimate the true risk for Type I error.

2. **NHST cannot tell us what we want to know.** We would like to know something like the probability that our research or alternative hypothesis is true, or the probability that the finding will replicate in future research, or how strong the effects were. In fact, NHST can tell us only the (theoretical) probability of obtaining the results in our data, given that H_0 is true (Cohen, 1994). NHST does not even do this well, given problems with its use in actual research practice.

3. Some philosophers of science argue that **progress in science requires us to discard faulty or incorrect evidence**. However, when researchers reject H_0, this is not "falsification" in that sense.[1]

4. **NHST is trivial because H_0 is always false.** Any nonzero difference (between μ_1 and μ_2) can be judged statistically significant if the sample size is sufficiently large (Kline, 2013).

5. **NHST requires us to think in terms of double negatives** (and people aren't very good at understanding double negatives). First, we set up a null hypothesis (of no treatment effect) that we almost always do not believe, and then we try to obtain evidence that would lead us to doubt this hypothesis. Double negatives are confusing and inconsistent with every day "psycho-logic" (Abelson & Rosenberg, 1958). In everyday reasoning, people have a strong preference to seek confirmatory evidence. People (including researchers) are confused by double negatives.

6. **NHST is misused in many research situations.** Assumptions and rules for proper use of NHST are stringent and are often violated in practice (as discussed in the next two sections). These violations often invalidate the inferences people want to make from p values.

Despite these criticisms, an argument can be made that NHST serves a valuable purpose when it is not misused. It can help assess whether results obtained in a study would be likely or unlikely to occur just because of sampling error when H_0 is true (Abelson, 1997; Garcia-Pérez, 2017). However, information about sampling error is also provided by CIs, in a form that may be less likely to lead to misunderstanding and yes/no thinking (Cumming, 2012).

1.5 COMMON MISUSES OF NHST

In actual practice, applications of NHST often do not conform to the ideal requirements for their use. Three sets of conditions are important for the proper use of NHST. I describe these as assumptions, rules, and handling of specific problems such as outliers. (These are fuzzy distinctions.)

In actual practice, it is difficult to satisfy all the requirements for p to be an accurate estimate of risk for Type I error. When these requirements are not met, values of p that appear in computer program results are biased; usually they underestimate the true risk for Type I error. When the true risk for Type I error is underestimated, both readers and writers of research reports may be overconfident that studies provide support for claims about findings. This can lead to publication and press-release distribution of false-positive results (Woloshin, Schwartz, Casella, Kennedy, & Larson, 2009). Inconsistent and even contradictory media reports of research findings may erode public trust and respect for science.

1.5.1 Violations of Assumptions

Most statistics textbooks precede the discussion of each new statistic with a list of formal mathematical assumptions about distribution shapes, independence of observations and residuals, and so forth. The list of assumptions for parametric analyses such as the independent-samples *t* test and one-way between-*S* ANOVA include:

- Data on quantitative variables are assumed to be normally distributed in the population from which samples were randomly drawn.
- Variances of scores in populations from which samples for groups were randomly drawn are assumed to be equal across groups (the homogeneity of variance assumption)
- Observations must be independent of one another. (Some textbooks do not explain this very important assumption clearly. See Chapter 2 in Volume I [Warner, 2020].)

For Pearson's *r* and bivariate regression, additional assumptions include:

- The relation between *X* and *Y* is linear.
- The variances of *Y* scores at each level of *X* are equal.
- Residuals from regression are uncorrelated with one another.

Advanced analyses often require additional assumptions.

Textbooks often provide information about evaluation of assumptions. However, most introductory data analysis exercises do not require students to detect or remedy violations of assumptions. The need for preliminary data screening and procedures for screening aren't clear in most introductory books. For NHST results to be valid, we need to evaluate whether assumptions are violated. However, journal articles often do not report whether assumptions were evaluated and whether remedies for violations were applied (Hoekstra, Kiers, & Johnson, 2012).

1.5.2 Violations of Rules for Use of NHST

I use the term *rules* to refer to other important guidelines about proper use of NHST. These are not generally included in lists of formal assumptions about distribution and independence of observations. These rules are often implicit; however, they are very important. These include the following:

- **Select the sample randomly from the (actual) population of interest** (Knapp, 2017). This is important whether you think about NHST in the traditional or classic manner, as a way to answer a yes/no question about the null hypothesis, or in terms of the New Statistics, with greater focus on CIs and less focus on *p* values. Bad practices in sampling limit generalizability of results and also compromise the logic of procedures of NHST.
- In practice, researchers often use convenience samples. When they want to generalize results, they imagine hypothetical populations similar to the sample in the study (invoking the idea of "proximal similarity" [Trochim, 2006] as justification for generalization beyond the sample). The use of convenience samples does not correspond closely to the situations the original developers of inferential statistics had in mind. For example, in industrial quality control, a population could be all the objects made by a factory in a month; the sample could be a random subset of

these objects. The logic of NHST inferential statistics makes more sense for random sampling. Studies based on accidental or convenience samples create much more difficult inference problems.

- **Select the statistical test and criterion for statistical significance (e.g., $\alpha < .05$, two tailed) prior to analysis.** This is important if you want to interpret *p* values as they have often been interpreted in the past, as a basis to make a yes/no decision about a null hypothesis. This rule is often violated in practice. For example, data analysts may use asterisks that appear next to correlations in tables and report that for one asterisk, $p < .05$; for two asterisks, $p < .01$; and for three asterisks, $p < .001$. Using asterisks to report a significance level separately for each correlation could be seen as implicitly setting the α criterion after the fact. On the other hand, many authorities recommend that instead of selecting specific α criteria, you should report an exact *p* value and not use the *p* value to make a yes/no decision about the believability of the null hypothesis. In other words, do not use *p* values as the basis to make statements such as "the result was statistically significant" or "reject H_0." Advocates of the New Statistics recommend that we should not rely on *p* values to make yes/no decisions.

- **Perform only one significance test** (or at most a small number of tests). The opposite of this is: Perform numerous statistical tests, and/or numerous variations of the same basic analysis, and then report only a few "statistically significant" results. This practice is often called *p*-hacking. Other names for *p*-hacking include data fishing, "the garden of forking paths" (Gelman & Loken, 2013), or my personal favorite, torturing the data until they confess (Mills, 1993).

Introductory statistics books usually discuss the problem of inflated risk for Type I error in the context of post hoc tests for ANOVA. They do not always make it clear that this problem is even more serious when people run dozens or hundreds of *t* tests or correlations.

1.5.3 Ignoring Artifacts and Data Problems That Bias *p* Values

Many artifacts that commonly appear in real data influence the magnitude of parameter estimates (such as *M*, *SD*, *r*, and *b*, among others) and *p* values. These include, but are not limited to:

- Univariate, bivariate, and multivariate outliers.
- Missing data that are not missing randomly.
- Measurement problems such as unreliability. For example, the obtained value of r_{xy} is attenuated (reduced) by unreliability of measures for *X* and *Y*.
- Mismatch of distribution shapes (for Pearson's *r* and regression statistics) that constrain the range of possible *r* values.

1.5.4 Summary

Consider an *F* ratio in a one-way between-*S* ANOVA. The logic for NHST goes something like this: If we formulate hypotheses and establish criteria for statistical significance and sample size prior to data collection, and if the null hypothesis is true, and if we take a random sample from the population of interest, and if all assumptions for the statistic are satisfied, and if we have not broken important rules for proper use of NHST, and if there are no artifacts such as outliers and missing values, then *p* should be an unbiased estimate of the likelihood of obtaining a value of *F* as large as, or larger than, the *F* ratio we obtained from our data. (Additional ifs could be added in many situations.)

This is a long conditional statement. The point is: Values of statistics such as *F* and *p* can provide the information described in ideal or imaginary situations in textbooks only when all of these conditions are satisfied. In actual research, one or many of these assumptions about conditions are violated. Therefore, statistics such as *F* and *p* rarely provide a firm basis for the conclusions described for ideal or imaginary research situations in textbooks. Problems with any of these (assumptions, rules, and artifacts) can result in biased *p* values that in turn may lead to false-positive decisions.

In real-life applications of statistics, it may be impossible to avoid all these problems. For all these reasons, I suggest that most *p* values should be taken with a very large grain of salt. *P* values are least likely to be misleading in simple experiments with a limited number of analyses, such as ANOVA with post hoc tests. They are highly likely to be misleading in studies that include large numbers of variables that are combined in different ways using many different analyses.

It is difficult to prioritize these problems; my guess is that violations of rules (such as running large numbers of significance tests and *p*-hacking) and neglect of sources of artifact (such as outliers) often create greater problems with *p* values in practice than violations of some of the formal assumptions about distributions of scores in populations (such as homogeneity of variance).

It requires some adjustment in thinking to realize that, to a very great extent, the numbers we obtain at the end of an analysis are strongly influenced by decisions made during data collection and analysis (Volume I [Warner, 2020]). Beginning students may think that final numerical results represent some "truth" about the world. We need to understand that with different data analysis decisions, we could have ended up with quite different answers. Greater transparency in reporting (Simmons, Nelson, & Simonsohn, 2011) helps readers understand the degree to which results may have been influenced by a data analyst's decisions.

1.6 THE REPLICATION CRISIS

Misuse and misinterpretation of statistics (particularly *p* values) is one of many factors that has contributed to rising concerns about the reproducibility of high-profile research findings in psychology. To evaluate reproducibility of research results, Brian Nosek and Jeff Spies founded the Center for Open Science in 2013 (Open Science Collaboration, 2015). Their aim was to increase openness, integrity, and reproducibility of scientific research. Participating scientists come from many fields, including astronomy, biology, chemistry, computer science, education, engineering, neuroscience, and psychology. Results reported for the first group of studies evaluated were disturbing. They conducted replications of 100 studies (both correlational and experimental) published in three psychology journals, using large samples (to provide adequate statistical power) and original materials if available. The average effect sizes were about half as large as the original results. Only 39 of the 100 replications yielded statistically significant outcomes (all original studies were "statistically significant"). This was not quite as bad as it sounds, because many original effect sizes associated with nonsignificant outcomes were within 95% CIs on the basis of replication effect sizes (Baker, 2015; Open Science Collaboration, 2015). These results attracted substantial attention and concern.

Failures to replicate have also been noted in biomedical research. Ioannidis (2005) examined 49 highly regarded medical studies from 13 prior years. He compared initial claims for intervention effectiveness with results in later studies with larger samples; 7 (16%) of the original studies were contradicted, and another 7 (16%) had smaller effects than the original study. Later studies have yielded even less favorable results. Begley and Ellis (2012) reported that biotechnology firm Amgen tried to confirm results from 53 landmark studies about issues such as new approaches to targeting cancers and alternative clinical uses for existing therapeutics. Findings were confirmed for only 6 (11%) studies. Baker and Dolgin (2017) noted that early results from the Cancer Reproducibility Project's examination of 6 cancer biology studies were mixed.

Do these replication failures indicate a "crisis"? That is debatable. Only a small subset of published studies were tested. Some of the original studies were chosen for replication because they reported surprising or counterintuitive results. Examination of *p* values is not the best way to assess whether results have been reasonably well replicated; *p* values are "fickle" and difficult to reproduce (Halsey, Curran-Everett, Vowler, & Drummond, 2015). It may be better to evaluate reproducibility using effect sizes or CIs instead of *p* values. Critics of the reproducibility projects argue that the replication methods and analyses were flawed (Gilbert, King, Pettigrew, & Wilson, 2016). It would be premature to conclude that large proportions of all past published research results would not replicate; however, concerns raised by failures to replicate should be taken seriously.

A failure to reproduce results does not necessarily mean that the original or past study was wrong. The replication study may be flawed, or the results may be context dependent (and might appear only in the specific circumstances in an earlier study, and not under the conditions in the replication study).

Concerns about reproducibility have led to a call for new approaches to reporting results, often called the New Statistics, along with a movement toward preregistration of study plans and Open Science, in which researchers more fully share information about study design and statistical analyses.

Many changes in research practice will be needed to improve reproducibility of research results (Wicherts et al., 2016). Misuse and misinterpretation of statistical significance tests (and *p* values) to make yes/no decisions about whether studies are "successful" have contributed to problems in replication. Some have even argued that NHST and *p* values are an inherently flawed approach to evaluation of research results (Krueger, 2001; Rozeboom, 1960). Cumming (2014) and others argue that a shift in emphasis (away from statistical significance tests and toward reports of effect size, CIs, and meta-analysis) is needed. However, many published papers still do not include effect size and CIs for important results (Watson, Lenz, Schmit, & Schmit, 2016).

1.7 SOME PROPOSED REMEDIES FOR PROBLEMS WITH NHST

1.7.1 Bayesian Statistics

Some authorities argue that we got off on the wrong foot (so to speak) when we adopted NHST in the early 20th century. Probability is a basic concept in statistical significance testing. The examples used to explain probability suggest that it is a simple concept. For example, if you draw 1 card at random from a deck of 52 cards with equal numbers of diamonds, hearts, spades, and clubs, what is the probability that the card will be a diamond? This example does not even begin to convey how complicated the notion of probability becomes in more complex situations (such as inference from sample to population).

NHST is based on a "frequentist" understanding of probability; this is not the only possible way to think about probability, and other approaches (such as Bayesian) may work better for some research problems. A full discussion of this problem is beyond the scope of this chapter; see Kruschke and Liddell (2018), Little (2006), Malakoff (1999), or Williamson (2013).

Researchers in a few areas of psychology use Bayesian methods. However, students typically receive little training in these methods. Whatever benefits this might have, a major shift toward the use of Bayesian methods in behavioral or social sciences seems unlikely to happen any time soon.

1.7.2 Replace $\alpha = .05$ with $\alpha = .005$

It has recently been suggested that problems with NHST could be reduced by setting the conventional α criterion to .005 instead of the current .05 (Benjamin et al., 2017). This would establish a more stringent standard for announcement of "new" findings. However, given the small effect sizes in many research areas, enormous sample sizes would be needed to have reasonable statistical power with $\alpha = .005$. This would be prohibitively costly. Bates (2017) and Schimmack (2017) argued that this approach is neither necessary nor sufficient and that it would make replication efforts even more unlikely. A change to this smaller α level is unlikely to be widely adopted.

1.7.3 Less Emphasis on NHST

The "new" statistics advocated by Cumming (2012, 2014) calls for a shift of focus. He recommended that research reports should focus more on

- confidence intervals,
- effect size information, and
- meta-analysis to combine effect size information across studies.

How "new" is the New Statistics? As noted by Cumming (2012) and others, experts have been calling for these changes for more than 40 years (e.g., Morrison & Henkel, 1970; Cohen, 1990, 1994; Wilkinson & Task Force on Statistical Inference, APA Board of Scientific Affairs, 1999). Cumming (2012, 2014) bolstered these arguments with further discussion of the ways that CIs (vs. *p* values) may lead data analysts to think about their data. Some argue that the New Statistics is not really "new" (Palij, 2012; Savalei & Dunn, 2011); CIs and significance tests are based upon the same information about sampling error. In practice, many readers may choose to convert CIs into *p* values so that they can think about them in more familiar terms. However, effect size reporting is critical; it provides information that is not obvious from examination of *p* values.

Unlike a shift to Bayesian approaches, or the use of $\alpha = .005$, including CIs and effect sizes in research reports would not be difficult or costly. In general, researchers have been slow to adopt these recommendations (Sharpe, 2013). The *Journal of Basic and Applied Social Psychology* (Trafimow & Marks, 2015) now prohibits publication of *p* values and related NHST results.

The following sections review the major elements of the New Statistics: CIs and effect size. CIs and effect size are both discussed in Volume I (Warner, 2020) for each bivariate statistic. A brief introduction to meta-analysis is also provided.

1.8 REVIEW OF CONFIDENCE INTERVALS

A confidence interval is an interval estimate for some unknown population characteristic or parameter (such as μ, the population mean) based on information from a sample (such as *M*, *SD*, and *N*). CIs can be set up for basic bivariate statistics using simple formulas. Unfortunately SPSS does not provide CIs for some statistics, such as Pearson's *r*. For more advanced statistics, CIs can be set up using methods such as bootstrapping, which is discussed in Chapter 15, on structural equation modeling, later in this book.

1.8.1 Review: Setting Up CIs

Consider an example of the CI for one sample mean, *M*. Suppose a data analyst has IQ scores for a sample of $N = 100$ cases, with these sample estimates: $M = 105$, $SD = 15$. In addition to reporting that mean IQ in the sample was M = 105, an interval estimate (a 95% CI) can be constructed, with lower and upper boundaries. The procedure used in this example can be used only when the sample statistic is known to have a normally shaped sampling distribution and when N is large enough that the standard normal or z distribution can be used to figure out what range of values lies within the center 95% of the distribution. (With smaller samples, t distributions are usually used.)

These are the steps to set up a CI:

- Decide on ***C*** **(level of confidence)** (usually this is 95%).
- Assuming that your sample statistic has a normally shaped sampling distribution, use the "critical values" from a z or standard normal distribution that correspond to the middle 95% of values. For a standard normal distribution, the middle 95% corresponds to the interval between $z_{\text{lower}} = -1.96$ and $z_{\text{upper}} = +1.96$. (Rounding these z values to −2 and +2 is reasonable when thinking about estimates.)
- Find the standard error (SE) for the sample statistic. The SE depends on sample size and standard deviation. For a sample mean, $SE_M = SD/\sqrt{N}$. Other sample statistics (such as r, b, and so forth) also have SEs that can be estimated.
- On the basis of $SD = 15$, and $N = 100$, we can compute the standard error of the sampling distribution for M: $SE_M = 15/\sqrt{100} = 15/10 = 1.5$.
- Now we combine SE_M with M and the z critical values that correspond to the middle 95% of the standard normal distribution to compute the CI limits:

$$\text{Lower limit} = M + z_{\text{lower}} \times SE_M = 105 - 1.96 * 1.5 = 105 - 2.94 = 102.06.$$

$$\text{Upper limit} = M + z_{\text{upper}} \times SE_M = 105 + 1.96 * 1.5 = 105 + 2.94 = 107.94.$$

This would be reported as "95% CI [102.06, 107.94]."

This procedure can be generalized and used with many other (but not all) sample statistics. To use this procedure, an estimate of the value of $SE_{\text{statistic}}$ is needed, and the sampling distribution for the statistic must be normal:

$$\text{Lower limit} = \text{Statistic} + z_{\text{lower}} \times SE_{\text{statistic.}} \quad (1.1)$$

$$\text{Upper limit} = \text{Statistic} + z_{\text{upper}} \times SE_{\text{statistic.}} \quad (1.2)$$

The statistic can be $(M_1 - M_2)$, r, or a raw-score regression slope b, for example. In more advanced analyses such as structural equation modeling, it is sometimes not possible to calculate the SE values for path coefficients directly, and it may be unrealistic to expect sampling distributions to be normal in shape. In these situations, Equations 1.1 and 1.2 cannot be used to set up CIs. The chapters that introduce structural equation modeling and logistic regression discuss different procedures to set up CIs for these situations.

1.8.2 Interpretation of CIs

It is incorrect to say that there is a 95% probability that the true population mean μ lies within a 95% CI. (It either does, or it doesn't, and we cannot know which.) We can make a long-range prediction that, if we have a population with known mean and standard deviation, and set a fixed sample size, and draw thousands of random samples from that population, that 95% of the CIs set up using this information will contain μ and the other 5% will not contain μ. Cumming and Finch (2005) provided other correct interpretations for CIs.

1.8.3 Graphing CIs

Upper and lower limits of CIs may be reported in text, tables, or graphs. One common type of graph is an error bar chart, as shown in Figure 1.1. (Bar charts can also be set up with error bars.) For either error bar or bar chart graphs, the graph may be rotated, such that error bars run from left to right instead of from bottom to top.

The data in Figure 1.1 are excerpted from an actual study. Undergraduates reported positive affect and the number of servings of fruit and vegetables they consumed in a typical day. Earlier research suggested that higher fruit and vegetable intake was associated with higher positive affect. Given the large sample size, number of servings could be treated as a group variable (i.e., the first group ate no servings of fruits and vegetables per day, the second group ate one serving per day, etc.) This was useful because past research suggested that the increase in positive affect might not be linear.

The vertical "whiskers" in Figure 1.1 show the 95% CI limits for each group mean. The horizontal line that crosses the Y axis at about 32.4 helps clarify that the CI for the zero servings of fruits and vegetables group did not overlap with the CIs for the groups of persons who ate three, four, or five servings per day.

In graphs of this type, the author must indicate whether the error bars correspond to a CI (and what level of confidence). Some graphs use similar-looking error bar markers to indicate the interval between $-1\ SE_M$ and $+1\ SE_M$ or the interval between $-1\ SD$ and $+1\ SD$.

Figure 1.1 Mean Positive Affect for Groups With Different Fruit and Vegetable Intake (With 95% CI Error Bars)

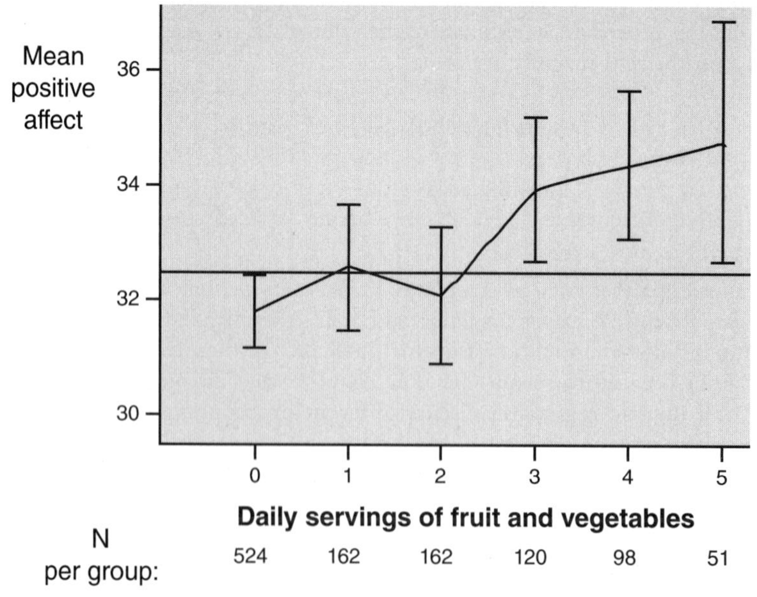

Source: Adapted from Warner, Frye, Morrell, and Carey (2017).

1.8.4 Understanding Error Bar Graphs

A reader can make two kinds of inferences from error bars in this type of graph (Figure 1.1). First, error bars can be used to guess which group means differed significantly. Cumming (2012, 2014) cautioned that analysts should not automatically convert CI information into p values for significance tests when they think about their results. However, if readers choose to do that, it is important to understand the way CIs and two-tailed p values are related. In general, if the CIs for two group means do not overlap in graphs such as Figure 1.1, the difference between means is statistically significant (assuming that the level of confidence corresponds to the α level, i.e., 95% confidence and $\alpha = .05$, two tailed). On the other hand, the difference between a pair of group means can be statistically significant even if the CIs for the means overlap slightly. Whether the difference is statistically significant depends on the amount of overlap between CIs (Cumming & Finch, 2005; Knezevic, 2008).

The nonoverlapping CIs for the zero-servings group and five-servings group indicates that if a t test were done to compare these two group means, using $\alpha = .05$, two tailed, this difference would be statistically significant. There is some overlap in the CIs for the two-servings and three-servings groups. This difference might or might not be statistically significant using $\alpha = .05$, two tailed.

The second kind of information a reader should look for is practical or clinical significance. Mean positive affect was about 34 for the five-servings group and 32 for the zero-servings group. Is that difference large enough to value or care about? Would a typical person be motivated to raise fruit and vegetable consumption from zero to five servings if that meant a chance to increase positive affect by two points? (Maybe there are easier ways to "get happy.")

Numbers on the scale for positive affect scores are meaningless unless some context is provided. In this example, the minimum possible score for positive affect was 10 points, and the maximum was 50 points. A 2-point difference on a 50-point rating scale does not seem like very much. Also note that this graph "lies with statistics" in a way that is very common in both research reports and the mass media. The Y axis begins at about 30 points rather than the actual minimum value of 10 points. How different would this graph look if the Y axis included the entire possible range of values from 10 to 50?

In the final analyses in our paper (Warner et al., 2017), fruit and vegetable intake uniquely predicted about 2% of the variance in positive affect after controlling for numerous other variables that included exercise and sleep quality. That 2% was statistically significant. However, on the basis of 2% of the variance and a two-point difference in positive affect ratings for the low versus high fruit and vegetable consumption groups, I would not issue a press release urging people to eat fruit and get happy. Other variables (such as gratitude) have much stronger associations with positive affect. (It may be of theoretical interest that consumption of fruits and vegetables, but not sugar or fat consumption, was related to positive affect. Fruit and vegetable consumption is related to other important outcomes such as physical health.)

The point is: Information about actual and potential range of scores for the outcome variable can provide context for interpretation of scores (even when they are in essentially meaningless units). Readers also need to remember that the selection of a limited range of values to include on the Y axis creates an exaggerated perception of group differences.

1.8.5 Why Report CIs Instead of, or in Addition to, Significance Tests?

Cumming (2012) and others suggest these possible advantages of focusing on CIs rather than p values:

1. **Reporting the CI can move us away from the yes/no thinking** involved in statistical significance tests (unless we use the CI only to reconstruct the statistical significance test).

2. **CIs make us aware of the lack of precision of our estimates** (of values such as means). Information about lack of precision is more compelling when scores on a predicted variable are in meaningful units. Consider systolic blood pressure, given in millimeters of mercury (mm Hg). If the 95% CI for systolic blood pressure in a group of drug-treated patients ranges from 115 mm Hg (not considered hypertensive) to 150 mm Hg (hypertensive), potential users of the drug will be able to see that mean outcomes are not very predictable. (On the other hand, if the CI ranges from 115 to 120 mm Hg, mean outcomes can be predicted more accurately.)

3. **CIs may be more stable across studies than *p* values.** In studies of replication and reproducibility, overlap of CIs across studies may be a better way to assess consistency than asking if studies yield the same result on the binary outcome judgment: significant or not significant. *P* values are "fickle"; they tend to vary across samples (Halsey et al., 2015). Asendorpf et al. (2013) recommended that evaluation of whether two studies produce consistent results should focus on CI overlap rather than on "vote counting" (i.e., noticing whether both studies had $p < .05$).

Data analysts hope that CIs will be relatively narrow, because if they are not, it indicates that estimates of mean have considerable sampling error. Other factors being equal, the width of a CI depends on these factors:

- As *SD* increases (other factors being equal), the width of the *CI* increases.
- As level of confidence increases (other factors being equal), the width of the *CI* increases.
- As *N* increases (other factors being equal), the width of the *CI* decreases.

Despite calls to include CIs in research reports, many authors still do not do so (Sharpe, 2013). This might be partly because, as Cohen (1994) noted, they are often "so embarrassingly large!"

1.9 EFFECT SIZE

Bivariate statistics introduced in Volume I (Warner, 2020) were accompanied by a discussion of one (or sometimes more than one) effect size indexes. For χ^2, effect sizes include Cramer's *V* and ϕ. Pearson's *r* and r^2 directly provide effect size information. For statistics such as the independent-samples *t* test, several effect sizes can be used; these include point biserial *r* (r_{pb}), **Cohen's *d***, η, and η^2. It is also possible to think about the $(M_1 - M_2)$ difference as information about practical or clinical effect size terms if the dependent variable is measured in meaningful units such as dollars, kilograms, or inches. For ANOVA, η and η^2 are commonly used. Rosnow and Rosenthal (2003) discussed additional, less widely used effect size indexes.

1.9.1 Generalizations About Effect Sizes

1. Effect size is independent of sample size. For example, the magnitude of Pearson's *r* does not systematically increase as *N* increases.[2]

2. Some effect sizes have a fixed range of possible values (*r* ranges from −1 to +1), but other effect sizes do not (Cohen's *d* is rarely higher than 3 in absolute value, but it does not have a fixed limit).

3. Many effect sizes are in unit-free (or standardized) terms. For example, the magnitude of Pearson's r is not related to the units in which X and Y are measured.
4. On the other hand, effect size information can be presented in terms of the original units of measurement (e.g., $M_1 - M_2$). This is useful when original units of measurement were meaningful (Pek & Flora, 2018).
5. Some effect sizes can be directly converted (at least approximately) into other effect sizes (Rosnow & Rosenthal, 2003).
6. Cohen's (1988) guidelines for verbal labeling of effect sizes are widely used; these appear in Table 1.1. Alternative guidelines based on Fritz, Morris, and Richler (2012) appear in Table 1.2.
7. The value of a test statistic (such as the independent-samples t test) depends on both effect size and sample size or df. This is explained further in the next section.
8. Many journals now call for reporting of effect size information. However, many published research reports still do not include this information.
9. Judgments about the clinical or practical importance of research results should be based on effect size information, not based on p values (Sullivan & Feinn, 2012).
10. If you read a journal article that does not include effect size information, there is usually enough information for you to compute an effect size yourself. (There should be!)
11. Computer programs such as SPSS often do not provide effect sizes; however, effect sizes can be computed from the information provided.

Table 1.1 Suggested Verbal Labels for Cohen's *d* and Other Common Effect Sizes

Verbal Label Suggested by Cohen (1988)	Cohen's d	r, r_{pb},[a] b, Partial r, R, or β	r^2, R^2, or η^2
Large effect	0.8	.371	.138
(In-between area)	0.7	.330	.109
	0.6	.287	.083
Medium effect	0.5	.243	.059
(In-between area)	0.4	.196	.038
	0.3	.148	.022
Small effect	0.2	.100	.010
(In-between area)	0.1	.050	.002
No effect	0.0	.000	.000

Source: Adapted from Cohen (1988).

a. Point biserial r is denoted r_{pb}. For an independent-samples t test, r_{pb} is the Pearson's r between the dichotomous variable that represents group membership and the Y quantitative dependent variable.

Table 1.2 Effect Size Interpretations

Research Question	Effect Sizes	Minimum Reportable Effect[a]	Moderate Effect	Large Effect
Difference between two group means	Cohen's d	.41	1.15	2.70
Strength of association: linear	r, r_{pb}, R, partial r, β, tau	.2	.5	.8
Squared linear association estimates	r^2, partial r^2, R^2, adjusted R^2, sr^2	.04	.25	.64
Squared association (not necessarily linear)	η^2 and partial η^2	.04	.25	.64
Risk estimates[b]	RR, OR	2.0	3.0	4.0

Source: Adapted from Fritz et al. (2012).

a. The minimum values suggested by Fritz et al. are much higher than the ones proposed by Cohen (1988).

b. Analyses such as logistic regression (in which the dependent variable is a group membership, such as alive vs. dead) provide information about relative or comparative risk, for example, how much more likely is a smoker to die than a nonsmoker? This may be in the form of relative risk (RR) and an odds ratio (OR). See Chapter 16.

12. In the upcoming discussion of meta-analysis, examples often focus on effect sizes such as Cohen's *d* that describe the difference between group means for treatment and control groups. However, raw or standardized regression slope coefficients can also be treated as effect sizes in meta-analysis (Nieminen, Lehtiniemi, Vähäkangas, Huusko, & Rautio, 2013; Peterson & Brown, 2005).

13. CIs can be set up for many effect size estimates (Kline, 2013; Thompson, 2002b). Ultimately, it would be desirable to report these along with effect size. In the short term, just getting everyone to report effect size for primary results is probably a more reasonable goal.

1.9.2 Test Statistics Depend on Effect Size Combined With Sample Size

Consider the independent-samples *t* test. M_1 and M_2 denote the group means, SD_1 and SD_2 are the group standard deviations, and n_1 and n_2 denote the number of cases in each group. One of the effect sizes used with the independent-samples *t* is Cohen's *d* (the standardized distance or difference between the sample means M_1 and M_2). The difference between the sample means is standardized (converted to a unit-free distance) by dividing $(M_1 - M_2)$ by the pooled standard deviation s_p:

$$Cohen's\ d = \frac{M_1 - M_2}{s_p}. \tag{1.3}$$

Formulas for s_p sometimes appear complicated; however, s_p is just the weighted average of SD_1 and SD_2, weighted by sample sizes n_1 and n_2.

Sample size information for the independent-samples *t* test can be given as $(\sqrt{df/2})$, where $df = (n_1 + n_2) - 2$. The formula for the independent-samples *t* test can be given as a function of effect size *d* and sample size, as shown by Rosenthal and Rosnow (1991):

$$t = d\frac{\sqrt{df}}{2}. \quad (1.4)$$

Examining Equation 1.4 makes it clear that if effect size d is held constant, the absolute value of t increases as the df (sample size) increases. Thus, even when an effect size such as d is extremely small, as long as it is not zero, we can obtain a value of t large enough to be judged statistically significant if sample size is made sufficiently large. Conversely, if the sample size given by df is held constant, the absolute value of t increases as d increases. This dependence of magnitude of the test statistic on both effect size and sample size holds for other statistical tests (I have provided only a demonstration for one statistic, not a proof).

This is the important point: A very large value of t, and a correspondingly very small value of p, can be obtained even when the effect size d is extremely small. A small p value does not necessarily tell us that the results indicate a large or strong effect (particularly in studies with very large N's).

Furthermore, both the value of N and the value of d depend on researcher decisions. For an independent-samples t test, other factors being equal, d often increases when the researcher chooses types of treatments and/or dosages of treatments that cause large differences in the response variable and when the researcher controls within-group error variance through standardization of procedures and recruitment of homogeneous samples. Some undergraduate students became upset when I explained this: "You mean you can make the results turn out any way you want?" Yes, within some limits. When we obtain statistics in samples, such as values of M or Cohen's d or p, these values depend on our design decisions. They are not facts of nature. See Volume I (Warner, 2020), Chapter 12, for further discussion.

1.9.3 Using Effect Size to Evaluate Theoretical Significance

Judgments about theoretical significance are sometimes made on the basis of the magnitude of standardized effect size indexes such as d or r. One way to think about the importance of research results is to ask, Given the effect size, how much does this variable add to our ability to predict some outcome of interest, or to "explain variance"? Is the added predictive information sufficient to be "worthwhile" from a theoretical perspective? Is it useful to continue to include this variable in future theories, or are its effects so trivial as to be negligible?

For example, if X and Y have $r_{xy} = .10$ and therefore, $r^2 = .01$, then only 1% of the variance in Y is linearly predictable from X. By implication, the other 99% of the variance is related to other variables (or is due to nonlinear associations or is inherently unpredictable). Is it worth expending a lot of energy on further study of a variable that predicts only 1% of the variance? When an effect size is this small, very large N's are needed in future studies in order to have sufficient statistical power (i.e., a reasonably high probability of obtaining a statistically significant outcome). Researchers need to make their own judgments as to whether it is worth pursuing a variable that predicts such a small proportion of variance.

There are two reasons why authors may not report effect sizes. One is that SPSS does not provide effect size information for some common statistics, such as ANOVA. This lack is easy to deal with, because SPSS does provide the information needed to calculate effect size information by hand, and the computations are simple. This information is provided for each statistic in Volume I (Warner, 2020). For example, an η^2 effect size for ANOVA can be obtained by dividing SS_{effect} by SS_{total}. There may be another reason. Cohen (1994) noted that CIs are often embarrassingly large; effect sizes may often be embarrassingly small. It just does not sound very impressive to say, "I have accounted for 1% of the variance."

A long time ago, Mischel (1968) pointed out that correlations between personality measures and behaviors tended to be no larger than $r = .30$. This triggered a crisis and disputes in personality research. Social psychologists argued that the power of situations was much greater than personality. Epstein and O'Brien (1985) argued that it is possible to obtain higher correlations in personality with broader assessments and that typical effect sizes in social psychology were not much higher. However, at the time, $r = .30$ seemed quite low. This may have been because earlier psychological research in areas such as behavior analysis and psychophysics tended to yield much larger effects (stronger correlations). I wonder whether Cohen's labeling of $r = .3$ as a medium to large effect was based on the observation that in many areas of psychology, effects much larger than this are not common. Nevertheless, accounting for 9% of the variance does not sound impressive.

Prentice and Miller (1992) pointed out that in some situations, even small effects may be impressive. Some behaviors are probably not easy to change, and a study that finds some change in this behavior can be impressive even if the amount of change is small. They cited this example: Physical attractiveness shows strong relationships with some responses (such as interpersonal attraction). It is impressive to note that even in the courtroom, attractiveness has an impact on behavior; unattractive defendants were more likely to be judged guilty and to receive more punishment. If physical attractiveness has effects in even this context, its effects may apply to a very wide range of situations.

Sometimes social and behavioral scientists have effect size envy, imagining that effect sizes in other research domains are probably much larger. In fact, effect sizes in much biomedical research are similar to those in psychology (Ferguson, 2009). Rosnow and Rosenthal (2003) cited an early study that examined whether taking low-dose aspirin could reduce the risk for having a heart attack. Pearson's r (or ϕ) between these two dichotomous variables was $r = .034$. The percentage of men who did not have heart attacks in the aspirin group (51.7%) was significantly higher than the percentage of men who did not have heart attacks in the placebo group (48.3%). Assuming that these results are generalizable to a larger population (and that is always a question), a 3.4% improvement in health outcome applied to 1 million men could translate into prevention of about 34,000 heart attacks. From a public health perspective, $r = .034$ can be seen as a large effect. From the perspective of an individual, the evaluation could be different. An individual might reason, I might change my risk for heart attack from 51.7% (if I do not take aspirin) to 48.3% (if I do take aspirin). From that perspective, the effect of aspirin might appear to be less substantial.

1.9.4 Use of Effect Size to Evaluate Practical or Clinical Importance (or Significance)

It is important to distinguish between statistical significance and practical or clinical significance (Kirk, 1996; Thompson, 2002a). We have clear guidelines how to judge statistical significance (on the basis of p values). What do we mean by clinical or practical significance, and how can we make judgments about this? In everyday use, the word *significant* often means "sufficiently important to be worthy of attention." When research results are reported as statistically significant, readers tend to think that the treatment caused effects large enough to be noticed and valued in everyday life. However, the term *statistically significant* has a specific technical meaning, and as noted in the previous section, a result that is statistically significant at $p < .001$ may not correspond to a large effect size.

For a study comparing group means, practical significance corresponds to differences between group means that are large enough to be valued (a large $M_1 - M_2$ difference). In a regression study, practical significance corresponds to large and "valuable" increases in an outcome variable as scores on the independent variable increase (e.g., a large raw-score regression slope b).

Standardized effect sizes such as Cohen's d are sometimes interpreted in terms of clinical significance. However, examining the difference between group means ($M_1 - M_2$) in their original units of measurement can be a more useful way to evaluate the clinical or practical importance of results (Pek & Flora, 2018). $M_1 - M_2$ provides understandable information when variables are measured in meaningful and familiar units. Age in years, salary in dollars or euros or other currency units, and body weight in kilograms or pounds are examples of variables in meaningful units. Everyday people can understand results reported in these terms.

For example, if a study that compared final body weight between treatment (1) and control (2) groups, with mean weights M_1 = 153 lb in the treatment group and M_2 = 155 lb in the control group, everyday folks (as well as clinicians) probably would not think that a 2-lb difference is large enough to be noticeable or valuable. Most people would not be very interested in this new treatment, particularly if it is expensive or difficult. On the other hand, if the two group means differed by 20 or 30 lb, probably most people would view that as a substantial difference. Similar comparisons can be made for other different treatment outcomes (such as blood pressure with vs. without drug treatment).

Unfortunately, when people read about new treatments in the media, reports often say that a treatment effect was "statistically significant" or even "highly statistically significant." Those phrases can mislead people to think that the difference between group means (for weight, blood pressure, or other outcomes) in the study was extremely large.

Here are examples of criteria that could be used to judge whether results of studies are clinically or practically significant, that is, whether outcomes are different enough to matter:

- **Are group means so far apart that one mean is above, and the other mean is below, some diagnostic cutoff value?** For example, is systolic blood pressure in a nonhypertensive range for the treatment group and a hypertensive range for a control group?
- **Would people care about an effect this size?** This is relatively easy to judge when the variable is money. Judge and Cable (2004) examined annual salaries for tall versus short persons. They reported these mean annual salaries (in U.S. dollars): tall men, $79,835; short men, $52,704; tall women, $42,425, short women, $32,613. As always in research, there are many reasons we should hesitate to generalize their results to other situations or apply them to ourselves individually. However, tall men earned mean salaries more than $47,000 higher than short women. I am a short woman, and this result certainly got my attention.

In economics, value or "mattering" is called utility. Systematic studies could be done to see what values people (clients, clinicians, and others) attach to specific outcomes. For a person who earns very little money, a $1,000 salary increase may have a lot of value. For a person who earns a lot of money, the same $1,000 increase might be trivial. Utility of specific outcomes might well differ across persons according to characteristics such as age and sex.

- **How large does a difference have to be for most people to even notice or detect it?** At a bare minimum, before we speak of an effect detected in a study as an important finding, it should be noticeable in everyday life (cf. Donlon, 1984; Stricker, 1997).

1.9.5 Uses for Effect Sizes

- Effect sizes should be included in research reports. Standardized effect sizes (such as Cohen's *d* or *r*) provide a basis for labeling strength of relationships between variables as weak, moderate, or strong. Standardized effect sizes can be compared with those found in other studies and in past research. Additional information, such as raw-score regression slopes and group means in original units of measurement, can help readers understand the real-world or clinical implications of findings (at least if the original units of measurement were meaningful).

- Effect size estimates from past research can be used to do statistical power analysis to make sample-size decisions for future research.
- Finally, effect size information can be combined and evaluated across studies using meta-analysis to summarize existing information.

1.10 BRIEF INTRODUCTION TO META-ANALYSIS

A meta-analysis is a summary of effect size information from past research. It involves evaluating the mean and variance of effect sizes combined across past studies. This section provides only a brief overview. For details about meta-analysis, see Borenstein, Hedges, Higgins, and Rothstein (2009) or Field and Gillett (2010).

1.10.1 Information Needed for Meta-Analysis

The following steps are involved in information collection:

1. **Clearly identify the question of interest.** For example, how does number of bystanders (*X*) predict whether a person offers help (*Y*)? What is the difference in mean depression scores (*Y*) between persons who do and do not receive cognitive behavioral therapy (CBT) (*X*)?
2. **Establish criteria for inclusion (vs. exclusion) of studies ahead of time.** Decide which studies to include and exclude. This involves many judgments. Poor-quality studies may be discarded. Studies that are retained must be similar enough in conception and design that comparisons make sense (you can't compare apples and oranges). Reading meta-analyses in your own area of interest can be helpful.
3. **Do a thorough search for past research about this question.** This should include published studies, located using library databases, and unpublished data, obtained through personal contacts.
4. **Create a data file that has at least the following information for each study:**
   a. Author names and year of publication for each study.
   b. Number in sample (and within groups).
   c. Effect size information (you may have to calculate this if it is not provided). The most common effect sizes are Cohen's *d* and *r*. However, other types of effect size may be used.[3]
   d. If applicable, group sizes, means, and standard deviations.
   e. Additional information to characterize studies. If the number of studies included in the meta-analysis is large, it may be possible to analyze these variables as possible "moderators," that is, variables that are related to different effect sizes. In studies of CBT, the magnitude of treatment effect might depend on number of treatment sessions, type of depression, client sex, or even the year when the study was done. There are also "study quality" and study type variables, for example, Was the study double blind or not? Was there a nontreatment control group? Was it a within-*S* or between-*S* design? It is a good idea to have more than one reader code this information and to check for interobserver reliability.

1.10.2 Goals of Meta-Analysis

- **Estimate mean effect size.** When effect sizes are averaged across studies they are usually weighted by sample size (or sometimes by other characteristics of studies).

- **Evaluate the variance of effect sizes across studies.** The variation among effect sizes indicates whether results of studies seem to be homogeneous (that is, they all tended to yield similar effect sizes) or heterogeneous (they yielded different effect sizes). If effect sizes are heterogeneous and the number of studies is reasonably large, a moderator analysis is possible.
- **Evaluate whether certain moderator variables are related to difference in effect sizes.** For example, are smaller effect sizes obtained in recent CBT studies than in those done many years ago?

The mechanics of doing a meta-analysis can be complex. For example, the analyst must choose between a fixed- and a random-effects model (for discussion, see Field & Gillett, 2010); a random-effects model is probably more appropriate in many situations. SPSS does not have a built-in meta-analysis procedure; Field and Gillet (2010) provide free downloadable SPSS syntax files on their website, and references to software created by others, including routines in R. See the following sources for guidelines about reporting meta-analysis: Liberati et al. (2009) and Rosenthal (1995).

1.10.3 Graphic Summaries of Meta-Analysis

Forest plots are commonly used to describe results from meta-analysis. Figure 1.2 shows a hypothetical forest plot. Suppose that three studies were done to compare depression scores between a group that has had CBT and a control group that has not had therapy. For each study, the effect size, Cohen's *d*, is the difference between posttest depression scores for the CBT and control groups (divided by the pooled within-group standard deviation). A 95% CI is obtained for Cohen's *d* for each study.

The vertical line down the center of the table is the "line of no effect" that corresponds to $d = 0$. This would be the expected result if population means did not differ between CBT and control conditions. In this example, a negative value of *d* means that the treatment group had a better outcome (i.e., lower depression after treatment) than the control group.

Figure 1.2 Hypothetical Forest Plot for Studies That Assess Posttreatment Depression in Therapy and Control Groups

Study	CBT Therapy Group *N*	CBT Therapy Group mean (*SD*)	Control Group *N*	Control Group mean (*SD*)	Cohen's *d* and 95% CI	Weight (%)	Cohen's *d* and 95% CI
Study 1, year	34	9.77 (2.93)	34	10.29 (3.43)		27.5	−0.52 [−2.04, 1.00]
Study 2, year	36	8.40 (1.90)	36	8.90 (3.00)		46.9	−0.50 [−1.66, 0.66]
Study 3, year	30	10.26 (2.96)	30	12.09 (3.24)		25.6	−1.83 [−3.40, −0.26]
Total (95% CI)	100		100			100.0	−0.85 [−1.64, −0.05]

Test for heterogeneity Chi-square = 2.03 *df* = 2 *p* = 036

Test for overall effect *z* = 2.09 *p* = .04

−4.0 −2.0 0 2.0 4.0

Favors Intervention Favors Control

Source: Adapted with permission from the Royal Australian College of General Practitioners from: Ried K. "Interpreting and understanding meta-analysis graphs: A practical guide." *Australian Family Physician*, 2006; 35(8):635–38. Available at www.racgp.org.au/afp/200608/10624.

Reading across the line for Study 1: Author names and year are provided, then N, mean, and SD for the CBT and control groups. The horizontal line to the right, with a square in the middle, corresponds to the 95% CI for Cohen's d for Study 1. The size of the square is proportional to total N for that study. The weight given to information from each study in a meta-analysis can be based on one or more characteristics of studies, such as sample size. The final column provides the exact numerical results that correspond to the graphic version of the 95% CI for Cohen's d for each study.

The row denoted "Total" shows the 95% CI for the weighted mean of Cohen's d across all three studies, first in graphic and then in numerical form. The "Total" row has a diamond-shaped symbol; the end points of the diamond indicate the 95% CI for the average effect size across studies. This CI did not include 0.

The values in the lower left of the figure answer two questions about the set of effect sizes across all studies. First, does the weighted mean of Cohen's d combined across studies differ significantly from 0? The test for the overall effect, $z = 2.09$, $p = .04$, indicates that the null hypothesis that the overall average effect was zero can be rejected using $\alpha = .05$, two tailed. The mean Cohen's d that describes difference of depression scores for CBT compared with control group was −.87. This suggests that average mean depression was almost 1 standard deviation lower for persons who received CBT. That would be labeled a large effect using Cohen's standards (Table 1.1); it lies in between "minimal reportable effect" and a moderate effect using the guidelines of Fritz et al. (2012) (Table 1.2).

Second, are the effect sizes sufficiently similar or close together that they can be viewed as homogeneous? The test for heterogeneity result was $\chi^2 = 2.03$, $df = 2$, $p = .36$. The null hypothesis of homogeneity is not rejected. If the χ^2 test result were significant, this would suggest that some studies yielded different effect sizes than others. If the meta-analysis included numerous studies, it would be possible to look for moderator variables that might predict which studies have larger and which have smaller effects. An actual meta-analyses of CBT effectiveness suggested that effects were larger for studies done in the early years of CBT and smaller in studies done in recent years (Johnsen & Friborg, 2015). In other words, the year when each study was done was a moderator variable; effect sizes were larger, on average, in earlier years than in more recent years.

1.11 RECOMMENDATIONS FOR BETTER RESEARCH AND ANALYSIS

Extensive recommendations have been made for improvements in data analysis and research practices. These could substantially improve understanding of results from individual studies, reduce p-hacking, reduce the number of false-positive results, and improve replicability of research results.

Cumming (2012) recommended focusing more on CIs and effect sizes (and less on p values) in reports and interpretations of research results. In addition, meta-analyses should be used to summarize effect size information across studies. When effect size information is not examined, small p values are sometimes misunderstood as evidence of effects strong enough to be "worthy of notice," in situations where treatment effects may be too small to be valued, and perhaps too small to even be noticed by everyday observation.

Use of language should be precise. It is unfortunate that the phrase "statistically significant" includes a word (*significant*) that means "noteworthy and important" in everyday use. Authors should try to convey accurate information about effect size in a way that distinguishes between statistical and practical significance. If you describe $p < .001$ as "highly significant," this leads many readers to think that the effect of a treatment or intervention is strong enough to be valuable in the real world and worthy of notice. However, p values depend on N, as well as effect size. A very weak treatment effect can have a very small p value if N is sufficiently large.

Data analysts need to avoid *p*-hacking, "undisclosed flexibility," and lack of transparency in research reports (Simmons et al., 2011). Authors also need to avoid HARKing: hypothesizing after results are known (Kerr, 1998). HARKing occurs when a researcher makes up an explanation for a result that was not expected. For a detailed *p*-hacking checklist (things to avoid) see Wicherts et al. (2016). When *p*-hacking occurs, reported *p* values can greatly understate the true risk for Type I error, and this often leads data analysts and readers to believe that evidence against the null hypothesis is much stronger than it actually is. This in turn leads to overconfidence about findings and perhaps publication of false-positive results.

The most extensive list of recommendations about changes need to improve replicability of research comes from Asendorpf et al. (2013). All of the following are based on their recommendations. The entire following list is an abbreviated summary of their ideas; see their paper for detailed discussion.

1.11.1 Recommendations for Research Design and Data Analysis

- Use larger sample sizes. Other factors being equal, this increases statistical power and leads to narrower CIs.
- Use reliable measures. When measures have low reliability, correlations between quantitative measures are attenuated (i.e., made smaller), and within-group *SS* terms in ANOVA become larger.
- Use suitable methods of statistical analysis.
- Avoid multiple **underpowered** studies. An underpowered study has too few cases to have adequate statistical power to detect the effect size. Consider error introduced by multiple testing in underpowered studies.

  The literature is scattered with inconsistent results because underpowered studies produce different sets of significant (or nonsignificant) relations between variables. Even worse, it is polluted by single studies reporting overestimated effect sizes, a problem aggravated by the **confirmation bias in publication** and a tendency to reframe studies post hoc to feature whatever results came out significant. (Asendorpf et al., 2013)

- Do not evaluate whether results of a replication are consistent with the original study by "vote counting" of NHST results (e.g., did both studies have $p < .05$?). Instead note whether the CIs for the studies overlap substantially and whether the sample mean for the original study falls within the CI for the sample mean in the replication study.

1.11.2 Recommendations for Authors

- Increase transparency of reporting (include complete information about sample size decisions, criteria used for statistical significance, all variables that were measured and all groups included, and all analyses that were conducted). Specify how possible sources of bias such as outliers and missing values were evaluated and remedied. If cases, variables, or groups are dropped from final analysis, explain how many were dropped and why.
- Preregister research plans and predictions. For resources in psychology, see "Preregistration of Research Plans" (n.d.).
- Publish materials, data, and details of analysis (e.g., on a webpage or in a repository; see "Recommended Data Repositories," n.d.).

- Publish working papers and engage in online research discussion forums to promote dialog among researchers working on related topics.
- Conduct replications and make it possible for others to conduct replications.
- Distinguish between exploratory and "confirmatory" analyses.

It is obvious that these are difficult for authors to do, particularly those at early stages in their careers. Publication of large numbers of studies that yield statistically significant results is a de facto requirement for getting hired, promoted, tenured, and grant-funded. Publication pressure can lower research quality (Sarawitz, 2016). Requirements to replicate studies and report more detail about data analysis decisions will make the process of publication far more time consuming. Efforts to adhere to these guidelines will almost certainly lead to publishing fewer papers. This could be good for the research field (Nelson, Simmons, & Simonsohn, 2012). Changes in individual researcher behavior can only occur if researchers are taught better practices and if institutions such as departments, universities, and grant-funding agencies provide incentives that encourage researchers to produce smaller numbers of high-quality studies instead of rewarding publication of large numbers of studies.

1.11.3 Recommendations for Journal Editors and Reviewers

- Promote good research practice by encouraging honest reporting of less-than-perfect results.
- Do not insist on "confirmatory" studies; this discourages honest reporting when analyses are exploratory.
- Publish null findings (those with $p > .05$) to minimize publication bias (provided that the studies are well designed). (Of course, a nil result should not be interpreted as evidence that the null hypothesis of no treatment effect is true. It is just a failure to find evidence that is inconsistent with the null hypothesis.)
- Notice when a research report presents an unlikely outcome and raise questions about it. For example, Asendorpf et al. (2013) noted, "If an article reports 10 successful replications . . . each with a power of .60, the probability that all of the studies could have achieved statistical significance is less than 1%," even if the finding is actually "true."
- Allow reviewers to discuss papers with authors.
- Journals may give badges to papers with evidence of adherence to good practice such as study preregistration. *Psychological Science* does this; other journals are beginning to as well.
- Require authors to make raw data available to reviewers and readers.
- Reserve space for publication of replication studies, including failures to replicate.

1.11.4 Recommendations for Teachers of Research Methods and Statistics

To a great extent, textbooks and instructors teach what researchers are doing, and researchers, reviewers, and journal editors do what they have been taught to do. This discourages change. Incorporating issues such as the limitations of p values, the importance of reporting CIs and effect size, the risk for going astray into p-hacking during lengthy data analysis, and so forth, will help future researchers take these issues into account.

- Students need to understand the limitations of information from statistical significance tests and the problems created by inadequate statistical power, running multiple analyses, and selectively reporting only "significant" outcomes. In other words, they need to learn how to avoid *p*-hacking. Some of these ideas might be introduced in early courses; these topics are essential in intermediate and advanced courses. Many technical books cover these issues, but most textbooks do not.
- Graduate courses should focus more on "getting it right" and less on "getting it published."
- Students need to know about a priori power analysis as a tool for deciding sample size (as opposed to the practice of continuing to collect data until $p < .05$ can be obtained, one of many forms of *p*-hacking). Some undergraduate statistics textbooks include an introduction to statistical power. Earlier chapters in this book provided basic information about power for each bivariate statistic.
- The problems with inflated risk for Type I error that are raised by multiple analyses and multiple experiments should be discussed.
- Transparency in reporting should be encouraged. Students need to work on projects that use real data set with the typical problems faced in actual research (such as missing values and outliers). Students should be required to report details about data screening and any remedies applied to data to minimize sources of artifact such as outliers.
- Students can reanalyze raw data from published studies or conduct replication studies as projects in research methods and statistics courses.
- Instructors should promote critical thinking about research designs and research reports.

1.11.5 Recommendations About Institutional Incentives and Norms

- Departments and universities should focus on quality instead of quantity of publications when making hiring, salary, and promotion decisions.
- Grant agencies should insist on replications.

1.12 SUMMARY

The title of an article in *Slate* describes the current situation: "Science Isn't Broken. It's Just a Hell of a Lot Harder Than We Gave It Credit For" (Aschwanden, 2015). Self-correction and quality control mechanisms for science (including peer review and replication) do not work perfectly, but they can be made to work better. Progress in science requires weeding out false-positive results as well as generating new findings. Unfortunately, while generating new findings is incentivized, weeding out false positives is not. *P*-hacking without active intention to deceive is probably the most common reason for false-positive results.

Attempts to identify false-positive results (whether in one's own work or in the work of others) can be painful. Ideally this will happen in a culture of cooperation and constructive commentary, rather than competition and attack. Public abuse of individual researchers whose work cannot be replicated is not a good way to move forward. All of us have (at least on occasion) complained about nasty reviews. We need to remember, when we become upset about the "them" who wrote those nasty reviews, that "them" is "us," and treat one another kindly. Criticism can be provided in constructive ways.

The stakes are high. Press releases of inconsistent or contradictory results in mass media may reduce public respect for, and trust in, science. This is turn may reduce support for research funding and higher education. If researchers make exaggerated claims on the basis of limited evidence, and claims are frequently contradicted, this provides ammunition for antiscience and anti-intellectual elements in our society.

Change in research practices does not have to be all or nothing. It is easy to report CIs and effect sizes (as suggested by Cumming, 2014, and others). Meta-analyses are becoming more common in many fields. We can make more thoughtful assessments of effect sizes and distinguish between statistical and practical or clinical importance (Kirk, 1996; Thompson, 2002a). The many additional recommendations listed in the preceding section may have to be implemented more gradually, as institutional support for change increases.

COMPREHENSION QUESTIONS

1. If Researcher B tries to replicate a statistically significant finding reported by Researcher A, and Researcher B finds a nonsignificant result, does this prove that Researcher A's finding was incorrect? Why or why not?
2. What needs to be considered when comparing an original study by Researcher A and a replication attempt by Researcher B?
3. Is psychology the only discipline in which failures to replicate studies have been reported? (If not, what other disciplines? Your answer might include examples that go beyond those in this chapter.)
4. What does a *p* value tell you about:
   a. Probability that the results of a study will replicate in the future?
   b. Effect size (magnitude of treatment effect)?
   c. Probability that the null hypothesis is correct?
   d. What does a *p* value tell you?
5. "NHST logic involves a double negative." Explain.
6. What does it mean to say that H_0 is always false?
7. In words, what does Cohen's *d* tell you about the magnitude of differences between two sample means? Does *d* have a restricted range? Can it be negative?
8. How does the value of the *t* ratio depend on the values of *d* and *df*?
9. How does the width of a CI depend on the level of confidence, *N*, and *SD*?
10. Review: What is the difference between SE_M and *SD*? Which will be larger?
11. Consider Equation 1.4. Which term provides information about effect size? Which term provides information about sample size?
12. Describe violations of assumptions or rules that can bias values of *p*. Don't worry whether to call something an assumption versus a rule versus an artifact; these concepts overlap.
13. What are the major alternatives that have been suggested to the use of $\alpha < .05$ (NHST)?
14. What is *p*-hacking? What common researcher practices can be described as *p*-hacking? What effect does *p*-hacking have on the believability of research results?
15. What is HARKing, and how can it be misleading?
16. How could *p*-hacking contribute to the problems that sometimes arise when people try to replicate research studies?
17. Is it correct to say that a study with $p < .001$ shows stronger treatment effects than a study that reports $p < .05$? Why or why not?
18. How does theoretical significance differ from practical or clinical significance? What kinds of information is useful in evaluating practical or clinical significance?
19. When people report CIs instead of *p* values, how might this lead them to think about data differently?
20. Can you tell from a graph or bar chart that shows 95% CIs for the means of two groups whether the *t* test that compares group means using $\alpha = .05$ would be statistically significant? Explain your reasoning.

21. If a computer program or research report does not provide effect size information, is there any way for you to figure it out?

22. Explain the difference between ($M_1 - M_2$) and Cohen's *d*. Which is standardized? What kind of information does each of these potentially provide about effect size?

23. In addition to reporting effect size in research reports, discuss two other uses for effect size.

24. What three questions does a meta-analysis usually set out to answer?

25. Find a forest plot (either using a Google image search or by looking at studies in your research area). Unless you already understand odds ratios, make sure that the outcome variable is quantitative (some forest plots provide information about odds ratios; we have not discussed those yet). To the extent that you can, evaluate the following: Does the plot include all the information you would want to have? What does it tell you about the magnitude of effect in each study? The magnitude of effect averaged across all studies?

26. Describe three changes (in the behavior of individual researchers) that could improve future research quality. Describe two changes (in the behavior of institutions) that could help individuals make these changes. Do any of these changes seem easy to you? Which changes do you think are the most difficult (or unlikely)?

27. Has this chapter changed your understanding or thinking about how you will conduct research and analyze data in future? If so, how?

NOTES

[1]The eminent philosopher of science Karl Popper (cited in Meehl, 1978) argued that to advance science, we need to look for evidence that might disconfirm our preferred hypotheses. NHST is not Popperian falsification. Meehl (1978) pointed out that NHST actually does the opposite. It is a search for evidence to disconfirm the null hypothesis (not evidence to disconfirm the research or alternative hypothesis). When we use NHST (with sufficiently large samples), our preferred alternative hypotheses are not in jeopardy. Meehl argued that NHST is not a good way to advance knowledge in the social and behavioral sciences. It does not pose real challenges to our theories and is not well suited to deal with the sheer complexity of research questions in social and behavioral sciences. We make progress not only by generating new hypotheses and findings but also by discarding incorrect ideas and faulty evidence. Selective reporting of small p values does not help us discard incorrect ideas.

[2]An exception is that if N, the number of data points, becomes very small, the size of a correlation becomes large. If you have only $N = 2$ pairs of X, Y values, a straight line will fit perfectly, and r will equal 1 or −1. For values of N close to 2, values of r will be inflated because of "overfitting."

[3]Odds ratios or relative risk measures, which can be obtained from logistic regression, are also common effect sizes in meta-analyses. See Chapter 16 later in this book.

DIGITAL RESOURCES

Find **free study tools** to support your learning, including **eFlashcards, data sets, and web resources**, on the accompanying website at **edge.sagepub.com/warner3e**.

CHAPTER 2

ADVANCED DATA SCREENING

Outliers and Missing Values

2.1 INTRODUCTION

Extensive data screening should be conducted prior to all analyses. Univariate and bivariate data screening are still necessary (as described in Volume I [Warner, 2020]). This chapter provides further discussion of outliers and procedures for handling missing data. It is important to formulate decision rules for data screening and handling prior to data collection and to document the process thoroughly.

During data screening, a researcher does several things:

- Correct errors.
- "Get to know" the data (for example, identify distribution shapes).
- Assess whether assumptions required for intended analyses are satisfied.
- Correct violations of assumptions, if possible.
- Identify and remedy problems such as outliers, skewness, and missing values.

The following section suggests ways to keep track of the data-screening process for large numbers of variables.

2.2 VARIABLE NAMES AND FILE MANAGEMENT

2.2.1 Case Identification Numbers

If there are no case identification numbers, create them. Often the original case numbers used to identify individuals during data collection in social sciences are removed to ensure confidentiality. Case numbers that correspond to row numbers in the SPSS file can be created using this command: COMPUTE idnumber = $casenum (where $casenum denotes row number in the SPSS file). The variable idnumber can be used to label individual cases in graphs and identify which cases have outliers or missing values.

2.2.2 Codes for Missing Values

Missing values are usually identified by leaving cells in the SPSS data worksheet blank. In recode or compute statements, a blank cell corresponds to the value $sysmis. In some kinds of research, it is useful to document different reasons for missing values (Acock, 2005). For example, a survey response can be missing because a participant refuses to answer or cannot remember the information; a physiological measure may be missed because of equipment malfunction. Different numerical codes can be used for each type of missing value. Be sure to use number codes for missing that cannot occur as valid score values. For example, number of tickets for traffic violations could be coded 888 for "refused to answer" and 999 for "could not remember." Archival data files sometimes use multiple codes for missing. Missing values are specified and labeled in the SPSS Data Editor Variable View tab.

2.2.3 Keeping Track of Files

It is common for data analysts to go through a multiple-step process in data screening; this is particularly likely when longitudinal data are collected. A flowchart may be needed to keep track of scores that are modified and cases that are lost due to attrition. The CONSORT (Consolidated Standards of Reporting Trials) protocol describes a way to do this (Boutron, 2017). Figure 2.1 shows a template for a CONSORT flow diagram.

Figure 2.1 Flowchart: CONSORT Protocol to Track Participant Attrition and Data Handling

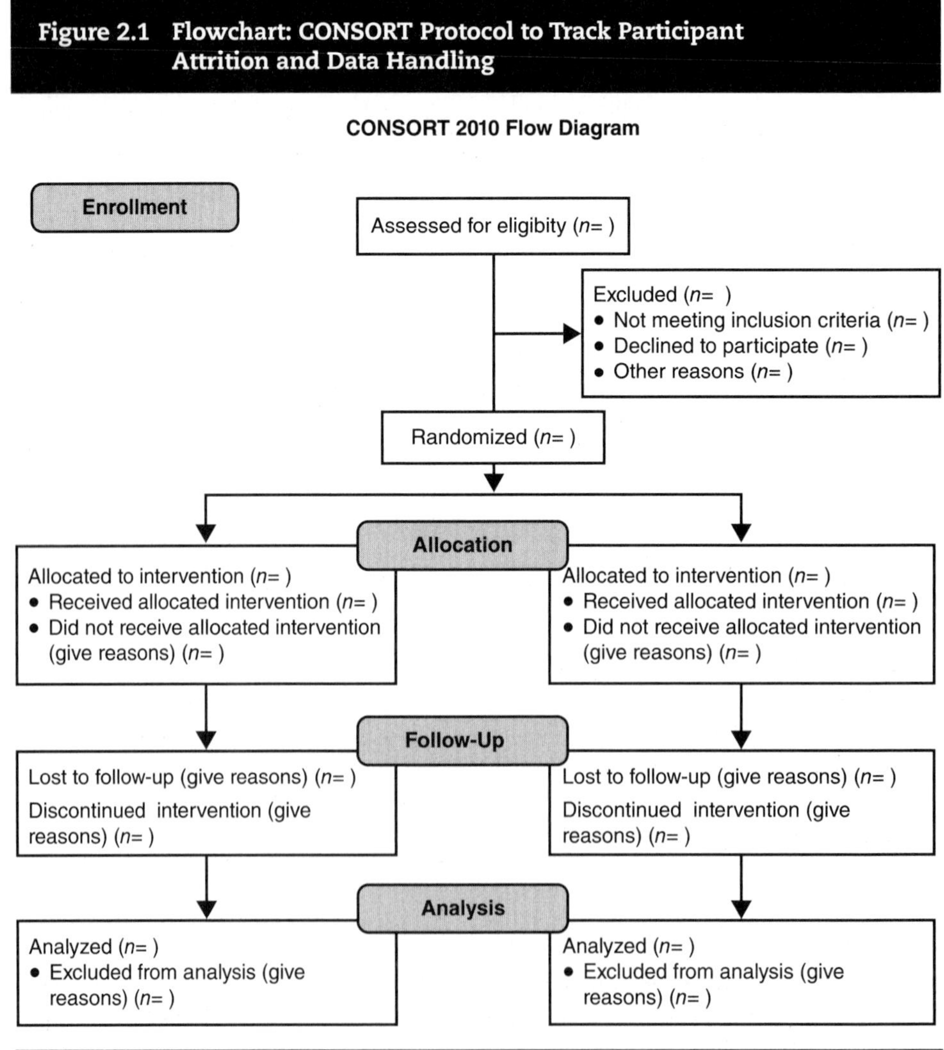

Source: http://www.consort-statement.org/consort-statement/flow-diagram.

It is important to retain the original data file and save modified data files at every step during this process. If you change your mind about some decisions, or discover errors, you may need to go back to earlier versions of files. Keep a detailed log that documents what was done to data at each step. Use of file names that include the date and time of file creation and/or words that remind you what was done at each step can be helpful when you need to locate the most recent version or backtrack to earlier versions. Naming a file "final" is almost never a good idea. (Files are time-stamped by computers, but these time stamps are not always adequate information.)

2.2.4 Use Different Variable Names to Keep Track of Modifications

If a variable will be transformed or recoded before use in later analysis, it is helpful to use an initial variable name indicating that this change has not yet been done. For example, an initial score for reaction time could be named raw_rt. The log-transformed version of the variable could be called log_rt. As another example, some self-report measures include reverse-worded questions. For example, most items in a depression scale might be worded such that higher degree of agreement indicates more depression (e.g., "I feel sad most of the time," rated on a scale from 1 = *strongly disagree* to 5 = *strongly agree*). Some items might be reverse worded (such that a high score indicates less depression; e.g., "Most days I am happy"). Before scores can be summed to create a total depression score, scoring for reverse-worded questions must be changed to make scores consistent (e.g., a score of 5 always indicates higher depression). The initial name for a reverse-worded question could be rev_depression1. The "rev" prefix would indicate that this item was worded in a direction opposite from other items. After recoding to change the direction of scoring, the new variable name could be depression1 (without the "rev" prefix). Then the total scale score could be computed by summing depression1, depression2, and so on. Avoid using the same names for variables before and after transformations or recodes, because this can lead to confusion.

2.2.5 Save SPSS Syntax

The Paste button in SPSS dialog boxes can be used to save SPSS commands generated by your menu selections into a syntax file. Save all SPSS syntax used to recode, transform, compute new variables, or make other modifications during the data preparation process. The syntax file documents what was done, and if errors are discovered, syntax can be edited to make corrections and all analyses can be done again. This can save considerable time.

Data screening is needed so that when the analyses of primary interest are conducted, the best possible information is available. If problems such as outliers and missing values are not corrected during data screening, the results of final analyses are likely to be biased.

2.3 SOURCES OF BIAS

Bias can be defined as over- or underestimation of statistics such as values of M, t ratios, and p values. Bias means that the sample statistic over- or underestimates the corresponding population parameters (e.g., M is systematically larger or smaller than μ). Bias can occur when assumptions for analyses are violated and when outliers or missing data are present.

Most of the statistics in this book (except for logistic regression) are special cases of the general linear model (GLM). Most GLM analyses were developed on the basis of the following assumptions. Some assumptions are explicit (i.e., assumed in derivations of statistics). There are also implicit assumptions and rules for the use of significance tests in practice (e.g., don't run hundreds of tests and report only those with $p < .05$; selected p values will greatly underestimate the risk for Type I decision error). Problems such as outliers and missing data often arise in real-world data. The actual practice of statistics is much messier than the ideal world imagined by mathematical statisticians. Here is a list of concerns that should be addressed in data screening. Some of these things are relatively easy to identify and correct, while others are more difficult.

- **Scores within samples must be independent of one another.** Whether this assumption is satisfied depends primarily on how data were collected (Volume I [Warner, 2020], Chapter 2). Scores in samples are not independent if participants can influence one another's behavior through processes such as persuasion, cooperation, imitation, or competition. See Kenny and Judd (1986, 1996) for discussion. When this assumption is violated, estimates of SD or SS_{within} are often too small; that makes estimates of t or F too large and results in inflated risk for Type I error. Violations of this assumption are a serious problem.

- **All relationships among variables are linear.** This is an extremely important assumption that we can check in samples by visual examination of scatterplots and by tests of nonlinearity. Nonlinear terms (such as X^2, in addition to X as a predictor of Y) can be added to linear regression models, but sometimes nonlinearity points to the need for other courses of action, such as nonlinear data transformation or analyses outside the GLM family.

- **Missing values** can lead to bias in the composition of samples and corresponding bias in estimation of statistics. Often, cases with missing values differ in some way from the cases with complete data. Suppose men are more likely not to answer a question about depression than women, or that students with low grades are more likely to skip questions about academic performance. If these cases are dropped, the sample becomes biased (the sample will underrepresent men and/or low-performing students). Later sections in this chapter discuss methods for evaluation of amount and pattern of missing values and replacement of missing values with estimated or imputed scores. Whether problems with missing values can be remedied depends on the reasons for missingness, as discussed in that section.

- **Residuals or prediction errors are independent of one another, are normally distributed, and have mean of 0 and equal variance for all values of predictor variables.** For regression analysis and related techniques such as time-series analysis, these assumptions can be evaluated using plots and descriptive statistics for residuals. Data analysts should beware the temptation to drop cases just because they have large residuals (Tabachnick & Fidell, 2018). This can amount to trimming the data to fit the model. Users of regression are more likely to focus on residuals than users of analysis of variance (ANOVA).

- **Some sample distribution shapes make *M* a poor description of central tendency.** For example, a bimodal distribution of ratings on a 1-to-7 scale, with a mode at the lowest and highest scores (as we would see for highly polarized ratings), is not well described by a sample mean (see Volume I [Warner, 2020], Chapter 5). We need to do something else with these data. If sample size is large enough, we may be able to treat each X score (e.g., $X = 1, X = 2, \ldots, X = 7$) as a separate group. With large samples (on the order of thousands) it may be better to treat some quantitative variables as categorical.

- **Some distribution types require different kinds of analysis.** For example, when a Y dependent variable is a count of behaviors such as occasions of drug use, the histogram for the distribution of Y may have a mode at 0. Analyses outside the GLM family, such as zero-inflated negative binomial regression, may be needed for this kind of dependent variable (see Appendix 2A). The remedy for this kind of problem is to choose an appropriate analysis.

- **Skewness of sample distribution shape.** Skewness can be evaluated by visual examination of histograms. SPSS provides a skewness index and its standard error; statistical significance of skewness can be assessed by examining $z = \text{skewness}/SE_{\text{skewness}}$, using the standard normal distribution to evaluate z. However, visual examination is often adequate and may provide insight into reasons for skewness that the skewness index by itself cannot provide. Skewness is not always a major problem. Sometimes sample skewness can be eliminated or reduced by removal or modification of outliers. If skewness is severe and not due to just a few outliers, transformations such as log may be useful ways to reduce skewness (discussed in a later section).

- Derivations of many statistical significance tests assume that **scores in samples are randomly selected from normally distributed populations.** This raises two issues. On one hand, some data analysts worry about the normality of their population distributions. I worry more about the use of convenience samples that were not selected from any well-defined population. The use of convenience samples can limit generalizability of results. On the other hand, Monte Carlo simulations that evaluate violations of this normally distributed population assumption for artificially generated populations of data and simple analyses such as the independent-samples *t* test often find that violations of this assumption do not seriously bias *p* values, provided that samples are not too small (Sawilowsky & Blair, 1992). There are significance tests, such as Levene *F*, to test differences between sample variances for *t* and *F* tests. However, tests that are adjusted to correct for violations of this assumption, such as the "equal variances not assumed" or Welch's *t*, are generally thought to be overly conservative. The issue here is that we often don't know anything about population distribution shape. For some simple analyses, such as independent-samples *t* and between-*S* ANOVA, violations of assumption of normal distribution in the population may not cause serious problems.

- **For more advanced statistical methods, violations of normality assumptions may be much more serious.** These problems can be avoided through the use of robust estimation methods that do not require normality assumptions (Field & Wilcox, 2017; Maronna, Martin, Yohai, & Salibián-Barrera, 2019).

- **Violation of assumption that all variables are measured without error** (that all measures are perfectly reliable and perfectly valid). This is almost never true in real data. Advanced techniques such as structural equation modeling include measurement models that take measurement error into account (to some extent).

- **Model must be properly specified.** A properly specified model in regression includes all the predictors of *Y* that should be included, includes terms such as interactions if these are needed, and does not include "garbage" variables that should not be statistically controlled. We can never be sure that we have a correctly specified model. Kenny (1979) noted that when we add or drop variables from a regression, we can have "bouncing betas" (regression slope estimates can change dramatically). The value of each beta coefficient depends on context (i.e., which other variables are included in the model). Significance tests for *b* coefficients vary depending on the set of variables that are controlled when assessing each predictor. Another way to say this is that we cannot obtain unbiased estimates of effects unless we control for the "right" set of variables. Decisions about which variables to control are limited by the variables that are available in the data set. Unfortunately, it is common practice for data analysts to add and/or drop control variables until they find that the predictor variable of interest becomes statistically significant.

Some of these assumptions (such as normally distributed scores in the population from which the sample was selected) cannot be checked. Some potential problems can be evaluated through screening of sample data.

2.4 SCREENING SAMPLE DATA

From a practical and applied perspective, what are the most important things to check for during preliminary data screening? First, remember that rules for identification and handling of problems with data, such as outliers, skewness, and missing values, should be established before you collect data. If you experiment with different rules for outlier detection and handling, run numerous analyses, and report selected results, the risk for committing Type I decision error increases, often substantially. Doing whatever it takes to obtain statistically significant values of *p* is called *p*-hacking (Wicherts et al., 2016), and this can lead to misleading results (Simmons, Nelson, & Simonsohn, 2011). Committing to decisions about data handling prior to data collection can reduce the temptation to engage in *p*-hacking.

2.4.1 Data Screening Needed in All Situations

- Individual scores should always be evaluated to make sure that all score values are plausible and accurate and that the ranges of scores in the sample (for important variables) corresponds at least approximately to the ranges of scores in the hypothetical population of interest. If a study includes persons with depression scores that range only from 0 to 10, we cannot generalize or extrapolate findings to persons with depression scores above 10. A frequency table provides information about range of scores.
- Missing values. Begin by evaluating how many missing data there are; frequency tables tell us how many missing values there are for each variable. If there are very few missing values (e.g., less than 5% of observations), missing data may not be a great concern, and it may be acceptable to let SPSS use default methods such as listwise deletion for handling missing data. If there are larger amounts of missing data, this raises concerns whether data are missing systematically. A later section in this chapter discusses the missing data problem further.
- Evaluate distribution skewness. When a distribution is asymmetrical, it often has a longer and thinner tail at one end than the other. It is possible that an appearance of skewness arises because of a few outliers. If this is the case, I recommend that you handle this as an outlier problem. A later section discusses possible ways to handle skewness if it is not due to just a few outliers. Some variables (such as income) predictably have very strong positive skewness. Sometimes nonlinear data transformations are used to reduce skewness.

2.4.2 Data Screening for Comparison of Group Means

- Make sure all groups have adequate n's. If we have at least $n = 30$ cases per group, and use two-tailed tests, violations of the population normality assumption and of assumptions about equal population variances do not seriously bias p value estimates (Sawilowsky & Blair, 1992). Some authorities suggest that even smaller values of n may be adequate. I believe that below some point (perhaps n of 20 per group), there is just not sufficient information to describe groups or to evaluate whether the group is similar to populations of interest. However, this is not an ironclad rule. In some kinds of research (such as neuroscience), it is reasonable for researchers to assume little variation among cases with respect to important characteristics such as brain structure and function, and recruiting and paying for cases can be expensive because of time-consuming procedures. For example, in behavioral neuroscience animal research, each case may require extensive training, then surgery, then extensive testing or evaluation or costly laboratory analysis of specimen materials. Procedures such as magnetic resonance imaging are very costly. Sometimes smaller *n*'s are all we can get.

- Check for outliers within groups. Outliers within groups affect estimates of both *M* and *SD*, and these in turn will affect estimates of *t* and *p*. The effect of outliers may be either to inflate or deflate the *t* ratio. Boxplots are a common way to identify outliers within groups.

- Examine distribution shapes in groups to evaluate whether *M* is a reasonable description of central tendency. Some distribution shapes (such as a bimodal distribution with modes at the extreme high and low ends of the distribution and distributions with large modes at 0) can make *M* a poor way to describe central tendency (Volume I [Warner, 2020], Chapter 5). If these distribution shapes are seen in sample data, the data analyst should consider whether comparison of means is a good way to evaluate outcomes.

2.4.3 Data Screening for Correlation and Regression

- Check that relations between all variables are linear if you plan to use linear correlation and linear regression methods. In addition, predictor variables should not be too highly correlated with one another. Visual examination of a scatterplot may be sufficient; regression can also be used to evaluate nonlinearity (discussed in Section 2.8).

- Check for outliers. Bivariate and multivariate outliers can inflate or deflate correlations among variables. Bivariate outliers can be detected by visual examination of an *X*, *Y* scatterplot. For more than two variables, you need to look for multivariate outliers (described later in this chapter).

- Evaluate whether *X* and *Y* have similar distribution shapes. It may be more important that *X* and *Y* have similar distribution shapes than that their sample distribution shapes are normal. When distribution shapes differ, the maximum obtainable value for *r* will have a limited range (not the full range from −1 to +1). This in turn will influence estimates for analyses that use *r* as a building block. Visual examination of histograms may be sufficient. See Appendix 10D in Volume I (Warner, 2020).

- Evaluate plots of residuals from regression to verify that they are (a) normally distributed and (b) not related to values of *Y* or *Y*′ (these are assumptions for regression). If you have only one predictor, screening raw scores on variables may lead to the same conclusions as screening residuals. Tabachnick and Fidell (2018) pointed out that when a researcher runs the final analysis of primary interest and then examines residuals, it can be tempting to remove or modify cases specifically because they cause poor fit in the final analysis. In other words, the data analyst may be tempted to trim the data (post hoc) to fit the model.

- Sample distributions that differ drastically from normal may alert you to the need for different kinds of analyses outside the GLM family (an example is provided in Appendix 2A).

2.5 POSSIBLE REMEDY FOR SKEWNESS: NONLINEAR DATA TRANSFORMATIONS

Nonlinear transformations of X (such as $1/X$, X^2, X^c for any value of c, base 10 or natural log of X, arcsine of X, and others) can change the shape of distributions (Tabachnick & Fidell, 2018). Although log transformations can potentially reduce positive or negative skewness in an otherwise normal distribution, they are not always appropriate or effective. In many situations, if distribution shape can be made reasonably normal by modifying or removing outliers, it may be preferable to do that. Log transformations make sense when at least one of the following conditions are met:

- The underlying distribution is exponential.
- It is conventional to use log transformations with this variable; readers and reviewers are familiar with it.
- Scores on the variable differ across orders of magnitude. Scores differ across orders of magnitude when the highest value is vastly larger than the smallest value. Consider the following example: The weight of an elephant can be tens of thousands of times greater than the weight of a mouse. Typical values for body weight for different species, given in kilograms, appear on the X axis in Figure 2.2.

Because of outliers (weight and metabolic rates for elephants), scores for body weight (and metabolism) of smaller species are crowded together in the lower left-hand corner of the graph, making it difficult to distinguish differences among most species. When the base 10 log is taken for both variables, as shown in Figure 2.3, scores for species are spread out more evenly on the X and Y axes. Differences among them are now represented in log units (orders of magnitude). In addition, the relation between log of weight and log of metabolic rate becomes linear (of course, this will not happen for all log-transformed variables).

Figure 2.2 Scatterplot of Metabolic Rate by Body Weight (Raw Scores)

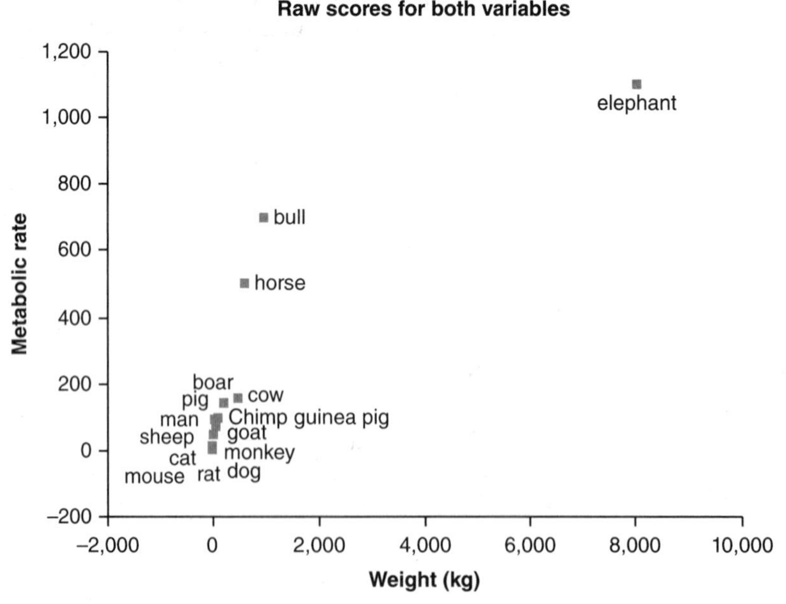

Source: Reprinted with permission from Dr. Tatsuo Motokawa.

Figure 2.3 Scatterplot of Log Metabolic Rate by Log Body Weight

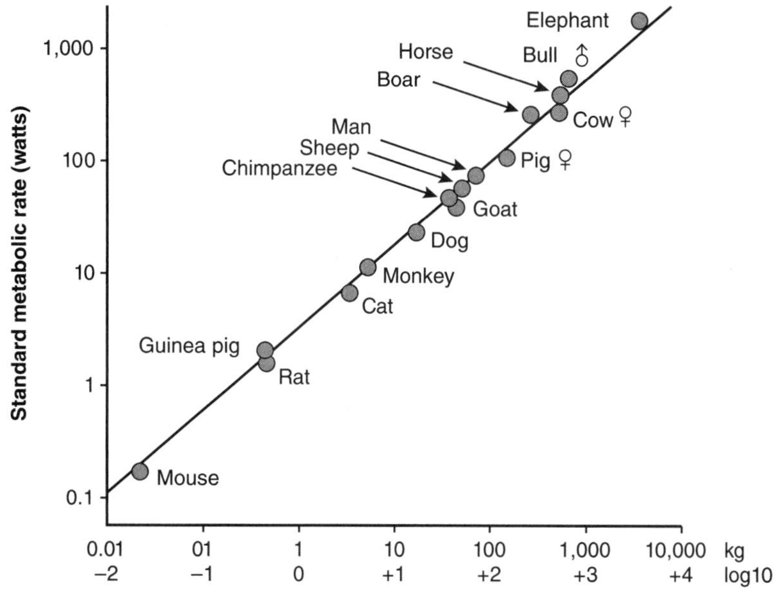

Source: Reprinted with permission from Dr. Tatsuo Motokawa.

Other transformations commonly used in some areas of psychology involve power functions, that is, replacing X with X^2, or X^c (where c is some power of X; the exponent c is not necessarily an integer value). Power transformations are used in psychophysical studies (e.g., to examine how perceived heaviness of objects is related to physical mass).

When individual scores for cases are proportions, percentages, or correlations, other nonlinear transformations may be needed. Data transformations such as arcsine (for proportions) or Fisher r to Z are used to correct problems with the shapes of sampling distributions that arise when the range of possible score values has fixed end points (–1 to +1 for correlation, 0 to 1.00 for proportion).

If you use nonlinear transformations to reduce skewness, examine a histogram for the transformed scores to see whether the transformation had the desired effect. In my experience, distributions of log-transformed scores often do not look any better than the raw scores. When X does not have a very wide range, the correlation of X with X^2, or X with log X, is often very close to 1. In these situations, the transformation does not have much effect on distribution shape.

2.6 IDENTIFICATION OF OUTLIERS

2.6.1 Univariate Outliers

Outliers can be a problem because many widely used statistics, such as the sample mean M, are not robust against the effect of outliers. In turn, other statistics that use M in computations (such as SD and r and t) can also be influenced by outliers. Outliers can bias estimates of parameters, effect sizes, standard errors, confidence intervals, and test statistics such as t and F ratios and their corresponding p values (Field, 2018).

It is often possible to anticipate which variables are likely to have outliers. If scores are ratings on 1-to-5 or 1-to-7 scales, extreme outliers cannot occur. However, many variables (such as income) have no fixed upper limit; in these situations, outliers are common. When you know ahead of time that some of your variables are likely to generate outliers, it's important to make decisions ahead of time. What rules will you use to identify scores as outliers, and what methods will you use to handle outliers? Outliers are sometimes obtained because of equipment malfunction or other forms of measurement error.

If groups will be compared, outlier evaluation should be done separately within each group (e.g., a separate boxplot within each group).

To review briefly:

- In boxplots, scores that lie outside the "whiskers" can be considered potential outliers (an open circle represents an outlier; an asterisk represents an extreme outlier). Boxplots are particularly appropriate for non-normally distributed data.
- Scores can be identified as outliers if they have z values greater than 3.29 in absolute value for the distribution within each group (Tabachnick & Fidell, 2018).

These are arbitrary rules; they are suggested here because they make sense in a wide range of situations. Aguinas, Gottfredson, and Joo (2013) provided numerous other possible suggestions for outlier identification.

2.6.2 Bivariate and Multivariate Outliers

Bivariate outliers affect estimates of correlations and regression slopes. In bivariate scatterplots it is easy to see whether an individual data point is far away from the cloud that contains most other data points. This distance can be quantified by computing a Mahalanobis distance. Mahalanobis distance can be generalized to situations with larger numbers of variables. A score with a large Mahalanobis distance corresponds to a point that is outside the cloud that contains most of the other data points, as shown in the three-dimensional plot for three variables in Figure 2.4. The most extreme multivariate outlier is shown as a filled circle near the top.

Figure 2.4 Multivariate Outlier for Combination of Three Variables

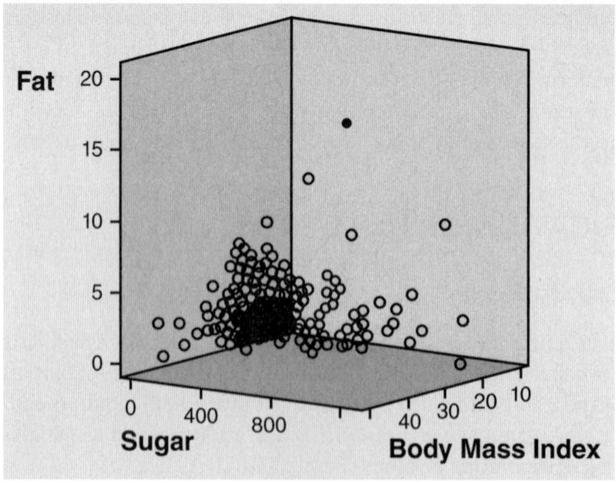

Source: Data selected and extensively modified from Warner, Frye, Morrell, and Carey (2017).

Note: Fat is the number of fat servings per day, and sugar is the number of sugar calories per day.

Figure 2.5 Linear Regression Dialog Boxes

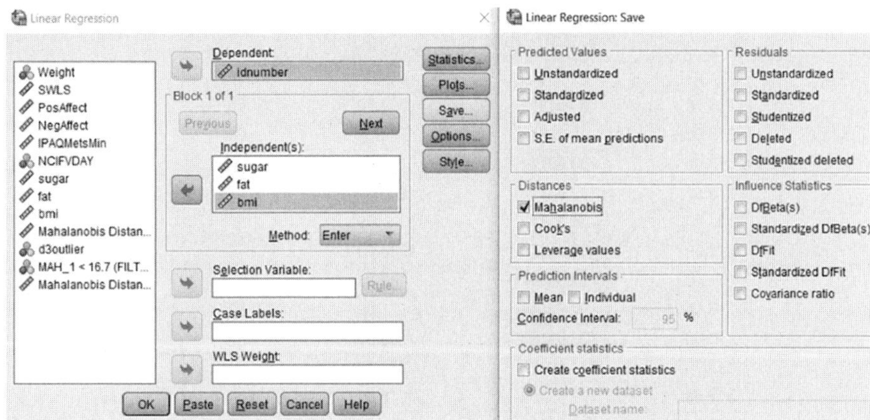

Mahalanobis distance can be obtained as a diagnostic when running analyses such as multiple regression and discriminant analysis. Tabachnick and Fidell (2018) suggested a method to obtain Mahalanobis distance for a set of variables without "previewing" the final regression analysis of interest. Their suggested method avoids the temptation to remove outliers that reduce goodness of fit for the final model. They suggested using the case identification number as the dependent variable in a linear regression and using the entire set of variables to be examined for multivariate outliers as predictors. (This works because multivariate outliers among predictors are unaffected by subject identification number; Tabachnick & Fidell, 2018).

Data in the file outlierfvi.sav are used to demonstrate how to obtain and interpret Mahalanobis distance for a set of hypothetical data. The initial menu selections are <Analyze> → <Regression> → <Linear>. This opens the Linear regression dialog box on the left-hand side of Figure 2.5. Idnumber is entered as the dependent variable. To examine whether there are multivariate outliers in a set of three variables, all three variables are entered as predictor variables in the Linear Regression dialog box. Click the Save button. This opens the Linear Regression: Save dialog box that appears on the right-hand side of Figure 2.5. Check the box for "Mahalanobis." Click Continue, then OK.

After the regression has been run, SPSS Data View (Figure 2.6) has a new variable named MAH_1. (The tag "_1" at the end of the variable name indicates that this is from the first regression analysis that was run.) This is the Mahalanobis distance score for each individual participant; it tells you the degree to which that person's combination of scores on fat, sugar, and body mass index (BMI) was a multivariate outlier, relative to the cloud that these scores occupied in three-dimensional space (shown previously in Figure 2.4). The file was sorted in descending order by values of MAH_1; the part of the file that appears in Figure 2.6 shows a subset of persons whose scores could be identified as multivariate outliers, because they had large values of Mahalanobis distance (many other cases not shown in Figure 2.6 had smaller values of Mahalanobis distance).

Mahalanobis distance has a χ^2 distribution with *df* equal to the number of predictor variables (Tabachnick & Fidell, 2018). The largest value was MAH_1 = 77.76 (for idnumber = 421). The critical value of chi squared with 3 *df*, using $\alpha = .001$, is 16.27. Using that value of χ^2 as a criterion, MAH_1 would be judged statistically significant for all cases listed in Figure 2.6. If the decision to use Mahalanobis distance as a criterion for the identification of outliers was made prior to data screening, scores for all three variables for the cases with significant values of Mahalanobis distance could be converted to missing values. If this results in fewer than 5% missing values, this small amount of missing data may not bias results. If more than 5% of cases have missing values, some form of imputation (described elsewhere in the chapter) could be used to replace the missing values with reasonable estimates.

Figure 2.6 SPSS Data View With Saved Mahalanobis Distance

idnumber	MAH_1	sugar	fat	bmi
421.00	77.76280	624.00	16.00	24.13
258.00	67.01817	338.00	11.50	21.25
213.00	40.08815	1248.00	.00	22.50
153.00	39.69702	.00	.00	43.16
278.00	39.51425	1040.00	9.00	17.16
134.00	34.02619	.00	4.50	18.47
210.00	33.44055	1248.00	3.00	21.92
151.00	33.32354	.00	.00	44.16
173.00	32.72449	.00	2.50	45.10
97.00	32.03592	966.00	4.50	31.20
147.00	23.09725	873.00	4.00	18.67
317.00	22.35106	1040.00	2.00	23.49
357.00	22.29595	104.00	8.00	21.14
125.00	22.11335	.00	3.50	23.12
425.00	21.08898	884.00	2.00	25.10
152.00	20.55901	244.00	3.50	39.06
112.00	20.04464	.00	.00	23.75
115.00	20.02894	602.00	8.00	21.77
80.00	19.83612	.00	2.00	39.53
145.00	18.91604	104.00	1.50	39.58
12.00	18.72081	936.00	1.00	21.63

Examination of scores for sugar and fat consumption and BMI for the case on the first row in Figure 2.6 indicates that this person had a BMI within normal range (the normal range for BMI is generally defined as 18.5 to 24.9 kg/m^2), even though this person reported consuming 16 servings of fat per day. (The value of 16 servings of fat per day was a univariate outlier.) Although this might be physically possible, this seems unlikely. In actual data screening, 16 servings of fat would have been tagged as a univariate outlier and modified at an earlier stage in data screening.

Several additional cases had statistically significant values for Mahalanobis distance. When there are numerous multivariate outliers, Tabachnick and Fidell (2018) suggested additional examination of this group of cases to see what might distinguish them from nonoutlier cases.

2.7 HANDLING OUTLIERS

2.7.1 Use Different Analyses: Nonparametric or Robust Methods

The most widely used parametric statistics (those covered in Volume I [Warner, 2020], and the present volume) that are part of the GLM are generally not robust against the effect of outliers. One way to handle outliers is to use different analyses. Many nonparametric statistics convert scores to ranks as part of computation; this gets rid of outliers. However, it would be incorrect to assume that use of nonparametric statistics makes everything simple. Statistics such as the Wilcoxon rank sum test do not require scores to be normally distributed, but they assume that the distribution shape is the same across groups, and in practice, data often violate that assumption.

Robust statistical techniques, often implemented using R (Field & Wilcox, 2017; Maronna et al., 2019) do not require the assumptions made for GLM. Robust methods are beyond the scope of this volume. They will likely become more widely used in the future.

2.7.2 Handling Univariate Outliers

Suppose that you have identified scores in your data file as univariate outliers (because they were tagged in a boxplot, because they had $z > 3.29$ in absolute value, or on the basis of other rules). Rules for identification and handling of outliers should be decided before data collection, if possible.

Here are the four most obvious choices for outlier handling; there are many other ways (Aguinas et al., 2013).

- Do nothing. Run the analysis with the outliers included.
- Discard all outliers. Removal of extreme values is often called truncation or trimming.
- Replace all outliers with the next largest score value that is not considered an outlier. The information in boxplots can be used to identify outliers and find the next largest score value that is not an outlier. This is called Winsorizing.
- Run the analysis with the outliers included, and also with the outliers excluded, and report both analyses. (Do not just report the version of the analysis that you liked better.)

No matter which of these guidelines you choose, you must document how many outliers were identified, using what rule, and what was done with these outliers. Try to avoid using different rules for different variables or cases. If you have a different story about each data point you remove, it will sound like *p*-hacking, and in fact, it will probably be *p*-hacking. (That said, there may be precedent or specific reasons for outlier handling that apply to some variables and not others.) Do not experiment with different choices for outlier elimination and modification and then report the version of analysis you like best. That is *p*-hacking; the reported *p* value will greatly underestimate the true risk for Type I decision error.

2.7.3 Handling Bivariate and Multivariate Outliers

Consider bivariate outliers first. If you have scores for height in inches (X) and body weight in pounds (Y), and one case has $X = 73$ and $Y = 110$, the univariate scores are not extreme. The combination, however, would be very unusual. Winsorizing might not get rid of the problem, but you could do other things (exclude the case, or run the analysis both with and without this case), as long as you can justify your choice on the basis of plans you made prior to data collection.

For multivariate outliers, it may be possible to identify which one or two variables make the case an outlier. In the previous example of a multivariate outlier, the extremely high value of fat (in row 1 of the data file that appears in Figure 2.6) seemed inconsistent with the normal BMI score. A decision might be made to replace the high fat score with a lower valid score value that is not an outlier. However, detailed evaluation of multivariate outliers to assess whether one or two variables are responsible may be too time consuming to be practical.

Some multivariate outliers may disappear when univariate outliers have been modified. However, multivariate outliers can arise even when none of the individual variables is a univariate outlier.

2.8 TESTING LINEARITY ASSUMPTION

If an association between variables is not linear, it can be described as nonlinear, curvilinear, or perhaps a polynomial trend. Visual examination of bivariate scatterplots may be sufficient to evaluate possible nonlinearity. It is possible to test whether departure from linearity is statistically significant using regression analysis to predict Y from X^2 and perhaps even X^3 (in addition to X). If adding X^2 to a regression equation that includes only X as a predictor leads to a significant increase in R^2, then the association can be called significantly nonlinear. The actual increase in R^2 would tell you whether nonlinearity predicts a trivial or large part of the variance in Y. For discussion of regression with two predictors, see Chapter 4.

If Y is a function of:

- Only X, then the X, Y function is a straight line; this represents a linear trend.
- X and X^2, then the X, Y function has one curve; this is a quadratic trend. It may resemble a U or inverted U shape.
- X, X^2, and X^3, then the function has two curves; this is a cubic trend.

Note that the number of curves in the X, Y function equals the highest power of X minus 1.

The bivariate regression model for a simple linear relationship is $Y' = b_0 + b \times X$. This can be expanded to include a quadratic term: $Y' = b_0 + b_1X + b_2X^2$. If the b_2 coefficient associated with the X^2 predictor variable is statistically significant, this indicates a significant departure from linearity. SPSS transform and compute commands are used to compute a new variable, Xsquared, that corresponds to X^2. (Similarly, we could compute $X^3 = X \times X \times X$; however, trends that are higher order than quadratic are not common in psychological data.)

The hypothetical data that correspond to the graphs in Figure 2.7 are in a file named linearitytest.sav (with $N = 13$ cases). Visual examination of the scatterplots in Figure 2.7 suggests a linear association of Y with X (left) and a quadratic association of Y with Q (right).

Let's first ask whether Y has a significantly nonlinear association with X for the scatterplot on the left of Figure 2.7. To do this, first compute the squared version of X. This can be done as follows using an SPSS compute statement. (If you are not familiar with compute statements, see Volume I [Warner, 2020], or an SPSS guide, or perform a Google search for this topic.)

```
COMPUTE Xsquared = X * X
```

(A better way to compute X^2 is $(X - M_X) \times (X - M_X)$, where M_X is the mean of X.[1])

Then run SPSS linear regression using X and the new variable that corresponds to the squared value of X (named Xsquared) as predictors.

```
REGRESSION
/MISSING LISTWISE
```

Figure 2.7 Linear Versus Quadratic Trend Data

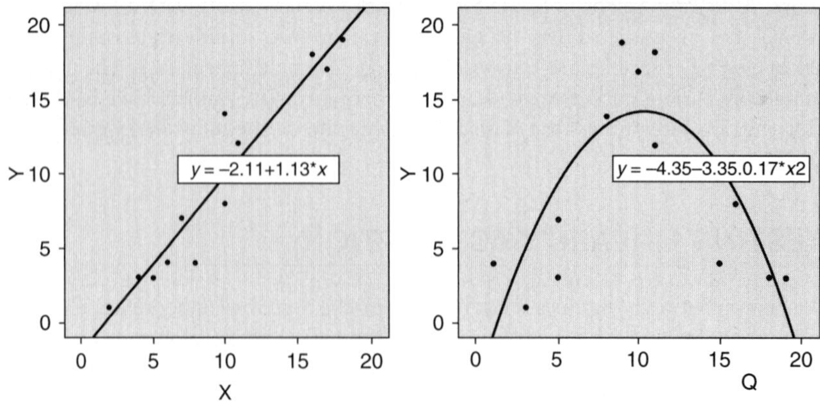

```
/STATISTICS COEFF OUTS R ANOVA
/CRITERIA=PIN(.05) POUT(.10)
/NOORIGIN
/DEPENDENT Y
/METHOD=ENTER X Xsquared
```

Partial results for this regression appear in Figure 2.8.

The X^2 term represents quadratic trend. If the b coefficient for this variable is statistically significant, the assumption of linearity is violated. In this situation, for X^2, $b = -.001$, $\beta = -.025$, $t(10) = -.048$, $p = .963$, two tailed (the df error term for the t test appears in a part of regression output that is not included here). The assumption of linearity is not significantly violated for X as a predictor of Y.

The same procedure can be used to ask whether there is a violation of the linearity assumption when Q is used to predict Y. First, compute a new variable named Qsquared; then run the regression using Q and Qsquared as predictors.

For Q^2, $b = -.173$, $\beta = -3.234$, $t(10) = -4.20$, and $p = .002$, two tailed (values from Figure 2.9). (While β coefficients usually range between -1 and $+1$, they can be far outside that range when squared terms or products between variables are used as predictors.) The linearity assumption was significantly violated for Q as a predictor of Y.

What can be done when the linearity assumption is violated? Sometimes a data transformation such as log will make the relation between a pair of variables more nearly linear; possibly the log of Q would have a linear association with the log of Y. However, this works in only a few situations. Another option is to incorporate the identified nonlinearity into later analyses, for example, include X^2 as a predictor in later regression analyses, so that the nonlinearity detected during data screening is taken into account.

Figure 2.8 Regression Coefficients for Quadratic Regression (Prediction of Y Scores From Scores on X and X^2)

Coefficients[a]

Model		Unstandardized Coefficients		Standardized Coefficients		
		B	Std. Error	Beta	t	Sig.
1	(Constant)	-2.200	2.276		-.967	.357
	X	1.214	.522	.980	2.325	.042
	Xsquared	-.001	.025	-.020	-.048	.963

a. Dependent Variable: Y

Figure 2.9 Regression Coefficients for Prediction of Y From Q and Q^2

Coefficients[a]

Model		Unstandardized Coefficients		Standardized Coefficients		
		B	Std. Error	Beta	t	Sig.
1	(Constant)	-4.350	3.970		-1.096	.299
	Q	3.552	.873	3.134	4.070	.002
	Qsquared	-.173	.041	-3.234	-4.199	.002

a. Dependent Variable: Y

2.9 EVALUATION OF OTHER ASSUMPTIONS SPECIFIC TO ANALYSES

Many analyses require additional evaluation of assumptions in addition to these preliminary assessments. In the past, you have seen that tests of homogeneity of variance were applied for independent-samples *t* tests and between-*S* ANOVA. Often, as pointed out by Field (2018), assumptions for different analyses are quite similar. For example, homogeneity of variance assumptions can be evaluated for the independent-samples *t* test, ANOVA, and regression. For advanced analyses such as multivariate analysis of variance, additional assumptions need to be evaluated. Additional screening requirements for new analyses are discussed when these analyses are introduced.

2.10 DESCRIBING AMOUNT OF MISSING DATA

2.10.1 Why Missing Values Create Problems

There are several reasons why missing data are problematic. Obviously, if your sample is small, missing values make the amount of information even smaller. There is a more subtle problem. Often, missing responses don't occur randomly. For example, people who are overweight may be more likely to skip questions about body weight. SPSS listwise deletion, the default method of handling missing data, just throws out the persons who did not answer this question. If you focus just on the subset of people who did answer a question, you may be looking at a different kind of sample (probably biased) than the original set of people recruited for the study.

Blank cells are often used to represent missing responses in SPSS data files. (Some archival data files use specific numerical values such as 99 or 77 to represent missing responses.)

SPSS does not treat these blanks as 0 when computing statistics such as means; it omits the cases with missing scores from computation. For many procedures there are two SPSS methods for handling missing values. Consider this situation: A researcher asks for correlations among all variables in this list: $X_1, X_2, \ldots, X_k$. If **listwise deletion** is chosen, then only the cases with valid scores for all of the *X* variables on the list are used when these correlations are calculated. If **pairwise deletion** is chosen, then each correlation (e.g., r_{12}, r_{13}, r_{23}) is computed using all the persons who have valid scores for that pair of variables. When listwise deletion is used, all correlations are based on the same *N* of cases. When pairwise deletion is used, if there are missing values, the *N*'s for different correlations will vary, and some of the *N*'s may be larger than the *N* reported using listwise deletion.

If the amount of missing data is less than 5%, use of listwise deletion may not cause serious problems (Graham, 2009). When the amount of missing data is larger, listwise deletion can yield a biased sample. For example, if students with low grades are dropped from a sample used in the analysis because they refused to answer some questions about grades, the remaining sample will mostly include students with higher grades. The sample will be biased and will not represent responses from students with lower grades.

I'll add another caution here. If you pay no attention to missing values, and you do a series of analyses with different variables, the total *N* will vary. For example, in your table of descriptive statistics, you may have 100 cases when you report *M* and *SD* for many variables. In a subsequent regression analysis, you may have only 85 cases. In an ANOVA, you might have only 50 cases. Readers are likely to wonder why *N* keeps changing. In addition, results can't be compared across these analyses because they are not based on data for the same set of cases. It is better to deal with the problem of missing values at the beginning and then work with the same set of cases in all subsequent analyses.

Missing value analysis involves two steps. First, we need to evaluate the amount and pattern of missing data. Then, missing values may be replaced with plausible scores prior to other analyses.

To illustrate procedures used with missing data, I used a subset of data obtained in a study by Warner and Vroman (2011). A subset of 240 cases and six variables with complete data

Figure 2.10 Number of Missing Values for Each of Six Variables (in Data File missingwb.sav)

Statistics

		Depression	Satisfactionw Life	NegativeAffect	Neuroticism	Sex	Socialdesirab ility
N	Valid	218	150	226	220	240	240
	Missing	22	90	14	20	0	0

was selected and saved in a file named nonmissingwb.sav. To create a corresponding file with specific patterns of missingness, I changed selected scores in this file to system missing and saved these data in the file named missingwb.sav.

2.10.2 Assessing Amount of Missingness Using SPSS Base

Initial assessments of amount of missing data do not require the SPSS Missing Values add-on module. Amount of missing data can be summarized three ways: for each variable, for the entire data set, and for each case or participant. To make an initial assessment, the SPSS frequencies was used; results appear in Figure 2.10.

For each variable: Four variables had some missing values; two variables did not have missing data (in other words, 4/6 = 66.7% of variables had at least one missing value). What number of cases (or percentage of values) were missing on each variable? This is also obtained from the frequencies procedure output in Figure 2.10. For example, out of 240 cases, depression had missing values on 22 cases (22/240 = 9.2%).

For the entire data set: Out of all possible values in the data set, what percentage were missing? The number of possible scores = number of variables × number of cases = 6 × 240 = 1,440. The number of missing values is obtained by summing the values in the "Missing" row in Figure 2.10: 22 + 90 + 14 + 20 + 0 + 0 = 146. Thus 146 of 1,440 scores are missing, for an overall missing data percentage of approximately 10%.

For each participant or case: Additional information is needed to evaluate the number of missing values for each case. To obtain this, create a dummy variable to represent missingness of scores on each variable (as suggested by Tabachnick & Fidell, 2018). The variable missingdepression corresponds to this yes/no question: Does the participant have a missing score on depression? Responses are coded 0 = no, 1 = yes. Dummy variables for missingness were created using the <Transform> → <Recode into Different Variables> procedure, as shown in Figure 2.11.

In the dialog box on top in Figure 2.11, specify the name of the existing (numerical) variable, in this example, depression. Create a name for the output variable in the right-hand side box (in this example, the output variable is named missingdepression). Click Change to move this new output variable name into the window under "Numeric Variable -> Output Variable." Then click Old and New Values. This leads to the second dialog box in Figure 2.11.

To define the first value of the dummy variable (a code of 1 if there is a missing value for depression), click the radio button to select the system missing value for depression as the old value; then enter the code for the new or output variable (1) into the "New Value" box on the right. Each participant who has a system missing value for depression is assigned a score of 1 on the new variable, missingdepression. Click Add to move this specification into the "Old --> New" box. To define the second value, select the radio button on the left for "All other values," and input 0 for "New Value" on the right; click Add. A participant with any other value, other than system missing, on depression is given a score of 0 on the new variable named missingdepression. Click Continue to return to the main dialog box, then click OK. The SPSS syntax that corresponds to these menu selections is:

Figure 2.11 Recode into Different Variables Dialog Box

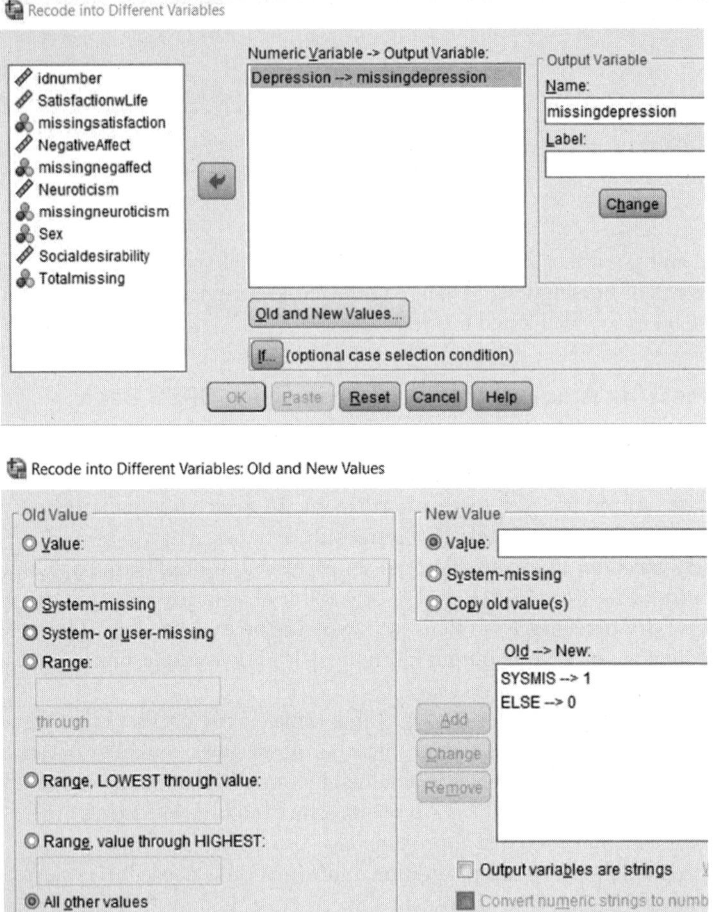

RECODE Depression (SYSMIS=1) (ELSE=0) INTO missingdepression

EXECUTE

The same operations can be used to create missingness variables for other variables (NegativeAffect, SatisfactionwLife, and Neuroticism). To find out how many variables had missing values for each participant, sum these new variables:

COMPUTE Totalmissing = missingdepression + missingsatisfaction + missingnegaffect + missingneuroticism

Then obtain a frequencies table for the new variable Totalmissing (see Figure 2.12). Only one person was missing values on all three variables. Most cases or participants were missing values on no variables (n = 116) or only one variable (n = 103).

2.10.3 Decisions Based on Amount of Missing Data

Amount of Missing Data in Entire Data Set

Graham (2009) stated that it may be reasonable to ignore the problem of missing values if the overall amount of missing data is below 5%. When there are very few missing data, the

Figure 2.12 Numbers of Participants or Cases Missing 0, 1, 2, and 3 Scores Across All Variables

Totalmissing

		Frequency	Percent	Valid Percent	Cumulative Percent
Valid	0	116	48.3	48.3	48.3
	1	103	42.9	42.9	91.3
	2	20	8.3	8.3	99.6
	3	1	.4	.4	100.0
	Total	240	100.0	100.0	

use of listwise deletion may be acceptable. In listwise deletion, cases that are missing values for any of the variables in the analysis are completely excluded. For example, if you run correlations among X_1, X_2, X_3, and X_4 using listwise deletion, a case is excluded if it is missing a value on any one of these variables. Pairwise deletion means that a case is omitted only for correlations that require a score that the case is missing; for example, if a case is missing a score on X_1, then that case is excluded for computation of r_{12}, r_{13}, and r_{14}, but retained for r_{23}, r_{34}, and r_{24}. Listwise and pairwise deletion are regarded as unacceptable for large amounts of missing data. Even with less than 5% missing, Graham still recommended using missing values imputation (discussed in upcoming sections) instead of listwise deletion.

Amount of Missing Data for Each Variable

Tabachnick and Fidell (2018) suggested that if a variable is not crucial to the analysis, that variable might be entirely dropped if it has a high proportion of missing values. Suppose that prior to data analysis, the analyst decided to discard variables with more than 33% missing values. Satisfaction with life was missing 38% of its values; it might be dropped using this preestablished rule. If a variable has numerous missing values, this may have been information that was not obtainable for many cases. (If the missing value were planned missing, the variable would not be dropped. For example, if only smokers are asked additional questions about amount of smoking, these variables would not be dropped simply because nonsmokers did not answer the questions.) It is not acceptable to drop variables after final analyses; dropping variables that influence outcomes such as *p* values at a late stage in the analysis can be a form of *p*-hacking. Any decision to drop a variable must be well justified.

Amount of Missing Data for Each Case

Analysts might also consider dropping cases with high percentages of missing values (as suggested by Tabachnick & Fidell, 2018). Completely dropping cases is equivalent to listwise deletion, and experts on missing values agree that listwise deletion is generally poor practice. However, it's worth considering the possibility that some participants may have provided really poor data. Some possible examples of extremely low quality survey data include the following: no answers for many questions, ridiculous or impossible responses (height 10 ft or 3 m), a series of identical ratings given for a long list of questions that assess different things (e.g., a string of scores such as 5, 5, 5, 5, 5, 5, 5 . . .), and inconsistent responses across questions (e.g., person responds "I have never smoked" to one question and then responds "I smoke an average of 10 cigarettes per day" to another question). These problems can arise because of poorly worded questions, or they may be due to lack of participant attention and effort or deliberate refusal to cooperate. If a decision is made to omit entire cases on the basis of data quality, be careful how this decision is presented, and make it clear that case deletions were thoughtful decisions, not (mindless, automatic) listwise deletion. Ideally, specific criteria for case deletion would be specified prior to data collection. However, participants can come up with types of poor data that are difficult to anticipate. In research other than surveys, analogous problems may arise.

The dummy variables used here to evaluate participant- or case-level missing data can also be used to evaluate patterns in missingness, as discussed in Section 2.13.

2.10.4 Assessment of Amount of Missingness Using SPSS Missing Values Add-On

The SPSS Missing Values add-on module can be used to obtain similar information about amount of missing data in a different format (without the requirement to set up dummy variables for missingness.)

The SPSS Missing Values add-on module provides two different procedures for analysis and **imputation of missing values**. Unfortunately, the menu options for these (at least up until SPSS Version 26) are confusing. (You can locate SPSS manuals by searching for "SPSS Missing Values manual" and locating the manual for the version number you are using.)

When you purchase a license for the Missing Values add-on, two new choices appear in the pull-down menu under <Analyze>. The first choice can be obtained by selecting these menu options <Analyze> → <Missing Value Analysis>. I have not used this procedure in this chapter, and I do not recommend it. The procedure that corresponds to these menu selections has an important limitation; it does not provide multiple imputation (only single imputation). Multiple imputation is strongly preferred by experts.

For all subsequent missing value analysis, I used these menu selections: <Analyze> → <Multiple Imputation>, as shown in Figure 2.13. The pull-down menu that appears when

Figure 2.13 Drop-Down Menu Selections to Open SPSS Missing Values Add-On Module

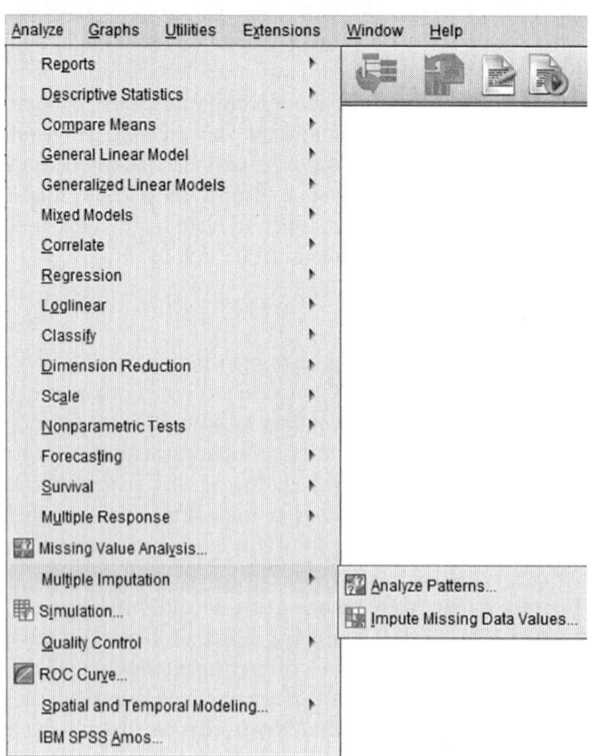

you click <Multiple Imputation> offers two choices: <Analyze Patterns> and <Impute Missing Data Values>. The procedures demonstrated in this chapter are run using these two procedures. First, descriptive information about the amount and pattern of missing data is obtained using the menu selections <Analyze> → <Multiple Imputation> → <Analyze Patterns>. Then the menu selections <Analyze> → <Multiple Imputation> → <Impute Missing Data Values> are used to generate multiple imputation of missing score values.

To obtain information about the amount of missing data, make these menu selections: <Analyze> → <Multiple Imputation> → <Analyze Patterns>, as shown in Figure 2.13. (The <Multiple Imputation> command appears in the <Analyze> menu only if you or your organization has purchased a separate license; it is not available in SPSS Base.)

In the Analyze Patterns dialog box (Figure 2.14), checkboxes can be used to select the kinds of information requested. I suggest that you include all variables in the "Analyze Across Variables" pane, not only the ones that you know have missing values. ("Analyze patterns" is a bit of a misnomer here; the information provided by this procedure is mainly for the amount of missing values rather than patterns of missingness.)

Only one part of the output is shown here (Figure 2.15). Figure 2.15 tells us that four of six of the variables (66.67%) had at least one missing value. One hundred sixteen of 240 of cases or participants (48.33%) had at least one missing value. Of the 2,400 values in the entire data set, 146 or 10.14% were missing. These graphics present information already obtained from SPSS Base. The Missing Values add-on module also generates graphics to show the co-occurrence of pairs or sets of missing variables (e.g., how many cases were missing scores on both depression and sex?). However, more useful ways to assess patterns of missingness are discussed in Sections 2.12 and 2.13.

Figure 2.14 Dialog Box for Analyze Patterns Procedure

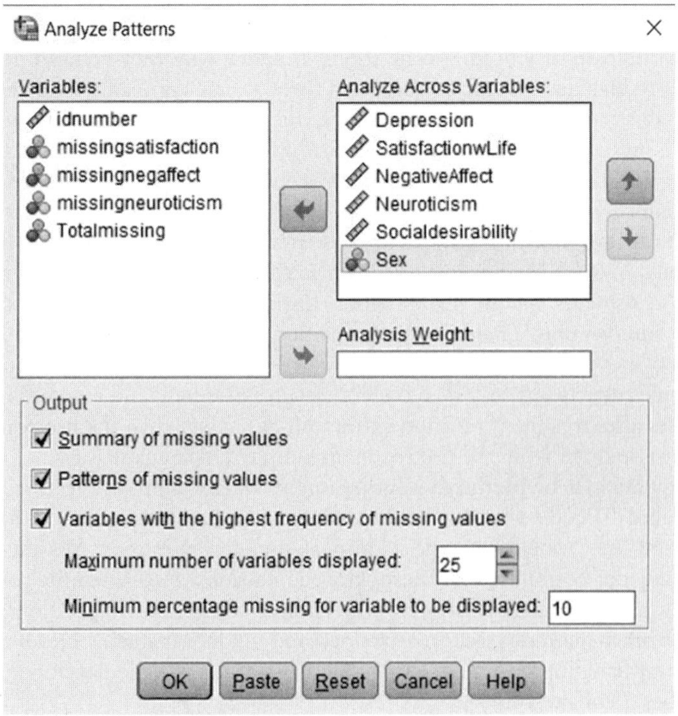

Analyze Patterns
Variables:
idnumber
missingsatisfaction
missingnegaffect
missingneuroticism
Totalmissing
Analyze Across Variables:
Depression
SatisfactionwLife
NegativeAffect
Neuroticism
Socialdesirability
Sex
Analysis Weight
Output
Summary of missing values
Patterns of missing values
Variables with the highest frequency of missing values
Maximum number of variables displayed: 25
Minimum percentage missing for variable to be displayed: 10
OK Paste Reset Cancel Help

Figure 2.15 Selected Output From Missing Values Analyze Patterns Procedure

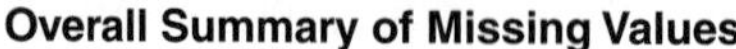

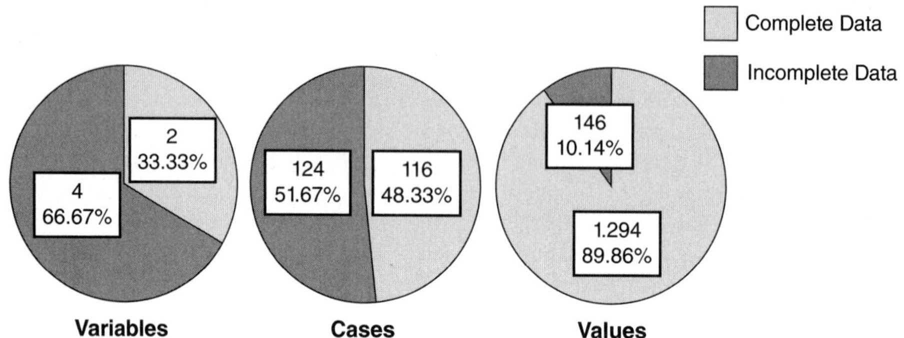

2.11 HOW MISSING DATA ARISE

Data can be missing for many reasons. Four common reasons are described; however, this does not exhaust the possibilities.

Refusal to participate: A researcher may initially contact 1,000 people to ask for survey participation. If only 333 agree to participate, no data are available for two thirds of the intended sample. Refusal to participate is unlikely to be random and can introduce substantial bias. There is nothing that can be done to replace this kind of missing data (the researcher could ask another 2,000 people to participate and obtain 666 more people). People who volunteer, or consent, to participate in research differ systematically from those who refuse (Rosenthal & Rosnow, 1975). It is essential to report numbers of person who refused to participate. It would also be useful to know why they refused. Refusal to participate leads to bias that cannot be corrected through later procedures such as imputation of missing values; imputation cannot replace this kind of lost data. The likelihood that the sample is not representative of the entire population that was contacted should be addressed in the discussion section when considering potential limitations of generalizability of results.

Attrition in longitudinal studies creates another kind of missingness. Imagine a longitudinal study in which participants are assigned (perhaps randomly) to different treatment conditions. Assessments may be made before treatment and at one or more times after the treatment or intervention. There is usually attrition. Participants may drop out of the treatment program, move and leave no contact information, die, or become unwilling or unable to continue. Some participants may miss one follow-up assessment and return for a later assessment. Samples after treatment or intervention can be smaller than the pretreatment sample, and they may also differ from the pretreatment sample in systematic ways.

Missing data may be planned: A survey might contain a funnel question, such as "Have you ever smoked?" People who say "yes" are directed to additional questions about smoking. People who say "no" would skip the additional smoking questions. Missing values would almost certainly not be imputed for these skipped questions. To shorten the time demands of a long survey, participants may be given only random subsets of the questions (and thus not have data for other questions, but in a planned and random manner). Development of better methods for handling planned missing data has encouraged the development of planned missing studies (Graham, 2009).

Missing values may have been used to replace outliers in previous data screening: One possible way to handle outliers (particularly when they are unbelievable or implausible) is to convert them to system missing values.

In an ideal situation, missing values would occur randomly, in ways that would not introduce bias in later data analysis. In actual data, missing values often occur in nonrandom patterns.

2.12 PATTERNS IN MISSING DATA

2.12.1 Type A and Type B Missingness

Patterns of missingness are usually described as one of these three types: missing completely at random, missing at random, and missing not at random (Rubin, 1976). To explain how these kinds of missingness differ, here is a distinction not found elsewhere in the missing values literature: I will refer to Type A and Type B missingness.

Consider **Type A missingness**. Suppose we have a *Y* variable (such as depression) that has missing values, and we also have data for other variables X_1, X_2, X_3, and so on (such as sex, neuroticism, and social desirability response bias). It is possible that missingness on *Y* is related to scores on one or more of the *X* variables; for example, men and people high in social desirability may be more likely to refuse to answer the depression questions than women and persons with low social desirability response bias. I will call this Type A missingness. The next few sections show that this kind of missingness can easily be detected and that state-of-the-art methods of replacement for missing values, such as **multiple imputation (MI)**, can correct for bias due to this type of missingness.

Now consider **Type B missingness**. It is conceivable that the likelihood of missing scores on *Y* (depression) depends on people's levels of depression. That is, people who would have had high scores on depression may be likely not to answer questions about depression. I will call this Type B missingness. Type B missingness is more difficult to identify than Type A missingness. (Sometimes it is impossible to identify Type B missingness.) Also, potential bias due to Type B missingness is more problematic and may not be correctable.

2.12.2 MCAR, MAR, and MNAR Missingness

The three patterns of missingness that appear widely in research on missing values were described by Rubin (1976). These are **missing completely at random (MCAR), missing at random (MAR)**, and **missing not at random (MNAR)**. Each of these patterns can be defined by the presence or absence of Type A and Type B missingness.

First consider MCAR missingness, as described by Schafer and Graham (2002): Assume that "variables X ($X_1, \dots X_p$) are known for all participants but Y is missing for some. If participants are independently sampled from the population . . . MCAR means that the probability that *Y* is missing for a participant does not depend on his or her own values of *X* or *Y*." Using the terms I suggest, MCAR does not have either Type A or Type B missingness.

The name MAR (missing at random) is somewhat confusing, because this pattern is not completely random. Schafer and Graham (2002) stated, "MAR means that the probability that *Y* is missing may depend on *X* but not *Y* . . . under MAR, there could be a relationship between missingness and *Y* induced by their mutual relationships to *X*, but there must be no residual relationship between them once *X* is taken into account." Using my terms, MAR may show Type A missingness (however, MAR must not show Type B missingness after corrections have been made for any Type A missingness).

The third and most troubling possible pattern is MNAR. Schafer and Graham (2002) stated that "MNAR means that the probability of missingness depends on Y. . . . Under MNAR, some residual dependence between missingness and *Y* remains after accounting for *X*." Using terms I suggest, MNAR has Type B missingness (and it may or may not also have Type A missingness).

MAR and MCAR patterns of missingness are called ignorable. This does not mean that we don't have to do anything about missing data if the pattern of missingness is judged to be MAR or MCAR. "Ignorable" means that, after state-of-the-art methods for replacement of missing values are used, results of analyses (such as *p* values) should not be biased.

MNAR (and Type B missingness, its distinguishing feature) are nonignorable forms of missingness. Even when state-of-the-art methods are used to impute scores for missing values in MNAR missing data, potential bias remains a problem that cannot be ignored. Discussion in a journal article must acknowledge the limitations imposed by this bias. For example, if we know that persons who are very depressed are likely to have missing data on the depression question, it follows that the people for whom we do have data represent a sample that is biased toward lower depression. Schlomer, Bauman, and Card (2010) urged researchers to consider the possible existence of MNAR and reasons why this might occur.

The degree to which missing values are problematic depends more on the pattern of missingness than the amount of missingness (Tabachnick & Fidell, 2018). MNAR is most problematic. Researchers should report information about pattern, as well as amount, of missing data. It is possible to find patterns in data that indicate problems with Type A missingness. However, it is impossible to prove that Type A and/or Type B missingness is absent.

2.12.3 Detection of Type A Missingness

Methods for detection of Type A missingness are discussed in the context of an empirical example in upcoming Section 2.13, including pairwise examination of variables and Little's test of MCAR. In this empirical example, Type A missingness occurs because missingness of depression scores is related to sex, neuroticism, socially desirable response bias, and other variables. The SPSS Missing Values add-on module provides all the necessary tests for Type A missingness. I will demonstrate that many of these tests can also be obtained using SPSS Base (the output from analysis using SPSS Base may be easier to understand). State-of-the-art methods for replacement of missing values are thought to correct most of the bias due to this type of missingness (Graham, 2009).

2.12.4 Detection of Type B Missingness

Unfortunately, evaluation of Type B missingness is difficult. It usually requires information that researchers don't have. Consider this example. If a question about school grade point average (GPA) is included in a survey, it is possible that students are more likely not to answer this if they have low GPAs. To evaluate whether Type B missingness is occurring, we need to know what the GPA scores would have been for the people who did not answer the question. Often there is no way to obtain this kind of information. In some situations, outside information can be helpful. Here are three examples of additional information that would help evaluate whether Type B missingness is occurring.

1. The researcher could follow up with the students who did not answer the GPA question and try again to obtain their answers. (Of course, if that information is obtained, it can be used to replace the missing value.)

2. The researcher could look for an independent source of data to find out what GPA answers would have been for people who did not answer the question. For example, universities have archival computer records of GPA data for all students. (Usually researchers cannot access this information.) If the researcher could obtain GPAs for all students, he or she could evaluate whether students who did not answer the question about GPA had lower GPA values than people who did answer the question. In this situation also, the values from archival data could be used to replace missing values in the self-report data.

3. An indirect way to assess Type B missingness would be to look at the distribution and range of GPA values in the sample of students and compare that with the distribution and range of GPA values for the entire university. Assume that the sample was drawn randomly from all students at the university. If the sample distribution for GPA contains a much lower proportion of GPAs below 2.0 than the university distribution, this would suggest that low-GPA students may have been less likely to report their GPAs than high-GPA students. This would indicate the presence of Type B missingness but would not provide a solution for it.

In the data set used as an empirical example, I know that neuroticism had Type B missingness (because, when I created my missing data file, I systematically turned higher scores on neuroticism into missing values). When I created Type B missingness for neuroticism, my new missing data file underrepresented people high in neuroticism, compared with the complete data set. Even after replacement of values using methods such as MI, generalization of findings to persons high on neuroticism would be problematic in this example.

Researchers often cannot identify, or correct for, Type B missingness. When Type B missingness is present (and probably it often is), researchers need to understand the bias this creates. Two types of bias may occur: Parameters may be over- or underestimated, and the sample may not be representative of, or similar to, the original population of interest. (For example, the sample may underrepresent certain types of persons, such as those highest on depression.) A researcher should address these problems and limitations in discussion of the study.

2.13 EMPIRICAL EXAMPLE: DETECTING TYPE A MISSINGNESS

To assess Type A missingness, we need to know whether missing versus nonmissing status for each variable is related to scores on other variables. This information can be obtained using the SPSS Missing Values add-on module. However, when first learning about missing values, doing a similar analysis in SPSS Base may make the underlying ideas clearer.

Figure 2.16 One-Way ANOVA Dialog Box: Assess Associations of Other Variables With Missingness on Negative Affect

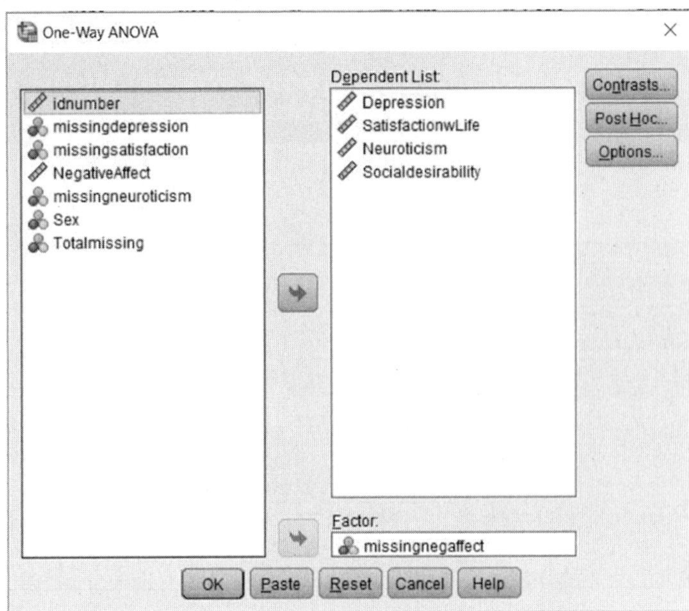

Earlier, in Section 2.10, a dummy "missingness" variable was created for each variable in the data set that had one or more missing values. These dummy variables can now be used to test Type A missingness. To see whether missingness on one variable (such as negative affect) is related to scores on other quantitative variables (such as response bias, negative affect, or neuroticism), means for those other quantitative variables are tested to see if they differ across the missing and nonmissing groups. It is convenient to use the SPSS one-way ANOVA procedure for comparison of means. To open the one-way ANOVA procedure, make the following menu selections: <Analyze> → <Compare Means> → <One-Way ANOVA>. The One-Way ANOVA dialog box in Figure 2.16 shows which variables were included. The Options button was used to select descriptive statistics (recall that means and other descriptive statistics are not provided unless requested explicitly). Selected results appear in Figure 2.17.

The groups (groups of persons missing or not missing negative affect scores) did not differ in mean satisfaction with life, $F(1, 148) = .106, p = .745$. Missingness on negative affect was related to scores on the other three variables; in other words, there is evidence of Type A missingness. The table of group means (not shown here) indicated that people in the missing negative affect group scored lower on neuroticism, higher in social desirability response bias, and lower on depression. Similar comparisons of means are needed for each of the other missingness dummy variables (e.g., ANOVAs to compare groups of missing vs. not missing status for Depression, SatisfactionwLife, etc.).

To evaluate whether missingness is related to a categorical variable such as sex, or to missingness on other variables, set up a contingency table using the SPSS crosstabs procedure.

The crosstabs results in Figure 2.18 indicate that sex was associated with missingness on depression; 22 of 112 men (almost 20%) of men were missing scores on depression; none of the women were missing depression scores. This was statistically significant, $\chi^2(1) = 27.68, p < .001$ (output not shown).

The SPSS Missing Values add-on module provides similar comparisons of group means and crosstabs (not shown here). An additional test available from the Missing Values add-on module is **Little's test of MCAR** (Little, 1988). Little's test essentially summarizes information from the individual tests for Type A missingness just described.

To obtain Little's test, open the Missing Values add-on module by selecting <Analyze> → <Missing Value Analysis> (not either of the two additional menu choices that appear to the right after selecting <Multiple Imputation>; refer back to Figure 2.13). The Missing Value Analysis dialog box appears as shown in Figure 2.19.

Figure 2.17 ANOVA Source Table: Comparison of Groups Missing Versus Not Missing Negative Affect Scores

ANOVA

		Sum of Squares	df	Mean Square	F	Sig.
Depression	Between Groups	690.064	1	690.064	6.310	.013
	Within Groups	23620.840	216	109.356		
	Total	24310.904	217			
SatisfactionwLife	Between Groups	1.263	1	1.263	.106	.745
	Within Groups	1760.505	148	11.895		
	Total	1761.768	149			
Neuroticism	Between Groups	201.893	1	201.893	5.991	.015
	Within Groups	7346.884	218	33.701		
	Total	7548.777	219			
Socialdesirability	Between Groups	686.018	1	686.018	104.892	.000
	Within Groups	1556.582	238	6.540		
	Total	2242.600	239			

Figure 2.18 Contingency Table for Missingness on Depression by Sex

Sex * missingdepression Crosstabulation

			missingdepression 0	missingdepression 1	Total
Sex	male	Count	90	22	112
		% within Sex	80.4%	19.6%	100.0%
	female	Count	128	0	128
		% within Sex	100.0%	0.0%	100.0%
Total		Count	218	22	240
		% within Sex	90.8%	9.2%	100.0%

Figure 2.19 Missing Value Analysis Dialog Box to Request Little's MCAR Test

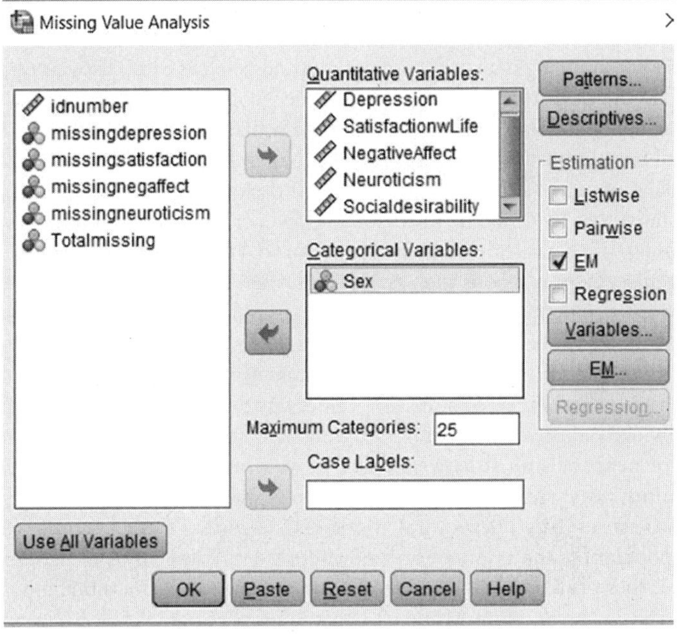

In the Missing Value Analysis dialog box, move all quantitative variables to the "Quantitative Variables" pane, and move any categorical variables into the separate "Categorical Variables" pane. Check the box for "EM" in the "Estimation" list, then click OK. (If you also want the *t* tests and crosstabs that were discussed earlier in SPSS Base, click the Descriptives button and use checkboxes in the Descriptives dialog box to request these; they are not included here.)

Little's MCAR test appears as a footnote to the "EM Means" table in Figure 2.20. This was statistically significant, $\chi^2(33) = 136.081, p < .001$. The null hypothesis is essentially that there is no Type A missingness for the entire set of variables. This null hypothesis is rejected (consistent with earlier ANOVA and crosstabs results showing that missingness was related to scores on other variables). This is additional evidence that Type A missingness is present. There is no similar empirical test for Type B missingness.

Figure 2.20 "EM Means" Table From SPSS Missing Values Analysis With Little's MCAR Test

EM Means[a]

Depression	SatisfactionwLife	NegativeAffect	Neuroticism	Socialdesirability
22.35	17.40	24.39	30.50	9.85

a. Little's MCAR test: Chi-Square = 136.081, DF = 33, Sig. = .000

2.14 POSSIBLE REMEDIES FOR MISSING DATA

There are essentially three ways to handle missing values. The first is to ignore them, that is, throw out cases with missing data using default methods such as SPSS listwise or pairwise deletion. (Somewhat different terms are used elsewhere; "complete case analysis" is synonymous with listwise deletion; "available data analysis" is equivalent to pairwise deletion; Pigott, 2001.) Listwise deletion is almost universally regarded as bad practice. However, Graham (2009) said that listwise deletion may yield acceptable results if the overall amount of missing data is less than 5%; he stated that "it would be unreasonable for a critic to argue that it was a bad idea" if an analyst chose to use listwise deletion in this situation. However, he recommended the use of missing data replacement methods such as MI even when there is less than 5% missing data.

One obvious problem with listwise deletion is reduction of statistical power because of a smaller sample size. A less obvious but more serious problem with listwise deletion is that discarding cases with missing scores can systematically change the composition of the sample. Recall that when I created a missing values pattern in the data set used as an example, I systematically deleted the cases with the highest scores on neuroticism (this created Type B missingness for neuroticism). If listwise deletion were used, subsequent analyses would not include any information about people who had the highest scores for neuroticism. That creates bias in two senses. First, if we want to generalize results from a sample to some larger hypothetical population, the sample now underrepresents some kinds of people in the population, Second, there is bias in estimation of statistics such as regression slopes, effect sizes, and p values (this is known from Monte Carlo studies that compared different methods for handling missing values in the presence of different types of pattern for missingness).

A second way to handle missing values is to replace them with simple estimates based on information in the data set. Missing scores on a variable could be replaced with the mean of that variable (for the entire data set or separately for each group). Missing values could be replaced with predicted scores from a regression analysis that uses other variables in the data set as predictors. These methods are not recommended (Acock, 2005), because they do not effectively reduce bias.

There are several state-of-the-art methods for replacement of missing values that involve more complex methods. Graham (2009) "fully endorses" multiple imputation. Monte Carlo work shows that MI is effective in reducing bias in many missing-values situations (but note that it cannot correct for bias due to Type B missingness). Graham and Schlomer et al. (2010) described other state-of-the-art procedures and the capabilities of several programs, including SAS, SPSS, Mplus, and others. They also described freely downloadable software for missing values.

The empirical example presented in the following section uses MI. Graham (2009) stated that MI performs well in samples as small as 50 (even with up to 18 predictors) and with as much as 50% missing data in the dependent variable. He explained that, contrary to some beliefs, it is acceptable to impute replacements for missing values on dependent variables. He suggested that a larger number of imputations than the SPSS default of 5 may be needed with larger amounts of missing data, possibly as many as 40 imputations.

2.15 EMPIRICAL EXAMPLE: MULTIPLE IMPUTATION TO REPLACE MISSING VALUES

To run MI using the SPSS Missing Values add-on module, start from the top-level menu. Choose <Analyze> → <Multiple Imputation>, then from the pop-up menu on the right, select <Impute Missing Data Values>. The resulting dialog box appears in Figure 2.21. All the variables of interest (both the variables with missing values and all other variables that will be used in later analyses) are included. Note that you can access a list of procedures that can be applied to imputed data in SPSS Help, as noted in this dialog box.

The number of imputations is set to 5 by default (note that a larger number of imputations, on the order of 40, is preferable for data sets with large percentages of missing data; Graham, 2009). A name for the newly created data set must be provided (in this example, Imputed Data).

MI does something comparable with replacement by regression. Each imputation estimates a different set of plausible values to replace each missing value for a variable such as depression; these plausible values are based on predictions from all other variables. The resulting data file (a subset appears in Figure 2.22) now contains six versions of the data: the original data and the five imputed versions. The first column indicates imputation number (0 for the original data).

Figure 2.21 Dialog Box for Impute Missing Data Values

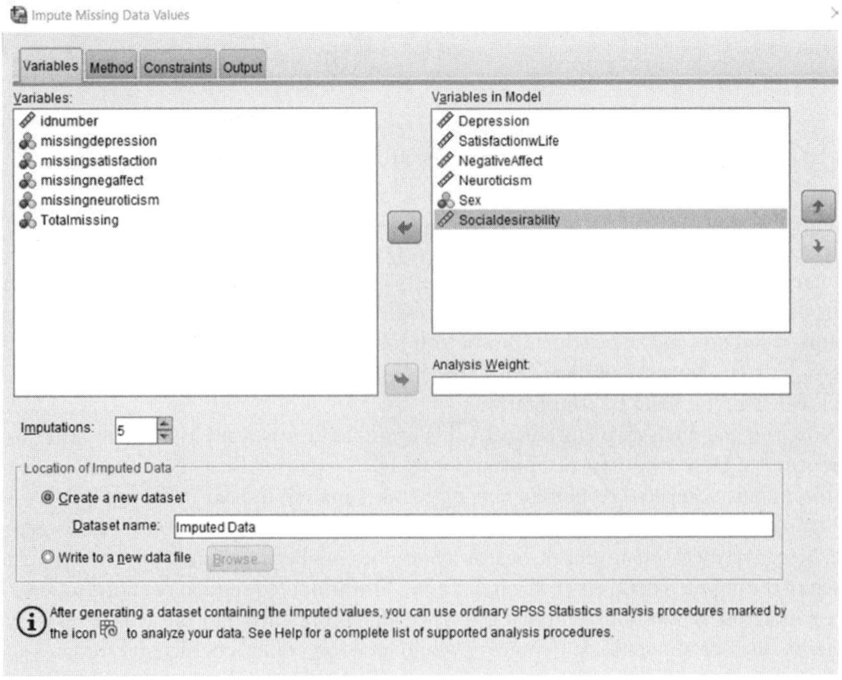

Figure 2.22 Selected Rows From Imputed Data Set

	Imputation_	idnumber	Depression	SatisfactionwLife	NegativeAffect	Neuroticism	Sex	Socialdesirabilit
235	0	235	18	16	.	26	1	1
236	0	236	.	20	18	23	1	1
237	0	237	0	20	10	24	1	1
238	0	238	26	18	37	37	2	1
239	0	239	30	9	32	26	2	
240	0	240	20	15	20	26	2	
241	1	1	30	18	24	32	2	
242	1	2	30	14	24	38	1	1
243	1	3	24	22	31	34	1	
244	1	4	44	17	31	35	2	
245	1	5	14	18	15	39	2	1
246	1	6	16	23	17	23	1	

Figure 2.23 Split File Command Used to Pool Results for Imputed Data File

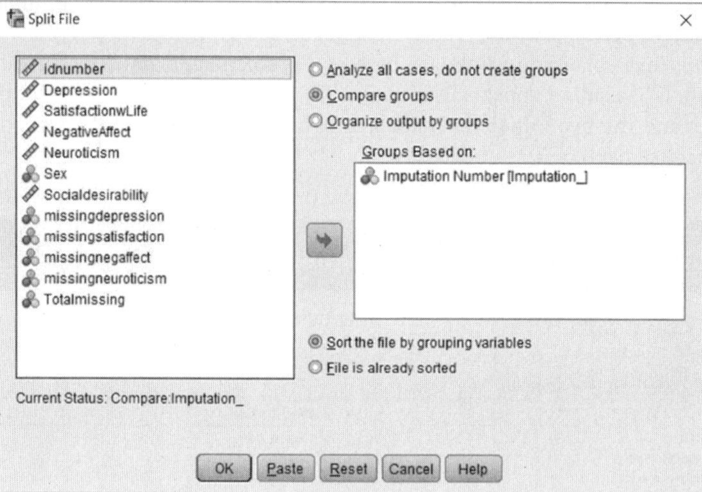

The final analysis of interest (for example, prediction of depression from the other five variables) is now run on all versions of the data (Imputations 0 through 5), and results are pooled (averaged) across data sets. Prior to the regression, select <Data> and <Split File> (not <Split into Files>).

In the Split File dialog box, move the variable Imputation Number into the pane under "Groups Based on" and select the radio button for "Compare groups." You should see a line that says Current Status: Compare:Imputation_ in the lower left corner. The SPSS syntax is: SPLIT FILE LAYERED BY Imputation_.

Now run the analysis of interest. In this example, it was a multiple regression to predict scores on Depression from SatisfactionwLife, NegativeAffect, Neuroticism, Sex, and Socialdesirability. Selected results for this regression analysis appear in Figure 2.24.

Figure 2.24 shows the regression coefficients (to predict Depression from SatisfactionwLife, NegativeAffect, Neuroticism, Sex, and Socialdesirability), separately for the original data, for each of the imputed data sets (1 through 5), and for the pooled results. We hope to see consistent results across all solutions, and that is usually what is obtained. For these data, results varied little across the five imputations. Reporting would focus on pooled coefficient estimates and the overall statistical significance of the regressions (in the ANOVA tables, not provided here).

Figure 2.24 Prediction of Depression From SatisfactionwLife, NegativeAffect, Neuroticism, Sex, and Socialdesirability Using Linear Regression: Original and Imputed Missing Values

Coefficients[a]

Imputation Number	Model		Unstandardized Coefficients B	Std. Error	Standardized Coefficients Beta	t	Sig.
Original data	1	(Constant)	14.365	6.686		2.149	.034
		SatisfactionwLife	-.832	.208	-.272	-4.003	.000
		NegativeAffect	.709	.119	.460	5.941	.000
		Neuroticism	.380	.139	.207	2.730	.007
		Sex	-1.840	1.362	-.090	-1.352	.179
		Socialdesirability	-.393	.298	-.091	-1.322	.189
1	1	(Constant)	14.954	4.441		3.367	.001
		SatisfactionwLife	-.780	.130	-.257	-5.981	.000
		NegativeAffect	.676	.073	.485	9.270	.000
		Neuroticism	.387	.089	.225	4.364	.000
		Sex	-2.295	.897	-.109	-2.559	.011
		Socialdesirability	-.348	.154	-.101	-2.263	.025
Pooled	1	(Constant)	12.420	4.888		2.541	.013
		SatisfactionwLife	-.851	.184		-4.625	.000
		NegativeAffect	.665	.090		7.397	.000
		Neuroticism	.428	.100		4.294	.000
		Sex	-1.442	1.105		-1.305	.200
		Socialdesirability	-.221	.185		-1.198	.238

a. Dependent Variable: Depression

2.16 DATA SCREENING CHECKLIST

Decisions about eligibility criteria, minimum group size, methods to handle outliers, plans for handling missing data, and so forth should be made prior to data collection. For longitudinal studies that compare treatment groups, **Consolidated Standards of Reporting Trials (CONSORT)** guidelines may be helpful (Boutron, 2017). Document what was done (with justification) at every step of the data-screening process. The following checklist for data screening and handling covers many research situations.

Some variation in the order of steps is possible. However, I believe that it makes sense to consider distribution shape prior to making decisions about handling outliers and to deal with outliers before imputing missing values. These suggestions are not engraved in stone. There are reasonable alternatives for most of the choices I have recommended.

1. Proofread the data set against original sources of data (if available). Replace incorrect scores with accurate data. Replace impossible score values with system missing.
2. Remove cases that do not meet eligibility criteria.
3. If group means will be compared, each group should have a minimum of 25 to 30 cases (Boneau, 1960). If some groups have smaller *n*'s, additional members for these groups might be obtained prior to other data analyses. Alternatively, groups with small *n*'s can be dropped, or combined with other groups (if that makes sense).
4. Assess distribution shapes by examining histograms. If groups will be compared, distribution shape should be assessed separately within each group. Some distribution shapes, such as Poisson, require different analyses than those covered in this book (see Appendix 2A).
5. Possibly apply data transformations (such as log or arcsine), but only if this makes sense. If distribution skewness is due to a few outliers, it may be preferable to deal with those outliers individually instead of transforming the entire set of scores.

6. Screen for univariate, bivariate, and multivariate outliers. Decide how to handle these (for example, convert extreme scores to less extreme values, or replace them with missing values).

7. Test linearity assumptions for associations between quantitative variables. If nonlinearity is detected, revisit the possibility of data transformations, or include terms such as X^2 in later analyses.

8. Assess amount and pattern of missing values. If there is greater than 50% missingness on a case or a variable, consider the possibility that these cases or variables provide such poor-quality data that they cannot be used. If cases or variables are dropped, this should be documented and explained.

9. Use multiple imputation to replace missing values (or use another state-of-the-art missing value replacement method, as discussed in Graham, 2009).

2.17 REPORTING GUIDELINES

At a minimum, the following questions should be answered. Some may require only a sentence or two; others may require more information. For additional suggestions about reporting, see Johnson and Young (2011), Recommendations 9 and 10, and Manly and Wells (2015).

In the "Introduction": What types of analyses were done and why were these chosen?

In the "Methods" section: Details about initial sample selection, measurements, group comparisons (if any), and other aspects of procedure.

In the "Results" section: Data screening and handling procedures should be described at the beginning of the "Results" section. This should address each of the following questions:

1. What were the final numbers of cases for final analysis, after any respondents were dropped because they declined to participate or did not meet eligibility criteria (or presented other problems)? For longitudinal studies, a CONSORT flowchart may be helpful (see Section 2.1).

2. If any variables were dropped from planned analyses because of poor measurement quality or if groups were omitted or combined because of small *n*'s, explain.

3. Explain rule(s) for outlier detection and the way outliers were handled, and note how many changes were made during outlier evaluation. Explain any data transformations.

4. Report the amount of missing values, such as the percentage of scores missing in the entire data set, the percentage missing for each variable, or the percentage of participants missing one or more scores.

5. Describe possible reasons for missing values.

6. Explain pattern in missing values. Type A missingness is present if Little's MCAR test is significant; details about the nature of missingness are found in the *t* tests and crosstabs that show how missingness dummy variables are related to other variables. It may not be possible to detect Type B missingness unless additional information is available beyond the data set; this possibility should be discussed. (Type A missingness is ignorable; Type B is problematic.)

7. Provide specific information about the imputation method used to replace missing values, including software, version, and commands; number of imputations; and any notable differences among results for different imputations and original data.

In the "Discussion" section: Be sure to explain the ways in which data problems, such as sample selection and missing values, may have (a) created bias in parameter estimates and (b) limited the generalizability of results.

2.18 SUMMARY

Before collecting data, researchers should decide on rules and procedures for data screening, outliers, and missing values, and then adhere to those rules. This information is required for preregistration of study plans. Open Science advocates preregistration as a way to improve completeness and transparency of reporting and calls for making data available for examination by other researches through publicly available data archives. Some journals offer special badges for papers that report preregistered studies. For further discussion, see Asendorpf et al. (2013) and Cumming and Calin-Jageman (2016).

In addition, professional researchers often seek research funding from federal grant agencies (e.g., the National Science Foundation, the National Institutes of Health). These agencies now require detailed plans for data handling in the proposals, for example, decisions about sample size on the basis of statistical power analysis, plans for identification and handling of outliers, and plans for management of missing data. A few professional journals (for example, *Psychological Science* and some medical journals) provide the opportunity to preregister detailed plans for studies including this information. Journal editors are beginning to require greater detail and transparency in reporting data screening than in the past. The requirement for detailed reporting of data handling is likely to increase.

For many decisions about outliers and missing value replacement, there is no one best option. This chapter suggests several options for handling outliers, but there are many others (Aguinas et al., 2013). This chapter describes the use of MI for replacement of missing values, but additional methods are available or may become available in the future.

The growing literature about missing values includes strong arguments for the use of MI and other state-of-the-art methods as ways to reduce bias. However, even state-of-the-art methods for replacement of missing values does not get rid of problems due to Type B missingness.

It is important to remember that many other common research practices may be even greater sources of bias. Use of convenience samples rather than random or representative samples limits the generalizability of findings. Practices such as *p*-hacking and hypothesizing after results are known to greatly inflate the risk for Type I error. Quality control during data collection is essential. Nothing that is done during data screening can make up for problems due to poor-quality data.

Numerous missing values situations are beyond the scope of this chapter, for example, imputation of missing values for categorical variables (Allison, 2002), attrition in longitudinal studies (Kristman, Manno, & Côté, 2005; Muthén, Asparouhov, Hunter, & Leuchter, 2011; Twisk & de Vente, 2002), missing data in multilevel or structural equation models, and missing values at the item level in research that uses multiple-item questionnaires to assess constructs such as depression (Parent, 2013).

Subsequent chapters assume that all appropriate data screening for generally required assumptions has been carried out. Additional data-screening procedures required for specific analyses will be introduced as needed.

APPENDIX 2A

Brief Note About Zero-Inflated Binomial or Poisson Regression

The following empirical example provides an illustration. Figure 2.25 is adapted from Atkins and Gallop (2007). The count variable in their study (on the *X* axis) is the number of steps each person has taken toward divorce, ranging from 0 to 10. The distribution is clearly non-normal; it has a mode at zero and positive skewness (a very small proportion of persons in the sample had taken 8 or more steps).

Figure 2.25 Four Models for Distribution Shape of Frequency Count Variable

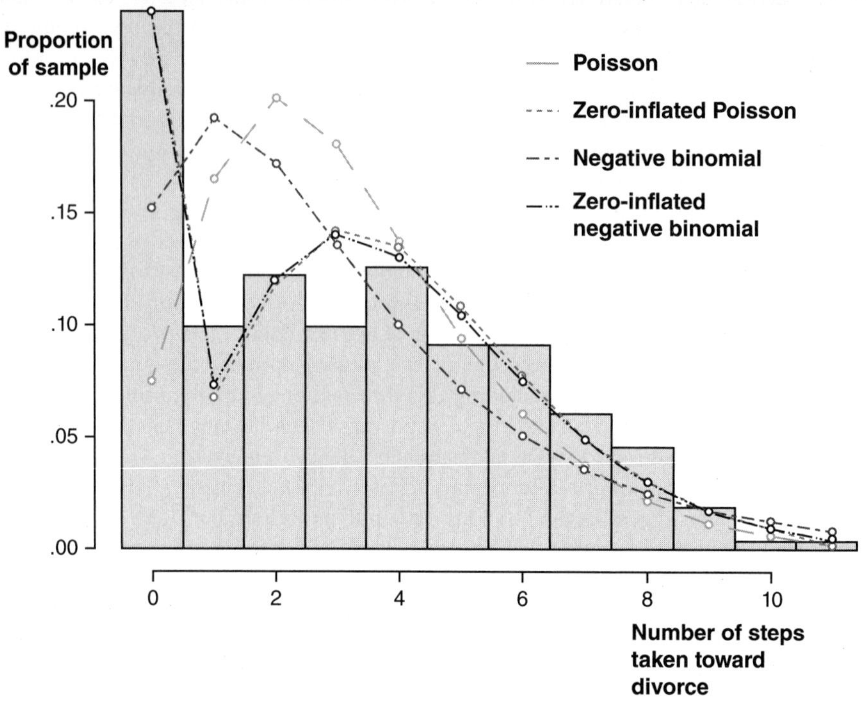

Source: Adapted from Atkins and Gallop (2007).

Note: Variable on the X axis is the number of steps or actions taken toward separation or divorce, ranging from 0 to 10.

Atkins and Gallop (2007) evaluated the fit of four mathematical distribution models to the empirical frequency distribution in Figure 2.25: Poisson, zero-inflated Poisson (ZIP), negative binomial, and zero-inflated negative binomial (ZINB). Quantitative criteria were used to evaluate model fit. They concluded that the ZIP model was the best fit for their data (results were very similar for the ZINB model). The regression analysis to predict number of steps toward divorce from other variables would be called zero-inflated Poisson regression; this is very different from linear regression.

It is possible to ask two questions about analyses in these models applied to behavior count variables. Consider illegal drug use as an example (e.g., Wagner, Riggs, & Mikulich-Gilbertson, 2015). First, we want to predict whether individuals use drugs or not. For those who do use drugs, a zero frequency of drug use in the past month is possible, but higher frequencies of use behaviors can occur. The set of variables that predicts frequency of drug use in this group may differ from the variables that predict use versus nonuse of drugs. This information would be missed if a data analyst applied ordinary linear regression.

The SPSS generalized linear models procedure can handle behavior count dependent variables. (Note that this is different from the GLM procedure used in Volume I [Warner, 2020].) For an online SPSS tutorial, see UCLA Institute for Digital Research & Education (2019). Atkins and Gallop (2007) provided extensive online supplemental material for their study. Note that count data should not be log transformed in an attempt to make them more nearly normally distributed (O'Hara & Kotze, 2010).

COMPREHENSION QUESTIONS

1. What can you look for in a histogram for scores on a quantitative variable?
2. What can you look for in a three-dimensional scatterplot?
3. What quantitative rule can be used to decide whether a univariate score is an outlier?
4. Are there situations in which can you justify deleting a case or participant completely? If so, what are they?
5. Under what conditions might you convert a score to system missing?
6. What is the point of running an analysis once with outliers included and once with outliers deleted?
7. What is a way to identify multivariate outliers using Mahalanobis distance?
8. Describe two distribution shapes (other than normal) that you might see in actual data (hint: any other distribution graph you have seen in this chapter, along with any strange things you might have seen in other data).
9. When can log transformations be used, and what potential benefits do these have? When should log transformations not be used?
10. If you have a dependent variable that represents a count of some behavior, would you expect data to be normally distributed? Why or why not? What types of distribution better describe this type of data? Can you use linear regression? What type of analysis would be preferable?
11. Which do authorities believe generally pose more serious problems in analysis: outliers or non-normal distribution shapes?
12. What problems arise when listwise deletion is used to handle missing values?

NOTE

[1]Chapter 7, on moderation, explains that when forming products between predictor variables, the correlation between X^2 and X can be reduced by using centered scores on X to compute the squared term. A variable is centered by subtracting out its mean. In other words, we can calculate $X^2 = (X - M_X) \times (X - M_X)$ where M_X is the mean of X. The significance of the quadratic trend is the same whether X is centered or not; however, judgments about whether there could also be a significant linear trend can change depending on whether X was centered before computing X^2.

DIGITAL RESOURCES

Find **free study tools** to support your learning, including **eFlashcards, data sets, and web resources,** on the accompanying website at **edge.sagepub.com/warner3e.**

CHAPTER 3

STATISTICAL CONTROL

What Can Happen When You Add a Third Variable?

3.1 WHAT IS STATISTICAL CONTROL?

Bivariate correlation (Pearson's r) is an index of the strength of the linear relationship between one independent variable (X_1) and one dependent variable (Y). This chapter moves beyond the two-variable research situation to ask, "Does our understanding of the nature and strength of the predictive relationship between a predictor variable, X_1, and a dependent variable, Y, change when we take a third variable, X_2, into account in our analysis? If so, how does it change?" This introduces one of the most important concepts in statistics: the idea of **statistical control**.

When we examine the association between each pair of variables (such as X_1 and Y) in the context of a larger analysis that includes one or more additional variables, our understanding of the way X_1 and Y are related often changes. For example, Judge and Cable (2004) reported that salary (Y) is predictable from height (X_1). An obvious question can be raised: Does this correlation occur because on average, women earn less than men, and are shorter than men? In this example, sex would be an X_2 control variable.

This chapter introduces concepts involved in statistical control using two simple analyses. The first is evaluation of an X_1, Y association for separate groups on the basis of X_2 scores. For example, Judge and Cable (2004) controlled for sex by reporting correlations between height (X_1) and salary (Y) separately for male and female groups. Second, the partial correlation between X_1 and Y, controlling for X_2, is another way of assessing whether controlling for X_2 changes the apparent relationship between other variables. Analyses in later chapters, such as multiple regression, implement statistical control by including control variables in the regression equation. It is easier to grasp statistical control concepts if we start with the simplest possible three-variable situation. You will then be able to see how this works in analyses that include more than three variables.

For a statistical analysis to make sense, it must be based on a properly specified model. "Properly specified" means that all the variables that should be included in the analysis are included and that no variables that should not be included in the analysis are included. *Model* refers to a specific analysis, such as a multiple regression equation to predict Y from X_1 and X_2. We can never know for certain whether we have a properly specified model. A well-developed theory can be a helpful guide when you choose variables. When researchers choose control variables, they usually choose them because other researchers have used these control variables in past research, or because the X_2 control variable is thought to be correlated with X_1 and/or Y. Later analyses (multiple regression) make it possible to include more than one control variable. When control (X_2) variables make sense, and are based on a widely accepted theory, then the description of the X_1, Y association when X_2 is controlled is usually preferred to a description of the association of X_1 and Y (r_{1Y}) by itself.

Statistical control can help rule out some potential rival explanations. Recall the requirements for making an inference that X causes Y. There must be a reasonable theory to explain how X might cause Y. Scores on X and Y must be statistically related, using whatever type of analysis is appropriate for the nature of the variables. X must come before Y in time (temporal precedence). We must be able to rule out all possible rival explanations (variables other than X that might be the real causes of Y). In practice, this last condition is difficult to achieve. Well-controlled experiments provide **experimental control** for rival explanatory variables (though methods such as holding variables constant or using random assignment of cases to group to try to achieve equivalence). Partial correlation and regression analyses provide forms of statistical control that can potentially rule out some rival explanatory variables. However, there are limitations to statistical control. We can statistically control only for variables that are measured and included in analyses, and there could always be additional variables that should have been considered as rival explanations, about which we have no information. (At least in theory, experimental methods such as random assignment of cases to groups should control for all rival explanatory variables, whether they are explicitly identified or not).

This chapter assumes familiarity with bivariate regression and correlation. Throughout this chapter, assumptions for the use of correlation and regression are assumed to be satisfied. Except where noted otherwise, all variables are quantitative. None of the following analyses would make any sense if assumptions for correlation (particularly the assumption of linearity) are violated. When we begin to examine statistical control, we need information about the nature of the association among all three pairs of variables: X_1 with Y, X_2 with Y, and X_1 with X_2. Subscripts are added to correlations to make it clear which variables are involved. A few early examples in this chapter use a categorical X_2 variable as the control variable; however, all examples used in this chapter assume that X_1 and Y are quantitative.

- The bivariate correlation between X_1 and Y is denoted r_{1Y} (or r_{Y1}, because correlation is symmetrical; that is, you obtain the same value whether you correlate X_1 with Y or Y with X_1).
- The bivariate correlation between X_2 and Y is denoted r_{2Y}.
- The bivariate correlation between X_1 and X_2 is denoted r_{12}.

In this chapter, X_1 denotes a predictor variable, Y denotes an outcome variable, and X_2 denotes a third variable that may be involved in some manner in the X_1, Y relationship. The X_2 variable can be called a **control variable** or a **covariate**. A control variable often (but not always) represents a rival explanatory variable. This chapter describes two methods of statistical control for one covariate, X_2, while examining the X_1, Y association. The first method is separating data into groups, on the basis of scores on the X_2 control variable, and then analyzing the X_1, Y association. The second method is obtaining a partial correlation between X_1 and Y controlling for X_2. Use of these methods can help understand how statistical control works, and these can be useful as forms of preliminary data screening. However, multiple regression and multivariate analyses are generally the way statistical control is done when data are analyzed and reported in journal articles.

You will see that when an X_2 variable is statistically controlled, the correlation between X_1 and Y can change in any way you can imagine. When the correlation between X_1 and Y is substantially different when we control for X_2, we need to explain why the relationship between X_1 and Y is different when X_2 is statistically controlled (than when X_2 is not controlled).

3.2 FIRST RESEARCH EXAMPLE: CONTROLLING FOR A CATEGORICAL X_2 VARIABLE

Suppose that X_1 is height, Y is vocabulary test score, and X_2 is grade level (Grade 1, 5, or 9). In this example, X_1 and Y are both quantitative variables. We assume that X_1 and Y are linearly related. X_2, grade level, is a convenient type of variable for the following examples because it can be treated as either a categorical variable that defines three groups (different grade levels) or as a quantitative variable that happens to have few different score values.

The analysis in this section includes two simple steps.

1. Find the bivariate correlation between X_1 and Y (ignoring X_2). This answers the question, How do X_1 and Y appear to be related when you do not control for X_2? Obtain an X_1, Y scatterplot as additional information about the relationship. You may also want to add case markers for values of X_2 to the plot, as discussed below.
2. Use the SPSS split file procedure to divide the data set into groups on the basis of the X_2 control variable (first, fifth, and ninth grade groups). Within each grade-level group, obtain an X_1, Y scatterplot and the X_1, Y correlation, r_{1Y}. You'll have values of r_{1Y} for the first grade group, the fifth grade group, and the ninth grade group. These r values within groups are statistically controlled to remove the effects of X_2, grade, because the value of X_2 is constant within each group.

Using these results, you can answer two questions:

- Do the values of r_{1Y} within the groups (first grade, fifth grade, and ninth grade) differ from the overall value of r_{1Y} obtained in Step 1? If so, how do they differ? Are they smaller or larger? (Unless you conduct statistical significance tests between correlations, as discussed in Appendix 10C in Chapter 10 in Volume I [Warner, 2020], these comparisons are only qualitative. Very large samples are required to have enough statistical power to judge differences between correlations significant; do not overinterpret small differences.)
- Do the values of r_{1Y} and the slopes in the scatterplots differ across these groups (i.e., between the first grade group, the fifth grade group, and the ninth grade group)? If these within-group correlations differ, this is possible preliminary evidence of an interaction between X_1 and X_2 as predictors of Y. If there is an interaction between X_1 and X_2 as predictors of Y, partial correlation and regression results will be misleading (unless the regression analysis includes interaction terms).

In the first hypothetical study, measures of height (X_1) and vocabulary (Y) were obtained for groups of schoolchildren in Grades 1, 5, and 9 (grade is the categorical X_2 control variable). Data for this example are in the file named heightvocabulary.sav. Before you start the analysis, you probably suspect that any correlation between height and vocabulary is silly or misleading (another word for this is *spurious*; spuriousness is discussed later in the chapter). Using these data, the following analyses were done.

First, before examining correlations it is a good idea to look at scatterplots. To obtain the scatterplot, make the following SPSS menu selections: <Graphs> → <Legacy Dialogs> → <Scatter/Dot>. In the first dialog box, click Simple Scatter, then Define. The Simple Scatterplot dialog box appears in Figure 3.1. Move the name of the dependent variable into the "Y Axis" box, the predictor or X_1 variable into the "X Axis" box, and the name of the control variable X_2 into the "Set Markers by" box. (This yields reasonable results only if X_2 has a small number of different values.) Then click OK.

Figure 3.1 SPSS Simple Scatterplot Dialog Box

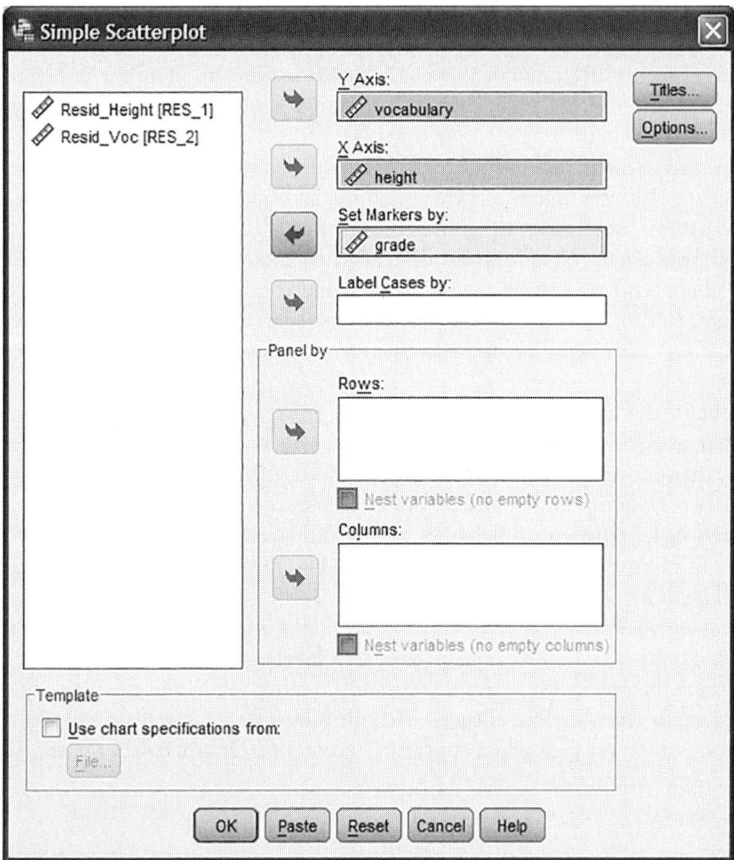

Figure 3.2 Scatterplot for Vocabulary With Height, With Fit Line at Total

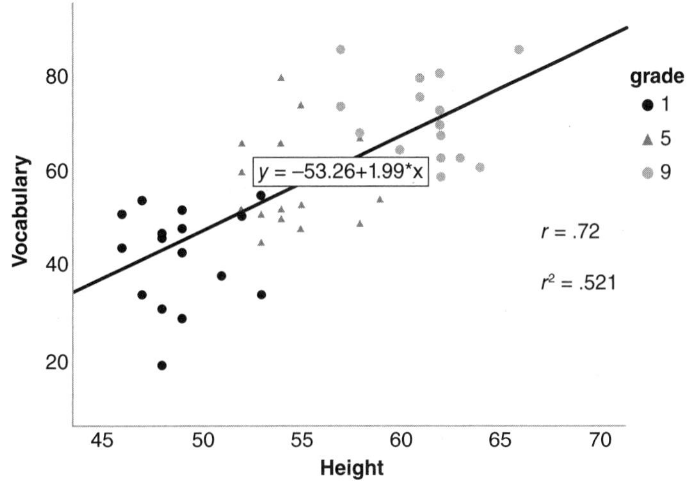

When the scatterplot appears, click on it twice to open it in the Chart Editor. Under the menu heading "Elements," click "Fit Line at Total" (this requests the best regression line for the total sample). The resulting scatterplot (with additional editing to improve appearance) appears in Figure 3.2. Case markers are used to identify group membership on the control variable (i.e., grade level). Scores for first graders appear as 1, scores for fifth graders appear as 5, and scores for ninth graders appear as 9 in this scatterplot. The three groups of scores show some separation across grade levels. Both height and vocabulary increase across grade levels.

If you examine the graph in Figure 3.2, you can see that the groups of scores for grade levels do not overlap very much. Figure 3.3 shows an exaggerated version of this scatterplot to make the pattern more obvious; circles are added to highlight the three separate groups of scores. If you focus on just one group at a time, such as Grade 5 (circled), you can see that within each group, there is no association between height and vocabulary. To confirm this, we can run the correlation (and/or bivariate regression) analysis separately within each group. You should also be able to see that height increases across grade levels and that vocabulary increases across grade levels.

Next obtain the bivariate correlation for the entire sample of $N = 48$ cases. From the top menu bar in SPSS, make the following menu selections: <Analyze> → <Correlate> → <Bivariate>, then move the names of the X_1 and Y variables into the "Variables" pane in the main bivariate correlation dialog box, as shown in Figure 3.4.

The zero-order correlation between height and vocabulary (not controlling for grade) that appears in Figure 3.5 is $r(46) = .716$, $p < .01$. The number in parentheses after r is usually the *df*. The *df* for a bivariate correlation = $N - 2$, where N is the number of cases. There is a strong, positive, linear association between height and vocabulary when grade level is ignored. What happens when we control for grade level?

To examine the grade-level groups separately, use the SPSS split file procedure to divide the data into grade levels. Select <Data> → <Split File> to open the Split File dialog box; the menu selections and first dialog box appear in Figure 3.6. (Do not select the similarly named <Split into Files> command.)

Figure 3.3 Exaggerated Group Differences Across Grade Levels

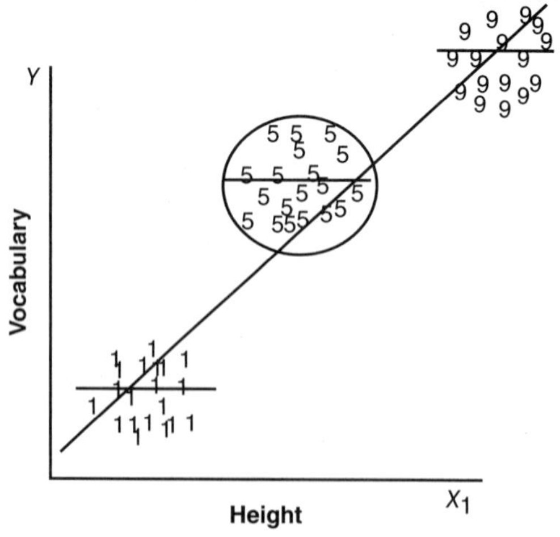

Figure 3.4 Bivariate Correlations Dialog Box

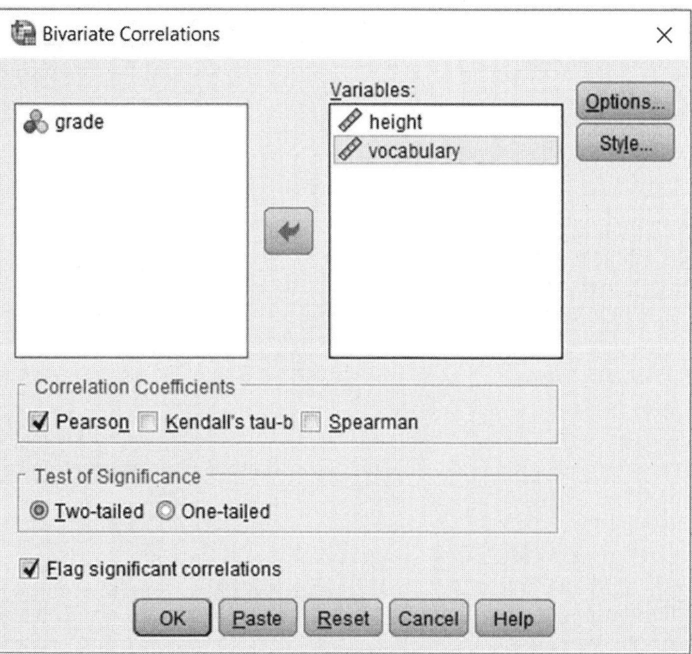

Figure 3.5 Correlation Between Height (X_1) and Vocabulary (Y) in Entire Sample

Correlations

		height	vocabulary
height	Pearson Correlation	1	.716**
	Sig. (2-tailed)		.000
	N	48	48
vocabulary	Pearson Correlation	.716**	1
	Sig. (2-tailed)	.000	
	N	48	48

**. Correlation is significant at the 0.01 level (2-tailed).

In the Split File dialog box in Figure 3.6, select the radio button for "Organize output by groups," then enter the name of the categorical control variable (grade) into the "Groups Based on" window, then click OK. All subsequent analysis will be reported separately for each grade level until you go back into the split file dialog box and make the selection "Analyze all cases, do not create groups."

Figure 3.6 Menu Selections and First Dialog Box for SPSS Split File Procedure

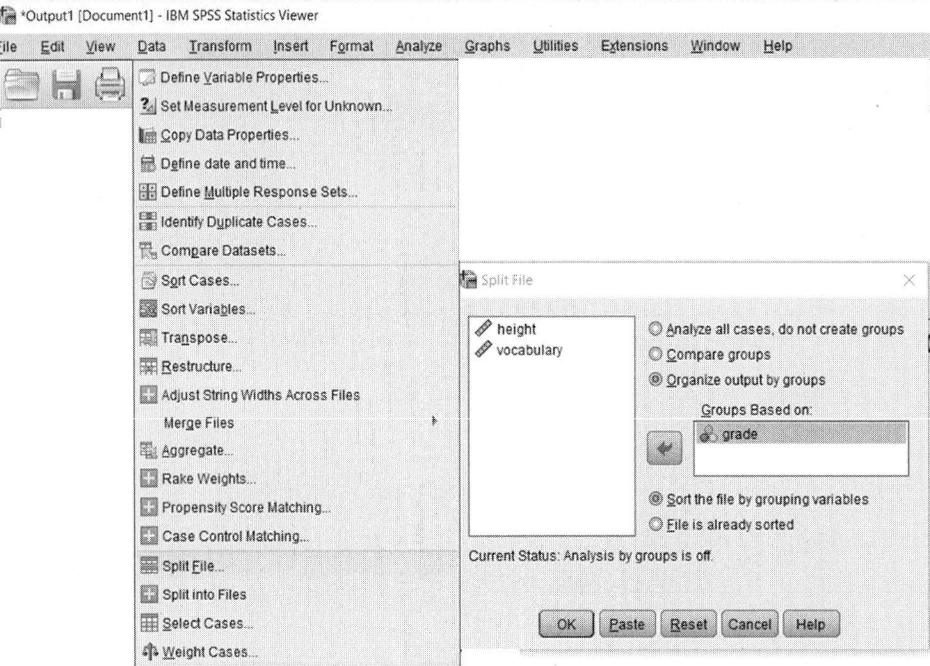

Now obtain the correlations between height (X_1) and vocabulary (Y) again, using the bivariate correlation procedure. The results for each of the three grade-level groups appear in Figure 3.7.

The within-group correlations in Figure 3.7 (r = .067 for Grade 1, r = .031 for Grade 5, and r = −.141 for Grade 9) did not differ significantly from 0; these correlations tell us how height and vocabulary are related when grade level is statistically controlled. Remember that the zero-order correlation between height and vocabulary, not controlling for grade level, was +.72 (p < .01) in Figure 3.5. Height and vocabulary appeared to be strongly, positively related when grade level was ignored; but when grade level was statistically controlled by looking at correlations within groups, height and vocabulary were not related. If you look separately at the clusters of data points for each grade, within each grade, the correlations between height and vocabulary do not differ significantly from zero. When we examine the height–vocabulary correlation separately with each grade-level group, we find out how height and vocabulary are related when height is held constant (by looking only within groups in which all members have the same grade-level score).

In this situation we can say that controlling for the X_2 control variable "explained away" or accounted for the seemingly positive correlation between height and vocabulary. We can conclude that the zero-order correlation between height and vocabulary was a **spurious correlation** (i.e., it was misleading). Later you will see that there are other possible

Figure 3.7 Correlations Between Height and Vocabulary Separately Within Each Grade

Correlations[a]

		height	vocabulary
height	Pearson Correlation	1	.067
	Sig. (2-tailed)		.806
	N	16	16
vocabulary	Pearson Correlation	.067	1
	Sig. (2-tailed)	.806	
	N	16	16

a. grade = 1

Correlations[a]

		height	vocabulary
height	Pearson Correlation	1	.031
	Sig. (2-tailed)		.909
	N	16	16
vocabulary	Pearson Correlation	.031	1
	Sig. (2-tailed)	.909	
	N	16	16

a. grade = 5

Correlations[a]

		height	vocabulary
height	Pearson Correlation	1	-.141
	Sig. (2-tailed)		.603
	N	16	16
vocabulary	Pearson Correlation	-.141	1
	Sig. (2-tailed)	.603	
	N	16	16

a. grade = 9

interpretations of situations in which controlling for an X_2 variable makes a correlation between X_1 and Y drop to zero.

3.3 ASSUMPTIONS FOR PARTIAL CORRELATION BETWEEN X_1 AND Y, CONTROLLING FOR X_2

Another way to evaluate the nature of the relationship between X_1 (height) and Y (vocabulary) while statistically controlling for X_2 (grade) is to compute a partial correlation between X_1 and Y, controlling for or partialling out X_2. The partial correlation between X_1 and Y controlling for X_2 is denoted $r_{1Y.2}$. The subscripts before the dot indicate which variables are being correlated. The subscripts after the dot indicate which variable(s) are being controlled. In this case we read $r_{1Y.2}$ as "the partial correlation between X_1 and Y, controlling for X_2."

For partial correlation to provide accurate information about the relationship between variables, the following assumptions about scores on X_1, X_2, and Y must be reasonably well satisfied. Detailed data screening procedures are not covered here; see Chapter 10 in Volume I (Warner, 2020) for review. Data screening should include the following:

1. Assess the types of variables. Partial correlation makes sense when X_1, X_2, and Y are all quantitative variables. (Under some circumstances, a dichotomous or dummy variable can be used in correlation analysis; for example, sex coded 1 = male and 2 = female can be correlated with height. However, you cannot use dichotomous variables as outcome or dependent variables in regression analysis.)

2. Ideally, scores on all variables should be approximately normally distributed. This can be assessed by examining histograms for all three variables. (The formal assumption is that scores are randomly sampled from normally distributed populations, and we have no way to test that assumption.)

3. There should not be extreme outliers or extreme bivariate outliers. Univariate outliers can be detected using boxplots (or other decision rules chosen prior to analysis). Bivariate outliers can be detected in scatterplots.

4. Examine scatterplots for all three pairs of variables. All three pairs of variables (X_1 with Y, X_2 with Y, and X_1 with X_2) must be linearly related. If they are not, use of Pearson correlation and partial correlation is not appropriate.

5. Other assumptions for use of Pearson's r (such as homogeneity of variance of Y across values of X_1) should be satisfied. Unfortunately, small samples usually don't provide enough information to evaluate these assumptions.

6. There must not be an interaction between X_1 and X_2 as predictors of Y (to say this another way, X_2 must not moderate the association between X_1 and Y). In the previous section, the SPSS split file procedure was used to divide the data set into groups on the basis of the categorical X_2 control variable, grade level. If the correlations or regression slopes for height and vocabulary had been different across groups, that would suggest a possible interaction between X_1 and X_2. Chapter 7, on moderation, explains how to test statistical significance for interactions. If an interaction is present, but not included in the analysis, partial correlations are misleading.

7. Factors that can artifactually influence the magnitudes of Pearson correlations must be considered whenever we examine other statistics that are based on these correlations. These are discussed in Appendix 10D at the end of the Chapter 10, on correlation, in Volume I (Warner, 2020). For example, if X_1 and Y both have low measurement reliability, the correlation between X_1 and Y will be attenuated or reduced, and any partial correlation that is calculated using r_{1Y} may also be inaccurate.

3.4 NOTATION FOR PARTIAL CORRELATION

To obtain a partial correlation between X_1 and Y controlling for X_2, we need the three bivariate or zero-order correlations among X_1, X_2, and Y. When we say that a correlation is "zero-order," we mean that the answer to the question "How many other variables were statistically controlled or partialled out when calculating this correlation?" is zero or none.

The following notation is used. Subscripts for r indicate which variables are involved in the analysis.

r_{Y1} or r_{1Y} denotes the zero-order bivariate Pearson correlation between Y and X_1.

r_{Y2} or r_{2Y} denotes the zero-order correlation between Y and X_2.

r_{12} or r_{21} denotes the zero-order correlation between X_1 and X_2.

For a **first-order partial correlation** between X_1 and Y, controlling for X_2, the term *first-order* tells us that only one variable (X_2) was statistically controlled when assessing how X_1 and Y are related. In a **second-order partial correlation**, the association between X_1 and Y is assessed while statistically controlling for two variables; for example, $r_{Y1.23}$ would be read as "the partial correlation between Y and X_1, statistically controlling for X_2 and X_3." Variables that follow the period in the subscript are control variables. In a kth-order partial correlation, there are k control variables. This chapter examines first-order partial correlation in detail; the conceptual issues involved in the interpretation of higher order partial correlations are similar.

The three zero-order correlations listed above (r_{1Y}, r_{2Y}, and r_{12}) provide information we can use to answer the question "When we control for, or take into account, a third variable called X_2, how does that change our description of the relation between X_1 and Y?" However, examination of separate scatterplots that show how X_1 and Y are related separately for each level of the X_2 variable provides additional, important information.

In the following examples, a distinction is made among three variables: an independent or predictor variable (denoted by X_1), a dependent or outcome variable (Y), and a control variable (X_2). The preliminary analyses in this chapter provide ways of exploring whether the nature of the relationship between X_1 and Y changes when you remove, partial out, or statistically control for the X_2 variable.

The following notation is used to denote the partial correlation between Y and X_1, controlling for X_2: $r_{Y1.2}$. The subscript 1 in $r_{Y1.2}$ refers to the predictor variable X_1, and the subscript 2 refers to the control variable X_2. When the subscript is read, pay attention to the position in which each variable is mentioned relative to the period in the subscript. The period within the subscript divides the subscripted variables into two sets. The variable or variables to the right of the period in the subscript are used as predictors in a regression analysis; these are the variables that are statistically controlled or partialled out. The variable or variables to the left of the period in the subscript are the variables for which the partial correlation is assessed while taking one or more control variables into account. Thus, in $r_{Y1.2}$, the subscript $Y1.2$ denotes the partial correlation between X_1 and Y, controlling for X_2.

In the partial correlation, the order in which the variables to the left of the period in the subscript are listed does not signify any difference in the treatment of variables; we could read either $r_{Y1.2}$ or $r_{1Y.2}$ as "the partial correlation between X_1 and Y, controlling for X_2." However, changes in the position of variables (before vs. after the period) do reflect a difference in their treatment. For example, we would read $r_{Y2.1}$ as "the partial correlation between X_2 and Y, controlling for X_1."

Another common notation for partial correlation is pr_1. The subscript 1 associated with pr_1 tells us that the partial correlation is for the predictor variable X_1. In this notation, it is implicit that the dependent variable is Y and that other predictor variables, such as X_2, are statistically controlled. Thus, pr_1 is the partial correlation that describes the predictive relation of X_1 to Y when one or more other variables are controlled.

3.5 UNDERSTANDING PARTIAL CORRELATION: USE OF BIVARIATE REGRESSIONS TO REMOVE VARIANCE PREDICTABLE BY X_2 FROM BOTH X_1 AND Y

To understand the partial correlation between X_1 and Y, controlling for X_2, it is helpful to do the following series of simple analyses. First, use bivariate regression to obtain residuals for the prediction of Y from X_2. This involves two steps. First, find the predicted value of Y (denoted Y') from the following bivariate regression:

$$Y' = b_0 + b_2X_2. \tag{3.1}$$

By definition, Y' represents the part of the Y scores that is predictable from X_2.

Then, to find the part of Y that is not predictable from X_2, we obtain the residual, that is, the difference between the original Y score and the predicted Y score. This residual is denoted Y^*.

$$\text{Residual for } Y = Y^* = (Y - Y').$$

Similar analyses are carried out to find the part of the X_1 score that is not predictable from X_2: First, do a bivariate regression to predict X_1 from X_2. The value of actual minus predicted X_1 scores, denoted X_1^*, is the part of the X_1 scores that is not related to X_2.

The partial correlation between X_1 and Y, controlling for or partialling out X_2, can be obtained by finding the correlation between X_1^* and Y^* (the parts of the X_1 and Y scores that are not related to X_2).

Consider this situation as an example. You want to know the correlation between X_1, everyday life stress, and Y, self-reported physical illness symptoms. However, you suspect that people high in the personality trait neuroticism (X_2) complain a lot about their everyday lives and also complain a lot about their health. Suppose you want to remove the effects of this complaining tendency on both X_1 and Y. To do that, you find the residuals from a regression that predicts Y, physical illness symptoms, from X_2, neuroticism. Call the residuals Y^*. You also find the residuals for prediction of X_1, everyday life stress, from X_2, neuroticism; these residuals are called X_1^*. When you find the correlation between X_1^* and Y^*, you can assess the strength of association between these variables when X_2 (neuroticism effects) has been completely removed from both variables.

Two control variables often used in personality research are neuroticism and social desirability response bias (social desirability measures assess a tendency to report answers that are more socially approved, instead of accurate answers). In ability measurement studies, a common control variable is verbal ability. (I received the Betty Crocker Homemaker of the Year award in high school because of my high score on a "homemaking skills" self-report test. However, scores on that test depended mostly on verbal and arithmetic ability. If they had partialled out verbal ability, this could have yielded the part of the homemaking test scores related to actual homemaking skills, and someone more deserving might have won the award. Did I mention that I flunked peach pie making?)

A partial correlation was obtained for the variables in the heightvocabulary.sav data examined in previous sections: height (X_1), vocabulary score (Y), and grade (X_2). Height increases with grade level; vocabulary increases with grade level. The fact that these scores both increase across grade levels may completely explain why they appear to be related. Figure 3.8 shows the SPSS Linear Regression dialog box to run the regression specified in Equation 3.1 (to predict X_1 from X_2—in this example, height from grade). Figure 3.9 shows the SPSS Data View worksheet after performing the regressions in Equations 3.1 (predicting

Figure 3.8 Bivariate Regression to Predict Height (X_1) From Grade in School (X_2)

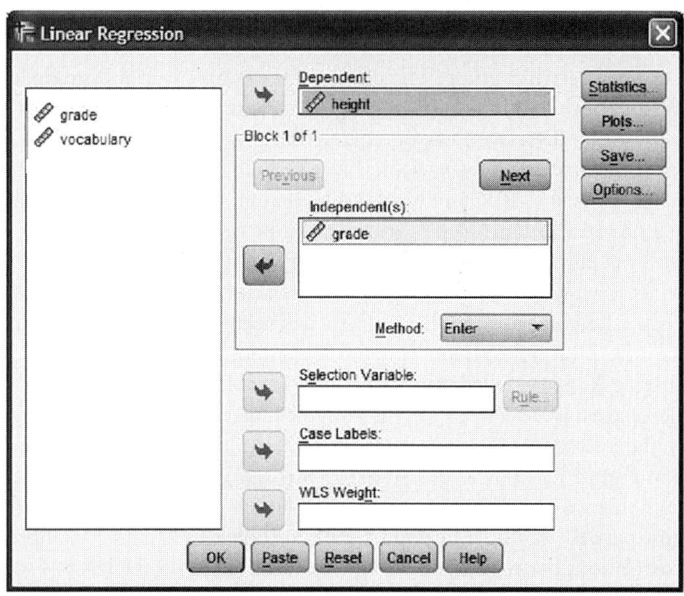

Figure 3.9 SPSS Data View Worksheet That Shows Scores on RES_1 and RES_2

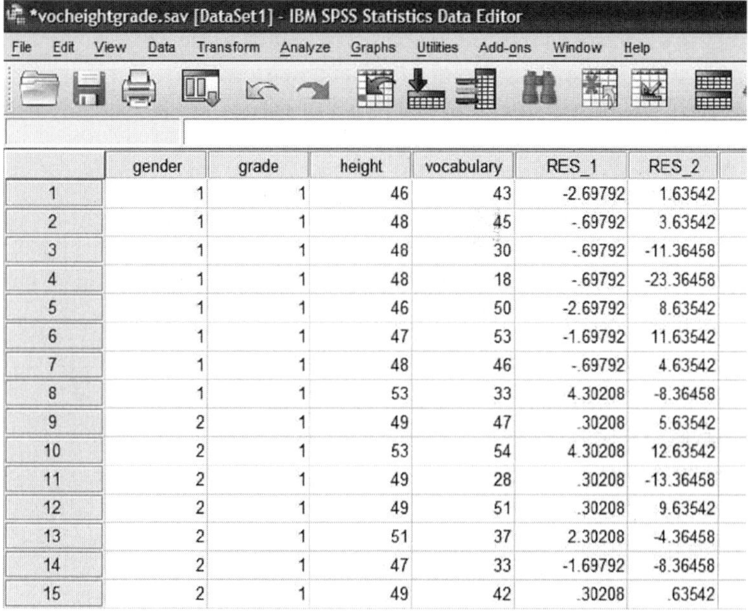

	gender	grade	height	vocabulary	RES_1	RES_2
1	1	1	46	43	-2.69792	1.63542
2	1	1	48	45	-.69792	3.63542
3	1	1	48	30	-.69792	-11.36458
4	1	1	48	18	-.69792	-23.36458
5	1	1	46	50	-2.69792	8.63542
6	1	1	47	53	-1.69792	11.63542
7	1	1	48	46	-.69792	4.63542
8	1	1	53	33	4.30208	-8.36458
9	2	1	49	47	.30208	5.63542
10	2	1	53	54	4.30208	12.63542
11	2	1	49	28	.30208	-13.36458
12	2	1	49	51	.30208	9.63542
13	2	1	51	37	2.30208	-4.36458
14	2	1	47	33	-1.69792	-8.36458
15	2	1	49	42	.30208	.63542

Note: RES_1 and RES_2 are the saved unstandardized residuals for the prediction of height from grade and the prediction of vocabulary from grade. In the text these are renamed Resid_Height and Resid_Voc.

height from grade) and 3.2 (predicting vocabulary score from grade). The residuals from these two separate regressions were saved as new variables and renamed.

Res_1, renamed Resid_Height, refers to the part of the scores on the X_1 variable, height, that was not predictable from or related to the control or X_2 variable, grade. Res_2, renamed Resid_Voc, refers to the part of the scores on the Y variable, vocabulary, that was not predictable from the control variable, grade. The effects of the control variable grade level (X_2) have been removed from both height and vocabulary by obtaining these residuals.

Finally, we use the bivariate Pearson correlation procedure (Figure 3.10) to obtain the correlation between these two new variables, Resid_Height and Resid_Voc. The correlation between these residuals, $r = -.012$ in Figure 3.11, corresponds to the value of the partial correlation between X_1 and Y, controlling for or partialling out X_2.

Note that X_2 is partialled out or removed from both variables (X_1 and Y). This partial $r = -.012$ tells us that X_1 (height) is not significantly correlated with Y (vocabulary) when variance that is predictable from grade level (X_2) has been removed from or partialled out of both the X_1 and the Y variables. Later you will learn about semipartial correlation, in which the control variable X_2 is partialled out of only one variable.

The value of partial r between height and vocabulary, controlling for grade (–.012) is approximately the average of the within-group correlations between height and vocabulary that appeared in Figure 3.7 $(.067 + .031 - .141)/3 = (-.043)/3 \approx -.012$. This correspondence is not close enough that we can use within-group correlations to compute an overall partial r, but it illustrates how partial r can be interpreted. Partial r between X_1 and Y is approximately the mean of the correlations between X_1 and Y for separate groups on the basis of scores for X_2.

Figure 3.10 Correlation Between Residuals for Prediction of Height From Grade (Resid_Height) and Residuals for Prediction of Vocabulary From Grade (Resid_Voc)

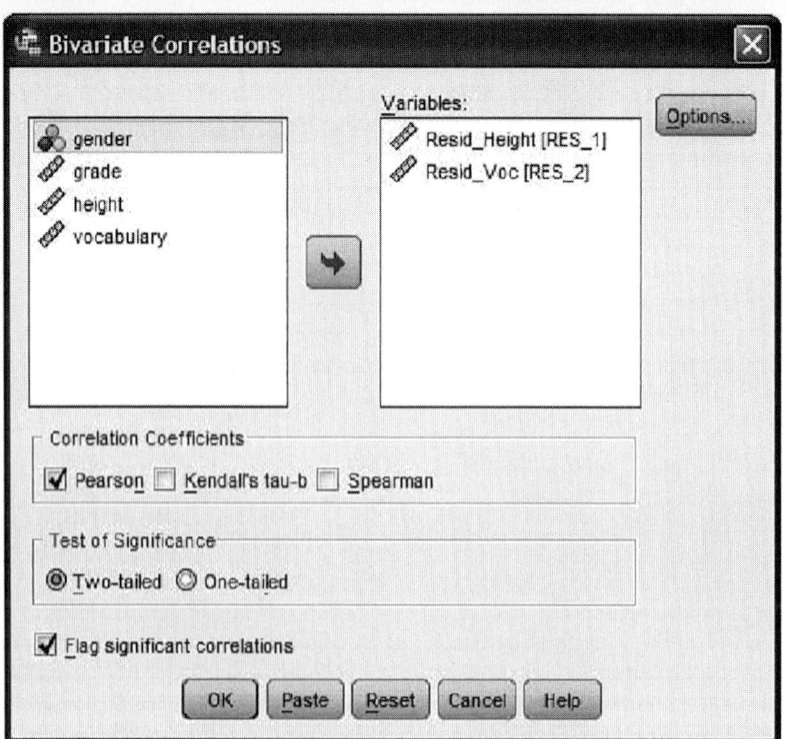

Figure 3.11 Correlation Between Residuals for Height and Vocabulary Using Grade as Control Variable

Correlations

		Resid_Height	Resid_Voc
Resid_Height	Pearson Correlation	1	-.012
	Sig. (2-tailed)		.937
	N	48	48
Resid_Voc	Pearson Correlation	-.012	1
	Sig. (2-tailed)	.937	
	N	48	48

Note: The control variable grade (X_2) was used to predict scores on the other variables (X_1, height, and Y, vocabulary). The variable Resid_Height contains the residuals from the bivariate regression to predict height (X_1) from grade (X_2). The variable Resid_Voc contains the residuals from the bivariate regression to predict vocabulary (Y) from grade (X_2). These residuals correspond to the parts of the X_1 and Y scores that are not related to or not predictable from grade (X_2).

In the preceding example, X_2 had only three score values, so we needed to examine only three groups. In practice, the X_2 variable often has many more score values (which makes looking at subgroups more tedious and less helpful; numbers of cases within groups can be very small).

3.6 PARTIAL CORRELATION MAKES NO SENSE IF THERE IS AN $X_1 \times X_2$ INTERACTION

The interpretation for partial correlation (as the mean of within-group correlations) does not make sense if assumptions for partial correlation are violated. If X_1 and X_2 interact as predictors of Y, partial correlation analysis will not help us understand the situation. A graduate student once brought me data that he didn't understand. He was examining predictors of job satisfaction for male and female MBA students in their first jobs. In his data, the control variable X_2 corresponded to sex, coded 1 = male, 2 = female. X_1 was a measure of need for power. Y was job satisfaction evaluations.

Figure 3.12 is similar to the data he showed me. Back in the late 1970s, when the data were collected, women who tried to exercise power over employees got more negative reactions than men who exercised power. This made management positions more difficult for women than for men. We can set up a scatterplot to show how evaluations of job satisfaction (Y) are related to MBA students' need for power (X_1). Case markers identify which scores belong to male managers (X_2 = m) and which scores belong to female managers (X_2 = f). Sex of manager was the X_2 or "controlled for" variable in this example.

If we look only at the scores for male managers (denoted by "m" in Figure 3.12), there was a positive correlation between need for power (X_1) and job satisfaction (Y). If we look only at the scores for female managers (denoted by "f" in Figure 3.12), there was a negative correlation between need for power (X_1) and job satisfaction (Y).

In this example, we could say that sex and need for power interact as predictors of job satisfaction evaluation; more specifically, for male managers, their job satisfaction evaluations increase as their need for power scores increase, whereas for female managers, their job satisfaction evaluations decrease as their need for power scores increase. We could also say that sex "moderates" the relationship between need for power and job satisfaction evaluation.

Figure 3.12 Interaction Between Sex (X_2) and Need for Power (X_1) as Predictors of Job Satisfaction Evaluation (Y)

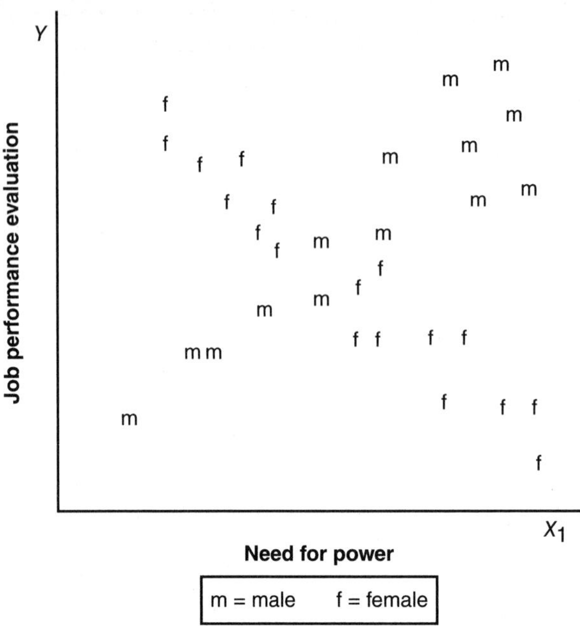

Note: Within the male group, X_1 and Y are positively correlated; within the female group, X_1 and Y are negatively correlated.

In this example, the slopes for the subgroups (male and female) had opposite signs. Moderation or interaction effects do not have to be this extreme. We can say that sex moderates the effect of X_1 on Y if the slopes to predict Y from X_1 are significantly different for men and women. The slopes do not actually have to be opposite in sign for an interaction to be present, and the regression lines within the two groups do not have to cross. Another type of interaction would be no correlation between X_1 and Y for women and a strong positive correlation between X_1 and Y for men. Yet another kind of interaction is seen when the b slope coefficient to predict Y from X_1 is positive for both women and men but is significantly larger in magnitude for men than for women.

The student who had these data found that the overall correlation between X_1 and Y was close to zero. He also found that the partial correlation between X_1 and Y, controlling for sex, was close to zero. When he looked separately at male and female groups, as noted earlier, he found a positive correlation between X_1 and Y for men and a negative correlation between X_1 and Y for women. (Robert Rosenthal described situations like this as "different slopes for different folks.")

If he had reported r_{1Y} (the correlation between need for power and job satisfaction, ignoring sex) near 0 and $r_{1Y.2}$ (the partial correlation between need for power and job satisfaction, controlling for sex) also near 0, this would not be an adequate description of his results. A reader would have no way to know from these correlations that there actually were correlations between need for power and job satisfaction but that the nature of the relation differed for men and women. One way to provide this information would be to report the correlation and regression separately for each sex. (Later you will see better ways to do this by including interaction terms in regression equations; see Chapter 7, on moderation.)

3.7 COMPUTATION OF PARTIAL *r* FROM BIVARIATE PEARSON CORRELATIONS

There is a simpler direct method for the computation of the partial *r* between X_1 and Y, controlling for X_2, on the basis of the values of the three bivariate correlations:

r_{Y1}, the correlation between Y and X_1,

r_{Y2}, the correlation between Y and X_2, and

r_{12}, the correlation between X_1 and X_2.

The formula to calculate the partial *r* between X_1 and Y, controlling for X_2, directly from the Pearson correlations is as follows:

$$pr_1 = r_{Y1.2} = \frac{(r_{1Y} - (r_{12} \times r_{2Y}))}{\sqrt{1-r_{12}^2}\sqrt{1-r_{2Y}^2}} \tag{3.2}$$

In the preceding example, where X_1 is height, Y is vocabulary, and X_2 is grade, the corresponding bivariate correlations were $r_{1Y} = +.716$, $r_{2Y} = +.787$, and $r_{12} = +.913$. If these values are substituted into Equation 3.2, the partial correlation $r_{Y1.2}$ is as follows:

$$\frac{+.716-(.913\times.787)}{\sqrt{1-.913^2}\sqrt{1-.787^2}} = \frac{.716-.71853}{\sqrt{.166431}\sqrt{.380631}}$$

$$= \frac{-.00253}{(.40796)\times(.61653)}$$

$$= \frac{-.00253}{.251692} \approx -.010$$

Within rounding error, this value of −.010 agrees with the value that was obtained from the correlation of residuals from the two bivariate regressions reported in Figure 3.11. In practice, it is rarely necessary to calculate a partial correlation by hand. If you read an article that reports only zero-order correlations, you could use Equation 3.2 to calculate partial correlations for additional information.

The most convenient way to obtain a partial correlation, when you have access to the original data, is the partial correlations procedure in SPSS. The SPSS menu selections <Analyze> → <Correlate> → <Partial>, shown in Figure 3.13, open the Partial Correlations dialog box, which appears in Figure 3.14. The names of the predictor and outcome variables (height and vocabulary) are entered in the pane that is headed "Variables." The name of the control variable, grade, is entered in the pane under the heading "Controlling for." (Note that more than one variable can be placed in this pane; that is, we can include more than one control variable.) The output for this procedure appears in Figure 3.15, where the value of the partial correlation between height and vocabulary, controlling for grade, is given as $r_{1Y.2} = -.012$; this partial correlation is not significantly different from 0 (and is identical to the correlation between Resid_Height and Resid_Voc reported in Figure 3.11).

Partial correlation is approximately (but not exactly) the mean of the X_1, Y correlations obtained by running an X_1, Y correlation for each score on the X_2 variable. In this example, where the X_2 variable is grade level, each grade level contained numerous cases. In situations where X_2 is a quantitative variable with many possible values, the same thing happens (essentially), but it is more difficult to imagine because some values of X_2 have few cases.

Figure 3.13 SPSS Menu Selections for Partial Correlation

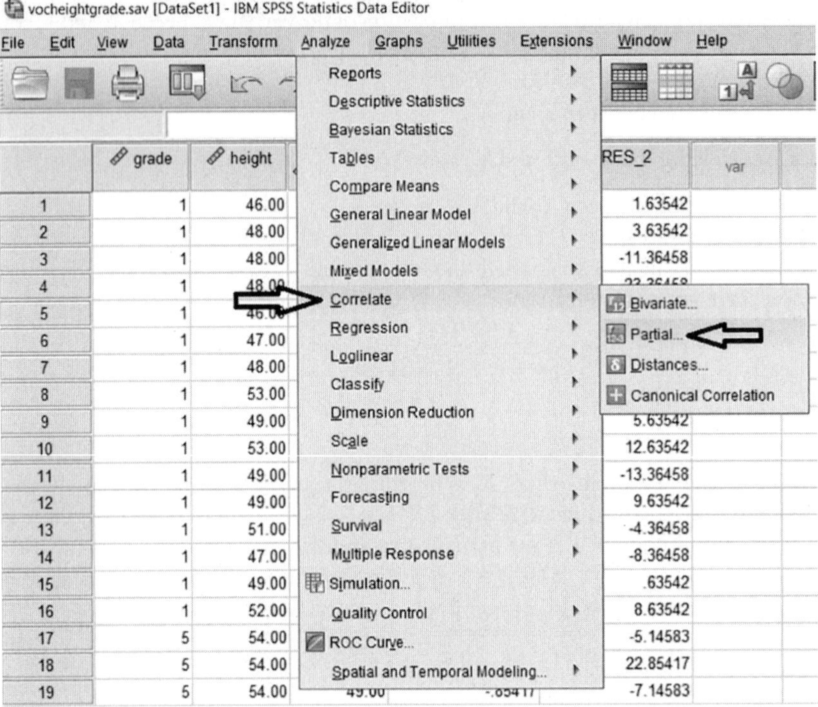

Figure 3.14 SPSS Dialog Box for the Partial Correlations Procedure

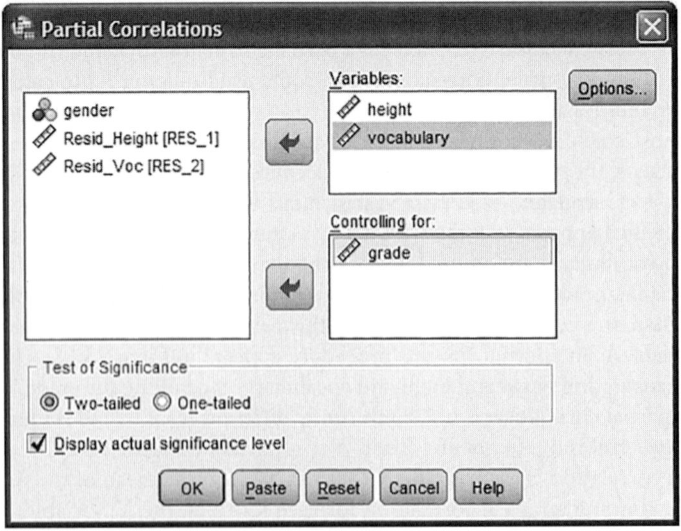

Figure 3.15 Output From SPSS Partial Correlations Procedure

Correlations

Control Variables			height	vocabulary
grade	height	Correlation	1.000	-.012
		Significance (2-tailed)	.	.938
		df	0	45
	vocabulary	Correlation	-.012	1.000
		Significance (2-tailed)	.938	.
		df	45	0

Note: First-order partial correlation between height (X_1) and vocabulary (Y), controlling for grade level (X_2). The partial correlation between height and vocabulary controlling for grade ($r = -.012$) is identical to the correlation between Resid_Height and Resid_Voc ($r = -.012$) that appeared in Figure 3.11.

3.8 SIGNIFICANCE TESTS, CONFIDENCE INTERVALS, AND STATISTICAL POWER FOR PARTIAL CORRELATIONS

3.8.1 Statistical Significance of Partial r

The null hypothesis that a partial correlation equals 0 can be tested by setting up a t ratio similar to the test for the statistical significance of an individual zero-order Pearson correlation. The SPSS partial correlations procedure provides this statistical significance test; SPSS reports an exact p value for the statistical significance of partial r. The degrees of freedom (df) for a partial correlation are $N - k$, where k is the total number of variables that are involved in the partial correlation, and N is the number of cases or participants.

3.8.2 Confidence Intervals for Partial *r*

Textbooks do not present detailed formulas for standard errors or confidence intervals for partial correlations. Olkin and Finn (1995) and Graf and Alf (1999) provided formulas for computation of the standard error for partial correlations; however, the formulas are complicated and not easy to work by hand. SPSS does not provide standard errors or confidence interval estimates for partial correlations.

3.8.3 Effect Size, Statistical Power, and Sample Size Guidelines for Partial *r*

Like Pearson's r (and r^2), the partial correlation $r_{Y1.2}$ and squared partial correlation $r^2_{Y1.2}$ can be interpreted directly as information about effect size or strength of association between variables. Effect size labels for values of Pearson's r can reasonably be used to describe effect sizes for partial correlations ($r = .10$ is small, $r = .30$ is medium, and $r = .50$ is large). Algina and Olejnik (2003) provided statistical power tables for correlation analysis with discussion of applications in partial correlation and multiple regression analysis. Later chapters discuss statistical power further in the situation where it is more often needed: multiple regression. In general, no matter what minimum sample sizes are given by statistical power tables, it is desirable to have large sample sizes for partial correlation, on the order of $N = 100$.

3.9 COMPARING OUTCOMES FOR $r_{Y1.2}$ AND r_{Y1}

When we compare the size and sign of the zero-order correlation between X_1 and Y with the size and sign of the partial correlation between X_1 and Y, controlling for X_2, several different outcomes are possible. The value of r_{1Y}, the zero-order correlation between X_1 and Y, can range from −1 to +1. The value of $r_{1Y.2}$, the partial correlation between X_1 and Y, controlling for X_2, can also potentially range from −1 to +1 (although in practice its actual range may be limited by the correlations of X_1 and Y with X_2). In principle, any combination of values of r_{1Y} and $r_{1Y.2}$ can occur (although some outcomes are much more common than others).

Here is a list of possible outcomes when r_{1Y} is compared with $r_{Y1.2}$:

1. Both r_{1Y} and $r_{Y1.2}$ are not significantly different from zero.
2. Partial correlation $r_{Y1.2}$ is approximately equal to r_{1Y}.
3. Partial correlation $r_{Y1.2}$ is not significantly different from zero, even though r_{1Y} differed significantly from 0.
4. Partial correlation $r_{Y1.2}$ is smaller than r_{1Y} in absolute value, but $r_{Y1.2}$ is significantly greater than 0.
5. Partial correlation $r_{Y1.2}$ is larger than r_{1Y} in absolute value, or opposite in sign from r_{Y1}.

Each of these is discussed further in later sections. Before examining possible interpretations for these five outcomes, we need to think about reasonable causal and noncausal hypotheses about associations among the variables X_1, Y, and X_2. Path models provide a way to represent these hypotheses. After an introduction to path models, we return to possible interpretations for the five outcomes listed above.

3.10 INTRODUCTION TO PATH MODELS

Path models are diagrams that represent *hypotheses* about how variables are related. These are often called "causal" models; that name is unfortunate because statistical analyses based on these models generally don't provide evidence that can be used to make causal inferences. However, some paths in these models represent causal hypotheses. The arrows shown in Table 3.1 represent three different hypotheses about how a pair of variables, X_1 and Y, may be related.

Given an r_{1Y} correlation from data, we can make only one distinction. If r_{1Y} does not differ significantly from 0, we prefer the model in row 1 (no association). If r_{1Y} does differ significantly from 0 (and whether it is positive or negative), we prefer one of the models in rows 2, 3, and 4. However, a significant correlation cannot tell us which of these three models is "true." First, we need a theory that tells us what potential causal connections make sense. Second, results from experiments in which the presumed causal variable is manipulated and other variables are controlled and observed provide stronger support for causal inferences. A noncausal association hypothesis should be the preferred interpretation of Pearson's r, unless we have other evidence for possible causality.

Table 3.1 Four Possible Hypothesized Paths Between Two Variables (X_1 and Y)

	Verbal Description of the Relationship Between X_1 and Y	Path Model for X_1 and Y	Model Is Consistent With These Values of r_{1Y} Correlation
1	X_1 and Y are not directly associated.	X_1 Y	r_{1Y} close to 0
2	X_1 and Y are associated but not in a causal way. They are correlated[a] (or confounded).	$X_1 \leftrightarrow Y$ (curved bidirectional arrow)	r_{1Y} significantly differs from 0
3	X_1 is hypothesized to cause Y	$X_1 \rightarrow Y$	r_{1Y} significantly differs from 0
4	Y is hypothesized to cause X_1	$Y \rightarrow X1$	r_{1Y} significantly differs from 0

a. The bidirectional arrow that represents a correlational relationship is often shown as a curve, but it can be straight.

3.11 POSSIBLE PATHS AMONG X_1, Y, AND X_2

When we take a third variable (X_2) into account, the number of possible models to represent relationships among variables becomes much larger. In Figure 3.16, there are three pairs of variables (X_1 and X_2, X_1 and Y, and X_2 and Y). Each rectangle can be filled in with almost any[1] of the four possible types of path (no relation, noncausal association, or one of the two arrows that correspond to a causal hypothesis). The next few sections of this chapter describe examples of causal models that might be proposed as hypotheses for the relationships among three variables. Conventionally, causal arrows point from left to right, or from top down. Locations of variables can be rearranged so that this can be done.

One of the most common (often implicit) models corresponds to the path model regression that represents X_1 and X_2 as correlated predictors of Y. A major difference between experimental and nonexperimental research is that experimenters can usually arrange for manipulated independent variables to be uncorrelated with each other. In nonexperimental research, we often work with predictors that are correlated, and analyses must take correlations between predictors into account. For the analysis in Figure 3.17, the bidirectional or noncausal path between X_1 and X_2 corresponds to the correlation r_{12}. The unidirectional arrows from X_1 to Y and from X_2 to Y can represent causal hypotheses, but when numerical results are obtained, it is usually better to interpret them as information about strength of predictive associations (and not make causal inferences). You will see later that regression with more than one predictor variable is used to obtain values (called path coefficients) that indicate the strength of association for each path. We haven't yet covered these methods yet, and for now, the "causal" path coefficients are denoted by question marks, indicating that you do not yet have methods to obtain these values. When

Figure 3.16 Blank Template for All Possible Paths Among Three Variables

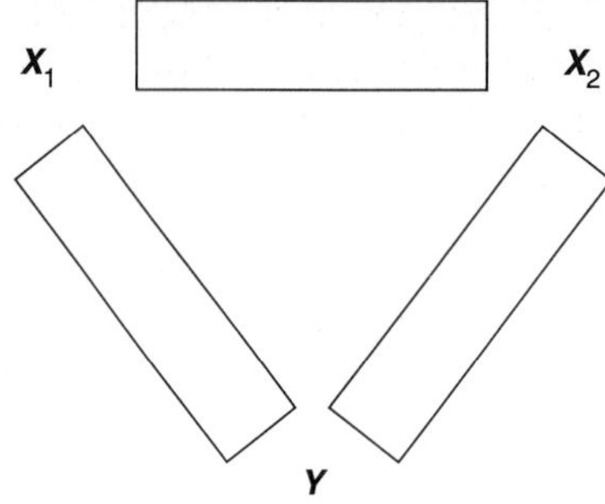

Figure 3.17 Path Model for Prediction of Y From Correlated Predictor (or Causal) Variables X_1 and X_2

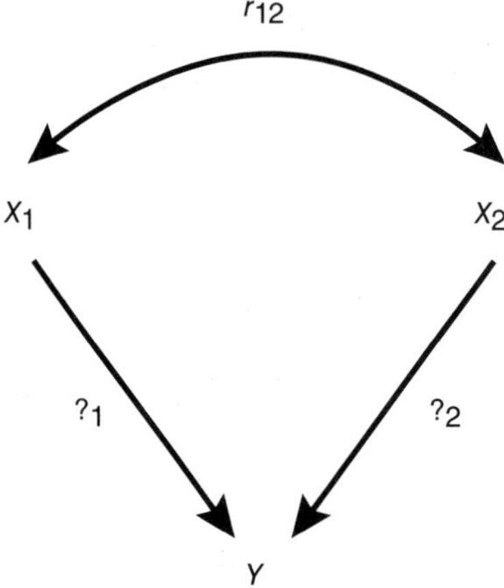

standardized path coefficients (β coefficients) are reported, their values can be interpreted like correlations. Values for these path coefficients will be estimated later using regression coefficients, which in turn depend on the set of correlations among all the variables in the path diagram or model.

Here is a hypothetical situation in which the model in Figure 3.17 might be used: Suppose we are interested in predicting Y, number of likes people receive on social media. Number of likes might be influenced or caused by X_1, quality of posts, and X_2, number of posts. In addition, X_1 and X_2 may be correlated or confounded; for instance, a person who posts frequently may also post higher quality material. Data for this situation can be analyzed using regression with two correlated predictor variables (discussed in the next chapter). Sometimes an X_2 variable is added to an analysis because a researcher believes that X_1 and X_2 together can predict more variance in the Y outcome than X_1 alone. In addition, researchers want to evaluate how adding X_2 to an analysis changes our understanding of the association between X_1 and Y. There are numerous possible outcomes for a three-variable analysis. The correlation for path r_{12} can either be nonsignificant or significant with either a positive or negative sign. The coefficient for any path marked with a question mark can be either significant or nonsignificant, positive or negative. A significant path coefficient $?_1$ indicates that X_1 is significantly related to Y, in the context of an analysis that also includes X_2 and paths that relate X_1 and Y to X_2.

Obtaining statistically significant paths for both the path $X_1 \rightarrow Y$ and the path $X_2 \rightarrow Y$ is not proof that X_1 causes Y or that X_2 causes Y. A model in which all paths are noncausal, as in Figure 3.18, would be equally consistent with statistically significant estimates for all path coefficients.

Path models are often more interesting and informative when at least one path is not statistically significant. Recall the example examining the association between height (X_1) and vocabulary (Y), controlling for grade level (X_2). When X_2 was included in the analysis, the association between height and vocabulary dropped to 0. Figures 3.19 and 3.20 depict two reasonable corresponding path models.

If $r_{1Y.2}$ is close to 0, we do not need a direct path between X_1 and Y in a model that includes X_2; for example, because the correlation of height and vocabulary dropped to zero when grade level was taken into account, we can drop the direct path between height and vocabulary from the model. We can conclude that the only reason we find a correlation between height and vocabulary is that each of them is correlated with (or perhaps caused by) grade level.

Unidirectional arrows in these models represent causal hypotheses. We cannot prove or disprove any of these hypotheses using data from nonexperimental research. However, we can interpret some outcomes for correlation and partial correlation as consistent with, or not consistent with, different possible models. This makes it possible, sometimes, to reduce the set of models that are considered plausible explanations for the relationships among variables. The next sections describe possible explanations to consider on the basis of comparison of r_{1Y} (not controlling for X_2) and $r_{1Y.2}$ (controlling for X_2). One of the few things a partial correlation tells us is whether it is reasonable to drop one (or for more complex situations, more than one) of the paths from the model. Beginning in the next chapter (regression with two predictor variables) we'll use coefficients from regression equations to make inferences about path model coefficients. This has advantages over the partial correlation approach.

Figure 3.18 Path Mode With Only Correlation Paths (No Causal Paths)

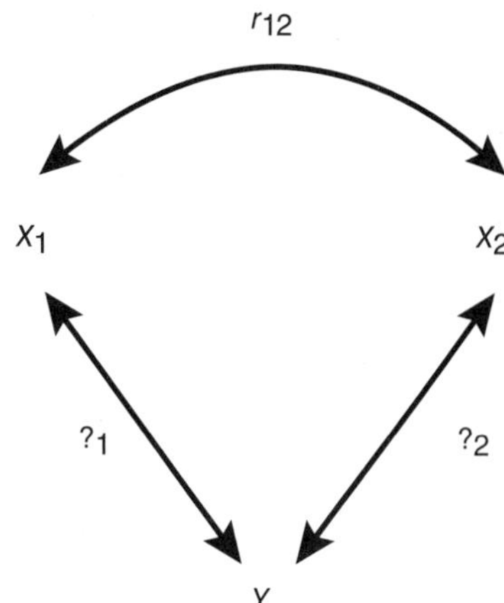

Figure 3.19 Path Model: Height and Vocabulary Are Both Correlated With Grade Level but Are Not Directly Related to Each Other

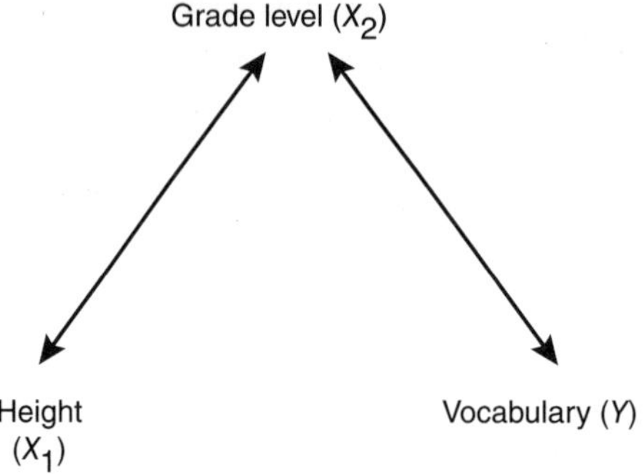

Figure 3.20 Height and Vocabulary Are Both Influenced or Caused by Maturation (Grade Level)

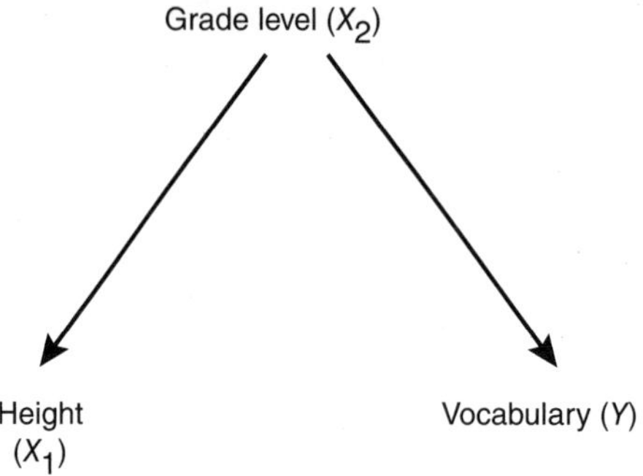

3.12 ONE POSSIBLE MODEL: X_1 AND Y ARE NOT RELATED WHETHER YOU CONTROL FOR X_2 OR NOT

One possible hypothetical model is that none of the three variables (X_1, X_2, and Y) is either causally or noncausally related to the others. This would correspond to a model that has no path (either causal or noncausal) between any pair of variables. If we obtain Pearson's r values for r_{12}, r_{1Y}, and r_{2Y} that are not significantly different from 0 (and all three correlations are too small to be of any practical or theoretical importance), those correlations are consistent with a model that has no paths among any of the three pairs of variables. The partial correlation between X_1 and Y, controlling for X_2, would also be 0 or very close to 0 in this situation. A researcher who obtains values close to 0 for all the bivariate (and partial) correlations would probably conclude that none of these variables are related to the others either causally or noncausally. This is usually not considered an interesting outcome. In path model terms, this would correspond to a model with no path between any pair of variables.

3.13 POSSIBLE MODEL: CORRELATION BETWEEN X_1 AND Y IS THE SAME WHETHER X_2 IS STATISTICALLY CONTROLLED OR NOT (X_2 IS IRRELEVANT TO THE X_1, Y RELATIONSHIP)

If the partial correlation between X_1 and Y, controlling for X_2, is approximately equal to the zero-order correlation between X_1 and Y, that is, $r_{1Y.2} \approx r_{1Y}$, we can say that the X_2 variable is "irrelevant" to the X_1, Y relationship. If a researcher finds a correlation he or she "likes" between an X_1 and a Y variable, and the researcher is asked to consider a rival explanatory variable X_2 he or she does not like, this may be the outcome the researcher wants. This outcome could correspond to a model with only one path, between X_1 and Y. The path could be either noncausal or causal; if causal, the path could be in the direction $X_1 \rightarrow Y$ or $Y \rightarrow X_1$). There would be no paths connecting X_2 with either X_1 or Y.

3.14 WHEN YOU CONTROL FOR X_2, CORRELATION BETWEEN X_1 AND Y DROPS TO ZERO

When $r_{1Y.2}$ is close to zero (but r_{1Y} is not close to zero), we can say that the X_2 control variable completely accounts for or explains the X_1, Y relationship. In this situation, the path model does not need a direct path from X_1 to Y, because we can account for or explain their relationship through associations with X_2. There are several possible explanations for this situation. However, because these explanations are all equally consistent with $r_{1Y.2} = 0$, information from data cannot determine which explanation is best. On the basis of theory, a data analyst may prefer one interpretation over others, but analysts should always acknowledge that other interpretations or explanations are possible (MacKinnon, Krull, & Lockwood, 2000).

If r_{1Y} differs significantly from 0, but $r_{1Y.2}$ does not differ significantly from zero:

- X_1 and X_2 could be interpreted as strongly correlated or confounded predictors (or causes) of Y; when $r_{1Y.2} = 0$, all the predictive information in X_1 may also be included in X_2. In that case, after we use X_2 to predict Y, we don't gain anything by adding X_1 as another predictor.
- The X_1, Y association may be spurious; that is, the r_{1Y} correlation may be nonzero only because X_1 and Y are both correlated with, or both caused by, X_2.
- The X_1, Y association may be completely mediated by X_2 (discussed below).
- The X_1, Y association might involve some form of suppression; that is, the absence of direct association between X_1 and Y may be clear only when X_2 is statistically controlled. (There are other forms of suppression in which r_{1Y} would not be close to 0.)

3.14.1 X_1 and X_2 Are Completely Redundant Predictors of Y

In the next chapter you will see that when X_1 and X_2 are used together as regression predictors of Y, it is possible for the contribution of X_1 to the prediction of Y (indexed by the unstandardized regression slope b) can be nonsignificant 0, even in situations where r_{1Y} is statistically significant. If all the predictive information available in X_1 is already available in X_2, this outcome is consistent with the path model in Figure 3.21. For example, suppose a researcher wants to predict college grades (Y) from X_1 (verbal SAT score) and X_2 (verbal SAT and math SAT scores). In this situation, X_1 could be completely redundant with X_2 as a predictor.

If the information in X_1 that is predictive of Y is also included in X_2, then adding X_1 as a predictor does not provide additional information that is useful to predict Y. Researchers try to avoid situations in which predictor variables are very highly correlated, because when this happens, any separate contributions of information from the predictor variables cannot be distinguished.

3.14.2 X_1, Y Correlation Is Spurious

In the height, vocabulary, and grade level example, the association between height and vocabulary became not statistically significant when grade level was controlled. There is no evidence of a direct association between height and vocabulary, so there is not likely to be a causal connection. Researchers are most likely to decide that a correlation is spurious when it is silly, or when there is no reasonable theory that would point to a direct association between X_1 and Y. There are several possible path models. For example, X_1 and Y might be correlated with each other only because they are both correlated with X_2 (as in the path model in Figure 3.22), or X_2 might be a common or shared cause of both X_1 and Y (as in Figure 3.23). Models in which one of the causal paths in Figure 3.23 changed to a correlational path are also possible.

Alternatively, we could propose that the maturation process that occurs from Grades 1 to 5 to 9 causes increases in both height and vocabulary. The hypothesis that X_1 and Y have a shared cause (X_2) corresponds to the path model in Figure 3.23. In this hypothetical situation, the only reason why height and vocabulary are correlated is that they share a common cause; when variance in height and vocabulary that can be explained by this shared cause is removed, these variables are not directly related.

Examples of spurious correlation intentionally involve foolish or improbable variables. For example, ice cream sales may increase as temperatures rise; homicide rates may also increase as temperatures rise. If we control for temperature, the correlation between ice cream sales and homicide rates drops to 0, so we would conclude that there is no direct relationship between ice cream sales and homicide but that the association of each of these

Figure 3.21 Outcome Consistent With Completely Redundant Predictor: All Predictive Information in X_1 Is Included in X_2

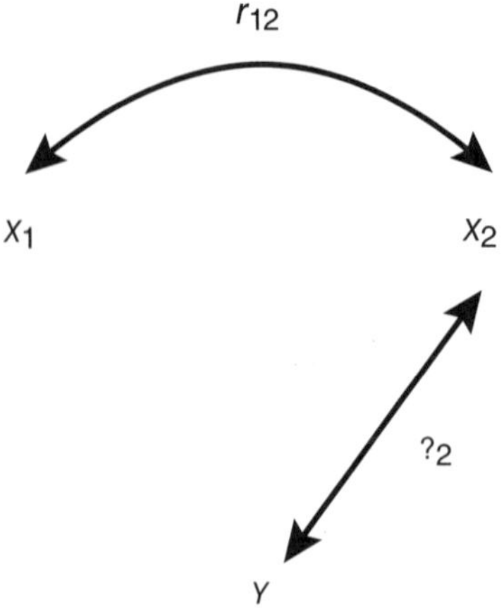

Figure 3.22 Path Model for One Kind of Spurious Association Between Height and Vocabulary

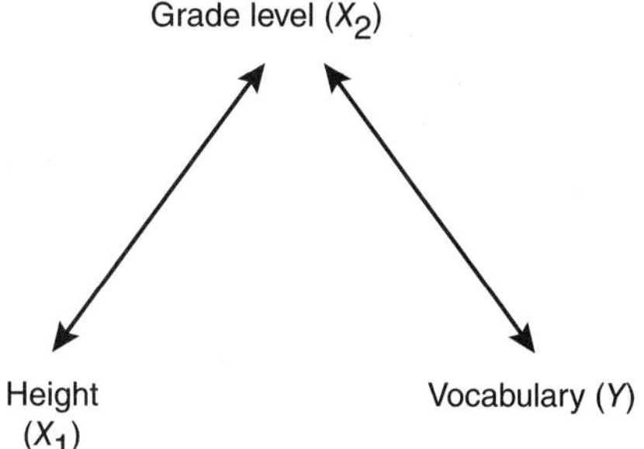

Figure 3.23 Path Model in Which Height and Vocabulary Have a Shared or Common Cause (Maturation)

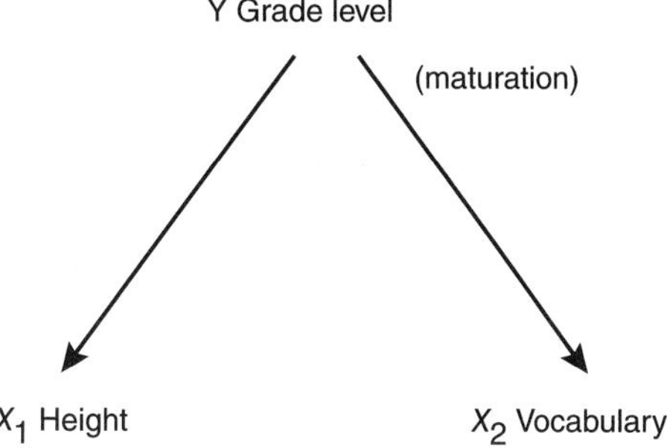

variables with outdoor temperature (X_2) creates the spurious or misleading appearance of a connection between ice cream consumption and homicide.

3.14.3 X_1, Y Association Is Completely Mediated by X_2

A mediation model involves a two-step causal sequence hypothesis. First, X_1 is hypothesized to cause X_2; then, X_2 is hypothesized to cause Y. If the X_1, Y association drops to 0 when we control for X_2, a direct path from X_1 to Y is not needed. For example, we might hypothesize that increases in age (X_1) cause increases in body weight (X_2); and then, increases in body weight (X_2) cause increases in blood pressure (Y). This is represented as a path model in Figure 3.24. If the X_1, Y association drops to 0 when we control for X_2, one possible inference is that the X_1 association with Y is completely mediated by X_2.

Consider an example that involves the variables age (X_1), body weight (X_2), and systolic blood pressure (Y). It is conceivable that blood pressure increases as people age but that this influence is mediated by body weight (X_2) and only occurs if weight changes with age and if blood pressure is increased by weight gain. The corresponding path model appears in Figure 3.24. The absence of a direct path from X_1 to Y is important; the absence of a direct path is what leads us to say that the association between X_1 and Y may be completely mediated. Methods for estimation of path coefficients for this model are covered in Chapter 9, on mediation.

Figure 3.24 Complete Mediation Model: Effects of Age (X_1) on Systolic Blood Pressure (Y) Are Completely Mediated by Body Weight (X_2)

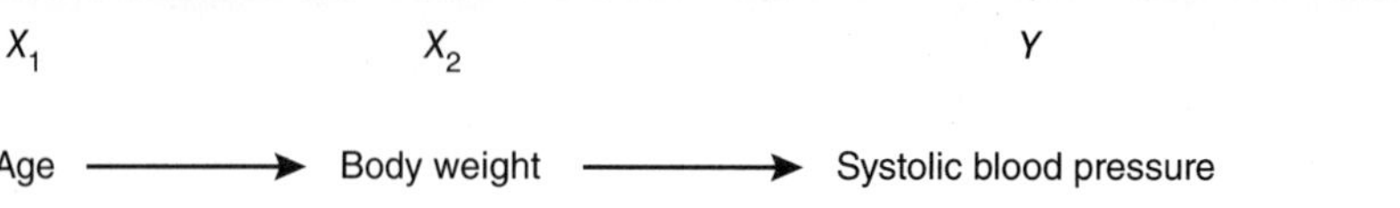

Note: Path diagrams for mediation often denote the mediating variable as M (instead of X_2).

3.14.4 True Nature of the X_1, Y Association (Their Lack of Association) Is "Suppressed" by X_2

The "true" nature of the X_1, Y association may be suppressed or disguised by X_2. In the shared cause example (Figure 3.23), the true nature of the height, vocabulary association is hidden when we look at their bivariate correlation; the true nature of the association (i.e., that there is no direct association) is revealed when we control for X_2 (grade level or maturation).

3.14.5 Empirical Results Cannot Determine Choice Among These Explanations

When we find a large absolute value for r_{1Y}, and a nonsignificant value of $r_{Y1.2}$, this may suggest any one of these explanations: spuriousness, completely redundant predictors, shared common cause, complete mediation, or suppression. Theories and common sense can help us rule out some interpretations as nonsense (for instance, it's conceivable that age might influence blood pressure; it would not make sense to suggest that blood pressure causes age). X_1 cannot cause Y if X_1 happens later in time than Y, and we would not hypothesize that X_1 causes Y unless we can think of reasons why this would make sense. Even when theory and common sense are applied, data analysts are often still in situations where it is not possible to decide among several explanations.

3.15 WHEN YOU CONTROL FOR X_2, THE CORRELATION BETWEEN X_1 AND Y BECOMES SMALLER (BUT DOES NOT DROP TO ZERO OR CHANGE SIGN)

This may be one of the most common outcomes when partial correlations are compared with zero-order correlations. The implication of this outcome is that the association between X_1 and Y can be only partly accounted for by a (causal or noncausal) path via X_2. A direct path (either causal or noncausal) between X_1 and Y is needed in the model, even when X_2 is included in the analysis.

We can consider the same potential explanations as for $r_{Y1.2} = 0$ and add the word *partly*.

1. The r_{1Y} correlation might be partly spurious (this language is not common).
2. X_1 and X_2 might be partly, but not completely, redundant predictors.
3. X_1 and X_2 might share common causes, but the shared causes might not be sufficient to completely explain their association.
4. The X_1, Y association might be partly mediated by X_2 (illustrated in Figure 3.25).
5. There may be partial suppression by X_2 of the true nature of the X_1, Y association.

Figure 3.25 Path Model for Partial Mediation of Effects of X_1 on Y by X_2

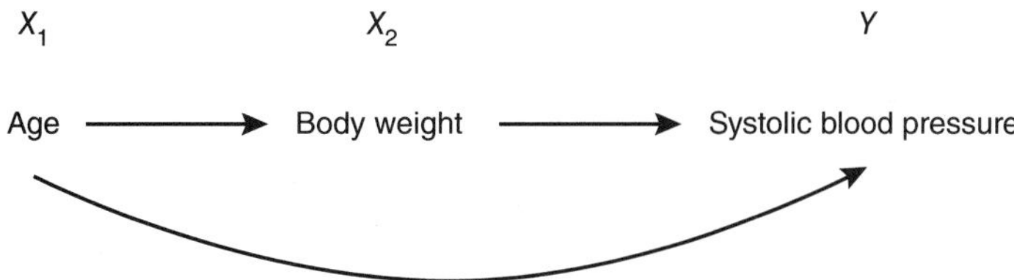

As in previous situations: empirical values of r_{1Y} and $r_{Y1.2}$, and of path coefficients from other analyses such as regression, cannot determine which among these explanations is more likely to be correct.

3.16 SOME FORMS OF SUPPRESSION: WHEN YOU CONTROL FOR X_2, $r_{1Y.2}$ BECOMES LARGER THAN r_{1Y} OR OPPOSITE IN SIGN TO r_{1Y}

The term **suppression** has been used to refer to different relations among variables. Cohen, Cohen, West, and Aiken (2013) suggested a broad definition: X_2 can be called a **suppressor variable** if it hides or suppresses the "true" association between X_1 and Y. Under that broad definition, any time $r_{1Y.2}$ differs significantly from r_{1Y}, we could say that X_2 acted as a suppressor variable. Implicitly we assume that the "true" nature of the X_1, Y association is seen only when we statistically control for X_2.

However, outcomes where $r_{Y1.2}$ is approximately equal to 0 or $r_{Y1.2}$ is less than r_{1Y}, described in the preceding sections, are common outcomes that are not generally seen as surprising or difficult to explain. Many authors describe outcomes as suppression only if they are surprising or difficult to explain, for example, when $r_{1Y.2}$ is larger than r_{1Y}, or when $r_{Y1.2}$ is opposite in sign from r_{1Y}. Paulhus, Robins, Trzesniewski, and Tracy (2004) explained different ways suppression has been defined and described three different types of suppression. Classical suppression occurs when an X_2 variable that is not predictive of Y makes X_1 a stronger predictor of Y (see Section 3.16.1 for a hypothetical example). These outcomes are not common in behavioral and social science research. However, it is useful to understand that they exist, particularly if you happen to find one of these outcomes when you compare $r_{1Y.2}$ to r_{1Y}. The following sections describe three specific types of suppression.

3.16.1 Classical Suppression: Error Variance in Predictor Variable X_1 Is "Removed" by Control Variable X_2

"Classical" suppression occurs if an X_2 variable that is not related to Y improves the ability of X_1 to predict Y when it is included in the analysis. This can happen if X_2 helps us remove irrelevant or error variance from X_1 scores.

Consider the following hypothetical situation. A researcher develops a written test of "mountain survival skills." The score on this test is the X_1 predictor variable. The researcher wants to demonstrate that scores on this test (X_1) can predict performance in an actual mountain survival situation (the score for this survival test is the Y outcome variable). The researcher knows that, to some extent, success on the written test depends on the level of verbal ability (X_2). However, verbal ability is completely uncorrelated with success in the actual mountain survival situation.

The diagram in Figure 3.26 uses overlapping circles to represent shared variance for all pairs of variables (as discussed in Chapter 10, on correlation, in Volume I [Warner, 2020]). For this hypothetical example, consider the problem of predicting skill in a mountain survival situation (Y) from scores on a written test of mountain survival skills (X_1). That prediction may not be good, because verbal skills (X_2) probably explain why some people do better on written tests (X_1). (I am guessing that a Daniel Boone–type hero might not have done well on such a test but would do very well outdoors.) It may be quite difficult to come up with questions that assess actual skill, independent of verbal ability. Verbal skills (X_2) probably have little or nothing to do with mountain survival (Y). From the point of view of a person who really wants to predict mountain survival skills, verbal ability is a nuisance or error variable.

We might think of the written test score as being made up of two parts:

- a part of the score that is relevant to survival skill and
- a part of the score that is related to verbal ability but is completely irrelevant to survival skill.

Figure 3.26 X_2 (a Measure of Verbal Ability) Is a Suppressor of Error Variance in the X_1 Predictor (a Written Test of Mountain Survival Skills); Y Is a Measure of Actual Mountain Survival

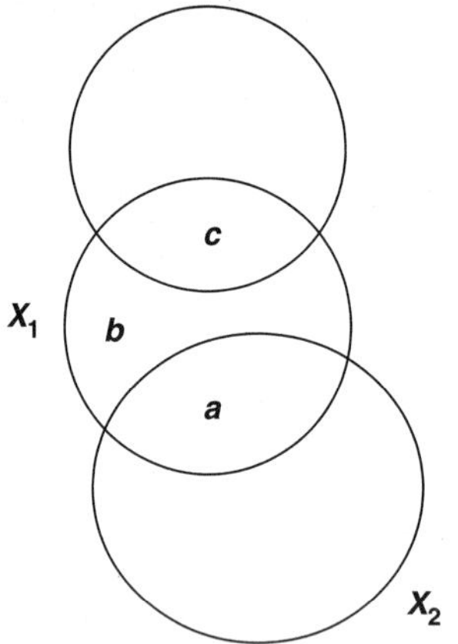

Total variance in X_1: $a + b + c = 1$.

Variance in X_1 that is predictive of Y: c.

Proportion of total variance in X_1 that is predictive of Y when X_2 is ignored: $c/(a + b + c)= c/1 =c$.

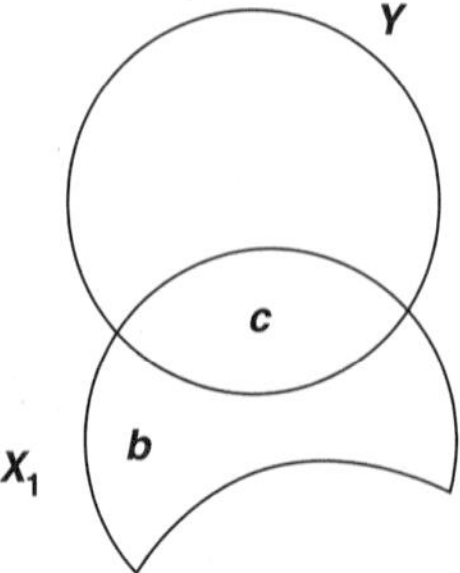

Total variance that remains in X_1 after X_2 is "partialled out" is $b + c$.

Variance in X_1 that is predictive of Y corresponds to area c.

Proportion of variance in X_1 that is predictive of Y when all variance in X_1 that is related to X_2 has been removed is: $c/(b + c)$.

Table 3.2 Correlations Among Measures in Hypothetical Mountain Survival Skills Study

	X_2 (Verbal Ability Control Variable)	Y (Actual Mountain Survival Performance)
X_1 (written test of mountain survival skills)	.25	—
X_2	.65	0

(There could be other components in this score, but let's imagine just these two.)

Table 3.2 has hypothetical correlations among the three variables. Using Equation 3.2 or an online calculator, the partial correlation based on these three bivariate correlations is $r_{1Y.2} = .33$. If verbal ability is not partialled out, the zero-order bivariate correlation between the written test survival skills, r_{1Y}, is .25. When verbal ability is partialled out, the correlation between the written test and survival skills, $r_{1Y.2}$, is .33. We can predict actual survival better when we use only the part of the test scores that is not related to verbal ability.

A path diagram is not helpful in understanding this situation. Use of overlapping circles is more helpful. In an overlapping circle diagram, each circle represents the total variance of one variable. Shared or overlapping area corresponds to shared variance. For variables X_1 and X_2, with $r = .65$, the proportion of overlap would be $r^2 = .42$. If two variables have a correlation of zero, their circles do not overlap. Figure 3.26 shows how partition of variance into explained and unexplained variance on the actual mountain survival task will work, given that X_2 isn't correlated at all with Y, but is highly correlated with X_1. (This is an uncommon outcome. You will more often see partition of variance that looks like the examples in the next chapter.)

The overlapping circle diagrams that appear in Figure 3.26 can help us understand what might happen in this situation. The top diagram shows that X_1 is correlated with Y and X_2 is correlated with X_1; however, X_2 is not correlated with Y (the circles that represent the variance of Y and the variance of X_2 do not overlap). If we ignore the X_2 variable, the squared correlation between X_1 and Y (r^2_{1Y}) corresponds to Area c in Figure 3.26. The total variance in X_1 is given by the sum of Areas $a + b + c$. In these circle diagrams, the total area equals 1.00; therefore, the sum $a + b + c = 1$. The proportion of variance in X_1 that is predictive of Y (when we do not partial out the variance associated with X_2) is equivalent to $c/(a + b + c) = c/1 = c$.

When we statistically control for X_2, we remove all the variance that is predictable from X_2 from the X_1 variable, as shown in the bottom diagram in Figure 3.26. The second diagram shows that after the variance associated with X_2 is removed, the remaining variance in X_1 corresponds to the sum of Areas $b + c$. The variance in X_1 that is predictive of Y corresponds to Area c. The proportion of the variance in X_1 that is predictive of Y after we partial out or remove the variance associated with X_2 now corresponds to $c/(b + c)$. Because $(b + c)$ is less than 1, the proportion of variance in X_1 that is associated with Y after removal of the variance associated with X_2 (i.e., $r^2_{Y1.2}$) is actually higher than the original proportion of variance in Y that was predictable from X_1 when X_2 was not controlled (i.e., r^2_{Y1}).

In this situation, the X_2 control variable suppresses irrelevant or **error variance** in the X_1 predictor variable. When we remove the verbal skills part of the written test scores by controlling for verbal ability, the part of the test score that is left becomes a better predictor of actual survival. It is not common to find a suppressor variable that makes some other predictor variable a better predictor of Y in actual research. However, sometimes a researcher can identify a factor that influences scores on the X_1 predictor and that is not related to or predictive of the scores on the outcome variable Y. In this example, verbal ability was one factor that influenced scores on the written test, but it was almost completely unrelated to actual mountain survival skills. Controlling for verbal ability (i.e., removing the variance associated with verbal ability from the scores on the written test) made the written test a better predictor of mountain survival skills.

If X_2 has a nearly 0 correlation with Y, and X_1 becomes a stronger predictor of Y when X_2 is statistically controlled, classical suppression is a possible explanation. It is better not to grasp at straws. If $r_{1Y} = .30$ and $r_{1Y.2} = .31$, $r_{Y1.2}$ is (a little) larger than r_{1Y}; however, this difference between r_{1Y} and $r_{Y1.2}$ may be too small to be statistically significant or to have a meaningful interpretation. It may be desirable to find variables that can make your favorite X_1 variable a stronger predictor of Y, but this does not happen often in practice.

3.16.2 X_1 and X_2 Both Become Stronger Predictors of Y When Both Are Included in Analysis

This outcome in which both X_1 and X_2 are more predictive of Y when the other variable has been statistically controlled has been described as cooperative, reciprocal, or mutual suppression. This can happen when X_1 and X_2 have opposite signs as predictors of Y, and X_1 and X_2 are positively correlated with each other. In an example provided by Paulhus et al. (2004), X_1 (self-esteem) and X_2 (narcissism) had relationships with opposite signs for the outcome variable Y (antisocial behavior). In their Sample 1, Paulhus et al. reported an empirical example in which the correlation between self-esteem and narcissism was +.32. Self-esteem had a negative zero-order relationship with antisocial behavior (–.27) that became more strongly negative when narcissism was statistically controlled (–.38). Narcissism had a positive association with antisocial behavior (.21) that became more strongly positive when self-esteem was statistically controlled (.33). In other words, each predictor had a stronger relationship with the Y outcome variable when controlling for the other predictor.

3.16.3 Sign of X_1 as a Predictor of Y Reverses When Controlling for X_2

Another possible form of suppression occurs when the sign of $r_{Y1.2}$ is opposite to the sign of r_{Y1}. This has sometimes been called negative suppression or net suppression; I prefer the term proposed by Paulhus et al. (2004), *crossover suppression*. In the following example, r_{1Y}, the zero-order correlation between crowding (X_1) and crime rate (Y_2) across neighborhoods is large and positive. However, when you control for X_2 (level of neighborhood socioeconomic status [SES]), the sign of the partial correlation between X_1 and Y, controlling for X_2, $r_{Y1.2}$, becomes negative. A hypothetical situation where this could occur is shown in Figure 3.27.

In this hypothetical example, the unit of analysis or case is "neighborhood"; for each neighborhood, X_1 is a measure of crowding, Y is a measure of crime rate, and X_2 is a categorical measure of income level (SES). X_2 (SES) is coded as follows: 1 = upper class, 2 = middle class, 3 = lower class. The pattern in this graph represents the following hypothetical situation. This example was suggested by correlations reported by Freedman (1975), but it illustrates a much stronger form of suppression than Freedman found in his data. For the hypothetical data in Figure 3.27, if you ignore SES and obtain the zero-order correlation between crowding and crime, you would obtain a large positive correlation, suggesting that crowding predicts crime. However, there are two confounds present: Crowding tends to be greater in lower SES neighborhoods (3 = low SES), and the incidence of crime also tends to be greater in lower SES neighborhoods.

Once you look separately at the plot of crime versus crowding within each SES category, however, the relationship becomes quite different. Within the lowest SES neighborhoods (SES code 3), crime is negatively associated with crowding (i.e., more crime takes place in "deserted" areas than in areas where there are many potential witnesses out on the streets). Freedman (1975) suggested that crowding, per se, does not "cause" crime; it just happens to be correlated with something else that is predictive of crime, namely, poverty or low SES.

Figure 3.27 Example: Crossover Suppression

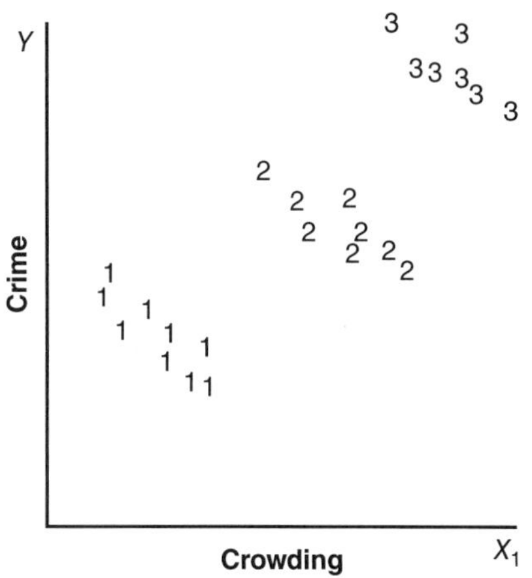

Note: On this graph, cases are marked by socioeconomic status (SES) level of the neighborhood (1 = high SES, 2 = medium SES, 3 = low SES). When SES is ignored, there is a large positive correlation between X_1 (neighborhood crowding) and Y (neighborhood crime). When the X_1, Y correlation is assessed separately within each level of SES, the relationship between X_1 and Y becomes negative. The X_2 variable (SES) suppresses the true relationship between X_1 (crowding) and Y (crime). Crowding and crime appear to be positively correlated when we ignore SES; when we statistically control for SES, it becomes clear that within SES levels, crowding and crime appear to be negatively related.

In fact, within neighborhoods matched in SES, Freedman reported that higher population density was predictive of *lower* crime rates.

3.17 "NONE OF THE ABOVE"

The foregoing sections describe some possible interpretations for comparisons of partial correlation outcomes with zero-order correlation outcomes. This does not exhaust all possibilities. Partial correlations can be misleading or difficult to interpret. Do not strain to explain results that don't make sense. Strange results may arise from sampling error, outliers, or problems with assumptions for correlations.

3.18 RESULTS SECTION

The first research example introduced early in the chapter examined whether height (X_1) and vocabulary (Y) are related when grade level (X_2) is statistically controlled. The results presented in Section 3.2 can be summarized briefly.

Results

The relation between height and vocabulary score was assessed for $N = 48$ students in three different grades in school: Grade 1, Grade 5, and Grade 9. The zero-order Pearson's r between height and vocabulary was statistically significant, $r(46) = .72$, $p < .001$, two tailed. A scatterplot of vocabulary scores by height (with individual points labeled by grade level) suggested that both vocabulary and height tended to increase with grade level. It seemed likely that the correlation between vocabulary and height was spurious, that is, attributable entirely to the tendency of both these variables to increase with grade level.

To assess this possibility, the relation between vocabulary and height was assessed controlling for grade. Grade was controlled for in two different ways. A first-order partial correlation was computed for vocabulary and height, controlling for grade. This partial r was not statistically significant, $r(45) = -.01$, $p = .938$. In addition, the correlation between height and vocabulary was computed separately for each of the three grade levels. For Grade 1, $r = .067$; for Grade 5, $r = .031$; and for Grade 9, $r = -.141$. None of these correlations was statistically significant, and the differences among these three correlations were not large enough to suggest the presence of an interaction effect (i.e., there was no evidence that the nature of the relationship between vocabulary and height differed substantially across grades).

When grade was controlled for, either by partial correlation or by computing Pearson's r separately for each grade level, the correlation between vocabulary and height became very small and was not statistically significant. This is consistent with the explanation that the original correlation was spurious. Vocabulary and height are correlated only because both variables increase across grade levels (and not because of any direct causal or noncausal association between height and vocabulary).

3.19 SUMMARY

Partial correlation can be used to provide preliminary exploratory information about relations among variables. When we take a third variable, X_2, into account, our understanding of the nature and strength of the association between X_1 and Y can change in several different ways.

This chapter outlines two methods to evaluate how taking X_2 into account as a control variable may modify our understanding of the way in which an X_1 predictor variable is related to a Y outcome variable. The first method involved dividing the data set into separate groups, on the basis of scores on the X_2 control variable (using the split file procedure in SPSS), and then examining scatterplots and correlations between X_1 and Y separately for each group. In the examples in this chapter, the X_2 control variables had a small number of possible score values (e.g., when sex was used as a control variable, it had just two values, male and female; when grade level in school and SES were used as control variables, they had just three score values). The number of score values on X_2 variables was kept small in these examples to make it easy to understand the examples. However, the methods outlined here are applicable in situations where the X_2 variable has a larger number of possible score values, as long as the assumptions for Pearson correlation and partial correlation are reasonably well met. Note, however, that if the X_2 variable has 40 possible different score values, and the total number of cases in a data set is only $N = 50$, it is quite likely that when any one score is selected (e.g., $X_2 = 33$), there may be only one or two cases with that value of X_2. When the n's within groups based on the value of X_2 become very small, it becomes impossible to evaluate assumptions such as linearity and normality within the subgroups, and estimates of the strength of association between X_1 and Y that are based on extremely small groups are not likely to be very reliable. The minimum sample sizes that are suggested for Pearson correlation and bivariate regression are on the order of $N = 100$. Sample sizes should be even larger for studies where an X_2 control variable is taken into account, particularly in situations where the researcher suspects the presence of an interaction or moderating variable; in these situations, the researcher needs to estimate a different slope to predict Y from X_1 for each score value of X_2.

We can use partial correlation to statistically control for an X_2 variable that may be involved in the association between X_1 and Y as a rival explanatory variable, a confound, a mediator, a suppressor, or in some other role. However, statistical control is generally a less effective method for dealing with extraneous variables than experimental control. Some methods of experimental control (such as random assignment of participants to treatment groups) are, at least in principle, able to make the groups equivalent with respect to hundreds of different participant characteristic variables. However, when we measure and statistically control for one specific X_2 variable in a nonexperimental study, we have controlled for only one of many possible rival explanatory variables. In a nonexperimental study, there may be dozens or hundreds of other variables that are relevant to the research question and whose influence is not under the researcher's control; when we use partial correlation and similar methods of statistical control, we are able to control statistically for only a few of these variables.

In this chapter, many questions were presented in the context of a three-variable research situation. For example, is X_1 confounded with X_2 as a predictor? When you control for X_2, does the partial correlation between X_1 and Y drop to 0? In multivariate analyses, we often take several additional variables into account when we assess each X_1, Y predictive relationship. However, the same issues that were introduced here in the context of three-variable research situations continue to be relevant for studies that include more than three variables.

A researcher may hope that adding a third variable (X_2) to the analysis will increase the ability to predict Y. A researcher may hope that adding an X_2 covariate will reduce or increase the strength of association between X_1 and Y. Alternatively, a researcher may hope that adding an X_2 covariate does not change the strength of association between X_1 and Y.

The next chapter (regression with two predictors) shows how a regression to predict Y from both X_1 and X_2 provides more information. This regression will tell us how much variance in Y can be predicted from X_1 and X_2 as a set. It will also tell us how well X_1 predicts Y when X_2 is statistically controlled and how well X_2 predicts Y when X_1 is statistically controlled.

COMPREHENSION QUESTIONS

1. When we assess X_1 as a predictor of Y, there are several ways in which we can add a third variable (X_2) and several "stories" that may describe the relations among variables. Explain what information can be obtained from the following two analyses:

   I. Assess the X_1, Y relation separately for each group on the X_2 variable.

   II. Obtain the partial correlation (partial r of Y with X_1, controlling for X_2).

   a. Which of these analyses (I or II) makes it possible to detect an interaction between X_1 and X_2? Which analysis assumes that there is no interaction?

   b. If there is an interaction between X_1 and X_2 as predictors of Y, what pattern would you see in the scatterplots in Analysis I?

2. Discuss each of the following as a means of illustrating the partial correlation between X_1 and Y, controlling for X_2. What can each analysis tell you about the strength and the nature of this relationship?

   I. Scatterplots showing Y versus X_1 (with X_2 scores marked in the plot).

   II. Partial r as the correlation between the residuals when X_1 and Y are predicted from X_2.

3. Explain how you might interpret the following outcomes for partial r:

   a. $r_{1Y} = .70$ and $r_{1Y.2} = .69$

   b. $r_{1Y} = .70$ and $r_{1Y.2} = .02$

   c. $r_{1Y} = .70$ and $r_{1Y.2} = -.54$

   d. $r_{1Y} = .70$ and $r_{1Y.2} = .48$

4. What does the term *partial* mean when it is used in connection with correlations?

NOTE

[1]Some issues with path models are omitted from this simple introduction. Analysis methods described here cannot handle path models with feedback loops, such as $X_1 \rightarrow X_2 \rightarrow Y \rightarrow X_1 \rightarrow X_2 \rightarrow Y$, and so on, or paths for both $X_1 \rightarrow X_2$ and $X_2 \rightarrow X_1$. There are real-world situations where these models would be appropriate; however, different analytic methods would be required.

DIGITAL RESOURCES

Find **free study tools** to support your learning, including **eFlashcards, data sets, and web resources**, on the accompanying website at **edge.sagepub.com/warner3e**.

CHAPTER 4

REGRESSION ANALYSIS AND STATISTICAL CONTROL

4.1 INTRODUCTION

Bivariate regression involves one predictor and one quantitative outcome variable. Adding a second predictor shows how statistical control works in regression analysis. The previous chapter described two ways to understand statistical control. In the previous chapter, the outcome variable was denoted Y, the predictor of interest was denoted X_1, and the control variable was called X_2.

1. We can control for an X_2 variable by dividing data into groups on the basis of X_2 scores and then analyzing the X_1, Y relationship separately within these groups. Results are rarely reported this way in journal articles; however, examining data this way makes it clear that the nature of an X_1, Y relationship can change in many ways when you control for an X_2 variable.
2. Another way to control for an X_2 variable is obtaining a partial correlation between X_1 and Y, controlling for X_2. This partial correlation is denoted $r_{1Y.2}$. Partial correlations are not often reported in journal articles either. However, thinking about them as correlations between residuals helps you understand the mechanics of statistical control. A partial correlation between X_1 and Y, controlling for X_2, can be understood as a correlation between the parts of the X_1 scores that are not related to X_2, and the parts of the Y scores that are not related to X_2.

This chapter introduces the method of statistical control that is most widely used and reported. This method involves using both X_1 and X_2 as predictors of Y in a multiple linear regression. This analysis provides information about the way X_1 is related to Y, controlling for X_2, and also about the way X_2 is related to Y, controlling for X_1. This is called "multiple" regression because there are multiple predictor variables. Later chapters discuss analyses with more than two predictors. It is called "linear" because all pairs of variables must be linearly related. The equation to predict a raw score for the Y outcome variable from raw scores on X_1 and X_2 is as follows:

$$Y' = b_0 + b_1X_1 + b_2X_2. \tag{4.1}$$

There is also a standardized (or unit-free) form of this predictive equation to predict z scores for Y from z scores on X_1 and X_2:

$$z'_Y = \beta_1 z_{X1} + \beta_2 z_{X2}. \quad (4.2)$$

Equation 4.2 corresponds to the path model in Figure 4.1.

The information from the sample that is used for this regression is the set of bivariate correlations among all predictors: r_{12}, r_{1Y}, and r_{2Y}. The values of the coefficients for paths from z_{X1} and z_{X2} to z_Y (denoted β_1 and β_2 in Figure 4.1) are initially unknown. Their values can be found from the set of three bivariate correlations, as you will see in this chapter. The β_1 path coefficient represents the strength of prediction of z_Y from z_{X1}, controlling for z_{X2}. The β_2 path coefficient represents the strength of prediction of z_Y from z_{X2}, controlling for z_{X1}. A regression analysis that includes z_{X1} and z_{X2} as predictors of z_Y, as shown in Equation 4.1, provides estimates for these β coefficients. In regression, the predictive contribution of each independent variable (e.g., z_{X1}) is represented by a β coefficient, and the strengths of associations are assessed while statistically controlling for all other independent variables (in this example, controlling for z_{X2}).

This analysis provides information that is relevant to the following questions:

1. How well does the entire set of predictor variables (X_1 and X_2 together) predict Y? Both a statistical significance test and an effect size are provided.
2. How much does each individual predictor variable (X_1 alone, X_2 alone) contribute to prediction of Y? Each predictor variable has a significance test to evaluate whether its b slope coefficient differs significantly from zero, effect size information (i.e., the percentage of variance in Y that can be predicted by X_1 alone, controlling for X_2), and the percentage of variance in Y that can be predicted by X_2 alone, controlling for X_1.

The b_1 and b_2 regression coefficients in Equation 4.1 are partial slopes. That is, b_1 represents the number of units of change in Y that are predicted for each one-unit increase in X_1 when X_2 is statistically controlled or partialled out of X_1. In many research situations, X_1 and X_2 are partly redundant (or correlated) predictors of Y; in such situations, we need to control for,

Figure 4.1 Path Model: Standardized Regression to Predict z_Y From Correlated Predictors z_{X1} and z_{X2}

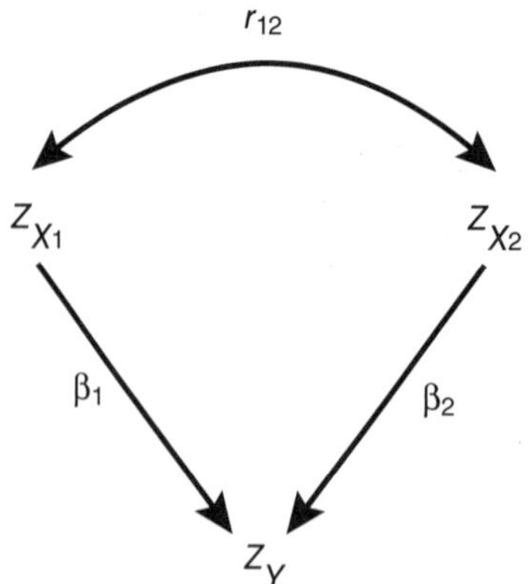

or partial out, the part of X_1 that is correlated with or predictable from X_2 to avoid "double counting" the information that is contained in both the X_1 and X_2 variables.

To understand why this is so, consider a trivial prediction problem. Suppose that you want to predict people's total height in inches (Y) from two measurements that you make using a yardstick: distance from hip to top of head (X_1) and distance from waist to floor (X_2). You cannot predict Y by summing X_1 and X_2, because X_1 and X_2 contain some duplicate information (the distance from waist to hip). The $X_1 + X_2$ sum would overestimate Y because it includes the waist-to-hip distance twice. When you perform a multiple regression of the form shown in Equation 4.1, the b coefficients are adjusted so that information included in both variables is not double counted. Each variable's contribution to the prediction of Y is estimated using computations that partial out other predictor variables; this corrects for, or removes, any information in the X_1 score that is predictable from the X_2 score (and vice versa).

To compute coefficients for the bivariate regression equation $Y' = b_0 + bX$, we need the correlation between X and Y (r_{XY}), as well as the means and standard deviations of X and Y. In regression analysis with two predictor variables, we need the means and standard deviations of Y, X_1, and X_2 and the correlation between each predictor variable and the outcome variable Y (r_{1Y} and r_{2Y}). We also need to know about (and adjust for) the correlation between the predictor variables (r_{12}).

Multiple regression is a frequently reported analysis that includes statistical control. Most published regression analyses include more than two predictor variables. Later chapters discuss analyses that include larger numbers of predictors. All techniques covered later in this book incorporate similar forms of statistical control for correlation among multiple predictors (and later, correlations among multiple outcome variables).

4.2 HYPOTHETICAL RESEARCH EXAMPLE

Suppose that a researcher measures age (X_1) and weight (X_2) and uses these two variables to predict blood pressure (Y). Data are in the file ageweightbp.sav. In this situation, it would be reasonable to expect that the predictor variables would be correlated with each other to some extent (e.g., as people get older, they often tend to gain weight). It is plausible that both predictor variables might contribute unique information toward the prediction of blood pressure. For example, weight might directly cause increases in blood pressure, but in addition, there might be other mechanisms through which age causes increases in blood pressure; for example, age-related increases in artery blockage might also contribute to increases in blood pressure. In this analysis, we might expect to find that the two variables together are strongly predictive of blood pressure and that each predictor variable contributes significant unique predictive information. Also, we would expect that both coefficients would be positive (i.e., as age and weight increase, blood pressure should also tend to increase).

Many outcomes are possible when two variables are used as predictors in a multiple regression. The overall regression analysis can be either significant or not significant, and each predictor variable may or may not make a statistically significant unique contribution. As we saw in the discussion of partial correlation, the assessment of the contribution of an individual predictor variable controlling for another variable can lead to the conclusion that a predictor provides useful information even when another variable is statistically controlled. Conversely, a predictor can become nonsignificant when another variable is statistically controlled. The same types of interpretations (e.g., spuriousness, possible mediated relationships) described for partial correlation outcomes can be considered possible explanations for multiple regression results. In this chapter, we will examine the two-predictor situation in detail; comprehension of the two-predictor situation is extended to regression analyses with more than two predictors in later chapters.

When we include two (or more) predictor variables in a regression, we sometimes choose one or more of the predictor variables because we hypothesize that they might be causes of the Y variable or at least useful predictors of Y. On the other hand, sometimes rival predictor variables are included in a regression because they are correlated with, confounded with, or redundant with a primary explanatory variable; in some situations, researchers hope to demonstrate that a rival variable completely "accounts for" the apparent correlation between the primary variable of interest and Y, while in other situations, researchers hope to show that rival variables do not completely account for any correlation of the primary predictor variable with the Y outcome variable. Sometimes a well-chosen X_2 control variable can be used to partial out sources of measurement error in another X_1 predictor variable (e.g., verbal ability is a common source of measurement error when written tests are used to assess skills that are largely nonverbal, such as playing tennis or mountain survival). An X_2 variable may also be included as a predictor because the researcher suspects that the X_2 variable may "suppress" the relationship of another X_1 predictor variable with the Y outcome variable.

4.3 GRAPHIC REPRESENTATION OF REGRESSION PLANE

For bivariate (one-predictor) regression, a two-dimensional graph (the scatterplot of Y values for each value of X) is sufficient. The regression prediction equation $Y' = b_0 + bX$ corresponds to a line on this scatterplot. If the regression fits the data well, most actual Y scores fall relatively close to the regression line. The b coefficient represents the slope of this line (for a one-unit increase in X, the regression equation predicts a b-unit increase in Y').

Figure 4.2 Three-Dimensional Graph of Multiple Regression Plane With X_1 and X_2 as Predictors of Y

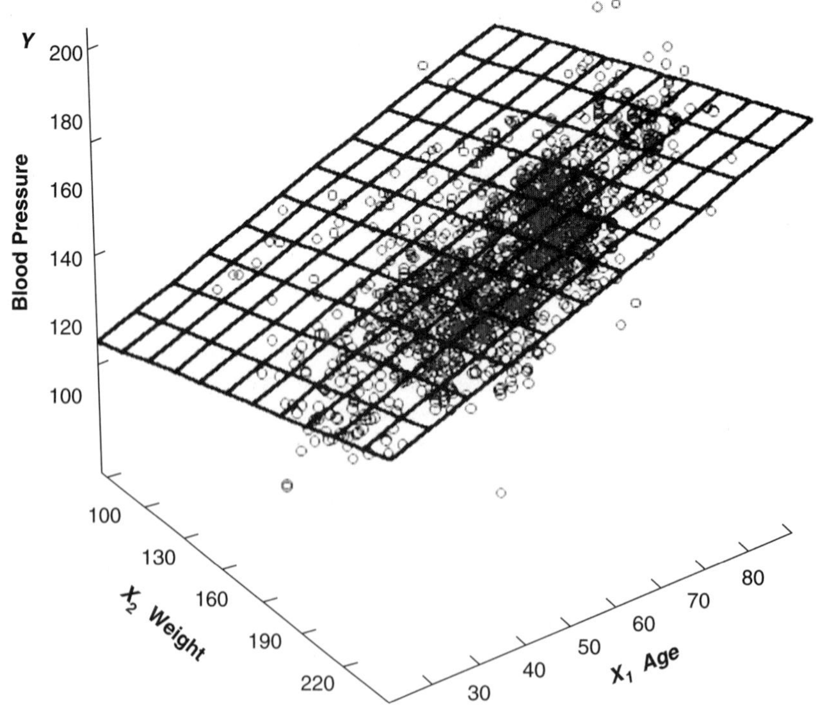

Source: Reprinted with permission from Palmer, M., http://ordination.okstate.edu/plane.jpg.

When we add a second predictor variable, X_2, we need a three-dimensional graph to represent the pattern on scores for three variables. Imagine a cube with X_1, X_2, and Y dimensions; the data points form a cluster in this three-dimensional space. For a good fit, we need a **regression plane** that has the actual points clustered close to it in this three-dimensional space. See Figure 4.2 for a graphic representation of a regression plane.

A more concrete way to visualize this situation is to imagine the X_1, X_2 points as locations on a tabletop (where X_1 represents the location of a point relative to the longer side of the table and X_2 represents the location along the shorter side). You could draw a grid on the top of the table to show the location of each subject's X_1, X_2 pair of scores on the flat plane represented by the tabletop. When you add a third variable, Y, you need to add a third dimension to show the location of the Y score that corresponds to each particular pair of X_1, X_2 score values; the Y values can be represented by points that float in space above the top of the table. For example, X_1 can be age, X_2 can be weight, and Y can be blood pressure. The regression plane can then be represented by a piece of paper held above the tabletop, oriented so that it is centered within the cluster of data points that float in space above the table. The b_1 slope represents the degree of tilt in the paper in the X_1 direction, parallel to the width of the table (i.e., the slope to predict blood pressure from age for a specific weight). The b_2 slope represents the slope of the paper in the X_2 direction, parallel to the length of the table (i.e., the slope to predict blood pressure from weight at some specific age).

Thus, the partial slopes b_1 and b_2, described earlier, can be understood in terms of this graph. The b_1 partial slope (in the regression equation $Y' = b_0 + b_1X_1 + b_2X_2$) has the following verbal interpretation: For a one-unit increase in scores on X_1, the best fitting regression equation makes a b_1-point increase in the predicted Y' score (controlling for or partialling out any changes associated with the other predictor variable, X_2).

4.4 SEMIPARTIAL (OR "PART") CORRELATION

The previous chapter described how to calculate and interpret a partial correlation between X_1 and Y, controlling for X_2. One way to obtain $r_{Y1.2}$ (the partial correlation between X_1 and Y, controlling for X_2) is to perform a simple bivariate regression to predict X_1 from X_2, run another regression to predict Y from X_2, and then correlate the residuals from these two regressions (X^*_1 and Y^*). This correlation is denoted by $r_{1Y.2}$, which is read as "the partial correlation between X_1 and Y, controlling for X_2." This partial r tells us how X_1 is related to Y when X_2 has been removed from or partialled out of both the X_1 and the Y variables. The squared partial r correlation, $r^2_{Y1.2}$, can be interpreted as the proportion of variance in Y that can be predicted from X_1 when all the variance that is linearly associated with X_2 is removed from both the X_1 and the Y variables.

Partial correlations are sometimes reported in studies where the researcher wants to assess the strength and nature of the X_1, Y relationship with the variance that is linearly associated with X_2 completely removed from both variables. This chapter introduces a slightly different statistic (the semipartial or **part correlation**) that provides information about the **partition of variance** between predictor variables X_1 and X_2 in regression in a more convenient form. A semipartial correlation is calculated and interpreted slightly differently from the partial correlation, and a different notation is used. The semipartial (or "part") correlation between X_1 and Y, controlling for X_2, is denoted by $r_{Y(1.2)}$. Another common notation for the semipartial correlation is sr_i, where X_i is the predictor variable. In this notation for semipartial correlation, it is implicit that the outcome variable is Y; the predictive association between X_i and Y is assessed while removing the variance from X_i that is shared with any other predictor variables in the regression equation. The parentheses around 1.2 indicate that X_2 is partialled out of only X_1. It is not partialled out of Y, which is outside the parentheses.

To obtain this semipartial correlation, we remove the variance that is associated with X_2 from only the X_1 predictor (and not from the Y outcome variable). For example, to obtain the semipartial correlation $r_{Y(1.2)}$, the semipartial correlation that describes the strength of the association between Y and X_1 when X_2 is partialled out of X_1, do the following:

1. First, run a simple bivariate regression to predict X_1 from X_2. Obtain the residuals (X^*_1) from this regression. X^*_1 represents the part of the X_1 scores that is not predictable from or correlated with X_2.
2. Then, correlate X^*_1 with Y to obtain the semipartial correlation between X_1 and Y, controlling for X_2. Note that X_2 has been partialled out of, or removed from, only the other predictor variable, X_1; the variance associated with X_2 has not been partialled out of or removed from Y, the outcome variable.

This is called a semipartial correlation because the variance associated with X_2 is removed from only one of the two variables (and not removed entirely from both X_1 and Y as in partial correlation analysis).

It is also possible to compute the semipartial correlation, $r_{Y(1.2)}$, directly from the three bivariate correlations (r_{12}, r_{1Y}, and r_{2Y}):

$$r_{Y(1.2)} = \frac{r_{1Y} - \left(r_{2Y} \times r_{12}\right)}{\sqrt{1 - r_{12}^2}}. \tag{4.3}$$

In many data sets, the partial and semipartial correlations (between X_1 and Y, controlling for X_2) yield similar values. The squared semipartial correlation has a simpler interpretation than the squared partial correlation when we want to describe the partitioning of variance among predictor variables in a multiple regression. The squared semipartial correlation between X_1 and Y, controlling for X_2—that is, $r^2_{Y(1.2)}$ or sr^2_1—is equivalent to the proportion of the total variance of Y that is predictable from X_1 when the variance that is shared with X_2 has been partialled out of X_1. It is more convenient to report squared semipartial correlations (instead of squared partial correlations) as part of the results of regression analysis.

4.5 PARTITION OF VARIANCE IN *Y* IN REGRESSION WITH TWO PREDICTORS

In multiple regression analysis, one goal is to obtain a partition of variance for the dependent variable Y (blood pressure) into variance that can be accounted for or predicted by each of the predictor variables, X_1 (age) and X_2 (weight), taking into account the overlap or correlation between the predictors. Overlapping circles can be used to represent the proportion of shared variance (r^2) for each pair of variables in this situation, as shown in Figure 4.3. Each circle has a total area of 1 (this represents the total variance of z_Y, for example). For each pair of variables, such as X_1 and Y, the squared correlation between X_1 and Y (i.e., r^2_{Y1}) corresponds to the proportion of the total variance of Y that overlaps with X_1, as shown in Figure 4.3.

The total variance of the outcome variable (such as Y, blood pressure) corresponds to the entire circle in Figure 4.3 with sections that are labeled a, b, c, and d. We will assume that the total area of this circle corresponds to the total variance of Y and that Y is given in z-score units, so the total variance or total area $a + b + c + d$ in this diagram corresponds to a value of 1.0. As in earlier examples, overlap between circles that represent different variables corresponds to squared correlation; the total area of overlap between X_1 and Y (which corresponds to the sum of Areas a and c) is equal to r^2_{1Y}, the squared correlation between X_1 and Y. One goal of multiple regression is to obtain information about the partition of variance in the outcome variable into the following components. Area d in the diagram corresponds

Figure 4.3 Partition of Variance of Y in a Regression With Two Predictor Variables, X_1 and X_2

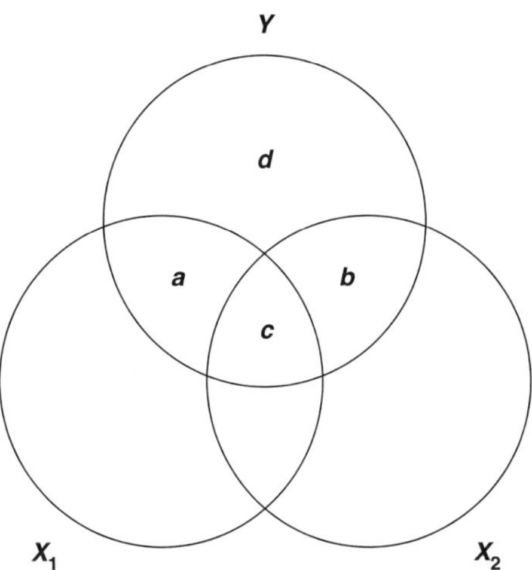

Note: The areas *a*, *b*, *c*, and *d* correspond to the following proportions of variance in Y, the outcome variable: Area *a* sr^2_1, the proportion of variance in Y that is predictable uniquely from X_1 when X_2 is statistically controlled or partialled out; Area *b* sr^2_2, the proportion of variance in Y that is predictable uniquely from X_2 when X_1 is statistically controlled or partialled out; Area *c*, the proportion of variance in Y that could be explained by either X_1 or X_2 (Area *c* can be obtained by subtraction, e.g., $c = 1 - [a + b + d]$); Area $a + b + c$ $R^2_{Y.12}$, the overall proportion of variance in Y predictable from X_1 and X_2 combined; Area *d* $1 - R^2_{Y.12}$, the proportion of variance in Y that is not predictable from either X_1 or X_2.

to the proportion of variance in Y that is not predictable from either X_1 or X_2. Area *a* in this diagram corresponds to the proportion of variance in Y that is uniquely predictable from X_1 (controlling for or partialling out any variance in X_1 that is shared with X_2). Area *b* corresponds to the proportion of variance in Y that is uniquely predictable from X_2 (controlling for or partialling out any variance in X_2 that is shared with the other predictor, X_1). Area *c* corresponds to a proportion of variance in Y that can be predicted by either X_1 or X_2. We can use results from a multiple regression analysis that predicts Y from X_1 and X_2 to deduce the proportions of variance that correspond to each of these areas, labeled *a*, *b*, *c*, and *d*, in this diagram.

We can interpret squared semipartial correlations as information about variance partitioning in regression. We can calculate zero-order correlations among all these variables by running Pearson correlations of X_1 with Y, X_2 with Y, and X_1 with X_2. The overall squared zero-order bivariate correlations between X_1 and Y and between X_2 and Y correspond to the areas that show the total overlap of each predictor variable with Y as follows:

$$a + c = r^2_{Y1},$$

$$b + c = r^2_{Y2}.$$

The squared partial correlations and squared semipartial r's can also be expressed in terms of areas in the diagram in Figure 4.3. The squared semipartial correlation between X_1 and Y, controlling for X_2, corresponds to Area *a* in Figure 4.3; the squared semipartial correlation sr^2_1 can be interpreted as "the proportion of the total variance of Y that is uniquely predictable from X_1." In other words, sr^2_1 (or $r^2_{Y(1.2)}$) corresponds to Area *a* in Figure 4.3.

The squared partial correlation has a somewhat less convenient interpretation; it corresponds to a ratio of areas in the diagram in Figure 4.3. When a partial correlation is calculated, the variance that is linearly predictable from X_2 is removed from the Y outcome variable, and therefore, the proportion of variance that remains in Y after controlling for X_2 corresponds to the sum of Areas a and d. The part of this remaining variance in Y that is uniquely predictable from X_1 corresponds to Area a; therefore, the squared partial correlation between X_1 and Y, controlling for X, corresponds to the ratio $a/(a + d)$. In other words, $pr^2{}_1$ (or $r^2{}_{Y1.2}$) corresponds to a ratio of areas, $a/(a + d)$.

We can reconstruct the total variance of Y, the outcome variable, by summing Areas a, b, c, and d in Figure 4.3. Because Areas a and b correspond to the squared semipartial correlations of X_1 and X_2 with Y, it is more convenient to report squared semipartial correlations (instead of squared partial correlations) as effect size information for a multiple regression. Area c represents variance that could be explained equally well by either X_1 or X_2.

In multiple regression, we seek to partition the variance of Y into components that are uniquely predictable from individual variables (Areas a and b) and areas that are explainable by more than one variable (Area c). We will see that there is more than one way to interpret the variance represented by Area c. The most conservative strategy is not to give either X_1 or X_2 credit for explaining the variance that corresponds to Area c in Figure 4.3. Areas a, b, c, and d in Figure 4.3 correspond to proportions of the total variance of Y, the outcome variable, as given in the table below the overlapping circles diagram.

In words, then, we can divide the total variance of scores on the Y outcome variable into four components when we have two predictors: the proportion of variance in Y that is uniquely predictable from X_1 (Area a, $sr^2{}_1$), the proportion of variance in Y that is uniquely predictable from X_2 (Area b, $sr^2{}_2$), the proportion of variance in Y that could be predicted from either X_1 or X_2 (Area c, obtained by subtraction), and the proportion of variance in Y that cannot be predicted from either X_1 or X_2 (Area d, $1 - R^2{}_{Y.12}$).

Note that the sum of the proportions for these four areas, $a + b + c + d$, equals 1 because the circle corresponds to the total variance of Y (an area of 1.00). In this chapter, we will see that information obtained from the multiple regression analysis that predicts scores on Y from X_1 and X_2 can be used to calculate the proportions that correspond to each of these four areas (a, b, c, and d). When we write up results, we can comment on whether the two variables combined explained a large or a small proportion of variance in Y; we can also note how much of the variance was predicted uniquely by each predictor variable.

If X_1 and X_2 are uncorrelated with each other, then there is no overlap between the circles that correspond to the X_1 and X_2 variables in this diagram and Area c is 0. However, in most applications of multiple regression, X_1 and X_2 are correlated with each other to some degree; this is represented by an overlap between the circles that represent the variances of X_1 and X_2. When some types of suppression are present, the value obtained for Area c by taking 1.0 – Area a – Area b – Area d can actually be a negative value; in such situations, the overlapping circle diagram may not be the most useful way to think about variance partitioning. The partition of variance that can be made using multiple regression allows us to assess the total predictive power of X_1 and X_2 when these predictors are used together and also to assess their unique contributions, so that each predictor is assessed while statistically controlling for the other predictor variable.

In regression, as in many other multivariable analyses, the researcher can evaluate results in relation to several different questions. The first question is, Are the two predictor variables together significantly predictive of Y? Formally, this corresponds to the following null hypothesis:

$$H_0 \colon R_{Y.12} = 0. \tag{4.4}$$

In Equation 4.4, an explicit notation is used for R (with subscripts that specifically indicate the dependent and independent variables). That is, $R_{Y.12}$ denotes the multiple R for a regression equation in which Y is predicted from X_1 and X_2. In this subscript notation, the variable to the left of the period in the subscript is the outcome or dependent variable; the numbers to the right of the period represent the subscripts for each of the predictor variables (in this example, X_1 and X_2). This explicit notation is used when it is needed to make it clear exactly which outcome and predictor variables are included in the regression.

In most reports of multiple regression, these subscripts are omitted, and it is understood from the context that $\boldsymbol{R}^2$ stands for the proportion of variance explained by the entire set of predictor variables that are included in the analysis. Subscripts on R and R^2 are generally used only when it is necessary to remove possible ambiguity. Thus, the formal null hypothesis for the overall multiple regression can be written more simply as follows:

$$H_0\text{: } R = 0. \tag{4.5}$$

Recall that multiple R refers to the correlation between Y and Y' (i.e., the correlation between observed scores on Y and the predicted Y' scores that are formed by summing the weighted scores on X_1 and X_2, $Y' = b_0 + b_1X_1 + b_2X_2$).

A second set of questions that can be addressed using multiple regression involves the unique contribution of each individual predictor. Sometimes, data analysts do not test the significance of individual predictors unless the F for the overall regression is statistically significant. Requiring a significant F for the overall regression before testing the significance of individual predictor variables used to be recommended as a way to limit the increased risk for Type I error that arises when many predictors are assessed; however, the requirement of a significant overall F for the regression model as a condition for conducting significance tests on individual predictor variables probably does not provide much protection against Type I error in practice.

For each predictor variable in the regression—for instance, for X_i—the null hypothesis can be set up as follows:

$$H_0\text{: } b_i = 0, \tag{4.6}$$

where b_i represents the unknown population raw-score slope[1] that is estimated by the sample slope. If the b_i coefficient for predictor X_i is statistically significant, then there is a significant increase in predicted Y values that is uniquely associated with X_i (and not attributable to other predictor variables).

It is also possible to ask whether X_1 is more strongly predictive of Y than X_2 (by comparing β_1 and β_2). However, comparisons between regression coefficients must be interpreted very cautiously; factors that artifactually influence the magnitude of correlations can also artifactually increase or decrease the magnitude of slopes.

4.6 ASSUMPTIONS FOR REGRESSION WITH TWO PREDICTORS

For the simplest possible multiple regression with two predictors, as given in Equation 4.1, the assumptions that should be satisfied are basically the same as the assumptions for Pearson correlation and bivariate regression. Ideally, all the following conditions should hold:

1. The Y outcome variable should be a quantitative variable with scores that are approximately normally distributed. Possible violations of this assumption can be assessed by looking at the univariate distributions of scores on Y. The X_1 and X_2 predictor variables should be normally distributed and quantitative, or one or

both of the predictor variables can be dichotomous (or dummy) variables. If the outcome variable, Y, is dichotomous, then a different form of analysis (binary logistic regression) should be used.

2. The relations among all pairs of variables (X_1, X_2), (X_1, Y), and (X_2, Y) should be linear. This assumption of linearity can be assessed by examining bivariate scatterplots for all possible pairs of these variables. Scatterplots should not have any extreme bivariate outliers.

3. There should be no interactions between variables, such that the slope that predicts Y from X_1 differs across groups that are formed on the basis of scores on X_2. An alternative way to state this assumption is that the regressions to predict Y from X_1 should be homogeneous across levels of X_2. This can be qualitatively assessed by grouping subjects on the basis of scores on the X_2 variable and running a separate X_1, Y scatterplot or bivariate regression for each group; the slopes should be similar across groups. If this assumption is violated and if the slope relating Y to X_1 differs across levels of X_2, then it would not be possible to use a flat plane to represent the relation among the variables as in Figure 4.2. Instead, you would need a more complex surface that has different slopes to show how Y is related to X_1 for different values of X_2. (Chapter 7, on moderation, demonstrates how to include interaction terms in regression models and how to test for the statistical significance of interactions between predictors.)

4. Variance in Y scores should be homogeneous across levels of X_1 (and levels of X_2); this assumption of homogeneous variance can be assessed in a qualitative way by examining bivariate scatterplots to see whether the range or variance of Y scores varies across levels of X. Formal tests of homogeneity of variance are possible, but they are rarely used in regression analysis. In many real-life research situations, researchers do not have a sufficiently large number of scores for each specific value of X to set up a test to verify whether the variance of Y is homogeneous across values of X.

As in earlier analyses, possible violations of these assumptions can generally be assessed reasonably well by examining the univariate frequency distribution for each variable and the bivariate scatterplots for all pairs of variables. Many of these problems can also be identified by graphing the standardized residuals from regression, that is, the $z_Y - z'_Y$ prediction errors. Some problems with assumptions can be detected by examining plots of residuals in bivariate regression; the same issues should be considered when examining plots of residuals for regression analyses that include multiple predictors. That is, the mean and variance of these residuals should be fairly uniform across levels of z'_Y, and there should be no pattern in the residuals (there should not be a linear or curvilinear trend). Also, there should not be extreme outliers in the plot of standardized residuals. Some of the problems that are detectable through visual examination of residuals can also be noted in univariate and bivariate data screening; however, examination of residuals may be uniquely valuable as a tool for the discovery of multivariate outliers. A multivariate outlier is a case that has an unusual combination of values of scores for variables such as X_1, X_2, and Y (even though the scores on the individual variables may not, by themselves, be outliers). A more extensive discussion of the use of residuals for the assessment of violations of assumptions and the detection and possible removal of multivariate outliers is provided in Chapter 4 of Tabachnick and Fidell (2018). Multivariate or bivariate outliers can have a disproportionate impact on estimates of b or β slope coefficients (just as they can have a disproportionate impact on estimates of Pearson's r). That is, sometimes omitting a few extreme outliers results in drastic changes in the size of b or β coefficients. It is undesirable to have the results of a regression analysis depend to a great extent on the values of a few extreme or unusual data points.

If extreme bivariate or multivariate outliers are identified in preliminary data screening, it is necessary to decide whether the analysis is more believable with these outliers included, with the outliers excluded, or using a data transformation (such as log of X) to reduce the

Figure 4.4 SPSS Dialog Box to Request Matrix Scatterplots

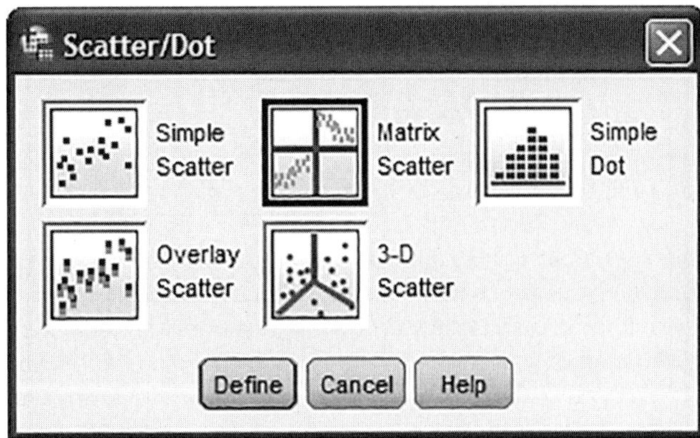

impact of outliers on slope estimates. If outliers are identified and modified or removed, the rationale and decision rules for the handling of these cases should be clearly explained in the write-up of results.

The hypothetical data for this example consist of data for 30 cases on three variables (in the file ageweightbp.sav): blood pressure (Y), age (X_1), and weight (X_2). Before running the multiple regression, scatterplots for all pairs of variables were examined, descriptive statistics were obtained for each variable, and zero-order correlations were computed for all pairs of variables using the methods described in previous chapters. It is also a good idea to examine histograms of the distribution of scores on each variable to assess whether scores on continuous predictor variables are reasonably normally distributed without extreme outliers.

A matrix of scatterplots for all possible pairs of variables was obtained through the SPSS menu sequence <Graph> → <Legacy Dialogs> → <Scatter/Dot>, followed by clicking on the "Matrix Scatter" icon, shown in Figure 4.4. The names of all three variables (age, weight, and blood pressure) were entered in the dialog box for matrix scatterplots, which appears in Figure 4.5. The SPSS output shown in Figure 4.6 shows the matrix scatterplots for all pairs of variables: X_1 with Y, X_2 with Y, and X_1 with X_2. Examination of these scatterplots suggested that relations between all pairs of variables were reasonably linear and there were no bivariate outliers. Variance of blood pressure appeared to be reasonably homogeneous across levels of the predictor variables. The bivariate Pearson correlations for all pairs of variables appear in Figure 4.7.

On the basis of preliminary data screening (including histograms of scores on age, weight, and blood pressure that are not shown here), it was judged that scores were reasonably normally distributed, relations between variables were reasonably linear, and there were no outliers extreme enough to have a disproportionate impact on the results. Therefore, it seemed appropriate to perform a multiple regression analysis on these data; no cases were dropped, and no data transformations were applied.

If there appear to be curvilinear relations between any variables, then the analysis needs to be modified to take this into account. For example, if Y shows a curvilinear pattern across levels of X_1, one way to deal with this is to recode scores on X_1 into group membership codes (e.g., if X_1 represents income in dollars, this could be recoded as three groups: low, middle, and high income levels); then, an analysis of variance (ANOVA) can be used to see whether means on Y differ across these groups (on the basis of low, medium, or high X scores). Another possible way to incorporate nonlinearity into a regression analysis is to include X^2 (and perhaps higher powers of X, such as X^3) as a predictor of Y in a regression equation of the following form:

$$Y' = b_0 + b_1X^1 + b_2X^2 + b_3X^3 + \cdots. \quad (4.7)$$

Figure 4.5 SPSS Scatterplot Matrix Dialog Box

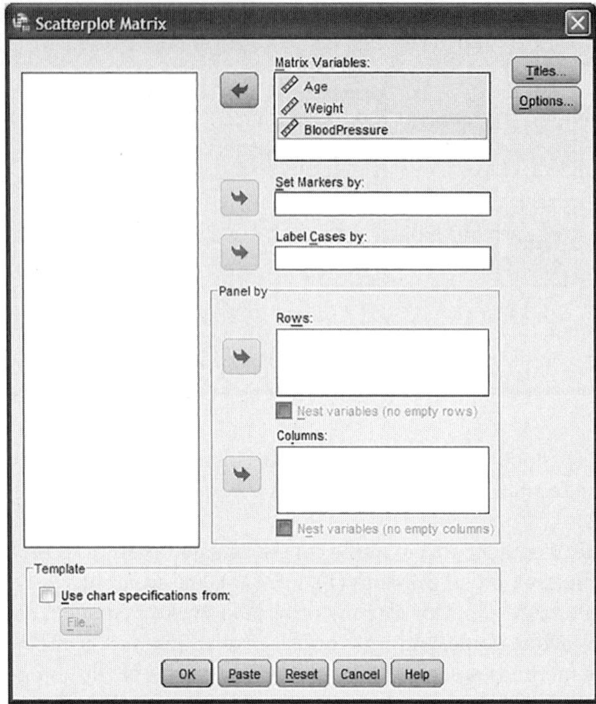

Note: This generates a matrix of all possible scatterplots between pairs of listed variables (e.g., age with weight, age with blood pressure, and weight with blood pressure).

Figure 4.6 Matrix of Scatterplots for Age, Weight, and Blood Pressure

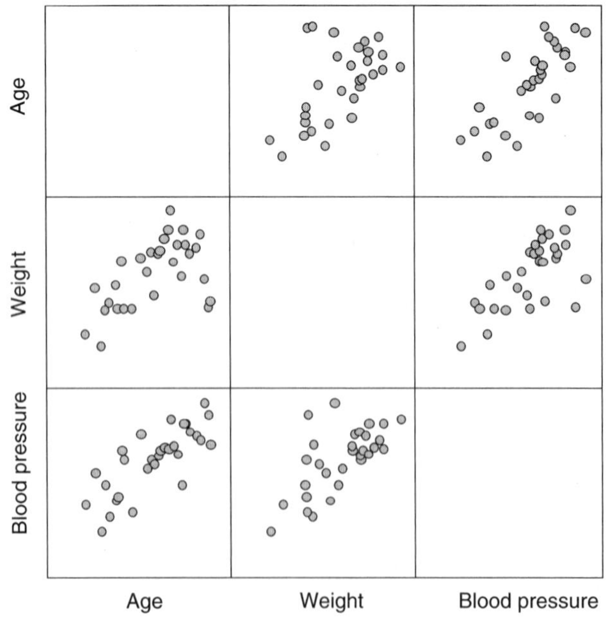

Figure 4.7 Bivariate Correlations Among Age, Weight, and Blood Pressure

Correlations

		Age	Weight	Blood Pressure
Age	Pearson Correlation	1	.563**	.782**
	Sig. (2-tailed)		.001	.000
	N	30	30	30
Weight	Pearson Correlation	.563**	1	.672**
	Sig. (2-tailed)	.001		.000
	N	30	30	30
BloodPressure	Pearson Correlation	.782**	.672**	1
	Sig. (2-tailed)	.000	.000	
	N	30	30	30

** Correlation is significant at the 0.01 level (2-tailed).

In practice, it is rare to encounter situations where powers of X higher than X^2, such as X^3 or X^4 terms, are needed. Curvilinear relations that correspond to a U-shaped or inverse U-shaped graph (in which Y is a function of X and X^2) are more common.

Finally, if an interaction between X_1 and X_2 is detected, it is possible to incorporate one or more interaction terms into the regression equation using methods that will be described in later chapters. A regression equation that does not incorporate an interaction term when there is in fact an interaction between predictors can produce misleading results. When we do an ANOVA, most programs automatically generate interaction terms to represent interactions among all possible pairs of predictors. However, when we do regression analyses, interaction terms are not generated automatically; if we want to include interactions in our models, we must add them explicitly. The existence of possible interactions among predictors is therefore easy to overlook when regression analysis is used.

4.7 FORMULAS FOR REGRESSION WITH TWO PREDICTORS

4.7.1 Computation of Standard-Score Beta Coefficients

The coefficients to predict z'_Y from z_{X1}, z_{X2} ($z'_Y = \beta_1 z_{X1} + \beta_2 z_{X2}$) can be calculated directly from the zero-order Pearson's r's among the three variables Y, X_1, and X_2, as shown in Equations 4.8 and 4.9. In a subsequent section, a simple path model is used to show how these formulas were derived:

$$\beta_1 = \frac{r_{Y1} - r_{12}r_{Y2}}{1 - r_{12}^2}, \tag{4.8}$$

and

$$\beta_2 = \frac{r_{Y2} - r_{12}r_{Y1}}{1 - r_{12}^2}. \tag{4.9}$$

4.7.2 Formulas for Raw-Score (*b*) Coefficients

Given the beta coefficients and the means (M_Y, M_{X1}, and M_{X2}) and standard deviations (SD_Y, SD_{X1}, and SD_{X2}) of Y, X_1, and X_2, respectively, it is possible to calculate the *b* coefficients for the raw-score prediction equation shown in Equation 4.1 as follows:

$$b_1 = \frac{SD_Y}{SD_{X1}} \times \beta_1, \tag{4.10}$$

and

$$b_2 = \frac{SD_Y}{SD_{X2}} \times \beta_2. \tag{4.11}$$

Note that these equations are analogous to Equation 4.1 for the computation of *b* from *r* (or β) in a bivariate regression, where $b = (SD_Y/SD_X)r_{XY}$. To obtain *b* from β, we need to restore the information about the scales on which *Y* and the predictor variable are measured (information that is not contained in the unit-free beta coefficient). As in bivariate regression, a *b* coefficient is a rescaled version of β, that is, rescaled so that the coefficient can be used to make predictions from raw scores rather than *z* scores.

Once we have estimates of the b_1 and b_2 coefficients, we can compute the intercept b_0:

$$b_0 = M_Y - b_1 M_{X1} - b_2 M_{X2}. \tag{4.12}$$

This is analogous to the way the intercept was computed for a bivariate regression, where $b_0 = M_Y - bM_X$. There are other by-hand computational formulas to compute *b* from the sums of squares and sums of cross products for the variables; however, the formulas shown in the preceding equations make it clear how the *b* and β coefficients are related to each other and to the correlations among variables. In a later section of this chapter, you will see how the formulas to estimate the beta coefficients can be deduced from the correlations among the variables, using a simple path model for the regression. The computational formulas for the beta coefficients, given in Equations 4.8 and 4.9, can be understood conceptually: They are not just instructions for computation. These equations tell us that the values of the beta coefficients are influenced not only by the correlation between each *X* predictor variable and *Y* but also by the correlations between the *X* predictor variables.

4.7.3 Formulas for Multiple *R* and Multiple R^2

The multiple *R* can be calculated by hand. First of all, you could generate a predicted *Y'* score for each case by substituting the X_1 and X_2 raw scores into the equation and computing *Y'* for each case. Then, you could compute Pearson's *r* between *Y* (the actual *Y* score) and *Y'* (the predicted score generated by applying the regression equation to X_1 and X_2). Squaring this Pearson correlation yields R^2, the multiple *R* squared; this tells you what proportion of the total variance in *Y* is predictable from X_1 and X_2 combined.

Another approach is to examine the ANOVA source table for the regression (part of the SPSS output). As in the bivariate regression, SPSS partitions SS_{total} for *Y* into $SS_{\text{regression}} + SS_{\text{residual}}$. Multiple R^2 can be computed from these sums of squares:

$$R^2 = \frac{SS_{regression}}{SS_{total}}. \tag{4.13}$$

A slightly different version of this overall goodness-of-fit index is called the "adjusted" or "shrunken" R^2. This is adjusted for the effects of sample size (N) and number of predictors. There are several formulas for adjusted R^2; Tabachnick and Fidell (2018) provided this example:

$$R^2_{adj} = 1 - \left(1 - R^2\right)\left(\frac{N-1}{N-k-1}\right), \tag{4.14}$$

where N is the number of cases, k is the number of predictor variables, and R^2 is the squared multiple correlation given in Equation 4.13. R^2_{adj} tends to be smaller than R^2; it is much smaller than R^2 when N is relatively small and k is relatively large. In some research situations where the sample size N is very small relative to the number of variables k, the value reported for R^2_{adj} is actually negative; in these cases, it should be reported as 0. For computations involving the partition of variance (as shown in Figure 4.14), the unadjusted R^2 was used rather than the adjusted R^2.

4.7.4 Test of Significance for Overall Regression: F Test for H_0: $R = 0$

As in bivariate regression, an ANOVA can be performed to obtain sums of squares that represent the proportion of variance in Y that is and is not predictable from the regression, the sums of squares can be used to calculate mean squares (MS), and the ratio $MS_{\text{regression}}/MS_{\text{residual}}$ provides the significance test for R. N stands for the number of cases, and k is the number of predictor variables. For the regression examples in this chapter, the number of predictor variables, k, equals 2.

$$F = \frac{SS_{regression} / k}{SS_{residual} / (N-k-1)}, \tag{4.15}$$

with (k, $N - k - 1$) degrees of freedom (df).

If the obtained F ratio exceeds the tabled critical value of F for the predetermined alpha level (usually $\alpha = .05$), then the overall multiple R is judged statistically significant.

4.7.5 Test of Significance for Each Individual Predictor: t Test for H_0: $b_i = 0$

Recall that many sample statistics can be tested for significance by examining a t ratio of the following form; this kind of t ratio can also be used to assess the statistical significance of a b slope coefficient.

$$t = \frac{\textit{Sample statistic} - \textit{Hypothesized population parameter}}{SE_{\textit{sample statistic}}}.$$

The output from SPSS includes an estimated standard error (SE_b) associated with each raw-score slope coefficient (b). This standard error term can be calculated by hand in the following way. First, you need to know SE_{est}, the standard error of the estimate, which can be computed as

$$SE_{est} = SD_Y \sqrt{\left(1 - R^2\right)} \times \sqrt{\frac{N}{N-2}}. \tag{4.16}$$

SE_{est} describes the variability of the observed or actual Y values around the regression prediction at each specific value of the predictor variables. In other words, it gives us some idea of the typical magnitude of a prediction error when the regression equation is used to generate a Y' predicted value. Using SE_{est}, it is possible to compute an SE_b term for each b coefficient, to describe the theoretical sampling distribution of the slope coefficient. For predictor X_i, the equation for SE_{bi} is as follows:

$$SE_{bi} = \frac{SE_{est}}{\sqrt{\Sigma\left(X_i - M_{Xi}\right)^2}}. \tag{4.17}$$

The hypothesized value of each b slope coefficient is 0. Thus, the significance test for each raw-score b_i coefficient is obtained by the calculation of a t ratio, b_i divided by its corresponding SE term:

$$t_i = \frac{b_i}{SE_{bi}} \text{ with } (N - k - 1)\ df. \tag{4.18}$$

If the t ratio for a particular slope coefficient, such as b_1, exceeds the tabled critical value of t for $N - k - 1$ df, then that slope coefficient can be judged statistically significant. Generally, a two-tailed or nondirectional test is used.

Some multiple regression programs provide an F test (with 1 and $N - k - 1$ df) rather than a t test as the significance test for each b coefficient. Recall that when the numerator has only 1 df, F is equivalent to t^2.

4.7.6 Confidence Interval for Each *b* Slope Coefficient

A confidence interval (CI) can be set up around each sample b_i coefficient, using SE_{bi}.

To set up a 95% CI, for example, use the t distribution table to look up the critical value of t for $N - k - 1$ df that cuts off the top 2.5% of the area, t_{crit}:

$$\text{Upper bound of 95\% CI} = b_i + t_{crit} \times SE_{bi}. \tag{4.19}$$

$$\text{Lower bound of 95\% CI} = b_i - t_{crit} \times SE_{bi}. \tag{4.20}$$

4.8 SPSS REGRESSION

To run the SPSS linear regression procedure and to save the predicted Y' scores and the unstandardized residuals from the regression analysis, the following menu selections were made: <Analyze> → <Regression> → <Linear>.

In the SPSS Linear Regression dialog box (which appears in Figure 4.8), the name of the dependent variable (blood pressure) was entered in the box labeled "Dependent"; the names of both predictor variables were entered in the box labeled "Independent(s)." CIs for the b slope coefficients and values of the part and partial correlations were requested in addition to the default output by clicking the Statistics button and checking the boxes for CIs and for part and partial correlations. Note that the value that SPSS calls a "part" correlation is called the "semipartial" correlation by most textbook authors. The part correlations are needed to calculate the squared part or semipartial correlation for each predictor variable and to work out the partition of variance for blood pressure. Finally the Plots button was clicked, and a graph of standardized residuals against standardized predicted scores was requested to evaluate whether assumptions for regression were violated. The resulting SPSS syntax was copied into the Syntax Editor by clicking the Paste button; this syntax appears in Figure 4.9.

Figure 4.8 SPSS Linear Regression Dialog Box for a Regression to Predict Blood Pressure From Age and Weight

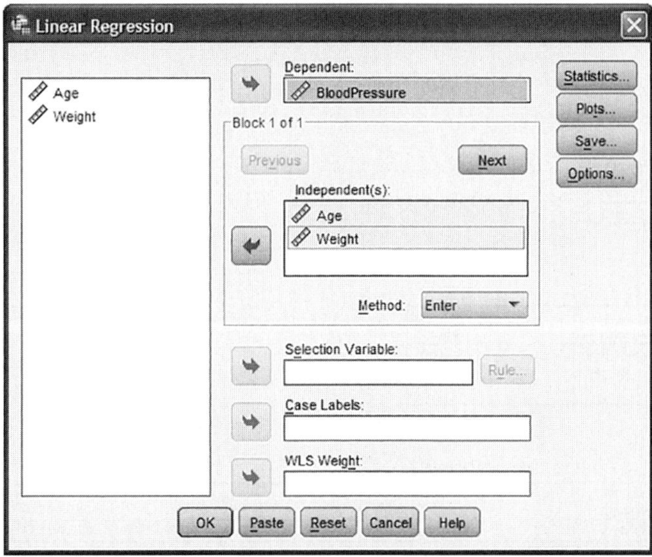

Figure 4.9 Syntax for the Regression to Predict Blood Pressure From Age and Weight (Including Part and Partial Correlations and a Plot of Standardized Residuals)

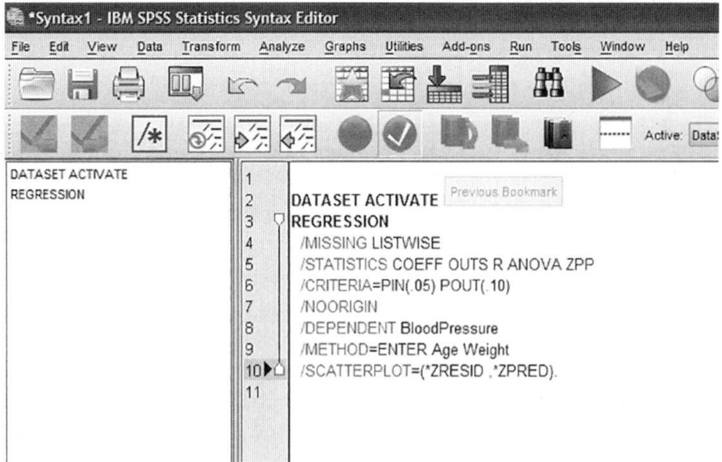

The resulting output for the regression to predict blood pressure from both age and weight appears in Figure 4.10, and the plot of the standardized residuals for this regression appears in Figure 4.11. The overall regression was statistically significant: $R = .83$, $F(2, 27) = 30.04$, $p < .001$. Thus, blood pressure could be predicted at levels significantly above chance from scores on age and weight combined. In addition, each of the individual predictor variables made a statistically significant contribution. For the predictor variable age, the raw-score regression coefficient b was 2.16, and this b slope coefficient differed significantly from 0, on

Figure 4.10 Output From SPSS Linear Regression to Predict Blood Pressure From Age and Weight

Variables Entered/Removed[b]

Model	Variables Entered	Variables Removed	Method
1	Weight, Age[a]	.	Enter

a. All requested variables entered.

b. Dependent Variable: BloodPressure

Model Summary[b]

Model	R	R Square	Adjusted R Square	Std. Error of the Estimate
1	.831[a]	.690	.667	36.692

a. Predictors: (Constant), Weight, Age

b. Dependent Variable: BloodPressure

ANOVA[b]

Model		Sum of Squares	df	Mean Square	F	Sig.
1	Regression	80882.13	2	40441.066	30.039	.000[a]
	Residual	36349.73	27	1346.286		
	Total	117231.9	29			

a. Predictors: (Constant), Weight, Age

b. Dependent Variable: BloodPressure

Coefficients[a]

Model		Unstandardized Coefficients		Standardized Coefficients			95% Confidence Interval for B		Correlations		
		B	Std. Error	Beta	t	Sig.	Lower Bound	Upper Bound	Zero-order	Partial	Part
1	(Constant)	-28.046	27.985		-1.002	.325	-85.466	29.373			
	Age	2.161	.475	.590	4.551	.000	1.187	3.135	.782	.659	.488
	Weight	.490	.187	.340	2.623	.014	.107	.873	.672	.451	.281

a. Dependent Variable: BloodPressure

Residuals Statistics[a]

	Minimum	Maximum	Mean	Std. Deviation	N
Predicted Value	66.13	249.62	177.27	52.811	30
Residual	-74.752	63.436	.000	35.404	30
Std. Predicted Value	-2.104	1.370	.000	1.000	30
Std. Residual	-2.037	1.729	.000	.965	30

a. Dependent Variable: BloodPressure

Figure 4.11 Plot of Standardized Residuals From Linear Regression to Predict Blood Pressure From Age and Weight

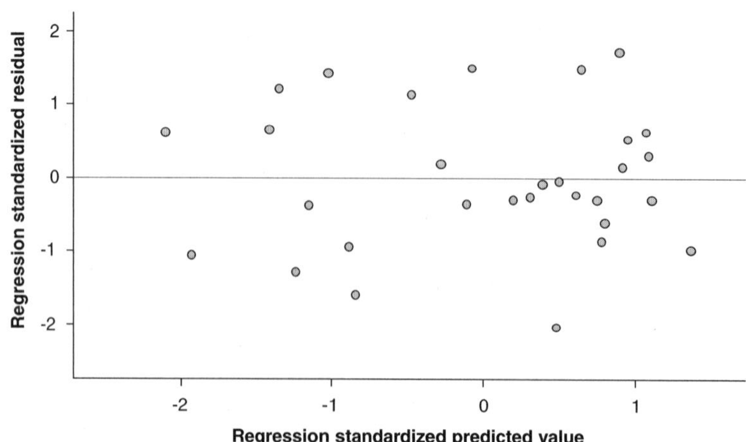

the basis of a t value of 4.55 with $p < .001$. The corresponding effect size for the proportion of variance in blood pressure uniquely predictable from age was obtained by squaring the value of the part correlation of age with blood pressure to yield $sr^2_{\text{age}} = .24$. For the predictor variable weight, the raw-score slope $b = .50$ was statistically significant: $t = 2.62$, $p = .014$; the corresponding effect size was obtained by squaring the part correlation for weight, $sr^2_{\text{weight}} = .08$. The pattern of residuals that is shown in Figure 4.11 does not indicate any problems with the assumptions. These regression results are discussed and interpreted more extensively in the model "Results" section that appears near the end of this chapter.

4.9 CONCEPTUAL BASIS: FACTORS THAT AFFECT THE MAGNITUDE AND SIGN OF β AND *b* COEFFICIENTS IN MULTIPLE REGRESSION WITH TWO PREDICTORS

It may be intuitively obvious that the predictive slope of X_1 depends, in part, on the value of the zero-order Pearson correlation of X_1 with Y. It may be less obvious, but the value of the slope coefficient for each predictor is also influenced by the correlation of X_1 with other predictors, as you can see in Equations 4.8 and 4.9. Often, but not always, we will find that an X_1 variable that has a large correlation with Y also tends to have a large beta coefficient; the sign of beta is often, but not always, the same as the sign of the zero-order Pearson's r. However, depending on the magnitudes and signs of the r_{12} and r_{2Y} correlations, a beta coefficient (like a partial correlation) can be larger, smaller, or even opposite in sign compared with the zero-order Pearson's r_{1Y}. The magnitude of a β_1 coefficient, like the magnitude of a partial correlation pr_1, is influenced by the size and sign of the correlation between X_1 and Y; it is also affected by the size and sign of the correlation(s) of the X_1 variable with other variables that are statistically controlled in the analysis.

In this section, we will examine a path diagram model of a two-predictor multiple regression to see how estimates of the beta coefficients are found from the correlations among all three pairs of variables involved in the model: r_{12}, r_{Y1}, and r_{Y2}. This analysis will make several things clear. First, it will show how the sign and magnitude of the standard-score coefficient β_i for each X_i variable are related to the size of r_{Yi}, the correlation of that particular predictor with Y, and also the size of the correlation of X_i and all other predictor variables included in the regression (at this point, this is the single correlation r_{12}).

Second, it will explain why the numerator for the formula to calculate β_1 in Equation 4.8 has the form $r_{Y1} - r_{12}r_{Y2}$. In effect, we begin with the "overall" relationship between X_1 and Y, represented by r_{Y1}; we subtract from this the product $r_{12} \times r_{Y2}$, which represents an indirect path from X_1 to Y via X_2. Thus, the estimate of the β_1 coefficient is adjusted so that it only gives the X_1 variable "credit" for any relationship to Y that exists over and above the indirect path that involves the association of both X_1 and Y with the other predictor variable X_2.

Finally, we will see that the formulas for β_1, pr_1, and sr_1 all have the same numerator: $r_{Y1} - r_{12}r_{Y2}$. All three of these statistics (β_1, pr_1, and sr_1) provide somewhat similar information about the nature and strength of the relation between X_1 and Y, controlling for X_2, but they are scaled slightly differently (by using different divisors) so that they can be interpreted and used in different ways.

Consider the regression problem in which you are predicting z scores on y from z scores on two independent variables X_1 and X_2. We can set up a path diagram to represent how two predictor variables are related to one outcome variable (Figure 4.12).

The path diagram in Figure 4.12 corresponds to this regression equation:

$$z'_Y = \beta_1 z_{X1} + \beta_2 z_{X2}. \tag{4.21}$$

Path diagrams represent hypothetical models (often called "causal" models, although we cannot prove causality from correlational analyses) that represent our hypotheses about the

Figure 4.12 Path Diagram for Standardized Multiple Regression to Predict z'_Y From z_{X1} and z_{X2}

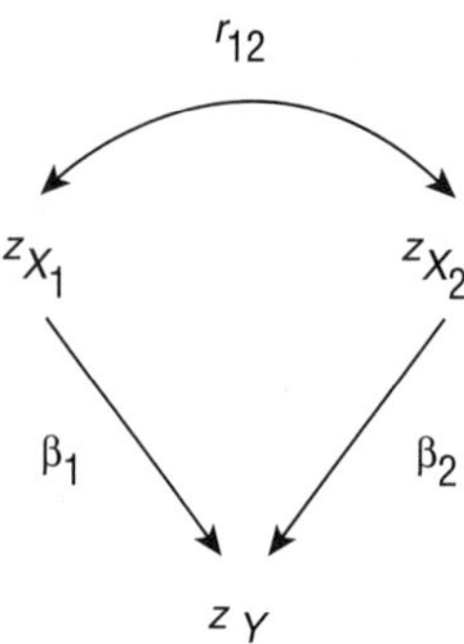

nature of the relations between variables. In this example, the path model is given in terms of z scores (rather than raw X scores) because this makes it easier to see how we arrive at estimates of the beta coefficients. When two variables in a path model diagram are connected by a double-headed arrow, it represents a hypothesis that the two variables are correlated or confounded (but there is no hypothesized causal connection between the variables). Pearson's r between these predictors indexes the strength of this confounding or correlation. A single-headed arrow ($X \rightarrow Y$) indicates a theorized causal relationship (such that X causes Y), or at least a directional predictive association between the variables. The "path coefficient" or regression coefficient (i.e., a beta coefficient) associated with it indicates the estimated strength of the predictive relationship through this direct path. If there is no arrow connecting a pair of variables, it indicates a lack of any direct association between the pair of variables, although the variables may be connected through indirect paths.

The path diagram that is usually implicit in a multiple regression analysis has the following general form: Each of the predictor (X) variables has a unidirectional arrow pointing from X to Y, the outcome variable. All pairs of X predictor variables are connected to each other by double-headed arrows that indicate correlation or confounding, but no presumed causal linkage, among the predictors. Figure 4.12 shows the path diagram for the standardized (z score) variables in a regression with two correlated predictor variables z_{X1} and z_{X2}. This model corresponds to a causal model in which z_{X1} and z_{X2} are represented as "partially redundant" or correlated causes or predictors of z_Y.

Our problem is to deduce the unknown path coefficients or standardized regression coefficients associated with the direct (or causal) path from each of the z_X predictors, β_1 and β_2, in terms of the known correlations r_{12}, r_{Y1}, and r_{Y2}. This is done by applying the tracing rule, as described in the following section.

4.10 TRACING RULES FOR PATH MODELS

The idea behind path models is that an adequate model should allow us to reconstruct the observed correlation between any pair of variables (e.g., r_{Y1}), by tracing the paths that lead from X_1 to Y through the path system, calculating the strength of the relationship for each path, and then summing the contributions of all possible paths from X_1 to Y.

Kenny (1979) provided a clear and relatively simple statement about the way in which the paths in this causal model can be used to reproduce the overall correlation between each pair of variables:

> The correlation between X_i and X_j equals the sum of the product of all the path coefficients [these are the beta weights from a multiple regression] obtained from each

of the possible tracings between X_i and X_j. The set of tracings includes all possible routes from X_i to X_j given that (a) the same variable is not entered twice and (b) a variable is not entered through an arrowhead and left through an arrowhead. (p. 30)

In general, the traced paths that lead from one variable, such as z_{X1}, to another variable, such as z'_Y, may include one direct path and also one or more indirect paths.

We can use the tracing rule to reconstruct exactly the observed correlation between any two variables from a path model from correlations and the beta coefficients for each path. Initially, we will treat β_1 and β_2 as unknowns; later, we will be able to solve for the betas in terms of the correlations.

Now, let's look in more detail at the multiple regression model with two independent variables (represented by the diagram in Figure 4.12). The path from z_{X1} to z_{X2} is simply r_{12}, the observed correlation between these variables. We will use the labels β_1 and β_2 for the coefficients that describe the strength of the direct, or unique, relationship of X_1 and X_2, respectively, to Y. β_1 indicates how strongly X_1 is related to Y after we have taken into account, or partialled out, the indirect relationship of X_1 to Y involving the path via X_2. β_1 is a partial slope: the number of standard deviation units of change in z_Y we predict for a 1-*SD* change in z_{X1} when we have taken into account, or partialled out, the influence of z_{X2}. If z_{X1} and z_{X2} are correlated, we must somehow correct for the redundancy of information they provide when we construct our prediction of Y; we don't want to double-count information that is included in both z_{X1} and z_{X2}. That is why we need to correct for the correlation of z_{X1} with z_{X2} (i.e., take into account the indirect path from z_{X1} to z_Y via z_{X2}) to get a clear picture of how much predictive value z_{X1} has that is *unique* to z_{X1} and not somehow related to z_{X2}.

For each pair of variables (z_{X1} and z_Y, z_{X2} and z_Y), we need to work out all possible paths from z_{Xi} to z_Y; if the path has multiple steps, the coefficients along that path are multiplied with each other. After we have calculated the strength of association for each path, we sum the contributions across paths. For the path from z_{X1} to z'_Y, in the diagram above, there is one direct path from z_{X1} to z'_Y, with a coefficient of β_1. There is also one indirect path from z_{X1} to z'_Y via z_{X2}, with two coefficients en route (r_{12} and β_2); these are multiplied to give the strength of association represented by the indirect path, $r_{12} \times \beta_2$. Finally, we should be able to reconstruct the entire observed correlation between z_{X1} and z_Y (r_{Y1}) by summing the contributions of all possible paths from z_{X1} to z'_Y in this path model. This reasoning based on the tracing rule yields the equation below:

$$\text{Total correlation} = \text{Direct path} + \text{Indirect path}.$$

$$r_{Y1} = \beta_1 + r_{12} \times \beta_2. \tag{4.22}$$

Applying the same reasoning to the paths that lead from z_{X2} to z'_Y, we arrive at a second equation of this form:

$$r_{Y2} = \beta_2 + r_{12} \times \beta_1. \tag{4.23}$$

Equations 4.22 and 4.23 are called the normal equations for multiple regression; they show how the observed correlations (r_{Y1} and r_{Y2}) can be perfectly reconstructed from the regression model and its parameter estimates β_1 and β_2. We can solve these equations for values of β_1 and β_2 in terms of the known correlations r_{12}, r_{Y1}, and r_{Y2} (these equations appeared earlier as Equations 4.8 and 4.9):

$$\beta_1 = \frac{r_{Y1} - r_{12} r_{Y2}}{1 - r_{12}^2},$$

and

$$\beta_2 = \frac{r_{Y2} - r_{12} r_{Y1}}{1 - r_{12}^2}.$$

The numerator for the betas is the same as the numerator of the partial correlation. Essentially, we take the overall correlation between X_1 and Y and subtract the correlation we would predict between X_1 and Y due to the relationship through the indirect path via X_2; whatever is left, we then attribute to the direct or unique influence of X_1. In effect, we "explain" as much of the association between X_1 and Y as we can by first looking at the indirect path via X_2 and only attributing to X_1 any additional relationship it has with Y that is above and beyond that indirect relationship. We then divide by a denominator that scales the result (as a partial slope or beta coefficient, in these two equations, or as a partial correlation, as in the previous chapter).

Note that if the value of β_1 is zero, we can interpret it to mean that we do not need to include a direct path from X_1 to Y in our model. If $\beta_1 = 0$, then any statistical relationship or correlation that exists between X_1 and Y can be entirely explained by the indirect path involving X_2. Possible explanations for this pattern of results include the following: that X_2 causes both X_1 and Y and the X_1, Y correlation is spurious, or that X_2 is a mediating variable, and X_1 influences Y only through its influence on X_2. This is the basic idea that underlies path analysis or so-called causal modeling: If we find that we do not need to include a direct path between X_1 and Y, then we can simplify the model by dropping a path. We will not be able to prove causality from path analysis; we can only decide whether a causal or theoretical model that has certain paths omitted is sufficient to reproduce the observed correlations and, therefore, is "consistent" with the observed pattern of correlations.

4.11 COMPARISON OF EQUATIONS FOR β, *b*, *pr*, AND *sr*

By now, you may have recognized that β, *b*, *pr*, and *sr* are all slightly different indexes of how strongly X_1 predicts Y when X_2 is controlled. Note that the (partial) standardized slope or β coefficient, the partial *r*, and the semipartial *r* all have the same term in the numerator: They are scaled differently by dividing by different terms, to make them interpretable in slightly different ways, but generally, they are similar in magnitude. The numerators for partial *r* (*pr*), semipartial *r* (*sr*), and beta (β) are identical. The denominators differ slightly because they are scaled to be interpreted in slightly different ways (squared partial *r* as a proportion of variance in *Y* when X_2 has been partialled out of *Y*; squared semipartial *r* as a proportion of the total variance of *Y*; and beta as a partial slope, the number of standard deviation units of change in *Y* for a one-unit *SD* change in X_1). It should be obvious from looking at the formulas that *sr*, *pr*, and β tend to be similar in magnitude and must have the same sign. (These equations are all repetitions of equations given earlier, and therefore, they are not given new numbers here.)

Standard-score slope coefficient β:

$$\beta_1 = \frac{r_{Y1} - r_{12} r_{Y2}}{1 - r_{12}^2}.$$

Raw-score slope coefficient *b* (a rescaled version of the β coefficient):

$$b_1 = \beta_1 \times \frac{SD_Y}{SD_{X1}}.$$

Partial correlation to predict Y from X_1, controlling for X_2 (removing X_2 completely from both X_1 and Y):

$$pr_1 \text{ or } r_{Y1.2} = \frac{r_{Y1} - r_{Y2}r_{12}}{\sqrt{\left(1-r_{Y2}^2\right)\times\left(1-r_{12}^2\right)}}.$$

Semipartial (or part) correlation to predict Y from X_1, controlling for X_2 (removing X_2 only from X_1, as explained in this chapter):

$$sr_1 \text{ or } r_{Y(1.2)} = \frac{r_{Y1} - r_{Y2}r_{12}}{\sqrt{\left(1-r_{12}^2\right)}}.$$

Because these equations all have the same numerator (and they differ only in that the different divisors scale the information so that it can be interpreted and used in slightly different ways), it follows that your conclusions about how X_1 is related to Y when you control for X_2 tend to be fairly similar no matter which of these four statistics (b, β, pr, or sr) you use to describe the relationship. If any one of these four statistics exactly equals 0, then the other three also equal 0, and all these statistics must have the same sign. They are scaled or sized slightly differently so that they can be used in different situations (to make predictions from raw vs. standard scores and to estimate the proportion of variance accounted for relative to the total variance in Y or only the variance in Y that isn't related to X_2).

The difference among the four statistics above is subtle: β_1 is a partial slope (how much change in z_Y is predicted for a 1-*SD* change in z_{X1} if z_{X2} is held constant). The partial r describes how X_1 and Y are related if X_2 is removed from both variables. The semipartial r describes how X_1 and Y are related if X_2 is removed only from X_1. In the context of multiple regression, the squared semipartial r (sr^2) provides the most convenient way to estimate effect size and variance partitioning. In some research situations, analysts prefer to report the b (raw-score slope) coefficients as indexes of the strength of the relationship among variables. In other situations, standardized or unit-free indexes of the strength of relationship (such as β, sr, or pr) are preferred.

4.12 NATURE OF PREDICTIVE RELATIONSHIPS

When reporting regression, it is important to note the signs of b and β coefficients, as well as their size, and to state whether these signs indicate relations that are in the predicted direction. Researchers sometimes want to know whether a pair of b or β coefficients differ significantly from each other. This can be a question about the size of b in two different groups of subjects: For instance, is the β slope coefficient to predict salary from years of job experience significantly different for male versus female subjects? Alternatively, it could be a question about the size of b or β for two different predictor variables in the same group of subjects (e.g., Which variable has a stronger predictive relation to blood pressure: age or weight?).

It is important to understand how problematic such comparisons usually are. Our estimates of β and b coefficients are derived from correlations; thus, any factors that artifactually influence the sizes of correlations such that the correlations are either inflated or deflated estimates of the real strength of the association between variables can also potentially affect our estimates of β and b. Thus, if women have a restricted range in scores on drug use (relative to men), a difference in Pearson's r and the beta coefficient to predict drug use for women versus men might be artifactually due to a difference in the range of scores on the outcome variable for the two groups. Similarly, a difference in the reliability of measures for the two groups could create an artifactual difference in the size of Pearson's r and regression coefficient estimates. It is probably never possible to rule out all possible sources of artifact that might explain the different sizes of r and β coefficients (in different samples or for different predictors). If a researcher wants to interpret a difference between slope coefficients as evidence for a difference in the strength of the association between variables, the researcher should demonstrate that the two groups do not differ in range of scores, distribution shape of scores, reliability of measurement, existence of outliers, or other factors that may affect the size of correlations. However, no matter how many possible sources of artifact are considered, comparison of slopes and correlations remains problematic. Later chapters describe use of dummy variables and interaction terms to test whether two groups, such as women versus men, have significantly different slopes for the prediction of Y from some X_i variable. More sophisticated methods that can be used to test equality of specific model parameters, whether they involve comparisons across groups or across different predictor variables, are available within the context of structural equation modeling (SEM) analysis using programs such as Amos.

4.13 EFFECT SIZE INFORMATION IN REGRESSION WITH TWO PREDICTORS

4.13.1 Effect Size for Overall Model

The effect size for the overall model—that is, the proportion of variance in Y that is predictable from X_1 and X_2 combined—is estimated by computation of an R^2. This R^2 is shown in the SPSS output; it can be obtained either by computing the correlation between observed Y and predicted Y' scores and squaring this correlation or by taking the ratio $SS_{\text{regression}}/SS_{\text{total}}$:

$$R^2 = \frac{SS_{regression}}{SS_{total}}. \tag{4.24}$$

Note that this formula for the computation of R^2 is analogous to the formulas given in earlier chapters for eta squared ($\eta^2 = SS_{\text{between}}/SS_{\text{total}}$ for an ANOVA; $R^2 = SS_{\text{regression}}/SS_{\text{total}}$ for multiple regression). R^2 differs from η^2 in that R^2 assumes a *linear* relation between scores on Y and scores on the predictors. On the other hand, η^2 detects differences in mean values of Y across different values of X, but these changes in the value of Y do not need to be a linear function of scores on X. Both R^2 and η^2 are estimates of the proportion of variance in Y scores that can be predicted from independent variables. However, R^2 (as described in this chapter) is an index of the strength of *linear* relationship, while η^2 detects patterns of association that need not be linear.

For some statistical power computations, such as those presented by Green (1991), a different effect size for the overall regression equation, called f^2, is used:

$$f^2 = R^2/(1 - R^2). \tag{4.25}$$

4.13.2 Effect Size for Individual Predictor Variables

The most convenient effect size to describe the proportion of variance in Y that is uniquely predictable from X_i is the squared semipartial correlation between X_i and Y, controlling for all other predictors. This semipartial (also called the part) correlation between each predictor and Y can be obtained from the SPSS regression procedure by checking the box for the part and partial correlations in the optional statistics dialog box. The semipartial or part correlation (sr) from the SPSS output can be squared by hand to yield an estimate of the proportion of uniquely explained variance for each predictor variable (sr^2).

If the part correlation is not requested, it can be calculated from the t statistic associated with the significance test of the b slope coefficient. It is useful to know how to calculate this by hand so that you can generate this effect size measure for published regression studies that don't happen to include this information:

$$sr_i^2 = \frac{t_i^2}{df_{residual}}\left(1-R^2\right), \tag{4.26}$$

where t_i is the ratio b_i/SE_{bi} for the X_i predictor variable, the df residual = $N - k - 1$, and R^2 is the multiple R^2 for the entire regression equation. The verbal interpretation of sr^2_i is the proportion of variance in Y that is uniquely predictable from X_i (when the variance due to other predictors is partialled out of X_i).

Some multiple regression programs do not provide the part or semipartial correlation for each predictor, and they report an F ratio for the significance of each b coefficient; this F ratio may be used in place of t^2_i to calculate the effect size estimate:

$$sr_i^2 = \frac{F}{df_{residual}}\left(1-R^2\right). \tag{4.27}$$

4.14 STATISTICAL POWER

Tabachnick and Fidell (2018) discussed a number of issues that need to be considered in decisions about sample size; these include alpha level, desired statistical power, number of predictors in the regression equation, and anticipated effect sizes. They suggested the following simple guidelines. Let k be the number of predictor variables in the regression (in this chapter, $k = 2$). The effect size index used by Green (1991) was f^2, where $f^2 = R^2/(1 - R^2)$; $f^2 = .15$ is considered a medium effect size. Assuming a medium effect size and $\alpha = .05$, the minimum desirable N for testing the significance of multiple R is $N > 50 + 8k$, and the minimum desirable N for testing the significance of individual predictors is $N > 104 + k$. Tabachnick and Fidell recommended that the data analyst choose the larger number of cases required by these two decision rules. Thus, for the regression analysis with two predictor variables described in this chapter, assuming the researcher wants to detect medium-size effects, a desirable minimum sample size would be $N = 106$. (Smaller N's are used in many of the demonstrations and examples in this textbook, however.) If there are substantial violations of assumptions (e.g., skewed rather than normal distribution shapes) or low measurement reliability, then the minimum N should be substantially larger; see Green for more detailed instructions. If N is extremely large (e.g., $N > 5{,}000$), researchers may find that even associations that are too weak to be of any practical or clinical importance turn out to be statistically significant.

To summarize, then, the guidelines described above suggest that a minimum N of about 106 should be used for multiple regression with two predictor variables to have reasonable power to detect the overall model fit that corresponds to approximately medium-size R^2 values. If more precise estimates of required sample size are desired, the guidelines given by Green (1991) may be used. In general, it is preferable to have sample sizes that are somewhat larger than the minimum values suggested by these decision rules. In addition to having a large enough sample size to have reasonable statistical power, researchers should also have samples large enough so that the CIs around the estimates of slope coefficients are reasonably narrow. In other words, we should try to have sample sizes that are large enough to provide reasonably precise estimates of slopes and not just samples that are large enough to yield "statistically significant" results.

4.15 ISSUES IN PLANNING A STUDY

4.15.1 Sample Size

A minimum N of at least 100 cases is desirable for a multiple regression with two predictor variables (the rationale for this recommended minimum sample size is given in Section 4.14 on statistical power). The examples presented in this chapter use fewer cases, so that readers who want to enter data by hand or perform computations by hand or in an Excel spreadsheet can replicate the analyses shown.

4.15.2 Selection of Predictor and/or Control Variables

The researcher should have some theoretical rationale for the choice of independent variables. Often, the X_1, X_2 predictors are chosen because one or both of them are implicitly believed to be "causes" of Y (although a significant regression does not provide evidence of causality). In some cases, the researcher may want to assess the combined predictive usefulness of two variables or to judge the relative importance of two predictors (e.g., How well do age and weight in combination predict blood pressure? Is age a stronger predictor of blood pressure than weight?). In some research situations, one or more of the variables used as predictors in a regression analysis serve as control variables that are included to control for competing causal explanations or to control for sources of contamination in the measurement of other predictor variables.

Several variables are often used to control for contamination in the measurement of predictor variables. For example, many personality test scores are related to social desirability; if the researcher includes a good measure of social desirability response bias as a predictor in the regression model, the regression may yield a better description of the predictive usefulness of the personality measure. Alternatively, of course, controlling for social desirability could make the predictive contribution of the personality measure drop to zero. If this occurred, the researcher might conclude that any apparent predictive usefulness of that personality measure was due entirely to its social desirability component.

After making a thoughtful choice of predictors, the researcher should try to anticipate the possible different outcomes and the various possible interpretations to which these would lead. Selection of predictor variables on the basis of "data fishing"—that is, choosing predictors because they happen to have high correlations with the Y outcome variable in the sample of data in hand—is not recommended. Regression analyses that are set up in this way are likely to report "significant" predictive relationships that are instances of Type I error. It is preferable to base the choice of predictor variables on past research and theory rather than on sizes of correlations. (Of course, it is possible that a large correlation that turns up unexpectedly may represent a serendipitous finding; however, replication of the correlation with new samples should be obtained.)

4.15.3 Collinearity (Correlation) Between Predictors

Although multiple regression can be a useful tool for separating the unique predictive contributions of correlated predictor variables, it does not work well when predictor variables are extremely highly correlated (in the case of multiple predictors, high correlations among many predictors are referred to as **multicollinearity**). In the extreme case, if two predictors are perfectly correlated, it is impossible to distinguish their predictive contributions; in fact, regression coefficients cannot be calculated in this situation.

To understand the nature of this problem, consider the partition of variance illustrated in Figure 4.13 for two predictors, X_1 and X_2, that are highly correlated with each other. When there is a strong correlation between X_1 and X_2, most of the explained variance cannot be

Figure 4.13 Diagram of Partition of Variance With Highly Correlated (Multicollinear) Predictors

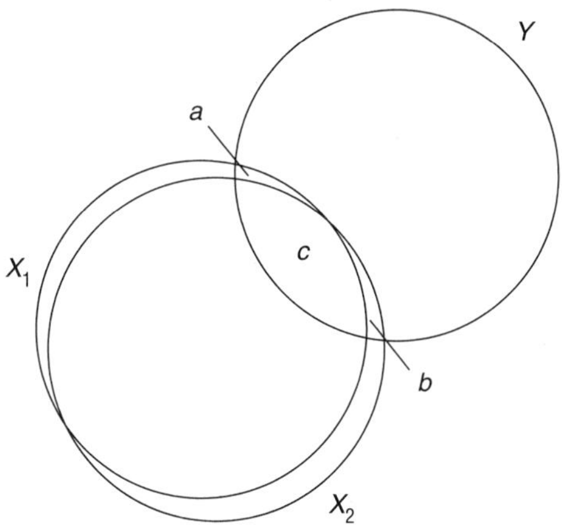

Note: Area c becomes very large and Areas *a* and *b* become very small when there is a large correlation between X_1 and X_2.

attributed uniquely to either predictor variable; in this situation, even if the overall multiple *R* is statistically significant, neither predictor may be judged statistically significant. The area (denoted as Area *c* in Figure 4.13) that corresponds to the variance in *Y* that could be predicted from either X_1 or X_2 tends to be quite large when the predictors are highly intercorrelated, whereas Areas *a* and *b*, which represent the proportions of variance in *Y* that can be uniquely predicted from X_1 and X_2, respectively, tend to be quite small.

Extremely high correlations between predictors (in excess of .9 in absolute value) may suggest that the two variables are actually measures of the same underlying construct (Berry, 1993). In such cases, it may be preferable to drop one of the variables from the predictive equation. Alternatively, sometimes it makes sense to combine the scores on two or more highly correlated predictor variables into a single index by summing or averaging them; for example, if income and occupational prestige are highly correlated predictors, it may make sense to combine these into a single index of socioeconomic status, which can then be used as a predictor variable.

Stevens (2009) identified three problems that arise when the predictor variables in a regression are highly intercorrelated (as shown in Figure 4.13). First, a high level of correlation between predictors can limit the size of multiple *R*, because the predictors are "going after much of the same variance" in *Y*. Second, as noted above, it makes assessment of the unique contributions of predictors difficult. When predictors are highly correlated with each other, Areas *a* and *b*, which represent their unique contributions, tend to be quite small. Finally, the error variances associated with each *b* slope coefficient (SE_b) tend to be large when the predictors are highly intercorrelated; this means that the CIs around estimates of *b* are wider and, also, that power for statistical significance tests is lower.

4.15.4 Ranges of Scores

As in correlation analyses, there should be a sufficient range of scores on both the predictor and the outcome variables to make it possible to detect relations between them. This, in turn, requires that the sample be drawn from a population in which the variables of interest show a reasonably wide range. It would be difficult, for example, to demonstrate strong age-related changes in blood pressure in a sample with ages that ranged only from 18 to 25 years; the relation between blood pressure and age would probably be stronger and easier to detect in a sample with a much wider range in ages (e.g., from 18 to 75).

4.16 RESULTS

The results of an SPSS regression analysis to predict blood pressure from both age and weight (for the data in ageweightbp.sav) are shown in Table 4.1. (Instead of presenting all information in one table, there could be separate tables for descriptive statistics, correlations among all variables, and regression results.) Description of results for regression should include bivariate correlations among all the predictor and outcome variables; mean and standard deviation for each variable involved in the analysis; information about the overall fit of the regression model (multiple R and R^2 and the associated F test); the b coefficients for the raw-score regression equation, along with an indication whether each b coefficient differs significantly from zero; the beta coefficients for the standard-score regression equation; and a squared part or semipartial correlation (sr^2) for each predictor that represents the proportion of variance in the Y outcome variable that can be predicted uniquely from each predictor variable, controlling for all other predictors in the regression equation. Confidence intervals should also be given for the b regression coefficients. Details about reporting are covered in the later chapter about regression with more than two predictor variables.

Table 4.1 Results of Standard Multiple Regression to Predict Blood Pressure (Y) From Age (X_1) and Weight (X_2)

Variables	Blood Pressure	Age	Weight	b	β	sr^2_{unique}
Age	+.78***			+2.161***	+.59	+.24
Weight	+.67***	+.56		+.490*	+.34	+.08
			Intercept = −28.05			
Mean	177.3	58.3	162.0			
SD	63.6	17.4	44.2			
						R^2 = .690
						R^2_{adj} = .667
						R = .831***

Note: The use of asterisks to denote whether p values are less than the most commonly selected α levels (.05, .01, and .001) is widespread, therefore you need to recognize it in research reports. I recommend that you replace asterisks with exact p values, below the corresponding statistic in parentheses, with a table footnote to explain what you have done. In addition, note that p values have not been adjusted for increased risk for Type I errors that arises when multiple tests are reported. In addition to that, if you ran numerous regression analysis and have reported only one, the analysis should be reported as exploratory. In this situation, p values should not be reported at all, or else footnoted to indicate that they are extremely inaccurate indications of risk for Type I error.

$^{*}p < .05$. $^{***}p < .001$.

The example given in the "Results" section below discusses age and weight as correlated or partly redundant predictor variables, because this is the most common implicit model when regression is applied. The following "Results" section reports more detail than typically included in journal articles.

Results

Initial examination of blood pressure data for a sample of N = 30 participants indicated that there were positive correlations between all pairs of variables. However, the correlation between the predictor variables age and weight, r = +.56, did not indicate extremely high multicollinearity.

For the overall multiple regression to predict blood pressure from age and weight, R = .83 and R^2 = .69. That is, when both age and weight were used as predictors, about 69% of the variance in blood pressure could be predicted. The adjusted R^2 was .67. The overall regression was statistically significant, $F(2, 27) = 30.04, p < .001$.

Age was significantly predictive of blood pressure when the variable weight was statistically controlled, $t(27) = 4.55, p < .001$. The positive slope for age as a predictor of blood pressure indicated that there was about a 2 mm Hg increase in blood pressure for each 1-year increase in age, controlling for weight. The squared semipartial correlation that estimated how much variance in blood pressure was uniquely predictable from age was sr^2 = .24. About 24% of the variance in blood pressure was uniquely predictable from age (when weight was statistically controlled).

Weight was also significantly predictive of blood pressure when age was statistically controlled, $t(27) = 2.62, p = .014$. The slope to predict blood pressure from weight was approximately b = +.49; in other words, there was about a 0.5 mm Hg increase in blood pressure for each 1-lb increase in body weight. The sr^2 for weight (controlling for age) was .08. Thus, weight uniquely predicted about 8% of the variance in blood pressure when age was statistically controlled.

The conclusion from this analysis is that the original zero-order correlation between age and blood pressure (r = .78 or r^2 = .61) was partly (but not entirely) accounted for by weight. When weight was statistically controlled, age still uniquely predicted 24% of the variance in blood pressure. One possible interpretation of this outcome is that age and weight are partly redundant as predictors of blood pressure; to the extent that age and weight are correlated with each other, they compete to explain some of the same variance in blood pressure. However, each predictor was significantly associated with blood pressure even when the other predictor variable was significantly controlled; both age and weight contribute uniquely useful predictive information about blood pressure in this research situation.

The predictive equations were as follows:

Raw-score version: Blood pressure′ = −28.05 + 2.16 × Age + .49 × Weight.

Standard-score version: $z_{\text{blood pressure}}' = .59 \times z_{\text{age}} + .34 \times z_{\text{weight}}$.

Although residuals are rarely discussed in the "Results" sections of journal articles, examination of plots of residuals can be helpful in detecting violations of assumptions or multivariate outliers; either of these problems would make the regression analysis less credible. The

Figure 4.14 Diagram of Partition of Variance for Prediction of Blood Pressure (Y) From Age (X_1) and Weight (X_2)

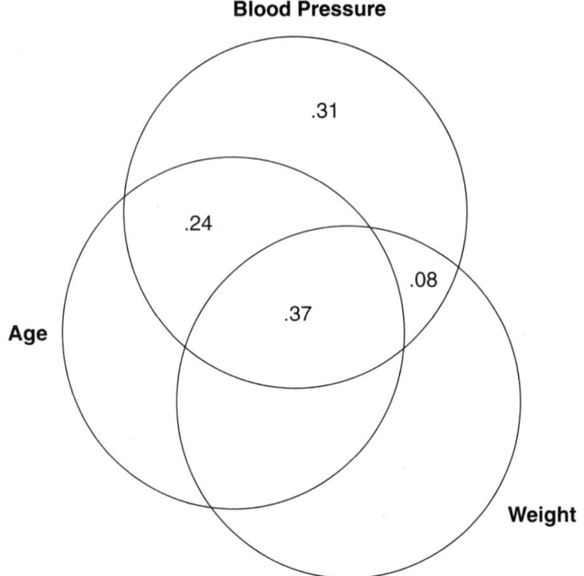

Note: Proportion of variance in blood pressure not predictable from age or weight = $1 - R^2 = .31$ = Area *d*. Proportion of variance in blood pressure uniquely predictable from age, controlling for weight = $sr^2_{age} = .24$ = Area *a*. Proportion of variance in blood pressure uniquely predictable from weight, controlling for age = $sr^2_{weight} = .08$ = Area *b*. Proportion of variance in blood pressure predictable by either age or weight = $1 - a - b - d = .37$ = Area *c*. Areas in the diagram do not correspond exactly to the proportions; the diagram is only approximate.

graph of standardized residuals against standardized predicted scores (in Figure 4.11) did not suggest any problem with the residuals. If all the assumptions for regression analysis are satisfied, the mean value of the standardized residuals should be 0 for all values of the predicted score, the variance of residuals should be uniform across values of the predicted score, the residuals should show no evidence of a linear or curvilinear trend, and there should be no extreme outliers.

Although this is not usually reported in a journal article, it is useful to diagram the obtained partition of variance so that you understand exactly how the variance in *Y* was divided. Figure 4.14 shows the specific numerical values that correspond to the variance components that were identified in Figure 4.2. Area *d* was calculated by finding $1 - R^2 = 1 - .69 = .31$, using the R^2 value of .69 from the SPSS output. Note that the unadjusted R^2 was used rather than the adjusted R^2; the adjusted R^2 can actually be negative in some instances. Numerical estimates for the proportions of unique variance predictable from each variable represented by Areas *a* and *b* were obtained by squaring the part correlations (also called the semipartial correlation) for each predictor. For age, the part or semipartial correlation in the SPSS output was $sr_{age} = .488$; the value of sr^2_{age} obtained by squaring this value was about .24. For weight, the part or semipartial correlation reported by SPSS was $sr_{weight} = .281$; therefore, $sr^2_{weight} = .08$. Because the sum of all four areas $(a + b + c + d)$ equals 1, once the values for Areas *a*, *b*, and *d* are known, a numerical value for Area *c* can be obtained by subtraction $(c = 1 - a - b - d)$.

In this example, 69% of the variance in blood pressure was predictable from age and weight in combination (i.e., $R^2 = .69$). This meant that 31% $(1 - R^2 = 1 - .69 = .31)$ of the variance in salaries could not be predicted from these two variables. Twenty-four percent of the variance in blood pressure was uniquely predictable from age (the part correlation for age was .488, so $sr^2_{\text{age}} = .24$). Another 8% of the variance in blood pressure was uniquely predictable from weight (the part correlation for weight was .281, so the squared part correlation for weight was about $sr^2_{\text{weight}} = .08$). Area *c* was obtained by subtraction of Areas *a*, *b*, and *d* from 1: $1 - .24 - .08 - .31 = .37$. Thus, the remaining 37% of the variance in blood pressure could be predicted equally well by age or weight (because these two predictors were confounded or redundant to some extent).

Note that it is possible, although unusual, for Area *c* to turn out to be a negative number; this can occur when one (or both) of the semipartial *r*'s for the predictor variables are larger in absolute value than their zero-order correlations with *Y*. When Area *c* is large, it indicates that the predictor variables are fairly highly correlated with each other and therefore "compete" to explain the same variance. If Area *c* turns out to be negative, then the overlapping circles diagram shown in Figure 4.3 may not be the best way to think about what is happening; a negative value for Area *c* suggests that some kind of suppression is present, and suppressor variables can be difficult to interpret.

4.17 SUMMARY

Regression with two predictor variables can provide a fairly complete description of the predictive usefulness of the X_1 and X_2 variables, although it is important to keep in mind that serious violations of the assumptions (such as nonlinear relations between any pair of variables and/or an interaction between X_1 and X_2) can invalidate the results of this simple analysis. Violations of these assumptions can often be detected by preliminary data screening that includes all bivariate scatterplots (e.g., X_1 vs. X_2, Y vs. X_1, and Y vs. X_2) and scatterplots that show the X_1, Y relationship separately for groups with different scores on X_2. Examination of residuals from the regression can also be a useful tool for identification of violation of assumptions.

Note that the regression coefficient *b* to predict a raw score on *Y* from X_1 while controlling for X_2 "partials out" or removes or controls only for the part of the X_1 scores that is *linearly* related to X_2. If there are nonlinear associations between the X_1 and X_2 predictors, then linear regression methods are not an effective way to describe the unique contributions of the predictor variables.

So far, you have learned several different analyses that can be used to evaluate whether X_1 is (linearly) predictive of *Y*. If you obtain a squared Pearson correlation between X_1 and *Y*, r^2_{1Y}, the value of r^2_{1Y} estimates the proportion of variance in *Y* that is predictable from X_1 when you do *not* statistically control for or partial out any variance associated with other predictor variables such as X_2. In Figure 4.3, r^2_{1Y} corresponds to the sum of Areas *a* and *c*. If you obtain a squared partial correlation between X_1 and *Y*, controlling for X_2 (which can be denoted either as pr^2_1 or $r^2_{Y1.2}$), $r^2_{Y1.2}$ corresponds to the proportion of variance in *Y* that can be predicted from X_1 when the variance that can be predicted from X_2 is removed from both *Y* and X_1; in Figure 4.3, $r^2_{Y1.2}$ corresponds to the ratio $a/(a + d)$. If you obtain the squared semipartial (or squared part) correlation between X_1 and *Y*, controlling for X_2, which can be denoted by either sr^2_1 or $r^2_{Y(1.2)}$, this value of $r^2_{Y(1.2)}$ corresponds to Area *a* in Figure 4.3, that is, the proportion of the total variance of *Y* that can be predicted from X_1 after any overlap with X_2 is removed from (only) X_1. Because the squared semipartial correlations can be used to deduce a partition of the variance (as shown in Figures 4.3 and 4.14), data analysts more often report squared semipartial correlations (rather than squared partial correlations) as effect size information in multiple regression.

The (partial) standard-score regression slope β_1 (to predict z_Y from z_{X1} while controlling for any linear association between z_{X1} and z_{X2}) can be interpreted as follows: For a one-unit increase in the standard-score z_{X1}, what part of a standard deviation increase is predicted in z_Y when the value of z_{X2} is held constant? The raw-score regression slope b_1 (to predict *Y* from

X_1 while controlling for X_2) can be interpreted as follows: For a one-unit increase in X_1, how many units of increase are predicted for the Y outcome variable when the value of X_2 is held constant?

In some circumstances, data analysts find it more useful to report information about the strength of predictive relationships using unit-free or standardized indexes (such as β or $r^2_{Y(1.2)}$). This may be particularly appropriate when the units of measurement for the X_1, X_2, and Y variables are all arbitrary or when a researcher wants to try to compare the predictive usefulness of an X_1 variable with the predictive usefulness of an X_2 variable that has completely different units of measurement than X_1. (However, such comparisons should be made very cautiously because differences in the sizes of correlations, semipartial correlations, and beta coefficients may be due partly to differences in the ranges or distribution shapes of X_1 and X_2 or the reliabilities of X_1 and X_2 or other factors that can artifactually influence the magnitude of correlations.)

In other research situations, it may be more useful to report the strength of predictive relationships by using raw-score regression slopes. These may be particularly useful when the units of measurement of the variables have some "real" meaning, for example, when we ask how much blood pressure increases for each 1-year increase in age.

Later chapters show how regression analysis can be extended in several different ways by including dummy predictor variables and interaction terms. In the general case, a multiple regression equation can have k predictors:

$$Y = b_0 + b_1X_1 + b_2X_2 + b_3X_3 + \cdots + b_kX_k. \tag{4.28}$$

The predictive contribution of each variable (such as X_1) can be assessed while controlling for all other predictors in the equation (e.g., $X_2, X_3, \ldots, X_k$). When we use this approach to variance partitioning—that is, each predictor is assessed controlling for all other predictors in the regression equation, the method of variance partitioning is often called **standard multiple regression** or **simultaneous multiple regression** (Tabachnick & Fidell, 2018). When variance was partitioned between two predictor variables X_1 and X_2 in this chapter, the "standard" method of partitioning was used; that is, sr^2_1 was interpreted as the proportion of variance in Y that was uniquely predictable by X_1 when X_2 was statistically controlled, and sr^2_2 was interpreted as the proportion of variance in Y that was uniquely predictable from X_2 when X_1 was statistically controlled.

COMPREHENSION QUESTIONS

1. Consider the hypothetical data in the file named sbpanxietyweight.sav. The research question is, How well can systolic blood pressure (SBP) be predicted from anxiety and weight combined? Also, how much variance in blood pressure is uniquely explained by each of these two predictor variables?

    a. As preliminary data screening, generate a histogram of scores on each of these three variables, and create a bivariate scatterplot for each pair of variables. Do you see evidence of violations of assumptions? For example, do any variables have non-normal distribution shapes? Are any pairs of variables related in a way that is not linear? Are there bivariate outliers?
    b. Run a regression analysis to predict SBP from weight and anxiety. As in the example presented in the chapter, make sure that you request the part and partial correlation statistics and a graph of the standardized residuals (*ZRESID) against the standardized predicted values (*ZPRED).
    c. Write up a "Results" section. What can you conclude about the predictive usefulness of these two variables, individually and combined?
    d. Does examination of the plot of standardized residuals indicate any serious violation of assumptions? Explain.
    e. Why is the *b* coefficient associated with the variable weight so much smaller than the *b* coefficient associated with the variable anxiety (even though weight accounted for a larger unique share of the variance in SBP)?
    f. Set up a table (similar to the one shown in Table 4.1) to summarize the results of your regression analysis.
    g. Draw a diagram (similar to the one in Figure 4.14) to show how the total variance of SBP is partitioned into variance that is uniquely explained by each predictor, variance that can be explained by either predictor, and variance that cannot be explained, and fill in the numerical values that represent the proportions of variance in this case.
    h. Were the predictors highly correlated with each other? Did they compete to explain the same variance? (How do you know?)
    i. Ideally, how many cases should you have to do a regression analysis with two predictor variables?

2. What is the null hypothesis for the overall multiple regression?

3. What null hypothesis is used to test the significance of each individual predictor variable in a multiple regression?

4. Which value in your SPSS output gives the correlation between the observed *Y* and predicted *Y* values?

5. Which value in your SPSS output gives the proportion of variance in *Y* (the dependent variable) that is predictable from X_1 and X_2 as a set?

6. Which value in your SPSS output gives the proportion of variance in *Y* that is uniquely predictable from X_1, controlling for or partialling out X_2?

7. Explain how the normal equations for a two-predictor multiple regression can be obtained from a path diagram that shows z_{X1} and z_{X2} as correlated predictors of z_Y, by applying the tracing rule.

8. The normal equations show the overall correlation between each predictor and *Y* broken down into two components, for example, $r_{1Y} = \beta_1 + r_{12}\beta_2$. Which of these components represents a direct (or unique) contribution of X_1 as a predictor of *Y*, and which one shows an indirect relationship?

9. For a regression (to predict Y from X_1 and X_2), is it possible to have a significant R but nonsignificant b coefficients for both X_1 and X_2? If so, under what circumstances would this be likely to occur?
10. What is multicollinearity in multiple regression, and why is it a problem?
11. How do you report effect size and significance test information for the entire regression analysis?
12. In words, what is this null hypothesis: H_0: $b = 0$?
13. How do you report the effect size and significance test for each individual predictor variable?
14. How are the values of b and β similar? How are they different?

NOTES

[1]Formal treatments of statistics use β to represent the population slope parameter in this equation for the null hypothesis. This notation is avoided in this textbook because it is easily confused with the more common use of β as the sample value of the standardized slope coefficient, that is, the slope to predict z'_Y from z_{X1}. In this textbook, β always refers to the *sample estimate* of a standard-score regression slope.

[2]It is possible to reconstruct the correlations (r_{Y1}, r_{Y2}) exactly from the model coefficients (β_1, β_2) in this example, because this regression model is "just identified"; that is, the number of parameters being estimated (r_{12}, β_1, and β_2) equals the number of correlations used as input data. In advanced applications of path model logic such as SEM, researchers generally constrain some of the model parameters (e.g., path coefficients) to fixed values, so that the model is "overidentified." For instance, if a researcher assumes that $\beta_1 = 0$, the direct path from z_{X1} to z_Y is omitted from the model. When constraints on parameter estimates are imposed, it is generally not possible to reproduce the observed correlations perfectly from the constrained model. In SEM, the adequacy of a model is assessed by checking to see how well the reproduced correlations (or reproduced variances and covariances) implied by the paths in the overidentified structural equation model agree with the observed correlations (or covariances). The tracing rule described here can be applied to standardized structural equation models to see approximately how well the structural equation model reconstructs the observed correlations among all pairs of variables. The formal goodness-of-fit statistics reported by SEM programs are based on goodness of fit of the observed variances and covariances rather than correlations.

DIGITAL RESOURCES

Find **free study tools** to support your learning, including **eFlashcards, data sets, and web resources**, on the accompanying website at **edge.sagepub.com/warner3e**.

CHAPTER 5

MULTIPLE REGRESSION WITH MULTIPLE PREDICTORS

5.1 RESEARCH QUESTIONS

The extension of multiple regression to situations in which there are more than two predictor variables is relatively straightforward. The raw-score version of a two-predictor multiple regression equation (as described in Chapter 4) is written as follows:

$$Y' = b_0 + b_1X_1 + b_2X_2. \quad (5.1)$$

The raw-score version of a regression equation with k predictor variables is written as follows:

$$Y' = b_0 + b_1X_1 + b_2X_2 + \cdots + b_kX_k. \quad (5.2)$$

In Equation 5.1, the b_1 slope represents the predicted change in Y for a one-unit increase in X_1, controlling for X_2. When there are more than two predictors in the regression, the slope for each individual predictor is calculated controlling for *all* other predictors; thus, in Equation 5.2, b_1 represents the predicted change in Y for a one-unit increase in X_1, controlling for $X_2, X_3, \ldots, X_k$ (i.e., controlling for all other predictor variables included in the regression analysis). For example, a researcher might predict 1st-year medical school grade point average (Y) from a set of several predictor variables such as college grade point average (X_1), Medical College Admissions Test (MCAT) physics score (X_2), MCAT biology score (X_3), quantitative evaluation of the personal goals statement on the application (X_4), a score on a self-reported empathy scale (X_5), and so forth. One goal of the analysis may be to evaluate whether this entire set of variables is sufficient information to predict medical school performance; another goal of the analysis may be to identify which of these variables are most strongly predictive of performance in medical school.

The standard-score version of a regression equation with k predictors is represented as follows:

$$z'_Y = \beta_1 z_{X1} + \beta_2 z_{X2} + \cdots + \beta_k z_{Xk}. \quad (5.3)$$

The beta coefficients in the standard-score version of the regression can be compared across variables to assess which of the predictor variables are more strongly related to the Y outcome variable when all the variables are represented in z-score form. (This comparison must be interpreted with caution, for reasons discussed in Volume I, Chapter 10 [Warner, 2020]; beta coefficients, like correlations, may be influenced by many types of artifacts, such as unreliability of measurement and restricted range of scores in the sample.)

We can conduct an overall or omnibus significance test to assess whether the entire set of all k predictor variables significantly predicts scores on Y; we can also test the significance of the slopes, b_i, for each individual predictor to assess whether each X_i predictor variable is significantly predictive of Y when all other predictors are statistically controlled.

The inclusion of more than two predictor variables in a multiple regression can serve the following purposes (M. Biderman, personal communication, July 12, 2011):

1. A regression that includes several predictor variables can be used to evaluate theories that include several variables that, according to theory, predict or influence scores on the outcome variable.
2. In a regression with more than two predictors, it is possible to assess the predictive usefulness of an X_i variable that is of primary interest while statistically controlling for more than one extraneous variable. As seen in Chapter 3, when we control for "other" variables, the apparent nature of the relation between X_i and Y can change in many different ways.
3. Sometimes a better prediction of scores on the Y outcome variable can be obtained by using more than two predictor variables. However, we should beware the "kitchen sink" approach to selection of predictors. It is not a good idea to run a regression that includes 10 or 20 predictor variables that happen to be strongly correlated with the outcome variable in the sample data; this approach increases the risk for Type I error. It is preferable to have a rationale for the inclusion of each predictor; each variable should be included (a) because a well-specified theory says it could be a "causal influence" on Y, (b) because it is known to be a useful predictor of Y, or (c) because it is important to control for the specific variable when assessing the predictive usefulness of other variables, because the variable is confounded with or interacts with other variables, for example.
4. When we use dummy predictor variables to represent group membership (as in Chapter 6), and the categorical variable has more than four levels, we need to include more than two dummy predictor variables to represent group membership.
5. In a regression with more than two predictor variables, we can use X^2 and X^3 (as well as X) to predict scores on a Y outcome variable; this provides us with a way to test for curvilinear associations between X and Y. (Scores on X should be centered before squaring X, that is, they should be transformed into deviations from the mean on X; see Aiken & West, 1991, for details.)
6. In addition, product terms between predictor variables can be included in regression to represent interactions between predictor variables (interaction is also called moderation).

Two new issues are addressed in this chapter. First, when we expand multiple regression to include k predictors, we need a general method for the computation of β and b coefficients that works for any number of predictor variables. These computations can be represented using matrix algebra; however, the reader does not need a background in matrix algebra to understand the concepts involved in the application of multiple regression. This chapter provides an intuitive description of the computation of regression coefficients; students who want to understand these computations in more detail will find a brief introduction to the matrix algebra for multiple regression in Appendix 5A at the end of this chapter. Second, there are several different methods for entry of predictor variables into multiple regression. These methods use different logic to partition the variance in the Y outcome variable among the individual predictor variables. Subsequent sections of this chapter describe these three major forms of order of entry in detail:

1. *Simultaneous or standard regression:* All the X predictor variables are entered in one step.
2. *Hierarchical regression (also called* ***sequential regression*** *or* ***user-determined order of entry in regression****):* X predictor variables are entered in a series of steps, with the order of entry determined by the data analyst.
3. ***Statistical regression*** *or* ***data-driven regression:*** The order of entry is based on the predictive usefulness of the individual X variables.

Both (2) and (3) are sometimes called "stepwise" regression. However, in this chapter, the term *stepwise* will be used in a much narrower sense, to identify one of the options for statistical or data-driven regression that is available in the SPSS regression program.

In this chapter, all three of these approaches to regression will be applied to the same data analysis problem. In general, the simultaneous approach to regression is preferable: It is easier to understand, and all the predictor variables are given equal treatment. In standard or simultaneous regression, when we ask, "What other variables were statistically controlled while assessing the predictive usefulness of the X_i predictor?" the answer is always "All the other X predictor variables." In a sense, then, all predictor variables are treated equally; the predictive usefulness of each X_i predictor variable is assessed controlling for *all* other predictors.

On the other hand, when we use hierarchical or statistical regression analysis, which involves running a series of regression equations with one or more predictor variables added at each step, the answer to the question "What other variables were statistically controlled while assessing the predictive usefulness of the X_i predictor variable?" is "Only the other predictor variables entered in the same step or in previous steps." Thus, the set of "statistically controlled variables" differs across the X_i predictor variables in hierarchical or statistical regression (analyses in which a series of regression analyses are performed). Predictor variables in sequential or statistical regression are treated "differently" or "unequally"; that is, the contributions for some of the X_i predictor variables are assessed controlling for none or few other predictors, while the predictive contributions of other variables are assessed controlling for most, or all, of the other predictor variables. Sometimes it is possible to justify this "unequal" treatment of variables on the basis of theory or temporal priority of the variables, but sometimes the decisions about order of entry are arbitrary.

Direct or standard or simultaneous regression (i.e., a regression analysis in which all predictor variables are entered in one step) usually, but not always, provides a more conservative assessment of the contribution made by each individual predictor. That is, usually the proportion of variance that is attributed to an X_i predictor variable is smaller when that variable is assessed in the context of a direct or standard or simultaneous regression, controlling for *all* the other predictor variables, than when the X_i predictor variable is entered in an early step in a hierarchical or statistical method of regression (and therefore is assessed controlling for only a subset of the other predictor variables).

The statistical or data-driven method of entry is not recommended, because this approach to order of entry often results in inflated risk for Type I error (variables that happen to have large correlations with the Y outcome variable in the sample, because of sampling error, tend to be selected earliest as predictors). This method is included here primarily because it is sometimes reported in journal articles. Statistical or data-driven methods of entry yield the largest possible R^2 using the smallest number of predictor variables within a specific sample, but they often yield analyses that are not useful for theory evaluation (or even for prediction of individual scores in different samples).

5.2 EMPIRICAL EXAMPLE

The hypothetical research problem for this chapter involves prediction of scores on a physics achievement test from the following predictors: intelligence quotient (IQ), emotional intelligence (EI), verbal SAT (VSAT) score, math SAT (MSAT) score, and gender (coded 1 = male, 2 = female). (Gender could be dummy coded, for example, 1 = male and 0 = female. However, the proportion of variance that is uniquely predictable by gender does not change when different numerical codes are used for the groups.) The first question is, How well are scores on physics predicted when this entire set of five predictor variables is included? The second question is, How much variance does each of these predictor variables uniquely account for? This second question can be approached in three different ways, using the standard, hierarchical, or statistical method of entry. The data set predictphysics.sav contains hypothetical data for 200 participants (100 male, 100 female) on these six variables, that is, the five predictor variables and the score on the dependent variable. (Note that because five subjects have missing data on physics score, the actual N in the regression analyses that follow is 195.)

5.3 SCREENING FOR VIOLATIONS OF ASSUMPTIONS

As the number of variables in analyses increases, it becomes increasingly time-consuming to do a thorough job of preliminary data screening. Detailed data screening will no longer be presented for the empirical examples from this point onward, because it would require a great deal of space; instead, there is only a brief description of the types of analyses that should be conducted for preliminary data screening.

First, for each predictor variable (and the outcome variable), you need to set up a histogram to examine the shape of the distribution of scores. Ideally, all quantitative variables (and particularly the Y outcome variable) should have approximately normal distribution shapes. If there are extreme outliers, the researcher should make a thoughtful decision whether to remove or modify these scores (see Chapter 2 for discussion of outliers). If there are dummy-coded predictors, the two groups should ideally have approximately equal n's, and in any case, no group should have fewer than 10 cases.

Second, a scatterplot should be obtained for every pair of quantitative variables. The scatterplots should show a linear relation between variables, homogeneous variance (for the variable plotted on the vertical axis) at different score values (of the variable plotted on the horizontal axis), and no extreme bivariate outliers.

Detection of possible multivariate outliers is most easily handled by examination of plots of residuals from the multiple regression and/or examination of information about individual cases (such as Mahalanobis D or **leverage** statistics) that can be requested and saved into the SPSS worksheet from the regression program. See Tabachnick and Fidell (2018) for further discussion of methods for detection and handling of multivariate outliers.

5.4 ISSUES IN PLANNING A STUDY

Usually, regression analysis is used in nonexperimental research situations, in which the researcher has manipulated none of the variables. In the absence of an experimental design, causal inferences cannot be made. However, researchers often select at least some of the predictor variables for regression analysis because they believe that these might be "causes" of the outcome variable. If an X_i variable that is theorized to be a "cause" of Y fails to account for a significant amount of variance in the Y variable in the regression analysis, this outcome may weaken the researcher's belief that the X_i variable has a causal connection with Y. On the other hand, if an X_i variable that is thought to be "causal" does uniquely predict a significant proportion of variance in Y even when confounded variables or competing causal variables are statistically controlled, this outcome may be interpreted as consistent with the possibility of causality. Of course, neither outcome provides proof for or against causality. An X_i variable may fail to be a statistically significant predictor of Y in a regression (even if it really is a cause) for many reasons: poor measurement reliability, restricted range, Type II error, a relation that is not linear, an improperly specified model, and so forth. On the other hand, an X_i variable that is not a cause of Y may significantly predict variance in Y because of some artifact; for instance, X_i may be correlated or confounded with some other variable that causes Y, or we may have an instance of Type I error. If there are measurement problems with any of the variables in the regression (poor reliability and/or lack of validity), of course, regression analysis cannot provide good-quality information about the predictive usefulness of variables.

As discussed in Chapter 4, the proportion of variance uniquely accounted for by X_i in a multiple regression, sr^2_i, is calculated in a way that adjusts for the correlation of X_i with all other predictors in the regression equation. We can obtain an accurate assessment of the proportion of variance attributable to X_i only if we have a correctly specified model, that is, a regression model that includes all the predictors that should be included and that does not include any predictors that should not be included. A good theory provides guidance about the set of variables that should be taken into account when trying to explain people's scores on a particular outcome variable. However, in general, we can never be sure that we have a correctly specified model.

What should be included in a correctly specified model? First, all the relevant "causal variables" that are believed to influence or predict scores on the outcome variable should be included. This would, in principle, make it possible to sort out the unique contributions of causes that may well be confounded or correlated with one another. In addition, if our predictor variables are "contaminated" by sources of measurement bias (such as general verbal ability or social desirability), measures of these sources of bias should also be included as predictors. In practice, it is not possible to be certain that we have a complete list of causes or a complete assessment of sources of bias. Thus, we can never be certain that we have a correctly specified model. In addition, a correctly specified model should include any moderator variables (see Chapter 7 for discussion of moderation or interaction in regression).

Usually, when we fail to include competing causes as predictor variables, the X_i variables that we do include in the equation may appear to be stronger predictors than they really are. For example, when we fail to include measures of bias (e.g., a measure of social desirability), this may lead to either over- or underestimation of the importance of individual X predictor variables. Finally, if we include irrelevant predictor variables in our regression, sometimes these take explained variance away from other predictors.

We must be careful, therefore, to qualify or limit our interpretations of regression results. The proportion of variance explained by a particular X_i predictor variable is specific to the sample of data and to the type of participants in the study; it is also specific to the context of the other variables that are included in the regression analysis. When predictor variables are added to (or dropped from) a regression model, the sr^2_i that indexes the unique variance explained by a particular X_i variable can either increase or decrease; the β_i that represents the partial slope for z_{Xi} can become larger or smaller (Kenny, 1979, called this "bouncing betas"). Thus, our judgment about the apparent predictive usefulness of an individual X_i variable is context dependent in at least three ways: It may be unique to the peculiarities of the particular sample of data, it may be limited to the types of participants included in the study, and it varies as a function of the other predictor variables that are included in (and excluded from) the regression analysis.

Past research (and well-developed theory) can be extremely helpful in deciding what variables ought to be included in a regression analysis, in addition to any variables whose possible causal usefulness a researcher wants to explore. Earlier chapters described various roles that variables can play. Regression predictors may be included because they are of interest as possible causes; however, predictors may also be included in a regression analysis because they represent competing causes that need to be controlled for, confounds that need to be corrected for, sources of measurement error that need to be adjusted for, moderators, or extraneous variables that are associated with additional random error.

The strongest conclusion a researcher is justified in drawing when a regression analysis is performed on data from a nonexperimental study is that a particular X_i variable is (or is not) significantly predictive of Y when a specific set of other X variables (that represent competing explanations, confounds, sources of measurement bias, or other extraneous variables) is controlled. If a particular X_i variable is still significantly predictive of Y when a well-chosen set of other predictor variables is statistically controlled, the researcher has a slightly stronger case for the possibility that X_i might be a cause of Y than if the only evidence is a significant zero-order Pearson correlation between X_i and Y. However, it is by no means proof of causality; it is merely a demonstration that, after we control for the most likely competing causes that we can think of, X_i continues to account uniquely for a share of the variance in Y.

Several other design issues are crucial, in addition to the appropriate selection of predictor variables. It is important to have a reasonably wide range of scores on the Y outcome variable and on the X predictor variables. As discussed in Volume I, Chapter 10 (Warner, 2020), a restricted range can artifactually reduce the magnitude of correlations, and restricted ranges or scores can also reduce the size of regression slope coefficients. Furthermore, we cannot assume that the linear regression equation will make accurate predictions for scores on X predictor variables that lie outside the range of X values in the sample. We need to have a sample size that is sufficiently large to provide adequate statistical power (see Section 5.12) and also large enough to provide reasonably narrow confidence intervals for the estimates of b slope coefficients; the larger the number of predictor variables (k), the larger the required sample size.

5.5 COMPUTATION OF REGRESSION COEFFICIENTS WITH *k* PREDICTOR VARIABLES

The equations for a standardized multiple regression with two predictors can be worked out using a path diagram to represent z_{X1} and z_{X2} as correlated predictors of z_Y, as shown in Figure 4.1. This results in the following equations to describe the way that the overall correlation between a specific predictor z_{X1} and the outcome z_Y (i.e., r_{1Y}) can be "deconstructed" into two components, a direct path from z_{X1} to z_Y (the strength of this direct or unique predictive relationship is represented by β_1) and an indirect path from z_{X1} to z_Y via z_{X2} (represented by $r_{12}\beta_2$):

$$r_{Y1} = \beta_1 + r_{12}\beta_2, \tag{5.4}$$

$$r_{Y2} = r_{12}\beta_1 + \beta_2. \tag{5.5}$$

We are now ready to generalize the procedures for the computation of β and b regression coefficients to regression equations that include more than two predictor variables. On a conceptual level, when we set up a regression that includes k predictor variables (as shown in Equations 5.1 and 5.2), we need to calculate the β_i and b_i partial slope coefficients that make the best possible prediction of Y from each X_i predictor variable (the beta coefficients are applied to z scores on the variables, while the b coefficients are applied to the raw scores in the original units of measurement). These partial slopes must control for or partial out any redundancy or linear correlation of X_i with all the other predictor variables in the equation (i.e., $X_1, X_2, \ldots, X_k$). When we had only two predictors, z_{X1} and z_{X2}, we needed to control for or partial out the part of the predictive relationship of z_{X1} with z_Y that could be accounted for by the path through the correlated predictor variable z_{X2}. More generally, the path model for a regression with several correlated predictor variables has the form shown in Figure 5.1.

When we compute a standardized partial slope β_1, which represents the unique predictive contribution of z_{X1}, we must "partial out" or remove all the indirect paths from z_{X1} to z_Y via each of the other predictor variables ($z_{X2}, z_{X3}, \ldots, z_{Xk}$). A formula to calculate an estimate of β_1 from the bivariate correlations among all the other predictors, as well as the other estimates of $\beta_2, \beta_3, \ldots, \beta_k$, is obtained by subtracting the indirect paths from the overall r_{1Y} correlation (as in Equation 5.3, except that in the more general case with k predictor variables, multiple indirect paths must be "subtracted out" when we assess the unique predictive relationship of each z_{Xi} variable with z_Y). In addition, the divisor for each beta slope coefficient takes into account the correlation between each predictor and Y and the correlations between all pairs of predictor variables. The point to understand is that the calculation of a β_i coefficient includes information about the magnitude of the Pearson correlation between X_i and Y, but the magnitude of the β_i coefficient is also adjusted for the correlations of X_i with all other X predictor variables (and the association of the other X predictors with the Y outcome variable). Because of this adjustment, a β_i slope coefficient can differ in size and/or in sign from the zero-order Pearson correlation of X_i with Y. Controlling for other predictors can greatly change our understanding of the strength and direction of the association between an individual X_i predictor variable and Y. For example, controlling for a highly correlated competing predictor variable may greatly reduce the apparent strength of the association between X_i and Y. On the other hand, controlling for a suppressor variable can actually make the predictive association between X_i and Y (represented by β_i) stronger than the zero-order correlation r_{1Y} or different in sign (see Chapter 3 for a discussion of suppressor variables).

Students who want to understand the computational procedures for multiple regression with more than two predictors in greater detail should see Appendix 5A at the end of this chapter for a brief introduction to matrix algebra and an explanation of the matrix algebra computation of the b and β slope coefficients for multiple regression with k predictor variables.

Figure 5.1 Path Model for Multiple Regression With k Variables ($X_1, X_2, \ldots, X_k$) as Correlated Predictors of Y

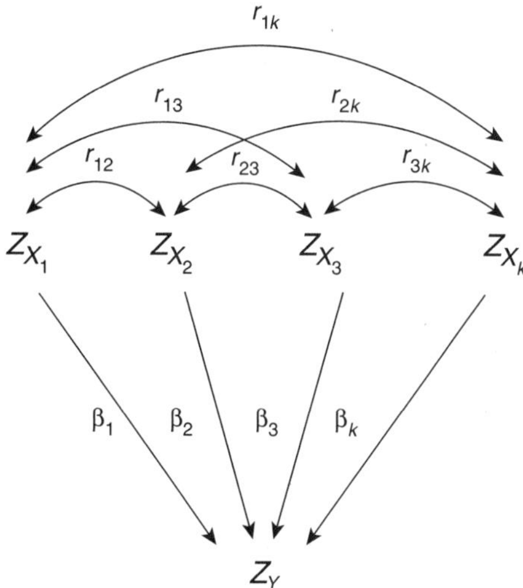

Once we have obtained estimates of the beta coefficients, we can obtain the corresponding *b* coefficients (to predict raw scores on *Y* from raw scores on the *X* predictor variables) by rescaling the slopes to take information about the units of measurement of the predictor and outcome variable into account, as we did earlier in Volume I, Chapter 11 (Warner, 2020), and Chapter 4 in the present volume:

$$b_i = \beta_i(SD_y/SD_i), \tag{5.6}$$

where SD_i is the standard deviation of X_i, the *i*th independent variable. The intercept b_0 is calculated from the means of the *X*'s and their *b* coefficients using the following equation:

$$b_0 = M_Y - b_1 M_{X1} - b_2 M_{X2} - \cdots - b_k M_{Xk}. \tag{5.7}$$

5.6 METHODS OF ENTRY FOR PREDICTOR VARIABLES

When regression analysis with two predictor variables was introduced in Chapter 4, we calculated one regression equation that included both predictors. In Chapter 4, only one regression analysis was reported, and it included both the X_1 and X_2 variables as predictors of *Y*. In Chapter 4, the predictive usefulness of X_1 was assessed while controlling for or partialling out any linear association between X_2 and X_1, and the predictive usefulness of X_2 was assessed while controlling for or partialling out any linear association between X_1 and X_2. The method of regression that was introduced in Chapter 4 (with all predictors entered at the same time) is equivalent to a method of regression that is called "standard" or "simultaneous" in this chapter. However, there are other ways to approach an analysis that involve multiple predictors. It is possible to conduct a regression analysis as a series of analyses and to enter just one predictor (or a set of predictors) in each step in this series of analyses. Doing a series of regression analyses makes it possible to evaluate how much additional variance is predicted by each X_i predictor variable (or by each set of predictor variables) when you control for only the variables that were entered in prior steps.

Unfortunately, the nomenclature that is used for various methods of entry of predictors into regression varies across textbooks and journal articles. In this textbook, three major approaches to method of entry are discussed. These are listed here and then discussed in more detail in later sections of this chapter:

1. *Standard or simultaneous or direct regression:* In this type of regression analysis, only one regression equation is estimated, all the X_i predictor variables are added at the same time, and the predictive usefulness of each X_i predictor is assessed while statistically controlling for any linear association of X_i with all other predictor variables in the equation.

2. *Sequential or hierarchical regression (user-determined order of entry):* In this type of regression analysis, the data analyst decides on an order of entry for the predictor variables on the basis of some theoretical rationale. A series of regression equations are estimated. In each step, either one X_i predictor variable or a set of several X_i predictor variables are added to the regression equation.

3. *Statistical regression (data-driven order of entry):* In this type of regression analysis, the order of entry of predictor variables is determined by statistical criteria. In Step 1, the single predictor variable that has the largest squared correlation with *Y* is entered into the equation; in each subsequent step, the variable that is entered into the equation is the one that produces the largest possible increase in the magnitude of R^2.

Unfortunately, the term *stepwise regression* is sometimes used in a nonspecific manner to refer to any regression analysis that involves a series of steps with one or more additional variables entered at each step in the analysis (i.e., to either Method 2 or 3 described above). It is sometimes unclear whether authors who label an analysis a stepwise regression are referring to a hierarchical or sequential regression (user determined) or to a statistical regression (data driven) when they describe their analysis as stepwise. In this chapter, the term *stepwise* is defined in a narrow and specific manner; *stepwise* refers to one of the specific methods SPSS uses for the entry of predictor variables in statistical regression. To avoid confusion, it is preferable to state as clearly as possible in simple language how the analysis was set up, that is, to make an explicit statement about order of entry of predictors (e.g., see Section 5.15).

5.6.1 Standard or Simultaneous Method of Entry

When all of the predictor variables are entered into the analysis at the same time (in one step), this corresponds to *standard* multiple regression (this is the term used by Tabachnick and Fidell, 2018). This method is also widely referred to as simultaneous or direct regression. In standard or simultaneous multiple regression, all the predictor variables are entered into the analysis in one step, and coefficients are calculated for just one regression equation that includes the entire set of predictors. The effect size that describes the unique predictive contribution of each X variable, sr^2_{unique}, is adjusted to partial out or control for any linear association of X_i with all the other predictor variables. This standard or simultaneous approach to multiple regression usually provides the most conservative assessment of the unique predictive contribution of each X_i variable. That is, usually (but not always) the proportion of variance in Y that is attributed to a specific X_i predictor variable is smaller in a standard regression analysis than when the X_i predictor is entered in an early step in a sequential or statistical series of regression equations. The standard method of regression is usually the simplest version of multiple regression to run and report.

5.6.2 Sequential or Hierarchical (User-Determined) Method of Entry

Another widely used method of regression involves running a series of regression analyses; at each step, one X_i predictor (or a set, group, or block of X_i predictors) selected by the data analyst for theoretical reasons is added to the regression analysis. The key issue is that the order of entry of predictors is determined by the data analyst (rather than by the sizes of the correlations among variables in the sample data). Tabachnick and Fidell (2018) called this method of entry, in which the data analyst decides on the order of entry of predictors, sequential or hierarchical regression. Sequential regression involves running a *series* of multiple regression analyses. In each step, one or more predictor variables are added to the model, and the predictive usefulness of each X_i variable (or set of X_i variables) is assessed by asking how much the R^2 for the regression model increases in the step when each predictor variable (or set of predictor variables) is first added to the model. When just one predictor variable is added in each step, the **increment in R^2, R^2_{inc}**, is equivalent to **incremental sr^2, sr^2_{inc}**, the squared part correlation for the predictor variable in the step when it first enters the analysis.

5.6.3 Statistical (Data-Driven) Order of Entry

In a statistical regression, the order of entry for predictor variables is based on statistical criteria. SPSS offers several different options for statistical regression. In forward regression, the analysis begins without any predictor variables included in the regression equation; in each step, the X_i predictor variable that produces the largest increase in R^2 is added to the regression equation. In backward regression, the analysis begins with all predictor variables included in the equation; in each step, the X_i variable is dropped, which leads to the smallest

Table 5.1 Summary of Nomenclature for Various Types of Regression

Common Names for the Procedure	What the Procedure Involves
Standard or simultaneous or direct regression	A single regression analysis is performed to predict Y from $X_1, X_2, \ldots, X_k$. The predictive contribution of each X_i predictor is assessed while statistically controlling for linear associations with all the other X predictor variables.
Sequential or hierarchical regression[a] (user-determined order of entry)	A series of regression analyses are performed. At each step, one or several X_i predictor variables are entered into the equation. The order of entry is determined by the data analyst on the basis of a theoretical rationale. The predictive usefulness of each X_i predictor variable is assessed while controlling for any linear association of X_i with other predictor variables that enter at the same step or at previous steps.
Statistical (or data-driven) regression[a]	A series of regression analyses are performed. X_i predictor variables are added to and/or dropped from the regression model at each step. An X_i predictor is added if it provides the maximum increase in R^2 (while controlling for predictors that are already in the model). An X_i predictor is dropped if removing it results in a nonsignificant reduction in R^2. Within SPSS, there are three types of statistical regression: forward, backward, and stepwise.

a. Some authors use the term *stepwise* to refer to either sequential or statistical regression. That usage is avoided here because it introduces ambiguity and is inconsistent with the specific definition of stepwise that is used by the SPSS regression program.

reduction in the overall R^2 for the regression equation. SPSS stepwise is a combination of the **forward method of entry** and **backward method of entry**; the analysis begins with no predictor variables in the model. In each step, the X_i predictor that adds the largest amount to the R^2 for the equation is added to the model, but if any X_i predictor variable no longer makes a significant contribution to R^2, that variable is dropped from the model. Thus, in an SPSS stepwise statistical regression, variables are added in each step, but variables can also be dropped from the model if they are no longer significant (after the addition of other predictors). The application of these three different methods of statistical regression (forward, backward, and stepwise) may or may not result in the same set of X_i predictors in the final model. As noted earlier, many writers use the term *stepwise* in a very broad sense to refer to any regression analysis where a series of regression equations are estimated with predictors added to the model in each step. In this chapter, stepwise is used to refer specifically to the type of variable entry just described here. Table 5.1 summarizes the preceding discussion about types of regression and nomenclature for these methods.

5.7 VARIANCE PARTITIONING IN STANDARD REGRESSION VERSUS HIERARCHICAL AND STATISTICAL REGRESSION

The three methods of entry of predictor variables (standard or simultaneous, sequential or hierarchical, and statistical) handle the problem of partitioning explained variance among predictor variables somewhat differently. Figure 5.2 illustrates this difference in variance partitioning.

In a standard or simultaneous entry multiple regression, each predictor is assessed controlling for all other predictors in the model; each X_i predictor variable gets credit only for variance that it shares uniquely with Y and not with any other X predictors (as shown in Figure 5.2a). In sequential or statistical regression, each X_i variable's contribution is assessed

Figure 5.2 Comparison of Partition of Variance in Standard Versus Hierarchical Regression

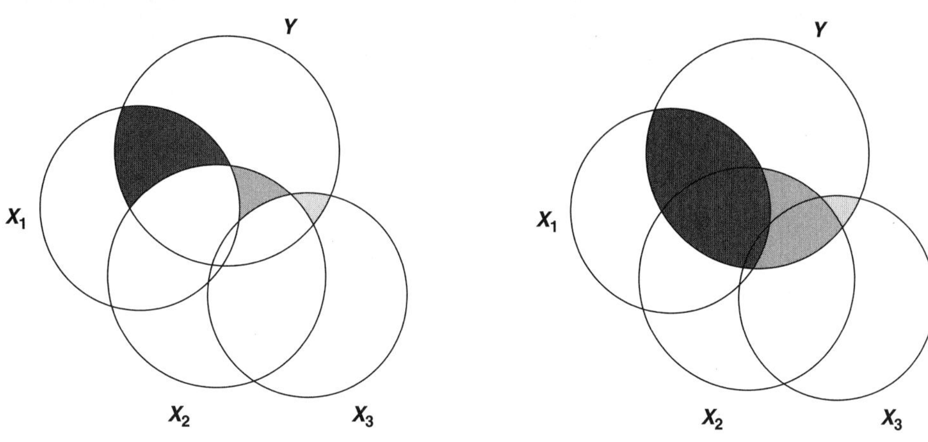

Note: Figure 5.2a shows partition of variance among three predictor variables in a standard regression. The contribution of each predictor is assessed controlling for all other predictors. The shaded areas correspond to the squared semipartial (or part) correlation of each predictor with Y (i.e., sr^2_{unique}). Figure 5.2b shows partition of variance among predictor variables in sequential or statistical regression. In this example, the predictor variables are added one at a time and three separate regressions are performed.

Step 1: $Y' = b_0 + b_1X_1$: Proportion of variance attributed to $X_1 = R^2$ for Step 1 regression (the black area in the diagram).

Step 2: $Y' = b_0 + b_1X_1 + b_2X_2$ (Step 2 model): Proportion of variance attributed to X_2 = increment in R^2 when X_2 is added to the model $= R^2_{Step2} - R^2_{Step1}$ = proportion of variance in Y that is uniquely predictable from X_2 (controlling for X_1) (dark gray area in the diagram).

Step 3: $Y' = b_0 + b_1X_1 + b_2X_2 + b_3X_3$ (Step 3 model): Proportion of variance attributed to X_3 = increment in R^2 when X_3 is added to the model $= R^2_{Step3} - R^2_{Step2}$ = proportion of variance in Y that is uniquely predictable from X_3, controlling for X_1 and X_2 (light gray area in the diagram).

controlling only for the predictors that enter in the same or earlier steps; variables that are entered in later steps are not taken into account (as shown in Figure 5.2b). In the example shown in Figure 5.2a, the standard regression, X_1 is assessed controlling for X_2 and X_3, X_2 is assessed controlling for X_1 and X_3, and X_3 is assessed controlling for X_1 and X_2. (Note that to keep notation simple in the following discussion, the number subscripts for predictor variables correspond to the step in which each variable is entered.)

On the other hand, in sequential or statistical regression, X_1 (the variable that enters in Step 1) is assessed controlling for none of the other predictors; X_2, the variable that enters in Step 2, is assessed only for variables that entered in prior steps (in this case, X_1); and X_3, the variable that enters in Step 3, is assessed controlling for all variables that entered in prior steps (X_1 and X_2) (see Figure 5.2b). In this example, the X_1 variable would get credit for a much smaller proportion of variance in a standard regression (shown in Figure 5.2a) than in sequential or statistical regression (shown in Figure 5.2b). Sequential or statistical regression essentially makes an arbitrary decision to give the variable that entered in an earlier step (X_1) credit for variance that could be explained just as well by variables that entered in later steps (X_2 or X_3). The decision about order of entry is sometimes arbitrary. Unless there are strong theoretical justifications, or the variables were measured at different points in time, it can be difficult to defend the decision to enter a particular predictor in an early step.

When a sequential (or hierarchical or user-determined) method of variable entry is used, it may be possible to justify order of entry on the basis of the times when the scores were obtained for predictors and/or the roles the predictor variables play in the theory. If the X_1, X_2, and X_3 predictors were measured a year, a month, and a week prior to the assessment of Y, respectively, then it might make sense to control for the X_1 variable before assessing what additional adjustments in the prediction of Y should be made on the basis of the later values of X_2 and X_3. Usually, our theories include some "preferred" predictor whose usefulness the researcher wants to demonstrate. When this is the case, the researcher can make a stronger case for the usefulness of her or his preferred predictor if she or he demonstrates that the preferred variable is still significantly predictive of Y when control or nuisance variables, or competing explanatory variables, have been controlled for or taken into account. It makes sense, in general, to include "control," "nuisance," or "competing" variables in the sequential regression in early steps and to include the predictors the researcher wants to subject to the most stringent test and to make the strongest case for in later steps. (Unfortunately, researchers sometimes do the opposite, that is, enter their "favorite" variables in early steps so that their preferred variables will get credit for larger shares of the variance.)

When statistical (data-driven) methods of variable entry are used (i.e., when predictor variables are entered into the model in order of their predictive usefulness), it is difficult to defend the resulting partition of variance. It is quite likely that this method of variable selection for predictors will result in an inflated risk for Type I error; that is, variables whose sample correlations overestimate their true population correlations with Y are likely to be included as predictors when statistical methods of variable entry are used. Tabachnick and Fidell (2018) pointed out that the significance tests in SPSS output are not adjusted to correct for the inflated risk for Type I error that arises when statistical regression methods are used.

The use of statistical regression is not recommended under any circumstances; this is a data-fishing technique that produces the largest R^2 possible from the minimum number of predictors, but it is likely to capitalize on chance, to result in a model that makes little sense, and to include predictors whose significance is due to Type I error. If the researcher cannot resist the temptation to use statistical regression despite this warning, then at least he or she should use the modified test procedures suggested by Wilkinson and Dallal (1981) to assess the statistical significance of the overall regression, as described later in the chapter.

5.8 SIGNIFICANCE TEST FOR AN OVERALL REGRESSION MODEL

In interpreting the results of a multiple regression analysis with k predictor variables, two questions are considered. First, is the overall multiple regression significantly predictive of Y? This corresponds to the null hypothesis that the multiple R (between the Y' calculated from the X's and the observed Y) equals 0. The test statistic for this omnibus test is the same as described in Chapter 4; it is an F ratio with k, $N - k - 1$ degrees of freedom (*df*), where N is the number of participants or cases and k is the number of predictor variables.

The null hypothesis for the overall test of the regression is

$$H_0: R = 0. \tag{5.8}$$

The F ratio that tests this null hypothesis can be calculated either from the sums of squares (*SS*) or from the overall R^2 for regression:

$$F = \frac{SS_{\text{regression}} / k}{SS_{\text{residual}} / (N - k - 1)}, \tag{5.9}$$

or, equivalently,

$$F = \frac{R^2 / k}{\left(1 - R^2\right) / (N - k - 1)} \tag{5.10}$$

SPSS provides an exact p value for the F ratio for the overall regression. If the obtained p value is smaller than the preselected alpha level, then the null hypothesis is rejected; the researcher concludes that Y scores can be predicted significantly better than chance when the entire set of predictor variables (X_1 through X_k) is used to calculate the predicted Y score. There is disagreement whether a statistically significant omnibus F test should be required before doing follow-up tests to assess the predictive contribution of individual predictor variables. Requiring a statistically significant overall F for the model might provide some protection against inflated risk for Type I error; however, it does not provide a guarantee of protection. When the omnibus F is significant, the researcher usually goes on to assess the predictive usefulness of each predictor variable (or sometimes the significance of sets or blocks of predictors considered as groups).

The omnibus test for the overall model is done the same way for standard (simultaneous) and sequential (user-determined) methods of entry. The researcher examines the F for the standard regression equation or for the equation in the final step of the sequential regression. If the p or significance value associated with this F ratio (in the SPSS output) is less than the predetermined alpha level (usually .05), then the overall multiple regression is judged to be statistically significant. This significance level reported by SPSS is accurate only if the researcher has run a single regression analysis. If the researcher has run a dozen different variations of the regression before deciding on a "best" model, then the p value in the SPSS output may seriously underestimate the real risk for Type I error that arises when a researcher goes "data fishing," in search of a combination of predictors that yields a large R^2 value.

In statistical methods of regression, the p values given in the SPSS output generally underestimate the true risk for Type I error. If the backward or stepwise method of entry is used, there is no easy way to correct for this inflated risk for Type I error. If "method = forward" is used, then the tables provided by Wilkinson and Dallal (1981), reproduced in Appendix 5B at the end of this chapter, can be used to look up appropriate critical values for multiple R^2. The value of critical R depends on the following factors: the desired alpha level (usually $\alpha = .05$); the number of "candidate" predictor variables, k (variables are counted whether or not they actually are entered into the analysis); the residual df, $N - k - 1$; and the ***F*-to-enter** that the user tells SPSS to use as a criterion in deciding whether to enter potential predictors into the regression equation. The table provides critical values of R^2 for F-to-enter values of 2, 3, and 4. For example, using the table in Appendix 5B, if there are $k = 20$ candidate predictor variables, $N = 221$ subjects, $N - k - 1 = 200$ df, $\alpha = .05$, and F-to-enter = 3.00, then the critical value from the table is $R^2 = .09$ (decimal points are omitted in the table). That is, the final regression equation described in the preceding example can be judged statistically significant at the .05 level if its multiple R^2 exceeds .09.

5.9 SIGNIFICANCE TESTS FOR INDIVIDUAL PREDICTORS IN MULTIPLE REGRESSION

The assessment of the predictive contribution of each individual X variable is handled differently in standard or simultaneous regression (in contrast to the sequential and statistical approaches). For standard regression, the researcher examines the t ratio (with $N - k - 1$ df) that assesses whether the b_i partial slope coefficient is statistically significant for each X_i predictor, for the output of the one regression model that is reported.

When you run either sequential or statistical regression, you actually obtain a series of regression equations. For now, let's assume that you have added just one predictor variable in each step (it is also possible to add groups or blocks of variables in each step). To keep the notation simple, let's also suppose that the variable designated X_1 happens to enter first, X_2 second, and so forth. The order of entry may either be determined arbitrarily by the researcher (the researcher tells SPSS which variable to enter in Step 1, Step 2, etc.) or be determined statistically; that is, the program checks at each step to see which variable would increase the R^2 the most if it were added to the regression and adds that variable to the regression equation.

In either sequential or statistical regression, then, a series of regression analyses is performed as follows:

Step 1:

Add X_1 to the model.

Predict Y from X_1 only.

Step 1 model: $Y' = b_0 + b_1X_1$.

X_1 gets credit for R^2_{inc} in Step 1.

$R^2_{inc} = R^2_{Step1} - R^2_{Step0}$ (assume that R^2 in Step 0 was 0).

Note that r^2_{1Y} is identical to R^2 and sr^2_1, in the Step 1 model, because at this point no other variable is controlled for when assessing the predictive usefulness of X_1. Thus, X_1 is assessed controlling for none of the other predictors.

R^2_{inc} or sr^2_{inc} for Step 1 corresponds to the black area in Figure 5.2b.

Step 2:

Add X_2 to the model.

Predict Y from both X_1 and X_2.

Step 2 model: $Y' = b_0 + b_1X_1 + b_2X_2$.

(Note that b_1 in the Step 2 equation must be reestimated controlling for X_2, so it is not, in general, equal to b_1 in Step 1.)

X_2 gets credit for the increase in R^2 that occurs in the step when X_2 is added to the model: $R^2_{inc} = R^2_{Step2} - R^2_{Step1}$.

Note that R^2_{inc} is equivalent to sr^2_2, the squared part correlation for X_2 in Step 2, and these terms correspond to the medium gray area in Figure 5.2b.

Step 3:

Add X_3 to the model.

Predict Y from X_1, X_2, and X_3.

Step 3 model: $Y' = b_0 + b_1X_1 + b_2X_2 + b_3X_3$.

X_3 gets credit for the proportion of incremental variance $R^2_{Step3} - R^2_{Step2}$ or sr^2_3 (the squared part correlation associated with X_3 in the step when it enters). This corresponds to the light gray area in Figure 5.2b.

Researchers do not generally report complete information about the regression analyses in all these steps. Usually, researchers report the b and beta coefficients for the equation in the *final* step and the multiple R and overall F ratio for the equation in the *final* step with all predictors included. To assess the statistical significance of each X_i individual predictor, the researcher looks at the t test associated with the b_i slope coefficient associated with X_i in the step when X_i first enters the model. If this t ratio is significant *in the step when* X_i *first enters the model*, this implies that the X_i variable added significantly to the explained variance, in the step when it first entered the model, controlling for all the predictors that entered in earlier steps. The R^2_{inc} for each predictor variable X_i on the step when X_i first enters the analysis provides effect size information (the estimated proportion of variance in Y that is predictable from X_i, statistically controlling for all other predictor variables included in this step).

It is also possible to report the overall F ratio for the multiple R for the regression equation at Step 1, Step 2, Step 3, and so forth, but this information is not always included.

Notice that it would not make sense to report a set of b coefficients such that you took your b_1 value from Step 1, b_2 from Step 2, and b_3 from Step 3; this mixed set of slope coefficients could not be used to make accurate predictions of Y. Note also that the value of b_1 (the slope to predict changes in Y from increases in X_1) is likely to change when you add new predictors in each step, and sometimes this value changes dramatically. As additional predictors are added to the model, b_1 usually decreases in absolute magnitude, but it can increase in magnitude or even change sign when other variables are controlled (see Chapter 3 to review discussion of ways that the apparent relationship of X_1 to Y can change when you control for other variables).

When reporting results from a standard or simultaneous regression, it does not matter what order the predictors are listed in your regression summary table. However, when reporting results from a sequential or statistical regression, predictors should be listed in the order in which they entered the model; readers expect to see the first entered variable on row 1, the second entered variable on row 2, and so forth.

The interpretation of $b_1, b_2, b_3, \ldots, b_k$ in a k-predictor multiple regression equation is similar to the interpretation of regression slope coefficients described in Chapter 4. For example, b_1 represents the number of units of change predicted in the raw Y score for a one-unit change in X_1, controlling for or partialling out all the other predictors $(X_2, X_3, \ldots, X_k)$. Similarly, the interpretation of β_1 is the number of standard deviations of change in predicted z_Y, for a 1-*SD* increase in z_{X1}, controlling for or partialling out any linear association between z_{X1} and all the other predictors.

For standard multiple regression, the null hypothesis of interest for each X_i predictor is

$$H_0: b_i = 0. \tag{5.11}$$

That is, we want to know whether the raw-score slope coefficient b_i associated with X_i differs significantly from 0. As in Chapter 4, the usual test statistic for this situation is a t ratio:

$$t = \frac{b_i}{SE_{bi}}, \text{ with } N - k - 1\,df. \tag{5.12}$$

SPSS and other programs provide exact (two-tailed) p values for each t test. If the obtained p value is less than the predetermined alpha level (which is usually set at $\alpha = .05$, two tailed), the partial slope b_i for X_i is judged statistically significant. Recall (from Volume I, Chapter 11 [Warner, 2020]) that if b_i is 0, then b_i, sr_i, and pr_i also equal 0. Thus, this t test can also be used to judge whether the proportion of variance that is uniquely predictable from X_i, sr^2_i, is statistically significant.

In sequential and statistical regression, the contribution of each X_i predictor variable is assessed in the step when X_i first enters the regression model. The null hypothesis is

$$H_0: R^2_{\text{inc}} = 0. \tag{5.13}$$

In words, this is the null hypothesis that the *increment* in multiple R^2 in the step when X_i enters the model equals 0. Another way to state the null hypothesis about the incremental amount of variance that can be predicted when X_i is added to the model is

$$H_0: sr^2_i = 0. \tag{5.14}$$

In words, this is the null hypothesis that the squared part correlation associated with X_i in the step when X_i first enters the regression model equals 0. When b, β, sr, and pr are calculated (see Chapter 4), they all have the same terms in the numerator. They are scaled differently (using different divisors), but if sr is 0, then b must also equal 0. Thus, we can use the t ratio associated with the b_i slope coefficient to test the null hypothesis H_0: $sr^2_i = 0$ (or, equivalently, H_0: $R^2_{\text{inc}} = 0$).

There is also an F test that can be used to assess the significance of the change in R^2 in a sequential or statistical regression, from one step to the next, for any number of added variables. In the example at the end of this chapter, it happens that just one predictor variable is added in each step. However, we can add a set or group of predictors in each step. To test the null hypothesis H_0: $R^2_{\text{inc}} = 0$ for the general case where m variables are added to a model that included k variables in the prior step, the following F ratio can be used:

Let $R^2_{\text{wo}} = R^2$ for the reduced model with only k predictors.[1]

Let $R^2_{\text{with}} = R^2$ for the full model that includes k predictors and m additional predictors.

Let N = number of participants or cases.

Let $R^2_{\text{inc}} = R^2_{\text{with}} - R^2_{\text{wo}}$ (note that R^2_{with} must be equal to or greater than R^2_{wo}).

The test statistic for H_0: $R^2_{\text{inc}} = 0$ is an F ratio with $m, N - k - m - 1$ df:

$$F_{\text{inc}} = \frac{R^2_{\text{inc}} / m}{(1 - R^2_{\text{with}}) / (N - k - m - 1)}. \tag{5.15}$$

When you enter just one new predictor X_i in a particular step, the F_{inc} for X_i equals the squared t ratio associated with X_i.

5.10 EFFECT SIZE

5.10.1 Effect Size for Overall Regression (Multiple *R*)

For all three methods of regression (standard, sequential, and statistical), the effect size for the overall regression model that includes all the predictors is indexed by multiple R, multiple R^2, and adjusted multiple R^2 (for a review of these, see Chapter 4). For standard regression, because there is just one regression equation, it is easy to locate this overall multiple R and R^2. For sequential and statistical regression, researchers always report multiple R and R^2 in the final step. Occasionally, they also report R and R^2 for every individual step.

5.10.2 Effect Size for Individual Predictor Variables (sr^2)

For standard or simultaneous regression, the most common effect size index for each individual predictor variable is sr^2_i; this is the squared part correlation for X_i. SPSS regression can report the part correlation (if requested); this value is squared by hand to provide an estimate of the unique proportion of variance predictable from each X_i variable. We will call this

sr^2_{unique} to indicate that it estimates the proportion of variance that each X predictor uniquely explains (i.e., variance that is not shared with *any* of the other predictors).

For either sequential or statistical regression, the effect size that is reported for each individual predictor is labeled either sr^2_{inc} or R^2_{inc} (i.e., the increase in R^2 in the step when that predictor variable first enters the model). When just one new predictor variable enters in a step, R^2_{inc} is equivalent to the sr^2_i value associated with X_i in the step when X_i first enters the model. If you request "*R* square change" as one of the statistics from SPSS, you obtain a summary table that shows the total R^2 for the regression model at each step and also the R^2 increment at each step of the analysis.

To see how the partition of variance among individual predictors differs when you compare standard regression with sequential or statistical regression, reexamine Figure 5.2. The sr^2_{unique} for variables X_1, X_2, and X_3 in a standard regression correspond to the black, dark gray, and light gray areas in Figure 5.2a, respectively. The sr^2_{inc} (or R^2_{inc}) terms for X_1, X_2, and X_3 in either a sequential or a statistical regression correspond to the black, dark gray, and light gray areas in Figure 5.2b, respectively. Note that in a standard regression, when predictors "compete" to explain the same variance in Y, none of the predictors gets credit for the explained variance that can be explained by other predictors. By contrast, in the sequential and statistical regressions, the contribution of each predictor is assessed controlling only for predictors that entered in earlier steps. As a consequence, when there is "competition" between variables to explain the same variance in Y, the variable that enters in an earlier step gets credit for explaining that shared variance. In sequential and statistical regressions, the total R^2 for the final model can be reconstructed by summing sr^2_1, sr^2_2, sr^2_3, and so forth. In standard or simultaneous regression, the sum of the sr^2_{unique} contributions for the entire set of X_i's is usually less than the overall R^2 for the entire set of predictors.

5.11 CHANGES IN *F* AND *R* AS ADDITIONAL PREDICTORS ARE ADDED TO A MODEL IN SEQUENTIAL OR STATISTICAL REGRESSION

Notice that as you add predictors to a regression model, an added predictor variable may produce a 0 or positive change in R^2; it cannot decrease R^2. However, the adjusted R^2 takes the relative sizes of k, number of predictor variables, and N, number of participants, into account; adjusted R^2 may go down as additional variables are added to an equation. The F ratio for the entire regression equation may either increase or decrease as additional variables are added to the model. Recall that

$$F = [R^2/df_{regression}]/[(1 - R^2)/df_{residual}], \tag{5.16}$$

where $df_{regression} = k$ and $df_{residual} = N - K - 1$.

As additional predictors are added to a regression equation, R^2 may increase (or remain the same), $df_{regression}$ increases, and $df_{residual}$ decreases. If R^2 goes up substantially, this increase may more than offset the change in $df_{regression}$, and if so, the net effect of adding an additional predictor variable is an increase in F. However, if you add a variable that produces little or no increase in R^2, the F for the overall regression may go down, because the loss of degrees of freedom for the residual term may outweigh any small increase in R^2. In general, F goes up if the added variables contribute a large increase in R^2, but if you add "garbage" predictor variables that use up a degree of freedom without substantially increasing R^2, the overall F for the regression can go down as predictor variables are added.

5.12 STATISTICAL POWER

According to Tabachnick and Fidell (2018), the ratio of N (number of cases) to k (number of predictors) has to be "substantial" for a regression analysis to give believable results. On the basis of work by Green (1991), they recommended a minimum $N > 50 + 8k$ for tests of multiple R and a minimum of $N > 104 + k$ for tests of significance of individual predictors. The larger of these two minimum N's should be used to decide how many cases are needed. Thus, for a multiple regression with $k = 5$ predictors, the first rule gives $N > 75$ and the second rule gives $N > 109$; at least 109 cases should be used.

This decision rule should provide adequate statistical power to detect medium effect sizes; however, if the researcher wants to be able to detect weak effect sizes, or if there are violations of assumptions such as non-normal distribution shapes, or if measurements have poor reliability, larger N's are needed. If statistical regression methods (such as stepwise entry) are used, even larger N's should be used. Larger sample sizes than these minimum sample sizes based on statistical power analysis are required to make the confidence intervals around estimates of b slope coefficients reasonably narrow. Note also that the higher the correlations among predictors, the larger the sample size that will be needed to obtain reasonably narrow confidence intervals for slope estimates.

On the other hand, in research situations where the overall sample size N is very large (e.g., $N > 10{,}000$), researchers may find that even effects that are too small to be of any practical or clinical importance may turn out to be statistically significant. For this reason, it is important to include information about effect size along with statistical significance tests.

Notice also that if a case has missing values on any of the variables included in the regression, the effective N is decreased. SPSS provides choices about handling data with missing observations. In pairwise deletion, each correlation is calculated using all the data available for that particular pair of variables. Pairwise deletion can result in quite different N's (and different subsets of cases) used for each correlation, and this inconsistency is undesirable. In listwise deletion, a case is entirely omitted from the regression if it is missing a value on any one variable. This provides consistency in the set of data used to estimate all correlations, but if there are many missing observations on numerous variables, listwise deletion of missing data can lead to a very small overall N.

5.13 NATURE OF THE RELATIONSHIP BETWEEN EACH X PREDICTOR AND Y (CONTROLLING FOR OTHER PREDICTORS)

It is important to pay attention to the sign associated with each b or β coefficient and to ask whether the direction of the relation it implies is consistent with expectations. It's also useful to ask how the partial and semipartial r's and b coefficients associated with a particular X_i variable compare with its zero-order correlation with Y, in both size and sign. As described in Chapter 3, when one or more other variables are statistically controlled, the apparent relation between X_i and Y can become stronger or weaker, become nonsignificant, or even change sign. The same is true of the partial slopes (and semipartial correlations) associated with individual predictors in multiple regression. The "story" about prediction of Y from several X variables may need to include discussion of the ways in which controlling for some Xs changes the apparent importance of other predictors.

It is important to include a matrix of correlations as part of the results of a multiple regression (not only correlations of each X_i with Y but also correlations among all the X predictors). The correlations between the X's and Y provide a baseline against which to evaluate whether including each X_i in a regression with other variables statistically controlled has made a difference in the apparent nature of the relation between X_i and Y. The correlations among the X's should be examined to see whether there were strong correlations among predictors (also called strong multicollinearity). When predictors are highly correlated with one another, they may compete to explain much of the same variance; also, when predictors are highly correlated, the researcher may find that none of the individual b_i slope coefficients are significant even when the overall R for the entire regression is significant. If predictors X_1 and X_2 are very highly correlated, the researcher may want to consider whether they are, in fact, both measures of the same thing; if so, it may be better to combine them (perhaps by averaging X_1 and X_2 or z_{X1} and z_{X2}) or drop one of the variables.

One type of information provided about multicollinearity among predictors is **tolerance**. The tolerance for a candidate predictor variable X_i is the proportion of variance in X_i that is not predictable from other X predictor variables that are already included in the regression equation. For example, suppose that a researcher has a regression equation to predict Y' from scores on X_1, X_2, and X_3: $Y' = b_0 + b_1X_1 + b_2X_2 + b_3X_3$. Suppose that the researcher is considering whether to add predictor variable X_4 to this regression. Several kinds of information are useful in deciding whether X_4 might possibly provide additional useful predictive information. One thing the researcher wants to know is, How much of the variance in X_4 is not already explained by (or accounted for) the other predictor variables already in the equation? To estimate the proportion of variance in X_4 that is not predictable from, or shared with, the predictor variables already included in the analysis—that is, X_1 through X_3—we could set up a regression to predict scores on the candidate variable X_4 from variables X_1 through X_3; the tolerance of candidate predictor variable X_4 is given by $1 - R^2$ for the equation that predicts X_4 from the other predictors X_1 through X_3. The minimum possible value of tolerance is 0; tolerance of 0 indicates that the candidate X_4 variable contains no additional variance or information that is not already present in predictor variables X_1 through X_3 (and that therefore X_4 cannot provide any "new" predictive information that is not already provided by X_1 through X_3). The maximum possible value of tolerance is 1.0; this represents a situation in which the predictor variable X_4 is completely uncorrelated with the other set of predictor variables already included in the model. If we are interested in adding the predictor variable X_4 to a regression analysis, we typically hope that it will have a tolerance that is not close to 0; tolerance that is substantially larger than 0 is evidence that X_4 provides new information not already provided by the other predictor variables.

5.14 ASSESSMENT OF MULTIVARIATE OUTLIERS IN REGRESSION

Examination of a histogram makes it possible to detect scores that are extreme univariate outliers; examination of bivariate scatterplots makes it possible to identify observations that are bivariate outliers (i.e., they represent unusual combinations of scores on X and Y, even though they may not be extreme on either X or Y alone); these are scores that lie outside the "cloud" that includes most of the data points in the scatterplot. It can be more difficult to detect multivariate outliers, as graphs that involve multiple dimensions are complex. Regression analysis offers several kinds of information about individual cases that can be used to identify multivariate outliers. For a more complete discussion, see Tabachnick and Fidell (2018).

First, SPSS can provide graphs of residuals (actual Y – predicted Y scores) against other values, such as Y' predicted scores. In a graph of standardized (z-score) residuals, about 99% of the values should lie between –3 and +3; any observations with standardized residual z-score values >3 in absolute value represent cases for which the regression made an unusually poor prediction. We should not necessarily automatically discard such cases, but it may be informative to examine these cases carefully to answer questions such as the following: Is the poor fit due to data entry errors? Was there something unique about this case that might explain why the regression prediction was poor for this participant? In addition, SPSS can provide saved scores on numerous case-specific diagnostic values such as "leverage" (slopes change by a large amount when cases with large leverage index values are dropped from the analysis; thus, such cases are inordinately influential). Mahalanobis D is another index available in SPSS that indicates the degree to which observations are multivariate outliers.

5.15 SPSS EXAMPLES

As an illustration of the issues in this chapter, the data in the SPSS file predictphysics.sav are analyzed using three different methods of multiple regression. The first analysis is standard or simultaneous regression; all five predictor variables are entered in one step. The second analysis is sequential regression; the five predictor variables are entered in a user-determined sequence, one in each step, in an order that was specified by using SPSS menu command selections. The third analysis is a statistical (data-driven) regression using "method = forward." In this analysis, a statistical criterion was used to decide the order of entry of variables. At each step, the predictor variable that would produce the largest increment in R^2 was added to the model; when adding another variable would not produce a statistically significant increment in R^2, no further variables were entered. Note that to use the Wilkinson and Dallal (1981) table (reproduced as Appendix 5B) to assess statistical significance of the overall final model obtained through forward regression, the user must specify a required minimum F-to-enter that matches one of the F values included in the Wilkinson and Dallal table (i.e., F-to-enter = 2.00, 3.00, or 4.00).

Details of data screening for these analyses are omitted. Prior to doing a multiple regression, the following preliminary screening should be done:

1. Histogram of scores on each predictor variable and Y: Check to see that the distribution shape is reasonably normal and that there are no extreme outliers or "impossible" score values.
2. Scatterplots between every pair of variables (e.g., all pairs of X variables and each X with Y). The scatterplots should show linear relations, homoscedastic variance, and no extreme bivariate outliers.

Three different multiple regressions were performed using the same data. Ordinarily, only one method is reported in a journal article; results from all three methods are reported here to illustrate how the nature of conclusions about the relative importance of predictors may differ depending on the method of entry that is chosen. Note that gender could have been dummy coded +1 and −1 or +1 and 0 (as in the examples of dummy variables that are presented in Chapter 6). However, the proportion of variance that is predictable from gender in a multiple regression is the same whether gender is coded +1, 0 or +1, +2, as in the examples that follow. The table included in this "Results" section is based on examples of summary tables in Tabachnick and Fidell (2018). It may be more convenient to break up the information into three separate tables: one for descriptive statistics on each variable (e.g., mean, SD), one for correlations among all variables, and one for regression results.

5.15.1 SPSS Menu Selections, Output, and Results for Standard Regression

Results for a Standard or Simultaneous Multiple Regression

Scores on a physics achievement test were predicted from the following variables: gender (coded 1 = male, 2 = female), EI, IQ, VSAT score, and MSAT score. The total N for this sample was 200; 5 cases were dropped because of missing data on at least one variable, and therefore for this analysis, $N = 195$. Preliminary data screening included examination of histograms of scores on all six variables and examination of scatterplots for all pairs of variables. Univariate distributions were reasonably normal, with no extreme outliers; bivariate relations were fairly linear, all slopes had the expected signs, and there were no bivariate outliers.

Standard multiple regression was performed; that is, all predictor variables were entered in one step. Zero-order, part, and partial correlations of each predictor with physics score were requested in addition to the default statistics. Results for this standard multiple regression are summarized in Table 5.2. See Figures 5.3 through 5.7 for SPSS menu selections and syntax and Figure 5.8 for the SPSS output.

To assess whether there were any outliers, the standardized residuals from this regression were plotted against the standardized predicted values (see the last panel in Figure 5.9). There was no indication of pattern, trend, or heteroscedasticity in this graph of residuals, nor were there any outliers; thus, it appears that the assumptions required for multiple regression were reasonably well met.

The overall regression, including all five predictors, was statistically significant, $R = .90$, $R^2 = .81$, adjusted $R^2 = .80$, $F(5, 189) = 155.83$, $p < .001$. Physics scores could be predicted quite well from this set of five variables, with approximately 80% of the variance in physics scores accounted for by the regression.

To assess the contributions of individual predictors, the t ratios for the individual regression slopes were examined. Three of the five predictors were significantly predictive of physics scores; these included gender, $t(189) = -5.58$, $p < .001$; VSAT score, $t(189) = -9.64$, $p < .001$; and MSAT score, $t(189) = 19.15$, $p < .001$. The nature of the predictive relation of gender was as expected; the negative sign for the slope for gender indicated that

Table 5.2 Results of Standard Multiple Regression to Predict Physics Score (Y) From Gender, IQ, EI, VSAT Score, and MSAT Score

	Physics Score	Gender	IQ	EI	VSAT Score	MSAT Score	b	β	sr^2_{unique}
Gender	−.37						−7.45***	−.21	.03
IQ	.34	.11					.13	.11	<.01
EI	−.39	.48	.04				−.03	−.02	<.01
VSAT score	−.13	.23	.64	.55			−.09***	−.56	.10
MSAT score	.69	−.03	.70	.04	.48		.36***	.88	.38
							Intercept = −45.20***		
Mean	79.6	—[a]	100.1	107.6	535.2	493.7			
SDA	17.6	—[a]	14.7	12.3	105.9	43.4			
									$R^2 = .805$
									$R^2_{adj} = .800$
									$R = .897$***

Note: Table format adapted from Tabachnick and Fidell (2018).

***$p < .001$.

a. Because gender was a dummy variable (coded 1 = male, 2 = female), mean and standard deviation are not reported. The sample included $n = 100$ men and $n = 100$ women.

Figure 5.3 Menu Selections for SPSS Linear Regression Procedure

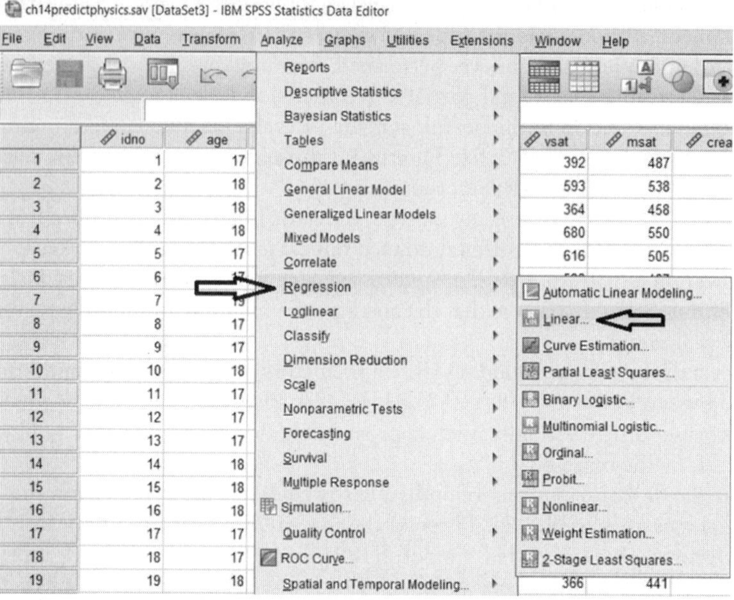

Figure 5.4 Regression Main Dialog Box for Standard or Simultaneous Linear Regression in SPSS (Method = Enter)

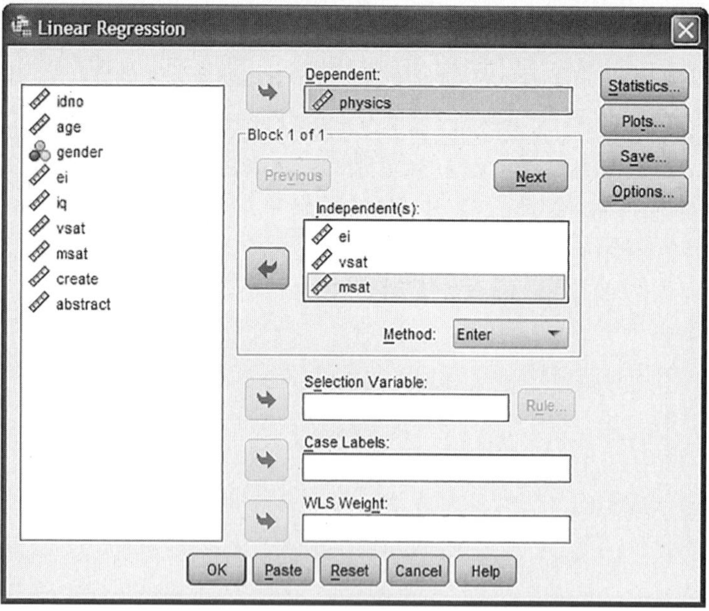

higher scores on gender (i.e., being female) predicted lower scores in physics. The predictive relation of MSAT score to physics score was also as predicted; higher scores on the MSAT predicted higher scores in physics. However, scores on the VSAT were negatively related to physics scores; that is, higher VSAT scores predicted lower scores in physics, which was contrary to expectations. The negative partial r for the prediction of physics score from VSAT score controlling for the other four predictors ($pr = -.57$) was stronger than the zero-order Pearson's r for the prediction of physics score from VSAT score without controlling for other variables ($r = -.13$), an indication of possible suppression effects; that is, it appears that the part of VSAT score that was unrelated to MSAT score and IQ was strongly predictive of poorer performance in physics. (See Chapter 3 for a review of suppression.) The

Figure 5.5 Statistics Requested for Standard Linear Regression in SPSS

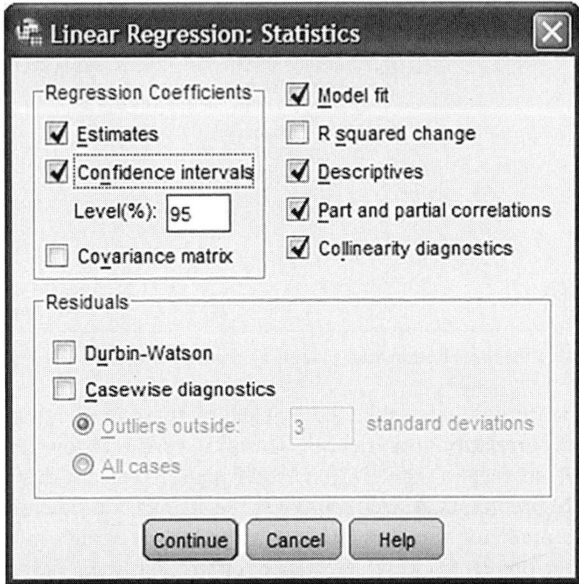

Figure 5.6 Request for Plot of Standardized Residuals From Linear Regression

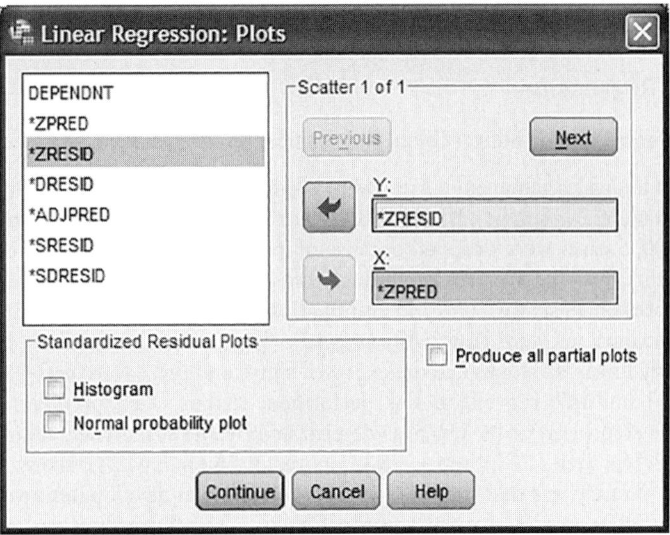

Figure 5.7 SPSS Syntax for Standard Regression

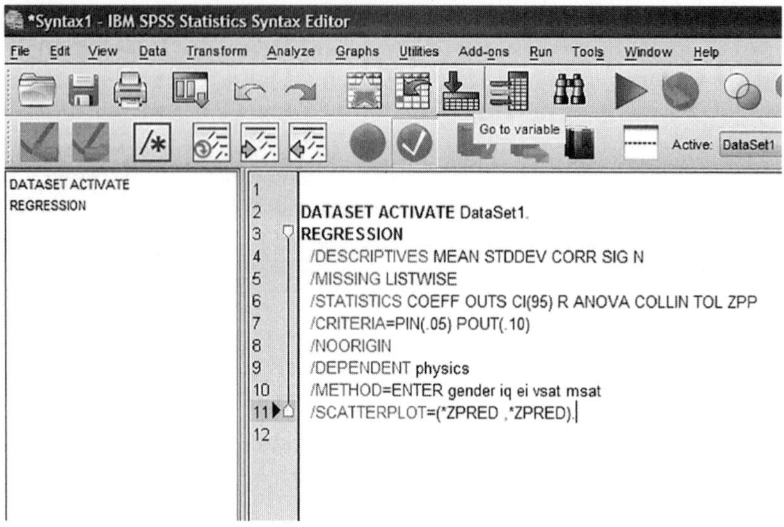

Note: Obtained by clicking the Paste button after making all the menu selections.

proportions of variance uniquely explained by each of these predictors (sr^2_{unique}, obtained by squaring the part correlation from the SPSS output) were as follows: sr^2 = .03 for gender, sr^2 = .096 for VSAT score, and sr^2 = .38 for MSAT score. Thus, in this sample and in the context of this set of predictors, MSAT score was the strongest predictor of physics score.

The other two predictor variables (EI and IQ) were not significantly related to physics score when other predictors were statistically controlled; their partial slopes were not significant. Overall, physics scores were highly predictable from this set of predictors; the strongest unique predictive contributions were from MSAT and VSAT scores, with a smaller contribution from gender. Neither EI nor IQ was significantly predictive of physics scores in this regression, even though these two variables had significant zero-order correlations with physics scores; apparently, the information that they contributed to the regression was redundant with other predictors.

5.15.2 SPSS Menu Selections, Output, and Results for Sequential Regression

Results for Sequential or Hierarchical Regression (User-Determined Order of Entry)

Scores on a physics achievement test were predicted from the following variables: gender (coded 1 = male, 2 = female), EI, IQ, VSAT score, and MSAT score. The total N for this sample was 200; 5 cases were dropped because of missing data on at least one variable, and therefore for this analysis, N = 195. Preliminary data screening included examination of histograms of scores on all six variables and examination of scatterplots for all pairs of variables. Univariate distributions were reasonably normal, with no extreme outliers; bivariate relations were fairly linear, all slopes had the expected signs, and there were no bivariate outliers.

Sequential multiple regression was performed; that is, each predictor variable was entered in one step in an order that was determined by the researcher, as follows: Step 1, gender; Step 2, IQ; Step 3, EI; Step 4, VSAT score; and Step 5, MSAT score. The rationale for this order of entry was that factors that emerge earlier in development were entered in earlier steps; VSAT score was entered prior to MSAT score arbitrarily. Zero-order, part, and

Figure 5.8 SPSS Output for Standard or Simultaneous Regression: Prediction of Physics Score From Gender, IQ, EI, VSAT Score, and MSAT Score

Descriptive Statistics

	Mean	Std. Deviation	N
physics	79.57	17.582	195
gender	1.50	.501	195
iq	100.09	14.687	195
ei	107.57	12.257	195
vsat	535.24	105.912	195
msat	493.67	43.367	195

Correlations

		physics	gender	iq	ei	vsat	msat
Pearson Correlation	physics	1.000	-.368	.344	-.394	-.129	.690
	gender	-.368	1.000	.108	.483	.234	-.028
	iq	.344	.108	1.000	.036	.641	.704
	ei	-.394	.483	.036	1.000	.548	.040
	vsat	-.129	.234	.641	.548	1.000	.484
	msat	.690	-.028	.704	.040	.484	1.000
Sig. (1-tailed)	physics	.	.000	.000	.000	.036	.000
	gender	.000	.	.066	.000	.001	.350
	iq	.000	.066	.	.308	.000	.000
	ei	.000	.000	.308	.	.000	.289
	vsat	.036	.001	.000	.000	.	.000
	msat	.000	.350	.000	.289	.000	.
N	physics	195	195	195	195	195	195
	gender	195	195	195	195	195	195
	iq	195	195	195	195	195	195
	ei	195	195	195	195	195	195
	vsat	195	195	195	195	195	195
	msat	195	195	195	195	195	195

Variables Entered/Removed[b]

Model	Variables Entered	Variables Removed	Method
1	msat, gender, ei, vsat, iq[a]	.	Enter

a. All requested variables entered.

b. Dependent Variable: physics

Model Summary[b]

Model	R	R Square	Adjusted R Square	Std. Error of the Estimate
1	.897[a]	.805	.800	7.870

a. Predictors: (Constant), msat, gender, ei, vsat, iq

b. Dependent Variable: physics

ANOVA[b]

Model		Sum of Squares	df	Mean Square	F	Sig.
1	Regression	48262.46	5	9652.492	155.833	.000[a]
	Residual	11706.91	189	61.941		
	Total	59969.37	194			

a. Predictors: (Constant), msat, gender, ei, vsat, iq

b. Dependent Variable: physics

Coefficients[a]

Model		Unstandardized Coefficients		Standardized Coefficients	t	Sig.	95% Confidence Interval for B		Correlations		
		B	Std. Error	Beta			Lower Bound	Upper Bound	Zero-order	Partial	Part
1	(Constant)	-45.201	9.125		-4.954	.000	-63.200	-27.202			
	gender	-7.453	1.335	-.212	-5.583	.000	-10.086	-4.819	-.368	-.376	-.179
	iq	.130	.070	.109	1.860	.064	-.008	.268	.344	.134	.060
	ei	-.032	.072	-.022	-.443	.658	-.173	.110	-.394	-.032	-.014
	vsat	-.093	.010	-.563	-9.643	.000	-.113	-.074	-.129	-.574	-.310
	msat	.357	.019	.881	19.155	.000	.320	.394	.690	.812	.616

a. Dependent Variable: physics

Note: All the predictors entered in one step.

Figure 5.9 Scatterplot to Assess Standardized Residuals From Linear Regression to Predict Physics Scores From Gender, IQ, EI, VSAT Score, and MSAT Score

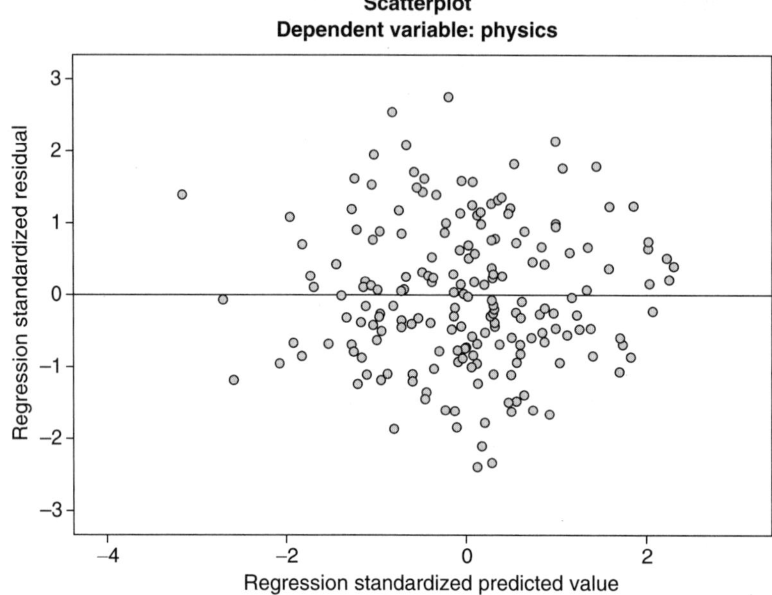

partial correlations of each predictor with physics were requested in addition to the default statistics. Results for this sequential multiple regression are summarized in Tables 5.3 and 5.4. (See Figures 5.10 through 5.12 for SPSS commands and Figure 5.13 for SPSS output.)

The overall regression, including all five predictors, was statistically significant, $R = .90$, $R^2 = .81$, adjusted $R^2 = .80$, $F(5, 189) = 155.83, p < .001$. Physics scores could be predicted quite well from this set of five variables, with approximately 80% of the variance in physics scores accounted for by the regression.

To assess the contributions of individual predictors, the *t* ratios for the individual regression slopes were examined for each variable in the step when it first entered the analysis. In Step 1, gender was statistically significant, $t(193) = -5.49, p < .001$; R^2_{inc} (which is equivalent to sr^2_{inc}) was .135. The nature of the relation of gender to physics score was as expected; the negative sign for the slope for gender indicated that higher scores on gender (i.e., being female) predicted lower scores in physics. IQ significantly increased the R^2 when it was entered in Step 2, $t(192) = 6.32$, $R^2_{\text{inc}} = .149$. (Note that the contribution of IQ, which is assessed in this analysis without controlling for EI, VSAT score, and MSAT score, appears to be much stronger than in the standard regression, where the contribution of IQ was assessed by controlling for all other predictors.) EI significantly increased the R^2 when it was entered in Step 3, $t(191) = -4.10, p < .001$, $R^2_{\text{inc}} = .058$. VSAT score significantly increased the R^2 when it was entered in Step 4, $t(190) = -5.27, p < .001$, $R^2_{\text{inc}} = .084$. MSAT score significantly increased the R^2 when it was entered in the fifth and final step, $t(189) = 19.16, p < .001$, $R^2_{\text{inc}} = .379$. Except for VSAT and EI, which were negatively related to physics score, the slopes of all predictors had the expected signs.

Overall, physics scores were highly predictable from this set of predictors; the strongest unique predictive contribution was from MSAT (even though this variable was entered in the last step). All five predictors significantly increased the R^2 in the step when they first entered.

Table 5.3 Results of Sequential Multiple Regression to Predict Physics Score (Y) From IQ, EI, VSAT Score, MSAT Score, and Gender

	Physics	MSAT Score	VSAT Score	Gender	IQ		b^a	α	sr^2_{inc}
MSAT score	.69						.36***	.88	.476
VSAT score	−.13	.48					−.09***	−.58	.280
Gender	−.37	−.03	.23				−7.73***	−.22	.043
IQ	.34	.70	.64	.11			.14*	.12	.006
EI (not entered)	−.39	.04	.55	.48	.04		—	—	—
						Intercept = −47.87***			
Mean[b]	79.6	493.7	535.2	—	100.1				
SD	17.6	43.4	105.9	—	14.7				
									R^2 = .805[c]
									R^2_{adj} = .800
									R^a = .897***

Note: Table format adapted from Tabachnick and Fidell (2018).

a. The significance values given in the SPSS output for all individual *b* coefficients were all <.001; however, forward regression leads to an inflated risk for Type I error, and therefore these *p* values may underestimate the true risk for Type I error in this situation. The significance of the multiple *R* for the final model was assessed using critical values that are appropriate for forward regression from Wilkinson and Dallal (1981) rather than the *p* values in the SPSS output.

b. Because gender was a dummy variable (coded 1 = male, 2 = female), mean and standard deviation are not reported. The sample included *n* = 100 men and *n* = 100 women.

c. In general, a multiple R would be smaller for a subset of four predictors than for five predictors. In this example, after rounding, the multiple R for four predictors was the same as the multiple R for five predictors, in Table 5.2, in the first three decimal places.

$^*p < .05$. $^{***}p < .001$.

Table 5.4 Summary of R^2 Values and R^2 Changes at Each Step in the Sequential Regression in Table 5.3

Predictors Included	R^2 for Model	*F* for Model	R^2 Change	*F* for R^2 Change
1. Gender	.135	*F*(1, 193) = 30.14***	.135	*F*(1, 193) = 30.14***
2. Gender, IQ	.284	*F*(2, 192) = 38.10***	.149	*F*(1, 192) = 39.97***
3. Gender, IQ, EI	.342	*F*(3, 191) = 33.08***	.058	*F*(1, 191) = 16.78***
4. Gender, IQ, EI, VSAT score	.426	*F*(4, 190) = 35.23**	.084	*F*(1, 190) = 27.76***
5. Gender, IQ, EI, VSAT score, MSAT score	.805	*F*(5, 189) = 155.83***	.379	*F*(1, 189) = 366.90***

$^{**}p < .01$. $^{***}p < .001$.

Figure 5.10 SPSS Dialog Box for Sequential or Hierarchical Regression

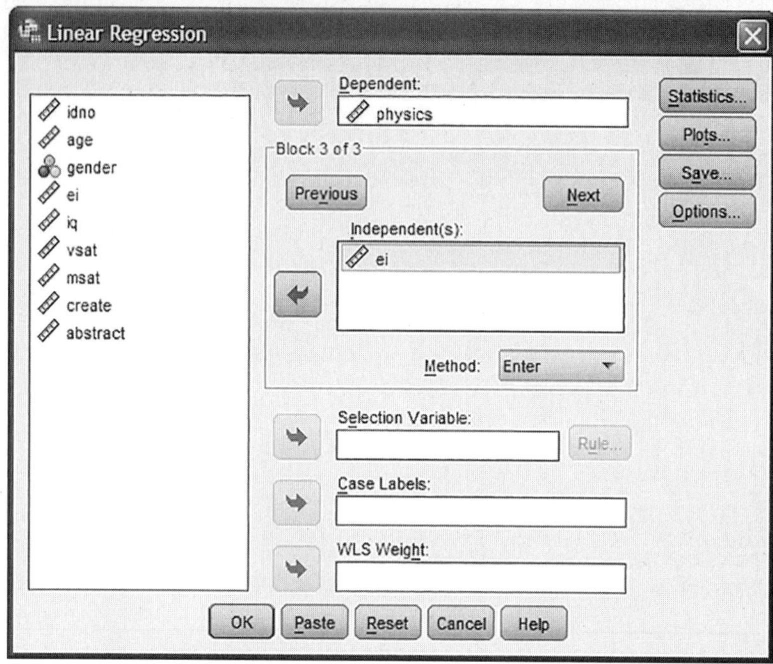

Note: One or more variables can be entered in each step. To enter gender in Step 1, place the name gender in the box, then click the Next button to go to the next step. In this example, only one variable was entered in each step. The dialog box is shown only for the step in which EI is entered as the third predictor.

Figure 5.11 Additional Statistics Requested for Sequential or Hierarchical Regression: R^2 Change

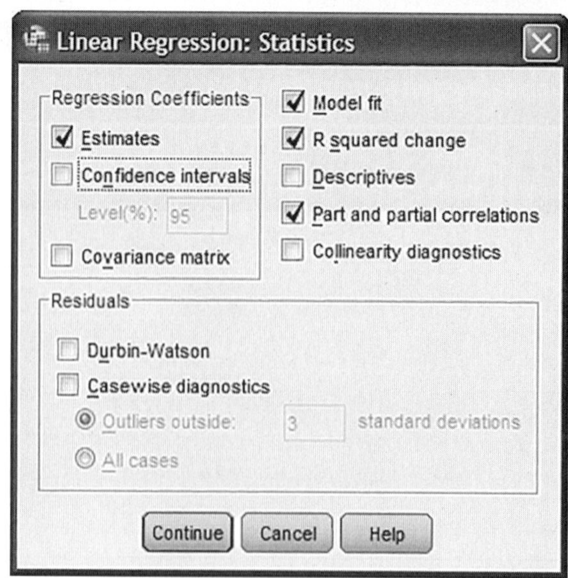

Figure 5.12 SPSS Syntax for the Sequential or Hierarchical Regression (User-Determined Order of Entry)

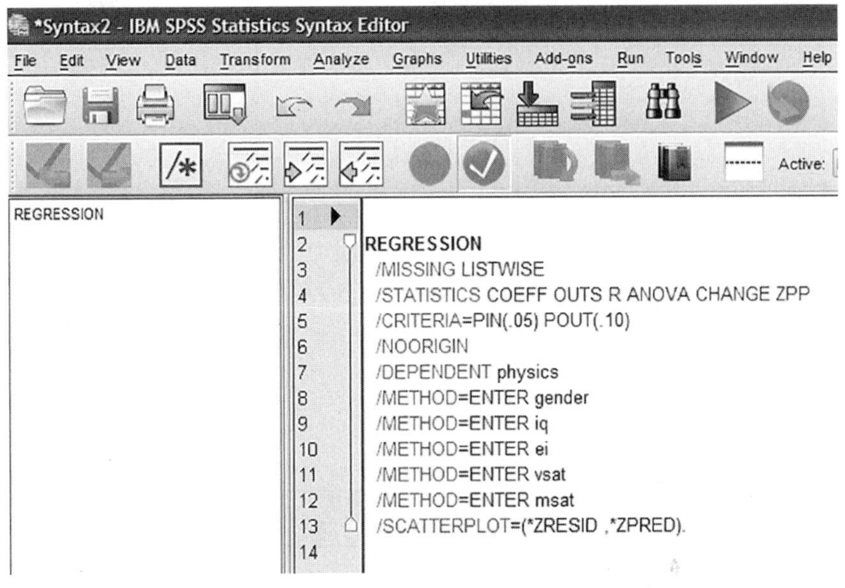

Note: Prediction of physics score from five variables each entered in one step: Step 1, gender; Step 2, IQ; Step 3, EI; Step 4, VSAT score; Step 5, MSAT score.

Figure 5.13 Selected Output From SPSS Sequential Regression

Variables Entered/Removed[b]

Model	Variables Entered	Variables Removed	Method
1	gender [a]	.	Enter
2	iq [a]	.	Enter
3	ei [a]	.	Enter
4	vsat [a]	.	Enter
5	msat [a]	.	Enter

a. All requested variables entered.

b. Dependent Variable: physics

Model Summary

					Change Statistics				
Model	R	R Square	Adjusted R Square	Std. Error of the Estimate	R Square Change	F Change	df1	df2	Sig. F Change
1	.368[a]	.135	.131	16.394	.135	30.143	1	193	.000
2	.533[b]	.284	.277	14.953	.149	39.975	1	192	.000
3	.585[c]	.342	.332	14.374	.058	16.778	1	191	.000
4	.653[d]	.426	.414	13.462	.084	27.759	1	190	.000
5	.897[e]	.805	.800	7.870	.379	366.897	1	189	.000

a. Predictors: (Constant), gender

b. Predictors: (Constant), gender, iq

c. Predictors: (Constant), gender, iq, ei

d. Predictors: (Constant), gender, iq, ei, vsat

e. Predictors: (Constant), gender, iq, ei, vsat, msat

(Continued)

Figure 5.13 (Continued)

Model		Sum of Squares	df	Mean Square	F	Sig.
1	Regression	8100.807	1	8100.807	30.143	.000[a]
	Residual	51868.56	193	268.749		
	Total	59969.37	194			
2	Regression	17039.07	2	8519.535	38.102	.000[b]
	Residual	42930.30	192	223.595		
	Total	59969.37	194			
3	Regression	20505.76	3	6835.253	33.082	.000[c]
	Residual	39463.61	191	206.616		
	Total	59969.37	194			
4	Regression	25536.37	4	6384.094	35.227	.000[d]
	Residual	34433.00	190	181.226		
	Total	59969.37	194			
5	Regression	48262.46	5	9652.492	155.833	.000[e]
	Residual	11706.91	189	61.941		
	Total	59969.37	194			

a. Predictors: (Constant), gender

b. Predictors: (Constant), gender, iq

c. Predictors: (Constant), gender, iq, ei

d. Predictors: (Constant), gender, iq, ei, vsat

e. Predictors: (Constant), gender, iq, ei, vsat, msat

f. Dependent Variable: physics

Coefficients[a]

Model		Unstandardized Coefficients		Standardized Coefficients	t	Sig.	Correlations		
		B	Std. Error	Beta			Zero-order	Partial	Part
1	(Constant)	98.876	3.707		26.675	.000			
	gender	-12.891	2.348	-.368	-5.490	.000	-.368	-.368	-.368
2	(Constant)	54.553	7.783		7.009	.000			
	gender	-14.365	2.154	-.410	-6.668	.000	-.368	-.434	-.407
	iq	.465	.074	.388	6.323	.000	.344	.415	.386
3	(Constant)	90.472	11.527		7.849	.000			
	gender	-9.700	2.363	-.277	-4.104	.000	-.368	-.285	-.241
	iq	.460	.071	.384	6.500	.000	.344	.426	.382
	ei	-.394	.096	-.275	-4.096	.000	-.394	-.284	-.240
4	(Constant)	53.387	12.887		4.143	.000			
	gender	-11.822	2.250	-.337	-5.255	.000	-.368	-.356	-.289
	iq	.858	.100	.716	8.537	.000	.344	.527	.469
	ei	.044	.123	.031	.363	.717	-.394	.026	.020
	vsat	-.087	.017	-.526	-5.269	.000	-.129	-.357	-.290
5	(Constant)	-45.201	9.125		-4.954	.000			
	gender	-7.453	1.335	-.212	-5.583	.000	-.368	-.376	-.179
	iq	.130	.070	.109	1.860	.064	.344	.134	.060
	ei	-.032	.072	-.022	-.443	.658	-.394	-.032	-.014
	vsat	-.093	.010	-.563	-9.643	.000	-.129	-.574	-.310
	msat	.357	.019	.881	19.155	.000	.690	.812	.616

a. Dependent Variable: physics

Excluded Variables[e]

Model		Beta In	t	Sig.	Partial Correlation	Collinearity Statistics Tolerance
1	iq	.388[a]	6.323	.000	.415	.988
	ei	-.283[a]	-3.826	.000	-.266	.767
	vsat	-.045[a]	-.656	.512	-.047	.945
	msat	.680[a]	14.858	.000	.731	.999
2	ei	-.275[b]	-4.096	.000	-.284	.767
	vsat	-.501[b]	-6.853	.000	-.444	.562
	msat	.821[b]	12.880	.000	.682	.494
3	vsat	-.526[c]	-5.269	.000	-.357	.303
	msat	.866[c]	15.469	.000	.747	.489
4	msat	.881[d]	19.155	.000	.812	.488

a. Predictors in the Model: (Constant), gender

b. Predictors in the Model: (Constant), gender, iq

c. Predictors in the Model: (Constant), gender, iq, ei

d. Predictors in the Model: (Constant), gender, iq, ei, vsat

e. Dependent Variable: physics

Note: Prediction of physics from five variables each entered in one step in a user-determined sequence: Step 1, gender; Step 2, IQ; Step 3, EI; Step 4, VSAT score; Step 5, MSAT score (descriptive statistics and correlations are omitted because they are the same as for the standard regression).

5.15.3 SPSS Menu Selections, Output, and Results for Statistical Regression

Results of Statistical Regression Using Method = Forward to Select Predictor Variables for Which the Increment in R^2 in Each Step Is Maximized

A statistical regression was performed to predict scores in physics from the following candidate predictor variables: gender (coded 1 = male, 2 = female), EI, IQ, VSAT score, and MSAT score. The total N for this sample was 200; 5 cases were dropped because of missing data on at least one variable, and therefore for this analysis, $N = 195$. Preliminary data screening included examination of histograms of scores on all six variables and examination of scatterplots for all pairs of variables. Univariate distributions were reasonably normal, with no extreme outliers; bivariate relations were fairly linear, all slopes had the expected signs, and there were no bivariate outliers.

Statistical multiple regression was performed using method = forward with the F-to-enter criterion value set at $F = 3.00$. That is, in each step, SPSS entered the one predictor variable that would produce the largest increase in R^2. When the F ratio for the R^2 increase due to additional variables fell below $F = 3.00$, no further variables were added to the model. This resulted in the following order of entry: Step 1, MSAT score; Step 2, VSAT score; Step 3, gender, and Step 4, IQ. EI did not enter the equation. See Figures 5.14 through 5.16 for SPSS commands and Figure 5.17 for SPSS output. Results of this sequential regression are summarized in Table 5.5.

Figure 5.14 SPSS Menu Selections for Statistical Regression (Method = Forward)

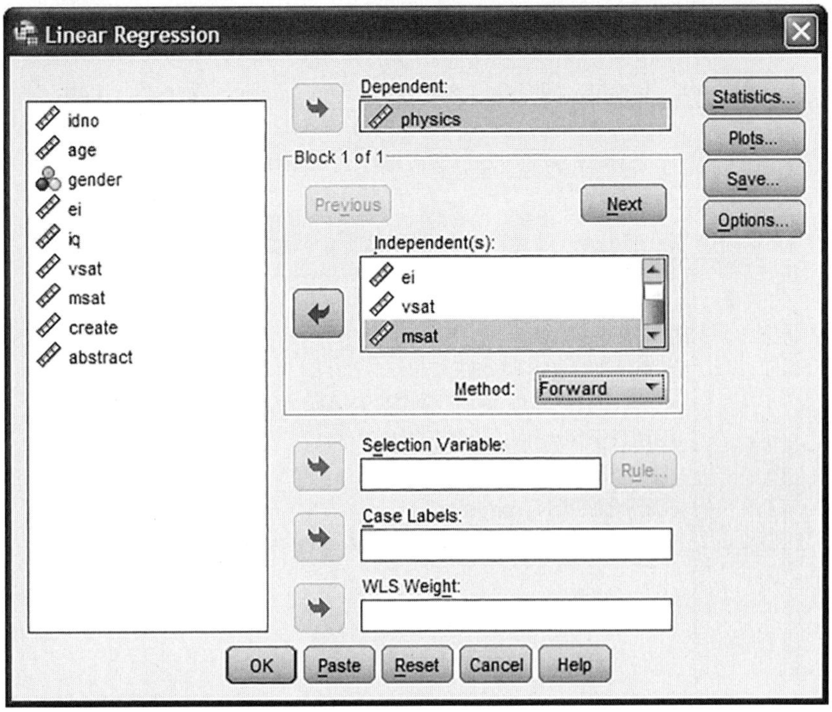

Note: Variables are entered by data-driven order of entry.

Figure 5.15 Selection of Criterion for F = 3.00 to Enter for Forward/ Statistical Regression

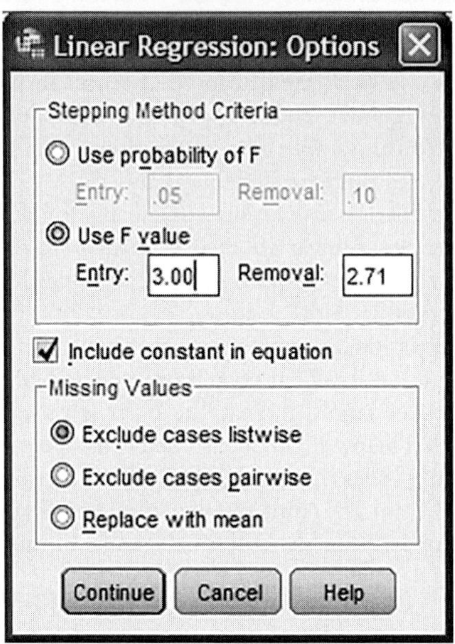

Figure 5.16 SPSS Syntax for Statistical Regression (Method = Forward, F-to-Enter Set at F = 3.00)

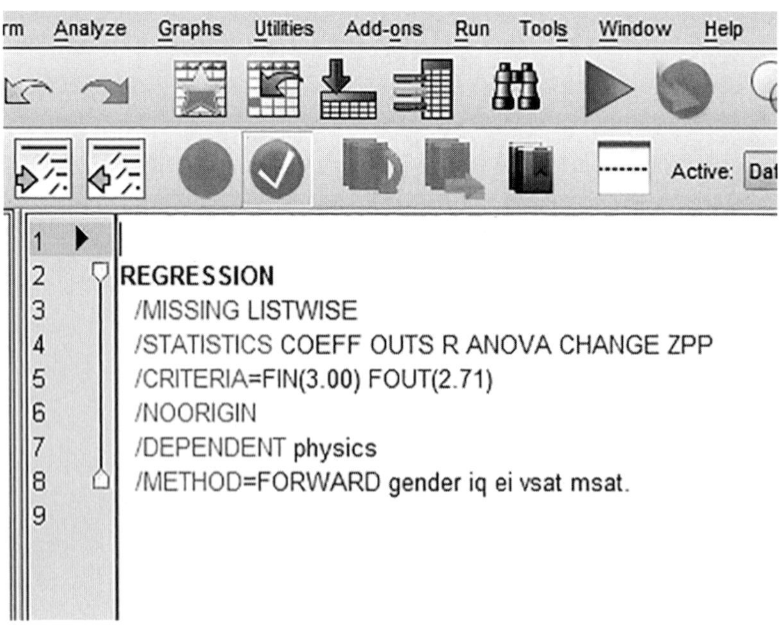

Figure 5.17 Selected Output From SPSS Statistical Regression (Method = Forward)

Model Summary

Model	R	R Square	Adjusted R Square	Std. Error of the Estimate	Change Statistics				
					R Square Change	F Change	df1	df2	Sig. F Change
1	.690[a]	.476	.473	12.757	.476	175.468	1	193	.000
2	.869[b]	.756	.753	8.732	.280	219.988	1	192	.000
3	.894[c]	.799	.796	7.945	.043	40.901	1	191	.000
4	.897[d]	.805	.800	7.854	.006	5.477	1	190	.020

a. Predictors: (Constant), msat

b. Predictors: (Constant), msat, vsat

c. Predictors: (Constant), msat, vsat, gender

d. Predictors: (Constant), msat, vsat, gender, iq

ANOVA[e]

Model		Sum of Squares	df	Mean Square	F	Sig.
1	Regression	28557.95	1	28557.952	175.468	.000[a]
	Residual	31411.42	193	162.753		
	Total	59969.37	194			
2	Regression	45330.62	2	22665.311	297.275	.000[b]
	Residual	14638.75	192	76.243		
	Total	59969.37	194			
3	Regression	47912.47	3	15970.825	253.003	.000[c]
	Residual	12056.90	191	63.125		
	Total	59969.37	194			
4	Regression	48250.29	4	12062.572	195.569	.000[d]
	Residual	11719.08	190	61.679		
	Total	59969.37	194			

a. Predictors: (Constant), msat

b. Predictors: (Constant), msat, vsat

c. Predictors: (Constant), msat, vsat, gender

d. Predictors: (Constant), msat, vsat, gender, iq

e. Dependent Variable: physics

Coefficients[a]

Model		Unstandardized Coefficients		Standardized Coefficients	t	Sig.	Correlations		
		B	Std. Error	Beta			Zero-order	Partial	Part
1	(Constant)	-58.543	10.467		-5.593	.000			
	msat	.280	.021	.690	13.246	.000	.690	.690	.690
2	(Constant)	-63.401	7.171		-8.841	.000			
	msat	.398	.017	.983	24.115	.000	.690	.867	.860
	vsat	-.100	.007	-.604	-14.832	.000	-.129	-.731	-.529
3	(Constant)	-50.271	6.841		-7.349	.000			
	msat	.382	.015	.943	25.077	.000	.690	.876	.814
	vsat	-.089	.006	-.535	-13.828	.000	-.129	-.707	-.449
	gender	-7.590	1.187	-.216	-6.395	.000	-.368	-.420	-.207
4	(Constant)	-47.871	6.839		-7.000	.000			
	msat	.357	.019	.880	19.200	.000	.690	.812	.616
	vsat	-.096	.007	-.581	-13.509	.000	-.129	-.700	-.433
	gender	-7.732	1.175	-.220	-6.582	.000	-.368	-.431	-.211
	iq	.145	.062	.121	2.340	.020	.344	.167	.075

a. Dependent Variable: physics

(Continued)

Figure 5.17 (Continued)

Excluded Variables[e]

Model		Beta In	t	Sig.	Partial Correlation	Collinearity Statistics Tolerance
1	gender	-.349[a]	-7.613	.000	-.482	.999
	iq	-.280[a]	-3.964	.000	-.275	.505
	ei	-.423[a]	-9.952	.000	-.583	.998
	vsat	-.604[a]	-14.832	.000	-.731	.766
2	gender	-.216[b]	-6.395	.000	-.420	.919
	iq	.103[b]	1.812	.072	.130	.387
	ei	-.161[b]	-3.722	.000	-.260	.633
3	iq	.121[c]	2.340	.020	.167	.386
	ei	-.066[c]	-1.470	.143	-.106	.527
4	ei	-.022[d]	-.443	.658	-.032	.412

a. Predictors in the Model: (Constant), msat

b. Predictors in the Model: (Constant), msat, vsat

c. Predictors in the Model: (Constant), msat, vsat, gender

d. Predictors in the Model: (Constant), msat, vsat, gender, iq

e. Dependent Variable: physics

Note: Variables are in data-driven order of entry.

Table 5.5 Results of Statistical (Method = Forward) Multiple Regression to Predict Physics Score (Y) From IQ, EI, VSAT Score, MSAT Score, and Gender

	Physics	MSAT	VSAT	Gender	IQ		b^a	β	sr^2_{inc}
MSAT score	.69						.36***	.88	.476
VSAT score	−.13	.48					−.09***	−.58	.280
Gender	−.37	−.03	.23				−7.73***	−.22	.043
IQ	.34	.70	.64	.11			.14*	.12	.006
EI (not entered)	−.39	.04	.55	.48	.04		—	—	—
						Intercept = −47.87***			
Mean[b]	79.6	493.7	535.2	—	100.1				
SD	17.6	43.4	105.9	—	14.7				
									$R^2 = .805^c$
									$R^2_{adj} = .800$
									$R^a = .897$***

Note: Table format adapted from Tabachnick and Fidell (2018).

a. The significance values given in SPSS output for all individual *b* coefficients were all <.001; however, forward regression leads to an inflated risk for Type I error, and therefore these *p* values may underestimate the true risk for Type I error in this situation. The significance of the multiple R for the final model was assessed using critical values that are appropriate for forward regression from Wilkinson and Dallal (1981) rather than the *p* values in the SPSS output.

b. Because gender was a dummy variable (coded 1 = male, 2 = female), mean and standard deviation are not reported. The sample included $n = 100$ men and $n = 100$ women.

c. In general, a multiple R would be smaller for a subset of four predictors than for five predictors. In this example, after rounding, the multiple R for four predictors was the same as the multiple R for five predictors, in Table 5.4, in the first three decimal places.

*$p < .05$. ***$p < .001$.

The overall regression, including four of the five candidate predictors, was statistically significant, $R = .90$, $R^2 = .81$, adjusted $R^2 = .80$, $F(4, 190) = 195.57$, $p < .001$. Physics scores could be predicted quite well from this set of four variables, with approximately 80% of the variance in physics scores accounted for by the regression. (Note that the F test in the SPSS output does not accurately reflect the true risk for Type I error in this situation. The critical values of R^2 from the Wilkinson and Dallal [1981] table in Appendix 5B provide a more conservative test of significance for the overall regression. For $k = 5$ candidate predictor variables, $df_{\text{residual}} = N - k - 1$ = between 150 and 200, F-to-enter set at $F = 3.00$, and $\alpha = .01$. The critical value of R^2 from the second page of the Wilkinson and Dallal table in Appendix 5B was R^2 between .05 and .04; thus, using the more conservative Wilkinson and Dallal test, this overall multiple regression would still be judged statistically significant.)

To assess the statistical significance of the contributions of individual predictors, the F ratio for R^2 increment was examined for each variable in the step when it first entered the analysis. In Step 1, MSAT score was entered; it produced an R^2 increment of .473, $F(1, 193) = 175.47$. In Step 2, VSAT score was entered; it produced an R^2 increment of .280, $F(1, 192) = 219.99$, $p < .001$. In Step 3, gender was entered; it produced an R^2 increment of .043, $F(1, 191) = 40.90$, $p < .001$. In Step 4, IQ was entered; it produced an R^2 increment of .006, $F(1, 190) = 5.48$, $p = .020$. (EI was not entered because its F ratio was below the criterion set for F-to-enter using forward regression.) Except for VSAT score, which was negatively related to physics score, the slopes of all predictors had signs that were in the predicted direction.

Overall, physics scores were highly predictable from this set of predictors; the strongest unique predictive contribution was from MSAT score (when this variable was entered in the first step). All four predictors significantly increased the R^2 in the step when they first entered.

5.16 SUMMARY

When more than two predictor variables are included in a regression, the basic logic remains similar to the logic in regression with two predictors: The slope and proportion of variance associated with each predictor variable are assessed controlling for other predictor variables. With k predictors, it becomes possible to assess several competing causal variables, along with nuisance or confounded variables; it also becomes possible to include additional terms to represent interactions.

There are three major methods of variable entry: standard, sequential, and statistical. In most applications, standard or simultaneous entry (all predictor variables entered in a single step) provides the simplest and most conservative approach for assessment of the contributions of individual predictors. When there is temporal priority among predictor variables (X_1 is measured earlier in time than X_2, for example), or when there is a strong theoretical rationale for order of entry, it can be useful to set up a regression as a sequential or hierarchical analysis in which the order of entry of predictor variables is determined by the data analyst. The statistical or data-driven methods of entry can be used in situations where the only research goal is to identify the smallest possible set of predictor variables that will generate the largest possible R^2 value, but the statistical approach to regression is generally not a good method of testing theoretical hypotheses about the importance of predictor variables; in the series of equations produced by data-driven or statistical regression, variables enter because they are strongly predictive of Y in the sample, but the variables that enter the regression may not have any meaningful relationship to the Y outcome variable or with other predictors. In addition, they may not provide accurate predictions for scores on Y outcome variables in different samples.

APPENDIX 5A

Use of Matrix Algebra to Estimate Regression Coefficients for Multiple Predictors

A matrix is a table of numbers or variables. An entire matrix of scores on X predictor variables is usually denoted by a boldface symbol (**X**), while individual values or elements are denoted by italic characters with subscripts to indicate which row and which column they come from. This discussion is limited to matrices that are two-dimensional; that is, they correspond to a set of rows and a set of columns. The matrix that is usually of interest in statistics is the matrix of data values; this corresponds to the SPSS worksheet. Each row of the matrix contains scores that belong to one participant or case. Each column of the matrix contains scores that correspond to values of one variable. The element X_{ij} is the score in row i and column j; it is the score for Participant i on the variable X_j. Suppose that N is the number of participants, and k is the number of variables. The matrix that contains data for N participants on k variables is described as an N-by-k matrix. The example below shows the notation for a matrix with k equal to three variables and four participants. The first subscript tells you what row of the matrix the element is in; the second subscript tells you the column. Thus, X_{41} is the score for Participant 4 on the variable X_1:

$$\mathbf{X} = \begin{array}{c} \\ P_1 \\ P_2 \\ P_3 \\ P_4 \end{array} \begin{array}{c} \begin{array}{ccc} X_1 & X_2 & X_3 \end{array} \\ \begin{bmatrix} X_{11} & X_{12} & X_{13} \\ X_{21} & X_{22} & X_{23} \\ X_{31} & X_{32} & X_{33} \\ X_{41} & X_{42} & X_{43} \end{bmatrix} \end{array}.$$

Several other matrices are common in statistics. One is a correlation matrix usually denoted by **R**; it contains the correlations among all possible pairs of k variables. For example, on the basis of the scores above for the matrix **X**, we could compute correlations between all possible pairs of variables (X_1 and X_2, X_1 and X_3, and X_2 and X_3) and then summarize the entire set of correlations in an **R** matrix:

$$\mathbf{R} = \begin{array}{c} \\ X_1 \\ X_2 \\ X_3 \end{array} \begin{array}{c} \begin{array}{ccc} X_1 & X_2 & X_3 \end{array} \\ \begin{bmatrix} 1.0 & r_{21} & r_{13} \\ r_{21} & 1.0 & r_{23} \\ r_{31} & r_{32} & 1.0 \end{bmatrix} \end{array}.$$

Note several characteristics of this matrix. All the diagonal elements equal 1 (because the correlation of a variable with itself is, by definition, 1.0). The matrix is "symmetrical" because each element below the diagonal equals one corresponding element above the diagonal; for example, $r_{12} = r_{21}, r_{13} = r_{31}$, and $r_{23} = r_{32}$. When we write a correlation matrix, we do not need to fill in the entire table. It is sufficient to provide just the elements below the diagonal to provide complete information about correlations among the variables. In general, if we have k predictor variables, we have $k \times (k - 1)/2$ correlations. In this example, with $k = 3$, we have $3 \times (2 - 1)/2 = 3$ correlations in the matrix **R**. For a set of k variables, **R** is a $k \times k$ matrix.

Another frequently encountered matrix is called a variance/covariance matrix; the population parameter version is usually denoted by $\mathbf{\Sigma}$, while the sample estimate version is generally denoted by **S**. This matrix contains the variance for each of the k variables in the diagonal, and the covariances for all possible pairs of variables are the off-diagonal elements. Like **R**,

S is symmetrical; that is, the covariance between X_1 and X_2 is the same as the covariance between X_2 and X_1:

$$\mathbf{S} = \begin{array}{c} \\ X_1 \\ X_2 \\ X_3 \end{array} \begin{array}{c} \begin{array}{ccc} X_1 & X_2 & X_3 \end{array} \\ \begin{bmatrix} \text{Var}(X_1) & \text{Cov}(X_1, X_2) & \text{Cov}(X_1, X_3) \\ \text{Cov}(X_2, X_1) & \text{Var}(X_2) & \text{Cov}(X_2, X_3) \\ \text{Cov}(X_3, X_1) & \text{Cov}(X_3, X_2) & \text{Var}(X_3) \end{bmatrix} \end{array}.$$

Recall that a correlation is essentially a standardized covariance (see Chapter 10 in Volume I [Warner, 2020]), that is,

$$r_{XY} = \text{Cov}(X, Y)/(SD_X \times SD_Y).$$

Thus, if we have **S**, it can easily be converted to an **R** matrix.

One additional matrix that is useful is a matrix that contains the sum of squares (*SS*) and the sum of cross products (SCP) for all the predictor variables:

$$SS(X) = \Sigma\,(X - M_X)^2,$$

$$\text{SCP}(X, Y) = \Sigma(X - M_X) \times (Y - M_Y).$$

The *SS* for a variable is equivalent to the SCP of that variable with itself. Thus, the **sum of cross products (SCP) matrix** for a set of variables X_1, X_2, and X_3 would have elements as follows:

$$\mathbf{SCP} = \begin{array}{c} \\ X_1 \\ X_2 \\ X_3 \end{array} \begin{array}{c} \begin{array}{ccc} X_1 & X_2 & X_3 \end{array} \\ \begin{bmatrix} SS(X_1) & \text{SCP}(X_1, X_2) & \text{SCP}(X_1, X_3) \\ \text{SCP}(X_2, X_1) & SS(X_2) & \text{SCP}(X_2, X_3) \\ \text{SCP}(X_3, X_1) & \text{SCP}(X_3, X_2) & SS(X_3) \end{bmatrix} \end{array}.$$

The covariance between two variables is just the SCP for those variables divided by *N*:

$$\text{Cov}(X, Y) = \text{SCP}(X, Y)/N.$$

Thus, we can easily convert an **SCP** matrix into an **S** matrix by dividing every element of the SCP matrix by *N* or *df*.

Another useful matrix is called the identity matrix, and it is denoted by **I**. The diagonal elements of an identity matrix are all ones, and the off-diagonal elements are all zeros, as in the following example of a 3 × 3 identity matrix:

$$\mathbf{I} = \begin{bmatrix} 1 & 0 & 0 \\ 0 & 1 & 0 \\ 0 & 0 & 1 \end{bmatrix}.$$

There are several special cases of matrix form. For example, a row vector is a matrix that has just one row and any number of columns. A column vector is a matrix that has just one column and any number of rows. A commonly encountered column vector is the set of scores on *Y*, the dependent variable in a regression. The score in each row of this vector corresponds to the Y score for one participant. If $N = 4$, then there will be four scores; in general, the column vector **Y** is a matrix with N rows and one column:

$$\mathbf{Y} = \begin{bmatrix} Y_1 \\ Y_2 \\ Y_3 \\ Y_4 \end{bmatrix}.$$

A matrix of size 1 × 1 (i.e., just one row and one column) is called a scalar; it corresponds to a single number (such as N).

5.A.1 Matrix Addition and Subtraction

It is possible to do arithmetic operations with matrices. For example, to add matrix **A** to matrix **B**, you add the corresponding elements (the elements in **A** that have the same combination of subscripts as the corresponding elements in **B**). Two matrices can be added (or subtracted) only if they are of the same size (i.e., they have the same number of rows and the same number of columns).

Here are three small matrices that will be used for practice in matrix operations:

$$\mathbf{A} = \begin{bmatrix} 5 & 3 \\ 2 & 1 \end{bmatrix}, \quad \mathbf{B} = \begin{bmatrix} 2 & 1 \\ 5 & 8 \end{bmatrix}, \quad \mathbf{C} = \begin{bmatrix} 4 & 2 & 6 \\ 7 & 3 & 9 \end{bmatrix}.$$

To add **B** to **A**, you simply add the corresponding elements:

$$\mathbf{A} + \mathbf{B} = \mathbf{D} = \begin{bmatrix} (5+2) & (3+1) \\ (2+5) & (1+8) \end{bmatrix} = \begin{bmatrix} 7 & 4 \\ 7 & 9 \end{bmatrix}.$$

To subtract matrix **B** from **A** (**A** – **B** = **E**), you need to subtract the corresponding elements, as follows:

$$\mathbf{A} + \mathbf{B} = \mathbf{E} = \begin{bmatrix} (5-2) & (3-1) \\ (2-5) & (1-8) \end{bmatrix} = \begin{bmatrix} 3 & 2 \\ -3 & -7 \end{bmatrix}.$$

It is not possible to add **A** to **C**, or subtract **C** from **A**, because the **A** and **C** matrices do not have the same dimensions; **C** has three columns, while **A** has only two columns.

5.A.2 Matrix Multiplication

Matrix multiplication is somewhat more complex than addition because it involves both multiplying and summing elements. It is only possible to multiply a **Q** matrix by a **W** matrix if the number of columns of **Q** equals the number of rows in **W**. In general, if you multiply an $m \times n$ matrix by a $q \times w$ matrix, n must equal q, and the product matrix will be of dimensions $m \times w$. Note also that the product **A** × **B** is not generally equal to the product **B** × **A**. The order in which the terms are listed makes a difference in how the elements are combined.

The method of matrix multiplication I learned from David Kenny makes the process easy. To multiply **A** × **B**, set up the computations as follows: Write the first matrix, **A**, on the left; write the second matrix, **B**, at the top of the page; put the product matrix (which we will call **D**) between them. Notice that we are multiplying a 2 × 2 matrix by another 2 × 2 matrix. The multiplication is possible because the number of columns in the first matrix (**A**) equals the number of rows in the second matrix (**B**). When we multiply an $m \times n$ matrix by a $q \times w$ matrix, the inner dimensions (n and q) must be equal, and these terms disappear; the product matrix **D** will be of size $m \times w$. Initially, the elements of the product matrix **D** are unknown (d_{11}, d_{12}, etc.):

$$\mathbf{B} = \begin{bmatrix} 2 & 1 \\ 5 & 8 \end{bmatrix}$$

$$\mathbf{A} = \begin{bmatrix} 5 & 3 \\ 2 & 1 \end{bmatrix} \begin{bmatrix} d_{11} & d_{12} \\ d_{21} & d_{22} \end{bmatrix}.$$

To find each element of **D**, you take the vector product of the row immediately to the left and the column immediately above. That is, you cross-multiply corresponding elements for each row combined with each column and then sum these products. For the problem above, the four elements of the **D** matrix are found as follows:

$$d_{11} = a_{11} \times b_{11} + a_{12} \times b_{21}$$

$$(5 \times 2) + (3 \times 5) = 25$$

$$d_{12} = a_{11} \times b_{12} + a_{12} \times b_{22}$$

$$(5 \times 1) + (3 \times 8) = 29$$

$$d_{21} = a_{21} \times b_{11} + a_{22} \times b_{21}$$

$$(2 \times 2) + (1 \times 5) = 9$$

$$d_{22} = a_{21} \times b_{12} + a_{22} \times b_{22}$$

$$(2 \times 1) + (1 \times 8) = 10.$$

Note that matrix multiplication is not commutative: **A** × **B** does not usually yield the same result as **B** × **A**.

To multiply two matrices, the number of columns in the first matrix must equal the number of rows in the second matrix; if this is not the case, then you cannot match the elements up one to one to form the vector product. In the sample matrices given above, matrix **A** was 2 × 2 and **C** was 2 × 3. Thus, we can multiply **A** × **C** and obtain a 2 × 3 matrix as the product; however, we cannot multiply **C** × **A**.

When you multiply any matrix by the identity matrix **I**, it is the matrix algebra equivalent of multiplication by 1. That is, **A** × **I** = **A**. As an example,

$$\mathbf{I} = \begin{bmatrix} 1 & 0 \\ 0 & 1 \end{bmatrix}$$

$$\mathbf{A} = \begin{bmatrix} 5 & 3 \\ 2 & 1 \end{bmatrix} \begin{bmatrix} (5*1)+(3*0)=5 & (5*0)+(1*3)=3 \\ (2*1)+(1*0)=2 & (2*0)+(1*1)=1 \end{bmatrix}$$

5.A.3 Matrix Inverse

This is the matrix operation that is equivalent to division. In general, dividing by a constant c is equivalent to multiplying by $1/c$ or c^{-1}. Both $1/c$ and c^{-1} are notations for the inverse of c. The matrix equivalent of 1 is the identity matrix, which is a square matrix with 1s in the diagonal and all other elements 0.

Suppose that we need to find the inverse of matrix **F**, and the elements of this **F** matrix are as follows:

$$\mathbf{F} = \begin{bmatrix} 2 & 6 \\ 4 & 1 \end{bmatrix}.$$

The inverse of **F**, denoted by $\mathbf{F}^{-1}$, is the matrix that yields **I** when we form the product $\mathbf{F} \times \mathbf{F}^{-1}$. Thus, one way to find the elements of the inverse matrix $\mathbf{F}^{-1}$ is to set up the following multiplication problem and solve for the elements of $\mathbf{F}^{-1}$:

$$\mathbf{F}^{-1} = \begin{bmatrix} f_{11} & f_{12} \\ f_{21} & f_{22} \end{bmatrix}$$

$$\mathbf{F} = \begin{bmatrix} 2 & 6 \\ 1 & 4 \end{bmatrix} \begin{bmatrix} 1 & 0 \\ 0 & 1 \end{bmatrix}.$$

When we carry out this multiplication, we find the following set of four equations in four unknowns:

$$2f_{11} + 6f_{21} = 1, 2f_{12} + 6f_{22} = 0,$$

$$1f_{11} + 4f_{21} = 0, 1f_{12} + 4f_{22} = 1.$$

We can solve for the elements of **F** by substitution:

$$\text{From } 1f_{11} + 4f_{21} = 0, \text{ we get } f_{11} = -4f_{21}.$$

Substituting this for f_{11} in the first equation, we get

$$2f_{11} + 6f_{21} = 1,$$

$$2(-4f_{21}) + 6f_{21} = 1,$$

$$-2f_{21} = 1,$$

$$f_{21} = -1/2.$$

Substituting this value for f_{21} back into the equation we had for f_{11} in terms of f_{21} yields

$$f_{11} = -4f_{21} = -4(-1/2) = +2.$$

Similarly, we can solve the other pair of equations:

$$2f_{12} + 6f_{22} = 0$$

can be rearranged to give

$$6f_{22} = -2f_{12},$$

$$f_{22} = -1/3 \times f_{12}.$$

Substituting this value for f_{22} (in terms of f_{12}) into the last equation, we have

$$1f_{12} + 4f_{22} = 1,$$

$$1f_{12} + 4(-1/3 \times f_{12}) = 1,$$

$$-1/3f_{12} = 1,$$

$$f_{12} = -3.$$

We can now find f_{22} by using this known value for f_{12}:

$$f_{22} = -1/3f_{12} = -1/3 \times (-3) = +1.$$

The inverse matrix $\mathbf{F}^{-1}$ is thus

$$\begin{bmatrix} 2 & -3 \\ -\frac{1}{2} & 1 \end{bmatrix}.$$

To check, we multiply $\mathbf{FF}^{-1}$ to make sure that the product is **I**:

$$\begin{bmatrix} 2 & -3 \\ -1/2 & 1 \end{bmatrix}.$$

$$\begin{bmatrix} 2 & 6 \\ 1 & 4 \end{bmatrix} \begin{bmatrix} 4-3=1 & -6+6=0 \\ 2-2=0 & -3+4=1 \end{bmatrix}.$$

Computational shortcuts and algorithms can be used to compute inverses for larger matrices. The method used here should make it clear that the inverse of **F**, $\mathbf{F}^{-1}$, is the matrix for which the $\mathbf{F} \times \mathbf{F}^{-1}$ product equals the identity matrix **I**. Returning to our original definition of the inverse of c as $1/c$, it should be clear that just as multiplying by $(1/c)$ is equivalent to dividing by c, multiplying a matrix by $\mathbf{F}^{-1}$ is essentially equivalent to "dividing" by the **F** matrix.

5.A.4 Matrix Transpose

The transpose operation involves interchanging the rows and columns of a matrix. Consider the **C** matrix given earlier:

$$\mathbf{C} = \begin{bmatrix} 4 & 2 & 6 \\ 7 & 3 & 9 \end{bmatrix}.$$

The transpose of **C**, denoted by **C′**, is found by turning column 1 into row 1, column 2 into row 2, and so forth. Thus,

$$\mathbf{C}' = \begin{bmatrix} 4 & 7 \\ 2 & 3 \\ 6 & 9 \end{bmatrix}.$$

If **C** is a 2 × 3 matrix, then **C′** will be a 3 × 2 matrix.

The transpose operation is useful because we sometimes need to "square" a matrix, that is, multiply a matrix by itself. For example, the **X** matrix is an $N \times k$ matrix that contains scores for N participants on k variables. If we need to obtain sums of squared elements, we cannot multiply $\mathbf{X} \times \mathbf{X}$ (the number of columns in the first matrix must equal the number of rows in the second matrix). However, we can multiply **X′X**; this will be a $k \times N$ matrix multiplied by an $N \times k$ matrix, and it will yield a $k \times k$ matrix in which the terms are sums of squared scores and sums of cross products of scores; these are the building blocks needed to obtain the SCP or variance/covariance matrix (**S**) for the k variables or the correlation matrix **R**.

5.A.5 Determinant

The **determinant of a matrix**, such as **A**, is denoted by |**A**|. For a 2 × 2 matrix, the determinant is found by subtracting the product of the elements on the minor diagonal (from upper right to lower left corner) from the product of the elements on the major diagonal (from upper left to lower right):

$$\left| \begin{bmatrix} a & b \\ c & d \end{bmatrix} \right| = ad - bc.$$

The computation of determinants for larger matrices is more complex but essentially involves similar operations, that is, forming products of elements along the major diagonals (upper left to lower right) and summing these products, as well as forming products of elements along minor diagonals (from lower left to upper right) and subtracting these products.

The determinant has a useful property: It tells us something about the amount of *nonshared* or nonredundant variance among the rows and/or columns of a matrix. If the determinant is 0, then at least one row is perfectly predictable from some linear combination of one or more other rows in the matrix. This means that the matrix is singular: One predictor variable in the data matrix is perfectly predictable from other predictor variables.

For example, consider the correlation matrix **R** for X_1 and X_2:

$$\mathbf{R} = \begin{bmatrix} 1 & r_{12} \\ r_{12} & 1 \end{bmatrix}.$$

The determinant of $\mathbf{R} = (1)(1) - (r_{12} \times r_{12}) = 1 - r^2_{12}$. Here, $1 - r^2_{12}$ is the proportion of variance in X_1 that is not shared with X_2 and vice versa. If $r_{12} = 1$, then the determinant of **R** will be 0; this would tell us that X_1 is perfectly predictable from X_2. When you have an **R** matrix for k variables, the determinant provides information about multicollinearities that are less obvious; for example, if $X_3 = X_1 + X_2$, the determinant of **R** for a matrix of correlations among X_1, X_2, and X_3 will be 0.

When a matrix has a determinant of 0, you *cannot* calculate an inverse for the matrix (because computation of the inverse, in most algorithms, involves dividing by the determinant).

Here is an example of a matrix **G**; let column 1 = X_1 and column 2 = X_2; note that $X_2 = 3 \times X_1$; that is, X_2 is perfectly predictable from X_1:

$$\mathbf{G} = \begin{bmatrix} 2 & 6 \\ 4 & 12 \end{bmatrix}.$$

The determinant of $\mathbf{G} = (2 \times 12) - (4 \times 6) = 24 - 24 = 0$.

Thus, it would not be possible to compute an inverse for this **G** matrix. When we do a multiple regression, the program computes a determinant for the matrices (such as **X′X** and **R**) that it calculates from the data. If a determinant of exactly 0 is found, this means that there is a perfect correlation between two predictors, or that one predictor variable is perfectly predictable from a weighted linear combination of other predictor variables. When this happens, a regression analysis cannot be performed. A zero determinant is reported using several different kinds of error message, such as "determinant = 0," "singular matrix," or "matrix not of full rank."

Before you attempt to compute an inverse for a matrix, you should ask two questions. First, is the matrix square (i.e., number of rows equal to number of columns)? You can only compute an inverse for a square matrix. Second, does the matrix have a determinant of 0? If the determinant equals 0, you cannot calculate an inverse.

In running multiple regression programs, information about the determinant of the **X** and **R** matrices is helpful in the assessment of multicollinearity, that is, the strength of correlation among predictors. Perfect multicollinearity means that at least one *X* variable can be predicted perfectly from one or more of the other *X* variables. A determinant near 0 indicates strong, but not perfect, multicollinearity; a near-zero determinant suggests that predictors are highly correlated. For reasons discussed in Chapter 4 (on bivariate regression), it is better to avoid situations in which the predictors are very highly correlated with one another. When there are correlations in excess of .9 among predictors, the *X* variables "compete" to explain the same variance, and sometimes no single variable is significant. Furthermore, as correlations among predictors increase, the width of the confidence intervals around estimates of *b* slope coefficients increases.

These matrix operations can be applied to the problem of finding *b* coefficient estimates in regression with *k* predictor variables.

Recall (from Chapter 4) that the normal equations for a regression to predict z_Y from z_{X1} and z_{X2} were as follows:

$$r_{Y1} = \beta_1 + r_{12}\beta_2, \quad (5.17a)$$

$$r_{Y2} = r_{12}\beta_1 + \beta_2. \quad (5.17b)$$

These equations were obtained by applying the tracing rule to a path diagram that represented z_{X1} and z_{X2} as correlated predictors of z_Y; these equations show how the observed correlation between each predictor variable and the *Y* outcome variable could be reconstructed from the direct and indirect paths that lead from each predictor variable to *Y*.

The next step is to rewrite these two equations using matrix notation. Once we have set up these equations in matrix algebra form, the computations can be represented using simpler notation, and the computational procedures can easily be generalized to situations with any number of predictors. The matrix notation that will be used is as follows:

$\mathbf{R}_{iY}$ is a column vector containing the correlation of each *X* predictor variable with *Y*; for the case with two independent variables, the elements of $\mathbf{R}_{iY}$ are

$$\begin{bmatrix} r_{Y1} \\ r_{Y2} \end{bmatrix}.$$

In general, $\mathbf{R}_{iY}$ is a column vector with k elements, each of which corresponds to the correlation of one X predictor variable with Y.

$\boldsymbol{\beta}$ is a column vector containing the beta coefficient, or standardized slope coefficient, for each of the predictors. For two predictors, the elements of $\boldsymbol{\beta}$ are as follows:

$$\begin{bmatrix} \beta_1 \\ \beta_2 \end{bmatrix}.$$

In general, the beta vector has k elements; each one is the beta coefficient for one of the z_{Xi} predictor variables.

$\mathbf{R}_{ii}$ is a matrix of correlations among all possible pairs of the predictors:

$$\mathbf{R}_{ii} = \begin{bmatrix} 1 & r_{12} \\ r_{12} & 1 \end{bmatrix}.$$

In the case of two predictors, $\mathbf{R}_{ii}$ is a 2×2 matrix; in the more general case of k predictors, this is a $k \times k$ matrix that includes the correlations among all possible pairs of predictor variables. The normal equations above (Equations 5.17a and 5.17b) can be written more compactly in matrix notation as

$$\mathbf{R}_{iY} = \mathbf{R}_{ii} \times \boldsymbol{\beta}.$$

To verify that the matrix algebra equation on the preceding line is equivalent to the set of two normal equations given earlier, multiply the $\mathbf{R}$ matrix by the beta vector:

$$\begin{bmatrix} \beta_1 \\ \beta_2 \end{bmatrix}.$$

$$\begin{bmatrix} 1 & r_{12} \\ r_{12} & 1 \end{bmatrix} \begin{bmatrix} \beta_1 + r_{12}\beta_2 \\ r_{12}\beta_1 + \beta_2 \end{bmatrix}.$$

When we had just two predictor variables, it was relatively easy to solve this set of two equations in two unknowns by hand using substitution methods (see Chapter 4). However, we can write a solution for this problem using matrix algebra operations; this matrix algebra version provides a solution for the more general case with k predictor variables.

To solve this equation—that is, to solve for the vector of beta coefficients in terms of the correlations among predictors and the correlations between predictors and Y—we just multiply both sides of this matrix equation by the inverse of $\mathbf{R}_{ii}$. Because $\mathbf{R}_{ii}\mathbf{R}_{ii}^{-1} = \mathbf{I}$, the identity matrix, this term disappears, leaving us with the following equation:

$$\boldsymbol{\beta} = \mathbf{R}^{-1}{}_{ii}\mathbf{R}_{iY}.$$

This matrix equation thus provides a way to calculate the set of beta coefficients from the correlations; when k, the number of predictor variables, equals 2, this is equivalent to the operations we went through to get the betas from the r's when we solved the normal equations by hand using substitution methods. This matrix notation generalizes to matrices of any size, that is, situations involving any number of predictors. It is not difficult to find the beta coefficients by hand when you only have two predictors, but the computations become cumbersome as the number of predictor variables increases. Matrix algebra gives us a convenient and compact notation that works no matter how many predictor variables there are.

Thus, no matter how many predictors we have in a multiple regression, the betas are found by multiplying the column vector of correlations between predictors and Y by the inverse of the correlation matrix among the predictors. Computer programs use algorithms for matrix inverse that are much more computationally efficient than the method of computing matrix inverse shown earlier in this appendix.

Note that if the determinant of $\mathbf{R}_{ii}$ is 0—that is, if any predictor is perfectly predictable from one or more other predictors—then we cannot do this computation (it would be equivalent to dividing by 0).

5.A.6 Using the Raw-Score Data Matrices for *X* and *Y* to Calculate *b* Coefficients

Another matrix representation of multiple regression involves calculating the b coefficients directly from the raw scores. Let $\mathbf{X}$ be the independent variable data matrix (each row contains one subject's raw scores, and each column contains scores on one of the independent variables). We will assume that prior to all the other computations in this section, each X variable and the Y variable are converted into deviations from their means. This X matrix is usually augmented by adding a column of ones; this column of ones will provide the information needed to estimate the intercept term, a. Let $\mathbf{Y}$ be the vector of raw scores of each subject on the dependent variable. To keep this very simple, let's just use two predictors and four subjects, but the method applies to any number of variables and subjects. The matrix computation that gives us the vector of b coefficients is

$$\mathbf{B} = (\mathbf{X'X})^{-1}\mathbf{X'Y},$$

where the $\mathbf{B}$ vector corresponds to the list of raw-score regression coefficients, including b_0, the intercept:

$$\begin{bmatrix} b_0 \\ b_1 \\ b_2 \\ \cdot \\ \cdot \\ \cdot \\ b_k \end{bmatrix}.$$

$\mathbf{B} = \mathbf{X'X}^{-1}\mathbf{X'Y}$ includes terms that are analogous to those in the earlier standard score equation, $\boldsymbol{\beta} = \mathbf{R}^{-1}_{ii}\mathbf{R}_{ii}$. $\mathbf{X'X}$, like $\mathbf{R}_{ii}$, contains information about covariances (or correlations) among the predictors. $\mathbf{X'Y}$, like $\mathbf{R}_{iY}$, contains information about the covariation (or correlation) between each predictor and Y. Each x_{ij} element is the deviation from the mean of the score of Subject i on variable j; X_i is the vector of scores of all subjects on variable i. Let's look at this product $\mathbf{X'X}$ (with an added column of ones in $\mathbf{X}$):

$$\begin{bmatrix} 1 & 1 & 1 & 1 \\ X_{11} & X_{21} & X_{31} & X_{41} \\ X_{12} & X_{22} & X_{32} & X_{42} \end{bmatrix} \begin{bmatrix} 1 & X_{11} & X_{12} \\ 1 & X_{21} & X_{22} \\ 1 & X_{31} & X_{32} \\ 1 & X_{41} & X_{42} \end{bmatrix} = \begin{bmatrix} N & \sum X_1 & \sum X_2 \\ \sum X_1 & X_{11}^2 + X_{21}^2 + X_{31}^2 + X_{41}^2 & X_{11}X_{12} + X_{21}X_{22} + X_{31}X_{32} + X_{41}X_{42} \\ \sum X_2 & X_{11}X_{12} + X_{21}X_{22} + X_{31}X_{32} + X_{41}X_{42} & X_{12}^2 + X_{22}^2 + X_{32}^2 + X_{42}^2 \end{bmatrix}.$$

The elements of this product matrix correspond to the following familiar terms:

$$\mathbf{X'X} = \begin{bmatrix} N & \sum X_1 & \sum X_2 \\ \sum X_1 & \sum X_1^2 & \sum X_1 X_2 \\ \sum X_2 & \sum X_1 X_2 & \sum X_2^2 \end{bmatrix}.$$

The diagonal elements of $\mathbf{X'X}$ contain the squared scores on each of the predictors or the basic information about the variance of each predictor. The off-diagonal elements of $\mathbf{X'X}$ contain the cross products of all possible pairs of predictors or the basic information about the covariance of each pair of predictors. The first row and column just contain the sums of the scores on each predictor, and the first element is N, the number of subjects. This is all the information that is needed to calculate the vector of raw-score regression coefficients, $\mathbf{B} = (b_0, b_1, b_2, \ldots, b_k)$. We can obtain the variance/covariance matrix $\mathbf{S}$ for this set of predictors by making a few minor changes in the way we compute $\mathbf{X'X}$. If we replace the scores on each X variable with deviations from the scores on the means of each variable, omit the added column of ones, form $\mathbf{X'X}$, and then divide each element of $\mathbf{X'X}$ by N, the number of scores, we obtain the variance/covariance matrix. The sample value of this matrix is usually denoted by $\mathbf{S}$; the corresponding population parameter matrix is $\mathbf{\Sigma}$:

$$\mathbf{S} = \begin{bmatrix} \text{Var}(X_1) & \text{Cov}(X_1, X_2) \\ \text{Cov}(X_1, X_2) & \text{Var}(X_2) \end{bmatrix}.$$

Similarly, $\mathbf{X'Y}$ involves cross-multiplying scores on the predictors with scores on the dependent variable (Y) to obtain a sum of cross products term for each predictor or X variable with Y:

$$\begin{bmatrix} y_1 \\ y_2 \\ y_3 \\ y_4 \end{bmatrix}$$

$$\begin{bmatrix} 1 & 1 & 1 & 1 \\ x_{11} & x_{21} & x_{31} & x_{41} \\ x_{12} & x_{22} & x_{32} & x_{42} \end{bmatrix} \begin{bmatrix} y_1 + y_2 + y_3 + y_4 \\ x_{11}y_1 + x_{21}y_2 + x_{31}y_3 + x_{41}y_4 \\ x_{12}y_1 + x_{22}y_2 + x_{32}y_3 + x_{42}y_4 \end{bmatrix}.$$

The elements of this product matrix correspond to

$$\mathbf{X'Y} = \begin{bmatrix} \sum Y \\ \sum X_1 Y \\ \sum X_2 Y \end{bmatrix}.$$

This $\mathbf{X'Y}$ product vector contains information that is related to the SCPs and could be used to calculate the SCP for each predictor variable with Y.

We can convert the variance/covariance matrix $\mathbf{S}$ to the correlation matrix $\mathbf{R}_{ii}$ in a few easy steps. Correlation is essentially a standardized covariance, that is, the X, Y covariance divided by SD_Y and SD_X. If we divide each row and each column through by the standard deviation for the corresponding variables, we can easily obtain $\mathbf{R}_{ii}$ from $\mathbf{S}$.

X′X and **X′Y** are the basic building blocks or "chunks" that are included in the computation of many multivariate statistics. Understanding the information that these matrices contain will help you to recognize what is going on as we look at additional multivariate techniques. Examination of the matrix algebra makes it clear that the values of β or *b* coefficients are influenced by the correlations among predictors as well as by the correlations between predictors and *Y*.

In practice, most computer programs find the *b* coefficients from **X′X** and **X′Y** and then derive other regression statistics from these, but it should be clear from this discussion that either the raw scores on the *X*'s and *Y* or the correlations among all the variables along with means and standard deviations for all variables are sufficient information to do a multiple regression analysis.

To summarize, computer programs typically calculate the raw-score regression coefficients, contained in the vector **B**, with elements $(b_0, b_1, b_2, \ldots, b_k)$ by performing the following matrix algebra operation: $\mathbf{B} = \mathbf{X'X}^{-1}\mathbf{X'Y}$. This computation cannot be performed if **X′X** has a zero determinant; a zero determinant indicates that at least one *X* variable is perfectly predictable from one or more other *X* variables. Once we have the **B** vector, we can write a raw score prediction equation as follows:

$$Y' = b_0 + b_1X_1 + b_2X_2 + \ldots + b_kX_k.$$

Each b_i coefficient has an associated standard error (SE_{bi}). This *SE* term may be used to set up a confidence interval for each b_i and to set up a *t* ratio to test the null hypothesis that each b_i slope equals 0. This is done using the same formulas reported earlier in Chapter 4.

The standardized slopes or beta coefficients can also be calculated for any number of predictors by doing the following matrix computation: $\boldsymbol{\beta} = \mathbf{R}^{-1}{}_{ii}\mathbf{R}_{iY}$. Once we have this beta vector, we can set up the equation to predict standard scores on *Y* from standardized or *z* scores on the predictors as follows:

$$z'_Y = \beta_1 z_{X1} + \beta_2 z_{X2} + \cdots + \beta_k z_{Xk}.$$

Once we have computed the β's, we can convert them to *b*'s (or vice versa).

It would be tedious to do the matrix algebra by hand (in particular, the computation of the inverse of **X′X** is time-consuming). You will generally obtain estimates of slope coefficients and other regression results from the computer program. Even though we will not do by-hand computations of coefficients from this point onward, it is still potentially useful to understand the matrix algebra formulas. First of all, when you understand the $(\mathbf{X'X})^{-1}\mathbf{X'Y}$ equation "as a sentence," you can see what information is taken into account in the computation of the *b* coefficients. In words, $(\mathbf{X'X})^{-1}\mathbf{X'Y}$ tells us to form all the sums of cross products between each pair of *X* predictor variables and *Y* (i.e., calculate **X′Y**) and then divide this by the matrix that contains information about all the cross products for all pairs of *X* predictor variables (i.e., *X′X*). The expression $(\mathbf{X'X})^{-1}\mathbf{X'Y}$ is the matrix algebra generalization of the computation for the single *b* slope coefficient in a regression of the form $Y' = b_0 + b_1X_1$. For a one-predictor regression equation, the estimate of the raw score slope $b = \text{SCP}/SS_X$, where SCP is the sum of cross products between *X* and *Y*, $\text{SCP} = \Sigma[(X - M_X) \times (Y - M_Y)]$, and SS_X is the sum of squares for *X*, $SS_X = \Sigma(X - M_X)^2$. The information contained in the matrix **X′X** is the multivariate generalization of SS_X. If we base all our computations on deviations of raw scores from the appropriate means, the *X′X* matrix includes the sum of squares for each individual X_i predictor and also the cross products for all pairs of *X* predictor variables; these cross products provide the information we need to adjust for correlation or redundancy among predictors. The information contained in the matrix **X′Y** is the multivariate generalization of SCP; these sums of cross products provide information that can be used to obtain a correlation between each X_i predictor and Y. Regression slope coefficients are obtained by dividing SCP by SS_X (or by premultiplying **X′Y** by the inverse of **X′X**, which is the matrix equivalent of dividing by **X′X**).

To compute b or β coefficients, we need to know about the covariance or correlation between each X predictor and Y, but we also need to take into account the covariances or correlations among all the X predictors. Thus, adding or dropping an X_i predictor can change the slope coefficient estimates for all the other X variables. The beta slope coefficient for a predictor can change (sometimes dramatically) as other predictors are added to (or dropped from) the regression equation.

Understanding the matrix algebra makes some of the error messages from computer programs intelligible. When SPSS or some other program reports "zero determinant," "singular matrix," or "matrix not of full rank," this tells you that there is a problem with the **X′X** matrix; specifically, it tells you that at least one X predictor variable is perfectly predictable from other X predictors. We could also say that there is perfect multicollinearity among predictors when the determinant of **X′X** is 0. For example, suppose that X_1 is VSAT score, X_2 is MSAT score, and X_3 is total SAT score (verbal score + math score). If you try to predict college grade point average from X_1, X_2, and X_3, the **X** data matrix will have a determinant of 0 (because "total SAT" X_3 is perfectly predictable from X_1 and X_2). To get rid of this problem, you would have to drop the X_3 variable from your set of predictors. Another situation in which perfect multicollinearity may arise involves the use of dummy variables as predictors (refer to Chapter 6). If you have k groups and you use k dummy variables, the score on the last dummy variable will be perfectly predictable from the scores on the first $k - 1$ dummy variables, and you will have a zero determinant for **X′X**. To get rid of this problem, you have to drop one dummy variable.

Another reason why it may be useful to understand the matrix algebra for multiple regression is that once you understand what information is contained in meaningful chunks of matrix algebra (such as **X′X**, which contains information about the variances and covariances of the X predictors), you will recognize these same terms again in the matrix algebra for other multivariate procedures.

APPENDIX 5B

Tables for Wilkinson and Dallal (1981) Test of Significance of Multiple R^2 in Forward Statistical Regression

Critical Values for Squared Multiple Correlation (R^2) in Forward Statistical Regression

$\alpha = .05$																	
		$N - k - 1$															
k	F	10	12	14	16	18	20	25	30	35	40	50	60	80	100	150	200
2	2	43	38	33	30	27	24	20	16	14	13	10	8	6	5	3	2
2	3	40	36	31	27	24	22	18	15	13	11	9	7	5	4	2	2
2	4	38	33	29	26	23	21	17	14	12	10	8	7	5	4	3	2
3	2	49	43	39	35	32	29	24	21	18	16	12	10	8	7	4	2
3	3	45	40	36	32	29	26	22	19	17	15	11	9	7	6	4	3
3	4	42	36	33	29	27	25	20	17	15	13	11	9	7	5	4	3
4	2	54	48	44	39	35	33	27	23	20	18	15	12	10	8	5	4

$\alpha = .05$																	
$N - k - 1$																	
k	***F***	**10**	**12**	**14**	**16**	**18**	**20**	**25**	**30**	**35**	**40**	**50**	**60**	**80**	**100**	**150**	**200**
4	3	49	43	39	36	33	30	25	22	19	17	14	11	8	7	5	4
4	4	45	39	35	32	29	27	22	19	17	15	12	10	8	6	5	3
5	2	58	52	47	43	39	36	31	26	23	21	17	14	11	9	6	5
5	3	52	46	42	38	35	32	27	24	21	19	16	13	9	8	5	4
5	4	46	41	38	35	52	29	24	21	18	16	13	11	9	7	5	4
6	2	60	54	50	46	41	39	33	29	25	23	19	16	12	10	7	5
6	3	54	48	44	40	37	34	29	25	22	20	17	14	10	8	6	5
6	4	48	43	39	36	33	30	26	23	20	17	14	12	9	7	5	4
7	2	61	56	51	48	44	41	35	30	27	24	20	17	13	11	7	5
7	3	59	50	46	42	39	36	31	26	23	21	18	15	11	9	7	5
7	4	50	45	41	38	35	32	27	24	21	18	15	13	10	8	6	4
8	2	62	58	53	49	46	43	37	31	28	26	21	18	14	11	8	6
8	3	57	52	47	43	40	37	32	28	24	22	19	16	12	10	7	5
8	4	51	46	42	39	36	33	28	25	22	19	16	14	11	9	7	5
9	2	63	59	54	51	47	44	38	33	30	27	22	19	15	12	9	6
9	3	58	53	49	44	41	38	33	29	25	23	20	16	12	10	7	6
9	4	52	46	43	40	37	34	29	25	23	20	17	14	11	10	7	6
10	2	64	60	55	52	49	46	39	34	31	28	23	20	16	13	10	7
10	3	59	54	50	45	42	39	34	30	26	24	20	17	13	11	8	6
10	4	52	47	44	41	38	35	30	26	24	21	18	15	12	10	8	6
12	2	66	62	57	54	51	48	42	37	33	30	25	22	17	14	10	8
12	3	60	55	52	47	44	41	36	31	28	25	22	19	14	12	9	7
12	4	53	48	45	41	39	36	31	27	25	22	19	16	13	11	9	7
14	2	68	64	60	56	53	50	44	39	35	32	27	24	18	15	11	8
14	3	61	57	53	49	46	43	37	32	29	27	23	20	15	13	10	8
14	4	43	49	46	42	40	37	32	29	26	23	20	17	13	11	9	7
16	2	69	65	61	58	55	53	46	41	37	34	29	25	20	17	12	9
16	3	61	58	54	50	47	44	38	34	31	28	24	21	17	14	11	8
16	4	53	50	46	43	40	38	33	30	27	24	21	18	14	12	10	8
18	2	70	67	63	60	57	55	49	44	40	36	31	27	21	18	13	9
18	3	62	59	55	51	49	46	40	35	32	30	26	23	18	15	12	9
18	4	54	50	46	44	41	38	34	31	28	25	22	19	15	13	11	8
20	2	72	68	64	62	59	56	50	46	42	38	33	28	22	19	14	10
20	3	62	60	56	52	50	47	42	37	34	31	27	24	19	16	12	9
20	4	54	50	46	44	41	37	35	32	29	26	23	20	16	14	11	8

Source: Reprinted with permission from Wilkinson and Dallal (1981).

Note: Decimals are omitted; *k* is the number of candidate predictors, *N* is sample size, and *F* is criterion *F*-to-enter.

Critical Values for Squared Multiple Correlation (R^2) in Forward Statistical Regression

$\alpha = .05$

		$N - k - 1$															
k	***F***	**10**	**12**	**14**	**16**	**18**	**20**	**25**	**30**	**35**	**40**	**50**	**60**	**80**	**100**	**150**	**200**
2	2	59	53	48	43	40	36	30	26	23	20	17	14	11	9	7	5
2	3	58	52	46	42	38	35	30	25	22	19	16	13	10	8	6	4
2	4	57	49	44	39	36	32	26	22	19	16	13	11	8	7	5	4
3	2	67	60	55	50	46	42	35	30	27	24	20	17	13	11	7	5
3	3	63	58	52	47	43	40	34	29	25	22	19	16	12	10	7	5
3	4	61	54	48	44	40	37	31	26	23	20	16	14	11	9	6	5
4	2	70	64	58	53	49	46	39	34	30	27	23	19	15	12	8	6
4	3	67	62	56	51	47	44	37	32	28	25	21	18	14	11	8	6
4	4	64	58	52	47	43	40	34	29	26	23	19	16	13	11	7	6
5	2	73	67	61	57	52	49	42	37	32	29	25	21	16	13	9	7
5	3	70	65	59	54	50	46	39	34	30	27	23	19	15	12	9	7
5	4	65	60	55	50	46	43	36	31	28	25	20	17	14	12	8	6
6	2	74	69	63	59	55	51	44	39	34	31	26	23	18	14	10	8
6	3	72	67	61	56	51	48	41	36	32	28	24	20	16	13	10	7
6	4	66	61	56	52	48	45	38	33	29	26	22	19	15	13	9	7
7	2	76	70	65	60	56	53	46	40	36	33	28	25	19	15	11	9
7	3	73	68	62	57	53	50	42	37	33	30	25	21	17	14	10	8
7	4	67	62	58	54	49	46	40	35	31	28	23	20	16	14	10	8
8	2	77	72	66	62	58	55	48	42	38	34	29	26	20	16	12	9
8	3	74	69	63	58	54	51	44	39	34	31	26	22	18	15	11	9
8	4	67	63	59	55	50	47	41	36	32	29	24	21	17	15	11	9
9	2	78	73	67	63	60	56	49	43	39	36	31	27	21	17	12	10
9	3	74	69	64	59	56	52	45	40	35	32	27	23	19	16	12	9
9	4	68	63	60	56	51	48	42	37	33	30	25	22	18	16	12	9
10	2	79	74	68	65	61	58	51	45	40	37	32	28	22	18	13	10
10	3	74	69	65	50	57	53	47	41	37	33	28	24	20	17	13	10
10	4	68	64	61	56	52	49	43	38	34	31	26	23	19	17	13	9
12	2	80	75	70	66	63	60	53	48	43	39	34	30	24	20	14	11
12	3	74	70	66	62	58	55	48	43	39	35	30	26	21	18	14	10
12	4	69	65	61	57	53	50	44	40	35	32	27	24	20	18	13	10
14	2	81	76	71	68	65	62	55	50	45	41	36	32	25	21	15	11
14	3	74	70	67	63	60	56	50	45	41	37	31	27	22	19	15	11
14	4	69	65	61	57	54	52	45	41	36	33	28	25	21	19	14	10
16	2	82	77	72	69	66	63	57	52	47	43	38	34	27	22	16	12
16	3	74	70	67	64	61	58	52	47	42	39	33	29	23	20	15	11

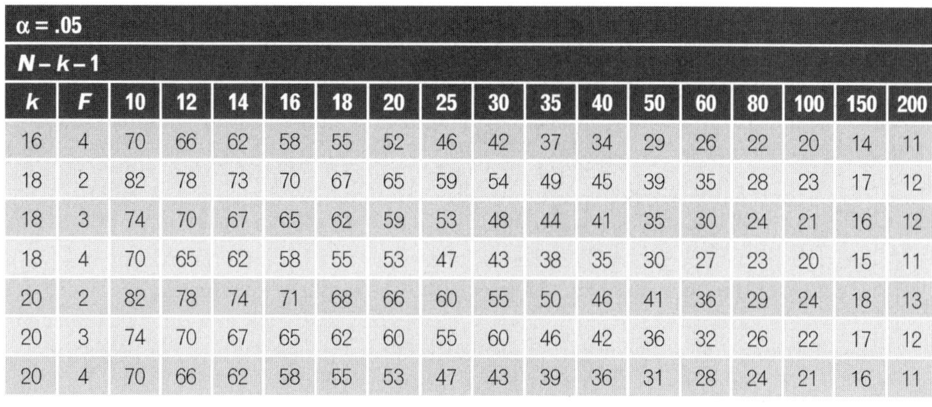

α = .05																	
N − k − 1																	
k	**F**	**10**	**12**	**14**	**16**	**18**	**20**	**25**	**30**	**35**	**40**	**50**	**60**	**80**	**100**	**150**	**200**
16	4	70	66	62	58	55	52	46	42	37	34	29	26	22	20	14	11
18	2	82	78	73	70	67	65	59	54	49	45	39	35	28	23	17	12
18	3	74	70	67	65	62	59	53	48	44	41	35	30	24	21	16	12
18	4	70	65	62	58	55	53	47	43	38	35	30	27	23	20	15	11
20	2	82	78	74	71	68	66	60	55	50	46	41	36	29	24	18	13
20	3	74	70	67	65	62	60	55	60	46	42	36	32	26	22	17	12
20	4	70	66	62	58	55	53	47	43	39	36	31	28	24	21	16	11

Source: Reprinted with permission from Wilkinson and Dallal (1981).

Note: Decimals are omitted; *k* is the number of candidate predictors, *N* is sample size, and *F* is criterion F-to-enter.

APPENDIX 5C

Confidence Interval for R^2

Information about confidence intervals for raw-score regression coefficients is not included in this chapter; that was presented in Volume I, Chapter 11 (Warner, 2020), and SPSS provides the 95% confidence intervals for values of regression coefficients if requested. An online calculator to obtain a confidence interval for R^2 (Figure 5.18) can be found at https://www.danielsoper.com/statcalc/calculator.aspx?id=28.

Figure 5.18 Online Calculator for Confidence Intervals for R^2

https://www.danielsoper.com/statcalc/calculator.aspx?id=28
R-square Confidence Interval Calculator
This calculator will compute the 99%, 95%, and 90% confidence intervals for an R² value (i.e., a squared multiple correlation), given the value of the R-square, the number of predictors in the model, and the total sample size.
Please enter the necessary parameter values, and then click 'Calculate'.
Number of predictors: 3
Observed R²: 0.25
Sample size: 30
Calculate!
Related Resources
x² Formulas
References
Related Calculators
Search

Source: Soper (2019).

COMPREHENSION QUESTIONS

1. Describe a hypothetical study for which multiple regression with more than two predictor variables would be an appropriate analysis. Your description should include one dependent variable and three or more predictors.
   a. For each variable, provide specific information about how the variable would be measured and whether it is quantitative, normally distributed, and so forth.
   b. Assuming that you are trying to detect a medium size R^2, what is the minimum desirable N for your study (on the basis of the guidelines in this chapter)?
   c. What regression method (e.g., standard, sequential, or statistical) would you prefer to use for your study, and why? (If sequential, indicate the order of entry or predictors that you would use and the rationale for the order of entry.)
   d. What pattern of results would you predict in the regression; that is, would you expect the overall multiple R to be significant? Which predictor(s) would you expect to make a significant unique contribution, and why?
2. What types of research situations often make use of multiple regression analysis with more than two predictors?
3. What is the general path diagram for multiple regression?
4. What are the different methods of entry for predictors in multiple regression?
5. What kind of reasoning would justify a decision to enter some variables earlier than others in a hierarchical regression analysis?
6. Suppose a researcher runs a "forward entry" statistical multiple regression with a group of 20 candidate predictor variables; the final model includes five predictors. Why are the p values shown in the SPSS output not a good indication of the true risk for Type I error in this situation? What correction can be used when you test the significance of the overall multiple R for a forward regression?
7. How can we obtain significance tests for sets or blocks of predictor variables?
8. What information do we need to assess linearity of associations and possible multicollinearity among predictors?
9. What information should be included in the report of a standard or simultaneous multiple regression (all predictors entered in one step)?
10. What information should be included in the report of a hierarchical or statistical multiple regression (i.e., a series of regressions with one variable or a group of variables entered in each step)?
11. What parts of the information that is reported for standard versus hierarchical regression are identical? What parts are different?
12. What is an R^2_{inc}? How is the R^2_{inc} for a single variable that enters in a particular step related to the sr^2_{inc} for that same single variable in the same step?
13. How can we identify disproportionately influential scores and/or multivariate outliers?
14. What is the correlation between Y and $(b_0 + b_1X_1 + b_2X_2 + \cdots + b_kX_k)$?
15. What is "tolerance?" What is the range of possible values for tolerance? Do we usually want the tolerance for a candidate predictor variable to be low or high?

SUGGESTED DATA ANALYSIS PROJECT FOR MULTIPLE REGRESSION

1. Select four independent variables and one dependent variable. They should all be quantitative.
2. Run a standard or simultaneous regression (i.e., all variables are entered in one step) with this set of variables.
3. Then run a hierarchical regression in which you arbitrarily specify an order of entry for the four variables.
4. Write up a "Results" section separately for each of these analyses using the model results sections and the tables in this chapter as a guide. Answer the following questions about your analyses:

   How does the equation in the last step of your hierarchical analysis (with all four variables entered) compare with your standard regression?

   Draw overlapping circle diagrams to illustrate how the variance is partitioned among the *X* predictor variables in each of these two analyses; indicate (by giving a numerical value) what percentage of variance is attributed (uniquely) to each independent variable. In other words, how does the variance partitioning in these two analyses differ? Which is the more conservative approach?

   Look at the equations for Steps 1 and 2 of your hierarchical analysis.

   a. Calculate the difference between the R^2 values for these two equations; show that this equals one of the squared part correlations (which one?) in your output.
   b. Evaluate the statistical significance of this change in R^2 between Steps 1 and 2. How does this F compare to the t statistic for the slope of the predictor variable that entered the model at Step 2?

   Consider the predictor variable that you entered in the first step of your hierarchical analysis. How does your evaluation of the variance due to this variable differ when you look at it in the hierarchical analysis compared with the way you look at it in the standard analysis? In which analysis does this variable look more "important," that is, appears to explain more variance? Why?

   Now consider the predictor variable that you entered in the last (fourth) step of your hierarchical analysis and compare your assessment of this variable in the hierarchical analysis (in terms of proportions of explained variance) with your assessment of this variable in the standard analysis.

   Look at the values of R and F for the overall model as they change from Steps 1 through 4 in the hierarchical analysis. How do R and F change in this case as additional variables are added in? In general, does R tend to increase or decrease as additional variables are added to a model? In general, under what conditions does F tend to increase or decrease as variables are added to a model?

   Suppose you had done a statistical (data-driven) regression (using the forward method of entry) with this set of four predictor variables. Which (if any) of the four predictor variables would have entered the equation and which one would have entered first? Why?

   If you had done a statistical regression (using a method of entry such as forward) rather than a hierarchical (user-determined order of entry), how would you change your significance testing procedures? Why?

NOTE

[1]The subscript wo stands for without; the first model predicts Y without (wo) the added m predictors, and the second model predicts Y with the added m predictors.

DIGITAL RESOURCES

Find **free study tools** to support your learning, including **eFlashcards, data sets, and web resources,** on the accompanying website at **edge.sagepub.com/warner3e.**

CHAPTER 6

DUMMY PREDICTOR VARIABLES IN MULTIPLE REGRESSION

6.1 WHAT DUMMY VARIABLES ARE AND WHEN THEY ARE USED

Previous examples of regression analysis have used scores on quantitative *X* variables to predict scores on a quantitative *Y* variable. However, it is possible to include group membership or categorical predictor variables as predictors in regression analysis. This can be done by creating dummy (dichotomous) predictor variables to represent information about group membership. A dummy or dichotomous predictor variable provides yes/no information for questions about group membership. For example, a simple dummy variable to represent sex corresponds to the following question: Is the participant female (0) or male (1)? Sex is an example of a two-group categorical variable that can be represented by a single dummy variable.

When we have more than two groups, we can use a set of dummy variables to provide information about group membership. For example, suppose that a study includes members of $k = 3$ political party groups. The categorical variable political party has the following scores: 1 = Democrat, 2 = Republican, and 3 = Independent. We might want to find out whether mean scores on a quantitative measure of political conservatism (*Y*) differ across these three groups. One way to answer this question is to perform a one-way analysis of variance (ANOVA) that compares mean conservatism (*Y*) across the three political party groups. In this chapter, we will see that we can also use regression analysis to evaluate how political party membership is related to scores on political conservatism.

However, we should not set up a regression to predict scores on conservatism from the *multiple-group* categorical variable political party, with party membership coded 1 = Democrat, 2 = Republican, and 3 = Independent. Multiple-group categorical variables usually do not work well as predictors in regression, because scores on a quantitative outcome variable, such as "conservatism," will not necessarily increase linearly with the score on the categorical variable that provides information about political party. The score values that represent political party membership may not be rank ordered in a way that reflects a **monotonic relationship** with changes in conservatism; as we move from Group 1 = Democrat to Group 2 = Republican, scores on conservatism may increase, but as we move from Group 2 = Republican to Group 3 = Independent, conservatism may decrease. Even if the scores that represent political party membership are rank ordered in a way that is monotonically associated with level of conservatism, the amount of change in conservatism between Groups 1 and 2 may not be equal to the amount of change in conservatism between Groups 2 and 3. In other words, scores on a multiple-group categorical predictor variable (such as political party coded 1 = Democrat, 2 = Republican, and 3 = Independent) are not necessarily *linearly* related to scores on quantitative variables.

If we want to use the categorical variable political party to predict scores on a quantitative variable such as conservatism, we need to represent the information about political party membership in a different way. Instead of using one categorical predictor variable with codes 1 = Democrat, 2 = Republican, and 3 = Independent, we can create two dummy or dichotomous predictor variables to represent information about political party membership, and we can then use these two dummy variables as predictors of conservatism scores in a regression. Political party membership can be assessed by creating dummy variables (denoted by D_1 and D_2) that correspond to two yes/no questions. In this example, the first dummy variable D_1, coded 1 = yes, 0 = no, corresponds to the following question: Is the participant a member of the Democratic Party? The second dummy variable D_2, also coded 1 = yes, 0 = no, corresponds to the following question: Is the participant a member of the Republican Party? We assume that group memberships for individuals are mutually exclusive and exhaustive; that is, each case belongs to only one of the three groups, and every case belongs to one of the three groups identified by the categorical variable. When these conditions are met, a third dummy variable is not needed to identify the members of the third group because, for Independents, the answers to the first two questions that correspond to the dummy variables D_1 and D_2 would be no. In general, when we have k groups or categories, a set of $k - 1$ dummy variables is sufficient to provide complete information about group membership. Once we have represented political party group membership by creating scores on two dummy variables, we can set up a regression to predict scores on conservatism (Y) from the scores on the two dummy predictor variables D_1 and D_2:

$$Y' = b_0 + b_1 D_1 + b_2 D_2. \tag{6.1}$$

In this chapter, we will see that the information about the association between group membership (represented by D_1 and D_2) and scores on the quantitative Y variable that can be obtained from the regression analysis in Equation 6.1 is equivalent to the information that can be obtained from a one-way ANOVA that compares means on Y across groups or categories. It is acceptable to use *dichotomous* predictor variables in regression and correlation analysis. This works because (as discussed in Volume I, Chapter 10 [Warner, 2020]) a dichotomous categorical variable has only two possible score values, and the only possible relationship between scores on a dichotomous predictor variable and a quantitative outcome variable is a linear one. That is, as you move from a score of 0 = female on sex to a score of 1 = male on sex, mean height or mean annual salary may increase, decrease, or stay the same; any change that can be observed across just two groups can be represented as linear. Similarly, if we represent political party membership using two dummy variables, each dummy variable represents a contrast between the means of two groups; for example, the D_1 dummy variable can represent the difference in mean conservatism between Democrats and Independents, and the D_2 dummy variable can represent the mean difference in conservatism between Republicans and Independents.

This chapter uses empirical examples to demonstrate that regression analyses that use dummy predictor variables (similar to Equation 6.1) provide information that is equivalent to the results of more familiar analyses for comparison of group means (such as ANOVA). There are several reasons why it is useful to consider dummy predictor variables as predictors in regression analysis. First, the use of dummy variables as predictors in regression provides a simple demonstration of the fundamental equivalence between ANOVA and multiple regression; ANOVA and regression are both special cases of a more general analysis called the general linear model (GLM). Second, researchers often want to include group membership variables (such as sex) along with other predictors in a multiple regression. Therefore, it is useful to examine examples of regression that include dummy variables along with quantitative predictors.

The computational procedures for regression remain the same when we include one or more dummy predictor variables. The most striking difference between dummy variables and quantitative variables is that the scores on dummy variables usually have small integer values (such as 1, 0, and −1). The use of small integers as codes simplifies the interpretation of the regression coefficients associated with dummy variables. When dummy variables are used as predictors in a multiple regression, the b raw-score slope coefficients provide information about differences between group means. The specific group means that are compared differ depending on the method of coding that is used for dummy variables, as explained in the following sections. Except for this difference in the interpretation of regression coefficients, regression analysis remains essentially the same when dummy predictor variables are included.

6.2 EMPIRICAL EXAMPLE

The hypothetical data for this example are provided by a study of predictors of annual salary in dollars for a group of N = 50 college faculty members; the complete data appear in the SPSS file salarysexcollege.sav. Predictor variables include the following: sex, coded 0 = female and 1 = male; years of job experience; college, coded 1 = liberal arts, 2 = sciences, and 3 = business; and an overall merit evaluation. Additional columns in the SPSS data worksheet in Figure 6.1, such as D_1, D_2, E_1, and E_2, represent alternative ways of coding group membership, which are discussed later in this chapter. All subsequent analyses in this chapter are based on the data in the file salarysexcollege.sav.

Figure 6.1 SPSS Data Worksheet for Hypothetical Faculty Salary Study

newestgendersalary [DataSet1] - SPSS Data Editor

File Edit View Data Transform Analyze Graphs Utilities Add-ons Window Help

	salary	years	gender	genyears	geneff	college	d1	d2	e1	e2	merit	Zmerit	Zyears	zyearmerit	Zsalary
1	31	0	0	0	-1	1	1	0	1	0	30	-.20458	-1.28239	.26	-1.56520
2	32	0	0	0	-1	3	0	0	-1	-1	10	-1.61545	-1.28239	2.07	-1.46263
3	33	1	0	0	-1	1	1	0	1	0	15	-1.26273	-1.09919	1.39	-1.36006
4	34	1	0	0	-1	2	0	1	0	1	29	-.27512	-1.09919	.30	-1.25750
5	33	2	0	0	-1	1	1	0	1	0	30	-.20458	-.91599	.19	-1.36006
6	40	2	0	0	-1	3	0	0	-1	-1	59	1.84119	-.91599	-1.69	-.64208
7	39	3	0	0	-1	2	0	1	0	1	55	1.55901	-.73279	-1.14	-.74465
8	51	3	0	0	-1	3	0	0	-1	-1	30	-.20458	-.73279	.15	.48618
9	54	3	0	0	-1	3	0	0	-1	-1	17	-1.12164	-.73279	.82	.79388
10	38	4	0	0	-1	1	1	0	1	0	11	-1.54490	-.54960	.85	-.84722
11	39	4	0	0	-1	2	0	1	0	1	45	.85358	-.54960	-.47	-.74465
12	43	4	0	0	-1	2	0	1	0	1	40	.50086	-.54960	-.28	-.33437
13	37	5	0	0	-1	1	1	0	1	0	55	1.55901	-.36640	-.57	-.94979
14	39	5	0	0	-1	1	1	0	1	0	30	-.20458	-.36640	.07	-.74465
15	41	6	0	0	-1	1	1	0	1	0	31	-.13403	-.18320	.02	-.53951
16	41	6	0	0	-1	1	1	0	1	0	30	-.20458	-.18320	.04	-.53951
17	42	7	0	0	-1	2	0	1	0	1	50	1.20629	.00000	.00	-.43694
18	46	9	0	0	-1	2	0	1	0	1	18	-1.05110	.36640	-.39	-.02667
19	49	12	0	0	-1	1	1	0	1	0	35	.14814	.91599	.14	.28104
20	54	15	0	0	-1	1	1	0	1	0	30	-.20458	1.46559	-.30	.79388
21	30	1	1	1	1	2	0	1	0	1	7	-1.82708	-1.09919	2.01	-1.66777
22	34	2	1	2	1	1	1	0	1	0	7	-1.82708	-.91599	1.67	-1.25750
23	36	2	1	2	1	2	0	1	0	1	30	-.20458	-.91599	.19	-1.05236
24	42	2	1	2	1	1	1	0	1	0	30	-.20458	-.91599	.19	-.43694
25	43	3	1	3	1	1	1	0	1	0	28	-.34566	-.73279	.25	-.33437
26	44	3	1	3	1	3	0	0	-1	-1	24	-.62784	-.73279	.46	-.23181
27	43	4	1	4	1	2	0	1	0	1	51	1.27684	-.54960	-.70	-.33437
28	45	4	1	4	1	2	0	1	0	1	22	-.76892	-.54960	.42	-.12924
29	46	4	1	4	1	1	1	0	1	0	7	-1.82708	-.54960	1.00	-.02667
30	47	4	1	4	1	2	0	1	0	1	35	.14814	-.54960	-.08	.07590
31	34	5	1	5	1	2	0	1	0	1	57	1.70010	-.36640	-.62	-1.25750
32	44	5	1	5	1	1	1	0	1	0	34	.07760	-.36640	-.03	-.23181
33	46	6	1	6	1	1	1	0	1	0	43	.71249	-.18320	-.13	-.02667
34	47	7	1	7	1	2	0	1	0	1	21	-.83947	.00000	.00	.07590

Data View Variable View

SPSS Processor is ready

The first research question that can be asked using these data is whether there is a significant difference in mean salary between men and women (ignoring all other predictor variables). This question could be addressed by conducting an independent-samples t test to compare male and female means on salary. In this chapter, a one-way ANOVA is performed to compare mean salary for male versus female faculty; then, salary is predicted from sex by doing a regression analysis to predict salary scores from a dummy variable that represents sex. The examples presented in this chapter demonstrate that ANOVA and regression analysis provide equivalent information about sex differences in mean salary. Examples or demonstrations such as the ones presented in this chapter do not constitute formal mathematical proof. Mathematical statistics textbooks provide formal mathematical proof of the equivalence of ANOVA and regression analysis.

The second question that will be addressed is whether there are significant differences in salary across the three colleges. This question will be addressed by doing a one-way ANOVA to compare mean salary across the three college groups and by using dummy variables that represent college group membership as predictors in a regression. This example demonstrates that membership in k groups can be represented by a set of $(k - 1)$ dummy variables. Data for this example are in the file salarysexcollege.sav.

6.3 SCREENING FOR VIOLATIONS OF ASSUMPTIONS

When we use one or more dummy variables as predictors in regression, assumptions are essentially the same as for any other regression analysis and for one-way ANOVA. As in other applications of ANOVA and regression, scores on the outcome variable Y should be quantitative and approximately normally distributed. If the Y outcome variable is categorical, logistic regression analysis should be used instead of linear regression; a brief introduction to binary logistic regression is presented in Chapter 16. Potential violations of the assumption of an approximately normal distribution shape for the Y outcome variable can be assessed by examining a histogram of scores on Y; the shape of this distribution should be reasonably close to normal. Methods for identification and handling outliers are discussed elsewhere.

The variance of Y scores should be fairly homogeneous across groups, that is, across levels of the dummy variables. The F tests used in ANOVA are fairly robust to violations of this assumption, unless the numbers of scores in the groups are small and/or unequal. When comparisons of means on Y across multiple groups are made using the SPSS t-test procedure, one-way ANOVA, or the GLM, the Levene test can be requested to assess whether the homogeneity of variance is seriously violated. SPSS multiple regression does not provide a formal test of the assumption of homogeneity of variance across groups. It is helpful to examine a graph of the distribution of scores within each group (such as a boxplot) to assess visually whether the scores of the Y outcome variable appear to have fairly homogeneous variances across levels of each X dummy predictor variable.

The issue of group size should also be considered in preliminary data screening (i.e., How many people are there in each of the groups represented by codes on the dummy variables?). For optimum statistical power and greater robustness to violations of assumptions, such as the homogeneity of variance assumption, it is preferred that there are equal numbers of scores in each group.[1] The minimum number of scores within each group should be large enough to provide a reasonably accurate estimate of group means. For any groups that include fewer than 10 or 20 scores, estimates of the group means may have confidence intervals that are quite wide. The guidelines about the minimum number of scores per group from Chapter 12 in Volume I (on the independent-samples t test [Warner, 2020]) and Chapter 13 in Volume I (on between-subjects [between-S] one-way ANOVA) can be used to judge whether the numbers in each group that correspond to a dummy variable, such as sex, are sufficiently large to yield reasonable statistical power and reasonable robustness against violations of assumptions.

For the hypothetical faculty salary data in the file salarysexcollege.sav, the numbers of cases within the groups are (barely) adequate. There were 20 female and 30 male faculty in the sample, of whom 22 were liberal arts faculty, 17 were sciences faculty, and 11 were business faculty. Larger group sizes are desirable in real-world applications of dummy variable analysis; relatively small numbers of cases were used in this example to make it easy for students to verify computations, such as group means, by hand.

All the SPSS procedures that are used in this chapter, including boxplot, scatterplot, Pearson correlation, one-way ANOVA, and linear regression, have been introduced and discussed in more detail in Volume I (Warner, 2020). Only the output from these procedures appears in this chapter. For a review of the menu selections and the SPSS dialog boxes for any of these procedures, refer to the chapter in which each analysis was first introduced.

To assess possible violations of assumptions, the following preliminary data screening was performed. A histogram was set up to assess whether scores on the quantitative outcome variable salary were reasonably normally distributed; this histogram appears in Figure 6.2. Although the distribution of salary values was multimodal, the salary scores did not show a distribution shape that was drastically different from a normal distribution. In real-life research situations, salary distributions are often positively skewed, such that there is a long tail at the upper end of the distribution (because there is usually no fixed upper limit for salary) and a truncated tail at the lower end of the distribution (because salary values cannot be lower than 0). Skewed distributions of scores can sometimes be made more nearly normal in

Figure 6.2 Histogram of Salary Scores

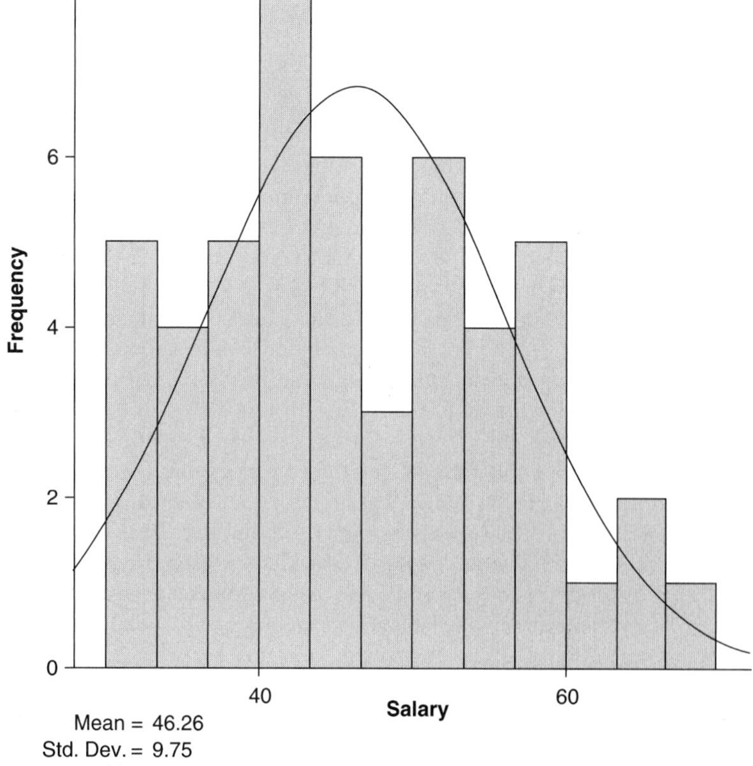

Figure 6.3 Boxplot of Salary Scores for Female and Male Groups

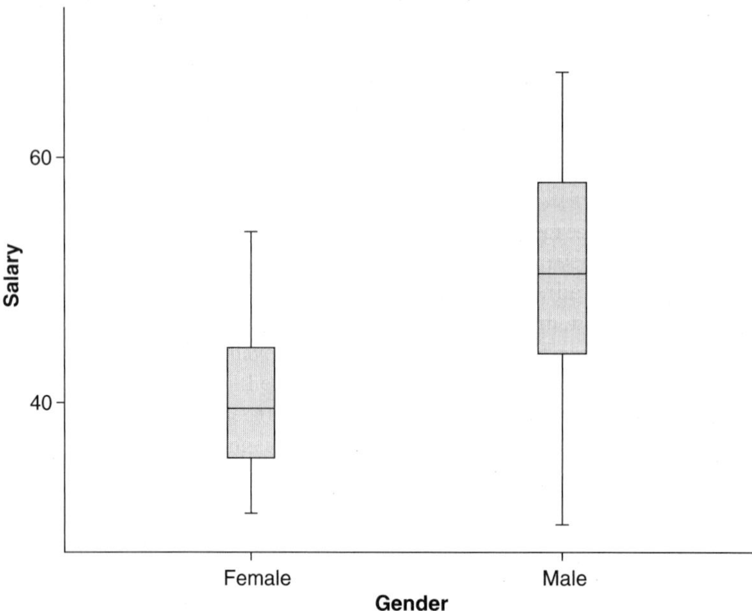

shape by taking the base 10 or natural logarithm of the salary scores. Logarithmic transformation was judged unnecessary for the artificial salary data that are used in the example in this chapter. Also, there were no extreme outliers in salary, as can be seen in Figure 6.2.

A boxplot of salary scores was set up to examine the distribution of outcomes for the female and male groups (Figure 6.3). A visual examination of this boxplot did not reveal any extreme outliers within either group; median salary appeared to be lower for women than for men. The boxplot suggested that the variance of salary scores might be larger for men than for women. This is a possible violation of the homogeneity of variance assumption; in the one-way ANOVA presented in a later section, the Levene test was requested to evaluate whether this difference between salary variances for the female and male groups was statistically significant (and the Levene test was not significant).

Figure 6.4 shows a scatterplot of salary (on the *Y* axis) as a function of years of job experience (on the *X* axis) with different case markers for female and male faculty. The relation between salary and years appears to be linear, and the slope that predicts salary from years appears to be similar for the female and male groups, although there appears to be a tendency for women to have slightly lower salary scores than men at each level of years of job experience; also, the range for years of experience was smaller for women (0–15) than for men (0–22). Furthermore, the variance of salary appears to be reasonably homogeneous across years in the scatterplot that appears in Figure 6.4, and there were no extreme bivariate outliers. Subsequent regression analyses will provide us with more specific information about sex and years as predictors of salary. We will use a regression analysis that includes the following predictors: years, a dummy variable to represent sex, and a product of sex and years. The results of this regression will help answer the following questions: Is there a significant increase in predicted salary associated with years of experience (when sex is statistically controlled)? Is there a significant difference in predicted salary between men and women (when years of experience is statistically controlled)?

Figure 6.4 Scatterplot of Salary by Years With Case Markers for Sex

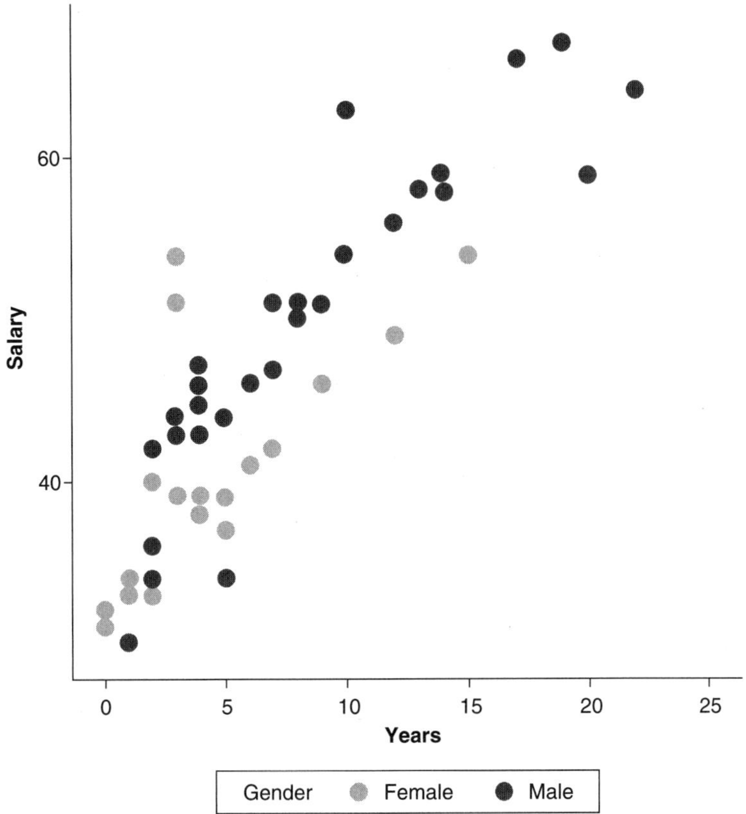

6.4 ISSUES IN PLANNING A STUDY

Essentially, when we use a dummy variable as a predictor in a regression, we have the same research situation as when we do a *t* test or ANOVA (both analyses predict scores on the *Y* outcome variable for two groups). When we use several dummy variables as predictors in a regression, we have the same research situation as in a one-way ANOVA (both analyses compare means across several groups). Therefore, the issues reviewed in planning studies that use *t* tests and one-way ANOVA are also relevant when we use a regression analysis as the method of data analysis. To put it briefly, if the groups that are compared received different "dosage" levels of some treatment variable, the dosage levels need to be far enough apart to produce detectable differences in outcomes. If the groups are formed on the basis of participant characteristics (such as age), the groups need to be far enough apart on these characteristics to yield detectable differences in outcome. Other variables that might create within-group variability in scores may need to be experimentally or statistically controlled to reduce the magnitude of error variance, as described in discussions of the independent-samples *t* test and one-way between-*S* ANOVA.

It is important to check that every group has a reasonable minimum number of cases. If any group has fewer than 10 cases, the researcher may decide to combine that group with one or more other groups (if it makes sense to do so) or exclude that group from the analysis. The statistical power tables that appeared in the one-way between-*S* ANOVA can be used to assess the minimum sample size needed per group to achieve reasonable levels of statistical power.

6.5 PARAMETER ESTIMATES AND SIGNIFICANCE TESTS FOR REGRESSIONS WITH DUMMY PREDICTOR VARIABLES

The use of one or more dummy predictor variables in regression analysis does not change any of the computations for multiple regression. The estimates of *b* coefficients, the *t* tests for the significance of individual *b* coefficients, and the overall *F* test for the significance of the entire regression equation are all computed using the methods described earlier for regression with multiple predictor variables. Similarly, sr^2 (the squared semipartial correlation), an estimate of the proportion of variance uniquely predictable from each dummy predictor variable, can be calculated and interpreted for dummy predictors in a manner similar to other types of regression. Confidence intervals for each *b* coefficient are obtained using the methods described earlier.

However, the *interpretation* of *b* coefficients when they are associated with dummy-coded variables is slightly different from the interpretation when they are associated with continuous predictor variables. Depending on the method of coding that is used, the b_0 (intercept) coefficient may correspond to the mean of one of the groups or to the grand mean of *Y*. The b_i coefficients for each dummy-coded predictor variable may correspond to contrasts between group means or to differences between group means and the grand mean.

6.6 GROUP MEAN COMPARISONS USING ONE-WAY BETWEEN-S ANOVA

6.6.1 Sex Differences in Mean Salary

Using the data in the file salarysexcollege.sav, a one-way between-*S* ANOVA was performed to assess whether mean salary differed significantly between female and male faculty. No other variables were taken into account in this analysis. A test of the homogeneity of variance assumption was requested (the Levene test).

The Levene test was not statistically significant, $F(1, 48) = 2.81, p = .10$. Thus, there was no statistically significant difference between the variances of salary for the female and male groups; the homogeneity of variance assumption was not violated. The mean salary for men (M_{male}) was 49.9 thousand dollars per year; the mean salary for women (M_{female}) was 40.8 thousand dollars per year. The difference between mean annual salary for men and women was statistically significant at the conventional $\alpha = .05$ level, $F(1, 48) = 13.02, p = .001$. The difference between the means ($M_{\text{male}} - M_{\text{female}}$) was +9.1 thousand dollars (i.e., on average, male faculty earned about 9.1 thousand dollars more per year than female faculty). The eta squared effect size for this sex difference was .21; in other words, about 21% of the variance in salaries could be predicted from sex. This corresponds to a large effect size. This initial finding of a sex difference in mean salary is not necessarily evidence of sex bias in salary levels. Within this sample, there was a tendency for female faculty to have fewer years of experience and for male faculty to have more years of experience, as well as salary increases as a function of years of experience. It is possible that the sex difference we see in this ANOVA is partly or completely accounted for by differences between men and women in years of experience. Subsequent analyses will address this by examining whether sex still predicts different levels of salary when "years of experience" is statistically controlled by including it as a second predictor of salary.

Note that when the numbers of cases in the groups are unequal, there are two different ways in which a grand mean can be calculated. An unweighted grand mean of salary can be obtained by simply averaging male and female mean salary, ignoring sample size. The

Figure 6.5 One-Way Between-S ANOVA: Mean Salary for Women and Men

Descriptives

salary

	N	Mean	Std. Deviation	Std. Error	95% Confidence Interval for Mean		Minimum	Maximum
					Lower Bound	Upper Bound		
Female	20	40.80	6.986	1.562	37.53	44.07	31	54
Male	30	49.90	9.714	1.774	46.27	53.53	30	67
Total	50	46.26	9.750	1.379	43.49	49.03	30	67

Test of Homogeneity of Variances

salary

Levene Statistic	df1	df2	Sig.
2.809	1	48	.100

ANOVA

salary

	Sum of Squares	df	Mean Square	F	Sig.
Between Groups	993.720	1	993.720	13.019	.001
Within Groups	3663.900	48	76.331		
Total	4657.620	49			

unweighted grand mean is (40.8 + 49.9)/2 = 45.35. However, in many statistical analyses, the estimate of the grand mean is weighted by sample size. The weighted grand mean in this example is found as follows:

$$[(n_{\text{male}} \times M_{\text{male}}) + (n_{\text{female}} \times M_{\text{female}})]/(n_{\text{male}} + n_{\text{female}})$$

$$= [(20 \times 40.8) + (30 \times 49.9)]/(20 + 30)$$

$$= 46.26.$$

The grand mean for salary reported in the one-way ANOVA in Figure 6.5 corresponds to the weighted grand mean. (This weighted grand mean is equivalent to the sum of the 50 individual salary scores divided by the number of scores in the sample.) In the regression analyses reported later in this chapter, the version of the grand mean that appears in the results corresponds to the unweighted grand mean.

6.6.2 College Differences in Mean Salary

In a research situation that involves a categorical predictor variable with more than two levels or groups and a quantitative outcome variable, the most familiar approach to data analysis is a one-way ANOVA. To evaluate whether mean salary level differs for faculty across the three different colleges in the hypothetical data set in the file salarysexcollege.sav, we can conduct a between-*S* one-way ANOVA using the SPSS one-way ANOVA procedure. The variable college is coded 1 = liberal arts, 2 = sciences, and 3 = business. The outcome variable, as in previous analyses, is annual salary in thousands of dollars. Orthogonal contrasts between colleges were also requested by entering custom contrast coefficients. The results of this one-way ANOVA appear in Figure 6.6.

Figure 6.6 One-Way Between-S ANOVA: Mean Salary Across Colleges

Descriptives

salary

	N	Mean	Std. Deviation	Std. Error	95% Confidence Interval for Mean Lower Bound	95% Confidence Interval for Mean Upper Bound	Minimum	Maximum
Liberal Arts	22	44.82	9.261	1.975	40.71	48.92	31	66
Sciences	17	44.71	9.777	2.371	39.68	49.73	30	67
Business	11	51.55	9.658	2.912	45.06	58.03	32	64
Total	50	46.26	9.750	1.379	43.49	49.03	30	67

Test of Homogeneity of Variances

salary

Levene Statistic	df1	df2	Sig.
.005	2	47	.995

ANOVA

salary

	Sum of Squares	df	Mean Square	F	Sig.
Between Groups	394.091	2	197.045	2.172	.125
Within Groups	4263.529	47	90.713		
Total	4657.620	49			

Contrast Coefficients

Contrast	college: Liberal Arts	Sciences	Business
1	1	-1	0
2	1	1	-2

Contrast Tests

		Contrast	Value of Contrast	Std. Error	t	df	Sig. (2-tailed)
salary	Assume equal variances	1	.11	3.076	.037	47	.971
		2	-13.57	6.515	-2.082	47	.043
	Does not assume equal variances	1	.11	3.086	.036	33.580	.971
		2	-13.57	6.591	-2.058	16.027	.056

The means on salary were as follows: 44.8 for faculty in liberal arts, 44.7 in sciences, and 51.6 in business. The overall F for this one-way ANOVA was not statistically significant, $F(2, 47) = 2.18$, $p = .125$. The effect size, eta squared, was obtained by taking the ratio $SS_{between}/SS_{total}$; for this ANOVA, $\eta^2 = .085$. This is a medium effect. In this situation, if we want to use this sample to make inferences about some larger population of faculty, we would not have evidence that the proportion of variance in salary that is predictable from college is significantly different from 0. However, if we just want to describe the strength of the association between college and salary within this sample, we could say that, for this sample, about 8.5% of the variance in salary was predictable from college. The orthogonal contrasts that were requested made the following comparison. For Contrast 1, the custom contrast coefficients were +1, −1, and 0; this corresponds to a comparison of the mean salaries between College 1 (liberal arts) and College 2 (sciences); this contrast was not statistically significant, $t(47) = .037$, $p = .97$. For Contrast 2, the custom contrast coefficients were +1, +1, and −2; this corresponds to a comparison of the mean salary for liberal arts and sciences faculty combined, compared with the business faculty. This contrast was statistically significant, $t(47) = -2.082$, $p = .043$. Business faculty had a significantly higher mean salary than the two other colleges (liberal arts and sciences) combined.

The next sections show that regression analyses with dummy predictor variables can be used to obtain the same information about the differences between group means. Dummy variables that represent group membership (such as female and male or liberal arts, sciences, and business colleges) can be used to predict salary in regression analyses. We will see that the information about differences among group means that can be obtained by doing regression with dummy predictor variables is equivalent to the information that we can obtain from one-way ANOVA. In future chapters, both dummy variables and quantitative variables are used as predictors in regression.

6.7 THREE METHODS OF CODING FOR DUMMY VARIABLES

The three coding methods that are most often used for dummy variables are given below (the details regarding these methods are presented in subsequent sections along with empirical examples):

1. Dummy coding of dummy variables
2. Effect coding of dummy variables
3. Orthogonal coding of dummy variables

In general, when we have *k* groups, we need only $(k - 1)$ dummy variables to represent information about group membership. Most dummy variable codes can be understood as answers to yes/no questions about group membership. For example, to represent college group membership, we might include a dummy variable D_1 that corresponds to the question, "Is this faculty member in liberal arts?" and code the responses as 1 = yes, 0 = no. If there are *k* groups, a set of $(k - 1)$ yes/no questions provides complete information about group membership. For example, if a faculty member reports responses of "no" to the questions, "Are you in liberal arts?" and "Are you in science?" and there are only three groups, then that person must belong to the third group (in this example, Group 3 is the business college).

The difference between dummy coding and effect coding is in the way in which codes are assigned for members of the last group, that is, the group that does not correspond to an explicit yes/no question about group membership. In dummy coding of dummy variables, members of the last group receive a score of 0 on all the dummy variables. (In effect coding of dummy variables, members of the last group are assigned scores of –1 on all the dummy variables.) This difference in codes results in slightly different interpretations of the *b* coefficients in the regression equation, as described in the subsequent sections of this chapter.

6.7.1 Regression With Dummy-Coded Dummy Predictor Variables

6.7.1.1 Two-Group Example With a Dummy-Coded Dummy Variable

Suppose we want to predict salary (*Y*) from sex; sex is a **dummy-coded dummy variable** with codes of 0 for female and 1 for male participants in the study. In a previous section, this difference was evaluated by doing a one-way between-*S* ANOVA to compare mean salary across female and male groups. We can obtain equivalent information about the magnitude of sex differences in salary from a regression analysis that uses a dummy-coded variable to predict salary. For this simple two-group case (prediction of salary from dummy-coded sex), we can write a regression equation using sex to predict salary (*Y*) as follows:

$$\text{Salary}' \text{ or } Y' = b_0 + b_1 \times \text{Sex}. \quad (6.2)$$

From Equation 6.2, we can work out two separate prediction equations: one that makes predictions of Y for women and one that makes predictions of Y for men. To do this, we substitute the values of 0 (for women) and 1 (for men) into Equation 6.2 and simplify the expression to obtain these two different equations:

$$Y' = b_0 \text{ (for women, sex = 0)}, \quad (6.3)$$

$$Y' = b_0 + b_1 \text{ (for men, sex = 1)}. \quad (6.4)$$

These two equations tell us that the constant value b_0 is the best prediction of salary for women, and the constant value $(b_0 + b_1)$ is the best prediction of salary for men. This implies that b_0 = mean salary for women, $b_0 + b_1$ = mean salary for men, and b_1 = the difference in mean salary between the male and female groups. The slope coefficient b_1 corresponds to the difference in mean salary for men and women. If the b_1 slope is significantly different from 0, it implies that there is a statistically significant difference in mean salary for men and women.

The results of the regression in Figure 6.7 provide the numerical estimates for the raw score regression coefficients (b_0 and b_1) for this set of data:

Figure 6.7 Regression to Predict Salary From Dummy-Coded Sex (or Gender)

Variables Entered/Removed[b]

Model	Variables Entered	Variables Removed	Method
1	gender[a]		Enter

a. All requested variables entered.

b. Dependent Variable: salary

Model Summary

Model	R	R Square	Adjusted R Square	Std. Error of the Estimate
1	.462[a]	.213	.197	8.737

a. Predictors: (Constant), gender

ANOVA[b]

Model		Sum of Squares	df	Mean Square	F	Sig.
1	Regression	993.720	1	993.720	13.019	.001[a]
	Residual	3663.900	48	76.331		
	Total	4657.620	49			

a. Predictors: (Constant), gender

b. Dependent Variable: salary

Coefficients[a]

Model		Unstandardized Coefficients		Standardized Coefficients	t	Sig.	Correlations		
		B	Std. Error	Beta			Zero-order	Partial	Part
1	(Constant)	40.800	1.954		20.884	.000			
	gender	9.100	2.522	.462	3.608	.001	.462	.462	.462

a. Dependent Variable: salary

$$\text{Salary}' = 40.8 + 9.1 \times \text{Sex}.$$

For women, with sex = 0, the predicted mean salary given by this equation is 40.8 + 9.1 × 0 = 40.8. Note that this is the same as the mean salary for women in the one-way ANOVA output in Figure 6.5. For men, with sex = 1, the predicted salary given by this equation is 40.8 + 9.1 = 49.9. Note that this value is equal to the mean salary for men in the one-way ANOVA in Figure 6.5.

The b_1 coefficient in this regression was statistically significant, $t(48) = 3.61, p = .001$. The F test reported in Figure 6.6 is equivalent to the square of the t test value for the null hypothesis that $b_1 = 0$ in Figure 6.7 ($t = +3.608$, $t^2 = 13.02$). Note also that the eta-squared effect size associated with the ANOVA ($\eta^2 = .21$) and the R^2 effect size associated with the regression are equal; in both analyses, about 21% of the variance in salaries was predictable from sex.

When we use a dummy variable with codes of 0 and 1 to represent membership in two groups, the value of the b_0 intercept term in the regression equation is equivalent to the mean of the group for which the dummy variable has a value of 0. The b_1 "slope" coefficient represents the difference (or contrast) between the means of the two groups. The slope, in this case, represents the change in the mean level of Y when you move from a code of 0 (female) to a code of 1 (male) on the dummy predictor variable.

Figure 6.8 shows a scatterplot of salary scores (on the Y axis) as a function of sex code (on the X axis). In this graph, the intercept b_0 corresponds to the mean on the dependent variable (salary) for the group that had a dummy variable score of 0 (women). The slope, b_1, corresponds to the difference between the means for the two groups, that is, the change in the predicted mean when you move from a code of 0 to a code of 1 on the dummy variable. That is, $b_1 = M_{\text{male}} - M_{\text{female}}$, the change in salary when you move from a score of 0 (female) to a score of 1 (male). The test of the statistical significance of b_1 is equivalent to the t test of the difference between the mean Y values for the two groups represented by the dummy variable in the regression.

Figure 6.8 Graph for Regression to Predict Salary From Dummy-Coded Sex (0 = Female, 1 = Male)

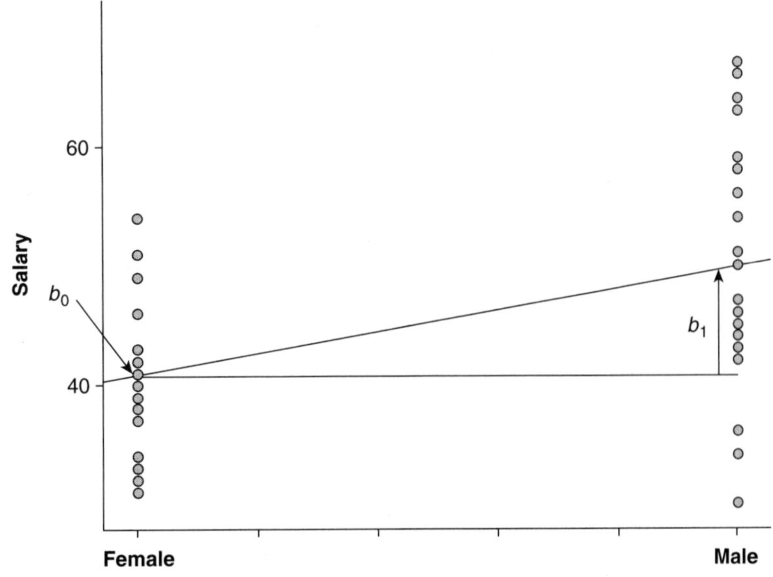

6.7.1.2 Multiple-Group Example With Dummy-Coded Dummy Variables

When there are multiple groups (number of groups = k), group membership can be represented by scores on a set of (k – 1) dummy variables. Each dummy variable essentially represents a yes/no question about group membership. In the preceding example, there are $k = 3$ college groups in the faculty data. In this example, we will use two dummy variables, denoted D_1 and D_2, to represent information about college membership in a regression analysis. D_1 corresponds to the following question: Is the faculty member from the liberal arts college (1 = yes, 0 = no)? D_2 corresponds to the following question: Is the faculty member from the sciences college (1 = yes, 0 = no)? For dummy coding, members of the last group receive a score of 0 on all the dummy variables. In this example, faculty from the business college received scores of 0 on both the D_1 and D_2 dummy-coded dummy variable. For the set of three college groups, the dummy-coded dummy variables that provide information about college group membership were coded as follows:

	D_1	D_2
Liberal arts	1	0
Sciences	0	1
Business	0	0

Now that we have created dummy variables that represent information about membership in the three college groups as scores on a set of dummy variables, mean salary can be predicted from college groups by a regression analysis that uses the dummy-coded dummy variables shown above as predictors:

$$Y' = b_0 + b_1D_1 + b_2D_2. \tag{6.5}$$

The results of the regression (using dummy-coded dummy variables to represent career group membership) are shown in Figure 6.9. Note that the overall F reported for the regression analysis in Figure 6.9 is identical to the overall F reported for the one-way ANOVA in Figure 6.6, $F(2, 47) = 2.17$, $p = .125$, and that the η^2 for the one-way ANOVA is identical to the R^2 for the regression ($\eta^2 = R^2 = .085$). Note also that the b_0 coefficient in the regression results in Figure 6.9 (b_0 = constant = 51.55) equals the mean salary for the group that was assigned score values of 0 on all the dummy variables (the mean salary for the business faculty was 51.55). Note also that the b_i coefficients for each of the two dummy variables represent the difference between the mean of the corresponding group and the mean of the comparison group whose codes were all 0; for example, $b_1 = -6.73$, which corresponds to the difference between the mean salary of Group 1, liberal arts ($M_1 = 44.82$) and mean salary of the comparison group, business ($M_3 = 51.55$); the value of the b_2 coefficient ($b_2 = -6.84$) corresponds to the difference between mean salary for the science ($M = 44.71$) and business ($M = 51.55$) groups. This regression analysis with dummy variables to represent college membership provided information equivalent to a one-way ANOVA to predict salary from college.

6.7.2 Regression With Effect-Coded Dummy Predictor Variables

6.7.2.1 Two-Group Example With an Effect-Coded Dummy Variable

We will now code the scores for sex slightly differently, using a method called "effect coding of dummy variables." In effect coding of dummy variables, a score value of 1 is used to represent a "yes" answer to a question about group membership; membership in the group

Figure 6.9 Regression to Predict Salary From Dummy-Coded College Membership

Variables Entered/Removed[b]

Model	Variables Entered	Variables Removed	Method
1	d2, d1[a]		Enter

a. All requested variables entered.

b. Dependent Variable: salary

Model Summary

Model	R	R Square	Adjusted R Square	Std. Error of the Estimate
1	.291[a]	.085	.046	9.524

a. Predictors: (Constant), d2, d1

ANOVA[b]

Model		Sum of Squares	df	Mean Square	F	Sig.
1	Regression	394.091	2	197.045	2.172	.125[a]
	Residual	4263.529	47	90.713		
	Total	4657.620	49			

a. Predictors: (Constant), d2, d1

b. Dependent Variable: salary

Coefficients[a]

Model		Unstandardized Coefficients		Standardized Coefficients			Correlations		
		B	Std. Error	Beta	t	Sig.	Zero-order	Partial	Part
1	(Constant)	51.545	2.872		17.949	.000			
	d1	-6.727	3.517	-.346	-1.913	.062	-.132	-.269	-.267
	d2	-6.840	3.685	-.336	-1.856	.070	-.116	-.261	-.259

a. Dependent Variable: salary

that does not correspond to a "yes" answer on any of the group membership questions is represented by a score of −1. In the following example, the **effect-coded dummy variable** "geneff" (Is the participant male?) is coded +1 for men and −1 for women. The variable geneff is called an effect-coded dummy variable because we used −1 (rather than 0) as the value that represents membership in the last group. Our overall model for the prediction of salary (Y) from sex, represented by the effect-coded dummy variable geneff, can be written as follows:

$$Y' = b_0 + b_1 \times \text{geneff}. \tag{6.6}$$

Substituting the values of +1 for men and −1 for women, the predictive equations for men and women become

$$Y' = b_0 + b_1 \text{ (for men, with geneff coded +1)}, \tag{6.7}$$

$$Y' = b_0 - b_1 \text{ (for women, with geneff coded −1)}. \tag{6.8}$$

From earlier discussions on t tests and ANOVA, we know that the best predicted value of Y for men is equivalent to the mean on Y for men, M_{male}; similarly, the best predicted value of Y for females is equal to the mean on Y for women, M_{female}. The two equations above, therefore, tell us that $M_{\text{male}} = b_0 + b_1$ and $M_{\text{female}} = b_0 - b_1$. What does this imply for the values of b_0 and b_1? The mean for men is b_1 units above b_0; the mean for women is b_1 units below b_0. With a little thought, you will see that the intercept b_0 must equal the grand mean on salary for both sexes combined.

Note that when we calculate a grand mean by combining group means, there are two different possible ways to calculate the grand mean. If the groups have the same numbers of scores, these two methods yield the same result, but when the groups have unequal numbers of cases, these two methods for computation of the grand mean yield different results. Whenever you do analyses with unequal numbers in the groups, you need to decide whether the unweighted or the weighted grand mean is a more appropriate value to report. In some situations, it may not be clear what default decision a computer program uses (i.e., whether the program reports the weighted or the unweighted grand mean), but it is possible to calculate both the weighted and unweighted grand means by hand from the group means; when you do this, you will be able to determine which version of the grand mean was reported in the SPSS output.

The unweighted grand mean for salary for men and women is obtained by ignoring the number of cases in the groups and averaging the group means together for men and women. For the male and female salary data, the **unweighted mean** is $(M_{\text{male}} + M_{\text{female}})/2 = (40.80 + 49.90)/2 = 45.35$. Note that the b_0 constant or intercept term in the regression in Figure 6.10 that uses effect-coded sex to predict salary corresponds to this unweighted grand mean of 45.35. When you run a regression to predict scores on a quantitative outcome variable from effect-coded dummy predictor variables, and the default methods of computation are used in SPSS, the b_0 coefficient in the regression equation corresponds to the unweighted grand mean, and effects (or differences between group means and grand means) are reported relative to this unweighted grand mean as a reference point. This differs slightly from the one-way ANOVA output in Figure 6.5, which reported the weighted grand mean for salary (46.26).

When effect-coded dummy predictor variables are used, the slope coefficient b_1 corresponds to the "effect" of sex; that is, $+b_1$ is the distance between the male mean and the grand mean, and $-b_1$ is the distance between the female mean and the grand mean. In Volume I, Chapter 13, on one-way ANOVA (Warner, 2020), the term used for these distances (group mean minus grand mean) was *effect*. The effect of membership in Group i in a one-way ANOVA is represented by α_i, where $\alpha_i = M_i - M_Y$, the mean of Group i minus the grand mean of Y across all groups.

This method of coding (+1 vs. –1) is called "effect coding," because the intercept b_0 in Equation 6.5 equals the unweighted grand mean for salary, M_Y, and the slope coefficient b_i for each effect-coded dummy variable E represents the effect for the group that has a code of 1 on that variable, that is, the difference between that group's mean on Y and the unweighted grand mean. Thus, when effect-coded dummy variables are used to represent group membership, the b_0 intercept term equals the grand mean for Y, the outcome variable, and each b_i coefficient represents a contrast between the mean of one group versus the unweighted grand mean (or the "effect" for that group). The significance of b_1 for the effect-coded variable geneff is a test of the significance of the difference between the mean of the corresponding group (in this example, men) and the unweighted grand mean. Given that geneff is coded –1 for women and +1 for men and given that the value of b_1 is significant and positive, the mean salary of men is significantly higher than the grand mean (and the mean salary for women is significantly lower than the grand mean).

Note that we do not have to use a code of +1 for the group with the higher mean and a code of –1 for the group with the lower mean on Y. The sign of the b coefficient can be either positive or negative; it is the combination of signs (on the code for the dummy variable and the b coefficient) that tells us which group had a mean that was lower than the grand mean.

Figure 6.10 Regression to Predict Salary From Effect-Coded Gender (or Sex)

Variables Entered/Removed[b]

Model	Variables Entered	Variables Removed	Method
1	geneff [a]		Enter

a. All requested variables entered.

b. Dependent Variable: salary

Model Summary

Model	R	R Square	Adjusted R Square	Std. Error of the Estimate
1	.462 [a]	.213	.197	8.737

a. Predictors: (Constant), geneff

ANOVA[b]

Model		Sum of Squares	df	Mean Square	F	Sig.
1	Regression	993.720	1	993.720	13.019	.001 [a]
	Residual	3663.900	48	76.331		
	Total	4657.620	49			

a. Predictors: (Constant), geneff

b. Dependent Variable: salary

Coefficients[a]

Model		Unstandardized Coefficients		Standardized Coefficients			Correlations		
		B	Std. Error	Beta	t	Sig.	Zero-order	Partial	Part
1	(Constant)	45.350	1.261		35.962	.000			
	geneff	4.550	1.261	.462	3.608	.001	.462	.462	.462

a. Dependent Variable: salary

The overall F result for the regression analysis that predicts salary from effect-coded sex (geneff) in Figure 6.10 is identical to the F value in the earlier analyses of sex and salary reported in Figures 6.5 and 6.7: $F(1, 48) = 13.02$, $p = .001$. The effect size given by η^2 and R^2 is also identical across these three analyses ($R^2 = .21$). The only difference between the regression that uses a dummy-coded dummy variable (Figure 6.7) and the regression that uses an effect-coded dummy variable to represent sex (Figure 6.10) is in the way in which the b_0 and b_1 coefficients are related to the grand mean and group means.

6.7.2.2 Multiple-Group Example With Effect-Coded Dummy Variables

If we used effect coding instead of dummy coding to represent membership in the three college groups used as an example earlier, group membership could be coded as follows:

	E_1	E_2
Liberal arts	1	0
Sciences	0	1
Business	–1	–1

That is, E_1 and E_2 still represent yes/no questions about group membership. E_1 corresponds to the following question: Is the faculty member in liberal arts (coded 1 = yes, 0 = no)? E_2 corresponds to the following question: Is the faculty member in sciences (coded 1 = yes, 0 = no)? The only change when effect coding (instead of dummy coding) is used is that members of the business college (the one group that does not correspond directly to a yes/no question) now receive a code of –1 on both the variables E_1 and E_2.

We can run a regression to predict scores on salary from the two effect-coded dummy variables E_1 and E_2:

$$\text{Salary}' \text{ or } Y' = b_0 + b_1E_1 + b_2E_2. \tag{6.9}$$

The results of this regression analysis are shown in Figure 6.11.

When effect coding is used, the intercept or b_0 coefficient is interpreted as an estimate of the (unweighted) grand mean for the Y outcome variable, and each b_i coefficient represents the effect for one of the groups, that is, the contrast between a particular group mean and the grand mean. (Recall that when dummy coding was used, the intercept b_0 was interpreted as the mean of the "last" group—namely, the group that did not correspond to a "yes" answer on any of the dummy variables—and each b_i coefficient corresponded to the difference between one of the group means and the mean of the "last" group, the group that is used as the reference group for all comparisons.) In Figure 6.11, the overall F value and the overall R^2 are the

Figure 6.11 Regression to Predict Salary From Effect-Coded College Membership

Variables Entered/Removed[b]

Model	Variables Entered	Variables Removed	Method
1	e2, e1 [a]	.	Enter

a. All requested variables entered.

b. Dependent Variable: salary

Model Summary

Model	R	R Square	Adjusted R Square	Std. Error of the Estimate
1	.291 [a]	.085	.046	9.524

a. Predictors: (Constant), e2, e1

ANOVA[b]

Model		Sum of Squares	df	Mean Square	F	Sig.
1	Regression	394.091	2	197.045	2.172	.125 [a]
	Residual	4263.529	47	90.713		
	Total	4657.620	49			

a. Predictors: (Constant), e2, e1

b. Dependent Variable: salary

Coefficients[b]

Model		Unstandardized Coefficients		Standardized Coefficients			Correlations		
		B	Std. Error	Beta	t	Sig.	Zero-order	Partial	Part
1	(Constant)	47.023	1.403		33.525	.000			
	e1	-2.205	1.828	-.179	-1.206	.234	-.238	-.173	-.168
	e2	-2.317	1.935	-.177	-1.197	.237	-.237	-.172	-.167

a. Dependent Variable: salary

same as in the two previous analyses that compared mean salary across college (in Figures 6.6 and 6.9): $F(2, 47) = 2.12$, $p = .125$, $R^2 = .085$. The *b* coefficients for Equation 6.9, from Figure 6.11, are as follows:

$$\text{Salary}' = 47.03 - 2.205 \times E_1 - 2.317 \times E_2.$$

The interpretation is as follows: The (unweighted) grand mean of salary is 47.03 thousand dollars per year. Members of the liberal arts faculty have a predicted annual salary that is 2.205 thousand dollars less than this grand mean; members of the sciences faculty have a predicted salary that is 2.317 thousand dollars less than this grand mean. Neither of these differences between a group mean and the grand mean is statistically significant at the $\alpha = .05$ level.

Because members of the business faculty have a score of −1 on both E_1 and E_2, the predicted mean salary for business faculty is

$$\text{Salary}' = 47.03 + 2.205 + 2.317 = 51.5 \text{ thousand dollars per year.}$$

We do not have a significance test to evaluate whether the mean salary for business faculty is significantly higher than the grand mean. If we wanted to include a significance test for this contrast, we could do so by rearranging the dummy variable codes associated with group membership, such that membership in the business college group corresponded to an answer of "yes" on either E_1 or E_2.

6.7.3 Orthogonal Coding of Dummy Predictor Variables

We can set up contrasts among group means in such a way that the former are orthogonal (the term *orthogonal* is equivalent to *uncorrelated* or *independent*). One method of creating orthogonal contrasts is to set up one contrast that compares Group 1 versus Group 2 and a second contrast that compares Groups 1 and 2 combined versus Group 3, as in the example below:

	Group		
	1 (Liberal Arts)	2 (Sciences)	3 (Business)
O_1	+1	−1	0
O_2	+1	+1	−2

The codes across each row should sum to 0. For each **orthogonally coded dummy variable**, the groups for which the code has a positive sign are contrasted with the groups for which the code has a negative sign; groups with a code of 0 are ignored. Thus, O_1 compares the mean of Group 1 (liberal arts) with the mean of Group 2 (sciences).

To figure out which formal null hypothesis is tested by each contrast, we form a weighted linear composite that uses these codes. That is, we multiply the population mean μ_k for Group *k* by the contrast coefficient for Group *k* and sum these products across the *k* groups; we set that weighted linear combination of population means equal to 0 as our null hypothesis.

In this instance, the null hypotheses that correspond to the contrast specified by the two O_i orthogonally coded dummy variables are as follows:

$$H_0 \text{ for } O_1\text{: } (+1)\mu_1 + (-1)\mu_2 + (0)\mu_3 = 0.$$

That is,

$$H_0 \text{ for } O_1\text{: } \mu_1 - \mu_2 = 0 \text{ (or } \mu_1 = \mu_2\text{).}$$

The O_2 effect-coded dummy variable compares the average of the first two group means (i.e., the mean for liberal arts and sciences combined) with the mean for the third group (business):

$$H_0 \text{ for } O_2\text{: } (+1)\mu_1 + (+1)\mu_2 + (-2)\mu_3, H_0 \text{ for } O_2\text{: } (\mu_1 + \mu_2) - 2\mu_3 = 0,$$

or

$$\frac{\mu_1 + \mu_2}{2} - \mu_3 = 0 \text{ or } \frac{\mu_1 + \mu_2}{2} = \mu_3.$$

We can assess whether the contrasts are orthogonal by taking the cross products and summing the corresponding coefficients. Recall that products between sets of scores provide information about covariation or correlation; see Volume I, Chapter 10 (Warner, 2020), for details. Because each of the two variables O_1 and O_2 has a sum (and therefore a mean) of 0, each code represents a deviation from a mean. When we compute the sum of cross products between corresponding values of these two variables, we are, in effect, calculating the numerator of the correlation between O_1 and O_2. In this example, we can assess whether O_1 and O_2 are orthogonal or uncorrelated by calculating the following sum of cross products. For O_1 and O_2, the sum of cross products of the corresponding coefficients is

$$(+1)(+1) + (-1)(+1) + (0)(-2) = 0.$$

Because this sum of the products of corresponding coefficients is 0, we know that the contrasts specified by O_1 and O_2 are orthogonal.

Of course, as an alternative way to see whether the O_1 and O_2 predictor variables are orthogonal or uncorrelated, SPSS can also be used to calculate a correlation between O_1 and O_2; if the contrasts are orthogonal, Pearson's r between O_1 and O_2 will equal 0 (provided that the numbers in the groups are equal).

For each contrast specified by a set of codes (e.g., the O_1 set of codes), any group with a 0 coefficient is ignored, and groups with opposite signs are contrasted. The direction of the signs for these codes does not matter; the contrast represented by the codes (+1, −1, and 0) represents the same comparison as the contrast represented by (−1, +1, 0). The b coefficients obtained for these two sets of codes would be opposite in sign, but the significance of the difference between the means of Group 1 and Group 2 would be the same whether Group 1 was assigned a code of +1 or −1.

Figure 6.12 shows the results of a regression in which the dummy predictor variables O_1 and O_2 are coded to represent the same orthogonal contrasts. Note that the t tests for significance of each contrast are the same in both Figure 6.6, where the contrasts were requested as an optional output from the SPSS one-way ANOVA procedure, and Figure 6.12, where the contrasts were obtained by using orthogonally coded dummy variables as predictors of salary. Only one of the two contrasts, the contrast that compares the mean salaries for liberal arts and sciences faculty with the mean salary for business faculty, was statistically significant.

Orthogonal coding of dummy variables can also be used to perform a trend analysis, for example, when the groups being compared represent equally spaced dosage levels along a continuum. The following example shows orthogonal coding to represent linear versus quadratic trends for a study in which the groups receive three different dosage levels of caffeine.

Figure 6.12 Regression to Predict Salary From Dummy Variables That Represent Orthogonal Contrasts

Variables Entered/Removed[b]

Model	Variables Entered	Variables Removed	Method
1	o2, o1[a]		Enter

a. All requested variables entered.

b. Dependent Variable: salary

Model Summary

Model	R	R Square	Adjusted R Square	Std. Error of the Estimate
1	.291 [a]	.085	.046	9.524

a. Predictors: (Constant), o2, o1

ANOVA[b]

Model		Sum of Squares	df	Mean Square	F	Sig.
1	Regression	394.091	2	197.045	2.172	.125 [a]
	Residual	4263.529	47	90.713		
	Total	4657.620	49			

a. Predictors: (Constant), o2, o1

b. Dependent Variable: salary

Coefficients[a]

Model		Unstandardized Coefficients		Standardized Coefficients			Correlations		
		B	Std. Error	Beta	t	Sig.	Zero-order	Partial	Part
1	(Constant)	47.023	1.403		33.525	.000			
	o1	.056	1.538	.005	.037	.971	-.013	.005	.005
	o2	-2.261	1.086	-.291	-2.082	.043	-.291	-.291	-.291

a. Dependent Variable: salary

	0 mg	150 mg	300 mg
O_1	–1	0	+1
O_2	+1	–2	+1

Note that the sum of cross products is again 0 [(–1)(+1) + (0)(–2) + (+1)(+1)], so these contrasts are orthogonal. A simple way to understand what type of trend is represented by each line of codes is to visualize the list of codes for each dummy variable as a template or graph. If you place values of –1, 0, and +1 from left to right on a graph, it is clear that these codes represent a linear trend. The set of coefficients +1, –2, and +1 or, equivalently, –1, +2, and –1 represent a quadratic trend. So, if b_1 (the coefficient for O_1) is significant with this set of codes, the linear trend is significant and b_1 is the amount of change in the dependent variable Y from 0 to 150 mg or from 150 to 300 mg. If b_2 is significant, there is a quadratic (curvilinear) trend.

6.8 REGRESSION MODELS THAT INCLUDE BOTH DUMMY AND QUANTITATIVE PREDICTOR VARIABLES

We can do a regression analysis that includes one (or more) dummy variables and one (or more) continuous predictors, as in the following example:

$$Y' = b_0 + b_1D + b_2X. \tag{6.10}$$

How is the b_1 coefficient for the dummy variable, D, interpreted in the context of this multiple regression with another predictor variable? The b_1 coefficient still represents the estimated difference between the means of the two groups; if sex was coded 0, 1 as in the first example, b_1 still represents the difference between mean salary Y for men and women. However, in the context of this regression analysis, this difference between means on the Y outcome variable for the two groups is assessed while statistically controlling for any differences in the quantitative X variable (such as years of job experience).

Numerical results for this regression analysis appear in Figure 6.13. From these results we can conclude that both sex and years are significantly predictive of salary. The coefficient

Figure 6.13 Regression to Predict Salary From Sex and Years

Variables Entered/Removed[b]

Model	Variables Entered	Variables Removed	Method
1	gender, years[a]		Enter

a. All requested variables entered.

b. Dependent Variable: salary

Model Summary

Model	R	R Square	Adjusted R Square	Std. Error of the Estimate
1	.880[a]	.775	.765	4.722

a. Predictors: (Constant), gender, years

ANOVA[b]

Model		Sum of Squares	df	Mean Square	F	Sig.
1	Regression	3609.469	2	1804.734	80.926	.000[a]
	Residual	1048.151	47	22.301		
	Total	4657.620	49			

a. Predictors: (Constant), gender, years

b. Dependent Variable: salary

Coefficients[a]

		Unstandardized Coefficients		Standardized Coefficients			Correlations		
Model		B	Std. Error	Beta	t	Sig.	Zero-order	Partial	Part
1	(Constant)	34.193	1.220		28.038	.000			
	years	1.436	.133	.804	10.830	.000	.866	.845	.749
	gender	3.355	1.463	.170	2.293	.026	.462	.317	.159

a. Dependent Variable: salary

to predict salary from sex (controlling for years of experience) was b_2 = 3.36, with $t(47)$ = 2.29, p = .026. The corresponding squared semipartial (or part) correlation for sex as a predictor of salary was $sr^2 = (.159)^2 = .03$. The coefficient to predict salary from years of experience, controlling for sex, was b = 1.44, $t(47)$ = 10.83, p <.001; the corresponding sr^2 effect size for years was $.749^2 = .56$. This analysis suggests that controlling for years of experience partly accounts for the observed sex differences in salary, but it does not completely account for sex differences; even after years of experience is taken into account, men still have an average salary that is about 3.36 thousand dollars higher than women's average salary at each level of years of experience. However, this sex difference is relatively small in terms of the proportion of explained variance. About 56% of the variance in salaries is uniquely predictable from years of experience (this is a very strong effect). About 3% of the variance in salary is predictable from sex (this is a medium-sized effect). Within this sample, for each level of years of experience, women are paid about 3.35 thousand dollars less than men who have the same number of years of experience; this is evidence of possible sex bias. Of course, it is possible that this remaining sex difference in salary might be accounted for by other variables. Perhaps more women are in the college of liberal arts, which has lower salaries, and more men are in the business and science colleges, which tend to receive higher salaries. Controlling for other variables such as college might help us to account for part of the sex difference in mean salary levels.

The raw-score b coefficient associated with sex in the regression in Figure 6.13 had a value of b = 3.355; this corresponds to the difference between the intercepts of the regression lines for men and women in Figure 6.14. For this set of data, it appears that sex differences in salary may be due to a difference in starting salaries (i.e., the salaries paid to faculty with 0 years of experience) and not due to differences between the annual raises in salary given

Figure 6.14 Graph of the Regression Lines to Predict Salary From Years of Experience Separately for Men and Women

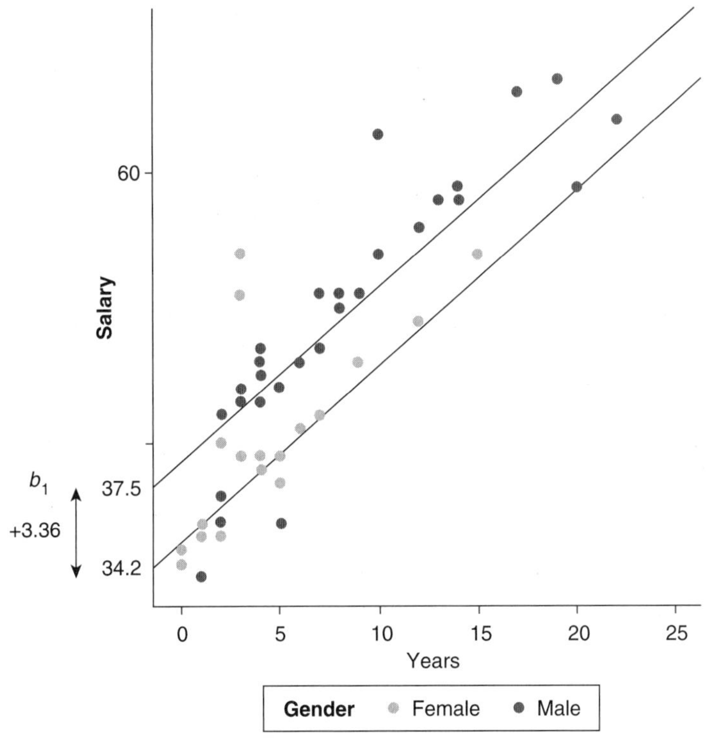

to male and female faculty. The model "Results" section at the end of the chapter provides a more detailed discussion of the results that appear in Figures 6.13 and 6.14.

The best interpretation for the salary data, on the basis of analyses that have been performed so far, appears to be the following: Salary significantly increases as a function of years of experience; there is also a sex difference in salary (such that men are paid significantly higher salaries than women) even when the effect of years of experience is statistically controlled. However, keep in mind that statistically controlling for additional predictor variables (such as college and merit) in later analyses could substantially change the apparent magnitude of sex differences in salary.

6.9 EFFECT SIZE AND STATISTICAL POWER

We can represent the effect size for the regression as a whole (i.e., the proportion of variance in Y is predictable from a set of variables that may include dummy-coded and/or continuous variables) by reporting multiple R and multiple R^2 as our overall effect size measures. We can represent the strength of the unique predictive contribution of an individual variable by reporting the estimate of sr^2_{unique} for each predictor variable, as in Chapter 4. The proportion of variance given by sr^2 can be used to describe the proportion of variance uniquely predictable from the contrast specified by a dummy variable, in the same manner in which it describes the proportion of variance uniquely predictable from a continuous predictor variable.

When only dummy variables are included in regression analysis, the regression is essentially equivalent to a one-way ANOVA (or a t test, if only two groups are being compared). Therefore, the tables presented in Volume I, Chapter 12 (independent-samples t test [Warner, 2020]) and Chapter 13 (one-way ANOVA), can be used to look up reasonable minimum sample sizes per group for anticipated effect sizes that are small, medium, or large. Whether the method used to make predictions and compare means across groups is ANOVA or regression, none of the groups should have a very small n. If $n < 20$ per group, nonparametric analyses may be more appropriate.

6.10 NATURE OF THE RELATIONSHIP AND/OR FOLLOW-UP TESTS

When dummy-coded group membership predictors are included in a regression analysis, the information that individual coefficients provide is equivalent to the information obtained from planned contrasts between group means in an ANOVA. The choice of the method of coding (dummy, effect, orthogonal) and the decision as to which group to code as the "last" group determine which set of contrasts the regression will include.

Whether we compare multiple groups by performing a one-way ANOVA or by using a regression equation with dummy-coded group membership variables as predictors, the written results should include the following: means and standard deviations for scores in each group, confidence intervals for each group mean, an overall F test to report whether there were significant differences in group means, planned contrasts or post hoc tests to identify which specific pairs of group means differed significantly, and a discussion of the direction of differences between group means. Effect-size information about the proportion of explained variance (in the form of an η^2 or sr^2) should also be included.

6.11 RESULTS

The hypothetical data showing salary scores for faculty (in the file salarysexcollege.sav) were analyzed in several different ways to demonstrate the equivalence between ANOVA and regression with dummy variables and to illustrate the interpretation of b coefficients for dummy variables in regression. The following example "Results" section reports the regression analysis for prediction of salary from years of experience and sex (as shown in Figure 6.13).

Results

To assess whether sex and years of experience significantly predict faculty salary, a regression analysis was performed to predict faculty annual salary in thousands of dollars from sex (dummy-coded 0 = female, 1 = male) and years of experience. The distribution of salary was roughly normal, the variances of salary scores were not significantly different for males and females, and scatter plots did not indicate nonlinear relations or bivariate outliers. No data transformations were applied to scores on salary and years, and all 50 cases were included in the regression analysis.

The results of this regression analysis (SPSS output in Figure 6.13) indicated that the overall regression equation was significantly predictive of salary; $R = .88$, $R^2 = .78$, adjusted $R^2 = .77$, $F(2, 47) = 80.93$, $p < .001$. Salary could be predicted almost perfectly from sex and years of job experience. Each of the two individual predictor variables was statistically significant. The raw score coefficients for the predictive equation were as follows:

$$\text{Salary}' = 34.19 + 3.36 \times \text{Sex} + 1.44 \times \text{Years}.$$

When controlling for the effect of years of experience on salary, the magnitude of the sex difference in salary was 3.36 thousand dollars. That is, at each level of years of experience, male annual salary was about 3.36 thousand dollars higher than female salary. This difference was statistically significant: $t(47) = 2.29$, $p = .026$.

For each 1-year increase in experience, the salary increase was approximately 1.44 thousand dollars for both females and males. This slope for the prediction of salary from years of experience was statistically significant: $t(47) = 10.83$, $p < .001$. The graph in Figure 6.14 illustrates the regression lines to predict salary for males and females separately. The intercept (i.e., predicted salary for 0 years of experience) was significantly higher for males than for females.

The squared semipartial correlation for years as a predictor of salary was $sr^2 = .56$; thus, years of experience uniquely predicted about 56% of the variance in salary (when sex was statistically controlled). The squared semipartial correlation for sex as a predictor of salary was $sr^2 = .03$; thus, sex uniquely predicted about 3% of the variance in salary (when years of experience was statistically controlled). The results of this analysis suggest that there was a systematic difference between salaries for male and female faculty and that this difference was approximately the same at all levels of years of experience. Statistically controlling for years of job experience, by including it as a predictor of salary in a regression that also used sex to predict salary, yielded results that suggest that the overall sex difference in mean salary was partly, but not completely, accounted for by sex differences in years of job experience.

6.12 SUMMARY

This chapter presented examples that demonstrated the equivalence of ANOVA and regression analyses that use dummy variables to represent membership in multiple groups. This discussion has presented demonstrations and examples rather than formal proofs; mathematical statistics textbooks provide formal proofs of equivalence between ANOVA and regression. ANOVA and regression are different special cases of the GLM.

If duplication of ANOVA using regression were the only application of dummy variables, it would not be worth spending so much time on them. However, dummy variables have important practical applications. Researchers often want to include group membership variables (such as sex) among the predictors that they use in multiple regression, and it is important to understand how the coefficients for dummy variables are interpreted.

An advantage of choosing ANOVA as the method for comparing group means is that the SPSS procedures provide a wider range of options for follow-up analysis, for example, post hoc protected tests. Also, when ANOVA is used, interaction terms are generated automatically for all pairs of (categorical) predictor variables or **factors**, so it is less likely that a researcher will fail to notice an interaction when the analysis is performed as a factorial ANOVA (as discussed in Volume I, Chapter 16 [Warner, 2020]) than when a comparison of group means is performed using dummy variables as predictors in a regression. ANOVA does not assume a linear relationship between scores on categorical predictor variables and scores on quantitative outcome variables. A quantitative predictor can be added to an ANOVA model (this type of analysis, called analysis of covariance [ANCOVA], is discussed in Chapter 8).

On the other hand, an advantage of choosing regression as the method for comparing group means is that it is easy to use quantitative predictor variables along with group membership predictor variables to predict scores on a quantitative outcome variable. Regression analysis yields equations that can be used to generate different predicted scores for cases with different score values on both categorical and dummy predictor variables. A possible disadvantage of the regression approach is that interaction terms are not automatically included in a regression; the data analyst must specifically create a new variable (the product of the two variables involved in the interaction) and add that new variable as a predictor. Thus, unless they specifically include interaction terms in their models (as discussed in Chapter 7), data analysts who use regression analysis may fail to notice interactions between predictors. A data analyst who is careless may also set up a regression model that is "nonsense"; for example, it would not make sense to predict political conservatism (Y) from scores on a categorical X_1 predictor variable that has codes 1 for Democrat, 2 for Republican, and 3 for Independent. Regression assumes a linear relationship between predictor and outcome variables; political party represented by just one categorical variable with three possible score values probably would not be linearly related to an outcome variable such as conservatism. To compare group means using regression in situations where there are more than two groups, the data analyst needs to create dummy variables to represent information about group membership. In some situations, it may be less convenient to create new dummy variables (and run a regression) than to run an ANOVA.

Ultimately, however, ANOVA and regression with dummy predictor variables yield essentially the same information about predicted scores for different groups. In many research situations, ANOVA may be a more convenient method to assess differences among group means. However, regression with dummy variables provides a viable alternative, and in some research situations (where predictor variables include both categorical and quantitative variables), a regression analysis may be a more convenient way of setting up the analysis.

COMPREHENSION QUESTIONS

1. Suppose that a researcher wants to do a study to assess how scores on the dependent variable heart rate (HR) differ across groups that have been exposed to various types of stress. Stress group membership was coded as follows:

   Group 1, no stress/baseline

   Group 2, mental arithmetic

   Group 3, pain induction

   Group 4, stressful social role play

   The basic research questions are whether these four types of stress elicited significantly different HRs overall and which specific pairs of groups differed significantly.

   a. Set up dummy-coded dummy variables that could be used to predict HR in a regression.

   Note that it might make more sense to use the "no-stress" group as the one that all other group means are compared with, rather than the group that happens to be listed last in the list above (stressful role play). Before working out the contrast coefficients, it may be helpful to list the groups in a different order:

   Group 1, mental arithmetic

   Group 2, pain induction

   Group 3, stressful social role play

   Group 4, no stress/baseline

   Set up dummy-coded dummy variables to predict scores on HR from group membership for this set of four groups.

   Write out in words which contrast between group means each dummy variable that you have created represents.

   b. Set up effect-coded dummy variables that could be used to predict HR in a regression.

   Describe how the numerical results for these effect-coded dummy variables (in 1b) differ from the numerical results obtained using dummy-coded dummy variables (in 1a). Which parts of the numerical results will be the same for these two analyses?

   c. Set up the coding for orthogonally coded dummy variables that would represent these orthogonal contrasts:

   Group 1 versus 2

   Groups 1 and 2 versus 3

   Groups 1, 2, and 3 versus 4

2. Suppose that a researcher does a study to see how level of anxiety (A_1 = low, A_2 = medium, A_3 = high) is used to predict exam performance (Y). Here are hypothetical data for this research situation. Each column represents scores on Y (exam scores).

   a. Would it be appropriate to do a Pearson correlation (and/or linear regression) between anxiety, coded 1, 2, 3 (for low, medium, high), and exam score? Justify your answer.
   b. Set up orthogonally coded dummy variables (O_1, O_2) to represent linear and quadratic trends, and run a regression analysis to predict exam scores from

A1, Low Anxiety	A2, Medium Anxiety	A3, High Anxiety
72	86	65
81	93	79
54	81	74
66	80	80
71	92	74

O_1 and O_2. What conclusions can you draw about the nature of the relationship between anxiety and exam performance?

c. Set up dummy-coded dummy variables to contrast each of the other groups with Group 2, medium anxiety; run a regression to predict exam performance (Y) from these dummy-coded dummy variables.

d. Run a one-way ANOVA on these scores; request contrasts between Group 2, medium anxiety, and each of the other groups. Do a point-by-point comparison of the numerical results for your ANOVA output with the numerical results for the regression in (2c), pointing out where the results are equivalent.

3. Why is it acceptable to use a dichotomous predictor variable in a regression when it is not usually acceptable to use a categorical variable that has more than two values as a predictor in regression?

4. Why are values such as +1, 0, and –1 generally used to code dummy variables?

5. How does the interpretation of regression coefficients differ for dummy coding of dummy variables versus effect coding of dummy variables? (Hint: In one type of coding, b_0 corresponds to the grand mean; in the other, b_0 corresponds to the mean of one of the groups.)

6. If you have k groups, why do you only need k – 1 dummy variables to represent group membership? Why is it impossible to include k dummy variables as predictors in a regression when you have k groups?

7. How does orthogonal coding of dummy variables differ from dummy and effect coding?

8. Write out equations to show how regression can be used to duplicate a t test or a one-way ANOVA.

NOTE

[1]Unequal numbers in groups make the interpretation of b coefficients for dummy variables more complex. For additional information about issues that should be considered when using dummy or effect codes to represent groups of unequal sizes, see Hardy (1993).

DIGITAL RESOURCES

Find **free study tools** to support your learning, including **eFlashcards, data sets, and web resources,** on the accompanying website at **edge.sagepub.com/warner3e.**

CHAPTER 7

MODERATION

Interaction in Multiple Regression

7.1 TERMINOLOGY

In factorial analysis of variance (ANOVA), when the effect of one predictor variable (Factor A) on the outcome variable *Y* differs within separate groups defined by Factor B, we call this an **interaction** between Factors A and B. Interaction can also be called **nonadditivity** because the presence of interaction means we need more than just additive main effects for Factors A and B to predict cell means for *Y*; an A × B interaction term is also needed. Most factorial ANOVA programs (including the SPSS general linear model [GLM] procedure) include all possible interactions among factors by default. For instance, if there are three factors, A, B, and C, the ANOVA source table includes *SS* terms for A, B, C, A × B, A × C, B × C, and A × B × C. This default can be overridden by specifying a custom model; that is, the user can delete some interaction terms from the factorial model. When this is done, *df* and *SS* terms for the deleted interaction term are pooled with error *df* and *SS*. Often, but not always, data analysts who use factorial designs have predictions about interactions.

Interactions between predictors can also be examined in multiple regression. In the context of regression, interaction is usually called **moderation**. Consider an example in which we want to predict interview skills (Y) from the categorical variable sex (X_1) and a quantitative measure of emotional intelligence (X_2). If we find that the slope that relates X_2 to Y is different for men than for women, we would say that sex is a **moderator variable**. The effect of X_2 on Y is moderated by, or changed by, or dependent upon sex.

By default, most regression analyses do not include interactions between predictors (i.e., they do not test hypotheses about moderation). Because of this, regression analysts can easily miss seeing interactions. Analysts can add interaction terms to regression equations by creating new predictor variables that are products between predictor variables such as X_1 and X_2, as discussed in this chapter. If a data analyst has theoretical predictions about interactions, terms should be added to the regression model to represent these interactions. If interactions are not predicted ahead of time, adding interaction terms to regression models in search of significant outcomes can be viewed and reported as exploratory, and is reasonable if the interactions make sense. However, if interaction terms are added to a regression model during an extended process of data torturing, in an attempt to obtain smaller p values for tests of the model, it is a form of p-hacking; this should be avoided.

In a classic paper, Baron and Kenny (1986) discussed moderation and mediation analyses. These terms should not be confused with each other; mediation is a completely different kind of hypothesis about associations among variables. An example of a mediation hypothesis is that first, X_1 causes X_2, and then, X_2 causes Y. If this model is consistent with regression results, one possible interpretation is that the effect of X_1 on Y may be mediated by X_2. Other interpretations are equally plausible, as discussed in Chapter 9, on mediation. Mediation is completely different from moderation.

When the term **effect** is used in statistics, it can imply causality if data come from an experiment. However, when data are not from an experiment, the term *effect* does not imply causality; it describes only the extent to which scores on an outcome variable Y can be predicted from one or more X variables. In most regression analysis situations, data come from a nonexperimental study. Results of regression analysis based on nonexperimental data should not be interpreted in causal terms. Regression results based on nonexperimental designs can be either consistent with, or inconsistent with, hypotheses about moderation and mediation.

A significant correlation between predictor variables X_1 and X_2 is *not* evidence of moderation or interaction (this seems to be a common misconception). An interaction can be present when predictors are correlated; however, it can also be present when the predictors are uncorrelated. The existence of correlation between predictors tells us nothing about potential interaction.

Hypotheses about moderation can be represented in path model form in two different ways. Figure 7.1 has an arrow from X_2 that points toward the X_1, Y path. This path diagram represents the idea that the coefficient for the X_1, Y path is modified by X_2. A second way to represent moderation appears in Figure 7.2. Moderation, or interaction between X_1 and X_2 as predictors of Y, can be assessed by including the product of X_1 and X_2 as an additional predictor variable in a regression model. The unidirectional arrows toward the outcome variable Y represent a regression model in which Y is predicted from X_1, X_2 and from a product term that represents an interaction between X_1 and X_2. In this model, the three predictors are shown as correlated with one another (these correlations are represented by the double-

Figure 7.1 Path Model That Represents Moderation of the X_1, Y Relationship by X_2

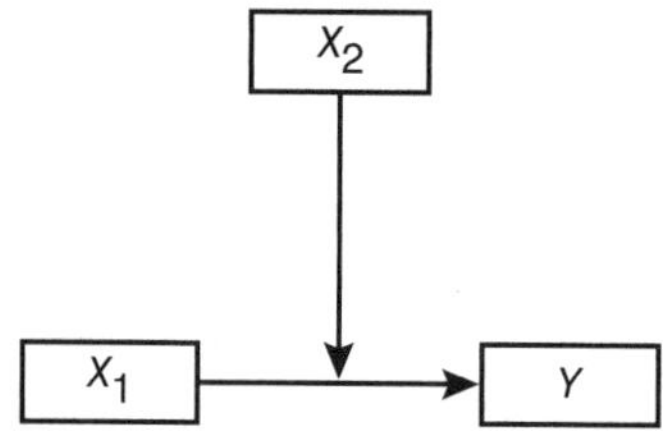

Figure 7.2 A Different Path Model That Represents Moderation of the X_1, Y Relationship by X_2

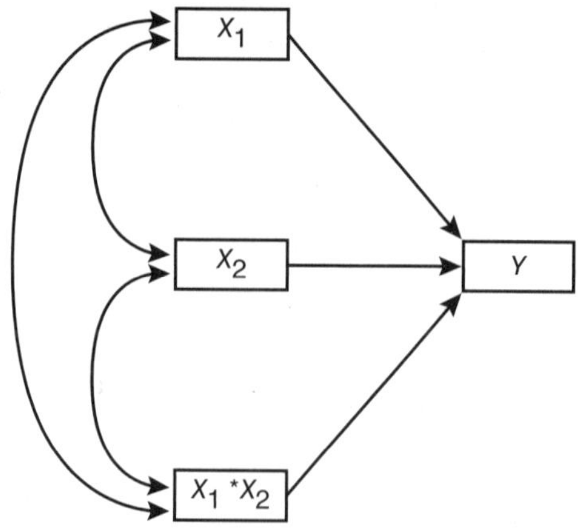

headed arrows). Predictors are often correlated in regression analyses, but as noted previously, those correlations are not evidence for or against the possible existence of moderation.

The path model in Figure 7.2 is a more convenient way to represent moderation; however, the diagram in Figure 7.1 is sometimes used.

The same assumptions are required for moderation analysis as for other regression analyses that include multiple predictor variables. The dependent variable Y must be quantitative, and ideally it should be approximately normally distributed with no extreme outliers. (Formally it is more correct to say that the residuals from regression are assumed to be normally distributed; however, the distribution of residuals often resembles the distribution for the Y variable.) All relations between variables are assumed to be linear. Other assumptions (e.g., homogeneity of variance of Y scores across levels of X_1 and other predictor variables) should also be met. Sample size requirements are similar to those for multiple regression with multiple predictors (also see Aiken & West, 1991).

This chapter is limited to consideration of the case for two predictors and their interaction. The generalization of the methods to larger number of predictors is reasonably straightforward. When we ask about the types of variables for X_1 and X_2, there are three possible combinations.

1. X_1 and X_2 can both be categorical variables.
2. One of the predictor variables can be categorical, the other can be quantitative. The categorical predictor variable is usually treated as the moderator variable.
3. Both X_1 and X_2 can be quantitative variables.

The following sections consider each of these three situations.

7.2 INTERACTION BETWEEN TWO CATEGORICAL PREDICTORS: FACTORIAL ANOVA

When both predictor variables are categorical, the most convenient way to analyze the data is usually factorial ANOVA (see Volume I, Chapter 16 [Warner, 2020]). In this situation, we might call categorical variable X_1 Factor A and categorical variable X_2 Factor B. In a factorial design, the values of a factor (or **levels of a factor**) can correspond to different types or amounts of treatment, or they may identify naturally occurring group memberships. A two-way factorial ANOVA provides F tests for a main effect for Factor A, a main effect for Factor B, and the A × B interaction. Eta squared (η^2) effect sizes can be obtained for each of these. Cell (group) means must be provided in some form to understand the nature of the interaction (and the nature of main effect group differences). The nature of an interaction can be illustrated as a line graph (in which lines connect the points that represent cell means), or as a bar graph (in which there is one bar for each group or cell, and the heights of bars represent group means). Equivalent information is provided by a table of cell means.

Factorial designs are common in experiments (in which a researcher manipulates one or more variables). However, they can also be used with categorical predictor variables that are not manipulated (i.e., that represent naturally occurring group membership).

As a brief review, consider results from a study conducted by Lyon and Greenberg (1991). (This data set is not provided.) They set up a 2 × 2 factorial study in which the first factor (family background of participant, i.e., whether the father was diagnosed with alcoholism) was assessed by self-report. This represents naturally occurring group membership. The second factor was experimentally manipulated. For the manipulation, a research assistant acted out a script; he asked the female participants to volunteer their time for another study to help him out.

Number of minutes volunteered was the dependent variable. The manipulated variable was the way the research assistant presented himself. In the nurturant ("Mr. Right") condition, he talked about the way he was supportive and helpful to other people. In the exploitive ("Mr. Wrong") condition, he talked about getting other people to write his papers and do his laundry. Women were randomly assigned to hear either the Mr. Right or Mr. Wrong performance. On the basis of theories about codependence, the researchers hypothesized that women who had fathers diagnosed with alcoholism would show a "codependent" response; that is, they would offer more time to help Mr. Wrong than Mr. Right. Conversely, women who did not have fathers diagnosed with alcoholism were expected to offer more time to Mr. Right.

In this study, there was one experimentally manipulated factor (nurturant vs. exploitive person) and one naturally occurring membership (whether the woman had a father diagnosed with alcoholism or not). In such situations, the naturally occurring group membership factor is often thought of as the moderator. That is, in describing the nature of the interaction, the focus was on the difference in the way women who did or did not have fathers diagnosed with

Figure 7.3 Plot of Cell Means for the Significant Interaction Between Type of Person Needing Help and Family Background of Female Participant as Predictors of Amount of Time Offered to Help

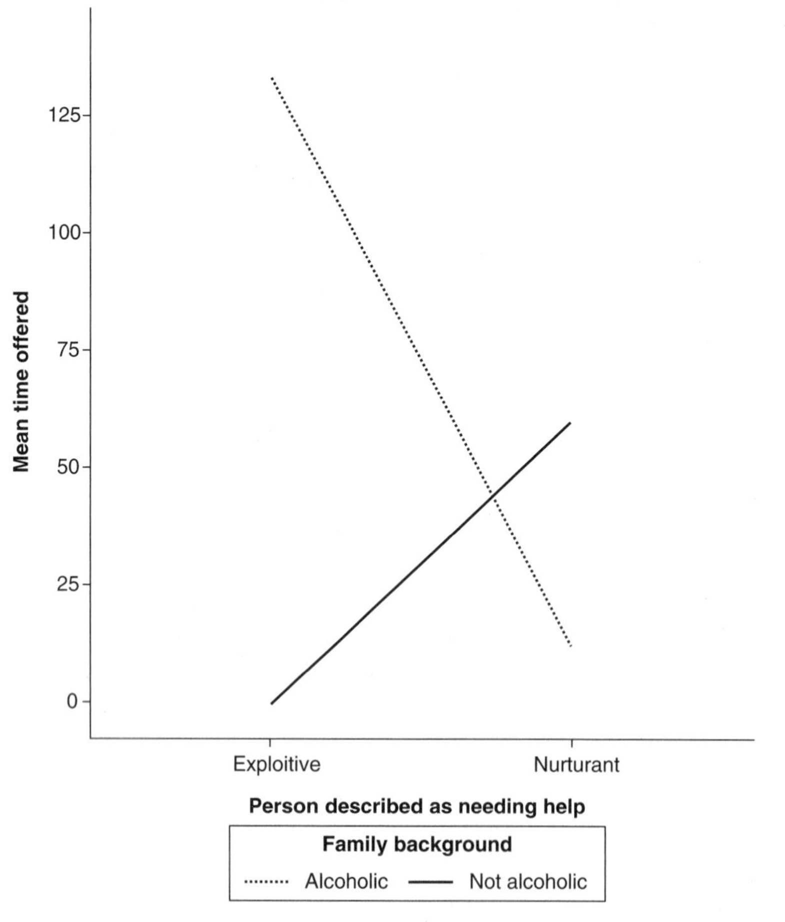

Source: Based on data in Lyon and Greenberg (1991).

alcoholism reacted to the Mr. Right and Mr. Wrong scenarios. The F for the interaction was statistically significant; the nature of the interaction can be seen in Figure 7.3.

The nature of the interaction in Figure 7.3 was consistent with their prediction. On average, women who had fathers diagnosed with alcoholism volunteered more time to help an exploitive person than a nurturant person. On the other hand, women who did not have fathers diagnosed with alcoholism volunteered more time to help a person who was described as nurturant (and in fact, none of them volunteered any time to help the exploitive person).

When an interaction is present, we can say that there are "different slopes for different folks" (as suggested by Robert Rosenthal). In this example, the "different folks" correspond to the naturally occurring family background groups (women who did vs. did not have fathers diagnosed with alcoholism). Women with fathers diagnosed with alcoholism offered more time in the Mr. Wrong condition; they offered less time to Mr. Right. On the other hand, women who did not have fathers diagnosed with alcoholism offered more time to Mr. Right and offered no time to Mr. Wrong. The effect of the manipulation (whether the research assistant was described as Mr. Right or Mr. Wrong) was different for the two types of folks (women from different family backgrounds). This interaction is **disordinal** (that is, the lines cross; the rank order of mean amount of time given to Mr. Right and Mr. Wrong was opposite for the two groups of women). Many interactions are not disordinal; none of the interactions in later graphs of this chapter show disordinal interactions.

When all predictor variables are categorical and the outcome variable is quantitative, and interactions between categorical predictors are of interest, it is usually most convenient to do the analysis as a factorial ANOVA (using the SPSS GLM procedure). It is possible to do the same analysis as a regression (using dummy variables to represent group membership on Factor A and Factor B and products between dummy variables to represent interactions). A regression analysis with dummy predictor variables yields the same results as a factorial ANOVA, although the information is presented in different ways in the output. Regression output usually does not directly include group means, although these can be calculated from the coefficients of the regression equation. The results for prediction of a quantitative Y variable from two categorical variables are the same for ANOVA and regression (provided that the regression includes an interaction term) because these analyses are both specific cases of the GLM.

The remainder of this chapter examines interactions that involve one categorical and one quantitative predictor or two quantitative predictors. It is convenient to use multiple regression analysis for these situations.

7.3 INTERACTION BETWEEN ONE CATEGORICAL AND ONE QUANTITATIVE PREDICTOR

The data set in the file firstexample.sav is used to demonstrate analysis of potential interactions between one categorical and one quantitative predictor variable. This hypothetical data set provides information to predict annual salary (given in thousands of dollars per year) from sex, college, and years of experience. For instance, if a faculty member's annual salary is given as 45.3, this corresponds to 45.3 × $1,000 = $45,300. The categorical variables used as predictors include sex (coded 0 = female, 1 = male), and college (coded 1 = liberal arts, 2 = business, and 3 = science and engineering). In all following examples, the dependent (Y) variable is salary.

In Chapter 6, on dummy variables, you saw that adding a categorical variable (such as sex) as a predictor in a regression equation provides information about mean differences in salary between groups defined by that categorical variable (i.e., men and women). In the following equation: $\text{Salary}' = b_0 + b_1\text{Sex} + b_2\text{Years}$, a statistically significant value for the b_1 coefficient tells us whether mean salaries differ for men versus women (after adjusting for the effect of the other predictor, years). The sign of the b_1 coefficient tells us which group had a higher mean or higher intercept. In other words, b_1 tells us about the difference between the intercepts of the regression lines that predict salary from years for the male versus female groups.

This equation, Salary′ = $b_0 + b_1$Sex + b_2Years, assumes that there is no interaction between sex and years as predictors of salary. Interaction can be understood as "different slopes for different folks." The sex, years, and salary example is a situation in which intercepts and slopes of lines to predict salary from years have unusually simple interpretations. The intercepts correspond to starting salaries (i.e., the predicted salaries for persons with 0 years of experience). Slopes within the male and female groups correspond to annual raises (i.e., for each additional year of experience, how much does salary tend to change?).

Prediction of salary from years is a nice example because the quantitative variable, years, can have a value of 0 (when a person is first hired). In this situation, the intercept b_0 (the predicted value of salary when years = 0) has a meaningful interpretation: It corresponds to average starting salary. In many real-world situations, it is not possible for anyone to have a score of 0 on the X predictor variable. In these cases, an intercept b_0 estimate is necessary to locate the regression line in a plot, but it has no meaningful interpretation.

Men and women may also have different average starting salaries. This can be called a "main effect" of sex on salary. In this regression analysis, men and women can have different intercepts (b_0 for women and $b_0 + b_1$ for men). If b_1 differs significantly from 0, they have significantly different starting salaries. (This was discussed in detail in Chapter 6.) (In later examples you will see that men and women may also receive different annual salary increases. In other words, there can be a different slope to predict salary from years for men than for women. That is called an interaction between sex and years as predictors of salary, or moderation of the effect of years on salary by sex.)

In the following example, the quantitative variable salary is predicted from the categorical variable sex and the quantitative variable years. (A new predictor to test significance of an interaction between sex and years will be introduced in a later section.) The imaginary data in the file firstexample.sav provides information to predict salary (Y, given in thousands of dollars) from sex and years.

Chapter 3 showed that you can get a preliminary look at within-group regression slopes by splitting the data file into groups (e.g., male and female groups) and examining the regression lines separately for each group. That is done in Section 7.4. This provides a preview look at data within each group. The preliminary look at data in Section 7.4 does not tell us whether any difference in slopes we might see by visual examination is large enough to be judged statistically significant. Examination of separate scatterplots also makes it possible to see potential bivariate outliers within each group and to assess whether the X_2, Y association is linear within each group.

7.4 PRELIMINARY DATA SCREENING: ONE CATEGORICAL AND ONE QUANTITATIVE PREDICTOR

Regressions that include interaction terms require assumptions like those for other multiple regressions. Scores on the Y' dependent variable must be quantitative; a histogram should show reasonably normally distributed scores without extreme outliers. Distribution shape for quantitative predictors is less critical, but an approximately normal distribution shape without extreme outliers is desirable. For categorical predictors, numbers of cases in each group should be reasonably large. (I prefer $n > 30$ per group, but people sometimes settle for smaller group sizes.) Associations among all quantitative variables (including predictor and dependent) should be linear (as assessed by scatterplots) and should not have extreme bivariate outliers.

To examine the association of an X_2 quantitative predictor and a Y outcome variable separately within groups, we set up scatterplots for Y and X_2 (separately for men and women). When we examine these separate scatterplots, we generally hope that relations between Y and X_2 are linear within both groups and that there are no bivariate outliers. (If relations within groups do not appear linear, more complicated analyses for assessment of nonlinear interactions is needed; see Aiken & West, 1991.) Evaluation of normality, outliers, and linearity is covered in Volume I (Warner, 2020) and in Chapter 2 in the present volume and are not repeated here. The following section shows how to examine associations of X_1 and Y separately within groups.

7.5 SCATTERPLOT FOR PRELIMINARY ASSESSMENT OF POSSIBLE INTERACTION BETWEEN CATEGORICAL AND QUANTITATIVE PREDICTOR

When one predictor is categorical, data screening should include examination of regression lines within subgroups (such as male and female groups) in scatterplots. Scatterplots can be used to evaluate whether these slopes are linear and can provide a preview of possible interaction. The data used in this section are in the file firstexample.sav. To obtain these separate regression lines, use the <Scatter/Dot> command. First, make the following menu selections: <Graphs> → <Legacy Dialogs> → <Scatter/Dot>. In the first dialog box, shown in Figure 7.4, select "Simple Scatter" and click Define.

In the next dialog box, shown in Figure 7.5, enter the regression dependent variable in the box for "Y Axis," the quantitative predictor years in the box for "X Axis," and the categorical moderator variable (sex) in the box for "Set Markers by." The categorical moderator variable has only two groups in this example, however, this procedure can also be used when categorical variables identify more than two groups. Then click OK. A scatterplot, similar to those seen in earlier regressions, appears (not shown here). The Chart Editor is used to obtain the separate regression lines within male and female subgroups. To open the Chart Editor, double-click on the scatterplot; the Chart Editor dialog box appears (as shown in Figure 7.6).

The scatterplot in Figure 7.6 is shown in the opened Chart Editor window. To open the Chart Editor, double-click on the scatterplot; the menu headings seen in Figure 7.6 appear. In Figure 7.6, I have already edited the lines and case markers to make them clearly distinguishable (unfilled triangles for each male case, filled circles for female cases).

To request separate regression lines for subgroups (men and women), make the following menu selections at the top-level menu bar in the Chart Editor window in Figure 7.6: <Elements> → <Fit Line at Subgroups>. The resulting regression lines appear in Figure 7.7 (this figure has been edited to improve appearance). Further chart editing procedures are not demonstrated here. There are many online tutorials (YouTube videos and PDF files) that

Figure 7.4 First Step in Obtaining Separate Regression Lines for Subgroups

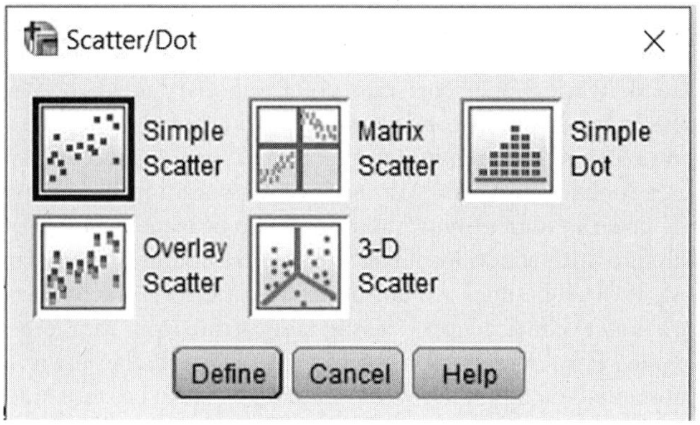

Figure 7.5 Simple Scatterplot Dialog Box to Identify Predictor, Outcome, and Marker Variables

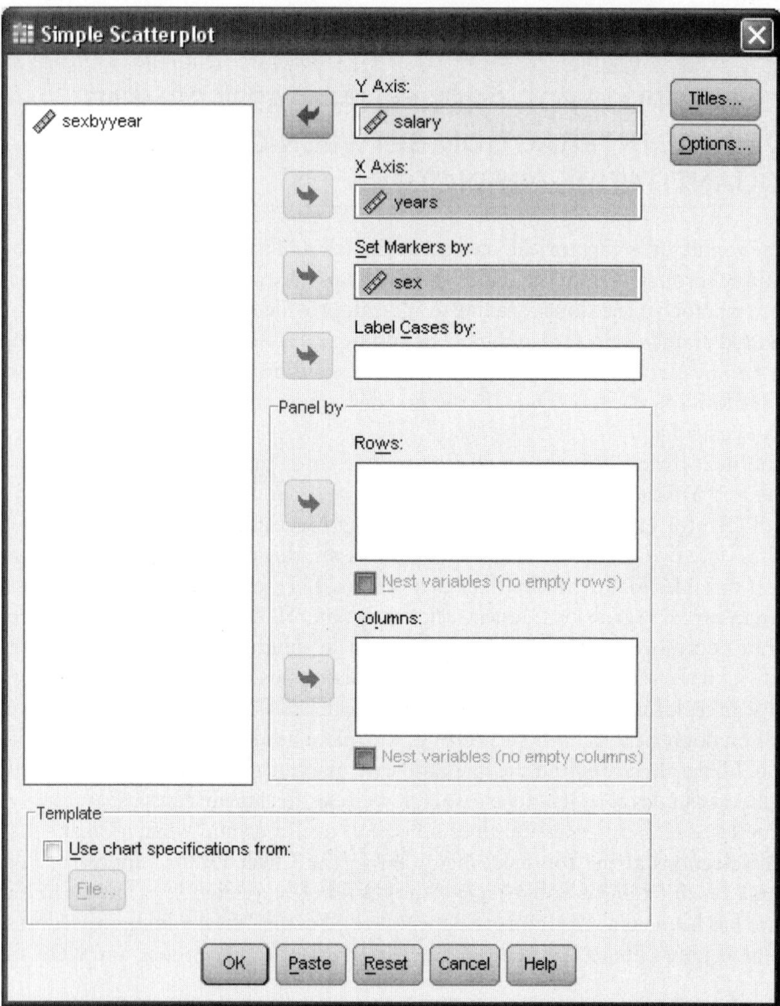

demonstrate additional editing features. Alternatively, you may prefer to use other programs to generate figures. Most graphs in this book were created using SPSS and then extensively edited to improve appearance.

The upper (solid, heavier) line represents predicted salary for men. The lower (dotted, lighter) line represents salary predictions for women. You should be able to see that, overall, women earn less on average than men in this hypothetical data set and that average predicted salary increases as a function of years. Also note that association between years and salary appears to be fairly linear in both groups; there appear to be some bivariate outliers. If curves had appeared, analysis of interaction would need to be take nonlinear interaction into account (see Aiken & West, 1991). The lines are not perfectly parallel. Perfectly parallel lines would indicate complete absence of interaction. The slight departure from parallel seen in Figure 7.7 may be due to sampling error. A regression analysis will tell us whether the observed departure from parallel is substantial enough, relative to sampling error, to be statistically significant.

Figure 7.6 Initial Scatterplot Shown in Chart Editor Dialog Box

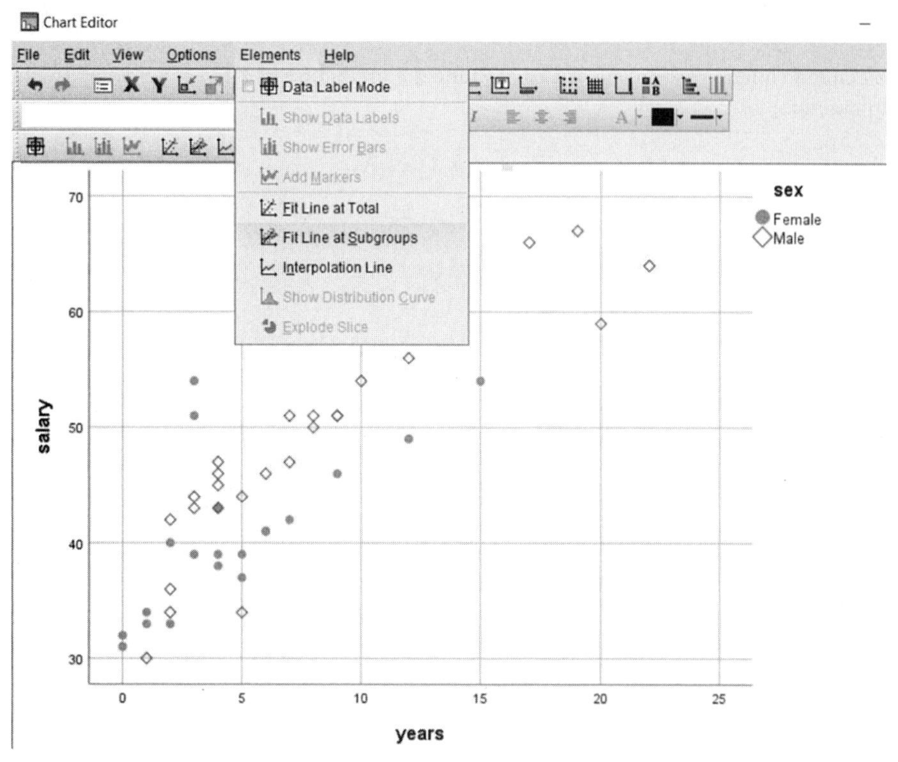

Figure 7.7 Separate Regression Lines to Predict Salary From Years Within Female and Male Subgroups in Data File firstexample.sav

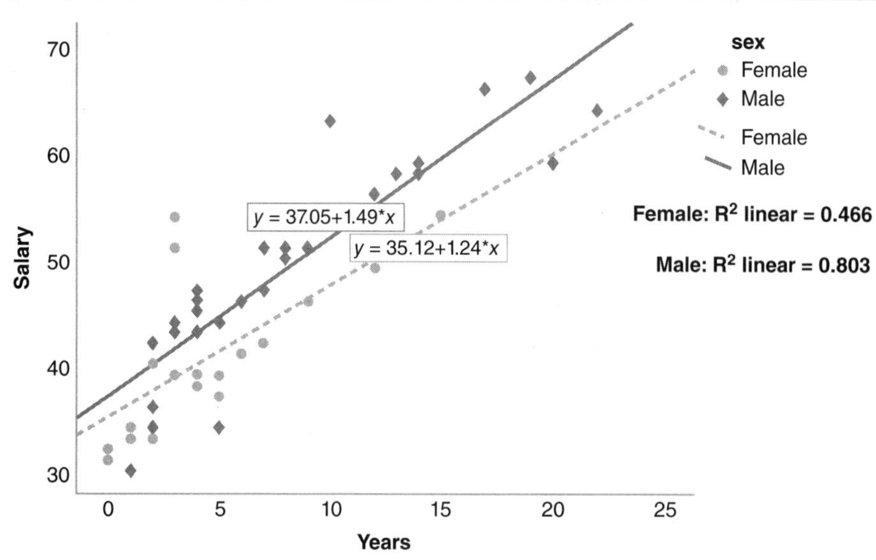

7.6 REGRESSION TO ASSESS STATISTICAL SIGNIFICANCE OF INTERACTION BETWEEN ONE CATEGORICAL AND ONE QUANTITATIVE PREDICTOR

The general equation for an analysis to examine the prediction of Y from X_1 and X_2 (without an interaction) is:

$$Y' = b_0 + b_1X_2 + b_2X_2. \tag{7.1}$$

With an added term to represent an interaction between X_1 and X_2, the equation becomes:

$$Y' = b_0 + b_1X_2 + b_2X_2 + b_3\ (X_1 \times X_2). \tag{7.2}$$

Note that when an interaction term (such as $X_1 \times X_2$) is included, the variables involved in that interaction (X_1, X_2) must also be included in the regression equation. The interaction term is created by forming the product of the two predictors; the new variable X1byX2 = $X_1 \times X_2$. In this example, we create a product variable called sexbyyears that is the product of sex by years. For the current example, the equation is: Salary′ = b_0 + b_1Sex + b_2Years + b_3(Sexbyyears).

To create this new interaction variable in SPSS, use the <Transform> → <Compute Variable> command, as shown in Figure 7.8. (Later you will see that if both X_1 and X_2 are quantitative, scores should be centered before they are multiplied. It is not necessary to center scores when one or both predictors are categorical; Aiken & West, 1991.)

Recall that in this data set (firstexample.sav), sex is coded 0 for female and 1 for male. By substituting those values of 0 and 1 into Equation 7.2, we can obtain separate prediction equations for men and women:

For women: Salary′ = $b_0 + b_1(0) + b_2$Years + b_3(0 × Years) = $b_0 + b_2$Years.

For men: Salary′ = $b_0 + b_1(1) + b_2$Years + b_3(1 × Years) = $(b_0 + b_1) + (b_2 + b_3)$Years.

Figure 7.8 Compute Variable Dialog Box to Compute Product (Interaction) Term

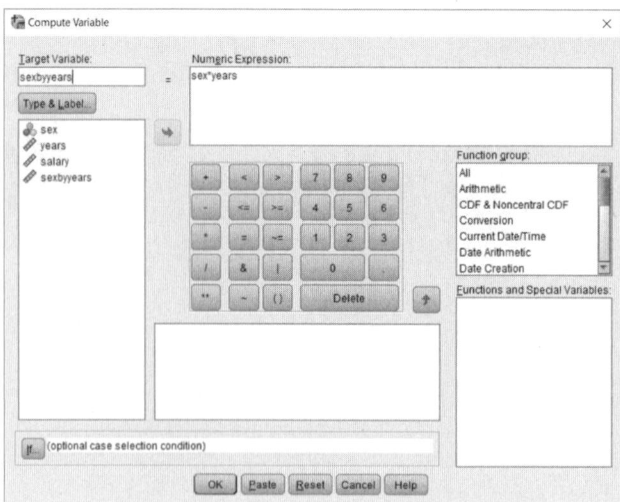

For women, the intercept for the regression line is b_0, and the slope is b_2. For men, the intercept is ($b_0 + b_1$), and the slope is ($b_2 + b_3$). A statistically significant value for b_1 indicates that the intercepts of the lines differ significantly for men versus women. If b_3 is statistically significant, it tells us that the slopes differ significantly for men versus women (in other words, that moderation or interaction is present).

Now let's examine a regression analysis that uses sex, years, and sexbyyears to predict salary in the data set firstexample.sav. All three variables are entered on one step as shown in Figure 7.9.

As in other uses of the SPSS regression procedure, click the Statistics button and request part and partial correlations (these provide effect size information). Part of the results appear in Figure 7.10. Note that the coefficient for b_3, which corresponds to a potential interaction, is not statistically significant; $b_3 = .258$, $t(46) = .808$, $p = .423$. This tells us that the lines in the graph in Figure 7.7 do not depart significantly from parallel. We can also say that we do not have different slopes for different folks; there is no interaction between sex and years; sex does not moderate the effect of years on salary.

Figure 7.9 Regression to Predict Salary From Sex, Years, and Sexbyyears

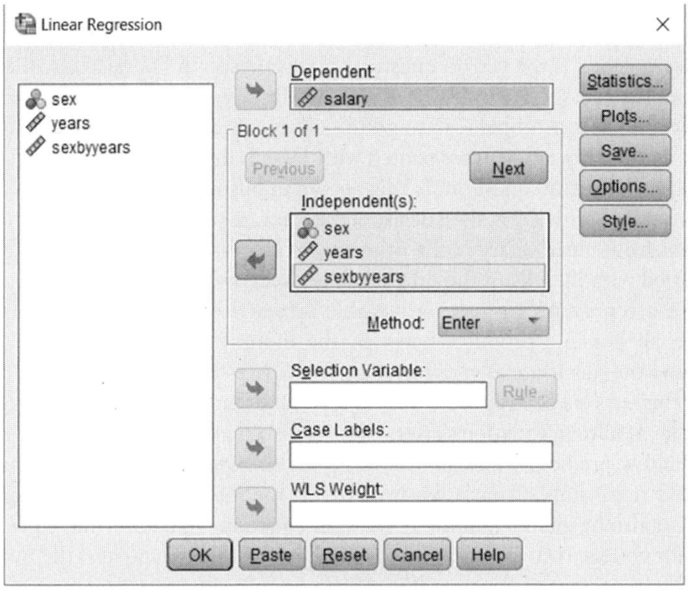

Figure 7.10 Regression Results for Prediction of Salary From Sex, Years, and Sexbyyears for Data in firstexample.sav

Coefficients[a]

Model		Unstandardized Coefficients B	Std. Error	Standardized Coefficients Beta	t	Sig.	Correlations Zero-order	Partial	Part
1	(Constant)	35.117	1.675		20.970	.000			
	sex	1.936	2.289	.098	.846	.402	.462	.124	.059
	years	1.236	.282	.692	4.383	.000	.866	.543	.304
	sexbyyears	.258	.320	.164	.808	.423	.816	.118	.056

a. Dependent Variable: salary

Upon finding that an interaction is not statistically significant, a data analyst may want to rerun the regression analysis using only sex and years as predictors. However, reporting a regression that includes a nonsignificant interaction term between sex and years would dramatize the absence of an interaction. If the original research questions included hypothesized interaction, an analysis that includes test for that interaction should be reported. On the other hand, if an interaction term was "thrown into" the analysis as part of an attempt to wring statistically significant results out of data, this is a form of *p*-hacking. In this situation it makes no sense to report a nonsignificant interaction term that was not predicted (particularly if the nature of the interaction is not interpretable).

7.7 INTERACTION ANALYSIS WITH MORE THAN THREE CATEGORIES

The file firstexample.sav includes the categorical variable college. The variable college represents college membership as 1 = liberal arts, 2 = business, and 3 = science and engineering. As discussed in Chapter 6, on dummy variables, when a categorical variable with k groups is used as a predictor in regression, $(k - 1)$ dummy variables are needed to represent group membership in order to use that categorical variable as a predictor in a regression equation. Dummy-coded dummy variables are used in this example. Each dummy variable can be thought of as the answer to a yes/no question.

In this example, D_1 is the yes/no question, Is this person in liberal arts (coded 1= yes, 0 = no)? If the coefficient associated with D_1 is statistically significant, it tells us that mean salary differs significantly between liberal arts and the other two colleges (business and science and engineering), and the b coefficient associated with D_1 tells us the magnitude of that difference.

D_2 is a different yes/no question: Is this person in business (coded 1 = yes, 0 = no)? If the coefficient associated with D_2 is statistically significant, that tells whether the mean salary in business differs from mean salary in the other two colleges, and the magnitude of the b coefficient associated with D_2 tells us the difference between means.

To set up a regression to assess a possible interaction between college and years, we need to form all possible products between the dummy variables that represent college (D_1 and D_2) and the quantitative variable years (e.g., $D_1 \times$ Years and $D_2 \times$ Years, called D1byyears and D2byyears). This can be done using the <Transform> → <Compute Variable> command. The resulting equation (interaction of a categorical variable with three groups with a quantitative predictor) has the following form. Note that when an interaction term is included in a regression, all variables involved in the interaction (in this example, D_1, D_2, and years) must also be included in the regression equation. The following equation includes main effects for college (D_1, D_2) and years as predictors of salary and also the two interaction terms needed to represent their interaction (D1byyears, D2byyears).

$$Y' = b_0 + b_1D_1 + b_2D_2 + b_3\text{Years} + b_4\text{D1byyears} + b_5\text{D2byyears}. \quad (7.3)$$

If we want a significance test for the overall college by years interaction, we need to find out how much predictive information the two combined interaction terms add to the analysis and whether this additional predicted variance is large enough to be statistically significant. This can be done using hierarchical regression (user-determined order of entry), which was covered in earlier chapters about multiple regression. In the following example (still using data in the file firstexample.sav), in Step 1, the variables that represent college and years are entered; in Step 2 (shown in Figure 7.11) the two interaction variables are added. In addition, request R squared change as an additional statistic (as shown in Figure 7.12).

Figure 7.11 Step 2 in Hierarchical Entry of Predictors: Interaction Terms Entered as Block 2 (Sex, D_1, and D_2 Were Entered in Block 1)

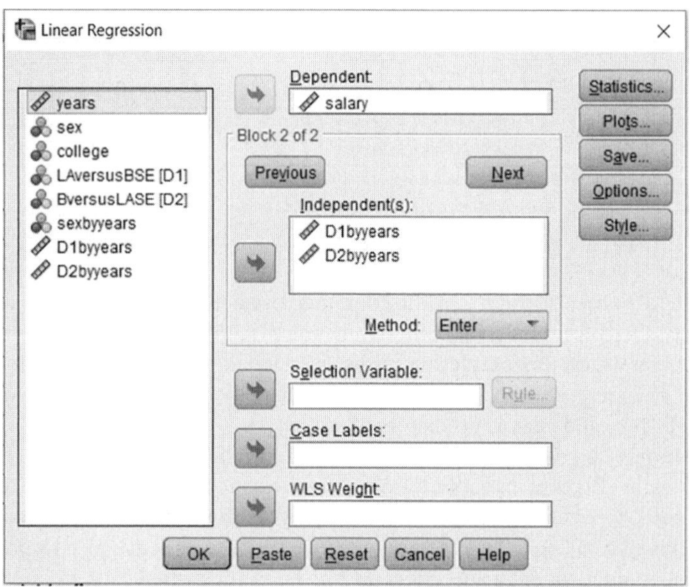

Figure 7.12 Request R Squared Change Table

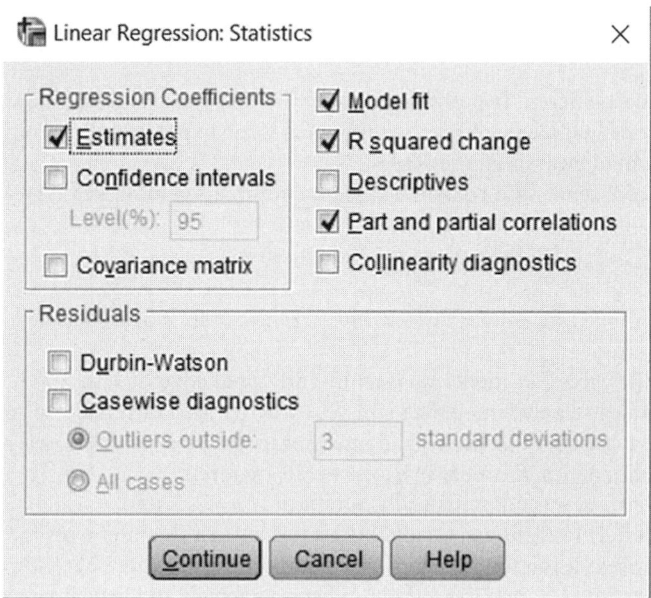

Figure 7.13 Results: R^2 Change Across Two Models: Step 1, Years and College; Step 2, Interaction Between Years and College

Model Summary

					Change Statistics				
Model	R	R Square	Adjusted R Square	Std. Error of the Estimate	R Square Change	F Change	df1	df2	Sig. F Change
1	.881[a]	.776	.761	4.766	.776	53.017	3	46	.000
2	.894[b]	.800	.777	4.604	.024	2.642	2	44	.082

a. Predictors: (Constant), BversusLASE, years, LAversusBSE

b. Predictors: (Constant), BversusLASE, years, LAversusBSE, D1byyears, D2byyears

For Model 1 in Step 1, the R^2 for the combined predictive contribution of years, D_1, and D_2 (college membership) was statistically significant, $F(3,46) = .776$, $p < .001$. The increase in R^2 for Model 2, in which the interaction terms were added to the model, was not statistically significant, R^2 change = .024, $F(2,44) = 2.642$, $p = .082$. There was no evidence of an interaction between college and year as predictors of salary.

Some authors (e.g., Aiken & West, 1991) refer to predictors such as years, D_1, and D_2 as representing "main effects" for years and college in the context of regressions that also include interaction terms between them. This is potentially confusing. In a regression without an interaction term (e.g., $Y' = b_0 + b_1D_1 + b_2D_2 + b_3\text{Years}$), it makes sense to interpret D_1 and D_2 as information about a college main effect. However, in the presence of an interaction term, D_1 and D_2 are better interpreted as information about the intercepts of different regression lines (Crawford, Jussim, & Pilanski, 2014; Whisman & McClelland, 2005). Interpretation of coefficients associated with dummy coded predictors such as D_1, D_2 in analyses that do not include interaction effects was discussed in the Chapter 6, on dummy variables, and that information is not repeated here.

7.8 EXAMPLE WITH DIFFERENT DATA: SIGNIFICANT SEX-BY-YEARS INTERACTION

Next we examine a different hypothetical data set that was constructed to show a sex-by-years interaction: secondexample.sav. A regression analysis is run to predict salary from sex, years, and sexbyyears. Results appear in Figure 7.14.

The raw score regression equation, based on information in Figure 7.14, is:

$$\text{Salary}' = b_0 + b_1\text{Sex} + b_2\text{Years} + b_3(\text{Sexbyyears})$$

$$= 34.582 - 1.5 \times \text{Sex} + 2.346 \times \text{Years} + 2.002 \times \text{Sexbyyears}.$$

The overall regression model was significantly predictive of salary, $F(3,56) = 144.586$, $p < .001$, with an extremely large effect size, $R^2 = .886$. (I used extremely large effect sizes in many examples to make it easy to see and interpret patterns of results. In real-world applications, in many disciplines, R^2 values in excess of .20 are rare.)

The effect for sex was not statistically significant, $b_1 = -1.5$, $t(56) = -.385$, $p = .702$, and the corresponding effect size was essentially 0. This tells us that the intercepts for the male-only and female-only regression lines do not differ significantly. The intercept for women is b_0, the intercept for men is $b_0 + b_1$, and the difference between female and male intercepts is b_1 (−1.5).

The effect for years was statistically significant, $b_2 = 2.346$, $t(56) = 11.101$, $p < .001$, $sr^2 = .25$. Because the sex-by-years interaction was also statistically significant, interpretation

Figure 7.14 SPSS Output for Regression: Sex, Years, and Sex × Years Interaction as Predictors of Salary (in Tens of Thousands of Dollars) for Data in secondexample.sav

Model Summary

Model	R	R Square	Adjusted R Square	Std. Error of the Estimate
1	.941[a]	.886	.880	7.854

a. Predictors: (Constant), sexbyyears, years, sex

ANOVA[a]

Model		Sum of Squares	df	Mean Square	F	Sig.
1	Regression	26756.614	3	8918.871	144.586	.000[b]
	Residual	3454.386	56	61.685		
	Total	30211.000	59			

a. Dependent Variable: salary

b. Predictors: (Constant), sexbyyears, years, sex

Coefficients[a]

Model		Unstandardized Coefficients		Standardized Coefficients			Correlations		
		B	Std. Error	Beta	t	Sig.	Zero-order	Partial	Part
1	(Constant)	34.582	2.773		12.470	.000			
	sex	-1.500	3.895	-.033	-.385	.702	.248	-.051	-.017
	years	2.346	.211	.658	11.101	.000	.808	.829	.502
	sexbyyears	2.002	.335	.530	5.980	.000	.684	.624	.270

a. Dependent Variable: salary

of the effect of years on salary differs by sex and should be interpreted taking sex into account. For the sex-by-years interaction, $b_3 = 2.002$, $t(56) = 5.98$, $p < .001$, $sr^2 = .073$.

Using codes female = 0 and male = 1, the female group is the reference group. For women, the intercept is b_0 and the slope is b_2. For men, the intercept is $(b_0 + b_1)$ and the slope is $(b_2 + b_3)$. If b_1 is statistically significant, the intercepts of the regression lines differ for men and women. If b_3 is statistically significant, the slopes of the regression lines differ for men and women (there is evidence of interaction or moderation). Values of b_1 and b_3 can be negative, positive, or 0. The values of b_1 and b_3 are independent (e.g., whether intercepts differ is unrelated to whether slopes differ).

Substituting in values of 0 and 1 for women and men, two separate regression equations are obtained:

$$\text{For women, Salary}' = b_0 + b_1(0) + b_2\text{Years} + b_3\ (0 \times \text{Years}) = b_0 + b_2\text{Years}.$$

$$\text{For women, Salary}' = 34.582 + 2.346 \times \text{Years}.$$

In words, women's average starting salary (at years = 0) is $34,582, and their average annual raise is $2,346. In other words, a female faculty member can calculate her predicted salary by substituting her years of experience; if she has 10 years of experience, the model predicts her salary = 34.582 + 2.346 × 10 = 58.042 (i.e., $58,042). Because the overall multiple R is so high for these hypothetical data, that predicted salary is probably close to actual salary for most members of the sample. In actual data, with much lower values of R, the estimates for many individuals will be incorrect by many thousands of dollars.

$$\text{For men, Salary}' = b_0 + b_1(1) + b_2\text{Years} + b_3(1 \times \text{Years}) = (b_0 + b_1) + (b_2 + b_3)\ \text{Years}.$$

$$\text{For men, Salary}' = (34.582 - 1.5) + (2.346 + 2.002) \times \text{Years} = 33.082 + 4.348.$$

Men's average starting salary is $33,082, and their average annual raise is $4,438.

A scatterplot with separate lines for female and male subgroups makes the nature of the interaction clear. The scatterplot in Figure 7.15 was created using the same procedures as in the previous example. Note that the regression equations for these lines correspond to the ones above that were deduced by substituting in scores of 0 and 1. Men and women have approximately equal salaries at year = 0 (the intercept). In the hypothetical data in secondexample.sav, the average annual raises received by men are much larger than those received by women, and the line to predict salary for years has a steeper slope for men than women. Using the Chart Editor, you can request separate regression lines by subgroup (in this example, male and female are the subgroups). In Figure 7.15 you can see that the intercept (starting salary) was almost exactly equal for the male and female groups. However, the slope (average salary increase for each additional year) was steeper for men than for women. Men tended to receive higher annual raises than women in this hypothetical example. Note that the regression equations for subgroups that appear in Figure 7.15 are the same as those deduced from the regression equation by substituting in values of 0 and 1 for sex. (The free interaction graphing program created by Daniel Soper can also be used to generate these lines; use of that program is demonstrated later in this chapter.)

7.9 FOLLOW-UP: ANALYSIS OF SIMPLE MAIN EFFECTS

We now have graphs of separate lines for men and women (in Figure 7.15). For each of these lines we may want to know: Is the multiple R to predict salary from years significant within each group? This is called an analysis of simple main effects (a term also used in factorial ANOVA). This is a common follow-up to evaluate significant interactions. It is possible, for example, that years is a significant predictor of salary for men but not for women.

Figure 7.15 Graph of Sex-by-Years Interaction in Prediction of Salary for Data in secondexample.sav

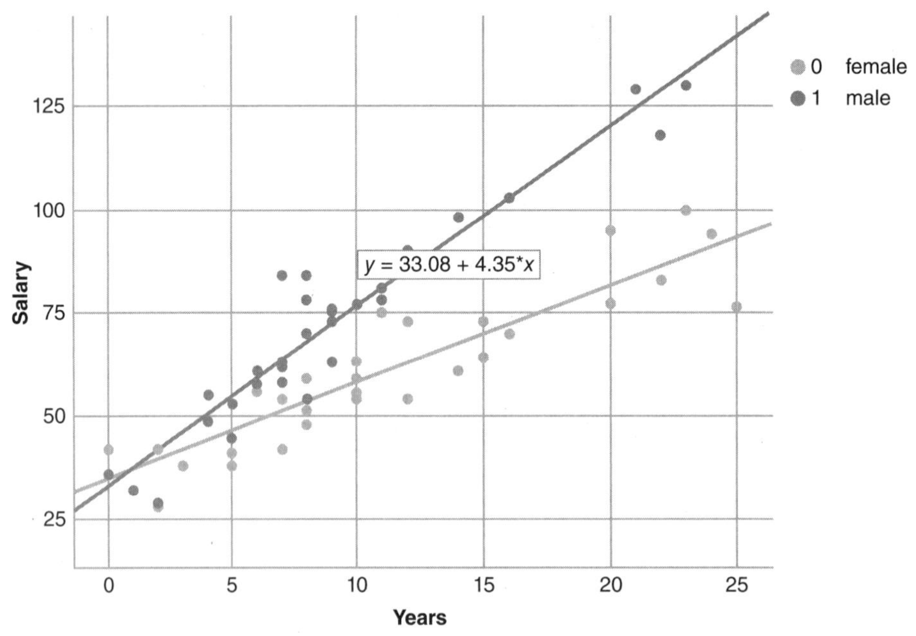

To obtain separate regressions for female and male subgroups (in the secondexample.sav data set), use the SPSS split file procedure: <Data> → <Split File>. (Note that this is different from the <Split into Files> menu option.) Within the Split File dialog box in Figure 7.16, select the radio button for "Organize output by groups"; then move the categorical moderator variable sex into the "Groups Based on" window, and click OK. Until this command is reversed (by going back into the Split File dialog box and choosing the radio button for "Analyze all cases, do not create groups"), all subsequent analyses are conducted separately for the male and female groups.

The next step is to run the regression to predict salary from years separately for each group. The variables sex and sexbyyears are not included in this set of regressions because, within each group, these would be constants. When this is done, the following output is obtained (Figures 7.17 and 7.18).

Within the female group, there was a statistically significant association between years and salary. Because there is only one predictor (years), we can either report F and R for the entire equation or t and sr^2 for the single predictor. The regression coefficients obtained here are the same as those obtained earlier using by-hand substitution of dummy variable codes into the regression equation that included the interaction term. For men, the regression to predict salary from years was also statistically significant, with $t(28) = 17.173, p < .001$ (Figure 7.18). The statistically significant interaction reported in Figure 7.14 tells us that these two groups have significantly different slopes to predict salary from years.

Interactions can be different from this example. For example, it is possible that a Y variable is significantly predictable from X_2 for one group (such as men) but not for the other group (women). Or the slope that relates X_2 to Y might be positive for one sex and negative for the other (this would be a disordinal interaction). In this example, the nature of the interaction was that men had a significantly larger positive slope to predict salary from years for women; in other words, they received higher salary increases per year (we know this because b_3 was statistically significant). However, starting salaries did not differ between groups (because b_1 was not significant).

Figure 7.16 SPSS Split File Dialog Box

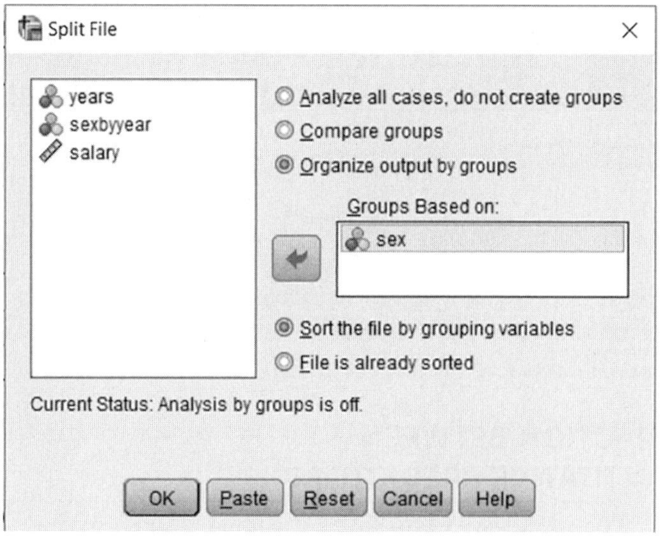

Figure 7.17 Simple Main Effects: Regression to Predict Salary From Years for Female Subgroup

Model Summary[a]

Model	R	R Square	Adjusted R Square	Std. Error of the Estimate
1	.899[b]	.807	.801	8.045

a. sex = female

b. Predictors: (Constant), years

ANOVA[a,b]

Model		Sum of Squares	df	Mean Square	F	Sig.
1	Regression	7601.620	1	7601.620	117.448	.000[c]
	Residual	1812.247	28	64.723		
	Total	9413.867	29			

a. sex = female

b. Dependent Variable: salary

c. Predictors: (Constant), years

Coefficients[a,b]

Model		Unstandardized Coefficients		Standardized Coefficients	t	Sig.	Correlations		
		B	Std. Error	Beta			Zero-order	Partial	Part
1	(Constant)	34.582	2.841		12.173	.000			
	years	2.346	.216	.899	10.837	.000	.899	.899	.899

a. sex = female

b. Dependent Variable: salary

Figure 7.18 Simple Main Effects: Prediction of Salary From Years for Male Subgroup

Model Summary[a]

Model	R	R Square	Adjusted R Square	Std. Error of the Estimate
1	.956[b]	.913	.910	7.658

a. sex = male

b. Predictors: (Constant), years

ANOVA[a,b]

Model		Sum of Squares	df	Mean Square	F	Sig.
1	Regression	17295.728	1	17295.728	294.908	.000[c]
	Residual	1642.139	28	58.648		
	Total	18937.867	29			

a. sex = male

b. Dependent Variable: salary

c. Predictors: (Constant), years

Coefficients[a,b]

Model		Unstandardized Coefficients		Standardized Coefficients	t	Sig.	Correlations		
		B	Std. Error	Beta			Zero-order	Partial	Part
1	(Constant)	33.082	2.666		12.408	.000			
	years	4.348	.253	.956	17.173	.000	.956	.956	.956

a. sex = male

b. Dependent Variable: salary

7.10 INTERACTION BETWEEN TWO QUANTITATIVE PREDICTORS

The statistical significance of an interaction between quantitative X_1 and X_2 predictors in a linear regression can be assessed by forming a new variable that is the product of the predictors and including this product term in a regression, along with the original predictor variables, as shown in Equation 7.4. A first approximation to a moderation model with quantitative X_1 and X_2 variables as predictors of Y could be given as:

$$Y' = b_0 + b_1X_1 + b_2X_2 + b_3(X_1 \times X_2), \tag{7.4}$$

where Y' is the predicted score on a quantitative Y outcome variable, and X_1 and X_2 are both quantitative predictor variables.

However, a problem arises when using the model in Equation 7.4. The product term $(X_1 \times X_2)$ is highly correlated with both X_1 and X_2. High correlations among predictors makes it difficult to distinguish their unique contributions. Aiken and West (1991) tell us that **centering** scores on X_1 and X_2 prior to calculating the $(X_1 \times X_2)$ product reduces this collinearity. Centering is done for each variable by subtracting the mean for that variable from each score (e.g., scores on X_1 are centered by subtracting the mean of X_1 from each individual X_1 score).

In practice, scores are centered for both quantitative predictors before calculating the product term to represent the interaction. This is a preview of the methods that are demonstrated in the next section.

1. Find the mean for each predictor variable (in this example, the means for X_1 and X_2). This information can be obtained from the SPSS <Descriptives> command.

2. Create new variables for centered scores. I use variable names that remind me which versions of the predictor have been centered. The <Transform> → <Compute Variable> command can be used to calculate these:

   $$X_1\_c = (X_1 - M_{X1}),$$

   $$X_2\_c = (X_2 - M_{X2}).$$

   Note that I have not centered scores on Y, the dependent variable. This is not necessary. (It isn't wrong, but it makes interpretation of results potentially confusing.)

3. Carry out an additional computation to obtain the product of the centered scores:

   $$\text{productcentered}X_1X_2 = X_1\_c \times X_2\_c.$$

4. Run a multiple linear regression with these three predictors: X_1, X_2, and productcenteredX_1X_2.

   Notice that I use the raw scores for X_1 and X_2 as predictors, rather than the centered scores. (It is not wrong to use the centered scores, but use of raw scores here makes interpretation of results easier.)

   Note that if you center Y or use the centered scores for X_1 and X_2 as predictors, interpretation of your results will differ from the examples that follow.

   In this example, based on Equation 7.4, the raw score for number of symptoms is predicted by:

   Y' = Intercept $b_0 + b_1 \times$ Healthy habits + $b_2 \times$ Age in years + $b_3 \times$ productcenteredX_1X_2.

5. Evaluate regression results (Figure 7.21) for the thirdexample.sav data with the following questions in mind:

   a. Is the overall model statistically significant, and what proportion of variance in Y scores can be predicted using X_1 and X_2 and their interaction? To answer this question, examine F in the ANOVA source table and multiple R^2.

b. Which of the three predictors (if any) are statistically significant when controlling for other predictors in the model; and what proportion of variance in Y can be uniquely predicted by each variable (i.e., what are the sr^2 effect sizes)? The p values (based on t values) are used to evaluate whether each raw score regression coefficient is statistically significant. Square each part correlation value to obtain sr^2, the proportion of variance predicted uniquely by each of the independent variables; this provides effect size information.

c. If the b_3 coefficient for the $X_1 \times X_2$ product term (productcenteredX_1X_2) is statistically significant, there is a statistically significant interaction between X_1 and X_2 as predictors of Y. If there is a statistically significant interaction, you need to be able to describe the nature of that interaction. To do this, set up a graph of predicted Y values for selected values of X_1 and X_2, as discussed in upcoming Section 7.13. In the following example, separate lines will be drawn for selected values of X_2 (age). Labels for the graph can be given in raw score units or standardized (z-score) units. On the basis of the pattern in this graph, you can say something about the nature of the interaction.

The following empirical example shows how to graph separate regression lines to predict symptoms from habits, separately for selected values of age, using SPSS linear regression and a separate graphics program. Instructions for the by-hand computation of selected predicted Y' values needed to create an interaction graph by hand, using procedures described by Aiken and West (1991), are provided in Appendix 7A at the end of this chapter.

7.11 SPSS EXAMPLE OF INTERACTION BETWEEN TWO QUANTITATIVE PREDICTORS

A hypothetical study involves assessing how age in years (X_2) interacts with number of healthy habits (X_1) (such as regular exercise, not smoking, and getting 8 hours of sleep per night) to predict number of physical illness symptoms (Y). The data for this hypothetical study are in the SPSS file named thirdexample.sav. Number of healthy habits ranged from 0 to 7; age ranged from 21 to 55. The equation that corresponds to this analysis is: Symptoms′ = b_0 + b_1Age + b_2Habits + b_3Agebyhabits.

Centered scores on age are obtained by finding the sample mean for age and subtracting this sample mean from scores on age to create a new variable, which in this example is named age_c (to represent centered age). Figure 7.19 shows the SPSS Compute Variable dialog box; age_c = age – 41.7253, where 41.7253 is the sample mean for age that was obtained by running descriptive statistics. Similarly, scores for habits_c were obtained by subtracting the mean number of habits from each individual score for number of habits.

After computing centered scores for both age and habits (age_c and habits_c, respectively), the compute procedure was used to form the interaction/product term agebyhabits = age_c × habits_c, as shown in Figure 7.20.

7.12 RESULTS FOR INTERACTION OF AGE AND HABITS AS PREDICTORS OF SYMPTOMS

The SPSS regression results for prediction of symptoms from age, habits, and the interaction of age and habits appear in Figure 7.21. The interaction was statistically significant, $b_3 = -.100$, $t(320) = -5.148$, $p < .001$. Age and habits were also statistically significant predictors of symptoms (i.e., older persons experienced a higher number of average symptoms compared

Figure 7.19 Creation of Centered Scores: Subtract Mean From Age to Create Age_c

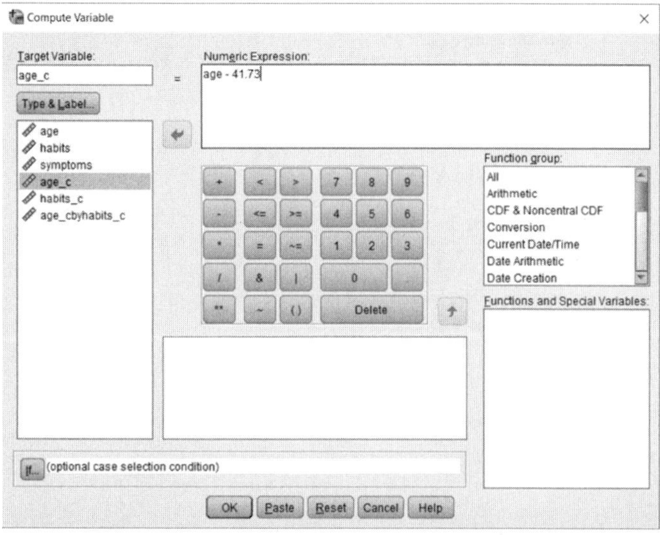

Figure 7.20 SPSS Compute Variable Dialog Box to Create Product Interaction Term

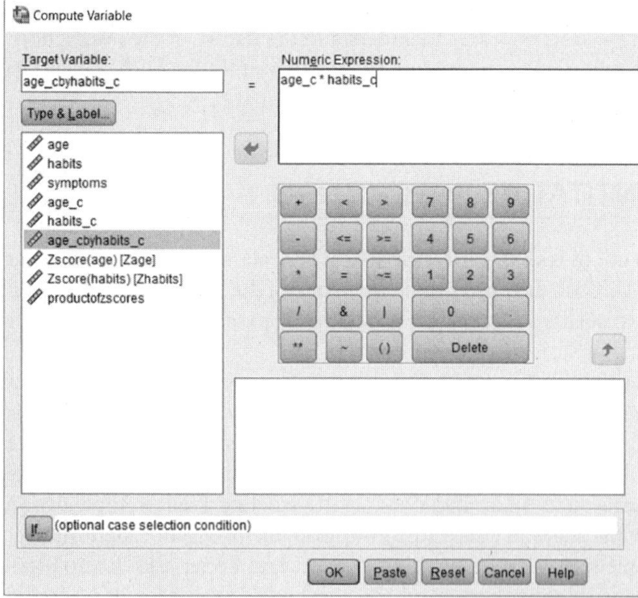

with younger persons; as number of healthy habits increased, number of symptoms decreased). The effect of habits on symptoms should be interpreted in light of the significant interaction between age and habits.

The overall regression model significantly predicted number of symptoms, $F(3, 320) = 286.93$, $p < .001$, $R^2 = .73$. Both age and number of healthy habits were significant predictors

Figure 7.21 Results for Regression That Includes Interaction Term Between Two Quantitative Variables for Data in the SPSS File Named thirdexample.sav

Model Summary

Model	R	R Square	Adjusted R Square	Std. Error of the Estimate
1	.854[a]	.729	.726	4.038

a. Predictors: (Constant), age_cbyhabits_c, age, habits

ANOVA[a]

Model		Sum of Squares	df	Mean Square	F	Sig.
1	Regression	14033.795	3	4677.932	286.928	.000[b]
	Residual	5217.127	320	16.304		
	Total	19250.923	323			

a. Dependent Variable: symptoms

b. Predictors: (Constant), age_cbyhabits_c, age, habits

Coefficients[a]

Model		Unstandardized Coefficients		Standardized Coefficients			Correlations		
		B	Std. Error	Beta	t	Sig.	Zero-order	Partial	Part
1	(Constant)	1.853	1.377		1.345	.180			
	age	.606	.025	.746	24.166	.000	.810	.804	.703
	habits	-1.605	.191	-.261	-8.394	.000	-.476	-.425	-.244
	age_cbyhabits_c	-.100	.019	-.153	-5.148	.000	-.006	-.277	-.150

a. Dependent Variable: symptoms

of symptoms. For age, b = .606, t(320) = 24.166, p < .001, sr^2 = .49. As age increased, number of symptoms also increased. For habits, b = –1.605, t(320) = –8.394, p < .001, sr^2 = .059. As number of healthy habits increased, symptoms decreased. These results should be interpreted in the context of the statistically significant interaction. For the interaction term, b = –.100, t(320) = –5.148, p < .001, sr^2 = .02. To understand the nature of the interaction, a graph of predicted values of symptoms as a function of selected values for both age and habits is needed.

7.13 GRAPHING INTERACTION FOR TWO QUANTITATIVE PREDICTORS

Graphing interactions is simple in factorial ANOVA; all we need is one value (a group mean) for each cell in the ANOVA. It is also simple when one of the variables involved in the interaction is categorical; all we need is a regression line separately for each group. Graphing is more complex when both predictor variables are quantitative.

The data analyst selects *representative* values for each of the predictor variables, uses the regression equation to generate predicted scores on the outcome variable for these selected values of the predictor variables, and then graphs the resulting predicted values. This interaction graph can be generated by hand, although procedures are somewhat tedious. The by-hand procedures used in Aiken and West (1991) are described in Appendix 7A.

The following example uses a free downloadable program called Interaction available at https://www.danielsoper.com/interaction/default.aspx. Soper asks you to participate in a brief online experiment that involves searching a webpage, then provides a free download of his program. The icon for the Soper Interaction program is an orange exclamation point. When you open the program, you see the screen view in Figure 7.22. Select <File> → < New Interaction Analysis> from the pull-down menu. In the second screen, Figure 7.23, you browse to find the data file. In the third screen view, Figure 7.24, you indicate whether you have missing values (which can be blanks or specific score values that have been designated as missing).

To assign roles to your variables (e.g., dependent, predictor, moderator) enter the variables in the appropriate windows in the Assign Model Variables screen (Figure 7.25).

Figure 7.22 Step 1 in Soper Interaction Program: Request New Interaction Analysis

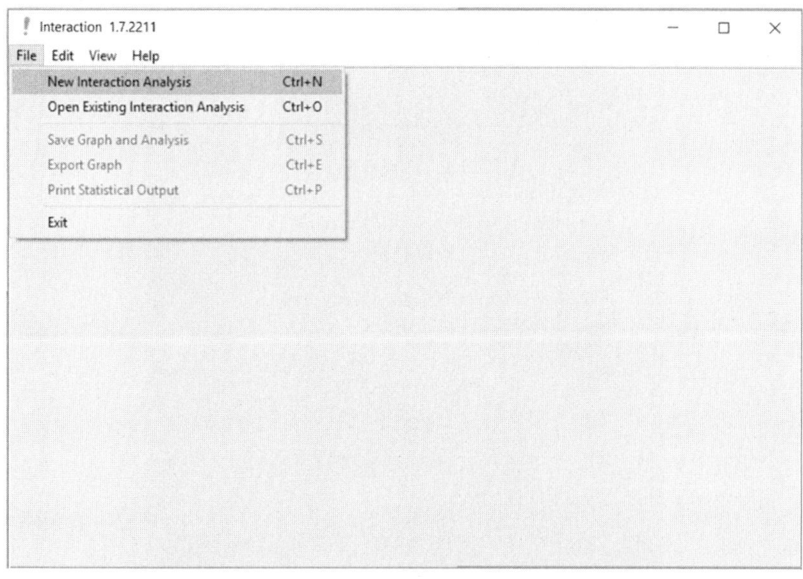

Source: Soper's Interaction program.

Figure 7.23 Step 2 in Soper Interaction Program: Select Data Source

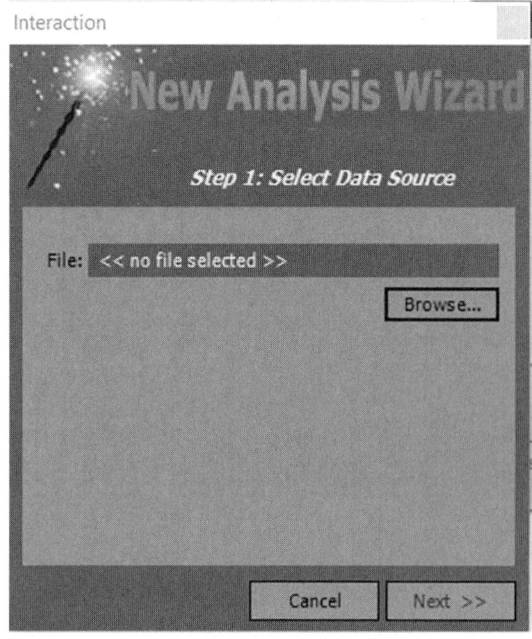

Source: Soper's Interaction program.

Figure 7.24 Step 3 in Soper Interaction Program: Identify Missing Values

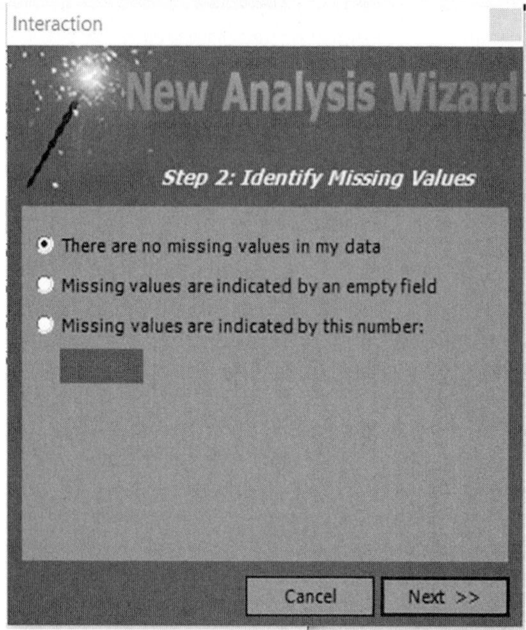

Source: Soper's Interaction program.

Figure 7.25 Step 4 in Soper Interaction Program: Assign Model Variables

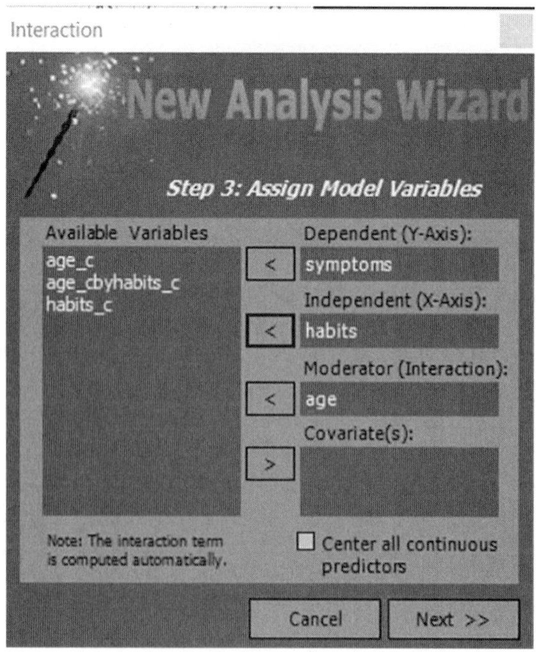

Source: Soper's Interaction program.

In this example it is obvious that symptoms is the dependent variable. You need to decide which of the X_1, X_2 predictors to define as the moderator. I chose age as the moderator in this variable because I wanted to see separate regression lines for different values of age. (You can also use this program if one or both of the predictors are categorical; if one predictor is categorical, like sex, you will almost certainly call that the moderator. You would obtain separate regression lines for each sex.)

Although we talked about the need to center scores on quantitative variables before computing the product, you do not need to check the box for "Center all continuous predictors" at this step. This command does not affect the way interaction analysis is conducted; it just changes the labels on the X axis on your scatterplot. If you check this box, tick-mark values of X will correspond to deviations from the mean of X. I prefer not to check the "Center all continuous predictors" box; this results in a graph with raw score values of X on the horizontal axis of the graph.

The next step (Figure 7.26) is to specify interaction levels. Here you tell the program for which selected values of age you want to see separate regression lines to predict symptoms from habits. These values of age are specified here in terms of z scores for age (distances of age from the mean). The most common choice is to graph lines for $z_{\text{moderator}}$ (z_{age}) = −1, 0, +1; or for $z_{\text{moderator}}$ (z_{age}) = −2, −1, 0, +1, +2). In other words, you will obtain lines for persons who are far below average in age, below average in age, average in age, above average in age, and far above average in age. (I do not recommend including z values of −3 and +3; when you consider values this far from the mean, there will be very few data points that correspond to the plotted regression lines.)

Figure 7.26 Step 5 in Soper Interaction Program: Specify Interaction Levels

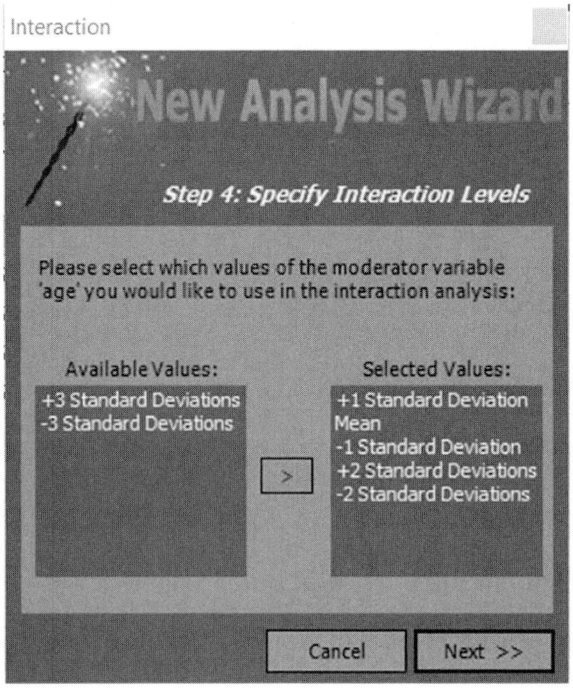

Source: Soper's Interaction program.

The selected z values for X_1 and X_2 are the same. If you decide to use z values of –2, –1, 0, +1, and +2 for age, these are also the levels of habit that are used to find the slope to predict symptoms from habits; for example, $z = -2$ is far below average in number of habits, $z = -1$ is below average, $z = 0$ (also called $z = M$ by Soper) is average, $z = +1$ is above average, and $z = +2$ is far above average in number of habits.

Click Next and then Finish to draw your graph and obtain results for analysis of simple main effects, that is, estimated coefficients and significance tests for each separate regression line. Results appear in Figures 7.27 and 7.28.

This graph makes the nature of the interaction clear. For the youngest persons (z age of –2) the slope to predict symptoms from habits is very shallow (it appears slightly negative, but in the analysis of simple main effects, it turns out that the regression coefficient is not statistically significant). In other words, the youngest persons have very few symptoms whether they practice many healthy habits or few. When you examine the line for the oldest adults ($z = +2$), the regression line has a substantial negative slope. Among older adults, those who practice very few healthy habits have high predicted numbers of symptoms; the more healthy habits they practice, the smaller the number of predicted symptoms. Intermediate values of z (–1, 0, and +1) suggest that the strength of association of habits with symptoms gradually increases as you move from younger to older age groups.

The z value labels on the X axis can be converted back into raw score values in years. Recall that $X = z \times SD + M$. For age, $M = 41.73$ and $SD = 9.505$. To convert z for age to raw score in years, years $= M + z \times SD = 41.73 + z \times 9.505$. The following ages in years (Table 7.1) could be used to label the lines in Figure 7.27.

For habits, $M = 3.14$ and $SD = 1.254$; z scores can be converted to raw score for number of habits using this equation: $3.14 + z \times 1.254$.

Statistical results for the analysis of simple main effects, and additional regression results, are obtained by clicking the "Statistical Output" tab (seen in Figure 7.27). Results are reported for analysis of the overall model with follow-up analysis of the simple main effects. Part of these results appear in Figure 7.28. (The simple main effects analysis (i.e., prediction of symptoms from habits within group) is shown only for the lowest age group, $z = -2$).

Figure 7.27 Interaction Graph for Age × Habits as Predictors of Symptoms (Generated by Soper Interaction Program)

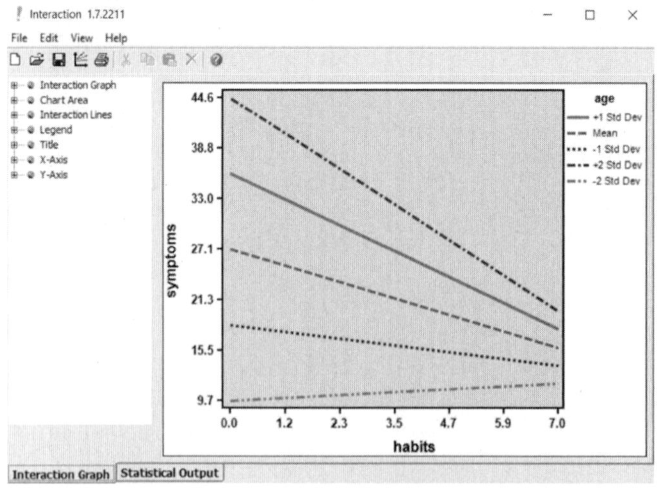

Source: Soper's Interaction program.

Figure 7.28 Selected Statistical Results Generated by Soper Program: Statistical Significance Tests for Overall Model and Simple Main Effects for Persons with *z* Age = −2, the Youngest Group

```
Monday, January 21, 2019  11:38:54 AM

Output generated by Interaction version 1.7.2211
Download the latest version at: http://www.danielsoper.com/Interaction
Copyright (c) 2006-2012 by Daniel S. Soper, Ph.D. All Rights Reserved.

******************************** MODEL SUMMARY ********************************

                                                  R: 0.853811095
                                           R Square: 0.728993386
                                  R Square Adjusted: 0.726452699
                      Standard Error of the Estimate: 4.037762144
   R Square Contribution of the Interaction Term(s): 0.022445620

                              RESEARCH MODEL: Y = B1X1
                                                + B2X2
                                                + B3X1X2
                                                + B0

                                       WHERE: Y = symptoms
                                             X1 = habits
                                             X2 = age
                                             B0 = Regression constant

****************************** END MODEL SUMMARY ******************************


************************** MODEL ANALYSIS OF VARIANCE **************************

               Sum of        Degrees       Mean
               Squares       of Freedom    Square        F             Significance
               ------------  ------------  ------------  ------------  ------------
   Regression: 14033.79543   3             4677.931811   286.9276642   0.000000001
     Residual: 5217.127403   320           16.30352313
        Total: 19250.92283   323

********************************** END ANOVA **********************************


***************************** INTERACTION LINE 5 *****************************
                                     Moderator: age
                        Level of the Moderator: -2 Std Dev
                                  Simple Slope: 0.290894604
                                     Intercept: 9.670284666
               Standard Error of Simple Slope: 0.390489143
                            Degrees of Freedom: 320
                                             t: 0.744949274
   Significance of Simple Slope (two-tailed): 0.456849145
   Significance of Simple Slope (one-tailed): 0.228424572

                                          Lower Bound   Upper Bound
                                          ------------  ------------
          95% CI around the Simple Slope: -0.47735567   1.059144886

********************************** END LINE 5 **********************************
```

$t(320) = .745$, $p = .46$, two-tailed

Source: Soper's Interaction program.

Table 7.1 Conversion of *z* Scores for Age to Raw Scores in Years

z	Computation for Corresponding Raw Score	Age in Years
−2	41.73 − 19.01	22.72
−1	41.73 − 9.505	32.22
0 or *M*	41.73 + 0	41.73
+1	41.73 + 9.505	51.23
+2	41.73 + 19.01	60.74

Note: The computation of the corresponding raw score on each line is as follows: grand mean − *z***SD*.

Table 7.2 Conversion of z Scores for Habits to Raw Score (Number of Habits)

z	Computation of Raw Score for Habits	Number of Habits
−2	3.14 − 2.508	.632
−1	3.14 − 1.254	1.886
0 or *M*	3.14	3.14
+1	3.14 + 1.254	4.394
+2	3.14 + 2.508	5.648

7.14 RESULTS SECTION FOR INTERACTION OF TWO QUANTITATIVE PREDICTORS

Following is an example of a "Results" section for the interaction of two quantitative predictors.

Results

A regression analysis was performed to assess whether healthy habits interact with age to predict number of symptoms of physical illness for a sample of $N = 320$ adults. Age ranged from 21 to 55 years, with $M = 41.73$, $SD = 9.505$. Number of healthy habits ranged from 0 to 7, with $M = 3.14$, $SD = 1.254$. Preliminary data screening did not suggest problems with assumptions of normality and linearity; there were no missing values or extreme outliers. Prior to forming a product term to represent an interaction between age and habits, scores on both variables were centered by subtracting sample means. The regression included age, habits, and an agebyhabits interaction term as predictors of symptoms.

The overall regression was statistically significant, $R = .854$, $R^2 = .729$, adjusted $R^2 = .726$, $F(3, 320) = 286.928$, $p < .001$. Unstandardized regression coefficients are reported, unless otherwise specified. There was a significant Age × Habits interaction, $b = -.100$, $t(320) = -5.148$, $p < .001$, $sr^2 = .0225$. There were also significant effects for age, $b = .606$, $t(320) = 24.166$, $p < .001$, $sr^2 = .494$, and for habits, $b = -1.605$, $t(320) = -8.394$, $p < .001$, $sr^2 = .0595$.

The nature of the interaction was examined in a plot (Figure 7.27) and through follow-up analysis of simple main effects. As age increased, the slope to predict symptoms from habits became more negative. Within the youngest group of persons (with z scores of −2) the b raw score slope to predict symptoms from age was not statistically significant, $t(320) = .745$, $p = .46$, two-tailed; 95% confidence interval (CI) [−.477, +1.06]. Within all other age levels (corresponding to z values of −1 or above) the slopes to predict symptoms from habit were statistically significant and negative. For the youngest persons in the sample, number of predicted symptoms was not associated with number of healthy habits. Among older participants, larger numbers of healthy habits predicted lower numbers of symptoms, and this association was strongest for the oldest adults.

Additional information should be included, possibly in other sections of the research report. Descriptive statistics for all variables in the analysis are often reported before other analyses, often as a table that includes *M*, *SD*, *CI* for *M*, and so forth.

If complete results are included for all five of the regression lines in the analysis of simple main effects, this would include all the information for the regression analysis discussed above. It would probably be more convenient to report this in table form.

7.15 ADDITIONAL ISSUES AND SUMMARY

Analysis of interactions in regression is not as difficult as it may seem at first glance, but there are several things you need to keep in mind to avoid confusion. The easiest way to become confused is by losing track of which variables are centered and which are not.

In examples in this chapter:

1. Scores for the Y outcome variable are not centered.
2. For categorical predictors with more than two groups, dummy variables are needed to represent group membership.
3. When either or both predictors in an interaction are categorical variables, centering scores on the two predictors is not necessary (Aiken & West, 1991).
4. When a statistically significant interaction is present, the main "story" in the "Results" section should be about the nature of the interaction.
5. When both variables involved in an interaction are quantitative, center their scores before computing the product that represents their interaction. However, when you set up the regression equation, do not center Y or X_1 or X_2. The predictive equation should be of this form: Y' (raw score, not centered) = $b_0 + b_1X_1$ (not centered) + b_2X_2 (not centered) + b_3 (product of centered X_1 and centered X_2 scores). Your results won't be incorrect if you center Y and/or X_1 and X_2, but they will be confusing to interpret.
6. You should examine whether regression lines in your scatterplot extend into areas of the graph where there are few or no data points. You should not extrapolate a linear regression into ranges of scores where you have no data (Bodner, 2016).
7. This chapter did not include computations of statistical significance tests for the separate regression lines for subgroups, analysis of nonlinear interactions, or analysis of three-way and higher order interactions. For information about these, see Aiken and West (1991) and Jaccard and Turissi (2003).

When an interaction in a regression analysis involves a dummy variable and a continuous variable, it is relatively easy to describe the nature of the interaction; there are "different slopes for different folks" (e.g., separate regression lines to predict salary as a function of years for men and women). When an interaction involves two quantitative predictors, slopes are obtained for selected values of each quantitative predictor, and these selected values are specified in terms of z scores. In both cases, a graph that shows the slope to predict Y from X_1, for selected values of X_2, can be very helpful in understanding the nature of the interaction.

Keep in mind that, as with any other result in statistics, a statistically significant interaction can be an instance of Type I error. If an interaction is statistically significant but was not predicted ahead of time, be skeptical; if you want to pursue it, do replication studies. If an interaction makes no sense, don't strain to come up with a story to explain it.

APPENDIX 7A

Graphing Interactions Between Quantitative Variables "by Hand"

It is easiest to work with z scores for X_1 and X_2 in this situation. (You can change numerical labels in your final graph back to the original raw score units at the end of the process.) Using your entire data file, do the following to obtain the regression equation you will use to generate predicted values to set up your graph.

1. Convert the quantitative predictors, X_1 and X_2, to z scores. This can be done using <Analyze> → <Descriptive Statistics> → <Descriptives> and checking the box for "Save standardized values as variables." The Descriptives dialog box appears in Figure 7.29. Do not convert scores on the dependent variable Y into z scores.

2. Use the SPSS <Transform> → <Compute Variable> command to create the product of z_{x1} and z_{x2}, as shown in Figure 7.30.

3. Run the regression to predict Y from z_{X1}, z_{X2}, and the product of these z scores. Results for this regression appear in Figure 7.31.

4. On the basis of the results in Figure 7.31, write out the equation to predict raw scores on Y (symptoms, in this example) from z scores on age (z_{age}) and habits (z_{habits}) as follows. *Note that when you write out this equation, you need to use the unstandardized regression coefficients* (because the dependent variable, symptoms, was not standardized).

$$\text{Symptoms}' = b_0 + b_1 z_{age} + b_2 z_{habits} + b_3(z_{age} \times z_{habits}).$$

   Equation with numerical values of coefficients:

$$\text{Symptoms}' = 22.101 + 5.759 \times z_{age} - 2.013 \times z_{habits} - 1.189 \times (\text{productofzscores}).$$

   Hold on to this equation.

5. You can now close the SPSS data file thirdexample.sav and create a new, much smaller data file in SPSS or Excel. I will call this new file worksheet.sav.

Figure 7.29 Saving *z* (Standardized) Scores for Predictor Variables

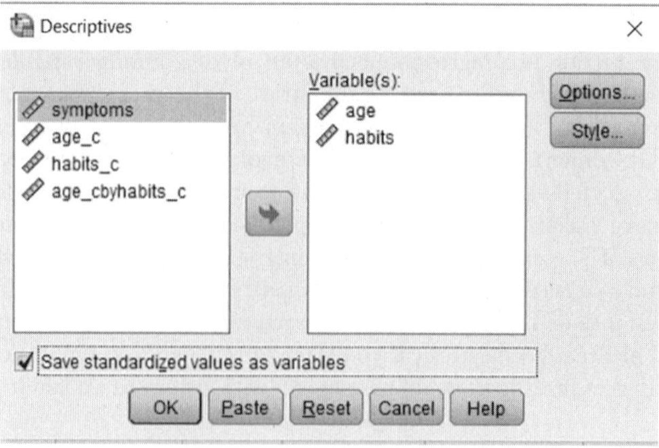

Figure 7.30 Computation of Interaction Term (Using *z* Scores for Age and Habits)

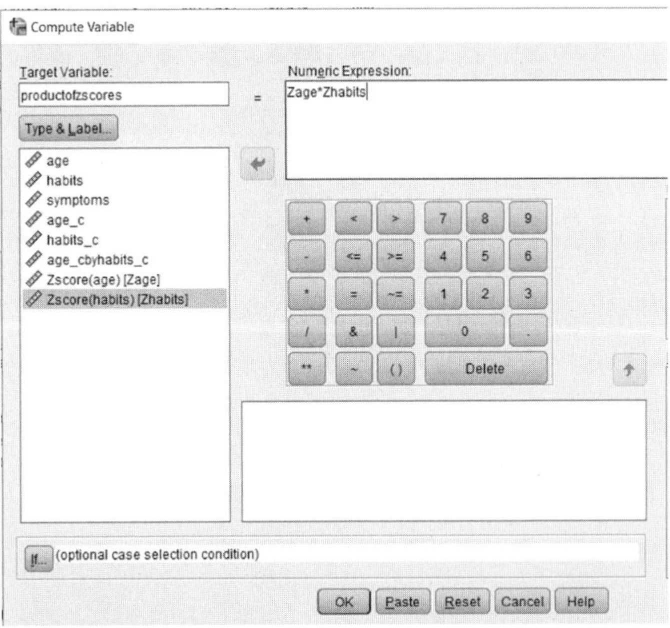

Figure 7.31 Regression Coefficients (From thirdexample.sav File) to Predict Symptoms From Age, Habits, and Their Interaction

Coefficients[a]

Model		Unstandardized Coefficients B	Unstandardized Coefficients Std. Error	Standardized Coefficients Beta	t	Sig.	Correlations Zero-order	Correlations Partial	Correlations Part
1	(Constant)	22.101	.236		93.493	.000			
	Zscore(age)	5.759	.238	.746	24.166	.000	.810	.804	.703
	Zscore(habits)	-2.013	.240	-.261	-8.394	.000	-.476	-.425	-.244
	productofzscores	-1.189	.231	-.153	-5.148	.000	-.006	-.277	-.150

a. Dependent Variable: symptoms

6. Enter the selected *z* scores for X_1 (age) and X_2 (habits) into the new file worksheet.sav. Column 1 is the selected *z* scores for age; column 2 is selected *z* scores for habits (I used the variable names selectedzage and selectedzhabits). Worksheet.sav in Figure 7.31 has one row for each of the possible 25 combinations (with 5 selected values for age and 5 selected values for habits, e.g., –2, –1, 0, +1, and +2, there are 5 × 5 possible combinations, so the worksheet file in Figure 7.32 has 25 rows).

7. Use the SPSS <Transform>→ <Compute Variable> command, as shown in Figure 7.33, to obtain the product of the variables selectedzage and selectedzhabits. This will create a new variable, productofzscores, in the third column of worksheet.sav (this step is not shown in the screenshot).

Figure 7.32 New SPSS File: worksheet.sav

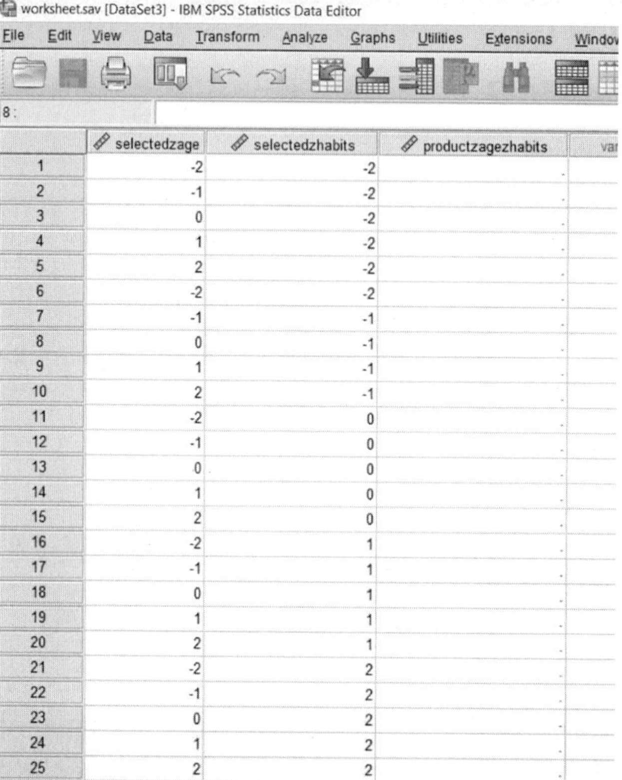

	selectedzage	selectedzhabits	productzagezhabits
1	-2	-2	.
2	-1	-2	.
3	0	-2	.
4	1	-2	.
5	2	-2	.
6	-2	-2	.
7	-1	-1	.
8	0	-1	.
9	1	-1	.
10	2	-1	.
11	-2	0	.
12	-1	0	.
13	0	0	.
14	1	0	.
15	2	0	.
16	-2	1	.
17	-1	1	.
18	0	1	.
19	1	1	.
20	2	1	.
21	-2	2	.
22	-1	2	.
23	0	2	.
24	1	2	.
25	2	2	.

Figure 7.33 Use of Transform Compute Statement to Compute Product Term for Interaction in worksheet.sav File

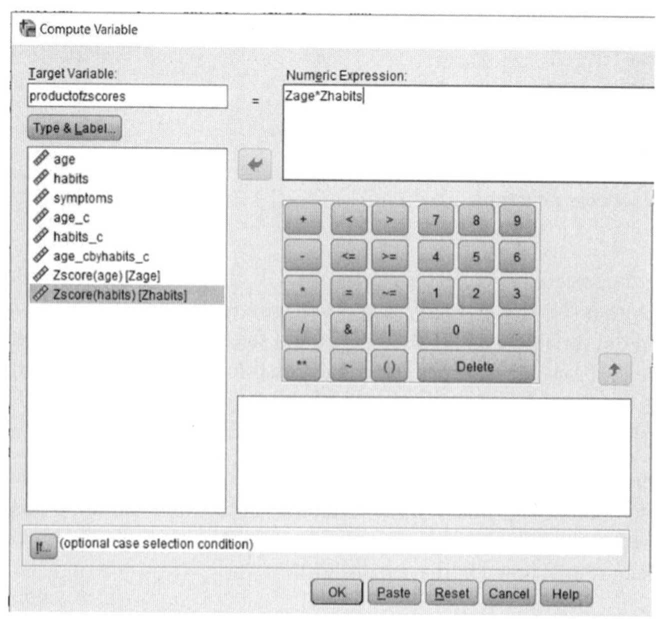

Figure 7.34 Computation of Predicted Symptoms Scores on the Basis of Selected z-Score Values of Age and Habits in worksheet.sav File

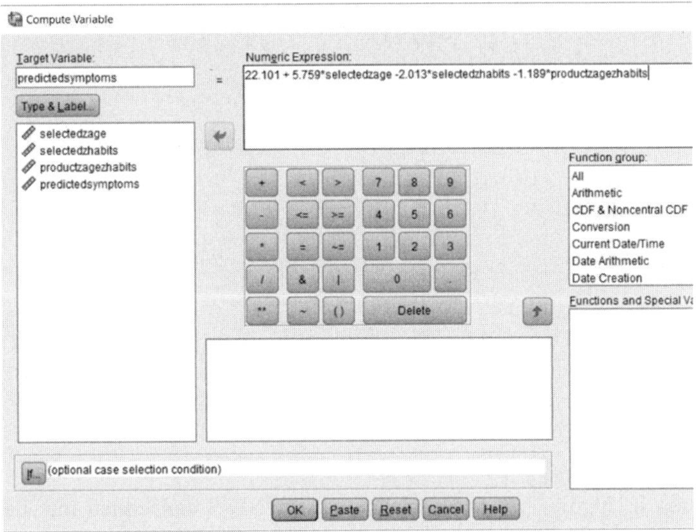

You now have the list of selected predictor values (in z-score terms, in columns 1, 2, and 3 of the file worksheet.sav) that will be substituted into the regression equation from step 4 to obtained predicted symptoms (in raw-score units). These predicted values of symptoms are the data points used to graph the regression lines.

8. To generate predicted values of symptoms for all possible combinations of values of z_{X1} and z_{X2}, use <Transform> → <Compute Variable> again, as shown in Figure 7.34 (still within the worksheet.sav file). You use the regression equation you obtained earlier to calculate a predicted Y value from the values of the first three columns in your worksheet file.

$$\text{Symptoms}' = 22.101 + 5.759 \times \text{selectedzage} - 2.013 \times \text{selectedzhabits} - 1.189 \times \text{productofzscores}.$$

After the <Compute Variable> command in Figure 7.34 has been executed, there will be a fourth column in worksheet.sav that has values for predicted symptoms (not shown).

9. Finally, request a line graph for predicted Y (symptom) values as a function of values of the other three variables in worksheet.sav (e.g., the selected values for X_1, for X_2, and for X_1X_2 (these predictors are all in z-score terms). In this example, X_1 (age) is the moderator variable.

Use <Graphs> → <Legacy Dialogs> → <Scatter/Dot> to open the first scatterplot dialog box, and select "Multiple."

In the Define Multiple Line: Summaries for Groups of Cases dialog box in Figure 7.36, select the radio button for "Other statistic (e.g., mean)," and enter the name of the outcome variable (in this example, predictedsymptoms). Enter the name of the moderator variable (in this example, selectedzage) in the "Define Lines by" box. Enter the name of the other predictor variable (in this example, selectedzhabit) in the "Category Axis" window. Then click OK.

Figure 7.35 Initial Line Charts Window: Request Multiple Lines

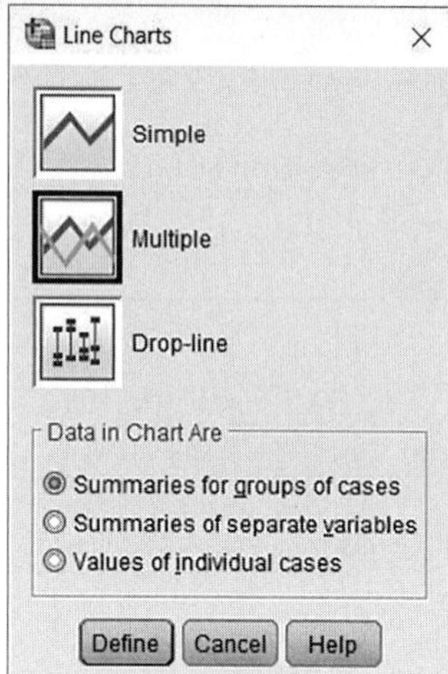

The graph in Figure 7.37 was generated using SPSS and edited in Chart Editor to improve its appearance. The lines are identical to those in the interaction graph obtained from the Soper Interaction program (except that in Soper's graph, the X axis labels are in raw scores, and in this graph, X axis labels are in z-score units).

As discussed earlier, z-score labels for both the predictor (habits, on the X axis) and the moderator variable (age, the labels for the five separate lines) can be converted back into raw score units that can be used to label the line graphs that show interactions, as shown in Tables 7.1 and 7.2.

Figure 7.36 Define Multiple Line Selections to Obtain Plots for Age by Habits Interaction Using Scores in worksheet.sav

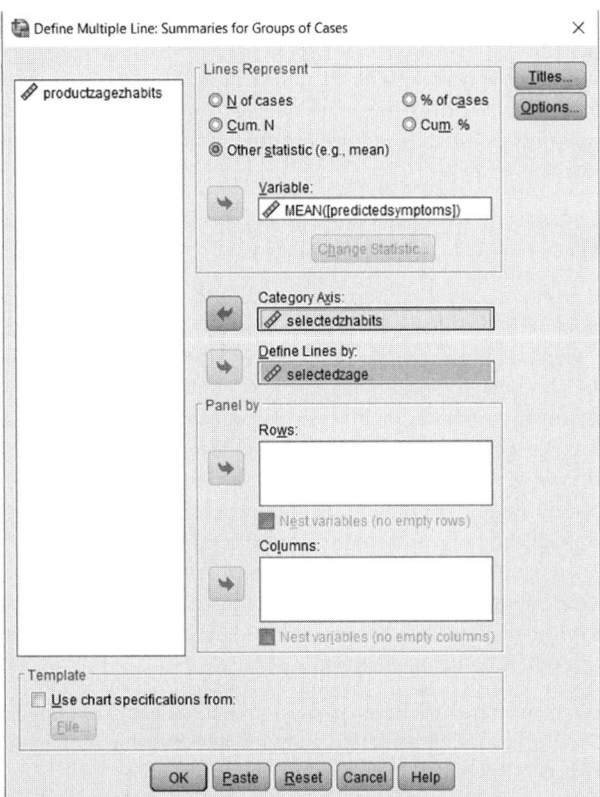

Figure 7.37 Graph of Age-by-Habits Interaction Generated Using Aiken and West (1991) Procedures (Edited in SPSS Chart Editor)

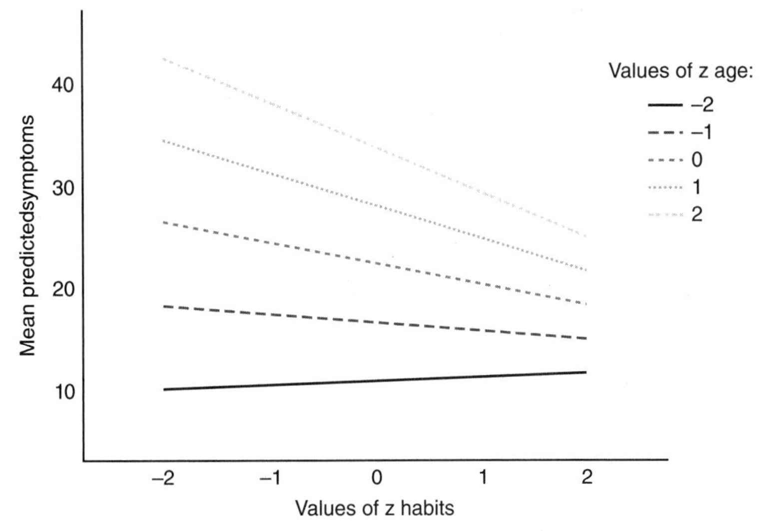

COMPREHENSION QUESTIONS

1. What preliminary graphs are needed to examine whether all associations between quantitative predictor and outcome variables are approximately linear? What graph would provide a preliminary assessment of possible moderation?

2. Briefly explain how the approach to interaction analysis differs for these combinations of types of predictor variables: two categorical independent variables, one categorical and one quantitative independent variable, and two quantitative independent variables. Assume a quantitative outcome variable.

3. How do you center scores on a quantitative predictor variable in SPSS? Why is centering usually used when both predictors are quantitative?

   The following questions involve interpretation of interaction analyses that use unpublished data in the file qualityofrelationships.sav. The main question in the study was the degree to which well-being outcomes (such as happiness, life satisfaction, and positive affect) could be predicted by social relationship variables such as social stress and social support, by personality variables such as extraversion and neuroticism, and by categorical variables such as sex (coded 1 = male, 2 = female in this data set) and dating (coded 1 = yes, 0 = no) and whether specific interactions were present. The literature on stress and social support often reports that social support has a "buffering" effect; that is, the happiness and well-being of people high in social support are not as negatively affected by stress, compared with people with lower social support. It has also been suggested that for women, social support is more strongly related to life satisfaction than for men. Other variables with self-explanatory names are included in the file, and you can use these if you want to run different interaction analyses.

4. Does neuroticism moderate the effects of social stress on life satisfaction? Specifically, does stress have a worse effect on life satisfaction for people who score high on neuroticism than low on neuroticism? All three of these variables are quantitative.

   Figure 7.38 shows the SPSS regression results for prediction of life satisfaction from neuroticism, social stress, and their interaction:

   Answer the following questions using the information in Figures 7.38, 7.39, and 7.40.

   - Was the overall model (including all three predictors) statistically significant, and what was the effect size?
   - Do you think I needed to center the scores on neuroticism and social stress before computing the product?
   - Was the interaction statistically significant? Report the numerical values for the interaction (e.g., *b*, *t*, *df*, *p*).
   - What additional information would you need to describe effect sizes (e.g., proportion of uniquely predicted variance for each predictor), and how would you obtain this information?
   - The interaction graph appears in Figure 7.39. Describe the nature of the interaction. Was this consistent with the prediction?
   - A selected part of the simple main effects analysis appears in Figure 7.40. Explain what is being examined in this analysis and interpret the results.

Figure 7.38 SPSS Output for Regression to Predict Life Satisfaction From Neuroticism, Social Stress, and Their Interaction

Model Summary

Model	R	R Square	Adjusted R Square	Std. Error of the Estimate
1	.449[a]	.201	.199	5.08172

a. Predictors: (Constant), neurotbystress, neuroticism, socialstress

ANOVA[a]

Model		Sum of Squares	df	Mean Square	F	Sig.
1	Regression	5485.767	3	1828.589	70.810	.000[b]
	Residual	21743.703	842	25.824		
	Total	27229.470	845			

a. Dependent Variable: lifesatisfaction

b. Predictors: (Constant), neurotbystress, neuroticism, socialstress

Coefficients[a]

Model		Unstandardized Coefficients B	Unstandardized Coefficients Std. Error	Standardized Coefficients Beta	t	Sig.
1	(Constant)	38.802	.897		43.256	.000
	neuroticism	-.115	.037	-.121	-3.096	.002
	socialstress	-.345	.038	-.355	-9.060	.000
	neurotbystress	-.012	.004	-.091	-2.943	.003

a. Dependent Variable: lifesatisfaction

Figure 7.39 Graph for Interaction Between Neuroticism and Social Stress as Predictors of Life Satisfaction

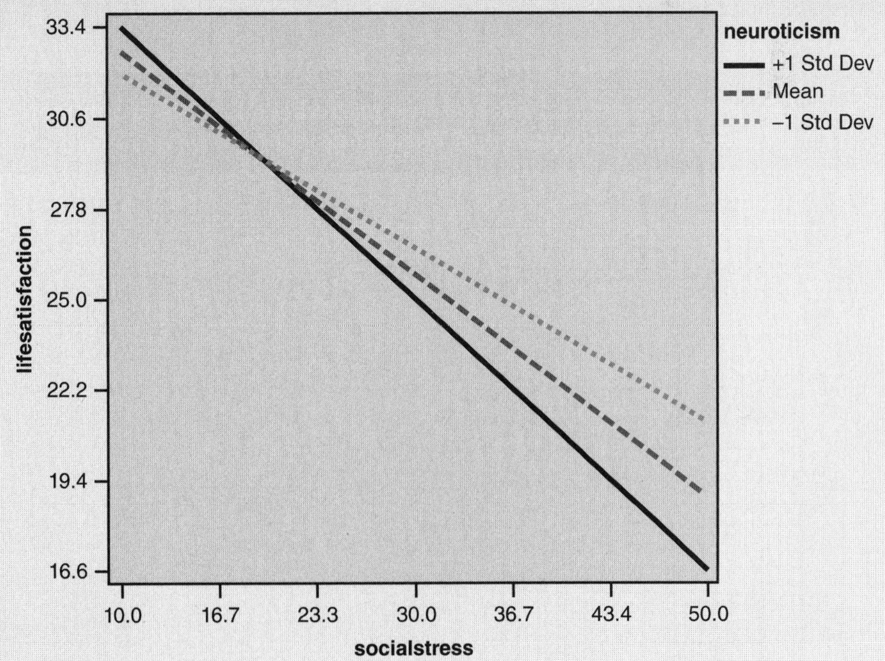

Figure 7.40 Results From Soper Interaction Program for Analysis of Simple Main Effects (One Regression Line Only)

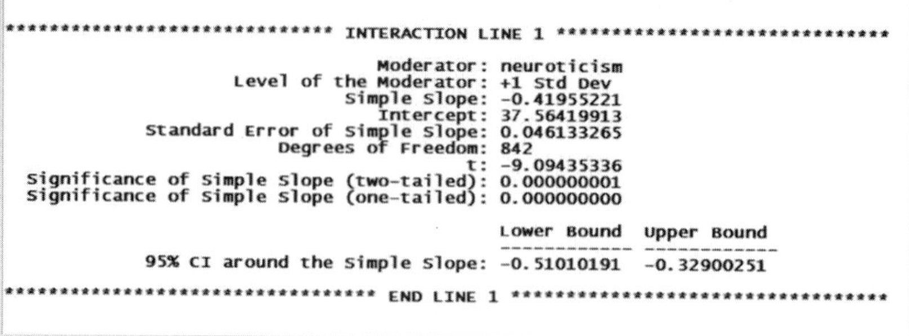

```
******************************* INTERACTION LINE 1 *******************************

                                     Moderator: neuroticism
                        Level of the Moderator: +1 Std Dev
                                  Simple Slope: -0.41955221
                                     Intercept: 37.56419913
                 Standard Error of Simple Slope: 0.046133265
                            Degrees of Freedom: 842
                                             t: -9.09435336
 Significance of Simple Slope (two-tailed): 0.000000001
 Significance of Simple Slope (one-tailed): 0.000000000

                                               Lower Bound   Upper Bound
                                               -----------   -----------
          95% CI around the Simple Slope:     -0.51010191   -0.32900251

********************************* END LINE 1 *********************************
```

Source: Soper's Interaction program.

5. It has been suggested that social support may be more important to women than to men for life satisfaction. This could mean that the regression line to predict life satisfaction for women has a steeper slope than the regression line for men. A regression was run to predict life satisfaction from sex, social support, and the product of sex and social support. Results are shown in Figures 7.41 and 7.42.

   Write as complete a "Results" section as you can, given the information provided. Identify any information you would want to have that is not included in the figures. What do you see in the scatterplot in Figure 7.42 that could be a problem?

6. On the basis of information in Figure 7.43, do you think that whether a person is dating moderates the effects of stress on happiness? Given that dating is coded 0 = no, 1 = yes, what can you say about the prediction of happiness from dating (e.g., are dating or nondating persons happier on average)?

Figure 7.41 SPSS Output From Regression to Predict Life Satisfaction From Sex, Social Support, and Their Interaction

Model Summary

Model	R	R Square	Adjusted R Square	Std. Error of the Estimate
1	.358[a]	.128	.125	5.30860

a. Predictors: (Constant), sexbysocialsupport, socialsupport, sex

ANOVA[a]

Model		Sum of Squares	df	Mean Square	F	Sig.
1	Regression	3481.823	3	1160.608	41.184	.000[b]
	Residual	23756.769	843	28.181		
	Total	27238.591	846			

a. Dependent Variable: lifesatisfaction

b. Predictors: (Constant), sexbysocialsupport, socialsupport, sex

Coefficients[a]

Model		Unstandardized Coefficients		Standardized Coefficients			Correlations		
		B	Std. Error	Beta	t	Sig.	Zero-order	Partial	Part
1	(Constant)	21.498	4.595		4.678	.000			
	sex	-5.840	2.702	-.468	-2.161	.031	.021	-.074	-.070
	socialsupport	.130	.116	.134	1.126	.261	.349	.039	.036
	sexbysocialsupport	.131	.067	.516	1.953	.051	.176	.067	.063

a. Dependent Variable: lifesatisfaction

Figure 7.42 Graph of Regressions to Predict Life Satisfaction From Social Stress and Sex

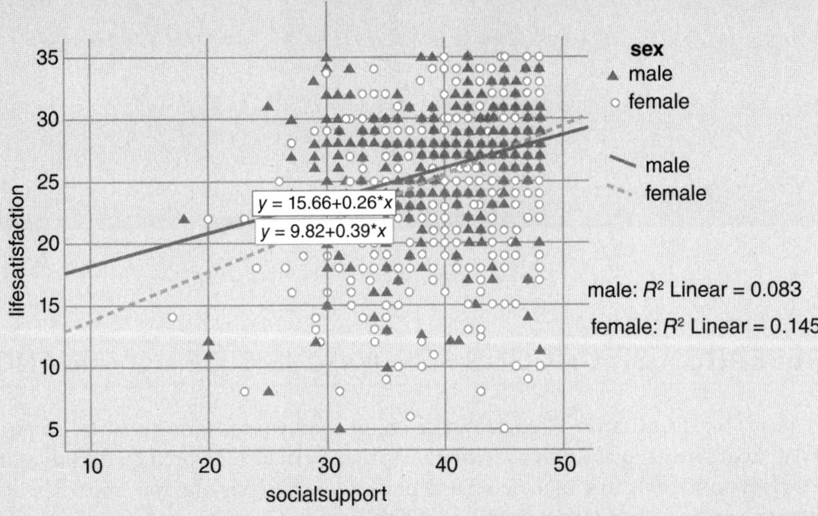

Figure 7.43 Selected SPSS Output: Coefficients for Regression to Predict Happiness From Dating, Social Stress, and Their Interaction

Coefficients[a]

Model		Unstandardized Coefficients B	Unstandardized Coefficients Std. Error	Standardized Coefficients Beta	t	Sig.	Correlations Zero-order	Correlations Partial	Correlations Part
1	(Constant)	6.995	.216		32.328	.000			
	dating	.633	.309	.315	2.050	.041	.082	.071	.064
	socialstress	-.065	.007	-.379	-8.732	.000	-.428	-.288	-.271
	datingbystress	-.017	.011	-.249	-1.596	.111	-.014	-.055	-.049

a. Dependent Variable: happiness

DIGITAL RESOURCES

Find **free study tools** to support your learning, including **eFlashcards, data sets, and web resources**, on the accompanying website at **edge.sagepub.com/warner3e**.

CHAPTER 8

ANALYSIS OF COVARIANCE

8.1 RESEARCH SITUATIONS FOR ANALYSIS OF COVARIANCE

Group means for quantitative Y scores can be compared by performing analysis of variance (ANOVA). Scores on a quantitative Y outcome variable can also be predicted from multiple linear regression models that include several predictors. Analysis of covariance (ANCOVA) combines one or more categorical predictor variables and one or more quantitative predictor variables, called covariates, to predict scores on a quantitative Y outcome variable. ANCOVA is, in a sense, a combination of ANOVA and regression. In ANCOVA, the typical research situation is as follows. A researcher has a set of treatment groups that correspond to levels of a factor (Factor A in the example in this chapter corresponds to three types of teaching methods). The researcher has preintervention quantitative measures of one or more individual participant characteristics, such as ability (X_c). The goal of the analysis is to assess whether scores on a quantitative Y outcome variable (such as scores on a final exam) differ significantly across levels of A when we statistically control for the individual differences among participants that are measured by the X_c covariate. The type of statistical control used in classic or traditional ANCOVA requires the absence of any interaction between the A treatment and the X_c covariate. We can thus think of ANCOVA as ANOVA with one or more added quantitative predictors called covariates. Because we can represent group membership predictor variables using dummy-coded predictor variables in regression, we can also understand ANCOVA as a special case of regression analysis in which the goal is to assess differences among treatment groups while controlling for scores on one or more covariates.

ANCOVA is often used in research situations where mean scores on a quantitative outcome variable are compared across groups that may not be equivalent in terms of participant characteristics (such as age or baseline level of mood). Comparison groups in a study are nonequivalent if the groups differ (prior to treatment) on one or more participant characteristics, such as age, motivation, or ability, that might possibly influence the outcome variable. **Nonequivalent control groups** (also called **nonequivalent comparison groups**) are often encountered in quasi-experimental research. A quasi-experiment is a study that resembles an experiment (up to a point); quasi-experimental designs typically involve comparison of behavior for groups that receive different types of treatment. However, in a quasi-experiment, the researcher generally has less control over the research situation than in a true experiment. Quasi-experiments are often conducted in field settings (such as schools), where researchers may have to administer interventions to already existing groups such as classrooms of students. In most quasi-experiments, the researcher does not have the ability to randomly assign individual participants to different treatment conditions. At best, the researcher may be able to randomly assign groups of students in different classrooms to different treatment conditions, but such quasi-experimental designs often result in a situation where prior to the treatment or intervention, the groups that are compared differ in participant characteristics, such as age, motivation, or ability. This nonequivalence can lead to a confound between the average participant characteristics for each group (such as motivation) and the type of treatment received.

One reason to use ANCOVA is that ANCOVA provides a way to assess whether mean outcome scores differ across treatment groups when a statistical adjustment is made to control for (or try to remove the effects of) different participant characteristics across groups. In most ANCOVA situations, the variable that we designate as the covariate X_c is a measure of some participant characteristic that differs across groups prior to treatment. In an ANCOVA, we want to evaluate whether scores on Y, the outcome variable of interest, differ significantly across groups when the Y scores are adjusted for group differences on the X_c covariate. Of course, this adjustment works well only if the assumptions for ANCOVA are not violated. The statistical adjustment for differences in participant characteristics between groups that can be made using ANCOVA is not typically as effective as experimental methods of controlling the composition of groups (such as random assignment of individuals to groups, creating matched samples, etc.). A second use of ANCOVA involves controlling for a covariate to remove one source of error variance from the outcome variable.

Even in a well-controlled laboratory experiment, a researcher sometimes discovers that groups that were formed by random assignment or more systematic matching techniques are not closely equivalent. If this nonequivalence is discovered early in the research process (e.g., before the intervention, the researcher notices that the mean age for participants is higher in the A_1 treatment group than in the A_2 treatment group), it may be possible to rerandomize and reassign participants to groups to get rid of this systematic difference between groups. Sometimes, even in a true experiment, the groups that are formed by random assignment of individual participants to conditions turn out not exactly equivalent with respect to characteristics such as age, because of unlucky randomization. If this problem is not corrected before the intervention is administered, even a true experiment can end up with a comparison of nonequivalent groups. ANCOVA is usually applied to data from quasi-experimental studies, but it may also be used to try to correct for nonequivalence of participant characteristics in the analysis of data from true experiments. However, when it is possible, it is preferable to correct the problem by redoing the assignment of participants to conditions to get rid of the nonequivalence between groups through experimental control; statistical control for nonequivalence, using ANCOVA, should be used only when the problem of nonequivalent groups cannot be avoided through design.

In ANCOVA, the null hypothesis that is of primary interest is whether means on the quantitative Y outcome variable differ significantly across groups after adjustment for scores on one or more covariates. For example, consider a simple experiment to assess the possible effects of food on mood. Some past research of possible drug-like effects on mood (Spring, Chiodo, & Bowen, 1987) suggests that the consumption of an all-carbohydrate meal has a calming effect on mood, while an all-protein meal may increase alertness. This hypothesis can be tested by doing an experiment. Suppose that participants are assigned to Group A_1 (an all-carbohydrate lunch) or Group A_2 (an all-protein lunch). We will consider two different possible covariates. The first is a measurement of a preexisting participant characteristic that may be linearly predictive of scores on the Y outcome variable. Before lunch, each participant rates "calmness" on a scale from 0 (*not at all calm*) to 10 (*extremely calm*); this score is X_c, the quantitative covariate. One hour after lunch, participants rate calmness again using the same scale; this second rating is Y, the outcome variable.

If we do a t test or an ANOVA to compare means on Y (calmness) across levels of A (carbohydrate vs. protein meal), the research question would be whether mean Y (calmness) is significantly higher in the carbohydrate group than in the protein group (when we *do not* take into account the individual differences in calmness before lunch or at baseline). When we do an ANCOVA to predict Y (calmness after lunch) from both A (type of food) and X_c (pretest level of calmness before lunch), the research question becomes whether mean calmness after lunch is significantly higher for the carbohydrate group than for the protein group when we control for or partial out any part of Y (the after-lunch calmness rating) that is linearly predictable from X_c (the before-lunch or baseline calmness rating).

The covariate in the preceding example (X_c) was a pretest measure of the same mood (calmness) that was used as the outcome or dependent variable. However, a covariate can be a measure of some other variable that is relevant in the research situation, either because it is confounded with treatment group membership or because it is linearly predictive of scores on the Y outcome variable, or both. For example, the covariate measured prior to the intervention in the study of food and mood might have been a measure of self-reported caffeine consumption during the 2-hour period prior to the study. Even if random assignment is used to place participants in the two different food groups, it is possible that because of unlucky randomization, one group might have a higher mean level of self-reported caffeine consumption prior to the lunch intervention. It is also possible that caffeine consumption is negatively correlated with self-reported calmness (i.e., participants may tend to report lower calmness when they have consumed large amounts of caffeine). Statistically controlling for caffeine consumption may give us a clearer idea of the magnitude of the effects of different types of food on mood for two reasons. First of all, statistically controlling for caffeine consumption as a covariate may help correct for any preexisting group differences or confounds between the amount of caffeine consumed and the type of food. Second, whether or not there is any confound between caffeine consumption and type of food, partialling out the part of the calmness scores that can be linearly predicted from caffeine consumption may reduce the error variance (MS_{within} in ANOVA or MS_{residual} if the analysis is a regression) and increase statistical power.

The regression equation for this simple ANCOVA is as follows: Let $A = 0$ for the protein group and $A = 1$ for the carbohydrate group; that is, A is a dummy-coded dummy variable. Let X_c be calmness ratings before lunch, and let Y be calmness ratings after lunch. The regression equation that corresponds to an ANCOVA to predict Y from A and X_c is as follows:

$$Y' = b_0 + b_1X_c + b_2A. \tag{8.1}$$

Controlling for an X_c covariate variable can be helpful in two different ways. First of all, it is possible that X_c is confounded with A. When we randomly assign participants to conditions in an experiment, we hope that random assignment will result in groups that are equivalent on all relevant participant characteristics. For example, in the study of food and mood, if we create treatment groups through random assignment, we hope that the resulting groups will be similar in gender composition, smoking behavior, level of stress, and any other factors that might influence calmness. In particular, if we have measured X_c, calmness before lunch, we can check to see whether the groups that we obtained are approximately equal on this variable. In addition, when we calculate the regression coefficient for the dummy variable for group membership, we can statistically control for any confound of X_c with A; that is, we can assess the relation of A to Y, controlling for or partialling out X_c. If A is still related to Y when we control for X_c, then we know that the group differences in mean Y are not entirely explained by a confound with the covariate.

Including a covariate can also be helpful in research situations where the covariate is not confounded with the treatment variable A. Consider a regression analysis in which we use only A to predict Y:

$$Y' = b_0 + b_1A. \tag{8.2}$$

The SS_{residual} term for Equation 8.1 will be equal to or less than the SS_{residual} for Equation 8.2. Therefore, the F ratio that is set up to evaluate the significance of the A variable for the analysis in Equation 8.1 (which includes X_c as a predictor) may have a smaller $MS_{residual}$ value than the corresponding F ratio for the analysis in Equation 8.2 (which does not include X_c as a predictor). If $SS_{residual}$ decreases more than $df_{residual}$ when the covariate is added to the analysis, then $MS_{residual}$ will be smaller for the ANCOVA in Equation 8.1 than for the ANOVA in Equation 8.2, and the F ratio for the main effect of A will tend to be larger for the analysis in Equation 8.1 than for the analysis in Equation 8.2. When we obtain a substantially smaller $SS_{residual}$ term by including a covariate in our analysis (as in Equation 8.1), we can say that the covariate X_c acts as a noise suppressor or provides **error variance suppression**.

When we add a covariate X_c to an analysis that compares Y means across levels of A, taking the X_c covariate into account while assessing the strength and nature of the association between Y and A can change our understanding of that relationship in two ways. Controlling for X_c can, to some degree, correct for a confound between X_c and type of A treatment; this correction may either increase or decrease the apparent magnitude of the differences between Y outcomes across levels of A. In addition, controlling for X_c can reduce the amount of error variance, and this can increase the power for the F test to assess differences in Y means across levels of A.

We will prefer ANCOVA over ANOVA in situations where

1. the covariate, X_c, is confounded with the group membership variable A, and we can assess the unique effects of A only when we statistically control for or partial out X_c, and/or
2. the covariate, X_c, is strongly predictive of Y, the outcome variable, and therefore including X_c as a predictor along with A (as in Equation 8.1) gives us a smaller error term and a larger F ratio for assessment of the main effect of A.

We can use ANCOVA only in situations where the assumptions for ANCOVA are met. All the usual assumptions for ANOVA and regression must be met for ANCOVA to be applied (i.e., scores on the Y outcome variable should be approximately normally distributed with homogeneous variances across levels of A, and scores on Y must be linearly related to scores for X_c). Additional assumptions are required for the use of ANCOVA, including the following. The covariate (X_c) must be measured prior to the administration of the A treatments, and there must be no treatment-by-covariate ($A \times X_c$) interaction. In addition, factors that cause problems in the estimation of regression models generally (such as unreliability of measurement, restricted ranges of scores, etc.) will also create problems in ANCOVA.

8.2 EMPIRICAL EXAMPLE

The hypothetical example for ANCOVA will be an imaginary quasi-experimental study to compare the effectiveness of three different teaching methods in a statistics course (Factor A). Group A_1 receives self-paced instruction (SPI), Group A_2 participates in a seminar, and Group A_3 receives lecture instruction. We will assume that the groups were not formed by random assignment; instead, each teaching method was carried out in a different preexisting classroom of students. The scores used in this example appear in Table 8.1.

The X_c covariate in this hypothetical research example corresponds to scores on a pretest of math ability. As we shall see, there are small differences among the A_1, A_2, and A_3 groups on this math pretest X_c variable. Thus, there is some confound between the mean level of math ability prior to the intervention in each group and the type of teaching method received by each group. ANCOVA can help adjust or correct for these differences in preexisting participant characteristics. The Y outcome variable in this hypothetical study corresponds to the score on a standardized final exam for the statistics course. If all the assumptions of ANCOVA are met, an ANCOVA in which we examine **adjusted means** on the Y outcome variable (i.e., Y scores that have been statistically adjusted to remove linear association with the X_c

Table 8.1 Data for Hypothetical ANCOVA Study

	A	X_c	Y	D_1	D_2	Int_1	Int_2
1	1	12	19	1	0	12	0
2	1	7	13	1	0	7	0
3	1	10	10	1	0	10	0
4	1	8	9	1	0	8	0
5	1	8	12	1	0	8	0
6	1	5	6	1	0	5	0
7	1	13	18	1	0	13	0
8	2	12	22	0	1	0	12
9	2	15	28	0	1	0	15
10	2	14	24	0	1	0	14
11	2	12	27	0	1	0	12
12	2	10	24	0	1	0	10
13	2	8	16	0	1	0	8
14	2	5	21	0	1	0	5
15	3	11	18	–1	–1	–11	–11
16	3	6	21	–1	–1	–6	–6
17	3	9	24	–1	–1	–9	–9
18	3	10	21	–1	–1	–10	–10
19	3	15	28	–1	–1	–15	–15
20	3	7	15	–1	–1	–7	–7
21	3	13	21	–1	–1	–13	–13

Source: Horton (1978, p. 175).

Note: Factor A is a categorical variable that represents which of three teaching methods each student received, Y is the final exam score, and X_c is the pretest on ability. Using an ANOVA program such as the GLM procedure in SPSS, the categorical variable A would be used to provide information about group membership, while the Y and X_c scores would be the dependent variable and covariate, respectively. It is also possible to run an ANCOVA using a linear regression program, but in this situation, we need to use dummy-coded variables to represent information about treatment group membership. Dummy-coded variables D_1 and D_2 provide an alternate way to represent treatment group membership. The variables Int_1 and Int_2 are the products of D_1 and D_2 with X_c, the covariate. To run the ANCOVA using linear regression, we could predict scores on Y from scores on D_1, D_2, and X_c. To test for the significance of a treatment-by-covariate interaction, we could assess whether the additional terms Int_1 and Int_2 are associated with a significant increase in R^2 for this regression analysis.

covariate) may provide us with clearer information about the nature of the effects of the three different teaching methods included in the study.

For this example, there are $n = 7$ participants in each of the three groups, for a total N of 21 (in real-life research situations, of course, larger numbers of scores in each group would be preferable). The covariate, X_c, is a math ability pretest. The outcome, Y, is the score on a standardized final exam. The scores for A, X_c, and Y and alternative forms of dummy-coded variables that represent treatment level appear in Table 8.1.

The research question is this: Do the three teaching methods (A) differ in mean final exam scores (Y) when pretest scores on math ability (X_c) are statistically controlled by doing an ANCOVA? What is the nature of the differences among group means? (That is, which pairs of groups differed significantly, and which teaching methods were associated with the highest mean exam scores?) As part of the interpretation, we will also want to consider the nature of the confound (if any) between A and X_c. (That is, did the groups differ in math ability?) Also, how did controlling for X_c change our understanding of the pattern of Y means?

(That is, how did the ANCOVA results differ from a one-way ANOVA comparing *Y* means across levels of *A*?)

It is usually more convenient to use the SPSS general linear model (GLM) procedure to perform ANCOVA. The ANCOVA example in this chapter was performed using the SPSS GLM procedure. However, if group membership is represented by dummy variables (such as D_1 and D_2), ANCOVA can also be performed using the SPSS regression procedure. The regression version of ANCOVA is not presented in this chapter.

8.3 SCREENING FOR VIOLATIONS OF ASSUMPTIONS

ANCOVA refers to a predictive model that includes a mix of both categorical and continuous predictor variables; however, several additional conventional limitations are set on an analysis before we can call it an ANCOVA:

1. As in other linear correlation and regression analyses, we assume that the X_c and Y quantitative variables have distribution shapes that are reasonably close to normal, that the relation between X_c and Y is linear, and that the variance of Y is reasonably homogeneous across groups on the A treatment variable.
2. The covariate X_c should not be influenced by the treatment or intervention; to ensure this, X_c should be measured or observed before we administer treatments to the groups. (If the covariate is measured after the treatment, then the covariate may also have been influenced by the treatment, and we cannot distinguish the effects of the intervention from the covariate.) In the hypothetical research example that involves comparison of teaching methods, the X_c covariate is measured prior to the teaching intervention, and therefore, the X_c covariate cannot be influenced by the teaching intervention.
3. In practice, we cannot expect ANCOVA adjustments to be really effective in separating the effects of A and X_c if they are very strongly confounded. ANCOVA makes sense only in situations where the correlation or confound between A and X_c is weak to moderate. Caution is required in interpreting ANCOVA results where the rank ordering of Y means across treatment groups changes drastically or where the distributions of X_c scores differ greatly across the A groups.
4. We assume that there is no treatment-by-covariate or $A \times X_c$ interaction; that is, we assume that the slope for the regression prediction of Y from X_c is the same within each group included in the A factor. Another term for this assumption of no **treatment-by-covariate interaction** is the **homogeneity of regression assumption**. Procedures to test whether there is a significant interaction are discussed below.
5. Measurement of the covariate should have high reliability.

If the assumption of no treatment-by-covariate interaction is violated, the data analyst may choose to report the results of an analysis that does include a treatment-by-covariate interaction; however, this analysis would no longer be called an ANCOVA (it might be described as a regression model that includes group membership, covariate, and group membership–by–covariate interaction terms).

Potential violations of these assumptions can be evaluated by preliminary data screening. At a minimum, preliminary data screening for ANCOVA should include the following:

1. Examination of histograms for Y and each of the X_c covariates; all distributions should be approximately normal in shape with no extreme outliers: The histograms

Figure 8.1 Preliminary Data Screening for ANCOVA: Normality of Distribution Shapes for Y (Quantitative Outcome Variable) and X_c (Quantitative Covariate)

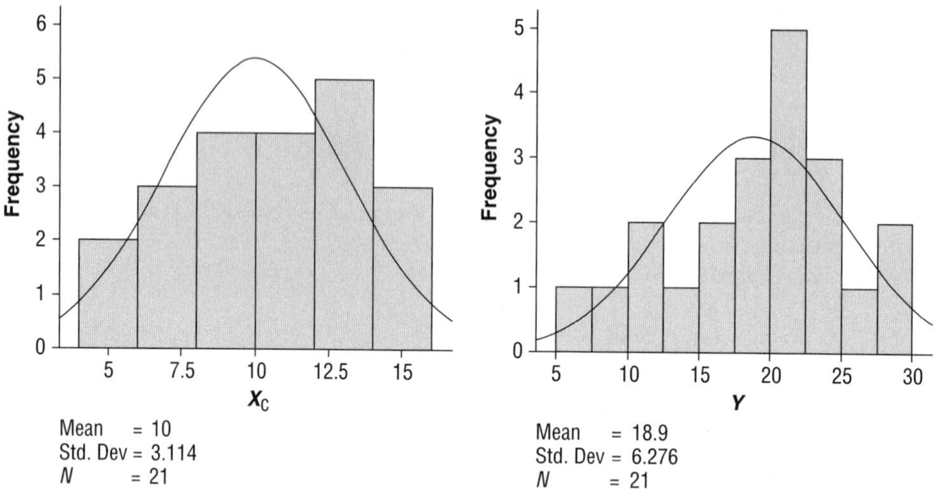

in Figure 8.1 and the scatterplot in Figure 8.2 indicate that for this small hypothetical data set, these assumptions do not appear to be severely violated.

2. Examination of scatterplots between Y and each X_c covariate and between all pairs of X_c covariates: All relations between pairs of variables should be approximately linear, with no extreme bivariate outliers.

3. Evaluation of the homogeneity of variance assumption for Y and assessment of the degree to which the X_c covariate is confounded with (i.e., systematically different across) levels of the A treatment factor: Both of these can be evaluated by running a simple one-way between-subjects ANOVA using A as the group membership variable and X_c and Y as separate outcome variables. Here, only selected results are reported (see Figure 8.3). Examination of the results in Figure 8.3 indicates that there is no statistically significant violation of the homogeneity of variance assumption for scores on Y.

4. This preliminary one-way ANOVA provides information about the nature of any confound between X_c and A. The mean values of X_c did not differ significantly across levels of A, $F(2, 18) = .61, p = .55, \eta^2 = .06$. (This eta squared effect size was obtained by dividing $SS_{between}$ by SS_{total} for the one-way ANOVA to predict X_c from A.) Thus, the confound between level of ability (X_c) and type of teaching method (A) experienced by each of the groups was relatively weak.

5. The one-way ANOVA to predict Y scores from type of teaching method (A) indicated that when the scores on the covariate are not taken into account, there are statistically significant differences in mean final exam scores across the three treatment groups: $F(2, 18) = 12.26, p < .001, \eta^2 = .58$. The nature of the differences in final exam scores across types of teaching group was as follows: Mean exam scores were highest for the seminar group, intermediate for the lecture group, and lowest for the SPI group.

Figure 8.2 Preliminary Data Screening for ANCOVA: Relationship Between Y and X_c

Correlations

		xc	y
xc	Pearson Correlation	1	.629**
	Sig. (2-tailed)		.002
	N	21	21
y	Pearson Correlation	.629**	1
	Sig. (2-tailed)	.002	
	N	21	21

**. Correlation is significant at the 0.01 level (2-tailed).

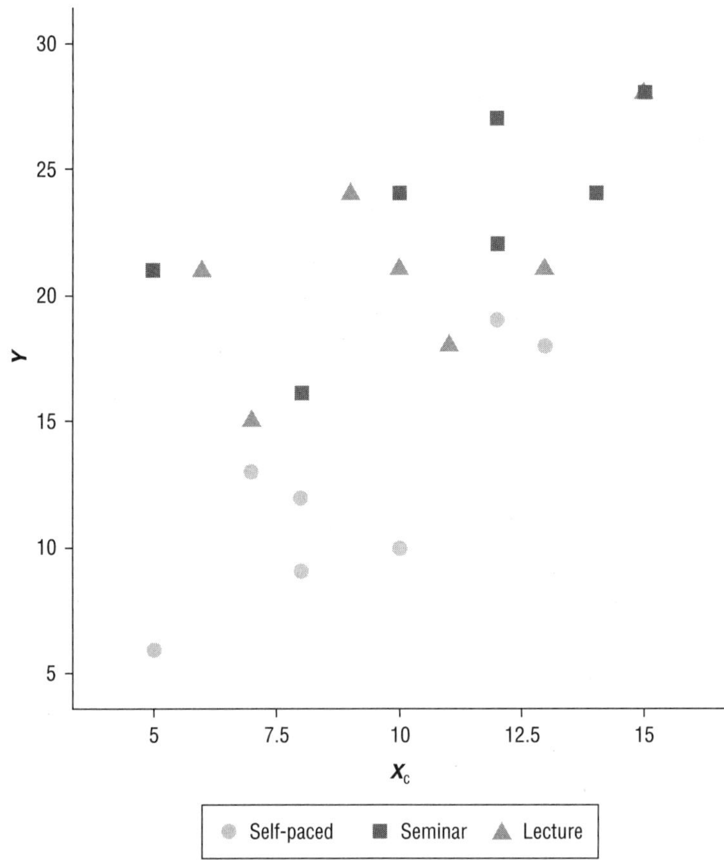

6. This one-way ANOVA to assess differences in Y means across A groups (without the X_c covariate) provides us with a basis for comparison. We can evaluate whether or how the inclusion of an X_c covariate in an ANCOVA model changes our understanding of the way in which mean Y scores differ across treatment groups, in comparison with the simple one-way ANOVA reported in Figure 8.3 (in which the

Figure 8.3 One-Way ANOVA to Evaluate Differences in Group Means Across Types of Teaching Methods (*A*) for the Covariate X_c and for the Unadjusted Outcome Scores on *Y*

Descriptives

		N	Mean	Std. Deviation	Std. Error	95% Confidence Interval for Mean Lower Bound	95% Confidence Interval for Mean Upper Bound	Minimum	Maximum
xc	Self Paced	7	9.00	2.828	1.069	6.38	11.62	5	13
	Seminar	7	10.86	3.485	1.317	7.63	14.08	5	15
	Lecture	7	10.14	3.185	1.204	7.20	13.09	6	15
	Total	21	10.00	3.114	.680	8.58	11.42	5	15
y	Self Paced	7	12.43	4.721	1.784	8.06	16.79	6	19
	Seminar	7	23.14	4.018	1.519	19.43	26.86	16	28
	Lecture	7	21.14	4.140	1.565	17.31	24.97	15	28
	Total	21	18.90	6.276	1.370	16.05	21.76	6	28

Test of Homogeneity of Variances

	Levene Statistic	df1	df2	Sig.
xc	.128	2	18	.881
y	.202	2	18	.819

ANOVA

		Sum of Squares	df	Mean Square	F	Sig.
xc	Between Groups	12.286	2	6.143	.608	.555
	Within Groups	181.714	18	10.095		
	Total	194.000	20			
y	Between Groups	454.381	2	227.190	12.265	.000
	Within Groups	333.429	18	18.524		
	Total	787.810	20			

X_c covariate is not statistically controlled). When we control for the X_c covariate in an ANCOVA reported later in this chapter, the apparent association between type of teaching method (*A*) and score on the final exam (*Y*) could become weaker or stronger; it is even possible that the rank ordering of adjusted means on *Y* can be different from the rank ordering of the original *Y* means.

7. Assessment of possible treatment-by-covariate interactions for each covariate: This can be requested in the SPSS GLM procedure by requesting a custom model that includes interaction terms. Researchers usually hope that the test of a treatment-by-covariate interaction in an ANCOVA research situation will not be statistically significant. If this interaction term is significant, it is evidence that an important assumption of ANCOVA is violated. If the treatment-by-covariate or $A \times X_c$ interaction term is nonsignificant, then this interaction term is dropped from the model, and an ANCOVA is performed without an interaction term between treatment and covariate. The ANCOVA that is used to generate adjusted group means (in the final "Results" section) generally does not include a treatment-by-covariate interaction. This simple ANCOVA corresponds to the type of ANCOVA model represented in Equation 8.1.

8.4 VARIANCE PARTITIONING IN ANCOVA

The most common approach to variance partitioning in ANCOVA is similar to the hierarchical method of variance partitioning discussed in Chapter 5, using multiple predictors in linear regression. In a typical ANCOVA, the researcher enters one (or more) *X* covariates as predictors of the *Y* outcome variable in Step 1. In Step 2, categorical predictors that represent treatment group membership are added, and the data analysts examine the results to see whether the categorical predictor variable(s) contribute significantly to our ability to predict scores on the *Y* outcome variable when scores on one or more covariates are statistically controlled.

We can use overlapping circles to represent the partition of SS_{total} for *Y* into sum of squares components for the covariate and for the treatment factor (just as we used overlapping circles in Chapter 5 to illustrate the partition of variance for *Y*, the outcome variable in a regression analysis, among correlated predictor variables). In the SPSS GLM procedure, we can obtain this type of variance partitioning (enter the covariate in Step 1, enter the treatment group membership variable in Step 2) by using the Univariate: Model dialog box in the GLM procedure to request the *SS* Type I method of computation for sums of squares and by listing the predictor variables in the Univariate: Model dialog box in the sequence in which they should enter the model. The variance partition that results from these analysis specifications is depicted by the overlapping circles in Figure 8.4a. We obtain a sum of squares for X_c; if we request *SS* Type I computation, this sum of squares for the covariate X_c is not adjusted for *A*, the treatment group variable. Then, we calculate an adjusted *SS* for *A* that tells us how strongly the *A* treatment variable is predictively related to scores on the *Y* outcome variable when the covariate X_c has been statistically controlled or partialled out. Figure 8.4a illustrates the partition of *SS* that is obtained when the *SS* Type I computation method is chosen; this is the most conventional approach to ANCOVA.

Figure 8.4 Variance Partitioning in an ANCOVA: (a) With One Treatment (*A*) and One Covariate (X_c) Using Type I *SS* (Hierarchical Approach) and (b) Using Type III *SS* (Standard Regression Approach)

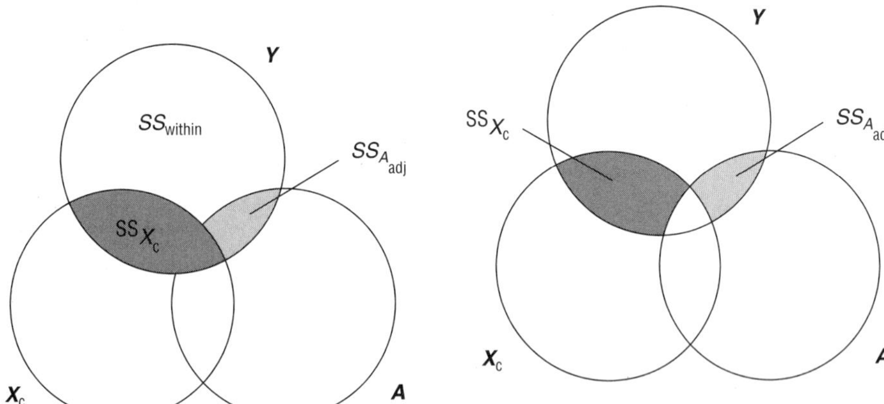

Note: In Figures 8.4a and 8.4b, the dark gray area corresponds to SS_{Xc}, the light gray area corresponds to SS_{Aadj}, and the sum of squares represents the remaining A main effect after statistically controlling for X_c, the covariate.

If you use the default method for computation of sums of squares in the GLM procedure, *SS* Type III, the *SS* term for each predictor is calculated by controlling for or adjusting for all the other predictors in the model. If the *SS* Type III method is used to compute the sums of squares in this situation, SS_A is adjusted for X_c and SS_{Xc} is adjusted for A, as shown in Figure 8.4b. Note that the SS_{Aadj} term is the same whether you use Type I or Type III SS; in either instance, SS_{Aadj} is adjusted for X_c. However, the *SS* term for the X_c covariate is estimated differently depending on whether Type I or Type III sum of squares is requested. Conventionally, when reporting ANCOVA, readers usually expect to see the Type I approach to variance partitioning, and therefore, the use of Type I *SS* is recommended. However, the primary interest in ANCOVA is usually the differences in group means rather than the proportion of variance that is predictable from the covariate, and the conclusion about that question is the same whether the data analyst selects Type I or Type III *SS*. More detailed information about the different methods of computation for sums of squares can be obtained by clicking the Help button in the SPSS GLM dialog box.

An ANCOVA model can include multiple covariates. An ANCOVA model can include more than one factor or treatment group variable, and in such cases, it will generally include interactions among factors. If you use the SS Type III method in GLM, then every predictor variable in the analysis is assessed controlling for every other predictor. If you use the *SS* Type I method (or sequential method of entry), the most conventional order of entry is as follows: all covariates first, all main effects next, all two-way interactions between factors after that, and so forth. When *SS* Type I is selected, predictor variables are entered into the analysis in the sequence in which they appear in the list provided by the data analyst. Note that each predictor is assessed by controlling for every other predictor that is entered in the same step or in previous steps.

8.5 ISSUES IN PLANNING A STUDY

When a researcher plans a study that involves comparisons of groups that are exposed to different treatments or interventions and if the researcher wants to make causal inferences, the preferred design is a well-controlled experiment. Random assignment of participants to groups and/or creation of matched samples are methods of experimental control for participant characteristics that are usually, although not always, effective in ruling out confounds between participant characteristics and type of treatment. Even in a well-designed experiment (in which participant characteristics are not confounded with treatments), it can be useful to measure one or more participant characteristics and to include them in the analysis as covariates so that the magnitude of the SS_{within} or SS_{residual} term used in the F test can be reduced.

In quasi-experiments, and sometimes even in randomized groups experiments, we may find that one or more participant characteristics are confounded with treatments. When such confounds are present, they make group differences difficult to interpret; we need to statistically control for or partial out the participant characteristics.

A common choice of covariate is a pretest measure on the same behavior that is assessed as the outcome. Thus, ANCOVA may be used to assess pretest–posttest differences across multiple treatment groups. (See Appendix 8A at the end of this chapter for a brief discussion of several alternative methods of analysis for pretest–posttest data.) However, any participant characteristic that has a strong correlation with Y may be useful as a noise suppressor in the data analysis, and any participant characteristic that is known to be confounded with treatment group membership, and thus is a rival explanation for any observed differences in outcomes, should be controlled as a covariate when comparing group means.

The statistical control of confounds using ANCOVA is not as powerful a method for dealing with participant characteristics as true experimental control (through random assignment of participants to conditions and/or holding participant characteristics constant by sampling homogeneous participants). Results of ANCOVA can be misleading if the covariate is not a reliable and valid measure of the construct of interest or if the basic assumptions for ANCOVA are violated.

8.6 FORMULAS FOR ANCOVA

ANCOVA can be understood as an ANOVA that compares adjusted Y means across groups (the adjustment involves regression of Y on one or more X_c covariates and subtracting any part of Y that is related to covariates). Note, however, that the regression slope that is used to make this adjustment is a partial or within-group slope (the slope for prediction of Y from X_c when you control for A, the treatment group variable).

It is somewhat easier to describe this adjustment if we describe ANCOVA as a regression analysis to predict Y from one or more covariates and from dummy variables that represent membership in treatment groups. An equation for regression analysis that is equivalent to an ANCOVA for the teaching method study described in Section 8.4 would look like this:

$$Y' = b_0 + b_1D_1 + b_2D_2 + b_3\,(X_c - M_{Xc}), \tag{8.3}$$

where Y is the outcome variable, final statistics exam score; X_c is the covariate, pretest on math ability; M_{Xc} is the grand mean of the covariate; and D_1, D_2 are the dummy variables that represent membership in the three teaching method groups. The b_3 regression coefficient is the partial regression slope that predicts Y from X_c while controlling for A group membership. The assumption of no treatment-by-covariate interaction is the assumption that the b_3 slope to predict Y from $X_c - M_{Xc}$ is the same within each of the A groups; the b_3 partial slope is, essentially, the average of the three within-group slopes to predict Y from $X_c - M_{Xc}$.

We can rewrite the previous equation as follows:

$$Y' - b_3(X_c - M_{Xc}) = b_0 + b_1D_1 + b_2D_2. \tag{8.4}$$

The new term on the left-hand side of Equation 8.4, $Y' - b_3\,(X_c - M_{Xc})$, corresponds to the adjusted Y score (Y^*). When we subtract $b_3\,(X_c - M_{Xc})$ from Y', we are removing from Y' the part of each Y score that is linearly predictable from X_c, to obtain an adjusted predicted Y^* score that is uncorrelated with X_c. The resulting equation,

$$Y^* = b_0 + b_1D_1 + b_2D_2, \tag{8.5}$$

corresponds to a one-way ANOVA to assess whether the adjusted Y^* means differ significantly across groups. The partition of the total variance of Y into the proportion of variance that is predictable from X_c and the proportion of variance that is predictable from A, statistically controlling for X_c, can be assessed by running appropriate regression analyses.[1]

The idea that students need to keep in mind is that ANCOVA is only a special case of regression in which one or more quantitative predictors are treated as covariates and one or more categorical variables represent membership in treatment groups. Using the matrix algebra for multiple regression (presented in Appendix 5A) with dummy variables to represent treatment group membership will yield parameter (slope) estimates for the ANCOVA model.

Matrix algebra is required to represent the computations of the various SS terms in ANCOVA (see Tabachnick & Fidell, 2018, for the matrix algebra). Only a conceptual description of the terms is presented here. In this simple example with only one covariate and one treatment factor, we need to obtain SS_{Xc}; this represents the part of SS_{total} for Y that is predictable from X_c, a covariate. We need to obtain an SS for A, usually called an adjusted SS_A, to remind us that we are adjusting for or partialling out the covariate; SS_{Aadj} represents the part of SS_{total} for Y that is predictable from membership in the groups on Factor A when the X_c

covariate is statistically controlled. Finally, we will also have SS_{within} or SS_{residual}, the part of the SS_{total} for Y that is not explained by either X_c or A.

The primary goal in ANCOVA is generally to assess whether the A main effects remain significant when one or more covariates are controlled. However, an F ratio may also be set up to assess the significance of each covariate, as follows:

$$F_{Xc} = (SS_{Xc}/df_{Xc})/(SS_{\text{residual}}/df_{\text{residual}}). \tag{8.6}$$

The null hypothesis that is usually of primary interest in ANCOVA is that the adjusted Y^* means are equal across levels of A:

$$H_0\text{: } \mu^*_{A1} = \cdots = \mu^*_{Ak}. \tag{8.7}$$

In words, Equation 8.7 corresponds to the null hypothesis that the adjusted Y^* means in the population are equal across levels of A; the means are adjusted by subtracting the parts of Y scores that are predictable from one or more X_c covariates. The corresponding F ratio to test the null hypothesis in Equation 8.7 is as follows:

$$F_A = (SS_{Aadj}/df_A)/(SS_{\text{residual}}/df_{\text{residual}}). \tag{8.8}$$

8.7 COMPUTATION OF ADJUSTED EFFECTS AND ADJUSTED Y* MEANS

The unadjusted effect (α_i) for level or Group i of an A factor in a between-S one-way ANOVA is sometimes included in discussions. The sample estimate of α_1 for Group 1 (the "effect" of being in Group 1) is obtained by finding the difference between the A_1 group mean on Y and the grand mean for Y.

$$\alpha_i = Y_{Ai} - Y_{\text{grand}}, \tag{8.9}$$

where Y_{Ai} is the mean of Y in Group A_i, and Y_{grand} is the grand mean of Y across all groups included in the study.

To do an ANCOVA, we need to compute adjusted effects, α^*_i, which have the X_c covariate partialled out:

$$\alpha^*_i = (Y_{Ai} - Y_{\text{grand}}) - b_3 \times (X_{\text{cAi}} - X_{\text{cgrand}}), \tag{8.10}$$

where X_{cAi} is the mean of the covariate X_c in Group A_i, X_{cgrand} is the mean of the X_c covariate across all groups in the study, and b_3 is the partial slope to predict Y from X_c in a regression that controls for A group membership by including dummy predictors, as in Equation 8.3.

Once we have adjusted effects, α^*_i, we can compute an estimated adjusted Y mean for each A_i group (Y^*_{Ai}) by adding the grand mean (Y_{grand}) to each adjusted effect estimate (α^*_i):

$$Y^*_{Ai} = \alpha^*_i + Y_{\text{grand}}. \tag{8.11}$$

8.8 CONCEPTUAL BASIS: FACTORS THAT AFFECT THE MAGNITUDE OF SS_{Aadj} AND SS_{residual} AND THE PATTERN OF ADJUSTED GROUP MEANS

When we control for a covariate X_c, several outcomes are possible for our assessment of the effects of the treatment group membership variable A. These outcomes are analogous to the outcomes described for partial correlation analysis in an earlier chapter.

1. When we control for X_c, the main effect of A (which was significant in an ANOVA) may become nonsignificant (in the ANCOVA), and/or the effect size for A may decrease.

2. When we control for X_c, the main effect of A (which was nonsignificant in an ANOVA) may become significant (in the ANCOVA), and/or the effect size for A may increase.

3. When we control for X_c, the conclusion about the statistical significance and effect size for A may not change.

4. When we control for X_c and calculate the adjusted group means using an ANCOVA, the rank ordering of Y means across levels of the A treatment group factor may change; in such cases, our conclusions about the nature of the relationship of A and Y may be quite different when we control for X_c than when we do not. This outcome is not common, but it can occur. Data analysts should be skeptical of adjusted means when adjustment for covariates drastically changes the rank ordering of treatment group means; when adjustments are this dramatic, there may be rather serious confounds between X_c and A, and in such situations, ANCOVA adjustments may not be believable.

8.9 EFFECT SIZE

When the SPSS GLM procedure is used to do an ANCOVA, effect sizes for predictor variables can be indexed by eta squared values, as in previous chapters in this book. Note that two versions of η^2 can be calculated for an ANCOVA. In ANCOVA, the *SS* term for a main effect, adjusted for the covariate, is called an adjusted *SS*. In an ANCOVA using *SS* Type I (or sequential regression), with one A factor and one X_c covariate, SS_{total} for Y is partitioned into the following sum of squares components:

$$SS_{\text{total}} = SS_{Xc} + SS_{Aadj} + SS_{\text{residual}}. \tag{8.12}$$

We can compute a "simple" η^2 ratio to assess the proportion of variance accounted for by the A main effect relative to the *total* variance in Y; this is the version of eta squared often used as an effect size one-way ANOVA.

$$\eta^2_A = SS_{Aadj} / SS_{\text{total}} = \frac{SS_{Aadj}}{SS_{Aadj} + SS_{Xc} + SS_{residual}}. \tag{8.13}$$

Alternatively, we can set up a "partial" η^2 to assess the proportion of variance in Y that is predictable from A when the variance associated with other predictors (such as the X_c covariates) has been partialled out or removed. When effect size information is requested, the SPSS GLM procedure reports a partial η^2 for each effect. For the simple ANCOVA example in this chapter, this corresponds to the partial η^2 in Equation 8.14:

$$\text{Partial } \eta^2_A = SS_{Aadj}/(SS_{Aadj} + SS_{\text{residual}}). \tag{8.14}$$

This partial η^2 tells us what proportion of the Y variance that remains after we statistically control for X_c is predictable from A. Note that the partial η^2 given by Equation 8.14 is usually larger than the simple η^2 in Equation 8.13 because SS_{Xc} is not included in the divisor for partial η^2.

More generally, partial η^2 for an effect in a GLM is given by the following equation:

$$\text{Partial } \eta^2_{\text{effect}} = SS_{\text{effect}}/(S_{\text{effect}} + SS_{\text{residual}}). \tag{8.15}$$

8.10 STATISTICAL POWER

Usually, researchers are not particularly concerned about statistical power for assessment of the significance of the X_c covariate(s). The covariates are usually included primarily because we want to adjust for them when we assess whether Y means differ across groups (rather than because of interest in the predictive usefulness of the covariates). We are more concerned with statistical power for assessment of main effects and interactions.

Note that including one or more covariates can increase statistical power for tests of main effects by reducing the size of SS_{residual} or SS_{within} and, therefore, increasing the size of F, but an increase in power is obtained only if the covariates are fairly strongly predictive of Y. Note also that if covariates are confounded with group membership, adjustment for covariates can move the group means farther apart (which will increase SS_{Aadj} and, thus, increase F_A), or adjustment for the covariate X_c can move the group means closer together (which will decrease SS_{Aadj} and reduce F_A). Thus, when one or more covariates are added to an analysis, main effects for group differences can become either stronger or weaker. It is also possible that controlling for covariates will change the rank ordering of Y means across levels of A, although this is not a common outcome. When a covariate is an effective noise suppressor, its inclusion in ANCOVA can improve statistical power. However, under some conditions, including a covariate decreases the apparent magnitude of between-group differences or fails to reduce within-group or error variance; in these situations, ANCOVA may be less powerful than ANOVA.

8.11 NATURE OF THE RELATIONSHIP AND FOLLOW-UP TESTS: INFORMATION TO INCLUDE IN THE "RESULTS" SECTION

First, you should describe what variables were included (one or more covariates, one or more factors, and the name of the Y outcome variable). You should specify which method of variance partitioning you have used (e.g., Type I vs. Type III *SS*) and the order of entry of the predictors. Second, you should describe the preliminary data screening and state whether there are violations of assumptions. Most of these assumptions (e.g., normally distributed scores on Y, linear relation between X_c and Y) are similar to those for regression. You should check to see whether there are homogeneous variances for Y across groups. In addition, you need to assess possible treatment-by-covariate interaction(s); this can be done by using the GLM (as in the example at the end of this chapter) or by adding an interaction to a regression model. If there are no significant interactions, rerun the ANCOVA with the treatment-by-covariate interaction terms omitted and report the results for that model. If there are significant interactions, blocking on groups on the basis of X_c scores may be preferable to ANCOVA.

In addition to checking for possible violations of assumptions, I also recommend running ANOVAs to see how Y varies as a function of treatment groups (not controlling for X_c); this gives you a basis for comparison, so that you can comment on how controlling for X_c changes your understanding of the pattern of Y means across groups. Also, run an ANOVA to see how X_c means vary across groups; this provides information about the nature of any confounds between group memberships and covariates.

Your "Results" section will typically include F ratios, *df* values, and effect size estimates for each predictor; effect sizes are typically indexed by partial η^2 when you use the GLM. If you include a source table for your ANCOVA, it is conventional to list the variables in the following order in the source table: first the covariate(s); then main effects; then, if ANCOVA involves a factorial design, two-way interactions and higher order interactions.

To describe the nature of the pattern of Y means across groups, you should include a table of both unadjusted and adjusted means for the Y outcome variable. The adjusted Y

means are corrected to remove any association with the covariate(s) or X_c variables; they are, in effect, estimates of what the means of Y might have been if the treatment groups had been exactly equal on the X_c covariates.

It is important to notice exactly how the ANCOVA (which assesses the pattern of Y means, controlling for the X_c covariates) differs from the corresponding ANOVA (which assessed the pattern of Y means without making adjustment for covariates). The overall F for main effects can become larger or smaller when you control for covariates depending on whether the adjustment to the Y means that is made using the covariates moves the Y means closer together, or farther apart, and whether $SS_{residual}$ or SS_{within} is much smaller in the ANCOVA than it was in the ANOVA.

If you have a significant main effect for a factor in your ANCOVA and the factor has more than two levels, you may want to do post hoc tests or planned contrasts to assess which pairs of group means differ significantly. If you have a significant interaction between factors in a factorial ANCOVA, you may want to present a graph of cell means, a table of cell means, or a table of adjusted effects to provide information about the nature of the interaction.

8.12 SPSS ANALYSIS AND MODEL RESULTS

To understand how Y, X_c, and A are interrelated, the following analyses were performed.

8.12.1 Preliminary Data Screening

A one-way ANOVA was performed to see whether Y differed across levels of A. This analysis provided an answer to the question, If you do *not* control for X_c, how does Y differ across levels of A, if at all? A one-way ANOVA was run to see whether X_c differed across levels of A. This analysis answered the question, To what extent did the A treatment groups begin with preexisting differences on X_c, the covariate, and what was the nature of any confound between X_c and A? These results were obtained using the SPSS one-way ANOVA procedure. The results for these ANOVAs (shown in Figure 8.3) indicated that the means on the covariate did not differ significantly across the three treatment groups; thus, there was a very weak confound between treatment and covariate (and that's good). To the extent that there was a confound between X_c (ability) and A (type of teaching method), students in Groups 2 and 3 (seminar and lecture) had slightly higher pretest math ability than students in Group 1 (the SPI group). Mean scores on Y, the final exam score, differed significantly across groups: Group 2 (seminar) and Group 3 (lecture) had higher mean final exam scores than Group 1 (SPI). We want to see whether controlling for the slight confound between teaching method and pretest math ability changes the Y differences across teaching groups.

To assess whether X_c and Y are linearly related, a scatterplot and a bivariate Pearson correlation between X_c and A were obtained. ANCOVA assumes that covariates are linearly related to the outcome variable Y. Unless there is a nonzero correlation between X_c and Y and/or some association between X_c and A, it is not useful to include X_c as a covariate. The scatterplot and Pearson's r were obtained using commands that were introduced in earlier chapters. The relation between X_c and Y appeared to be fairly linear (see the scatterplot in Figure 8.2). The correlation between X_c and Y was strong and positive, $r(19) = .63$, $p = .02$. Thus, X_c seemed to be sufficiently correlated with Y to justify its inclusion as a covariate. The preceding analyses provide two kinds of information: information about possible violations of assumptions (such as homogeneity of variance and linearity of association between X_c and Y) and information about all the bivariate associations between pairs of variables in this research situation. That is, How does the mean level of X_c differ across groups on the basis of Factor A? Is Y, the outcome variable, strongly linearly related to X_c, the covariate? How do mean Y scores differ across levels of A when the covariate X_c is not statistically controlled? It is useful to compare the final ANCOVA results with the original ANOVA to predict Y from A, so that we can see whether and how controlling for X_c has changed our understanding of the nature of the association between Y and A.

8.12.2 Assessment of Assumption of No Treatment-by-Covariate Interaction

Another assumption of ANCOVA is the no treatment-by-covariate interaction; that is, the slope relating Y to X_c should be the same within each of the A_1, A_2, and A_3 groups. In the scatterplot in Figure 8.2, the points were labeled with different markers for members of each group. Visual examination of the scatterplots does not suggest much difference among groups in the slope of the function to predict Y from X_c. Of course, it would be desirable to have much larger numbers of scores in each of the three groups when trying to assess the nature of the association between X_c and Y separately within each level of A. To examine the Y by X_c regression separately within each group, you could use the SPSS <Data> → <Split File> command to run a separate regression to predict Y from X_c for each level of the A factor, that is, for groups A_1, A_2, and A_3; results of separate regressions are not reported here.

To test for a possible violation of the assumption that there is no treatment-by-covariate ($A \times X_c$) interaction, a preliminary GLM analysis was conducted with an interaction term included in the model. Figure 8.5 shows the SPSS menu selections for the GLM procedure. From the top-level menu in the SPSS Data View worksheet, the following menu selections were made: <Analyze> → <General Linear Model> → <Univariate>. (This ANCOVA is "univariate" in the sense that there is only one Y outcome variable.) Figure 8.6 shows the main dialog box for the SPSS univariate GLM procedure. The names of the dependent variable (Y), treatment variable or factor (A), and covariate (X_c) were entered in the appropriate boxes in the main dialog box.

To add an interaction term to the ANCOVA model, it is necessary to request a custom model. Interactions between factors and covariates are not included by default; however, they

Figure 8.5 SPSS Menu Selections for the SPSS GLM Univariate Procedure

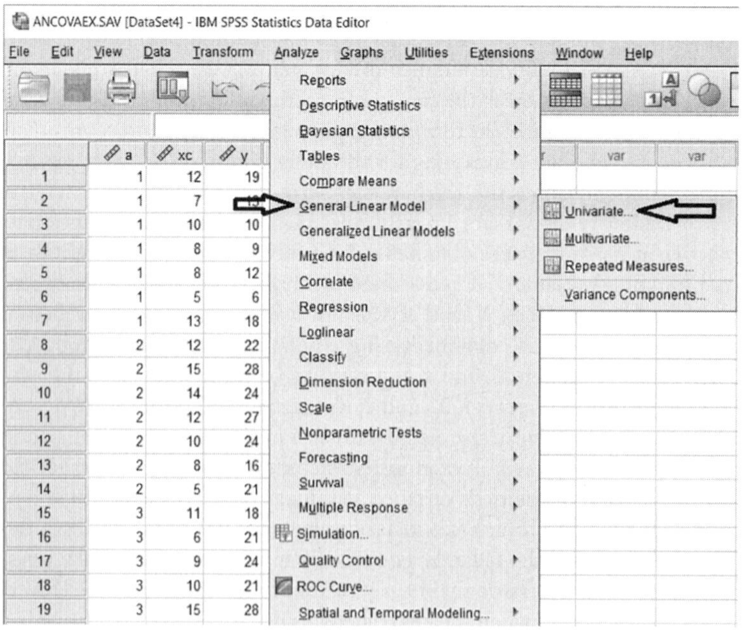

Figure 8.6 SPSS Univariate Dialog Box

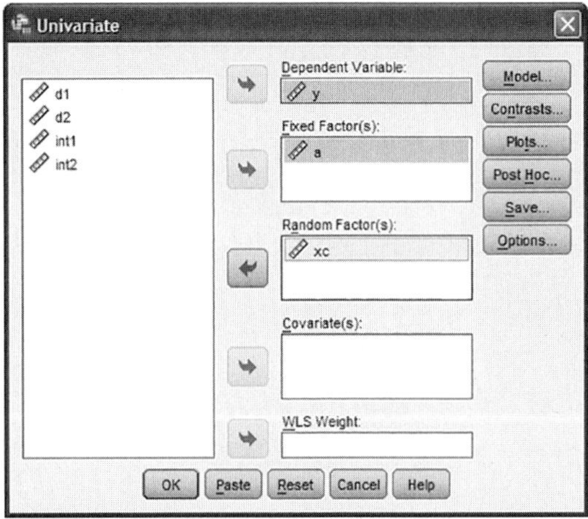

Note: A is the grouping variable, Y the outcome variable, and X_c the covariate.

can be specifically requested in the Univariate: Model window. Clicking the Model button in the main GLM dialog box, seen in Figure 8.6, opens up the Univariate: Model dialog box, which appears in Figure 8.7.

To specify a preliminary ANCOVA that included an $X_c \times A$ interaction term, the following selections were made within the Univariate: Model dialog box. First, among the three radio buttons at the top of the dialog box in Figure 8.7, select the "Build Terms" radio button in the center. Both X_c and A were highlighted in the left-hand pane, which included all possible factors and covariates, and "Interaction" was chosen from the pull-down menu in the "Build Term(s)" box. Clicking the right arrow created an $A \times X_c$ interaction term and placed it in the list of terms to be included in the custom model in the right-hand pane in Figure 8.7. Note that it was also necessary to move the terms that represent the main effect of A and the effect of X_c into this list to test for the interaction while controlling for main effects. The pull-down menu "Sum of Squares" offers a choice of four methods of computation for sums of squares (Types I, II, III, and IV). For this preliminary analysis, select "Type III"; this ensures that the interaction will be assessed while statistically controlling for both X_c and A. Click the Continue button to return to the main GLM dialog box, then OK to run the analysis. (Additional statistics such as adjusted group means and model parameter estimates were not requested for this preliminary analysis.)

The output that appears in the lower part of Figure 8.7 includes a test for the $A \times X_c$ interaction term (in the row labeled "a * xc"). This interaction was not statistically significant, $F(2, 15) = .68, p = .52$. Because this interaction was not statistically significant, there was no evidence that the "homogeneity of regression" or "no treatment-by-covariate interaction" assumption for ANCOVA was violated. Therefore, the next step will be to conduct and report a final ANCOVA that does *not* include an interaction term between A and X_c.

If this interaction between A and X_c had been statistically significant, it would mean that the slope to predict Y from X_c differed significantly across the A groups and that, therefore, it would be misleading to use the same regression slope to create adjusted group means for all the A groups (which is what ANCOVA does, in effect). At this point, the data analyst can report a model that includes an interaction term between treatment and covariate, but that analysis would not be called an ANCOVA. Horton (1978) mentioned the possibility of "ANCOVA with heterogeneous slopes"; however, most authorities require the absence of a treatment-by-covariate interaction before they will call the analysis an ANCOVA.

Figure 8.7 Custom Model Specification: Interaction Term Added to the Model (to Test for Possible $A \times X_c$ Interaction, Which Would Be a Violation of a Basic Assumption for ANCOVA)

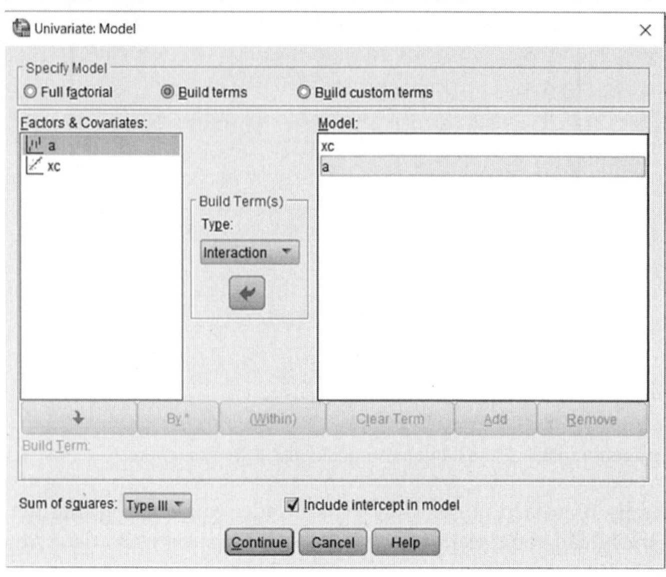

Tests of Between-Subjects Effects

Dependent Variable: y

Source	Type III Sum of Squares	df	Mean Square	F	Sig.
Corrected Model	633.192[a]	5	126.638	12.286	.000
Intercept	139.316	1	139.316	13.516	.002
a * xc	14.108	2	7.054	.684	.520
a	74.671	2	37.335	3.622	.052
xc	174.702	1	174.702	16.949	.001
Error	154.617	15	10.308		
Total	8293.000	21			
Corrected Total	787.810	20			

a. R Squared = .804 (Adjusted R Squared = .738)

Note: SS Type III is used here because we want to assess whether the $A \times X_c$ interaction is statistically significant, controlling for the effects of both A and X_c. Output for GLM that includes A, X_c, and $A \times X_c$ interaction term in the model. Because the $A \times X_c$ interaction term was not statistically significant (using the $\alpha = .05$ level of significance as the criterion), this interaction term is dropped from the model; the ANCOVA without an interaction term is reported in the next few figures.

8.12.3 Conduct Final ANCOVA Without Interaction Term Between Treatment and Covariate

Given that, for these hypothetical data, there was no significant interaction between treatment and covariate, the next step is to run a new GLM analysis that does not include this interaction term. To be certain that this interaction term is excluded from this new model, it is a good idea to click the Reset button for GLM so that all the menu selections made in the earlier analysis are cleared. We identify the variables in the main Univariate dialog box (as shown in Figure 8.6). To request additional descriptive statistics, click the Statistics button in the main dialog box; the Univariate: Options dialog box appears in Figure 8.8. In the following example, these were the requested statistics: First, the terms that represent the overall (grand) mean and the mean for each level of Factor *A* (a) were highlighted and moved into the pane under the heading "Display Means for." Checkboxes were used to request "Estimates of effect size" (i.e., partial η^2 values) and "Parameter estimates" (i.e., coefficients for the regression model that predicts *Y* from scores on X_c and dummy variables to represent group membership). The box labeled "Compare main effects" was checked to request selected comparisons between pairs of group means on the *A* factor. From the pull-down menu directly below this, the option for Bonferroni-corrected adjustments was selected. The Continue button was clicked to return to the main Univariate dialog box.

To request the use of Type I *SS* computational methods, and to make sure that X_c (the covariate) was entered before *A* (the treatment factor), the Model button was clicked. Within the

Figure 8.8 Options Selected for the Final ANCOVA Model: Display Means for Each Level of A (Type of Teaching Method)

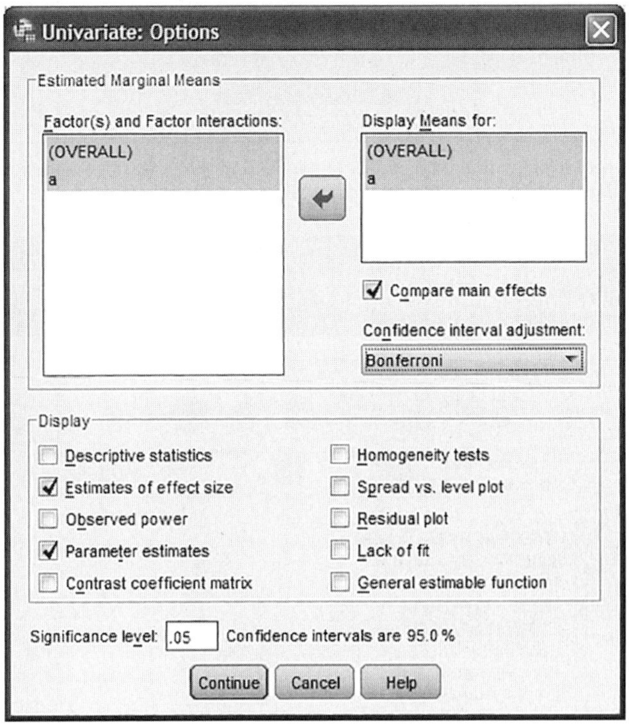

Note: Do Bonferroni-corrected contrasts between group means. Report estimates of effect size (partial η^2) and parameter estimates for the regression version of the ANCOVA.

Univariate: Model dialog box, which appears in Figure 8.9, *SS* Type I was selected from the pull-down menu for type of sums of squares. Under "Specify Model," the "Custom" radio button was clicked, then the terms X_c and A were highlighted and moved into the "Model" pane. The order in which the terms appear in this list is the order in which the predictors will enter the model. That is, for the analysis specified in Figure 8.9, the X_c covariate will be entered in the first step, and then an adjusted *SS* for A will be calculated controlling or adjusting for X_c. Clicking Continue returns us to the main dialog box. The Paste button was used to place the SPSS syntax that was generated by these menu selections into an SPSS syntax window, which appears in Figure 8.10.

Figure 8.9 Custom Model for Final Version of ANCOVA

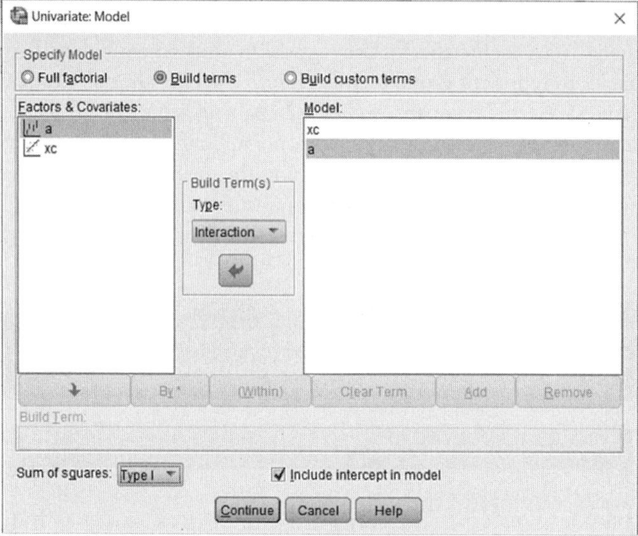

Note: Select Type I method of computation of SS. List variables in the order of entry (first X_c, then A) so that the SS for A is calculated while partialling out or controlling for X_c.

Figure 8.10 SPSS Syntax for the Final ANCOVA Analysis Specified by the Menu Selections Above

These menu selections correspond to an ANCOVA to assess whether mean scores on Y differ significantly across levels of A when the X_c covariate is statistically controlled. Results for this analysis appear in Figures 8.11 and 8.12. The result that is typically of primary interest in an ANCOVA is the F test for the A main effect, controlling for the covariate X_c. This appears in the "Tests of Between-Subjects Effects" panel in Figure 8.11, in the row for a. The mean Y scores differed significantly across levels of the A factor when the covariate X_c was statistically controlled, $F(2, 17) = 17.47$, $p < .001$. The "adjusted" Y means (i.e., adjusted to remove the association with the covariate X_c) are given in Figure 8.12, in the panel labeled "Estimates." Table 8.2 summarizes the original unadjusted Y means (from the one-way ANOVA that was reported in Figure 8.3) along with the adjusted Y means obtained from the GLM ANCOVA reported in Figures 8.11 and 8.12.

An example "Results" section for the foregoing ANCOVA follows.

Results

A quasi-experimental study was performed to assess whether three different teaching methods produced different levels of performance on a final exam in statistics. Group A_1 was given SPI, Group A_2 participated in a seminar, and Group A_3 received lecture instruction. Because the assignment of participants to groups was not random, there was a possibility that there might be preexisting differences in math ability; to assess

Figure 8.11 Output From the Final ANCOVA Analysis

Between-Subjects Factors

a	Value Label	N
1	Self Paced	7
2	Seminar	7
3	Lecture	7

Tests of Between-Subjects Effects

Dependent Variable: y

Source	Type I Sum of Squares	df	Mean Square	F	Sig.	Partial Eta Squared
Corrected Model	619.085[a]	3	206.362	20.792	.000	.786
Intercept	7505.190	1	7505.190	756.191	.000	.978
xc	311.938	1	311.938	31.430	.000	.649
a	307.146	2	153.573	15.473	.000	.645
Error	168.725	17	9.925			
Total	8293.000	21				
Corrected Total	787.810	20				

a. R Squared = .786 (Adjusted R Squared = .748)

Parameter Estimates

Dependent Variable: y

Parameter	B	Std. Error	t	Sig.	95% Confidence Interval Lower Bound	95% Confidence Interval Upper Bound	Partial Eta Squared
Intercept	11.486	2.653	4.330	.000	5.890	17.083	.524
xc	.952	.234	4.074	.001	.459	1.445	.494
[a=1]	-7.626	1.705	-4.473	.000	-11.223	-4.029	.541
[a=2]	1.320	1.692	.780	.446	-2.250	4.890	.035
[a=3]	0[a]	.	.	.	.	.	.

a. This parameter is set to zero because it is redundant.

Figure 8.12 Estimated (Adjusted for the Covariate X_c) Means on the Y Outcome Variable Across Levels of the A Teaching Method Factor

1. Grand Mean

Dependent Variable: y

Mean	Std. Error	95% Confidence Interval	
		Lower Bound	Upper Bound
18.905[a]	.687	17.454	20.355

a. Covariates appearing in the model are evaluated at the following values: xc = 10.00.

Estimates

Dependent Variable: y

a	Mean	Std. Error	95% Confidence Interval	
			Lower Bound	Upper Bound
Self Paced	13.381[a]	1.213	10.820	15.941
Seminar	22.327[a]	1.207	19.779	24.874
Lecture	21.007[a]	1.191	18.494	23.520

a. Covariates appearing in the model are evaluated at the following values: xc = 10.00.

Pairwise Comparisons

Dependent Variable: y

(I) a	(J) a	Mean Difference (I-J)	Std. Error	Sig.[a]	95% Confidence Interval for Difference[a]	
					Lower Bound	Upper Bound
Self Paced	Seminar	-8.946*	1.739	.000	-13.563	-4.329
	Lecture	-7.626*	1.705	.001	-12.153	-3.099
Seminar	Self Paced	8.946*	1.739	.000	4.329	13.563
	Lecture	1.320	1.692	1.000	-3.173	5.813
Lecture	Self Paced	7.626*	1.705	.001	3.099	12.153
	Seminar	-1.320	1.692	1.000	-5.813	3.173

Based on estimated marginal means

*. The mean difference is significant at the .05 level.

a. Adjustment for multiple comparisons: Bonferroni.

Table 8.2 Adjusted and Unadjusted Group Means for Final Statistics Exam Scores

		Unadjusted Mean of Y	Mean of Y, Adjusted for X_c (Y_i^*)	Ability at Pretest (X_c)
Group 1	Self-paced instruction	12.43	13.38	9.00
Group 2	Seminar	23.14	22.32	10.86
Group 3	Lecture	21.14	21.00	10.14

Note: Ability at pretest is the covariate.

this, students were given a pretest on math ability (X_c), and this score was used as a covariate. The dependent variable, Y, was the score on the final exam.

Preliminary data screening was done; scores on X_c and Y were reasonably normally distributed, with no extreme outliers. The scatterplot for X_c and Y showed a linear relation with no bivariate outliers. Scores on the math pretest (X_c) did not differ significantly across groups, $F(2, 18) = .608$, $p = .55$; however, the SPI group (A_1) scored slightly lower on the math pretest than the other two groups.

To assess whether there was an interaction between treatment and covariate, a preliminary ANCOVA was run using the SPSS GLM procedure with a custom model that included an $X_c \times A$ interaction term. This interaction was not statistically significant, $F(2, 15) = .684$, $p = .52$. This indicated no significant violation of the homogeneity of regression (or no treatment-by-covariate interaction) assumption. (Examination of the X_c-by-Y association within each of the A groups, as marked in Figure 8.2, suggested that the X_c-by-Y relation was reasonably linear within each group; however, the sample size within each group, $n = 7$, was extremely small in this example. Much better assessments could be made for linearity and possible treatment by covariate interaction if the numbers of scores were much larger for each group.) The assumption of no treatment-by-covariate interaction that is required for ANCOVA appeared to be satisfied; the final ANCOVA reported, therefore, does not include this interaction term.

When math ability at pretest (X_c) was not statistically controlled, the difference in final statistics exam scores (Y) was statistically significant, $F(2, 18) = 12.26$, $p < .001$. The main effect for type of instruction in the final ANCOVA using math ability scores (X_c) as a covariate was also statistically significant: $F(2, 17) = 17.47$, $p < .001$. The strength of the association between teaching method and final exam score was $\eta^2 = .58$ when math scores were not used as a covariate versus partial $\eta^2 = .645$ when math scores were used as a covariate. The effects of teaching method on final exam score appeared to be somewhat stronger when baseline levels of math ability (X_c) were statistically controlled.

Adjusted and unadjusted group means for the final exam scores are given in Table 8.2. The rank ordering of the group means was not changed by adjustment for the covariate; however, after adjustment, the means were slightly closer together. Nevertheless, the F for the main effect of instruction in the ANCOVA was larger than the F when X_c was not statistically controlled, because controlling for the variance associated with pretest math ability substantially reduced the size of the within-group error variance.

After adjustment for pretest math scores, students who received either lecture or seminar forms of instruction scored higher ($M_3 = 21.00$ and $M_2 = 22.32$) than students who received SPI ($M_1 = 13.38$). The parameter estimates in Figure 8.11 that correspond to the slope coefficients for the dummy variable for the A_1 group and the A_2 group provide information about contrasts between the adjusted Y means of the A_1 versus A_3 group and the adjusted Y means of the A_2 versus A_3 group, respectively. After adjustment for the X_c covariate, the difference between mean final exam scores for the A_1 group was 7.626 points lower than for the A_3 group, and this difference was statistically significant, $t(17) = -4.473$, $p < .001$. After adjustment for the X_c ability covariate, the difference between the mean final exam scores for the A_2 group was only 1.32 points lower than for the A_3 group, and this difference was not statistically significant, $t(17) = .780$, $p = .446$.

To summarize, after adjustment for the scores on the X_c ability pretest, mean final exam scores were significantly lower for the SPI group than for the lecture group, while adjusted mean final exam scores did not differ significantly between the seminar and lecture groups.

(Additional post hoc tests, not reported here, could be run to make other comparisons, such as SPI vs. seminar group means.)

8.13 ADDITIONAL DISCUSSION OF ANCOVA RESULTS

When the SPSS GLM procedure is used to run ANCOVA, it reports both unadjusted and adjusted Y means. We can compute the "unadjusted" effect for Y for each group by hand by subtracting the grand mean of Y (18.90) from each Y group mean. This gives the following:

Unadjusted α_i Effect for Group	
	$Y_{A_i} - Y_{grand} = \alpha_{Y_i}$
Group 1	12.43 – 18.90 = –6.47
Group 2	23.14 – 18.90 = 4.24
Group 3	21.14 – 18.90 = 2.24

Part of these unadjusted effects may be associated with preexisting group differences on X_c, the covariate. To see what the magnitude of these preexisting differences on X_c are, we compute the α_{Xc} effect for each group.

	$X_{c_i} - X_{Cgrand} = \alpha_{Xc_i}$
Group 1	9.00 – 10.00 = –1.00
Group 2	10.86 – 10.00 = + .86
Group 3	10.14 – 10.00 = + .14

Note that the computation of effects becomes a little more complicated when the numbers of scores in the groups are unequal, because we then need to choose whether to use the weighted or unweighted grand mean as the basis for computing deviations.

Now, we will need to know the raw score slope that relates scores on Y to scores on X_c while controlling for or taking into account membership on the A grouping variable. The SPSS GLM procedure provides this slope if parameter estimates are requested as part of the GLM output; see the last panel in Figure 8.11. The raw-score partial slope relating scores on Y to scores on X_c, controlling for A, is $b = +.952$. (It can be demonstrated that this partial slope approximately equals the average of the three within-group regression slopes to predict Y from X_c; however, the three separate within-group regressions to predict Y from X_c are not presented here.)

Now let's calculate the adjusted α^*_{Yi} effect for each of the three groups:

Adjusted Y Deviation or Effect (Removing Part of the Y Mean That Is Related to X_c)	
	$\alpha_{Y_i} - b(\alpha_{Xc_i}) = \alpha^*_{Y_i}$
Group 1	–6.47 – .952 (–1.00) = –5.528
Group 2	4.24 – .952 (+.86) = 3.421
Group 3	2.24 – .952 (+.14) = 2.107

Notice that these adjusted effects are essentially predictions of how far each Y group mean would have been from the grand mean if parts of the Y score that are linearly associated with X_c were completely removed or partialled out.

To construct the predicted or adjusted group means on Y, Y^*_i, adjusted for X_c, the covariate, you add the grand mean of Y to each of these adjusted effects. This gives us the following:

Adjusted Effect + Grand Mean = Adjusted Mean for Group A_i	
	$\alpha^*_i + Ygrand = Y^*_{A_i}$
Group 1	–5.528 + 18.90 = 13.372
Group 2	3.421 + 18.90 = 22.321
Group 3	2.107 + 18.90 = 21.007

Within rounding error, these agree with the values given for the adjusted or estimated means in the second panel, headed "Estimates," in Figure 8.12.

Finally, the partitions of sums of squares for the ANOVA results (prediction of Y from A) and the ANCOVA results (prediction of Y from A, controlling for X_c) are graphically illustrated in Figures 8.13 and 8.14. The inclusion of the X_c covariate resulted in the following changes in the values of the SS terms. Before adjustment for the covariate, the value of SS_A was 454.38; this numerical value comes from the one-way ANOVA to predict Y from A that was represented in Figure 8.3, and the partition of SS for Y in this one-way ANOVA is shown graphically in Figure 8.13. After controlling for the X_c covariate, the adjusted SS_{Aadj} was reduced to 307.146; this adjusted SS for A is obtained from the ANCOVA results in Figure 8.11, and the new partition of SS in the ANCOVA is shown graphically in Figure 8.14. By comparing the partition of the total SS for Y between the one-way ANOVA (as shown in Figure 8.13) and the ANCOVA (as shown in Figure 8.14), we can evaluate how statistically controlling for the X_c covariate has changed the partition of SS. The slight reduction in the SS for A when X_c was statistically controlled is consistent with the observation that after adjustment for X_c, the adjusted Y means were slightly closer together than the original (unadjusted) Y means. See Table 8.2 for a side-by-side comparison of the unadjusted and adjusted Y means. However, the inclusion of X_c in the analysis also had a substantial impact on the magnitude of SS_{within}. For the one-way ANOVA to predict Y from A, SS_{within} was 333.43. For the ANCOVA to predict Y from A while controlling for X_c, SS_{within} was (substantially) reduced to 168.73. Thus, X_c acted primarily as a noise suppressor in this situation; controlling for X_c reduced the magnitude of SS_A by a small amount; however, controlling for X_c resulted in a much greater reduction in the magnitude of SS_{within}. As a consequence, the F ratio for the main effect of A was larger in this ANCOVA than in the simple ANOVA, and the partial η^2 effect size (.645), which represents the strength of the association between A and Y, was larger in the ANCOVA than the simple η^2 (.577) in the ANOVA. Note, however, that including an X_c covariate can result in quite a different outcome, such as a substantial reduction in SS_A and not much reduction in SS_{within}.

8.14 SUMMARY

ANCOVA is likely to be used in the following research situations.

1. It is used to correct for nonequivalence in subject characteristics (either on a pretest measure or on another variable related to Y). In these situations, the ANCOVA

Figure 8.13 Partition of *SS* in ANOVA to Predict *Y* From *A* for Data in Table 8.1 (Ignoring the Covariate X_c)

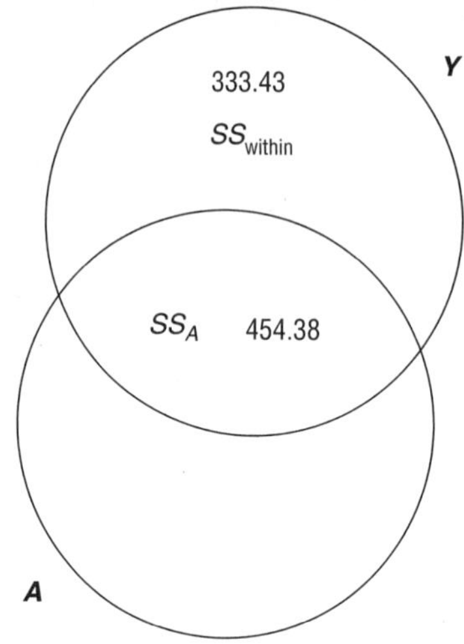

Note: Areas in the diagrams are not exactly proportional to the magnitudes of SS terms. $SS_{total} = SS_A + SS_{within} = 454.38 + 333.43 = 787.81$.

Figure 8.14 Partition of *SS* in ANCOVA to Predict *Y* From *A* and X_c Using Type I *SS*, With X_c Entered Prior to *A*

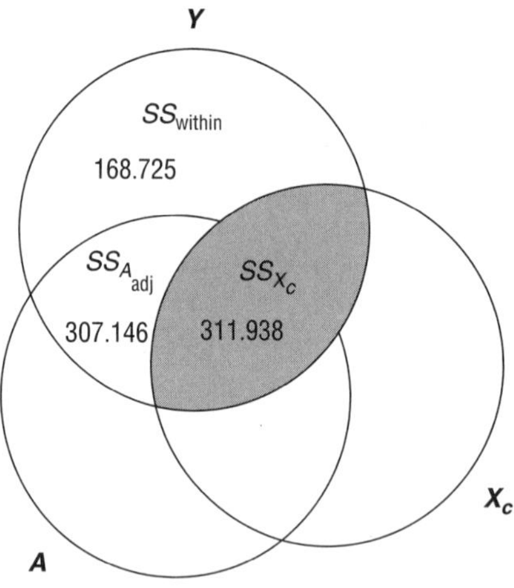

Note: $SS_{total} = SS_{Xc} + SS_{Aadj} + SS_{within} = 311.938 + 307.146 + 168.725 = 787.809$.

"adjusts" the group means for the *Y* outcome variable to the values they might have had if means on the covariate (X_c) had been equal across groups. When considering this application of ANCOVA, relevant questions include the following. How far apart were the groups on the covariate? Are the adjusted group means closer together, or farther apart, than the unadjusted group means? Researchers should interpret large adjustments to *Y* means cautiously, particularly if the adjustments result in larger group differences or a different rank ordering of groups.

2. It can be used to analyze pretest–posttest data. Several other methods that can be used to analyze pretest–posttest scores are briefly discussed in Appendix 8A.
3. It may be used to reduce error variance due to individual differences among subjects (i.e., a covariate may be included as noise suppressor).

Conventional restrictions on ANCOVA include the following:

1. The covariate X_c is a normally distributed quantitative variable, and it must be linearly related to Y.
2. There should be no treatment-by-covariate or $A \times X_c$ interaction (you need to test for this; if there is a significant interaction, you should not call your analysis an ANCOVA).
3. The covariate should not be influenced by the treatment (this is not usually a problem if the covariate is measured *before* treatment is administered but often is a problem if the covariate is measured after the treatment).
4. We assume that the covariate is reliably measured; otherwise, adjustments made using the covariates could be misleading.

The ANCOVA example presented here had only one factor and one covariate. ANCOVA may have more than one factor and may include two-way and higher order interactions among factors. ANCOVA may also include multiple covariates and/or multiple outcome variables. However, it should be apparent from this discussion that keeping track of how the variance is actually being partitioned becomes more complex as you add more variables to your design. Also, with multiple factors and/or multiple covariates, it is necessary to check for a large number of possible treatment-by-covariate interactions.

In some situations, ANCOVA provides a reasonable method for trying to remove or control for preexisting group differences in participant characteristics that are measured as covariates. But be wary: Unless you are certain that all the assumptions of ANCOVA are met, the adjusted scores on Y that you obtain from ANCOVA may be misleading. As Tabachnick and Fidell (2018) stated, "In all uses of ANCOVA, however, adjusted means must be interpreted with some caution. The mean Y score after adjustment for covariates may not correspond to any situation in the real world."

If you expect to do a great deal of quasi-experimental research involving nonequivalent control groups, Cook and Campbell's (1979) book *Quasi-Experimentation* is an extremely valuable resource on design and analysis.

APPENDIX 8A

Alternative Methods for the Analysis of Pretest–Posttest Data

Often it is desirable to obtain pretest observations on the outcome behavior (e.g., in research on the effects of stress on blood pressure, we nearly always include a prestress or baseline measure of blood pressure). Although **pretest–posttest designs** are extremely common, there is continuing controversy about the best way to analyze such data. Here are the most common methods for the analysis of data from a simple pretest–posttest design (e.g., blood pressure measures for N subjects at two points in time, before and after treatment):

1. *Analysis of gain scores or difference scores or change scores:* For each of the N subjects, simply compute the **change score** (or **gain score** or **difference score**) (posttest

blood pressure – pretest blood pressure); use this change score in any further analysis. For instance, the **paired-samples *t* test** (or **correlated-samples *t* test** or **direct-difference *t* test**) assesses whether the mean change differs significantly from 0; ANOVAs may be performed to see if change scores differ across groups such as male and female; correlational or regression analyses may be done to see if change scores are predictable from continuous variables such as weight, IQ, trait measures, and so on.

2. *ANCOVA using the posttest score as the outcome or* Y *variable and the pretest score as the covariate:* Differences among groups are assessed statistically controlling for, or partialling out, the pretest scores; in effect, the adjusted Y scores are estimates of the outcomes we might have obtained if the groups had been equal on the pretest or covariate variable.

3. *Repeated-measures ANOVA in which the pretest–posttest measures are represented as two levels of a within-subjects trials factor:* Interactions between this within-subjects trials factor and other between-subjects factors such as sex may be used to assess whether between-subjects factors predict different changes in response across time.

4. *Multiple regression, in which the* X_c *pretest variable is entered as a predictor of* Y *and pretest levels on* X_c *are statistically controlled when assessing whether other* X_i *predictors are significantly related to* Y*:* In effect, the researcher controls for or partials out the part of Y that can be predicted from pretest scores on X_c and treats the remaining part of Y as an index of change; partial slopes for other X_i predictors in the multiple regression are interpreted as predictors of change in Y. For instance, if X_c is depression at Time 1, Y is depression at Time 2, and X_1, X_2, and X_3 correspond to social support, coping skills, and stress, respectively, we would set up a regression to predict Y from X_c, X_1, X_2, and X_3 and interpret the b coefficients associated with X_1, X_2, and X_3 as predictors of change in level of depression from Time 1 to Time 2.

5. *Growth curve analysis:* This is a relatively new method for analysis of repeated measures (usually, studies that use growth curve analysis involve more than two repeated measures, but two measures can be done as a special case). It has become particularly popular as a way of assessing individual participant patterns of change over time in developmental psychology research. Typically, participants have two or several measures on a particular cognitive or emotional test per subject. If there are only two scores, all you can do to assess change is compute a change score (equivalent to a linear slope); X_c, the pretest score, is the intercept for the growth function, and d, the difference score, is its slope, and with only two points, the function must be linear. If there are more than two scores, you can begin to fit curves and estimate parameters such as intercept and slope for various types of growth curves; these growth curve parameters can be used as dependent variables in a multivariate analysis. Thus, growth curve analysis represents yet another way to look at pretest–posttest data.

8.A.1 Potential Problems With Gain or Change Scores

There is a potential problem with the use of change scores or gain scores, so ANCOVA is often viewed as a better means of controlling for pretest differences.

Here is the issue:

Let Y = posttest score and X_c = pretest score.

If we set up a bivariate regression to predict Y from X (as we implicitly do in an ANCOVA), we will obtain an equation like this: $Y' = b_0 + b_{Xc}$. In situations where X_c is a pretest on the same measurement as Y, the slope b is generally less than 1.

To obtain the part of the Y scores that is not related to or predictable from X_c—that is, an estimate of the change from pretest to posttest that is not correlated with the pretest score X_c—we could use the residuals from this simple bivariate regression:

$$\text{Adjusted change score } Y^* = Y - Y' \text{ or } Y - (b_0 + bX).$$

Note that Y^* or $Y - Y'$ is uncorrelated with X_c because of the way it is constructed, by using regression to divide each Y score into two components, one that is related to $X_c(Y')$ and the other that is not related to X_c $(Y - Y')$.

On the other hand, when we compute a simple gain or change score $d = Y - X_c$, we are calculating something like a regression residual, except that we essentially assume that $b = 1$. By construction, $Y - Y'$ is uncorrelated with X_c, and this estimate of Y' usually involves a value of b that is less than 1 (because pretest and posttest scores usually have similar standard deviations and are less than perfectly correlated). Subtracting $b \times X_c$, as in the regression or ANCOVA, will yield an index of change that is uncorrelated with X_c, by construction. When we subtract $1 \times X_c$ (as in the computation of a simple difference score), it is often an "overcorrection." When Y and X_c are repeated measures on the same variable, b is almost always less than 1. Often, although not always, simple change scores or gain scores that are computed by taking the $Y - X_c$ difference are artifactually negatively correlated with pretest levels of X_c. We can, of course, run a Pearson correlation between d and X_c to assess empirically whether this artifactual negative correlation is actually present.

Researchers often use simple gain scores or change scores even though they may be artifactually correlated with pretest levels; in many research areas (such as studies of cardiovascular reactivity), it has become conventional to do so. However, editors and reviewers sometimes point out potential problems with simple change scores; they may recommend ANCOVA or repeated-measures ANOVA as alternative methods of analysis for pretest–posttest data.

Apart from the possible artifactual correlation between d, a simple change score, and X_c, the pretest score, there is another issue to consider in the analysis of pretest–posttest data. Sometimes the nature of the differences between groups in a pretest–posttest design is quite different depending on the type of analysis that is used. For example, whether or not the between-group differences are statistically significant may depend on whether the researcher did an ANOVA on gain scores or an ANCOVA, controlling for pretest scores as a covariate. The rank ordering of groups on outcomes may even differ depending on which analysis was used; this phenomenon is called **Lord's paradox** (Lord, 1967). While it is not common for the nature of the effects in the study to appear entirely different for ANOVA versus ANCOVA, the researcher needs to be aware of this as a possibility.

It may be informative to analyze pretest–posttest data in several different ways (e.g., an ANOVA on change scores, an ANCOVA using X_c as the covariate, and a repeated-measures ANOVA). If all the different analyses yield essentially the same results, the researcher can then report the analysis that seems the clearest and mention (perhaps in a footnote) that alternative analyses led to the same conclusions. However, if the analyses yield radically different outcomes (e.g., group differences being significant in one analysis and not in the other or the apparent rank ordering of outcome changes across analyses), then the researcher faces a dilemma. Expert advice may be needed to decide which of the analyses provides the best assessment of change; otherwise, the results of the study may have to be judged inconclusive.

If you use change scores, it is a good idea to run a correlation between the change scores and pretest scores (X_c) to see if there is a negative correlation. Ideally, you might want

d and X_c to be unrelated. Sometimes, there is no negative correlation between simple change scores and pretest, and in this case, you can dismiss one of the potential problems by showing empirically that no artifactual negative correlation between X_c and d exists. On the other hand, it is possible that there is a relationship between initial status on the pretest variable and the amount of change; if this makes sense, then generating an adjusted change score that is forced to be uncorrelated with pretest level may not make sense.

Multilevel modeling provides another way to assess within-subject change; it can be set up with repeated-measures scores nested within subject and condition to evaluate within-person patterns of change (Adelson & Owen, 2012; Grimm & Ram, 2016).

Each of the various methods for the analysis of pretest–posttest data has both advantages and potential drawbacks. No single method is always the correct choice in all situations. Researchers need to think carefully about the choice among these options, particularly because the different analyses do not necessarily lead to the same conclusions.

COMPREHENSION QUESTIONS

1. What is a quasi-experimental design? What is a nonequivalent control group?
2. Why is ANCOVA often used in situations that involve nonequivalent control groups?
3. Is the statistical correction for nonequivalence among groups likely to be as effective a means of controlling for the covariate as experimental methods of control (such as creating equivalent groups by random assignment)?
4. What is a covariate?
5. What information is needed to judge whether scores on the covariate (X_c) differ across groups? What information is needed to assess whether scores on the covariate X are strongly linearly related to Y (the dependent variable)?
6. Is there any point in controlling for X as a covariate if X is not related to group membership and also not related to Y?
7. The usual or conventional order of entry in an ANCOVA is as follows:

   Step 1: one or more covariates
   Step 2: one or more group membership variables

   (If the design is factorial, interactions are often entered after main effects.) Why is this order of entry preferred to group variables in Step 1 and covariates in Step 2?
8. Running an ANCOVA is almost (but not exactly) like running the following two-step analysis:

   First, predict Y (the dependent variable) from X_1 (the covariate), and take the residuals from this regression; this residual is denoted X^*.
   Second, run an ANOVA on X^* to see if the mean on X^* differs across levels of A, the group membership variable.

   How is the two-step analysis described here different from what happens when you run an ANCOVA?
9. When we say that the group means in an ANCOVA are "adjusted," what are they adjusted for?
10. When SPSS reports "predicted" group means, what does this mean?
11. What is the homogeneity of regression assumption? Why is a violation of this assumption problematic? How can this assumption be tested?
12. When adjusted group means are compared with unadjusted group means, which of the following can occur? (Answer yes or no to each.)
    - The adjusted group means can be closer together than the unadjusted group means.
    - The adjusted group means can be different in rank order across groups from the unadjusted means.
    - The adjusted group means can be farther apart than the unadjusted group means.
13. Draw two overlapping circles to represent the shared variance between A (the treatment or group membership variable) and Y (the outcome variable). Now draw three overlapping circles to represent the shared variance among Y (a quantitative outcome variable), X (a quantitative covariate), and A (a group membership variable).

Use different versions of these diagrams to illustrate that the effect of including the covariate can be primarily to reduce the SS_{error} or primarily to reduce SS_A.

Data Analysis Project for ANCOVA

For this assignment, use the small batch of data provided below. Enter the data into an SPSS worksheet—one column for each variable (group membership, A; dependent variable, Y; and covariate, X_c).

To answer all the following questions, you will need to do these analyses:

1. Run a one-way ANOVA to compare mean levels of Y across groups on the A factor (without the covariate) and also a one-way ANOVA to compare mean levels of X_c, the covariate, across groups on the A factor (to see if scores on the covariate differ across treatment groups).

2. Create histograms for scores on X_c and Y and a scatterplot and bivariate Pearson's r to assess the association between X_c and Y.

3. Run a preliminary ANCOVA using the GLM procedure, with A as the group membership factor, Y as the dependent variable, and X_c as the covariate. In addition, use the Univariate: Model dialog box to create an interaction term between treatment (A) and covariate (X_c). The purpose of this preliminary ANCOVA is to test whether the assumption of the no treatment-by-covariate interaction is violated. This is not the version of the analysis that is generally reported as the final result.

4. Run your final ANCOVA using the GLM procedure. This is the same as in Section 8.12.3, except that you drop the $A \times X_c$ interaction term. Assuming that you have no significant treatment-by-covariate interaction, this is the ANCOVA for which you would report results. Use Type I *SS*, and make sure that the covariate precedes the group membership predictor variable in the list of predictors in the Univariate: Model dialog box in the GLM procedure.

Data for the Preceding Questions

These are hypothetical data. We will imagine that a three-group quasi-experimental study was done to compare the effects of three treatments on the aggressive behavior of male children. X_c, the covariate, is a pretest measure of aggressiveness: the number of aggressive behaviors performed by each child when the child is first placed in a neutral playroom situation. This measure was taken prior to exposure to the treatment. Children could not be randomly assigned to treatment groups, so the groups did not start out exactly equivalent on aggressiveness. The dependent variable, Y, is a posttest measure: the number of aggressive behaviors performed by each child after exposure to one of the three treatments. Treatment A consisted of three different films. The A_1 group saw a cartoon animal behaving aggressively. The A_2 group saw a human female model behaving aggressively. The A_3 group saw a human male model behaving aggressively. The question is whether these three models elicited different amounts of aggressive behavior when controlling (or not) for individual differences in baseline aggressiveness. The scores are given below.

Let's further assume that there was very good interrater reliability on these frequency counts of behaviors and that they are interval/ratio level of measure, normally distributed, and independent observations.

Write up your "Results" section using APA style; hand in your SPSS output. In addition, answer the following questions about your results:

1. Using overlapping circles, show how SS_{total} was partitioned in the ANOVA (to predict Y from A) versus the ANCOVA (to predict Y from A, controlling for X_c). Indicate what *SS* values correspond to each slice of the circles in these diagrams.

A_1		A_2		A_3	
X_c	Y	X_c	Y	X_c	Y
3	8	13	18	14	26
7	12	17	22	10	22
10	16	10	16	8	20
8	14	9	14	4	16
15	20	4	8	2	14
8	12	6	10	8	18
15	18	2	6	6	12

2. Compare the unadjusted *SS* terms from your ANOVA (for error and the effect of treatment) with the adjusted *SS* terms from your ANCOVA. Was the effect of including the covariate primarily to decrease error variance, primarily to take variance away from the treatment variable, or both? Does the effect of treatment look stronger or weaker when you control for the covariate?

3. Note the pattern of differences across treatment groups on X_c, the covariate or pretest. To what extent were there large differences in aggressiveness before the treatment was even administered? What was the nature of the confound: Did the most aggressive boys get a treatment that was highly effective or ineffective at eliciting aggression?

4. Look at the adjusted and unadjusted deviations that you got when you ran ANCOVA using the ANOVA program. Find and report the within-group slope that relates Y to X_c (for this, you will have to look at your regression output). Show how one of the adjusted means can be calculated.

5. After adjustment, were the group means closer together or farther apart? Is there any change in the rank order of group means when you compare adjusted versus unadjusted means? Do you think these adjustments are believable?

6. We make a number of assumptions to do ANCOVA: that the treatment does not affect the covariate, that there is no treatment-by-covariate interaction, and that the covariate is reliably measured. How well do you think these assumptions were met in this case? Why would it be a problem if these assumptions were not met?

7. Only if you have covered the appendix material that discusses other possible methods of analysis for pretest–posttest designs, briefly describe three other ways you could analyze these data.

NOTE

[1]The partitioning of variance in this situation can be worked out by representing the ANCOVA as a multiple regression problem (following Keppel & Zedeck, 1989, pp. 458–461) in which scores on Y are predicted from scores on both A (a treatment group variable) and X_c (a covariate). First, we

need to figure out the proportion of variance in Y that is predictable from X_c, the covariate. This can be done by finding R^2 for the following regression analysis: $Y = b_0 + bX_c$. Next, as in any other sequential or hierarchical regression analysis, we want to assess how much additional variance in Y is predictable from A when we control for X_c. To assess this R^2 increment, we can run a regression to obtain $R^2_{Y\cdot XA}$, $Y' = a + b_1X_c + b_2D_1 + b_2D_2$, where D_1 and D_2 are dummy-coded variables that represent membership in the A treatment groups. The proportion of variance uniquely predictable from A, adjusting or controlling for $X_c(R^2_{Aadj})$, corresponds to the difference between these two R^2 terms: $R^2_{Aadj} = R^2_{Y\cdot XA} - R^2_{Y\cdot X}$. Finally, to obtain the adjusted error variance or within-group variance, we calculate $1 - R^2_{Y\cdot XA}$. Values for SS_{Xc}, SS_{Aadj}, and SS_{residual} can be obtained by multiplying the SS_{total} for Y by these three proportions. When the SS terms are obtained in this manner, they should sum to $SS_{\text{total}} = SS_{Xc} + SS_{Aadj} + SS_{\text{residual}}$.

DIGITAL RESOURCES

Find **free study tools** to support your learning, including **eFlashcards, data sets, and web resources**, on the accompanying website at **edge.sagepub.com/warner3e**.

CHAPTER 9

MEDIATION

9.1 DEFINITION OF MEDIATION

Mediation involves a set of causal hypotheses. An initial causal variable X_1 may influence an outcome variable Y through a mediating variable X_2. (Some books and websites use different notations for the three variables; for example, on Kenny's mediation webpage, http://www.davidakenny.net/cm/mediate.htm, the initial causal variable is denoted X, the outcome as Y, and the mediating variable as M.) Mediation occurs if the effect of X_1 on Y is partly or entirely "transmitted" by X_2. A mediated causal model involves a causal sequence; first, X_1 causes or influences X_2; then, X_2 causes or influences Y. X_1 may have additional direct effects on Y that are not transmitted by X_2. A mediation hypothesis can be represented by a diagram of a **causal model**. Note that the term *causal* is used because the path diagram represents hypotheses about possible causal influence; however, when data come from nonexperimental designs, we can only test whether a hypothesized causal model is consistent or inconsistent with a particular causal model. That analysis falls short of proof that any specific causal model is correct.

9.1.1 Path Model Notation

Path model notation was introduced earlier and it is briefly reviewed here. We begin with two variables (X and Y). Arrows are used to correspond to paths that represent different types of relations between variables. The absence of an arrow between X and Y corresponds to an assumption that these variables are not related in any way; they are not correlated or confounded, and they are not directly causally connected. A unidirectional arrow corresponds to the hypothesis that one variable has a causal influence on the other—for example, $X \rightarrow Y$ corresponds to the hypothesis that X causes or influences Y; $Y \rightarrow X$ corresponds to the hypothesis that Y causes or influences X. A bidirectional or double-headed arrow represents a noncausal association, such as correlation or confounding of variables that does not arise from any causal connection between them. In path diagrams, these double-headed arrows may be shown as curved lines.

If we consider only two variables, X and Y, there are four possible models: (a) X and Y are not related in any way (this is denoted in a path diagram by the absence of a path between X and Y), (b) X causes Y ($X \rightarrow Y$), (c) Y causes X ($Y \rightarrow X$), and (d) X and Y are correlated but not because of any causal influence (XY). When a third variable is added, the number of possible relationships among the variables X_1, X_2, and Y increases substantially. One theoretical model corresponds to X_1 and X_2 as correlated causes of Y. For this model, the appropriate analysis is a regression to predict Y from both X_1 and X_2. Another possible hypothesis is that X_2 may be a moderator of the relationship between X_1 and Y; this is also described as an interaction between X_2 and X_1 as predictors of Y. Statistical significance and nature of interaction can be assessed using procedures described in Chapter 7, on moderation.

9.1.2 Circumstances in Which Mediation May Be a Reasonable Hypothesis

Because a mediated causal model includes the hypothesis that X_1 causes or influences X_2 and the hypothesis that X_2 causes or influences Y, it does not make sense to consider mediation analysis in situations where one or both hypotheses would be nonsense. For X_1 to be hypothesized as a cause of X_2, X_1 should occur before X_2, and there should be a plausible mechanism through which X_1 could influence X_2. For example, suppose we are interested in a possible association between height and salary (a few studies suggest that taller people earn higher salaries). It is conceivable that height influences salary (perhaps employers have a bias that leads them to pay tall people more money). It is not conceivable that a person's salary changes his or her height.

9.2 HYPOTHETICAL RESEARCH EXAMPLE

This hypothetical correlational study examines associations among three variables as an illustration of a mediation hypothesis: X_1, age; X_2, body weight; and Y, systolic blood pressure (SBP). The data are in an SPSS file named ageweightbp.sav. Note that for research applications of mediation analysis, much larger sample sizes should be used.

For these variables, it is plausible to hypothesize the following causal connections. Blood pressure tends to increase as people age. As people age, body weight tends to increase (this could be due to lower metabolic rate, reduced activity level, or other factors). Other factors being equal, increased body weight makes the cardiovascular system work harder, and this can increase blood pressure. It is possible that at least part of the age-related increase in blood pressure might be mediated by age-related weight gain. Figure 9.1 is a path model that represents this mediation hypothesis for this set of three variables.

To estimate the strength of association that corresponds to each path in Figure 9.1, a series of three ordinary least squares (OLS) linear regression analyses can be run. Note that a variable is dependent if it has one or more unidirectional arrows pointing toward it. We run a regression analysis for each dependent variable (such as Y), using all variables that have unidirectional arrows that point toward Y as predictors. For the model in Figure 9.1, the first regression predicts Y from X_1 (blood pressure from age). The second regression predicts X_2 from X_1 (weight from age). The third regression predicts Y from both X_1 and X_2 (blood pressure predicted from both age and weight).

9.3 LIMITATIONS OF "CAUSAL" MODELS

Path models similar to Figure 9.1 are called "causal" models because each unidirectional arrow represents a hypothesis about a possible causal connection between two variables. However, the data used to estimate the strength of relationship for the paths are almost always from nonexperimental studies, and nonexperimental data cannot prove causal hypotheses. If the path coefficient between two variables such as X_2 and Y (this coefficient is denoted b in Figure 9.1) is statistically significant and large enough in magnitude to indicate a change in the outcome variable that is clinically or practically important, this result is consistent with the possibility that X_2 might cause Y, but it is not proof of a causal connection. Numerous other situations could yield a large path coefficient between X_2 and Y. For example, Y may cause X_2; both Y and X_2 may be

Figure 9.1 Hypothetical Mediation Example: Effects of Age on Systolic Blood Pressure (SBP)

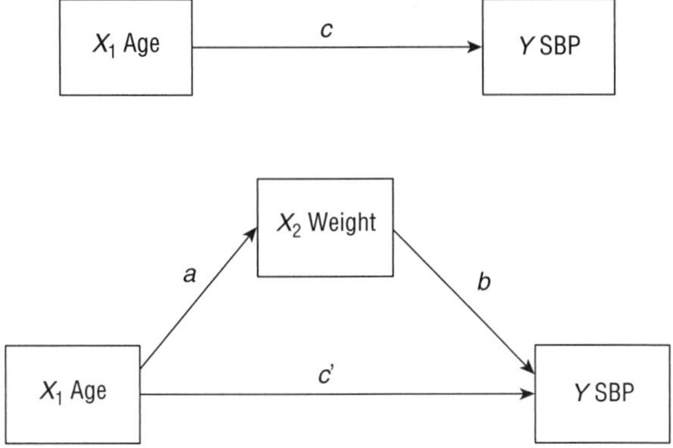

Note: Top panel: The total effect of age on SBP is denoted by c. Bottom panel: The path coefficients (a, b, and c') that estimate the strength of hypothesized causal associations are estimated by unstandardized regression coefficients. The product $a \times b$ estimates the strength of the mediated or indirect effect of age on SBP, that is, how much of the increase in SBP that occurs as people age is due to weight gain. The c' coefficient estimates the strength of the direct (also called partial) effect of age on SBP, that is, any effect of age on SBP that is not mediated by weight. The coefficients in this bottom panel decompose the total effect (c) into a direct effect (c') and an indirect effect ($a \times b$). When OLS regression is used to estimate unstandardized path coefficients, $c = (a \times b) + c'$; the total relationship between age and SBP is the sum of the direct relationship between age and SBP and the indirect or mediated effect of age on SBP through weight.

caused by some third variable, X_3; X_2 and Y may actually be measures of the same variable; the relationship between X_2 and Y may be mediated by other variables, X_4, X_5, and so on; or a large value for the b path coefficient may be due to sampling error.

9.3.1 Reasons Why Some Path Coefficients May Be Not Statistically Significant

If the path coefficient between two variables is not statistically significantly different from zero, there are also several possible reasons. If the b path coefficient in Figure 9.1 is close to zero, this could be because there is no causal or noncausal association between X_2 and Y. However, a small path coefficient could also occur because of sampling error or because assumptions required for regression are severely violated.

9.3.2 Possible Interpretations for Statistically Significant Paths

A large and statistically significant b path coefficient is *consistent* with the hypothesis that X_2 causes Y, but it is not proof of that causal hypothesis. Replication of results (such as values of $\boldsymbol{a}$, $\boldsymbol{b}$, and $\boldsymbol{c}'$ path coefficients in Figure 9.1) across samples increases confidence that findings are not due to sampling error. For predictor variables and/or hypothesized mediating variables that can be experimentally manipulated, experimental studies can be done to provide stronger evidence whether associations between variables are causal (MacKinnon, 2008). By itself, a single mediation analysis only provides preliminary nonexperimental evidence to evaluate whether the proposed causal model is plausible (i.e., consistent with the data).

9.4 QUESTIONS IN A MEDIATION ANALYSIS

Researchers typically ask two questions in a mediation analysis. The first question is whether there is a statistically significant mediated path from X_1 to Y via X_2 (and whether the part of the Y outcome variable score that is predictable from this path is large enough to be of practical importance). Recall from the discussion of the **tracing rule** in Chapter 4 that when a path from X to Y includes more than one arrow, the strength of the relationship for this multiple-step path is obtained by multiplying the coefficients for each included path. Thus, the strength of the mediated relationship (the path from X_1 to Y through X_2 in Figure 9.1) is estimated by the product of the $a \times b$ (ab) coefficients. The null hypothesis of interest is H_0: $ab = 0$. Note that the *unstandardized* regression coefficients are used for this significance test. Later sections in this chapter describe test statistics for this null hypothesis. If this mediated path is judged to be nonsignificant, the mediation hypothesis is not supported, and the data analyst would need to consider other explanations.

If there is a significant mediated path (i.e., the ab product differs significantly from zero), then the second question in the mediation analysis is whether there is also a significant direct path from X_1 to Y; this path is denoted c' in Figure 9.1. If c' is not statistically significant (or too small to be of any practical importance), a possible inference is that the effect of X_1 on Y is completely mediated by X_2. If c' is statistically significant and large enough to be of practical importance, a possible inference is that the influence of X_1 on Y is only partially mediated by X_2 and that X_1 has some additional effect on Y that is not mediated by X_2. In the hypothetical data used for the example in this chapter (in the SPSS file ageweightbp.sav), we will see that the effects of age on blood pressure are only partially mediated by body weight.

Of course, it is possible that there could be additional mediators of the effect of age on blood pressure; for example, age-related changes in the condition of arteries might also influence blood pressure. Models with multiple mediating variables are discussed briefly later in the chapter.

9.5 ISSUES IN DESIGNING A MEDIATION ANALYSIS STUDY

A mediation analysis begins with a minimum of three variables. Every unidirectional arrow that appears in Figure 9.1 represents a hypothesized causal connection and must correspond to a plausible theoretical mechanism. A model such as age → body weight → blood pressure seems reasonable; processes that occur with advancing age, such as slowing metabolic rate, can lead to weight gain, and weight gain increases the demands on the cardiovascular system, which can cause an increase in blood pressure. However, it would be nonsense to propose a model of the following form: blood pressure → body weight → age, for example; there is no reasonable mechanism through which blood pressure could influence body weight, and weight cannot influence age in years.

9.5.1 Types of Variables in Mediation Analysis

Usually all three variables (X_1, X_2, and Y) in a mediation analysis are quantitative. A dichotomous variable can be used as a predictor in regression (Chapter 6), and therefore it is acceptable to include an X_1 variable that is dichotomous (e.g., treatment vs. control) as the initial causal variable in a mediation analysis; OLS regression methods can still be used in this situation. However, both X_2 and Y are dependent variables in mediation analysis; if one or both of these variables are categorical, then **logistic regression** is needed to estimate regression coefficients, and this complicates the interpretation of outcomes (see MacKinnon, 2008, Chapter 11).

It is helpful if scores on the variables can be measured in meaningful units because this makes it easier to evaluate whether the strength of influence indicated by path coefficients is large enough to be clinically or practically significant. For example, suppose that we want to predict annual salary in dollars (Y) from years of education (X_1). An unstandardized regression coefficient is easy to interpret. A student who is told that each additional year of education predicts a $50 increase in annual salary will understand that the effect is too weak to be of any practical value, while a student who is told that each additional year of education predicts a $5,000 increase in annual salary will understand that this is enough money to be worth the effort. Often, however, measures are given in arbitrary units (e.g., happiness rated on a scale from 1 = *not happy at all* to 7 = *extremely happy*). In this kind of situation, it may be difficult to judge the practical significance of a half-point increase in happiness.

As in other applications of regression, measurements of variables are assumed to be reliable and valid. If they are not, regression results can be misleading.

9.5.2 Temporal Precedence or Sequence of Variables in Mediation Studies

Hypothesized causes must occur earlier in time than hypothesized outcomes (temporal precedence, as discussed in Volume I, Chapter 2 [Warner, 2020]). It seems reasonable to hypothesize that "being abused as a child" might predict "becoming an abuser as an adult"; it would not make sense to suggest that being an abusive adult causes a person to have experiences of abuse in childhood. Sometimes measurements of the three variables X_1, X_2, and Y are all obtained at the same time (e.g., in a one-time survey). If X_1 is a retrospective report of experiencing abuse as a child, and Y is a report of current abusive behaviors, then the requirement for temporal precedence (X_1 happened before Y) may be satisfied. In some studies, measures are obtained at more than one point in time; in these situations, it would be preferable to measure X_1 first, then X_2, and then Y; this may help establish temporal precedence. When all three variables are measured at the same point in time and there is no logical reason to believe one of them occurs earlier in time than the others, it may not be possible to establish temporal precedence.

9.5.3 Time Lags Between Variables

When measures are obtained at different points in time, it is important to consider the time lag between measures. If this time lag is too brief, the effects of X_1 may not be apparent yet when Y is measured (e.g., if X_1 is initiation of treatment with either placebo or Prozac, a drug that typically does not have full antidepressant effects until about 6 weeks, and Y is a measure of depression and is measured one day after X_1, then the full effect of the drug will not be apparent). Conversely, if the time lag is too long, the effects of X_1 may have worn off by the time Y is measured. Suppose that X_1 is receiving positive feedback from a relationship partner and Y is relationship satisfaction, and Y is measured 2 months after X_1. The effects of the positive feedback (X_1) may have dissipated over this period of time. The optimal time lag will vary depending on the variables involved; some X_1 interventions or measured variables may have immediate but not long-lasting effects, while others may require a substantial time before effects are apparent.

9.6 ASSUMPTIONS IN MEDIATION ANALYSIS AND PRELIMINARY DATA SCREENING

Unless the types of variables involved require different estimation methods (e.g., if a dependent variable is categorical, logistic regression methods are required), the coefficients (a, b, and c') associated with the paths in Figure 9.1 can be estimated using OLS regression. All of the assumptions required for regression (see Volume I, Chapter 11 [Warner, 2020], and Chapter 5 in the present volume) are also required for mediation analysis. Because preliminary data screening was presented in greater detail earlier, data-screening procedures are reviewed here only briefly. For each variable, histograms or other graphic methods can be used to assess whether scores on all quantitative variables are reasonably normally distributed, without extreme outliers. If the X_1 variable is dichotomous, both groups should have a reasonably large number of cases. Scatterplots can be used to evaluate whether relationships between each pair of variables appear to be linear (X_1 with Y, X_1 with X_2, and X_2 with Y) and to identify bivariate outliers.

Baron and Kenny (1986) suggested that a mediation model should not be tested unless there is a significant relationship between X_1 and Y. In more recent treatments of mediation, it has been pointed out that in situations where one of the path coefficients is negative, there can be significant mediated effects even when X_1 and Y are not significantly correlated (Hayes, 2009). This can be understood as a form of suppression. If none of the pairs of variables in the model are significantly related to one another in bivariate analyses, however, there is not much point in testing mediated models.

MacKinnon, Krull, and Lockwood (2000) explained that the patterns of results obtained from analysis of models such as the models presented here for partial or complete mediation cannot prove mediation hypotheses. The patterns of results that might be interpreted as evidence of mediation are equally consistent with the outcomes expected for suppression and confounded predictor. Empirical results do not make it possible to distinguish which of these explanations is "better."

9.7 PATH COEFFICIENT ESTIMATION

The most common way to obtain estimates of the path coefficients that appear in Figure 9.1 is to run the following series of regression analyses. These steps are similar to those recommended by Baron and Kenny (1986), except that, as suggested in recent treatments of mediation (MacKinnon, 2008), a statistically significant outcome on the first step is not considered a requirement before going on to subsequent steps.

Step 1: First, a regression is run to predict Y (SBP) from X_1 (age). (SPSS procedures for this type of regression were provided in Volume I, Chapter 11 [Warner, 2020], and Chapter 4 in the present volume.) The raw or unstandardized regression coefficient from this regression corresponds to path **c**. This step is sometimes omitted; however, it provides information that can help evaluate how much controlling for the X_2 mediating variable reduces the strength of association between X_1 and Y. Figure 9.2 shows the regression coefficients part of the output. The unstandardized regression coefficient for the prediction of Y (BloodPressure—note that there is no space within the SPSS variable name) from X_1 (age) is $c = 2.862$; this is statistically significant, $t(28) = 6.631, p < .001$. (The N for this data set is 30; therefore, the df for this t ratio is $N - 2 = 28$.) Thus, the overall effect of age on blood pressure is statistically significant.

Step 2: Next a regression is performed to predict the mediating variable (X_2, weight) from the causal variable (X_1, age). The results of this regression provide the path coefficient for the path denoted a in Figure 9.1 and also the standard error of a (s_a) and the t test for the statistical significance of the a path coefficient (t_a). The coefficient table for this regression appears in Figure 9.3. For the hypothetical data, the unstandardized a path coefficient was 1.432, with $t(28) = 3.605, p = .001$.

Step 3: Finally, a regression is performed to predict the outcome variable Y (blood pressure) from both X_1 (age) and X_2 (weight). (Detailed examples of regression with two predictor variables appeared in Chapter 4.) This regression provides estimates of the unstandardized coefficients for path b (and s_b and t_b) and also path c' (the direct or remaining effect of X_1 on Y when the mediating variable has been included in the analysis). See Figure 9.1 for the corresponding path diagram. From Figure 9.4, path

Figure 9.2 Regression Coefficient to Predict Blood Pressure (Y) From Age (X_1)

Coefficients[a]

Model		Unstandardized Coefficients		Standardized Coefficients		
		B	Std. Error	Beta	t	Sig.
1	(Constant)	10.398	26.222		.397	.695
	Age	2.862	.432	.782	6.631	.000

a. Dependent Variable: BloodPressure

Note: The raw-score slope in this equation, 2.862, corresponds to coefficient c in the path diagram in Figure 9.1.

Figure 9.3 Regression Coefficient to Predict Weight (Mediating Variable X_2) From Age (X_1)

Coefficients[a]

Model		Unstandardized Coefficients		Standardized Coefficients		
		B	Std. Error	Beta	t	Sig.
1	(Constant)	78.508	24.130		3.254	.003
	Age	1.432	.397	.563	3.605	.001

a. Dependent Variable: Weight

Note: The raw-score slope from this equation, 1.432, corresponds to the path labeled a in Figure 9.1.

$b = .49$, $t(27) = 2.623$, $p = .014$; path $c' = 2.161$, $t(27) = 4.551$, $p < .001$. These unstandardized path coefficients are used to label the paths in a diagram of the causal model (top panel of Figure 9.5). These values are also used later to test the null hypothesis H_0: $ab = 0$. In many research reports, particularly when the units in which the variables are measured are not meaningful or not easy to interpret, researchers report the standardized path coefficients (these are called beta coefficients in the SPSS output); the bottom panel of Figure 9.5 shows the standardized path coefficients. Sometimes the estimate of the c coefficient appears in parentheses, next to or below the c' coefficient, in these diagrams.

In addition to examining the path coefficients from these regressions, the data analyst should pay some attention to how well the X_1 and X_2 variables predict Y. From Figure 9.4, $R^2 = .69$, adjusted $R^2 = .667$, and this is statistically significant, $F(2, 27) = 30.039$, $p < .001$. These two variables do a good job of predicting variance in blood pressure.

9.8 CONCEPTUAL ISSUES: ASSESSMENT OF DIRECT VERSUS INDIRECT PATHS

When a path that leads from a predictor variable X to a dependent variable Y involves other variables and multiple arrows, the overall strength of the path is estimated by multiplying the

Figure 9.4 Regression Coefficient to Predict Blood Pressure (Y) From Age (X_1) and Mediating Variable Weight (X_2)

Model Summary

Model	R	R Square	Adjusted R Square	Std. Error of the Estimate
1	.831[a]	.690	.667	36.692

a. Predictors: (Constant), Weight, Age

ANOVA[b]

Model		Sum of Squares	df	Mean Square	F	Sig.
1	Regression	80882.132	2	40441.066	30.039	.000[a]
	Residual	36349.735	27	1346.286		
	Total	117231.867	29			

a. Predictors: (Constant), Weight, Age

b. Dependent Variable: BloodPressure

Coefficients[a]

Model		Unstandardized Coefficients		Standardized Coefficients	t	Sig.
		B	Std. Error	Beta		
1	(Constant)	-28.046	27.985		-1.002	.325
	Age	2.161	.475	.590	4.551	.000
	Weight	.490	.187	.340	2.623	.014

a. Dependent Variable: BloodPressure

Note: The raw-score slope for *a* in this equation, 2.161, corresponds to the path labeled *c′* in Figure 9.1; the raw-score slope for weight in this equation, .490, corresponds to the path labeled *b*.

coefficients for each leg of the path (as discussed in the introduction to the tracing rule in Chapter 4).

9.8.1 The Mediated or Indirect Path: *ab*

The strength of the indirect or mediated effect of age on blood pressure through weight is estimated by multiplying the *ab* path coefficients. In many applications, one or more of the variables are measured in arbitrary units (e.g., happiness may be rated on a scale from 1 to 7). In such situations, the unstandardized regression coefficients may not be very informative, and research reports often focus on standardized coefficients.[1] The standardized (β) coefficients for the paths in the age, weight, and blood pressure hypothetical data appear in the bottom panel of Figure 9.5. Throughout the remainder of this section, all path coefficients are given in standardized (β-coefficient) form.

When the path from X to Y has multiple parts or arrows, the overall strength of the association for the entire path is estimated by multiplying the coefficients for each part of the path. Thus, the unit-free index of strength of the **mediated effect** (the effect of age on blood pressure, through the mediating variable weight) is given by the product of the standardized

Figure 9.5 Path Coefficients for the Mediation Analysis of Age, Weight, and SBP

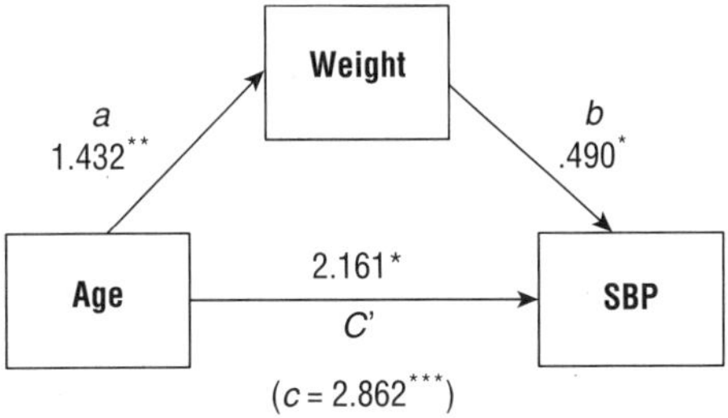

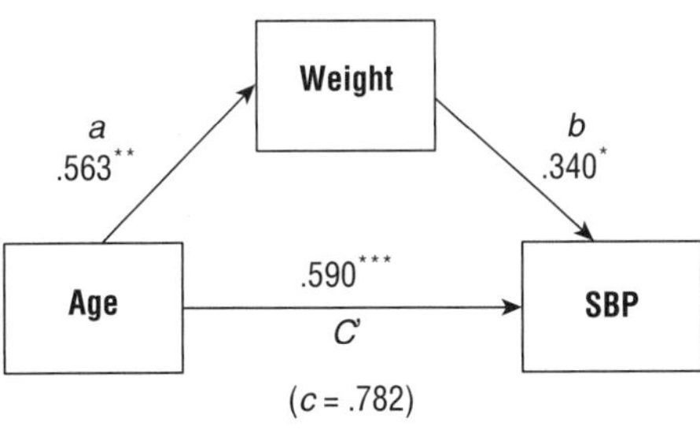

$^{*}p < .05$, $^{**}p < .01$, and $^{***}p < .001$, all two tailed.

estimates of the path coefficients, *ab*. For the standardized coefficients, this product = (.563 × .340) = .191. The strength of the direct or nonmediated path from age to SBP corresponds to *c*′; the standardized coefficient for this path is .590. In other words, for a 1-*SD* increase in z_{Age}, we predict a .191 increase in z_{SBP} through the mediating variable z_{Weight}. In addition, we predict a .590 increase in z_{SBP} due to direct effects of z_{Age} (effects that are not mediated by z_{Weight}); this corresponds to the *c*′ path. The total effect of z_{Age} on z_{SBP} corresponds to path *c*, and the standardized coefficient for path *c* is .782 (the beta coefficient to predict z_{SBP} from z_{Age} in Figure 9.5).

9.8.2 Mediated and Direct Path as Partition of Total Effect

The mediation analysis has partitioned the total effect of age on blood pressure (*c* = .782) into a **direct effect** (*c*′ = .590) and a mediated effect (*ab* = .191). (Both of these are given in terms of standardized or unit-free path coefficients.) It appears that mediation through weight, while statistically significant, explains only a small part of the total effect of age on blood pressure in this hypothetical example. Within rounding error, $c = c' + ab$, that is, the **total effect** is the sum of the direct and mediated effects. These terms are additive when OLS regression is used to obtain estimates of coefficients; when other estimation methods such as maximum likelihood are used (as in structural equation modeling [SEM] programs), these equalities may not hold. Also note that if there are missing data, each regression must be performed on the same set of cases in order for this additive association to work.

Note that even if the researcher prefers to label and discuss paths using standardized regression coefficients, information about the unstandardized coefficients is required to carry out additional statistical significance tests (to find out whether the product *ab* differs significantly from zero, for example).

9.8.3 Magnitude of Mediated Effect

When variables are measured in meaningful units, it is helpful to think through the magnitude of the effects in real units, as discussed in this paragraph. (The discussion in this paragraph is helpful primarily in research situations in which units of measurement have some real-world practical interpretation.) All of the path coefficients in the rest of this paragraph are unstandardized regression coefficients. From the first regression analysis, the *c* coefficient for the total effect of age on blood pressure was $c = 2.862$. In simple language, for each 1-year increase in age, we predict an increase in blood pressure of 2.862 mm Hg. On the basis of the *t*-test result in Figure 9.2, this is statistically significant. Taking into account that people in wealthy countries often live to age 70 or older, this implies substantial age-related increases in blood pressure; for example, for a 30-year increase in age, we predict an increase of 28.62 mm Hg in blood pressure, and that is sufficiently large to be clinically important. This tells us that the total effect of age on SBP is reasonably large in terms of clinical or practical importance. From the second regression, we find that the effect of age on weight is $a = 1.432$; this is also statistically significant, on the basis of the *t* test in Figure 9.3. For a 1-year increase in age, we predict almost 1.5 lb in weight gain. Again, over a period of 10 years, this implies a sufficiently large increase in predicted body weight (about 14.32 lb) to be of clinical importance. The last regression (in Figure 9.4) provides information about two paths, *b* and c'. The *b* coefficient that represents the effect of weight on blood pressure was $b = .49$; this was statistically significant. For each 1-lb increase in body weight, we predict almost a half-point increase in blood pressure. If we take into account that people may gain 30 or 40 lb over the course of a lifetime, this would imply weight-related increases in blood pressure on the order of 15 or 20 mm Hg. This also seems large enough to be of clinical interest. The **indirect effect** of age on blood pressure is found by multiplying $a \times b$, in this case, $1.432 \times .49 = .701$. For each 1-year increase in age, a .7 mm Hg increase in blood pressure is predicted through the effects of age on weight. Finally, the direct effect of age on blood pressure when the mediating variable weight is statistically controlled or taken into account is represented by $c' = 2.161$. Over and above any weight-related increases in blood pressure, we predict about a 2.2-unit increase in blood pressure for each additional year of age. Of the total effect of age on blood pressure (a predicted 2.862 mm Hg increase in SBP for each 1-year increase in age), a relatively small part is mediated by weight (.701), and the remainder is not mediated by weight (2.161). (Because these are hypothetical data, this outcome does not accurately describe the importance of weight as a mediator in real-life situations.) The mediation analysis partitions the total effect of age on blood pressure ($c = 2.862$) into a direct effect ($c' = 2.161$) and a mediated effect ($ab = .701$). Within rounding error, $c = c' + ab$, that is, the total effect *c* is the sum of the direct (c') and mediated (*ab*) effects.

9.9 EVALUATING STATISTICAL SIGNIFICANCE

Several methods to test the statistical significance of mediated models have been proposed. The four most widely used procedures are briefly discussed: Baron and Kenny's (1986) causal-steps approach, joint significance tests for the *a* and *b* path coefficients, the **Sobel test** (Sobel, 1982) for H_0: $ab = 0$, and the use of bootstrapping to obtain confidence intervals (CIs) for the *ab* product that represents the mediated or indirect effect.

9.9.1 Causal-Steps Approach

Fritz and MacKinnon (2007) reviewed and evaluated numerous methods for testing whether mediation is statistically significant. A subset of these methods is described here.

Their review of mediation studies conducted between 2000 and 2003 revealed that the most frequently reported method was the causal-steps approach described by Baron and Kenny (1986). In Baron and Kenny's initial description of this approach, in order to conclude that mediation may be present, several conditions were required: first, a significant total relationship between X_1, the initial cause, and Y, the final outcome variable (i.e., a significant path c); significant a and b paths; and a significant ab product using the Sobel test or a similar method, as described in Section 9.9.3. The decision of whether to call the outcome partial or complete mediation then depends on whether the c' path that represents the direct path from X_1 to Y is statistically significant; if c' is not statistically significant, the result may be interpreted as complete mediation; if c' is statistically significant, then only partial mediation may be occurring. Kenny has also noted elsewhere (http://www.davidakenny.net/cm/mediate.htm) that other factors, such as the sizes of coefficients and whether they are large enough to be of practical significance, should also be considered and that, as with any other regression analysis, meaningful results can be obtained only from a **correctly specified model**.

This approach is widely recognized, but it is not the most highly recommended procedure at present for two reasons. First, there are (relatively rare) cases in which mediation may occur even when the original X_1, Y association is not significant. For example, if one of the paths in the mediation model is negative, a form of suppression may occur such that positive direct and negative indirect effects tend to cancel each other out to yield a small and nonsignificant total effect. (If a is negative, while b and c' are positive, then when we combine a negative ab product with a positive c' coefficient to reconstitute the total effect c, the total effect c can be quite small even if the separate positive direct path and negative indirect paths are quite large.) MacKinnon, Fairchild, and Fritz (2007) referred to this as **inconsistent mediation**; the mediator acts as a **suppressor variable**. See Chapter 3 for further discussion and an example of inconsistent mediation. Second, among the methods compared by Fritz and MacKinnon (2007), this approach had relatively low statistical power.

9.9.2 Joint Significance Test

Fritz and MacKinnon (2007) also discussed a joint significance test approach to testing the significance of mediation. The data analyst simply asks whether the a and b coefficients that constitute the mediated path are both statistically significant; the t tests from the regression results are used. (On his mediation webpage at http://www.davidakenny.net/cm/mediate.htm, Kenny suggests that if this approach is used, and if an overall risk for Type I error of .05 is desired, each test should use α = .025, two tailed, as the criterion for significance.) This approach is easy to implement and has moderately good statistical power compared with the other test procedures reviewed by Fritz and MacKinnon. However, it is not the most frequently reported method; journal reviewers may prefer better known procedures.

9.9.3 Sobel Test of H_0: $ab = 0$

Another method to assess the significance of mediation is to examine the product of the a, b coefficients for the mediated path. (This is done as part of Baron and Kenny's [1986] causal-steps approach.) The null hypothesis, in this case, is H_0: $ab = 0$. To set up a z-test statistic, an estimate of the standard error of this ab product (**SE_{ab}**) is needed. Sobel (1982) provided the following approximate estimate for SE_{ab}:

$$SE_{ab} \approx \sqrt{b^2 s_a^2 + a^2 s_b^2}, \tag{9.1}$$

where

a and b are the raw (unstandardized) regression coefficients that represent the effect of X_1 on X_2 and the effect of X_2 on Y, respectively;

s_a is the standard error of the *a* regression coefficient;

s_b is the standard error of the *b* regression coefficient.

Using the standard error from Equation 9.1 as the divisor, the following z ratio for the Sobel test can be set up to test the null hypothesis H_0: $ab = 0$:

$$z = ab/SE_{ab}. \tag{9.2}$$

The *ab* product is judged to be statistically significant if z is greater than +1.96 or less than −1.96. This test is appropriate only for large sample sizes. The Sobel test is relatively conservative, and among the procedures reviewed by Fritz and MacKinnon (2007), it had moderately good statistical power. It is sometimes used in the context of Baron and Kenny's (1986) causal-steps procedure and sometimes reported without the other causal steps. The Sobel test can be done by hand; Preacher and Leonardelli (2008) provide an online calculator at http://quantpsy.org/sobel/sobel.htm to compute this z test given either the unstandardized regression coefficients and their standard errors or the t ratios for the *a* and *b* path coefficients. Their program also provides z tests on the basis of alternative methods of estimating the standard error of *ab* suggested by the **Aroian test** (Aroian, 1947) and **Goodman test** (Goodman, 1960).

The Sobel test was carried out for the hypothetical data on age, weight, and blood pressure. (Note again that the N in this demonstration data set is too small for the Sobel test to yield accurate results; these data are used only to illustrate the use of the techniques.) For these hypothetical data, $a = 1.432$, $b = .490$, $s_a = .397$, and $s_b = .187$. These values were entered into the appropriate lines of the calculator provided at Preacher's webpage; the results appear in Figure 9.6. Because $z = 2.119$, with $p = .034$, two tailed, the *ab* product that represents the effect of age on blood pressure mediated by weight can be judged statistically significant.

Note that the z tests for the significance of *ab* assume that values of this *ab* product are normally distributed across samples from the same population; it has been demonstrated empirically that this assumption is incorrect for many values of *a* and *b*. Because of this, authorities on mediation analysis (MacKinnon, Preacher, and their colleagues) now recommend bootstrapping methods to obtain CIs for estimates of *ab*.

9.9.4 Bootstrapped Confidence Interval for *ab*

Bootstrapping has become widely used in situations where the analytic formula for the standard error of a statistic is not known and/or there are violations of assumptions of normal distribution shape (Iacobucci, 2008). A sample is drawn from the population (with replacement), and values of *a*, *b*, and *ab* are calculated for this sample. This process is repeated many times (bootstrapping procedures typically allow users to request from 1,000 up to 5,000 different samples). The value of *ab* is tabulated across these samples; this provides an empirical sampling distribution that can be used to derive a value for the standard error of *ab*. Results of such bootstrapping indicate that the distribution of *ab* values is often asymmetrical, and this asymmetry should be taken into account when setting up CI estimates of *ab*. This CI provides a basis for evaluation of the single estimate of *ab* obtained from analysis of the entire data set. Bootstrapped CIs do not require that the *ab* statistic have a normal distribution across samples. If this CI does not include zero, the analyst concludes that there is statistically significant mediation. Some bootstrapping programs include additional refinements, such as bias correction (see Fritz & MacKinnon, 2007). Most SEM programs, such as Amos, can provide bootstrapped CIs. A detailed example is presented in Chapter 15, on structural equation modeling.

Figure 9.6 Sobel Test Results for H_0: $ab = 0$, Using Calculator Provided by Preacher and Leonardelli at http://quantpsy.org/sobel/sobel.htm

Input:			Test statistic:	Std. Error:	p-value:
a	1.43 2	Sobel test:	2.119	0.330	0.034
b	.490	Aroian test:	2.068	0.339	0.038
s_a	.397	Goodman test:	2.175	0.322	0.029
s_b	.187	Reset all			

Input:			Test statistic:	p-value:
t_a	3.60 5	Sobel test:	2.120	0.033
t_b	2.62 3	Aroian test:	2.069	0.038
		Goodman test:	2.176	0.029
		Reset all		

Note: This test is recommended for use only with large N samples. The data set used for this example has N = 30; this was used only as a demonstration.

9.10 EFFECT SIZE INFORMATION

Effect size information is usually given in unit-free form (Pearson's r and r^2 can both be interpreted as effect sizes). The raw or unstandardized path coefficients from mediation analysis can be converted to standardized slopes; alternatively, we can examine the correlation between X_1 and X_2 to obtain effect-size information for the a path, as well as the partial correlation between X_2 and Y (controlling for X_1) to obtain effect-size information for the b path. There are potential problems with comparisons among standardized regression or path coefficients. For example, if the same mediation analysis involving the same set of three variables is conducted in two different samples (e.g., a sample of women and a sample of men), these samples may have different standard deviations on variables such as the predictor X_1 and the outcome variable Y. Suppose that the male and female samples yield b and c' coefficients that are very similar, suggesting that the amount of change in Y as a function of X_1 is about the same across the two groups. When we convert raw-score slopes to standardized slopes, this may involve multiplying and dividing by different standard deviations for men and women, and different standard deviations within these groups could make it appear that the groups have different relationships between variables (different standardized slopes but similar unstandardized slopes).

Unfortunately, both raw score (b) and standardized (β) regression coefficients can be influenced by numerous sources of artifact that may operate differently in different groups. Appendix 10C in Volume I (Warner, 2020) reviewed numerous factors that can artifactually influence the size of r (such as outliers, curvilinearity, different distribution shapes for X and Y, unreliability of measurement of X and Y, etc.). Chapter 11 in Volume I demonstrated that β coefficients can be computed from bivariate correlations and that b coefficients are rescaled versions of β. When Y is the outcome and X is the predictor, $b = \beta \times (SD_Y/SD_X)$. Both b and β coefficients can be influenced by many of the same problems as correlations. Therefore, if we try to compare regression coefficients across groups or samples, differences in regression coefficients across samples may be due partly to artifacts. Considerable caution is required whether we want to compare standardized or unstandardized coefficients.

Despite concerns about potential problems with standardized regression slopes (as discussed by Greenland et al., 1991), data analysts often include standardized path coefficients in reports of mediation analysis, particularly when some or all variables are not measured in meaningful units. In reporting results, authors should make it clear whether standardized or unstandardized path coefficients are reported. Given the difficulties just discussed, it is a good idea to include both types of path coefficients.

9.11 SAMPLE SIZE AND STATISTICAL POWER

Assuming the hypothesis of primary interest is H_0: $ab = 0$, how large does sample size need to be to have an adequate level of statistical power? Answers to questions about sample size depend on several pieces of information: the alpha level, desired level of power, the type of test procedure, and the population effect sizes for the strength of the association between X_1 and X_2, as well as X_2 and Y. Often, information from past studies can help researchers make educated guesses about effect sizes for correlations between variables. In the discussion that follows, $\alpha = .05$ and desired power of .80 are assumed. We can use the correlation between X_1 and X_2 as an estimate of the effect-size index for a and the partial correlation between X_2 and Y, controlling for X_1, as an estimate of the effect size for b. On the basis of recommendations about verbal labels for effect size given by Cohen (1988), Fritz and MacKinnon (2007) designated a correlation of .14 as small, a correlation of .39 as medium, and a correlation of .59 as large. They reported statistical power for combinations of small (S), medium (M), and large (L) effect sizes for the a and b paths. For example, if a researcher plans to use the Sobel test and expects that both the a and b paths correspond to medium effects, the minimum recommended sample size from Table 9.1 would be 90.

A few cautions are in order: Sample sizes from this table may not be adequate to guarantee significance, even if the researcher has not been overly optimistic about anticipated effect size. Even when the power table suggests that fewer than 100 cases might be adequate for statistical power for the test of H_0: $ab = 0$, analysts should keep in mind that small samples lead to more sampling error in estimates of path coefficients. For most studies that test mediation models, minimum sample sizes of 150 to 200 would be advisable if possible.

9.12 ADDITIONAL EXAMPLES OF MEDIATION MODELS

Several variations of the basic mediation model in Figure 9.1 are possible. For example, the effect of X_1 on Y could be mediated by multiple variables instead of just one (see Figure 9.7). Mediation could involve a multiple-step causal sequence. Mediation and moderation can both occur together. The following sections provide a brief introduction to each of these research situations; for more extensive discussion, see MacKinnon (2008).

9.12.1 Multiple Mediating Variables

In many situations, the effect of a causal variable X_1 on an outcome Y might be mediated by more than one variable. Consider the effects of personality traits (such as extraversion

Table 9.1 Empirical Estimates of Sample Size Needed for Power of .80 When Using $\alpha = .05$ as the Criterion for Statistical Significance in Three Different Types of Mediation Analysis

ab Effect Size[a]	Joint Significance[b]	Sobel[c]	Bootstrapped Confidence Interval[d]
SS	530	667	558
SM	403	422	406
SL	403	412	398
MS	405	421	404
MM	74	90	78
ML	58	66	59
LS	405	410	401
LM	59	67	59
LL	36	42	36

Source: Adapted from Fritz and MacKinnon (2007, Table 3, p. 237).

Note: These power estimates may be inaccurate when measures of variables are unreliable, assumptions of normality are violated, or categorical variables are used rather than quantitative variables.

[a]*SS indicates that both a and b are small effects, SM indicates that a is small and b is medium, and SL indicates that a is small and b is large.*

[b]*Joint significance test:* Requirement that the *a* and *b* coefficients each are statistically significant.

[c]A *z test for* H_0: *ab* using a method to estimate SE_{ab} proposed by Sobel (1982).

[d]*Without bias correction.*

and neuroticism) on happiness. Extraversion is moderately positively correlated with happiness. Tkach and Lyubomirsky (2006) suggested that the effects of trait extraversion on happiness may be at least partially mediated by behaviors such as social activity. For example, people who score high on extraversion tend to engage in more social activities, and people who engage in more social activities tend to be happier. They demonstrated that, in their sample, the effects of extraversion on happiness were partially mediated by engaging in social activity, but there was still a significant direct effect of extraversion on happiness. Their mediation analyses examined only one behavior at a time as a potential mediator. Multiple mediators can easily be examined using SEM programs such as Amos, discussed later in this chapter.

9.12.2 Multiple-Step Mediated Paths

It is possible to examine a mediation sequence that involves more than one intermediate step, as in the sequence $X_1 \rightarrow X_2 \rightarrow X_3 \rightarrow Y$. If only partial mediation occurs, additional paths would need to be included in this type of model; for further discussion, see Taylor, MacKinnon, and Tein (2008).

9.12.3 Mediated Moderation and Moderated Mediation

It is possible for moderation (described in another chapter) to co-occur with mediation in two different ways. Mediated moderation occurs when two initial causal variables (let's

Figure 9.7 Path Model for Multiple Mediating Variables Showing Standardized Path Coefficients

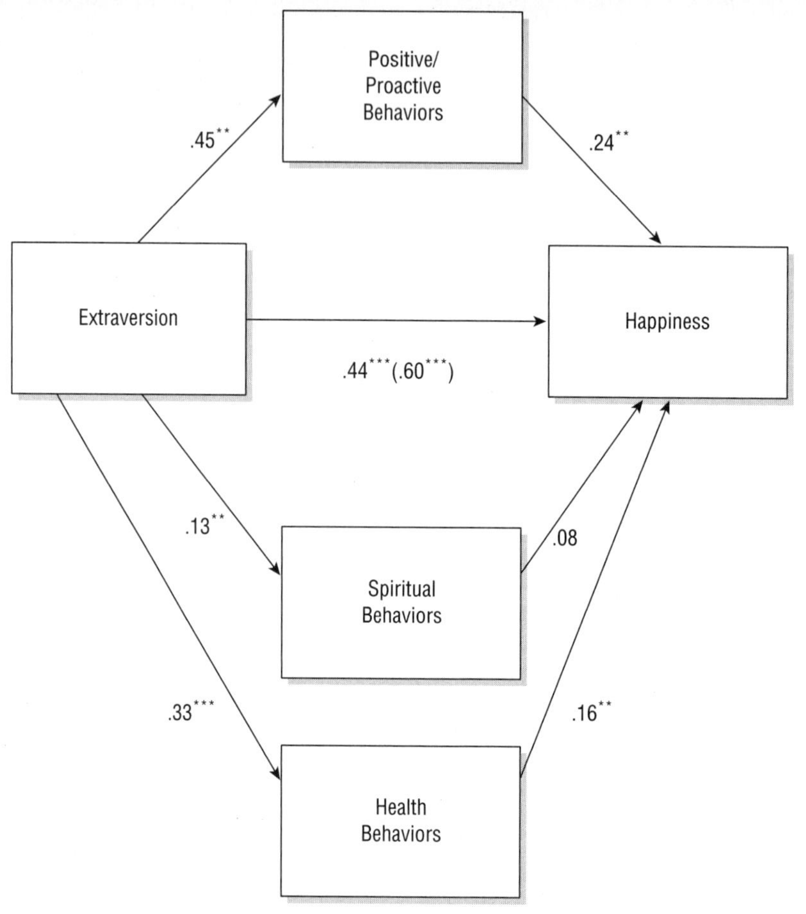

Source: Adapted from Warner and Vroman (2011).

Note: Coefficient estimates and statistical significance tests were obtained using the Indirect.sps script (output not shown). The effect of extraversion on happiness was partially mediated by behaviors. Positive/proactive behaviors ($a_1 \times b_1$) and health behaviors ($a_3 \times b_3$) were significant mediators; spiritual behaviors did not significantly mediate effects of extraversion on happiness.

call these variables A and B) have an interaction ($A \times B$), and the effects of this interaction involve a mediating variable. In this situation, A, B, and the $A \times B$ interaction are included as initial causal variables, and the mediation analysis is conducted to assess the degree to which a potential mediating variable explains the impact of the $A \times B$ interaction on the outcome variable. The PROCESS macros provided by Andrew Hayes (2017, 2019) are extremely useful for assessment of models with moderated mediation, or mediated moderation.

Moderated mediation occurs when you have two different groups (e.g., men and women), and the strength or signs of the paths in a mediation model for the same set of variables differ across these two groups. Many SEM programs, such as Amos, make it possible to compare path models across groups and to test hypotheses about whether one, or several, path coefficients differ between groups (e.g., men vs. women). Further discussion can be found in

Edwards and Lambert (2007); Muller, Judd, and Yzerbyt (2005); and Preacher, Rucker, and Hayes (2007). Comparison of models across groups using the Amos SEM program was demonstrated by Byrne (2016).

9.13 NOTE ABOUT USE OF STRUCTURAL EQUATION MODELING PROGRAMS TO TEST MEDIATION MODELS

SEM programs such as LISREL, EQS, Mplus, and Amos make it possible to test models that include multiple-step paths (e.g., mediation hypotheses) and to compare results across groups (to test moderation hypotheses). In addition, SEM programs make it possible to include multiple indicator variables for some or all of the constructs; in theory, this makes it possible to assess multiple indicator measurement reliability. Most SEM programs now also provide bootstrapping; most analysts now view SEM programs as the preferred method for assessment of mediated models. More extensive discussion of other types of analyses that can be performed using SEM is beyond the scope of this book; for further information, see Byrne (2009) or Kline (2016).

There are two reasons why it is worthwhile to learn how to use Amos (or other SEM programs) to test mediated models. First, it is now generally agreed that bootstrapping is the preferred method to test the statistical significance of indirect effects in mediated models; bootstrapping may be more robust to violations of assumptions of normality. Second, once a student has learned to use Amos (or other SEM programs) to test simple mediation models similar to the example in this chapter, the program can be used to add additional predictor and/or mediator variables, as shown in Figure 9.11.

9.14 RESULTS SECTION

For the hypothetical data in this chapter, a "Results" section could read as follows. Results presented here are based on the output from linear regression (Figures 9.2–9.4) and the Sobel test result in Figure 9.6. (Results would include slightly different numerical values if the Amos output is used.)

Results

A mediation analysis was performed using Baron and Kenny's (1986) causal-steps approach; in addition, a bootstrapped CI for the *ab* indirect effect was obtained using procedures described by Preacher and Hayes (2008). The initial causal variable was age, in years; the outcome variable was systolic blood pressure (SBP), in millimeters of mercury; and the proposed mediating variable was body weight, in pounds. [Note: The sample N, mean, standard deviation, minimum and maximum scores for each variable, and correlations among all three variables would generally appear in earlier sections.] Refer to Figure 9.1 for the path diagram that corresponds to this mediation hypothesis. Preliminary data screening suggested that there were no serious violations of assumptions of normality or linearity. All coefficients reported here are unstandardized, unless otherwise noted; $\alpha = .05$, two tailed, is the criterion for statistical significance.

The total effect of age on SBP was significant, $c = 2.862$, $t(28) = 6.631$, $p < .001$; each 1-year increase in age predicted approximately a 3-point increase in SBP. Age was significantly predictive of the hypothesized mediating variable, weight; $a = 1.432$, $t(28) = 3.605$, $p = .001$. When controlling for age, weight was significantly predictive of SBP, $b = .490$, $t(27) = 2.623$, $p = .014$. The estimated direct effect of age on SBP, controlling for weight, was $c' = 2.161$, $t(27) = 4.551$, $p < .001$.

SBP was predicted quite well from age and weight, with adjusted $R^2 = .667$ and $F(2, 27) = 30.039$, $p < .001$.

The indirect effect, *ab*, was .701. This was judged to be statistically significant using the Sobel test, $z = 2.119$, $p = .034$. [Note: The Sobel test should be used only with much larger sample sizes than the *N* of 30 for this hypothetical data set.] Using the SPSS script for the indirect procedure (Preacher & Hayes, 2008), bootstrapping was performed; 5,000 samples were requested; a bias-corrected and accelerated CI was created for *ab*. For this 95% CI, the lower limit was .0769 and the upper limit was 2.0792.

Several criteria can be used to judge the significance of the indirect path. In this case, both the *a* and *b* coefficients were statistically significant, the Sobel test for the *ab* product was significant, and the bootstrapped CI for *ab* did not include zero. By all these criteria, the indirect effect of age on SBP through weight was statistically significant. The direct path from age to SBP (c') was also statistically significant; therefore, the effects of age on SBP were only partly mediated by weight.

The upper diagram in Figure 9.5 shows the unstandardized path coefficients for this mediation analysis; the lower diagram shows the corresponding standardized path coefficients.

Comparison of the coefficients for the direct versus indirect paths ($c' = 2.161$ vs. $ab = .701$) suggests that a relatively small part of the effect of age on SBP is mediated by weight. There may be other mediating variables through which age might influence SBP, such as other age-related disease processes.

9.15 SUMMARY

This chapter demonstrates how to assess whether a proposed mediating variable (X_2) may partly or completely mediate the effect of an initial causal variable (X_1) on an outcome variable (Y). The analysis partitions the total effect of X_1 on Y into a direct effect, as well as an indirect effect through the X_2 mediating variable. The path model represents causal hypotheses, but readers should remember that the analysis cannot prove causality if the data are collected in the context of a nonexperimental design. If controlling for X_2 completely accounts for the correlation between X_1 and Y, this could happen for reasons that have nothing to do with mediated causality; for example, this can occur when X_1 and X_2 are highly correlated with each other because they measure the same construct. A mediation analysis should be undertaken only when there are good reasons to believe that X_1 causes X_2 and that X_2 in turn causes Y. In addition, it is highly desirable to collect data in a manner that ensures temporal precedence (i.e., X_1 occurs first, X_2 occurs second, and Y occurs third).

These analyses can be done using OLS regression; however, use of SPSS scripts provided by Preacher and Hayes (2008) provides bootstrapped estimates of CIs, and most analysts now believe this provides better information than statistical significance tests that assume normality. SEM programs provide even more flexibility for assessment of more complex models.

If a mediation analysis suggests that partial or complete mediation may be present, additional research is needed to establish whether this is replicable and real. If it is possible to manipulate or block the effect of the proposed mediating variable experimentally, experimental work can provide stronger evidence of causality (MacKinnon, 2008).

COMPREHENSION QUESTIONS

1. Suppose that a researcher first measures a Y outcome variable, then measures an X_1 predictor and an X_2 hypothesized mediating variable. Why would this not be a good way to collect data to test the hypothesis that the effects of X_1 on Y may be mediated by X_2?
2. Suppose a researcher wants to test a mediation model that says that the effects of math ability (X_1) on science achievement (Y) are mediated by sex (X_2). Is this a reasonable mediation hypothesis? Why or why not?
3. A researcher believes that the prediction of Y (job achievement) from X_1 (need for power) is different for men versus women (X_2). Would a mediation analysis be appropriate? If not, what other analysis would be more appropriate in this situation?
4. Refer to Figure 9.1. If a, b, and ab are all statistically significant (and large enough to be of practical or clinical importance), and c' is not statistically significant and/or not large enough to be judged practically or clinically important, would you say that the effects of X_1 on Y are partially or completely mediated by X_2?
5. What pattern of outcomes would you expect to see for coefficient estimates in Figure 9.1; for example, which coefficients would need to be statistically significant and large enough to be of practical importance, for the interpretation that X_2 only partly mediates the effects of X_1 on Y? Which coefficients (if any) should be not statistically significant if the effect of X_1 on Y is only partly mediated by X_2?
6. In Figure 9.1, suppose that you initially find that path c (the total effect of X_1 on Y) is not statistically significant and too small to be of any practical or clinical importance. Does it follow that there cannot possibly be any indirect effects of X_1 on Y that are statistically significant? Why or why not?
7. Using Figure 9.1 again, consider this equation: $c = (a \times b) + c'$. Which coefficients represent direct, indirect, and total effects of X_1 on Y in this equation?
8. A researcher believes that the a path in a mediated model (see Figure 9.1) corresponds to a medium unit-free effect size and the b path in a mediated model also corresponds to a medium unit-free effect size. If assumptions are met (e.g., scores on all variables are quantitative and normally distributed), and the researcher wants to have power of about .80, what sample size would be needed for the Sobel test (according to Table 9.1)?
9. Give an example of a three-variable study for which a mediation analysis would make sense. Be sure to make it clear which variable is the proposed initial predictor, mediator, and outcome.
10. Briefly comment on the difference between the use of a bootstrapped CI (for the unstandardized estimate of ab) versus the use of the Sobel test. What programs can be used to obtain the estimates for each case? Which approach is less dependent on assumptions of normality?

NOTES

[1]For discussion of potential problems with comparisons among standardized regression coefficients, see Greenland, Maclure, Schlesselman, Poole, and Morgenstern (1991). Despite the problems they and others have identified, research reports still commonly report standardized

regression or path coefficients, particularly in situations where variables have arbitrary units of measurement.

DIGITAL RESOURCES

Find **free study tools** to support your learning, including **eFlashcards, data sets, and web resources**, on the accompanying website at **edge.sagepub.com/warner3e**.

CHAPTER 10

DISCRIMINANT ANALYSIS

10.1 RESEARCH SITUATIONS AND RESEARCH QUESTIONS

Up to this point, the analyses that have been discussed have involved the prediction of a score on a quantitative *Y* outcome variable. For example, in multiple regression (MR) we set up an equation to predict the score on a quantitative *Y* outcome variable (such as heart rate) from a combination of scores on several predictor variables (such as gender, anxiety, smoking, caffeine consumption, and level of physical fitness). However, in some research situations, the *Y* outcome variable of primary interest is categorical. For example, in medical research, we may want to predict whether a patient is more likely to die (Outcome Group 1) or survive (Outcome Group 2) on the basis of the patient's scores on several relevant predictor variables. This chapter discusses discriminant analysis (DA) as one possible way to approach the problem of predicting group membership on a categorical *Y* outcome variable from several *X* predictors. As in MR, these *X* predictors are often quantitative variables; the *X* predictors in DA may also include dummy-coded variables that represent information about group membership.

Like MR, DA involves finding optimal weighted linear combinations of scores on several *X* predictor variables, that is, sums or composites of *X* scores that make the best possible predictions for scores on a *Y* outcome variable. The difference between MR and DA is in the type of outcome variable. For MR, the *Y* outcome variable is quantitative; for DA, the *Y* outcome variable is categorical. Because DA involves a different type of outcome variable, different types of information are needed to evaluate the adequacy of prediction. In MR, we assess the accuracy of prediction of scores on a quantitative *Y* outcome variable by looking at the magnitude of multiple *R* (the correlation between actual *Y* and predicted *Y'* scores) and the magnitude of SS_{residual} (residual sum of squares). For DA, we will examine other types of information, such as the percentage of cases for which the predicted group membership agrees with the actual group membership.

In both MR and DA, the goal is to find an "optimal weighted linear combination" of scores on a set of *X* predictor variables, that is, a combination of raw scores or *z* scores for the *X* variables that makes the best possible prediction of scores on the outcome variable. A "weighted linear combination" of scores on a set of *X* variables is a sum of scores on several *X* variables. A combination of scores on *X* variables is linear as long as we use only addition and subtraction in forming the composites; if we use nonlinear arithmetic operations, such as division, multiplication, X^2, and so forth, then the resulting composite is a nonlinear combination of scores on the *X* variables. In MR, the weight associated with each *X* predictor variable corresponds to its regression slope or regression coefficient. The weights for an optimal weighted linear combination in MR can be given in raw score form; that is, *b* slope coefficients or weights are applied to raw scores on *X* variables to estimate the raw score on a Y variable. Alternatively, the optimal predictive weights may be reported in standardized form as beta coefficients that are applied to z scores on X variables to generate predicted z scores on a Y outcome variable. In the MR analyses described in earlier chapters, the optimal weighted linear combination of scores on X can be obtained from a raw-score version of the multiple regression equation, as shown in Equation 10.1:

$$Y' = b_0 + b_1X_1 + b_2X_2 + \cdots + b_kX_k. \tag{10.1}$$

In the standardized version of MR, we set up an optimal weighted linear combination of standardized or z scores on the X predictor variables:

$$z'_Y = \beta_1 z_{X1} + \beta_2 z_{X2} + \cdots + \beta_k z_{Xk}. \tag{10.2}$$

The magnitude and sign of the β_i coefficient for each z_{Xi} predictor variable indicate how an increase in scores on each z_{Xi} variable that is included in the composites influences the overall composite predicted score (z'_Y). In other words, each β_i coefficient tells us how much "weight" is given to z_{Xi} when forming the sum of the z scores across several predictors.

Let's consider a different example involving a weighted linear composite. In this example with the specific weighted linear composite U given in Equation 10.3,

$$U = +.8 \times z_{X1} - .4 \times z_{X2} + .03 \times z_{X3}, \tag{10.3}$$

the score on the z_{X1} variable has a stronger association with the value of the composite U than the score on the z_{X3} variable because z_{X1} has a much larger beta value or weight associated with it. High scores on the linear composite U, described by Equation 10.3, will be strongly and positively associated with high scores on z_{X1} (because z_{X1} has a large positive weight of +.80). High scores on U will be moderately associated with low scores on z_{X2} because z_{X2} has a moderate negative weight (–.40) in the formula used to compute the composite variable U. Scores on the new linear composite function U will have a correlation close to 0 with scores on z_{X3} because the weight associated with the z_{X3} term in the computation of U (.03) is extremely small.

In MR, the goal of the analysis is to find regression slopes or weights that yield an **optimal weighted linear composite** of scores on the X predictor variables; this composite is "optimal" in the sense that it has the maximum possible correlation with the quantitative Y outcome variable in MR and the minimum possible value of SS_{residual}. By analogy, in DA, the goal of the analysis is to find the DA function coefficients that yield the optimal weighted linear composites of scores on the X predictor variables; the optimal weighted linear composites obtained in DA, which are called **discriminant functions**, are optimal in the sense that they have the largest possible SS_{between} and the smallest possible SS_{within} for the groups defined by the Y categorical outcome variable. Another sense in which the discriminant function scores obtained from DA may be optimal is that these scores can be used to classify cases into groups with the highest possible percentage of correct classifications into groups.

DA can be performed to answer one or both of the following types of questions (Huberty, 1994). First of all, DA can be used to make *predictions about group membership outcomes*. For example, DA could be used to predict whether an individual patient is more likely to be a member of Group 1 (patients who are having heart attacks) or Group 2 (patients whose chest pain is not due to heart attacks) on the basis of each individual patient's scores on variables that can be assessed during the initial intake interview and lab tests, such as age, gender, high-density lipoprotein and low-density lipoprotein cholesterol levels, triglyceride level, smoking, intensity of reported chest pain, and systolic blood pressure (SBP). In the initial study to develop the predictive model, we need data that include actual group membership (e.g., based on later more definitive tests, "Is the patient a member of Group 1, heart attack, or Group 2, chest pain that is not due to heart attack?"). We can use a sample of patients for whom the correct diagnosis is known (on the basis of subsequent diagnostic work that takes longer to obtain) to develop a statistical formula that makes the best possible prediction of ultimate diagnosis using the information available during the initial intake interview. Once we have this statistical formula, we may be able to use it to try to predict the ultimate diagnosis for new groups of incoming patients for whom we have only the initial intake information. Similarly, an admissions officer at a university might do an initial model development study to see what combination of variables (verbal SAT score, math SAT score, high school grade point average, high school class rank, etc.) best predicts whether applicants to a college subsequently succeed (graduate from the college) or not (drop out or fail). Once a statistical prediction model has been developed using data for students whose college outcomes are known, this model can be applied to the information obtained from the applications of new incoming students to try to predict the group outcome (success or failure) for these new individuals whose actual group outcomes are not yet known.

Second, *DA can be used to describe the nature of differences between groups.* If we can identify a group of patients who have had heart attacks (Group 1) and compare them with a group of patients who have not had heart attacks (Group 2), we can determine what pattern of scores on variables such as age, high-density lipoprotein and low-density lipoprotein cholesterol levels, triglyceride level, smoking, anxiety, hostility, and SBP best describes the differences between these two groups of patients. In studies that involve DA, researchers may be interested in the accurate prediction of group membership, a description of the nature of differences between the groups, or both kinds of information.

DA is one of several analyses that may be used to predict group membership on a categorical outcome variable. When all the variables involved in an analysis are categorical (both the outcome and all the predictor variables), log-linear analysis is more appropriate. Another approach to the prediction of group membership outcomes that has become increasingly popular in recent years is called logistic regression. Logistic regression involves evaluating the **odds** of different group outcomes; for example, if a patient is 5 times as likely to die as to survive, the odds of death are described as 5:1. Logistic regression is sometimes preferred to DA because it requires less restrictive assumptions about distributions of scores. In addition, results derived from logistic regression analysis can be interpreted as information about the changes in the likelihood of negative outcomes as scores on each predictor variable increase.

DA can provide useful information about the nature of group differences. Discriminant functions can be used to describe the dimensions (i.e., combinations of scores on variables) on which groups differ. In DA, research questions that can be addressed include the following: Can we predict group membership at levels above chance, on the basis of participants' scores on the discriminating variables? What pattern of scores on the predictor variables predicts membership in each group? Which predictor variables provide unique information about group membership when correlations with other predictor variables are statistically controlled?

In some applications of DA, different types of **classification errors** have different consequences or costs. For instance, when a patient who is at high risk for heart disease is erroneously predicted to be a member of the low-risk group, the patient may not receive the necessary treatment. Different costs and consequences arise when a healthy patient is incorrectly predicted to belong to a high-risk category (which may lead to unnecessary diagnostic tests or treatment). One type of information that can be obtained as part of DA is a table that shows the numbers and types of classification errors (i.e., participants for whom the predicted group membership is not the same as the actual group membership).

In DA, as in MR, it is assumed that the predictor (or discriminating) variables are likely to be correlated with each other to some degree. Thus, it is necessary to statistically control for these intercorrelations when we form optimal weighted linear combinations of the discriminating variables (i.e., discriminant functions).

A discriminant function is an optimal weighted linear combination of scores on the X predictor variables. The equation for a discriminant function may be given in terms of the raw scores on the X predictor variables or in terms of z scores or standard scores. In a situation where the Y outcome variable corresponds to just two groups, the equation for a single standardized discriminant function D_1 that is used to predict group membership is written as follows (where p is the number of discriminating variables):

$$D_1 = d_{11}z_1 + d_{12}z_2 + \cdots + d_{1p}z_p, \tag{10.4}$$

where $z_1, z_2, \ldots, z_p$ are the z-score or standard-score versions of the predictor variables X_1, $X_2, \ldots, X_p$. These X's are usually quantitative predictor variables, although it is possible for one or more of these predictors to be dichotomous or dummy variables. The equation for the discriminant function D_1 in Equation 10.4 is analogous to the standardized regression in Equation 10.2, except that the outcome variable for DA corresponds to a prediction of group membership, whereas the outcome variable in regression is a score on a quantitative (Y) variable. Later, we will see that when the Y outcome variable corresponds to k groups, we may be able to calculate up to $(k - 1)$ different discriminant functions. (The number of discriminant functions that can be obtained is also limited by the number of predictor variables, p or, more formally, by the number of predictor variables that are not completely predictable from other predictor variables.)

As a hypothetical example, suppose that we want to predict membership in one of the following three diagnostic groups (Group 1 = no diagnosis, Group 2 = neurotic, Group 3 = psychotic). Suppose that for each patient, we have scores on the following quantitative measures: X_1, anxiety; X_2, depression; X_3, delusions; and X_4, hallucinations. In our initial study, we have patients whose diagnoses (none, neurotic, psychotic) are known; we want to develop formulas that tell us how well these group memberships can be predicted from scores on the X's. If we can develop predictive equations that work well to predict group membership, the equations may help us characterize how the members of the three groups differ; also, in future clinical work, if we encounter a patient whose diagnosis is unknown, for whom we have scores on X_1 through X_4, we may be able to use the equations to predict a diagnosis for that patient.

If the categorical variable that is predicted in the DA has just two groups, only one discriminant function is calculated. In general, if there are k groups and p discriminating variables, the number of different discriminant functions that will be obtained is the minimum of these two values: $(k - 1)$ = (number of groups − 1) and p, the number of discriminating variables. The units or the specific values of scores on the discriminant function D do not matter; the D values are generally scaled to have a mean of 0 and a within-group variance of 1, similar to standardized z scores. What does matter is the relation of discriminant scores to group membership. When we have just two groups on the Y outcome variable, the goal of DA is to find the weighted linear combination of scores on the standardized predictor variables (D_1 in Equation 10.4) that has the largest possible variance between groups and the smallest possible variance within groups. If we take the new variate or function called D_1 (where the value of D_1 is given by Equation 10.4) as the outcome variable and perform a one-way analysis of variance (ANOVA) on these scores across groups on the Y variable, then DA provides an optimal set of weights, that is, the set of weights that yields values of D_1 scores for which the F ratio in this ANOVA is maximized.

If there are more than two groups, for example, if $k = 3$, more than one discriminant function may be needed to differentiate between group members; for instance, in a DA with three groups, the first discriminant function might distinguish members of Group 1 from Groups 2 and 3, while the second discriminant function might distinguish members of Group 3 from Groups 1 and 2.

Boundary decision rules can be applied to the values of discriminant functions to make predictions about group membership. For example, if there are just two groups, there will be only one discriminant function, D_1. The boundary that divides the D_1 values for most Group 1 members from those of most members of Group 2 might fall at $D_1 = 0$, with lower scores on D_1 for members of Group 1. In that case, our decision rule would be to classify persons with D_1 scores less than 0 as members of Group 1 and to classify persons with D_1 scores greater than 0 as members of Group 2. When there are two or more discriminant functions, the decision rule depends jointly on the values of several discriminant functions. For two discriminant functions, the decision rule can be shown graphically in the form of a **territorial map** with D_1 (scores on the first discriminant function) on the *X* axis and D_2 (scores on the second discriminant function) on the *Y* axis. Each case can be plotted in this two-dimensional map on the basis of its score on D_1 (along the horizontal axis) and its score on D_2 (along the vertical axis). If the DA is successful in creating discriminant functions that have scores that differ clearly across groups, then the cases that belong to each group will tend to fall close together in the territorial map, and members of different groups will tend to occupy different regions in the space given by this map. We can set up boundaries to represent how the members of Groups 1, 2, and 3 are spatially distributed in this map (on the basis of their combinations of scores on D_1 and D_2). When DA is successful, we can draw boundaries to separate this map into a separate spatial region that corresponds to each group. When a participant has scores on D_1 and D_2 that place him or her in the part of the territorial map that is occupied mostly by members of Group 1, then we will classify that individual as a member of Group 1, as in the example that follows.

Figure 10.1 shows a hypothetical territorial map in which the members of three groups (no diagnosis, neurotic, and psychotic) can be differentiated quite well by using scores on two discriminant functions. In this imaginary example, suppose that the first discriminant function, D_1, is a composite of scores on the variables in which there are large positive **discriminant function coefficients** for the variables anxiety and depression (and near-zero coefficients for delusions and hallucinations). From Figure 10.1, we can see that members of the "neurotic" group scored high on D_1; if D_1 is a function that involves large positive weights or coefficients for the predictor variables anxiety and depression, then the diagram in Figure 10.1 tells us that the neurotic group scored high on a composite of anxiety and depression compared with the other two groups. High scores on this first discriminant function D_1 correspond to high scores on anxiety and depression, and they are not related to people's scores on delusions or hallucinations; therefore, we might interpret D_1 as a dimension that represents how much each person exhibits "everyday psychopathologies." On the other hand, suppose that the second discriminant function, D_2, has large coefficients for delusions and hallucinations. Also, suppose that the psychotic group scored high on D_2 (and thus high on a composite score for delusions and hallucinations) relative to the other two groups. We might interpret the combination of variables that is represented by D_2 as an index of "major psychopathologies" or "delusional disorders." In Figure 10.1, members of the three groups have D_1 and D_2 scores that correspond to clearly separated regions in the territorial map. In contrast, Figure 10.2 shows a second hypothetical territorial map in which the groups cannot be well discriminated because the discriminant function scores overlap substantially among all three groups.

Sometimes a discriminant function that combines information about scores on two or more variables can accurately distinguish between members of two groups even though the groups do not differ significantly on either of the two individual variables. Consider the hypothetical data shown in Figure 10.3, which illustrate a problem in physical anthropology. Suppose that we want to classify a set of skulls as belonging to apes (A) or humans (H) on the basis of measurements such as jaw width (X_1) and forehead height (X_2). Scores on X_1 and X_2 are shown as a scatterplot in Figure 10.3. The two groups (whose scores are marked in the scatterplot using A for ape and H for human) do not differ substantially on their scores on

Figure 10.1 Hypothetical Example of a Discriminant Analysis Study

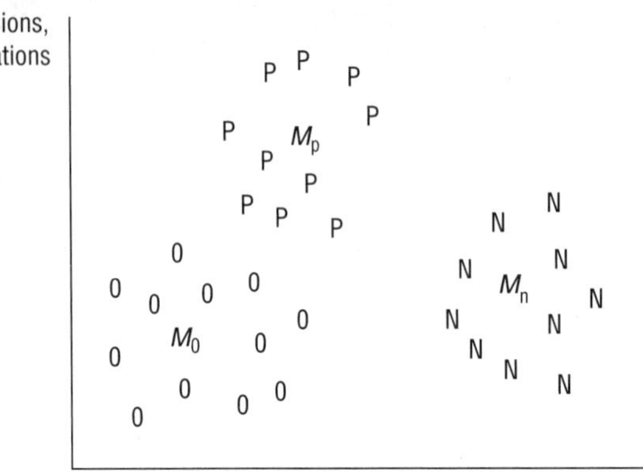

Note: There were three diagnostic groups (0 = no diagnosis, N = neurotic, and P = psychotic) and four discriminating variables (X_1, anxiety; X_2, depression; X_3, delusions; and X_4, hallucinations).

Figure 10.2 Hypothetical Example: Discriminant Analysis With Substantial Overlap in Values of Both D_1 and D_2 Among Groups (and, Therefore, Poor Prediction of Group Membership From Discriminant Function Scores)

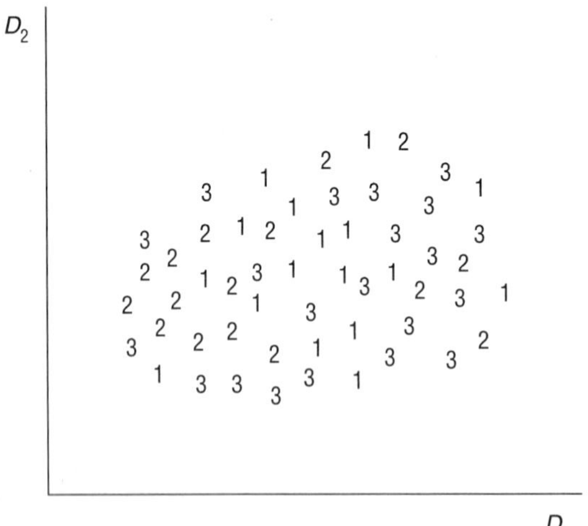

either X_1 or X_2 alone. However, a weighted combination of scores on X_1 and X_2 can be used to create a new function; this new discriminant function D_1 is represented by the dotted line in Figure 10.3. The boundary between the A and H groups appears as a solid line in Figure 10.3. This boundary corresponds to a decision rule that is based on scores on the discriminant function. In this example, if the score on the discriminant function D_1, which is represented by the dotted line, is greater than 0, we would classify the case as a member of group H

Figure 10.3 Hypothetical Example of a Situation in Which Discrimination Between Groups A and H Is Better for a Combination of Scores on Two Variables (X_1, X_2) Than for Either Variable Alone

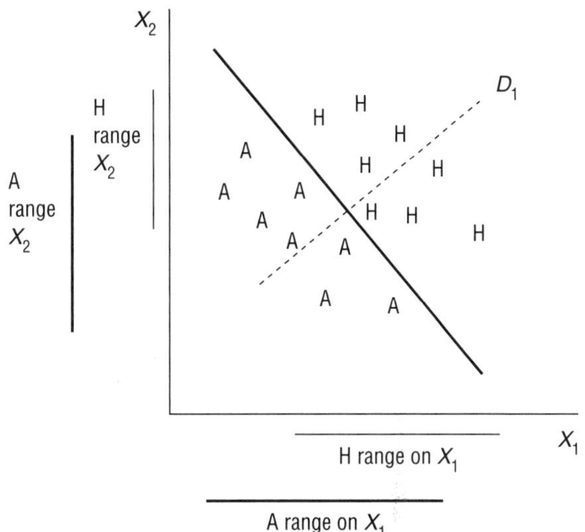

(human); if the score on D_1 is less than 0, we would classify the case as a member of group A (ape); this decision rule would provide perfectly accurate group classification. An alternative view of this graph appears in Figure 10.4. Suppose that we use the following decision rule to predict group membership on the basis of D_1 scores: For cases with scores on $D_1 > 0$, classify as H; for scores on $D_1 < 0$, classify as A. It is apparent from the graph that appears in Figure 10.4 that scores on this new function D_1 differentiate the two groups quite well.

In some research situations, DA provides a similar description of the way in which individual variables differ across groups compared with a series of univariate ANOVAs on the individual variables (Huberty & Morris, 1989). However, a DA takes the intercorrelations among the p predictor variables into account when it makes predictions for group membership or provides a description about the nature of group differences, whereas a series of univariate ANOVAs that examine each of the p predictor variables in isolation to see how it differs across groups does not take these intercorrelations into account, and it will usually inflate the risk for Type I error. Multivariate analyses that consider patterns of scores can sometimes yield significant differences between groups even when none of the univariate differences is significant, as in the skull classification example in Figure 10.3. DA may have greater statistical power (compared with univariate ANOVAs for each of the p predictor variables), but it does not always provide better power.

The research questions for a DA may include any or all of the following:

1. First of all, there is a test of significance for the overall model, that is, an omnibus test to assess whether, when all the discriminating variables and discriminant functions are considered, the model predicts group membership at levels significantly better than chance. For the overall model, there is a multivariate goodness-of-fit statistic called **Wilks' lambda (Λ)**. Wilks' Λ can be interpreted as the proportion of variance in one or several discriminant functions that is not associated with (or predictive of) group membership. Thus, Wilks' Λ might be better thought of as a "badness-of-fit" measure, because it is an estimate of the proportion of unexplained variance; thus,

Figure 10.4 Group Classification Decision Rule for Data in Figure 10.3

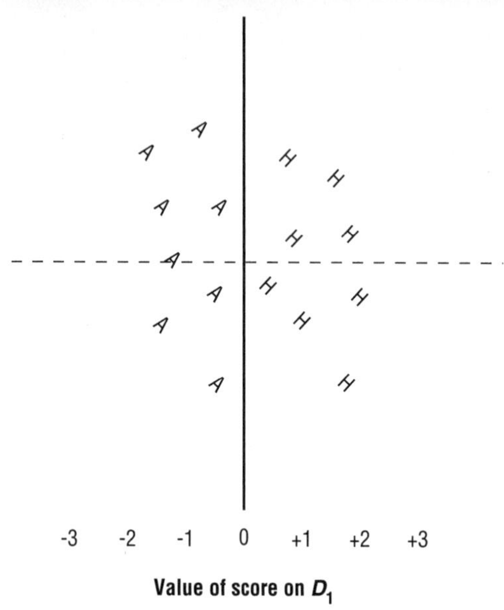

Note: Figure 10.4 is a rotated version of Figure 10.3. D_1 (dotted line) is the score on one discriminant function. If the score on $D_1 > 0$, classify as H. If the score on $D_1 < 0$, classify as A.

it is analogous to $(1 - \eta^2)$ in an ANOVA or $(1 - R^2)$ in a regression. There are three different *df* terms associated with a Wilks' Λ on the basis of the following three numbers: *N*, the overall number of cases; *k*, the number of groups; and *p*, the number of predictor or discriminating variables. It would be cumbersome to look up critical values for Λ in a table that needs to be indexed by three different *df* terms. The SPSS discriminant procedure converts Wilks' Λ to χ^2. This conversion from Λ to χ^2 and the associated *df* are only approximate; note that in some programs, including the SPSS general linear model (GLM) procedure, Wilks' Λ is converted to an *F* ratio. After Λ has been converted into a more familiar test statistic, tables of critical values for these familiar distributions can be used to assess the significance of the overall DA model. SPSS provides *p* values for the chi-square that corresponds to Wilks' Λ. One way to understand the null hypothesis for DA is as follows: For the population, means on the discriminant functions are equal across groups. Another way to understand the null hypothesis is to say that, in the population, group membership is not related to (i.e., cannot be predicted from) scores on the discriminant functions.

2. The second major research question in DA is the following: How many discriminant functions are useful in differentiating among groups? When we have only $k = 2$ groups, we can only obtain one discriminant function. However, when there are more than two groups, there may be more than one discriminant function; the number of discriminant functions for *k* groups is usually the minimum of $(k - 1)$ and *p*, where *p* is the number of predictor variables. (This assumes that the correlation matrix for the *p* variables does not have a determinant of 0; that is, none of the *p* predictor variables is perfectly predictable from one or more other predictor variables.) The discriminant functions are extracted in such a way that scores on the first discriminant function, D_1, have the largest possible between-group differences.

Scores on the second discriminant function, D_2, must be uncorrelated with scores on D_1, and D_2 scores must predict the largest possible between-group variance that is not explained by D_1.

The discriminant functions are numbered (1, 2, 3, etc.). They are rank ordered from the most predictive to the least predictive of group membership. Scores on the discriminant functions D_1, D_2, D_3, and so forth, are uncorrelated with one another. If we have $k = 5$ groups and at least four predictor variables, for example, we will obtain $k - 1 = 4$ different discriminant functions. Our research question will be the following: How many of these four discriminant functions are useful in predicting group membership?

To answer this question, we conduct a **dimension reduction analysis**; that is, we evaluate how many discriminant functions are needed to achieve reasonably accurate predictions of group membership. SPSS DA actually tests the significance of *sets* of discriminant functions rather than the significance of each individual discriminant function; the tests are set up in a manner that may seem counterintuitive at first. For instance, if we have four discriminant functions (denoted D_1 through D_4), SPSS will provide significance tests for the following sets of functions:

Set 1: D_1, D_2, D_3, D_4

Set 2: D_2, D_3, D_4 (i.e., all discriminant functions after, or not including, the first)

Set 3: D_3, D_4

Set 4: D_4

The proportion of between-groups variance associated with each set of discriminant functions typically decreases as we reduce the number of discriminant functions included in the sets. If Set 2 is significant, then Set 1 is usually also significant because Set 1 includes the same functions as Set 2. In practice, it is helpful to read this list from the bottom up. If Set 4 is not significant, then we can drop D_4 from our interpretation. If Set 3 is also not significant, then we can drop D_3 also from our interpretation. If Set 2 is significant, then we may decide that we need to include the first two discriminant functions in our interpretation and discussion of group differences. Usually, we hope that we can account for between-group differences using a relatively small set of discriminant functions. Each discriminant function can be thought of as a "dimension" along which groups may differ. When we make the decision to include only the first one or two discriminant functions from a larger set of discriminant functions in our interpretation of DA, it is called "dimension reduction"; that is, we decide that we only need the information about group differences that is provided by scores on the first one or two discriminant functions (and that later functions do not add much additional information about the nature of group differences or do not lead to much improvement in the prediction of group membership). Sometimes, but not always, the discriminant functions can be given meaningful interpretations or labels as descriptors of underlying "dimensions" on which groups differ.

3. Another possible question is related to the evaluation of the contribution of individual predictors or discriminating variables: Which variables are useful in discriminating among groups or predicting group membership? SPSS discriminant analysis (unlike SPSS regression) does not provide direct statistical significance tests for the contribution of individual predictor variables. DA provides three pieces of information about the contributions of individual predictors; together, these can be used to decide which predictor variables provide important and/or unique information about group differences.

First of all, as described above, the researcher decides how many of the discriminant functions should be interpreted. Then, only for the discriminant functions that are retained for interpretation, the researcher asks the following question: Which predictor variable(s) had "large" correlations (in absolute value) with each function? The magnitudes of correlations between predictor variables and discriminant functions are judged relative to some arbitrary cutoff value; for example, a correlation in excess of .5 (in absolute value) could be interpreted as an indication that a predictor is strongly related to a discriminant function. (The choice of a cutoff value such as .50 is arbitrary.)

Second, the standardized discriminant function coefficient for each variable can be examined to assess which predictor variables were given the greatest weight in computing each discriminant function score when intercorrelations among the discriminating variables are taken into account. The standardized discriminant function coefficients (denoted d in Equation 10.4) are analogous to beta weights or coefficients in an MR. To see how scores on a particular predictor are related to group membership, we need to consider two things: First of all, is a high score on X_i associated with a high or low score on D? Second, does each group tend to score high or low on discriminant function D? Together, these two pieces of information help us understand how X_i scores differ across groups in the context of this multivariate analysis.

Third, the univariate ANOVAs show how scores on each individual discriminating variable differ across groups when each variable's correlations with other predictors are *not* taken into account. For each discriminating variable, there is an F ratio that indicates whether means differed significantly across groups; note, however, that the p values that SPSS reports for these univariate F's are not corrected for inflated risk for Type I error.

If all these three pieces of information provide a consistent story about the strength and nature of the relationship between group membership and X_i scores, then it is relatively easy to write up a description of how groups differ on the X's and/or what pattern of X scores leads us to predict that an individual is a member of each group. However, recall that the zero-order correlation of an individual X_i with Y and the partial slope (β_i) to predict z_Y from z_{Xi}, while controlling for other X's, can be quite different. The same kind of difference can arise in DA; that is, the discriminant function coefficient d_i for z_{Xi} can have a sign that implies a different association between group membership and X_i scores than found in a simple one-way ANOVA. When the results of the DA (in which each X_i predictor variable's association with group membership is assessed while controlling for all other X predictors) differ qualitatively from the results of univariate ANOVAs (in which means on each X_i variable across groups are assessed separately for each X variable, without taking intercorrelations among the X's into account), the researcher needs to think very carefully about descriptions of group differences.

4. Which groups can (or cannot) be discriminated? As in an ANOVA, if there are more than two groups in a DA, our null hypothesis is essentially that the means on the discriminant functions are equal across all groups. If we find that the overall DA is significant, and there are more than two groups, follow-up tests are needed to assess which of the pairwise contrasts between groups are statistically significant. A significant overall DA usually implies that at least one group differs significantly from the others (just as a significant one-way ANOVA usually implies at least one significant contrast between group means). However, a significant overall DA analysis does not necessarily imply that all groups can be distinguished. The SPSS

DA program provides several kinds of information about group differences. The contingency table at the end of the DA output summarizes information about the numbers of persons who were correctly classified as members of each group and also indicates the numbers of misclassifications. This table makes it possible to see whether there were certain groups whose members were frequently misclassified into other groups. It is also possible to examine the territorial map (described later) to see if the clusters of scores on Discriminant Functions 1 and 2 were clearly separated in the territorial map (as in Figure 10.1 or Figure 10.3) or mixed together (as in Figure 10.2).

5. What meaning (if any) can we attach to each discriminant function? (This question is optional and is likely to be of greater interest in more theoretical studies.) It is sometimes useful to name or label a discriminant function; this name should be based on a careful consideration of which *X* variables had high correlations with this discriminant function and which groups show the largest differences on the discriminant function. The names given to discriminant functions are analogous to the names given to factors or components in a factor analysis. It may make sense, particularly in theory-driven studies, to think of the discriminant functions as "dimensions" along which the groups differ. In the hypothetical study for which results are shown in Figure 10.1, we might call D_1 (which is highly correlated with anxiety and depression) the "everyday psychological distress" dimension; the "neurotic" group had high scores on this discriminant function and thus high scores on anxiety and depression, compared with both the "no-diagnosis" and "psychotic" groups. The second discriminant function D_2 was highly correlated with delusions and hallucinations; this might be called a "major psychopathology" or "delusional disorder" dimension; on this dimension, the psychotic group scored higher than either the no-diagnosis or neurotic group.

   However, sometimes in practice, the variables that are given most weight in forming the discriminant function cannot be easily interpreted. If it does not make sense to think of the measures that make the strongest contribution to a particular discriminant function as different measures of the same underlying construct or dimension, an interpretation should not be forced. In research where the selection of predictor variables is guided by theory, it is more likely that the discriminant functions will be interpretable as meaningful "dimensions" along which groups differ.

6. What types of classification errors are most common? A classification error occurs when the predicted group membership for an individual case does not correspond to the actual group membership. Do we need to modify decision rules to reduce the occurrence of certain types of costly prediction errors? (This question is optional and is most likely to be of interest in clinical prediction applications.) For example, DA may be used as a statistical method of making medical diagnoses (on the basis of past records of initial symptom patterns and final diagnosis or medical outcome) or in hiring or admission decisions (on the basis of the history of applicant credential information and success or failure in the job or academic program). In such applied situations, a goal of research may be to generate a formula that can be used to make predictions about future individual cases for which the ultimate medical, job, or educational outcome is not yet known.

   Consider the problem of deciding whether a patient is having a heart attack (Group 1) or not having a heart attack (Group 2), on the basis of initial information such as the patient's age, gender, blood pressure, serum cholesterol, description of location and intensity of pain, and so forth. A physician needs to make an initial guess about diagnosis on the basis of incomplete preliminary information (a definitive diagnosis of heart attack requires more invasive tests that take more time). There are two possible kinds of classification errors: a false positive (the physician thinks the patient is having a heart attack when the patient is not really having a heart attack) and a false negative (the physician thinks the patient is not having a heart attack when the patient really is having a heart attack). In clinical situations, these two types of errors have different types of costs and consequences associated with them. A false positive results in unnecessarily alarming the patient, running unnecessary tests, or perhaps administering unnecessary treatment; there may be some discomfort and even risk, as well as financial cost, associated with this type of error. Usually, in medicine, a false-negative error is seen as more costly; if a patient who really has a life-threatening condition is mistakenly judged to be healthy, necessary care may not be given, and the patient may get worse or even die (and the physician may possibly be sued for negligence). Thus, in medical decisions, there may be a much greater concern with avoiding false negatives than avoiding false positives.

There are a number of ways in which we may be able to reduce the occurrence of false negatives, for example, by finding more useful predictor variables. However, an alternative way of reducing false negatives involves changing the cutoff boundaries for classification into groups. For example, if we decide to classify people as hypertensive when they have SBP > 130 mm Hg (instead of SBP > 145 mm Hg), lowering the cutoff to 130 mm Hg will decrease the false-negative rate, but it will also increase the false-positive rate. The classification of individual cases and the costs of different types of classification errors are not usually major concerns in basic or theory-driven research, but these are important issues when DA is used to guide practical or clinical decision making.

10.2 INTRODUCTION TO EMPIRICAL EXAMPLE

As an example consider the data in the file talent.sav. This is a subset of cases and variables from the Project Talent study (Lohnes, 1966). High school students took aptitude and ability tests, and they were asked to indicate their choices of future career. For this analysis, students who selected the following types of future careers were selected: Group 1, business; Group 2, medicine; and Group 3, teaching or social work. Six of the achievement and aptitude test scores were used as predictors of career choice: abstract (reasoning), reading, English, math, mechanic (mechanical aptitude), and office (skills). Later in this chapter, the results of DA are presented. All six predictors were entered in one step (this is analogous to a standard or simultaneous MR). In addition, one-way ANOVA was performed to show how the means on the saved scores on Discriminant Function 1 differed across the three groups. A simple unit-weighted composite variable was also created (raw scores on office skills minus raw scores on math); one-way ANOVA was performed to assess how the means of the simple unit composite variable (the difference between office skills and math scores) varied across groups.

10.3 SCREENING FOR VIOLATIONS OF ASSUMPTIONS

It may be apparent from the previous section that DA involves some of the same issues as ANOVA (it assesses differences in means on the discriminant functions across groups) and some issues similar to those in MR (it takes intercorrelations among predictors into account when constructing the optimal weighted composite for each discriminant function). The assumptions about data structure required for DA are similar to those for ANOVA and regression but also include some new requirements. Just as in regression, the quantitative X predictor variables should be approximately normally distributed, with no extreme outliers, and all pairs of X predictor variables should be linearly related, with no extreme outliers (this can be assessed by an examination of scatterplots). Just as in ANOVA, the numbers in the groups should exceed some minimum size. (If the number of scores within a group is smaller than the number of predictor variables, for example, it is not possible to calculate independent estimates of all the values of within-group variances and covariances. Larger sample sizes are desirable for better statistical power.)

One additional assumption is made, which is a multivariate extension of a more familiar assumption. When we conduct a DA, we need to compute a sum of cross products (SCP) matrix. (See Appendix 5A in Chapter 5 for an introduction to matrix algebra.) Let $\mathbf{SCP}_i$ stand for the sample **SCP** matrix for the scores within group i. The elements of the sample **SCP** matrix are given below; an **SCP** matrix is computed separately within each group and also for the entire data set with all groups combined.

$$\mathbf{SCP} = \begin{bmatrix} SS(X_1) & \text{SCP}(X_1, X_2) & \cdots & \text{SCP}(X_1, X_p) \\ \text{SCP}(X_2, X_1) & SS(X_2) & \cdots & \text{SCP}(X_2, X_p) \\ \text{SCP}(X_p, X_1) & \text{SCP}(X_p, X_2) & \cdots & SS(X_p) \end{bmatrix}. \quad (10.5)$$

The diagonal elements of each **SCP** matrix are the sums of squares for $X_1, X_2, \ldots, X_p$. Each SS term is found as follows: $SS(X_i) = \Sigma(X_i - M_{Xi})^2$. The off-diagonal elements of each **SCP** matrix are the sums of cross products for all possible pairs of X variables. For X_i and X_j, $\text{SCP} = \Sigma[(X_i - M_{Xi}) \times (X_j - M_{Xj})]$. Recall that in a univariate ANOVA, there was an assumption of homogeneity of variance; that is, the variance of the Y scores was assumed to be approximately equal across groups. In DA, the analogous assumption is that the elements of the variance/covariance matrix are homogeneous across groups.

Note that a sample **SCP** matrix can be used to obtain a sample estimate of the variance/matrix **S** for each group of data. To obtain the elements of **S** from the elements of **SCP**, we divide each element of **SCP** by the appropriate n or df term to convert each SS into a variance and each individual SCP element into a covariance, as shown below.

$$\mathbf{S} = \begin{bmatrix} \text{Var}(X_1) & \text{Cov}(X_1, X_2) & \cdots & \text{Cov}(X_1, X_p) \\ \text{Cov}(X_2, X_1) & \text{Var}(X_2) & \cdots & \text{Cov}(X_2, X_p) \\ \text{Cov}(X_p, X_1) & \text{Cov}(X_p, X_2) & \cdots & \text{Var}(X_p) \end{bmatrix}.$$

A sample variance/covariance matrix **S** is an estimate of a corresponding population variance/covariance matrix denoted by $\mathbf{\Sigma}$. Each diagonal element of the $\mathbf{\Sigma}$ matrix corresponds to a population variance for each of the predictors—for example, $\sigma^2_1, \sigma^2_2, \ldots, \sigma^2_k$—and the off-diagonal elements of the $\mathbf{\Sigma}$ matrix correspond to the population covariance for all possible pairs of X variables.

An assumption for DA is that these $\mathbf{\Sigma}_i$ population variance/covariance matrices are homogeneous across groups. This assumption about homogeneity includes the assumption that the variances for $X_1, X_2, \ldots, X_p$ are each homogeneous across groups and, in addition, the assumption that covariances between all pairs of the X's are also homogeneous across groups. The assumption of **homogeneity of variance/covariance matrices** across groups can be written as follows:

$$H_0: \mathbf{\Sigma}_1 = \mathbf{\Sigma}_2 = \cdots = \mathbf{\Sigma}_k. \tag{10.6}$$

In words, this null hypothesis corresponds to the assumption that corresponding elements of these $\mathbf{\Sigma}$ matrices are equal across groups; in other words, the variances of each X variable are assumed to be equal across groups, and in addition, the covariances among all possible pairs of X variables are also assumed to be equal across groups. Box's M test (available as an optional test in SPSS) tests whether this assumption is significantly violated.

Box's M test can be problematic for at least two reasons. First, it is very sensitive to nonnormality of the distribution of scores. Also, when the number of cases in a DA is quite large, Box's M may indicate statistically significant violations of the homogeneity of the variance/covariance assumption, even when the departures from the assumed pattern are not serious enough to raise problems with the DA. On the other hand, when the number of cases in the DA is very small, Box's M may not be statistically significant at conventional alpha levels (such as $\alpha = .05$), even when the violation of assumptions is serious enough to cause problems with the DA. As with other preliminary tests of assumptions, because of the degrees of freedom involved, Box's M is more sensitive to violations of the assumption of homogeneity in situations where this violation may have less impact on the validity of the results (i.e., when the overall N is large); also, Box's M is less sensitive to violations of the assumption of homogeneity in situations where these violations may cause more problems (i.e., when the overall N is small). To compensate for this, when the overall number of cases is very large, it may be preferable to evaluate the statistical significance of Box's M using a smaller alpha level (such as $\alpha = .01$ or $\alpha = .001$). When the overall number of cases involved in the DA is quite small, it may be preferable to evaluate the significance of Box's M using a higher alpha level (such as $\alpha = .10$).

Complete information about the **SCP** matrices within each of the groups can be requested as an optional output from the DA procedure. Examination of these may make it possible to locate the variables or groups that create problems. When there are serious violations of the assumption of multivariate normality and the homogeneity of variances and covariances across groups, it may be possible to remedy these violations. Identification and removal of outliers can be helpful; decisions about procedures should be made prior to data collection and handling of outliers should be documented in the research report. If prior plans included these situations, data analysts might consider dropping individual cases with extreme scores (because these may have inflated variances or covariances in the groups that they belonged to), dropping individual variables (which have unequal variances or covariances across groups), or dropping a group (if that group has variances and/or covariances that are quite different from those in other groups). However, decisions about outliers should be based on plans made in advance. Dropping cases for which an analysis makes poor predictions may result in pruning the data to fit the model and can be a form of p-hacking.

10.4 ISSUES IN PLANNING A STUDY

A study that involves DA requires design decisions about the number and size of groups and also requires the selection of discriminating variables. The goal of DA may be the development of a statistical model that makes accurate predictions of group memberships, the development of a model that provides theoretically meaningful information about the nature of differences between groups, or both. In the first situation, the selection of discriminating variables may be driven more by practical considerations (e.g., what information is inexpensive to obtain), while in the second situation, theoretical issues may be more important in the selection of variables.

As in other one-way analyses, it is helpful, although not essential, to have equal numbers of cases in the groups. If the numbers in the groups are unequal, the researcher must choose between two different methods of classification of scores into predicted groups. The default method of classification in SPSS DA assumes that the model should classify the same proportion into each group regardless of the actual numbers; thus, if there are $k = 4$ groups, approximately one fourth or 25% of the cases will be classified as members of each group. An alternative method available in SPSS (called "priors = size") uses the proportions of cases in the sample as the basis for the classification rule; thus, if there were four groups with $n = 40$, 25, 25, and 10, then approximately 40% of the cases would be classified into Group 1, 25% into Group 2, and so forth.

The number of cases within each group should exceed the number of predictor variables (Tabachnick & Fidell, 2018). To obtain adequate statistical power, larger numbers of cases than this minimum sample size are desirable. If n is less than the number of variances and covariances that need to be calculated within each group, there are not enough degrees of freedom to calculate independent estimates of these terms, and the SCP matrix will be singular (i.e., it will have a determinant of 0). In practice, as long as a pooled covariance matrix is used, SPSS can still carry out a DA even if some groups have a singular **SCP** matrix or **S** variance/covariance matrix. When the within-group numbers are very small, it is difficult to assess violations of the assumption about the homogeneity of variance/covariance matrices across groups (i.e., Box's M test does not have good statistical power to detect violations of this assumption when the sample sizes within groups are small).

The second major design issue is the selection of discriminating variables. In applied research, it may be important to choose predictor variables that can be measured easily at low cost. In basic or theory-driven research, the choice of predictors may be based on theory. As in any multivariate analysis, the inclusion of "garbage" variables (i.e., variables that are unrelated to group membership and/or are not theoretically meaningful) can reduce statistical power and also make the contributions of other variables difficult or impossible to interpret.

10.5 EQUATIONS FOR DISCRIMINANT ANALYSIS

A DA produces several types of information about patterns in the data, and much of this information can be understood by analogy to the information we obtain from ANOVA and MR. The omnibus test statistic in DA that summarizes how well groups can be differentiated (using all discriminating variables and all discriminant functions) is called Wilks' Λ. Wilks' Λ is interpreted as the proportion of variance in discriminant function scores that is *not* predictable from group membership. Thus, Wilks' Λ is analogous to $(1 - \eta^2)$ in ANOVA or $(1 - R^2)$ in MR.

DA (like ANOVA) involves a partition of the total variances and covariances (for a set of X discriminating variables considered jointly as a set) into between-group and within-group variances and covariances. In one-way ANOVA, the partition of variance was as follows: $SS_{\text{total}} = SS_{\text{between}} + SS_{\text{within}}$. In DA, where **SCP** corresponds to a matrix that contains sums of squares and sums of cross products for all the predictor X variables, the corresponding matrix algebra expression is

$$\mathbf{SCP}_{\text{total}} = \mathbf{SCP}_{\text{between}} + \mathbf{SCP}_{\text{within}}. \tag{10.7}$$

The elements of $\mathbf{SCP}_{\text{total}}$ consist of the total sums of squares for each of the X's (in the diagonal elements) and the total sums of cross products for all pairs of X variables (in the off-diagonal elements) for the entire data set with all groups combined. Next, within each group (let the group number be indicated by the subscript i), we obtain $\mathbf{SCP}_i$. The elements of $\mathbf{SCP}_i$ within each group consist of the sums of squares of the X's just for scores within Group i and the sums of cross products for all pairs of X's just for scores within Group i. We obtain a reasonable estimate of the error term for our analysis, $\mathbf{SCP}_{\text{within}}$, by pooling, combining, or averaging these $\mathbf{SCP}_i$ matrices across groups; this pooling or averaging only makes sense if the assumption of homogeneity of $\boldsymbol{\Sigma}$ across groups is satisfied. We can compute the overall within-group variance/covariance matrix $\mathbf{S}_{\text{within}}$ from the **SCP** matrices within groups as follows:

$$\mathbf{SCP}_{\text{within}} = (\mathbf{SCP}_1 + \mathbf{SCP}_2 + \cdots + \mathbf{SCP}_k)/(n_1 + n_2 + \cdots + n_k - k), \tag{10.8}$$

where k is the number of groups. Note that this is analogous to the computation of SS_{within} in a one-way ANOVA, as a sum of $SS_1, SS_2, \ldots, SS_k$, or to the computation of s^2_{pooled} for a t test, where s^2_{pooled} is the weighted average of s^2_1 and s^2_2.

In univariate ANOVA, to judge the relative magnitude of explained versus unexplained variance, an F ratio is examined ($F = MS_{\text{between}}/MS_{\text{within}}$). However, in DA we consider a ratio of the determinant of $\mathbf{SCP}_{\text{within}}$ to the determinant of $\mathbf{SCP}_{\text{total}}$. It is necessary to summarize the information about variance contained in each of these **SCP** matrices; this is done by taking the determinant of each matrix. One formula for the computation of Wilks' Λ is

$$\Lambda = |\mathbf{SCP}_{\text{within}}|/|\mathbf{SCP}_{\text{total}}|. \tag{10.9}$$

Recall that the determinant of a matrix is a single value that summarizes the variance of the X's while taking into account the correlations or covariances among the X's (so that redundant information isn't double counted). If there is just one discriminating variable, note that Wilks' Λ becomes

$$\Lambda = SS_{\text{within}}/SS_{\text{total}}. \tag{10.10}$$

Wilks' Λ is equivalent to the proportion of unexplained, or within-groups, variance in a one-way ANOVA. In a one-way ANOVA, Λ is equivalent to $(1 - \eta^2)$, and $\eta^2 = SS_{\text{between}}/SS_{\text{total}}$.

Note that an F ratio has two *df* terms because its distribution varies as a function of the degrees of freedom for within groups and the degrees of freedom for between groups. A Wilks' Λ statistic has three different *df* terms, which are based on the three factors that are involved in a DA:

N, the total number of subjects or cases;

k, the number of groups; and

p, the number of discriminating variables.

In principle, one could look up critical values for Wilks' Λ, but it is unwieldy to use tables that depend on three different *df* terms. It is more convenient to convert Wilks' Λ into a more familiar test statistic. The SPSS discriminant analysis procedure converts the Λ to an approximate chi-square with $df = p \times (k - 1)$:

$$\chi^2 = -\{N - [(p + k)/2] - 1\}\ln \Lambda, \tag{10.11}$$

where N is the number of cases, p the number of variables, and k the number of groups.

Finally, we convert Wilks' Λ to a familiar effect size index, η^2:

$$\eta^2 = 1 - \Lambda. \tag{10.12}$$

To summarize, all the preceding equations provide information about the omnibus test for DA. If Wilks' Λ is reasonably small, and if the corresponding omnibus F is statistically significant, we can conclude that group membership can be predicted at levels significantly better than chance when all the discriminant functions in the analysis are considered as a set.

Another question that may be of interest is the following: Which of the discriminating variables provide useful information about group membership? To answer this question, we need to look at sizes and signs of the discriminant function coefficients. The discriminant function coefficients for discriminant function D_1 were given in Equation 10.4. More generally, there can be several discriminant functions, denoted $D_1, D_2, \ldots, D_{k-1}$. For each discriminant function D_i, there are separate discriminant function coefficients, as shown in the following expression: $D_i = d_{i1}z_1 + d_{i2}z_2 + \ldots + d_{ip}z_p$. That is, each discriminant function D_i is constructed from a different weighted linear combination of the *z* scores $z_1, z_2, \ldots, z_p$. For each discriminant function that differs significantly across groups (i.e., it is significantly predictive of group membership), the data analyst may want to interpret or discuss the meaning of the discriminant function by describing which variables were given the largest weights in computation of the discriminant function. These *d* coefficients are analogous to beta coefficients in an MR; they provide the optimal weighted linear composite of the *z* scores on the *X* predictor variables; that is, the *d* coefficients are the set of weights associated with the *z* scores on the predictor variables for which the value of each new discriminant function, D_i, has the largest possible SS_{between} and the smallest possible SS_{within}. Other factors being equal, an *X* variable that has a *d* coefficient that is large (in absolute value) is judged to provide unique information that helps predict group membership when other discriminating variables are statistically controlled.

The discriminant function coefficients for a DA are obtained using a type of matrix algebra computation that we have not encountered before: the eigenvalue/eigenvector problem (see Appendix 10A). An eigenvector is a vector or list of coefficients; these correspond to the weights that are given to predictor variables (e.g., discriminant function coefficients, factor loadings) when we form different weighted linear combinations of variables from a set of *X*'s. The elements of the eigenvector must be rescaled to be used as discriminant function coefficients. Each eigenvector has one corresponding eigenvalue. In DA, the **eigenvalue (λ)** for a particular discriminant function corresponds to the value of the goodness-of-fit measure $SS_{\text{between}}/SS_{\text{within}}$. In other words, the eigenvalue that corresponds to a discriminant function gives us the ratio of SS_{between} to SS_{within} for a one-way ANOVA on scores on that discriminant function.

When properly rescaled, an eigenvector gives the *d* coefficients for one standardized discriminant function; like beta coefficients in an MR, these are applied to *z* scores. The eigenvalue (λ) yields an effect size measure (the **canonical correlation [r_c]**) that describes how strongly its corresponding discriminant function scores are associated with group membership. The term *canonical* generally indicates that the correlation is between weighted linear composites that include multiple predictor and/or multiple outcome variables. The canonical correlation (r_c) between scores on a discriminant function and group membership is analogous to η in a one-way ANOVA. The subscript *i* is used to identify which of the discriminant functions each canonical correlation, r_c, is associated with; for example, r_{c2} is the canonical correlation that describes the strength of association between scores on D_2 and group membership. The eigenvalue λ_i for each discriminant function can be used to calculate a canonical correlation r_{ci}. For example, this canonical correlation r_{c1} describes how strongly scores on D_1 correlate with group membership:

$$r_{c1} = \sqrt{\frac{\lambda_1}{1+\lambda_1}}. \tag{10.13}$$

The matrix algebra for the eigenvalue problem generates solutions with the following useful characteristics. First, the solutions are rank ordered by the size of the eigenvalues; that is, the first solution corresponds to the discriminant function with the largest possible λ_1

(and r_{c1}), the second corresponds to the next largest possible λ_2 (and r_{c2}), and so on. In addition, scores on the discriminant functions are constrained to be uncorrelated with each other; that is, the scores on D_1 are uncorrelated with the scores on D_2 and with scores on all other discriminant functions.

Because the discriminant function scores are uncorrelated with one another, it is relatively easy to combine the explained variance across discriminant functions and, therefore, to summarize how well an entire set of discriminant functions is associated with or predictive of group memberships. One way to decide how many discriminant functions to retain for interpretation is to set an alpha level (such as $\alpha = .05$), look at the p values associated with tests of sets of discriminant functions, and drop discriminant functions from the interpretation if they do not make a statistically significant contribution to the prediction of group membership, as discussed earlier in the section about "dimension reduction." Thus, in some research situations, it may be possible to discriminate among groups adequately using only the first one or two discriminant functions.

The notation used for the standardized discriminant function coefficients for each discriminant function (the subscript i is used to denote Discriminant Functions 1, 2, . . . , i, and the second subscript on each d coefficient indicates which of the X predictor variables that coefficient is associated with) is as follows:

$$D_i = d_{i1}z_1 + d_{i2}z_2 + \cdots + d_{ip}z_p. \tag{10.14}$$

These d coefficients, like beta coefficients in an MR, tend to lie between −1 and +1 (although they may lie outside this range). To provide a more concrete example of the optimal weighted linear combination represented by Equation 10.14, in the empirical example at the end of the chapter, we will compute and save discriminant function scores that are calculated from the formula for D_1 and then conduct a one-way ANOVA to see how scores on this new variable D_1 vary across groups. Comparison of the output of this one-way ANOVA with the DA results help make it clear that the DA involves comparison of the means for "new" outcome variables—that is, scores on one or more discriminant functions across groups.

The d coefficients can be interpreted like beta coefficients; each beta coefficient tells us how much weight is given to the z score on each X discriminating variable when all other X variables are statistically controlled. SPSS can also report raw score discriminant function coefficients; these are not generally used to interpret the nature of group differences, but they may be useful if the goal of the study is to generate a prediction formula that can be used to predict group membership easily from raw scores for future cases. SPSS also reports **structure coefficients**; that is, for each discriminating variable, its correlation with each discriminant function is reported. Some researchers base their discussion of the importance of discriminating variables on the magnitudes of the standardized discriminant function coefficients, while others prefer to make their interpretation on the basis of the pattern of structure coefficients. See Appendix 10B for discussion of additional equations in DA.

10.6 CONCEPTUAL BASIS: FACTORS THAT AFFECT THE MAGNITUDE OF WILKS' Λ

If the groups in a DA have high between-groups variance (and relatively low within-groups variance) on one or more of the discriminating variables, it is possible that the overall Wilks' Λ (the proportion of variance in discriminant scores that is not associated with group membership) will be relatively small. However, other factors may influence the size of Wilks' Λ, including the strength and signs of correlations among the discriminating variables and whether combinations of variables differentiate groups much more effectively than any single variable (for a more detailed and technical discussion of this issue, see Bray & Maxwell, 1985). Figure 10.1 shows a graphic example in which three groups are clearly separated in a territorial map (scores on D_1 and scores on D_2). This clear spatial separation corresponds to a situation in which Wilks' Λ will turn out to be relatively small. On the other hand, Figure 10.2 illustrates a hypothetical situation in which there is no clear spatial separation among the three groups. The overlap among the groups in Figure 10.2 corresponded to a research situation in which Wilks' Λ would be quite large.

10.7 EFFECT SIZE

One question that can be answered using DA is the following: How well can group membership be predicted overall, using all discriminating variables and all discriminant functions? To answer this question, we look at the overall Wilks' Λ to find the corresponding effect size, η^2:

$$\eta^2 = 1 - \Lambda. \tag{10.15}$$

For each discriminant function, we can compute an effect size index (a canonical correlation, r_c) from its eigenvalue, λ_i:

$$r_{c1} = \sqrt{\frac{\lambda_1}{1+\lambda_1}}. \tag{10.16}$$

There is no effect size index for each individual discriminating variable that can be interpreted as a proportion of variance (i.e., nothing like the sr^2_{unique} effect size estimate for each predictor in a regression). It is possible to evaluate the relative importance of the discriminating variables by looking at their structure coefficients, their standardized discriminant function coefficients, and possibly also the univariate *F*'s (from a one-way ANOVA on each discriminating variable). However, these comparisons are qualitative, and the cutoff values used to decide which *X* predictors make "strong" predictive contributions are arbitrary.

10.8 STATISTICAL POWER AND SAMPLE SIZE RECOMMENDATIONS

Because the size of Wilks' Λ (and its corresponding χ^2 or F test of significance) is potentially influenced by such a complex set of factors (including not only the strength of the univariate associations of individual predictors with group membership but also the signs and magnitudes of correlations among predictors), it is difficult to assess the sample size requirements for adequate statistical power. Stevens (2009) cited Monte Carlo studies that indicate that estimates of the standardized discriminant function coefficients and the structure coefficients are unstable unless the sample size is large. Accordingly, Stevens recommended that the total number of cases for a DA should be at least 20 times as large as p, the number of discriminating variables. Another issue related to sample size is the requirement that the number of cases within each group be larger than the number of predictor variables, p (Tabachnick & Fidell, 2018).

Another set of guidelines for minimum sample size in a k group situation with p outcome variables is provided in the next chapter in a power table adapted from Stevens (2009) that provides minimum suggested sample sizes per group to achieve a power of .70 or greater using $\alpha = .05$ and three to six discriminating variables. Recommended sample size increases as the expected effect size becomes smaller. A more detailed treatment of this problem is provided by Lauter (1978).

10.9 FOLLOW-UP TESTS TO ASSESS WHAT PATTERN OF SCORES BEST DIFFERENTIATES GROUPS

If the overall DA model is not significant (i.e., if the chi-square associated with the overall Wilks' Λ for the set of all discriminant functions is nonsignificant), then it does not make sense to do follow-up analyses to assess which pairs of groups differed significantly and/or which discriminating variables provided useful information. If the overall model is significant, however, these follow-up analyses may be appropriate.

If there are only two groups in the DA, no further tests are required to assess contrasts. However, if there are more than two groups, post hoc comparisons of groups may be useful to identify which pairs of groups differ significantly in their scores on one or more discriminant functions. This can be done more easily using the multivariate ANOVA (MANOVA) procedure introduced in the next chapter. Analysts may also want to evaluate which among the X predictor variables are predictive of group membership. A qualitative evaluation of the discriminant function coefficients and structure coefficients can be used to answer this question; variables with relatively large discriminant coefficients (in absolute value) are more predictive of group membership in a multivariate context, where correlations with the other X predictor variables are taken into consideration. It is also possible to examine the one-way ANOVA in which the means on each X variable are compared across groups (not controlling for the intercorrelations among the X variables), but these ANOVA results need to be interpreted with great caution for two reasons. First, the significance or p values reported for these ANOVAs by SPSS are not adjusted to limit the risk for Type I error that arises when a large number of significance tests are performed. The easiest way to control for inflated risk for Type I error that is likely to arise in this situation, where many significance tests are performed, is to use the Bonferroni correction. That is, for a set of q significance tests, to obtain an experiment-wise alpha (EW_α) of .05, set the per comparison alpha (PC_α) at $.05/q$. Second, these tests do not control for the intercorrelations among predictor variables; an individual X predictor variable can have a quite different relationship with an outcome variable (both in magnitude and in sign) when you do control for other predictor variables than when you do not control for those other variables.

To assess the predictive usefulness of individual variables, the analyst may look at three types of information, including the magnitude and sign of the standardized discriminant function coefficients, the magnitude and sign of the structure coefficients, and the significance of the univariate F for each variable, along with the unadjusted group means for variables with significant overall F's. The conclusions about the importance of an individual X_i predictor variable (and even about the nature of the relation between X_i and group membership, e.g., whether membership in one particular group is associated with high or low scores on X_i) may differ across these three sources of information. It is fairly easy to draw conclusions and summarize results when these three types of information appear to be consistent. When they are apparently inconsistent, the researcher must realize that the data indicate a situation where X_i's role as a predictor of group membership is different when other X predictors are statistically controlled (in the computation of discriminant function coefficients) compared with the situation when other X predictors are ignored (as in the computation of the univariate ANOVAs). Review discussions of statistical control in previous chapters, if necessary, to understand why a variable's predictive usefulness may change when other variables are controlled and to get an idea of possible explanations (such as spuriousness, confounding, and suppression) that may need to be considered.

Note that if none of the individual X's is a statistically significant predictor of group membership, the optimal linear combinations of X's that are formed for use as discriminant functions may not differ significantly across groups either. However, it is sometimes possible to discriminate significantly between groups using a combination of scores on two or more predictors even when no single predictor shows strong differences across groups (as in the example that appeared in Figures 10.3 and 10.4). On the other hand, usually, when there is a nonsignificant overall result for the DA, it generally means that none of the individual predictors was significantly related to group membership. However, there can be situations in which one or two individual predictors are significantly related to group membership, and yet, the overall DA is not significant. This result may occur when other "garbage" variables that are added to the model decrease the error degrees of freedom and reduce statistical power.

In clinical or practical prediction situations, it may be important to report the specific equation(s) used to construct discriminant function scores (from standardized *z* scores and/or from raw scores) so that readers can use the predictive equation to make predictions in future situations where group membership is unknown. (An additional type of discriminant function coefficient that can be obtained, the Fisher linear discriminant function, is described in Appendix 10B. The Fisher coefficients are not generally used to interpret the nature of group differences, but they provide a simpler method of prediction of group membership for each individual score, particularly in research situations where there are more than two categories or groups.) For example, a personnel director may carry out research to assess whether job outcome (success or failure) can be predicted from numerous tests that are given to job applicants. On the basis of data collected from people who are hired and observed over time until they succeed or fail at the job, it may be possible to develop an equation that does a reasonably good job of predicting job outcomes from test scores. This equation may then be applied to the test scores of future applicants to predict whether they are more likely to succeed or fail at the job in the future. See Appendix 10B for a further discussion of the different types of equations used to predict group membership for individual cases.

10.10 RESULTS

High school seniors were given a battery of achievement and aptitude tests; they were also asked to indicate their choices of future career. Three groups were included: Group 1, business careers (n = 98); Group 2, medicine (n = 27); and Group 3, teaching or social work (n = 39). Scores on the following six tests were used as discriminating variables: English, reading, mechanical reasoning, abstract reasoning, math, and office skills. The valid number for this analysis was N = 164. For better statistical power, it would be preferable to have larger numbers in the groups than the numbers in the data set used in this example.

The SPSS menu selections to access the DA procedure appear in Figure 10.5. From the top-level menu in the SPSS Data View, make the following selections: <Analyze> → <Classify> → <Discriminant>. The name of the categorical group membership outcome variable, cargroup, is placed in the box under the heading "Grouping Variable." The Define Range button was used to open up an additional window (not shown here) where the range of score codes for the groups included in the analysis was specified; scores that ranged from 1 to 3 on cargroup were included in this analysis. The decision to enter all predictor variables in one step was indicated by clicking the radio button marked "Enter independents together" in the DA dialog box in Figure 10.6. The Statistics button was clicked so that additional optional statistics could be requested.

From the list of optional statistics, the following selections were made (see Figure 10.7): "Means," "Univariate ANOVAs," "Box's M," and "Within-groups correlation." The univariate ANOVAs provide information about the differences in means across groups separately for each independent variable. This information may be useful in characterizing group differences, but note that the significance tests associated with these univariate ANOVAs are not corrected for the inflated risk for Type I error that arises when multiple tests are done; the researcher may want to use the Bonferroni correction when assessing these univariate differences. Box's M test provides a significance test of the assumption that the variance/covariance

Figure 10.5 SPSS Menu Selections for Discriminant Analysis Procedure

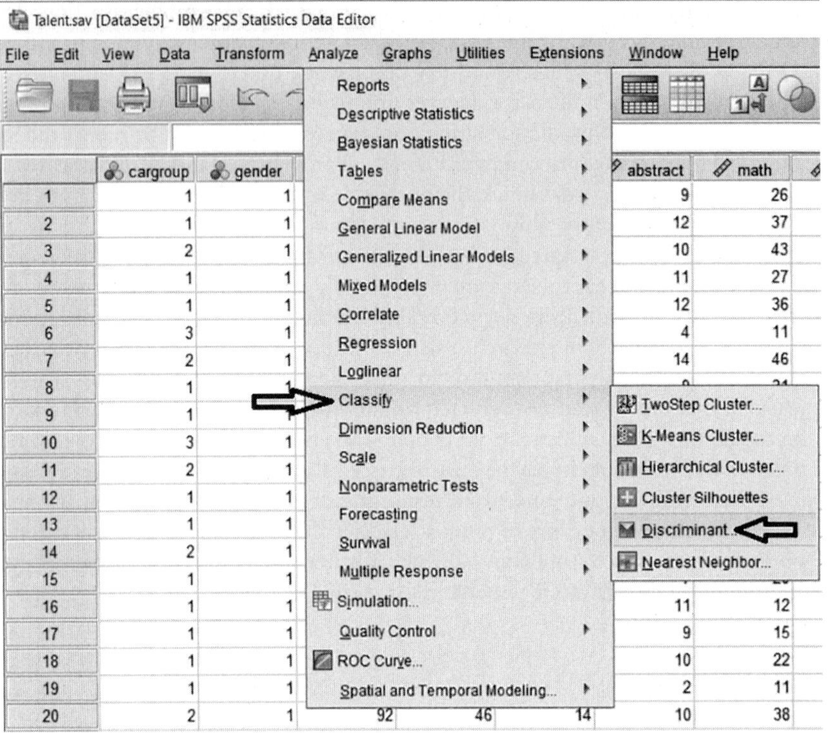

Figure 10.6 Main SPSS Dialog Box for Discriminant Analysis Procedure

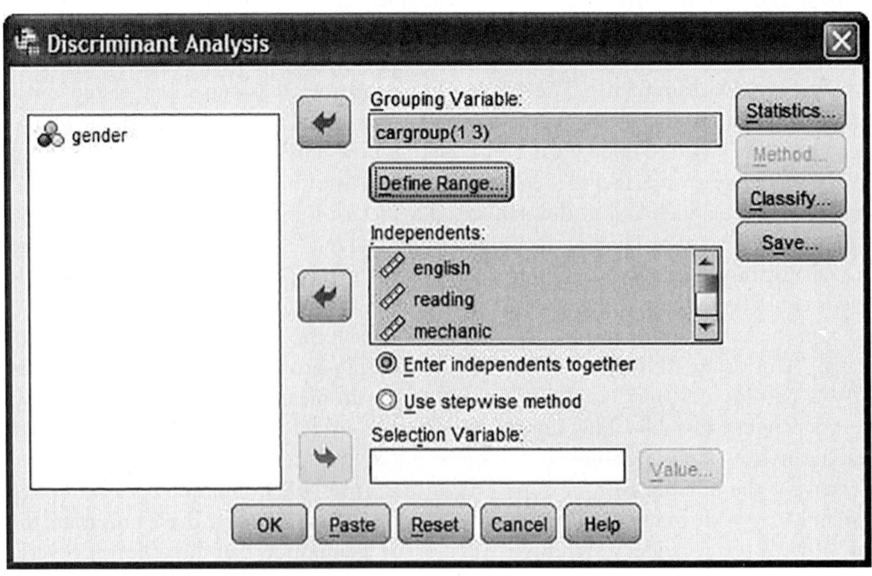

Figure 10.7 Statistics for Discriminant Analysis

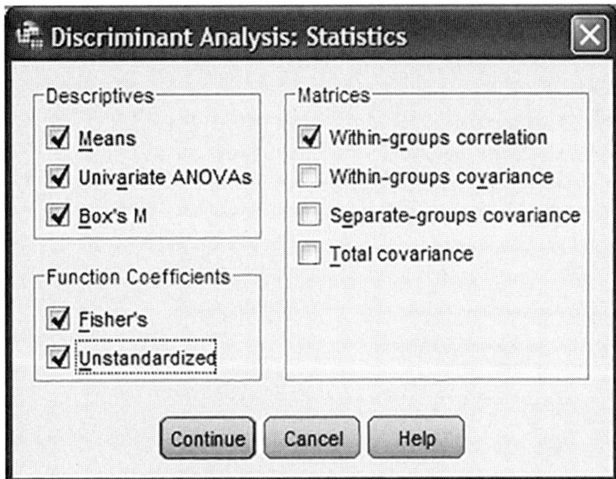

matrix Σ is homogeneous across groups. If Box's M indicates serious problems with the assumption of homogeneity of the variance/covariance matrices, it may be useful to request the separate-groups covariance matrices (these do not appear here). Examination of these matrices may help identify the variables and/or groups for which the inequalities are greatest.

Back in the main dialog box for DA, the Classify button was clicked to open up a dialog box to request additional information about the method used for group classification and about the accuracy of the resulting predicted group memberships. From the classification method options in Figure 10.8, the selections made were as follows: To use prior probabilities as a basis for the classification rule (instead of trying to classify equal proportions of cases into each of the three groups), the radio button "Compute from group sizes" was selected. That is, instead of trying to classify 33% of the participants into each of the three career groups (which would yield poor results, because there were so many more participants in Group 1 than in Groups 2 and 3), the goal was to classify about 98/164 = 60% into Group 1, 27/164 = 16% into Group 2, and 39/164 = 24% into Group 3. Under the heading "Display," the option "Summary table" was checked. This requests a contingency table that shows actual versus predicted group membership; from this table, it is possible to judge what proportion of members of each group was correctly classified and, also, to see what types of classification errors (e.g., classifying future medicine aspirants as business career aspirants) were most frequent. Under the heading "Use Covariance Matrix," the option "Within-groups" was chosen; this means that the $\mathbf{SCP}_{\text{within}}$ matrix was pooled or averaged across groups. (If there are serious inequalities among the **SCP** matrices across groups, it may be preferable to select "Separate-groups," i.e., to keep the **SCP** matrices separate instead of pooling them into a single estimate.) Under the heading "Plots," "Territorial map" was selected. This provides a graphic representation of how the members of the groups differ in their scores on Discriminant Function 1 (D_1 scores on the X axis) and Discriminant Function 2 (D_2 scores on the Y axis). A territorial map is useful because it helps us understand the nature of the differences among groups on each discriminant function; for example, what group(s) differed most clearly in their scores on D_1? Finally, the dialog box for the <Save> command was opened, and a checkbox was used to request that the discriminant function scores should be saved as new variables in the SPSS worksheet (see Figure 10.9). The SPSS syntax that resulted from these menu selections is shown in Figure 10.10.

Figure 10.8 Classification Methods for Discriminant Analysis

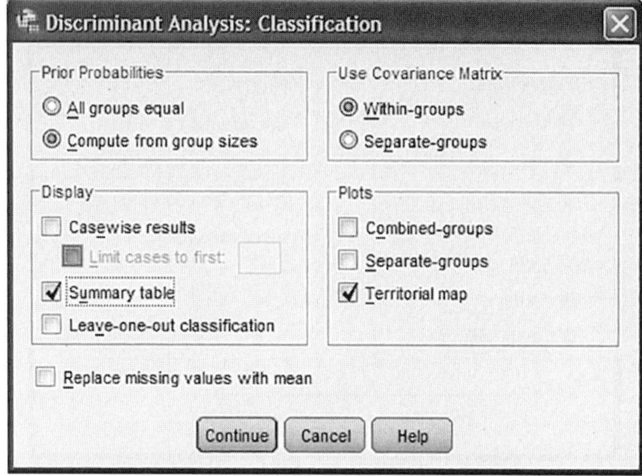

Figure 10.9 Command to Save the Computed Scores on Discriminant Functions 1 and 2 as New Variables in the SPSS Worksheet

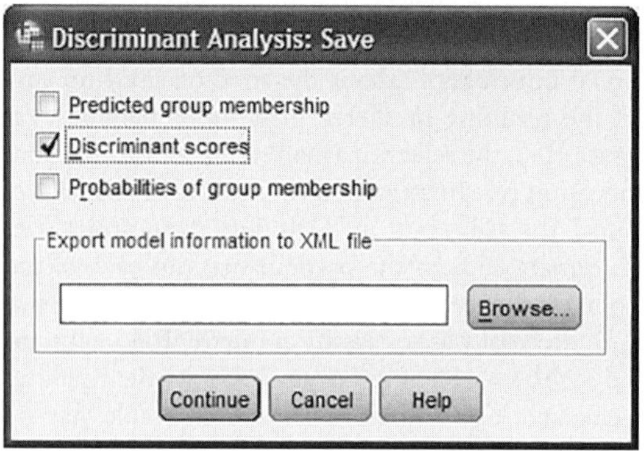

The output from the DA specified by the previous menu selections appears in Figures 10.11 through 10.23. The first two parts of the output report the simple univariate ANOVA results. That is, a univariate ANOVA was done to assess whether the means on each of the individual variables (English, reading, mechanical reasoning, abstract reasoning, math, and office skills) differed across the three levels of the career group choice outcome variable. The mean and standard deviation for each variable in each group appear in Figure 10.11. The F tests (which assess whether means differ significantly across the three groups, separately for each variable) appear in Figure 10.12. From the information in Figures 10.11 and 10.12, we can evaluate the group differences on each variable in isolation from the other variables that are included in the DA. For example, if we do not statistically control for the other five predictor variables, what is the nature of differences on English test scores across the three groups? When English scores are assessed in isolation, there is no significant difference in mean English scores across the three career groups, $F(2, 161) = .219, p = .803$.

Figure 10.10 **SPSS Syntax for Discriminant Analysis to Predict Cargroup (1 = Business, 2 = Medical, 3 = Teaching/Social Work) From Scores on English, Reading, Mechanical Reasoning, Abstract Reasoning, Math, and Office Skills, With All Predictor Variables Entered in One Step (Using Group Size as the Basis for Prior Probability of Group Membership)**

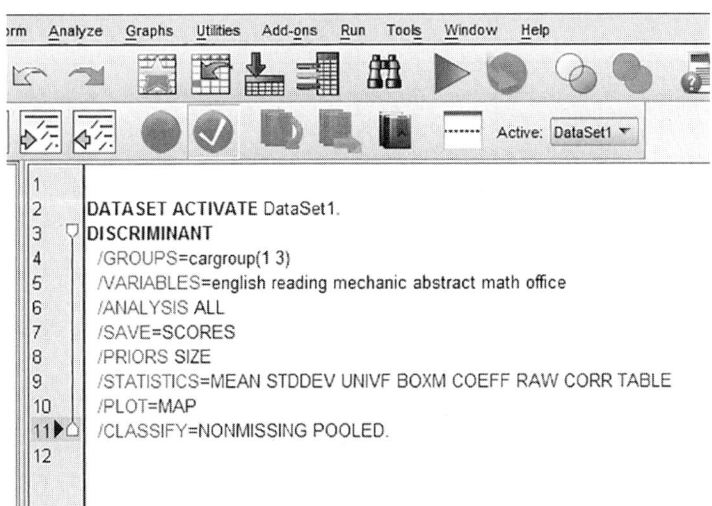

Reporting multiple significance tests (as in Figure 10.12) leads to an increased risk for Type I error, and the p values that appear in the SPSS output are not adjusted in any way to correct for this problem. It is advisable to make an adjustment to the alpha levels that are used to assess the individual F's to avoid an increased risk for Type I error. If we want an overall risk for Type I error of $\alpha = .05$ for the entire set of six F tests that appear in Figure 10.12, we can limit the risk for Type I error by using the Bonferroni correction. For each individual F test, an appropriate PC_α level that is used to assess statistical significance can be obtained by taking the overall experiment-wise Type I error rate, $EW_\alpha = .05$, and dividing this $EW\alpha$ by the number of significance tests (q) that are performed. In this example, the number of significance tests included in Figure 10.12 is $q = 6$, so the Bonferroni-corrected PC_α level for each test would be $EW\alpha/q = .05/6 = .008$. In other words, when we assess the statistical significance of each univariate F test included in Figure 10.12, we require that the p value obtained be less than .008 before we judge a univariate F test to be statistically significant. By this criterion, only two of the individual variables (math and office skills) showed statistically significant differences across the three career groups, with $F(2, 161) = 10.53$, $p < .001$, and $F(2, 161) = 23.32$, $p < .001$, respectively. Wilks' Λ for this test is .884; from this we can compute $\eta^2 = 1 - \Lambda$, so $\eta^2 = .116$. That is, almost 12% of the variance in math scores is associated with between-group differences. Returning to Figure 10.11 to find the means on math for these three groups, we find that mean math scores were higher for Group 2, medicine ($M_2 = 26.37$), and Group 3, teaching or social work ($M_3 = 25.59$), than for Group 1, business ($M_1 = 20.05$). In the absence of any other information and without statistically controlling for the correlation between math and other predictors, we would tend to predict that a student with low scores on math might choose business as a future career (Group 1) rather than medicine or teaching or social work.

To summarize the results so far, if we conduct a univariate ANOVA for each of the six variables, only two of the variables (math and office skills) showed significant differences

Figure 10.11 Selected Output From a Discriminant Analysis (All Predictor Variables Entered in One Step); Univariate Means on Each Variable for Each Group

Group Statistics

cargroup		Mean	Std. Deviation	Valid N (listwise) Unweighted	Weighted
business	english	88.44	8.688	98	98.000
	reading	32.44	8.489	98	98.000
	mechanic	9.12	3.681	98	98.000
	abstract	9.08	2.477	98	98.000
	math	20.05	7.159	98	98.000
	office	28.84	8.281	98	98.000
medicine	english	87.15	9.326	27	27.000
	reading	34.15	10.524	27	27.000
	mechanic	10.22	5.213	27	27.000
	abstract	9.70	2.614	27	27.000
	math	26.37	9.958	27	27.000
	office	18.37	9.199	27	27.000
teach_social	english	88.72	13.330	39	39.000
	reading	34.90	8.926	39	39.000
	mechanic	9.49	3.094	39	39.000
	abstract	9.69	3.229	39	39.000
	math	25.59	8.696	39	39.000
	office	20.62	8.583	39	39.000
Total	english	88.29	10.024	164	164.000
	reading	33.30	8.961	164	164.000
	mechanic	9.39	3.846	164	164.000
	abstract	9.33	2.695	164	164.000
	math	22.41	8.503	164	164.000
	office	25.16	9.601	164	164.000

across career groups. Looking back at the group means in Figure 10.11, we can see that Group 1 (business careers) scored lower than Groups 2 and 3 (medicine, teaching or social work) on math skills and higher than Groups 2 and 3 on office skills. If we used only our univariate test results, therefore, we could make the following preliminary guesses about group membership: Students with low scores on math and high scores on office skills may be more likely to be in Group 1, business. However, subsequent DA will provide us with more information about the nature of the differences among groups when intercorrelations among the p variables are taken into account; it will also provide us with summary information about our ability to predict group membership on the basis of patterns of scores on the entire set of p predictor variables.

These univariate results are sometimes, although not always, included as part of the DA results. They provide a basis for comparison; the discriminant function coefficients tell us how much information about group membership is contributed by each predictor variable

Figure 10.12 Univariate ANOVA Results From the Discriminant Analysis

Tests of Equality of Group Means

	Wilks's Lambda	F	df1	df2	Sig.
english	.997	.219	2	161	.803
reading	.985	1.196	2	161	.305
mechanic	.989	.880	2	161	.417
abstract	.987	1.028	2	161	.360
math	.884	10.529	2	161	.000
office	.775	23.319	2	161	.000

Figure 10.13 Pooled Within-Groups Correlation Matrix

Pooled Within-Groups Matrices

		english	reading	mechanic	abstract	math	office
Correlation	english	1.000	.647	.212	.492	.537	-.006
	reading	.647	1.000	.380	.489	.586	-.140
	mechanic	.212	.380	1.000	.378	.551	-.259
	abstract	.492	.489	.378	1.000	.529	-.018
	math	.537	.586	.551	.529	1.000	-.217
	office	-.006	-.140	-.259	-.018	-.217	1.000

when we statistically control for correlations with all other predictors; the univariate results tell us how each variable is related to group membership when we do *not* control for other predictors. It can be useful to examine whether the nature of group differences changes (or remains essentially the same) when we control for correlations among predictors.

Figure 10.13 shows the correlations among all predictor variables; these correlations were calculated separately within each of the three groups, then pooled (or averaged) across groups. This information is analogous to the information obtained from the matrix of correlations among predictors in an MR. We do not want to see extremely high correlations among predictor variables because this indicates that the contributions of individual predictors can't be distinguished. On the other hand, if the predictors are almost entirely uncorrelated with one another, the multivariate analysis (DA) may not yield results that are very different from a set of univariate results. In rare situations where predictor variables are completely uncorrelated, it may be appropriate to just look at them individually using univariate methods. DA is most appropriate when there are weak to moderate correlations among predictors. In this case, none of the correlations among predictors exceeded .6 in absolute value; thus, there was no serious problem with multicollinearity among predictors.

Figure 10.14 shows the results of Box's M test. Recall that the null hypothesis that this statistic tests corresponds to the assumption of homogeneity of the **SCP** matrices across groups:

$$H_0: \boldsymbol{\Sigma}_1 = \boldsymbol{\Sigma}_2 = \boldsymbol{\Sigma}_3.$$

We would prefer that this test be nonsignificant. If Box's M is significant, it indicates a violation of an assumption that may make the results of DA misleading or uninterpretable. In this instance, Box's $M = 63.497$, and the approximate F for Box's $M = F(42, 20{,}564) = 1.398$,

Figure 10.14 Box's *M* Test of Equality of Variance/Covariance Matrices

Log Determinants

cargroup	Rank	Log Determinant
business	6	19.295
medicine	6	20.257
teach_social	6	20.501
Pooled within-groups	6	20.130

The ranks and natural logarithms of determinants printed are those of the group covariance matrices.

Test Results

Box's M		63.497
F	Approx.	1.398
	df1	42
	df2	20563.91
	Sig.	.045

Tests null hypothesis of equal population covariance matrices.

$p = .045$. If the alpha level used to evaluate the significance of Box's M is set at $\alpha = .01$ (because the *df* associated with Box's M test are so high), this value of Box's M indicates no significant violation of the assumption of homogeneous variance/covariance matrices.

As in earlier situations where we tested for violations of assumptions, there is a dilemma here. The larger the number of cases, the less problematic the violations of assumptions tend to be. However, the larger the number of cases, the more statistical power we have for the detection of violations of assumptions. That is, in studies with relatively large numbers, we have a better chance of judging violations of assumptions to be statistically significant, and yet, these violations may create serious problems when N is large. When N is rather large, it is reasonable to use a relatively small alpha level to judge the significance of tests of assumptions; in this case, with $N = 164$, $\alpha = .01$ was used as the criterion for significance of Box's M.

If Box's M test suggests a serious violation of the assumption of the equality of the Σ's, the data analyst has several options. The application of log transformations to non-normally distributed variables (as recommended by Tabachnick & Fidell, 2018) may be an effective way of reducing differences in the variances of X across groups and the impact of multivariate outliers. An alternative statistical analysis that makes less restrictive assumptions about patterns in the data (such as binary logistic regression) may be preferred in situations where Box's M suggests serious violations of assumptions. In this example, none of these potential remedies was applied; students may wish to experiment with the data to see whether any of these changes would correct the violation of the assumption of homogeneity of variance and covariance across groups.

The data included three groups based on the choice of future career (Group 1 = business, Group 2 = medicine, Group 3 = teaching or social work). There were six discriminating variables: scores on achievement and aptitude tests, including English, reading, mechanical reasoning, abstract reasoning, math, and office skills. The goal of the analysis was to assess whether future career choice was predictable from scores on some or all of these tests. The first DA was run with all six predictor variables entered in one step. The main results for the overall DA (with all the six predictor variables entered in one step) appear in Figure 10.15. In this example, with k = number of groups = 3 and p = number of discriminating variables = 6, there were two discriminant functions (the number of discriminant functions

Figure 10.15 Summary of Canonical Discriminant Functions

Eigenvalues

Function	Eigenvalue	% of Variance	Cumulative %	Canonical Correlation
1	.448[a]	98.4	98.4	.556
2	.007[a]	1.6	100.0	.085

a. First 2 canonical discriminant functions were used in the analysis.

Wilks's Lambda

Test of Function(s)	Wilks's Lambda	Chi-square	df	Sig.
1 through 2	.685	59.879	12	.000
2	.993	1.156	5	.949

that can be calculated is usually determined by the minimum of $k - 1$ and p). The panel headed "Wilks's Lambda" contains significance tests for sets of discriminant functions. The first row, "1 through 2," tests whether group membership can be predicted using D_1 and D_2 combined; that is, it provides information about the statistical significance of the entire model, using all discriminant functions and predictor variables. For this omnibus test, $\Lambda = .685$ (and thus, $\eta^2 = 1 - \Lambda = .315$). That is, about 32% of the variance on the first two discriminant functions (considered jointly) is associated with between-groups differences. This was converted to a chi-square statistic, $\chi^2(12) = 59.88, p < .001$; thus, the overall model, including both Discriminant Functions 1 and 2, significantly predicted group membership.

To answer the question about whether we need to include all the discriminant functions in our interpretation or just a subset of the discriminant functions, we look at the other tests of subsets of discriminant functions. In this example, there is only one possibility to consider; the second row of the Wilks' Λ table tests the significance of Discriminant Function 2 alone. For D_2 alone, $\Lambda = .993$, $\chi^2(5) = 1.16, p = .949$. Thus, D_2 alone did not contribute significantly to the prediction of group membership. We can limit our interpretation and discussion of group differences to the differences that involve D_1. Note that we do not have a direct significance test for D_1 alone in this set of DA results.

Additional information about the relative predictive usefulness of D_1 and D_2 is presented in the panel headed "Eigenvalues." For D_1, the eigenvalue $\lambda_1 = .448$ (recall that this corresponds to $SS_{\text{between}}/SS_{\text{within}}$ for a one-way ANOVA comparing groups on D_1 scores). This eigenvalue can be converted to a canonical correlation:

$$r_c = \sqrt{\frac{.448}{1+.448}} = .556.$$

This r_c is analogous to an η correlation between group membership and scores on a quantitative variable in one-way ANOVA. If r_c is squared, we obtain information analogous to η^2, that is, the proportion of variance in scores on D_1 that is associated with between-group differences. It is clear that in this case, D_1 is strongly related to group membership ($r_{c1} = .556$), while D_2 is very weakly (and nonsignificantly) related to group membership ($r_{c2} = .085$). The total variance that can be predicted by D_1 and D_2 combined is obtained by summing the eigenvalues (.448 + .007 = .455). The manner in which this predicted variance should be apportioned between D_1 and D_2 is assessed by dividing each eigenvalue by this sum of eigenvalues, that is, by the variance that can be predicted by

D_1 and D_2 combined (.455); .448/.455 = .985 or 98.5% of that variance is due to D_1. It is clear, in this case, that D_1 significantly predicts group membership; D_2 provides very little additional information, and thus, D_2 should not be given much importance in the interpretation.

In the following written results, only the results for D_1 are included. To see what information was included in D_1, we look at the table in Figure 10.16. The values in this table correspond to the *d* coefficients in Equation 10.4. Column 1 gives the coefficients that provide the optimal weighted linear combination of *z* scores that are used to construct scores on D_1; column 2 provides the same information for D_2. These *d* coefficients are analogous to beta coefficients in an MR; they tell us how much weight or importance was given to each *z* score in calculating the discriminant function when we control for or partial out correlations with other predictors. No tests of statistical significance are provided for these coefficients. Instead, the data analyst typically decides on an (arbitrary) cutoff value; a variable is viewed as useful if its standardized discriminant function coefficient exceeds this arbitrary cutoff (in absolute value). In this example, the arbitrary cutoff value used was .5. By this standard, just two of the six predictor variables were given a substantial amount of weight when computing D_1: math (d_{1math} = –.794) and office skills ($d_{1office}$ = +.748). In some data analysis situations, it might be reasonable to choose a lower "cutoff" value when deciding which variables make an important contribution, for example, discriminant coefficients that are above .4 or even .3 in absolute value. In this situation, use of such low cutoff values would result in judging English and mechanical reasoning as "important" even though neither of these variables had a significant univariate *F*.

It is important to understand that, in DA as in MR, the strength of the association and even the direction of the association between scores on each predictor variable and the outcome variable can be quite different when the association is evaluated in isolation than when controlling for other predictor variables included in the analysis. In some situations, the same variables are "important" in both sets of results; that is, the magnitude and sign of discriminant function coefficients are consistent with the magnitude and direction of differences between group means in univariate ANOVAs.

In this research example, the combination of variables on which the three groups showed maximum between-group differences was D_1; high scores on D_1 were seen for students who had high scores on office skills (which had a positive coefficient) and low scores on math (which had a negative coefficient for D_1).

Figure 10.17 reports another type of information about the relationship between each individual predictor variable and each discriminant function coefficient; this is called the "structure matrix." The values in this table provide information about the correlation between each discriminating variable and each discriminant function. For example, for D_1 (Discriminant

Figure 10.16 Standardized Canonical Discriminant Function Coefficients

Standardized Canonical Discriminant Function Coefficients

	Function	
	1	2
english	.420	.281
reading	.015	.533
mechanic	.443	-.937
abstract	-.116	-.027
math	-.794	.243
office	.748	.098

Function 1), the two variables that had the largest correlations with this discriminant function were office skills and math; the correlation between scores on office skills and scores on D_1 was +.80, while the correlation between scores on math and scores on D_1 was −.54. These correlations do not have formal significance tests associated with them; they are generally evaluated relative to some arbitrary cutoff value. In this example, an arbitrary cutoff of .5 (in absolute value) was used. By this criterion, the two variables that were highly correlated with D_1 were office skills (with a correlation of +.80) and math (with a correlation of −.54). Note that the asterisks next to these structure coefficients do *not* indicate statistical significance. For each variable, the structure coefficient marked with an asterisk was just the largest correlation when correlations of variables were compared across discriminant functions.

Figure 10.18 reports the coefficients that are used to construct discriminant function scores from raw scores on the X_i predictor variables (see Appendix 10B for further explanation). These coefficients are analogous to the raw score slope coefficients in a raw score MR,

Figure 10.17 Structure Matrix: The Correlation of Each Predictor Variable With Each Discriminant Function (e.g., Scores on the Office Skills Test Correlate +.80 With Scores on Discriminant Function 1)

Structure Matrix

	Function	
	1	2
office	.803*	.213
math	-.540*	.155
abstract	-.168*	.144
mechanic	-.138	-.577*
english	.035	.544*
reading	-.172	.474*

Pooled within-groups correlations between discriminating variables and standardized canonical discriminant functions. Variables ordered by absolute size of correlation within function.

*. Largest absolute correlation between each variable and any discriminant function

Note: The asterisks do not indicate statistical significance.

Figure 10.18 Raw Score Discriminant Function Coefficients

Canonical Discriminant Function Coefficients

	Function	
	1	2
english	.042	.028
reading	.002	.060
mechanic	.115	-.243
abstract	-.043	-.010
math	-.099	.030
office	.088	.012
(Constant)	-4.415	-3.035

Unstandardized coefficients

as shown in Equation 10.1. While these coefficients are not usually interpreted as evidence of the strength of the predictive association of each X_i variable, they may be used to generate discriminant function scores that in turn may be used to predict group membership on the basis of raw scores.

Figure 10.19 reports the Fisher classification coefficients. These are generally used in research situations in which the primary goal is prediction of group membership. Note that these differ from the discriminant functions discussed earlier; we obtain one Fisher classification function for each of the *k* groups. In situations where the number of groups is larger than two, the Fisher classification coefficients provide a more convenient way of predicting group membership (and generating estimates of probabilities of group membership on the basis of scores on the X_i predictor variables). See Appendix 10B for further discussion. The Fisher coefficients are not generally used to interpret or describe the nature of group differences.

Figure 10.21 reports the "prior probability" of classification into each of the three groups. In this example, because the sizes of groups (numbers of cases in groups) were used as the basis for the classification procedure, the "prior probability" of classification into each group is equivalent to the proportion of cases in the sample that are in each group.

At this point, we have three types of information about the predictive usefulness of these six variables: the univariate *F*'s, the standardized canonical discriminant function coefficients, and the structure coefficients. In this example, we reach the same conclusion about the association between scores on predictor variables and choice of career, no matter which of these three results we examine. Among the six variables, only the scores on math and office skills are related to group membership; scores on the other four variables are not related to group

Figure 10.19 Fisher's Linear Discriminant Functions (One Function for Each Group)

Classification Function Coefficients

	cargroup		
	business	medicine	teach_social
english	1.186	1.120	1.138
reading	-.312	-.323	-.306
mechanic	1.008	.869	.833
abstract	-.308	-.242	-.256
math	-.405	-.262	-.279
office	.394	.261	.286
(Constant)	-52.700	-47.289	-48.660

Fisher's linear discriminant functions

Figure 10.20 Means on Discriminant Functions 1 and 2 (Centroids) for Each Group

Functions at Group Centroids

	Function	
cargroup	1	2
business	.541	-.008
medicine	-.954	-.147
teach_social	-.697	.123

Unstandardized canonical discriminant functions evaluated at group means

Figure 10.21 Prior Probabilities for Groups (Based on Number of Cases in Each Group Because "Priors = Size" Was Selected)

Prior Probabilities for Groups

cargroup	Prior	Cases Used in Analysis	
		Unweighted	Weighted
business	.598	98	98.000
medicine	.165	27	27.000
teach_social	.238	39	39.000
Total	1.000	164	164.000

membership. Just as beta coefficients may differ in sign from zero-order Pearson correlations between a predictor and Y, and just as group means in an analysis of covariance may change rank order when we adjust for a covariate, we can find in a DA that the nature of group differences on a particular variable looks quite different in the univariate ANOVA results compared with that in the DA results. That is, the relation between English and group membership might be quite different when we do not control for the other five predictors (in the univariate ANOVAs) than when we do control for the other five predictors (in the computation of the standardized discriminant function coefficients and structure coefficients). In situations like this, the discussion of results should include a discussion of how these two sets of results differ.

So far, we know that a high score on D_1 represents high scores on office skills and low scores on math. How are scores on D_1 related to group membership? This can be assessed by looking at the values of functions at group centroids (in Figure 10.20) and at the plot of the territorial map in Figure 10.22. A **centroid** is the mean of scores on each of the discriminant functions for members of one group. For the first discriminant function, the business group has a relatively high mean on D_1 (.541), while the other two groups have relatively low means (–.954 for medicine and –.697 for teaching or social work). In words, then, the business group had relatively high mean scores on D_1; members of this group scored high on office skills and low on math relative to the other two groups. Note that the means (or centroids) for D_2 were much closer together than the means on D_1 (centroids on D_2 were –.008, –.147, and +.123). This is consistent with earlier information (such as the small size of r_{c2}) that the groups did not differ much on D_2 scores.

The territorial map (in Figure 10.22) graphically illustrates the decision or classification rule that is used to predict group membership for individuals on the basis of their scores on the discriminant functions D_1 and D_2. Scores on D_1 are plotted on the horizontal axis; scores on D_2 are plotted on the vertical axis. The strings of ones correspond to the boundary of the region for which we classify people as probable members of Group 1, the strings of twos correspond to the boundary of the region for which we classify people as probable members of Group 2, and so forth. Roughly, anyone with a positive score on D_1 was classified as a probable member of Group 1, business. The other two groups were differentiated (but not well) by scores on D_2; higher scores on D_2 led us to classify people as members of Group 3, teaching or social work; lower scores on D_2 led us to classify people as members of Group 2, medicine. However, as we shall see from the table of predicted versus actual group membership in the last panel of the DA output, members of these two groups could not be reliably differentiated in this analysis.

Together with earlier information, this territorial map helps us interpret the meaning of D_1. High scores on D_1 corresponded to high scores on office skills and low scores on math achievement, and high scores on D_1 predicted membership in the business career group.

Figure 10.22 Territorial Map

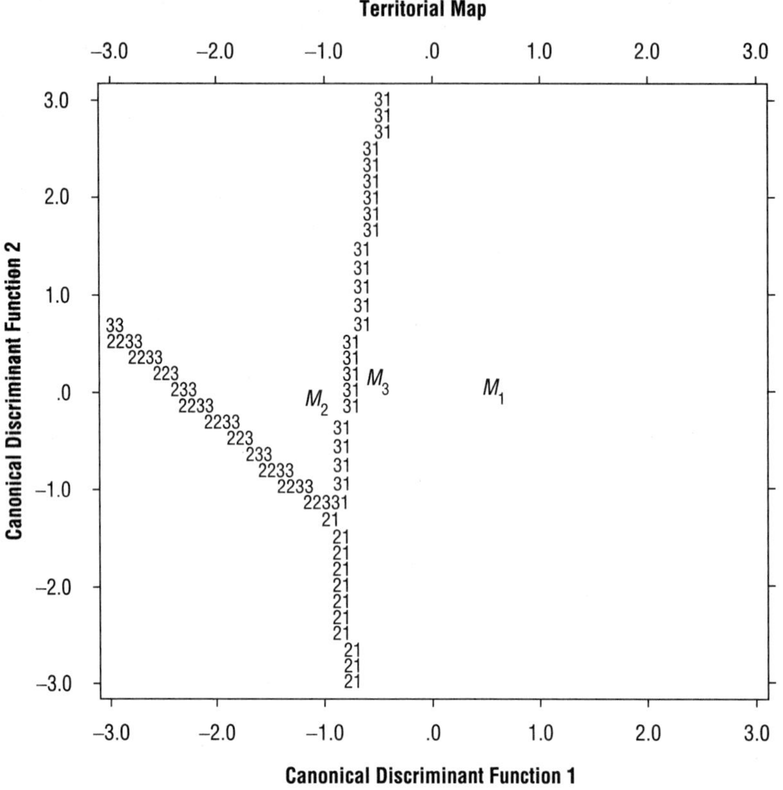

Note: The centroids M_1, M_2, and M_3 are not part of the SPSS output; these were added by the author.

The centroids of the groups appear as asterisks in the territorial map produced by SPSS; in Figure 10.22, these asterisks were replaced by the symbols M_1, M_2, and M_3 (to denote the centroids for Groups 1, 2, and 3, respectively). If the DA does a good job of discriminating among groups, these centroids should be far apart, and each centroid should fall well within the region for which people are classified into a particular group. In this example, the centroid for Group 1 is far from the centroids for Groups 2 and 3; this tells us that members of Group 1 could be distinguished from Groups 2 and 3 fairly well. However, the centroids of Groups 2 and 3 were relatively close together, and the centroid for Group 2 was not even within the region for Group 2 classification, which tells us that Group 2 members are not identified well at all by this DA. Note that in some DA solutions, it is possible that none of the cases is classified into a particular group, and when this happens, there is no region in the territorial map that corresponds to that group.

Finally, the classification results displayed in Figure 10.23 summarize the association between actual group ("Original" in the table in Figure 10.23) membership and predicted group membership. If the classification on the basis of discriminant function scores was very accurate, most cases should fall into the diagonal cells of this table (i.e., the actual and predicted group membership should be identical). In fact, only 68% of the participants were correctly classified overall. Members of Group 1 were most often classified correctly (91 of 98, or about 93%, were correctly classified). However, this analysis did not do a good job of

Figure 10.23 Summary Table of Original (Actual) Versus Predicted Group Membership

Classification Results[a]

		cargroup	Predicted Group Membership business	medicine	teach_social	Total
Original	Count	business	91	2	5	98
		medicine	15	6	6	27
		teach_social	21	3	15	39
	%	business	92.9	2.0	5.1	100.0
		medicine	55.6	22.2	22.2	100.0
		teach_social	53.8	7.7	38.5	100.0

a. 68.3% of original grouped cases correctly classified.

identifying members of Groups 2 and 3; in fact, the majority of members of these groups were misclassified. Thus, even though the overall model was statistically significant (on the basis of the chi-square for D_1 and D_2 combined), group membership was not predicted well at all for members of Groups 2 and 3. Only two of the six variables (scores on office skills and math) were useful in differentiating groups; high scores on office skills and low scores on math predicted a preference for business, but these scores provided no information about the choice between medicine and teaching. The following "Results" section illustrates how the information from the preceding DA could be written up for a journal article.

Results

A discriminant function analysis was done with all predictor variables entered in one step to assess how well future career choice could be predicted from scores on six aptitude and achievement tests. The three groups being compared were Group 1, business (n = 98); Group 2, medicine (n = 27); and Group 3, teachers and social workers (denoted by teach_social in the SPSS output) (n = 39). The six discriminating variables were scores on the following: English, reading, mechanical reasoning, abstract reasoning, math, and office skills. Preliminary screening indicated that scores on all six variables were approximately normally distributed.

A pooled within-groups correlation matrix was calculated to assess whether there was multicollinearity among these predictor variables; see Figure 10.13. No correlation exceeded .7 in absolute value, so multicollinearity was not a problem. Most test scores were modestly positively intercorrelated (with r's of the order of +.2 to +.6), except that scores on the office skills test were rather weakly correlated with scores on the other five variables. Using $\alpha = .01$, Box's M test for possible violation of the homogeneity of within-group variance/covariance matrices was judged to be nonsignificant.

Because there were three groups, two discriminant functions were created. Discriminant Function 2 had a canonical correlation of .0853; thus, it was weakly related to group membership. The chi-square for Discriminant Function 2 alone was not statistically significant, $\chi^2(5) = 1.156$, $p = .949$. Discriminant Function 1 had a canonical correlation of .556; thus, it was moderately related to group membership. The test for the combined predictive value of Discriminant Functions 1 and 2 combined was statistically significant: $\chi^2(12) = 59.879$, $p < .001$. Because the second discriminant function was not statistically significant, the coefficients for this function were not interpreted. Overall, the prediction of group membership was not very good; Wilks' Λ, which is analogous to $1 - \eta^2$ or the percentage of variance in the discriminant scores that is *not* explained by group membership, was .69. Thus, only 31% of the variance in discriminant scores was due to between-group differences.

Overall, 68% of the subjects were correctly classified into these three career groups by the DA (see Figure 10.23). This is not a very high rate of correct classification. Given that such a large proportion of the sample chose business careers, we could have achieved a 60% correct classification rate (98 of 164) simply by classifying all the cases into the business group. When group sizes are unequal, the correct rate of classification that would be expected by chance depends on both the number of groups and the group sizes, as discussed in Tabachnick and Fidell (2018).

From Figure 10.23, it is evident that the discriminant function did a good job of classifying actual business career group subjects (91 of 98 were correctly predicted to be in the business career group) but did a very poor job of predicting group membership for the other two groups. In fact, most persons in Groups 2 and 3 were classified incorrectly as members of other groups. The univariate *F*s and standardized discriminant function coefficients for the six discriminating variables are shown in Table 10.1.

If an arbitrary cutoff of .50 is used to decide which of the standardized discriminant coefficients are large, then only two of the predictor variables have large coefficients for standardized Discriminant Function 1: math and office skills. Because the second discriminant function alone was not statistically significant, coefficients on Discriminant Function 2 were not interpreted.

Table 10.1 reports summary information about individual predictor variables, evaluated using both univariate methods (individual ANOVAs) and multivariate methods (DA).

The nature of the pattern suggested by the results in Table 10.1 (along with the pattern of univariate group means that appeared in Figure 10.11) was as follows. High scores on office skills and low scores on math corresponded to high scores on the first discriminant function. Office skills and math were also the only individual variables that had significant univariate *F*s. The nature of these mean differences among groups was as follows: Members of Group 1, business careers, had higher mean scores on the office skills test (M_1 = 28.83) than members of Group 2, medicine (M_2 = 18.37) or Group 3, teaching (M_3 = 20.62). The business career group had much lower math scores (M_1 = 20.05) than the medicine group (M_2 = 26.37) or the teaching group (M_3 = 25.59). Thus, high scores on office skills and low scores on math were associated with the choice of a business career.

Finally, the territorial map (Figure 10.22) indicated that the first discriminant function separates Group 1 (business) from Groups 2 and 3 (medicine, teaching). People with high scores on the first discriminant function (and, therefore, high scores on office skills and low scores on math) were predicted to be in the office group. The second discriminant function tended to separate the medicine and teaching groups, but these two groups were not well discriminated by the second function. This is evident because the group centroids for Groups 2 and 3 on Discriminant Function 2 were quite close together and because the proportion of correct classifications of persons into Groups 2 and 3 was quite low.

A discriminant function that primarily represented information from the office skills and math scores predicted membership in the business career group at

Table 10.1 Summary of Information About Individual Predictor Variables for Discriminant Analysis: Differences Among Three Career Choice Groups (Business, Medicine, Teaching) on Six Achievement and Aptitude Tests

	Standardized Discriminant Function Coefficients[a]			
Variable	Function 1	Function 2	Univariate F	p Values[b] for Univariate F
English	.42	.28	0.22	*ns*
Reading	.02	.53	1.20	*ns*
Mechanic	.44	−.94	0.88	*ns*
Abstract	−.12	−.03	1.03	*ns*
Math	−.79	.24	10.53	<.001
Office	.75	.10	23.32	<.001

Note: To achieve an overall risk for Type I error of .05 for the entire set of six *F* tests, each individual *F* ratio was judged significant only if its *p* value was less than .05/6, that is, for $p < .008$ (using Bonferroni-corrected PCα levels).

[a]*Some authors present structure coefficients to show how each variable is correlated with each discriminant function (in addition to or instead of the standardized discriminant function coefficients).*

[b]*To correct for the inflated risk for Type I error, Bonferroni-corrected alpha levels were used to assess the significance of each univariate F.*

levels slightly better than chance. However, none of the variables was effective in discriminating between Group 2, medicine, and Group 3, teaching; the DA did not make accurate group membership predictions for members of these two groups. A reanalysis of these data could correct for this problem in a number of different ways; for example, it might be possible to identify some other variable that does discriminate between people who choose medicine versus teaching and social work as careers.

10.11 ONE-WAY ANOVA ON SCORES ON DISCRIMINANT FUNCTIONS

One final analysis is included to help students understand what a DA does. This is not information that is typically reported in a journal article along with other DA results. One of the commands included in the preceding DA was the <Save> command; this saved the computed discriminant function scores for each person for both D_1 and D_2 as new variables in the SPSS worksheet. A second new variable was computed for each subject in this data set; this was a simple unit-weighted composite of scores on the two variables that, according to the DA results, were the most useful variables for the prediction of group membership:

$$\text{Unit-weighted composite} = U = (+1 \times \text{Office skills}) + (-1 \times \text{Math}). \quad (10.17)$$

Figure 10.24 Creation of a Unit-Weighted Composite Variable: (+1) × Office Skills + (−1) × Math in the SPSS Compute Variable Dialog Box

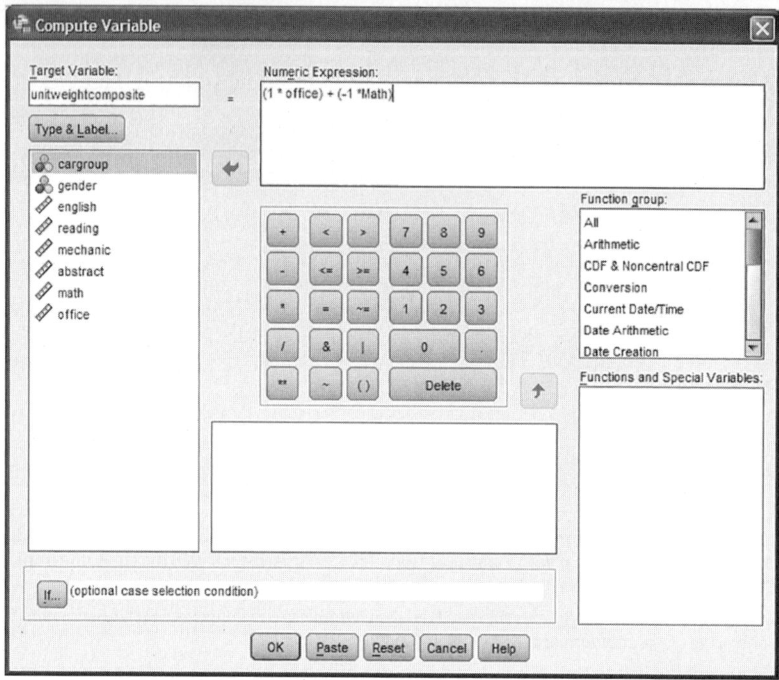

The simple unit-weighted composite computed using Equation 10.17 is just the difference between each person's raw scores on office skills and math. Figure 10.24 shows how the scores on this new variable named "unitweightcomposite" or U were calculated using the SPSS compute variable procedure.

One-way ANOVA was performed to evaluate how the three career groups differed in mean scores on the following variables: D_1, the saved score on Discriminant Function 1 from the preceding DA, and the unit-weighted composite or U, computed from Equation 10.17 above. The results of this one-way ANOVA appear in Figure 10.25.

For D_1, there was a statistically significant difference in means across groups; $F(2, 161) = 36.10, p < .001$. The corresponding value of eta squared (proportion of variance that is associated with between-group differences) was estimated as .309 or 31%. Refer to Figure 10.15 to the table headed "Wilks's Lambda" and examine the value of Λ for the overall model; for the set of both discriminant functions D_1 and D_2, $\Lambda = .685$. (However, the second discriminant function did not make a significant contribution to discrimination between groups; therefore, most of the explained variance is associated with D_1 in this example.) Recall that $1 - \Lambda$ is equivalent to eta squared; in this preceding DA (see Figure 10.15), $1 - \Lambda = 1 - .685 = .315$. In the DA results, about 31.5% of the variance in the scores on the two discriminant functions considered jointly was not associated with group membership. In this particular example, the proportion of variance due to between-group differences on the D_1 variable in the one-way ANOVA (.3096) was slightly smaller than the value of $1 - \Lambda$ (.315) for the overall DA. The proportion of explained variance or eta squared in the one-way ANOVA for differences in D_1 across groups would have been closer to the value of $1 - \Lambda$ if the second discriminant function had an even weaker predictive power. The eta squared for the ANOVA on D_1 scores would have been much smaller than $1 - \Lambda$ if the second discriminant function had a much greater

Figure 10.25 One-Way ANOVA to Compare Scores Across Three Career Groups

Descriptives

		N	Mean	Std. Deviation	Std. Error	95% Confidence Interval for Mean		Minimum	Maximum
						Lower Bound	Upper Bound		
Discriminant Scores from Function 1 for Analysis 1	business	98	.5405178	.93837212	.0947899	.3523861	.7286495	-2.42326	2.20927
	medicine	27	-.9544909	1.223373	.2354383	-1.4384412	-.4705407	-3.37797	1.16265
	teach_social	39	-.6974227	.98240739	.1573111	-1.0158824	-.3789631	-2.70931	1.11760
	Total	164	.0000000	1.196107	.0934003	-.1844305	.1844305	-3.37797	2.20927
unitweightcomposite	business	98	8.7857	12.03367	1.21558	6.3731	11.1983	-31.00	32.00
	medicine	28	-8.3929	15.84260	2.99397	-14.5360	-2.2497	-34.00	20.00
	teach_social	40	-4.5250	12.88208	2.03684	-8.6449	-.4051	-38.00	20.00
	Total	166	2.6807	14.86466	1.15372	.4028	4.9587	-38.00	32.00

ANOVA

		Sum of Squares	df	Mean Square	F	Sig.
Discriminant Scores from Function 1 for Analysis 1	Between Groups	72.200	2	36.100	36.100	.000
	Within Groups	161.000	161	1.000		
	Total	233.200	163			
unitweightcomposite	Between Groups	9162.925	2	4581.462	27.359	.000
	Within Groups	27295.15	163	167.455		
	Total	36458.08	165			

Note: Two different dependent variables are reported here. For the first analysis, the dependent variable was saved discriminant scores from Function 1 (D_1). For the second analysis, the dependent variable was the unit-weighted composite computed as follows: U = office skills – math scores.

predictive power; this is the case because the ANOVA, unlike the DA, only took scores on D_1 (and not scores on D_2) into account.

Now, examine the ANOVA results in Figure 10.25 for the dependent variable U, that is, the unweighted linear composite of scores (just the difference between office skills and math raw scores). For this ANOVA, $\eta^2 = .251$. About 25% of the variance in a simple unit-weighted linear composite of scores on just two of the variables, office skills and math, was associated with membership in the three career groups. This was less than the 31.5% of variance on D_1 and D_2 combined that was associated with group membership in the DA. This empirical example illustrates (but does not prove) that an optimally weighted linear composite of scores (derived from DA) provides a better prediction of group membership than a simple unit-weighted composite of the two "best" predictor variables.

In other words, the DA obtained the *optimal* weighted linear combination of scores on the X variables. For these data, the score on D_1 that had the largest possible differences between groups (and the minimum possible within-group variance) corresponded to $D_1 = +.42z_{\text{English}} + .01z_{\text{reading}} + .44z_{\text{mechanic}} - .12z_{\text{abstract}} - .79z_{\text{math}} + .75z_{\text{office}}$. (These standardized discriminant function coefficients were obtained from Figure 10.16.) These coefficients created the D_1 function that had the minimum possible Λ and the largest possible η^2 and F.

However, it is worth noting that we are able to show significant (although much smaller) differences between the groups by examining scores on the unit-weighted composite, which is obtained by assigning a weight of +1 to any variable with a large positive discriminant coefficient, a weight of −1 to any variable with a large negative discriminant function coefficient, and a weight of 0 to any variable with a small discriminant function coefficient. Although the between-group differences for this unit-weighted composite U (which in this example was +1 × office skills − 1 × math) were smaller than the between-group differences for the discriminant function D_1, in this case, there was still a significant difference between groups on this simpler unit-weighted composite.

This last analysis was presented for two reasons. First of all, it may help us understand what the DA actually does through a more concrete example using a simpler, more familiar analysis (one-way ANOVA). Second, there may be research situations in which it is important to present complex results in a simpler way, and in those situations, the DA may serve as a guide to the construction of simpler unit composite variables that may be adequate, in some situations, to describe group differences. Note that the discriminant function coefficients are optimized to predict group membership only for the sample(s) used to estimate these coefficients, and they may work less well when applied to new data.

10.12 SUMMARY

In DA, scores on several quantitative variables are used to predict group membership; the researcher is usually interested in the amount of weight given to each predictor variable. When assumptions that are required for DA (such as the assumption of homogeneity of the variance/covariance matrix across groups) are violated, other analyses that involve the prediction of group membership from scores on several quantitative predictors (such as binary logistic regression) may be preferred. Logit regression also provides information about relative risk or comparative probabilities of group membership that can be very useful in clinical prediction situations.

As in MR, it is possible to do a stepwise entry of predictor variables into a DA on the basis of the predictive usefulness of variables. This is not recommended, particularly in situations where the goal of the analysis is to obtain a theoretically meaningful description of the nature of group differences.

It may be helpful to look back and see how the information that can be obtained from DA compares with the information that can be obtained from analyses discussed earlier

Table 10.2 Information Provided by Discriminant Analysis in Comparison With ANOVA and Multiple Regression

Type of Information	ANOVA	Multiple Regression	Discriminant Analysis
Overall effect size for entire model	η^2: Proportion of variance in *Y* outcome scores that is predictable from group membership	Multiple *R* and multiple R^2: Proportion of variance in *Y* outcome variable predictable from the *X* predictor variables	Wilks' Λ: Proportion of variance in discriminant function scores that is not associated with or predictive of group membership (equivalent to $1 - \eta^2$)
Overall significance test for entire model	$F = MS_{between}/MS_{within}$	$F = MS_{regression}/MS_{residual}$	Wilks' Λ is usually converted to an approximate χ^2 or *F*
Strength and nature of predictive contribution of each individual X_i predictor variable	η^2	For each X_i predictor, examine the values of the *b* and β coefficients (and sr^2)	For each X_i predictor, examine the value of the raw-score or standardized-score discriminant function coefficients or the structure coefficients
Statistical significance of each X_i individual predictor variable	*F* for each factor	For each X_i predictor, examine the value of the *t* ratio associated with its *b* raw-score slope coefficient	No significance test for contribution of each X_i predictor provided by SPSS

(ANOVA and MR). To highlight the similarities, Table 10.2 outlines the types of information that each of these analyses provide. In the next chapter we will see that MANOVA is another analysis that is even more closely related to DA. MANOVA is essentially DA in reverse; that is, in MANOVA, researchers usually speak of predicting a set of scores on a list of outcome variables for different groups (e.g., treatment groups in an experiment). In DA, researchers usually speak of predicting group membership from scores on a set of predictor variables. However, both DA and MANOVA fundamentally deal with the same data analysis problem: examining how scores on a categorical variable are related to or predictable from scores on several quantitative variables. DA, however, is generally limited to the consideration of just one categorical outcome variable, while MANOVA may include several categorical predictor variables or factors. When we discuss MANOVA we shall see that DA provides a useful follow-up analysis that provides more detailed information about the nature of group differences found in MANOVA.

APPENDIX 10A

The Eigenvalue/Eigenvector Problem

Discriminant analysis requires the use of matrix algebra to solve the eigenvalue/eigenvector problem. In a sense, when we solve for eigenvectors and eigenvalues for a set of variables, we are "repackaging" the information about the variances and covariances of the variables. We do this by forming one (or more) weighted linear composites of the variables, such that these weighted linear composites are orthogonal to each other; in DA, each weighted linear composite is maximally predictive of group membership.

In DA, we begin with the following two matrices:

B is the between-groups **SCP** matrix for the set of *X* predictors. Computations for this matrix are not described in this chapter and can be found in Stevens (2009, Chapter 5).

W is the pooled within-groups **SCP** matrix for the set of *X* predictors. **W** is obtained by summing $\mathbf{SCP}_1 + \mathbf{SCP}_2 + \cdots + \mathbf{SCP}_k$, using the notation from an earlier section of this chapter.

The matrix algebra equivalent to division is multiplication by the inverse of a matrix, so $\mathbf{BW}^{-1}$ is the ratio of the between- versus within-groups **SCP** matrices. Our goal is to "repackage" the information about the between- versus within-groups variances and covariances of *X* variables in the form of one or several discriminant functions. This is accomplished by finding the eigenvectors and eigenvalues for $\mathbf{BW}^{-1}$ and rescaling the eigenvectors so that they can be used as coefficients to construct the standardized discriminant function scores; the corresponding eigenvalue for each discriminant function, λ, provides information about the ratio $SS_{\text{between}}/SS_{\text{within}}$ for each discriminant function.

In DA, the matrix product $\mathbf{BW}^{-1}$ is the matrix for which we want to solve the eigenproblem. Each eigenvector will be a list of *p* values (one for each dependent variable). When properly scaled, the eigenvector elements will be interpretable as weights or coefficients for these variables—discriminant function coefficients, for instance. Each eigenvector has a corresponding constant associated with it called an eigenvalue. The eigenvalue, when properly scaled, will be an estimate of the amount of variance that is explained by the particular weighted linear combination of the dependent variables that you get by using the (rescaled) eigenvector elements to weight the dependent variable. When we extract solutions, we first extract the one solution with the largest possible eigenvalue; then, we extract the solution that has the second largest eigenvalue, which is orthogonal to the first solution, and so on, until all the possible solutions have been extracted.

Here is the matrix algebra for the eigenproblem in DA. Let $\mathbf{BW}^{-1}$ be the matrix for which you want to find one or more eigenvectors, each with a corresponding eigenvalue. We need to find one or more solutions to the equation

$$(\mathbf{BW}^{-1} - \lambda\mathbf{I})\mathbf{V} = 0.$$

Note that this equation implies that $\mathbf{BW}^{-1}\mathbf{V} = \lambda\mathbf{I} \times \mathbf{V}$, where λ corresponds to the one (or more) eigenvalues and **V** corresponds to the matrix of eigenvectors. This equation is the eigenproblem. Tabachnick and Fidell (2018) provided a worked numerical example for a 2 × 2 matrix; however, the matrix denoted **D** in their example is replaced by the matrix $\mathbf{BW}^{-1}$ as shown above in computation of a DA.

When we solve the eigenproblem, more than one solution may be possible; each solution consists of one eigenvector and its corresponding eigenvalue. In general, if there are *k* groups and *p* discriminating variables, the number of different solutions is the minimum of $(k - 1)$ and *p*. If there is perfect multicollinearity among the *p* predictor variables, then we must replace *p*, the number of predictors, by the rank of the matrix that represents the scores on the *p* predictor variables; the rank of the matrix tells us, after taking multicollinearity among predictors into account, how many distinguishable independent variables remain. Thus, for a design with four groups and six discriminating variables, there are potentially three solutions, each one consisting of an eigenvector (the elements of which are rescaled to obtain the *d* coefficients for a discriminant function) and its corresponding eigenvalue (λ), which can be used to find a canonical correlation that tells us how strongly scores on the discriminant function are associated with group membership. For more extensive treatment of the eigenproblem and its multivariate applications, see Tatsuoka (1988, pp. 135–162).

APPENDIX 10B

Additional Equations for Discriminant Analysis

Two additional equations that are involved in discriminant function analysis are the following. These are used primarily in situations where the researcher's main interest is in the correct classification of cases (rather than a description of the nature of group differences or theory testing about possible "causal" variables).

The equation for Discriminant Function 1 can be given in terms of z scores on the X predictor variables, as in Equation 10.4: $D_1 = d_{11}z_1 + d_{12}z_2 + \cdots + d_{1p}z_p$. This equation is analogous to the standard-score version of an MR prediction equation; for example, $z'_Y = \beta_1 z_{X1} + \beta_2 z_{X2} + \cdots + \beta_k z_{Xk}$.

We can also obtain a raw-score version of the equation for each discriminant function:

$$D_1 = b_0 + b_1X_1 + b_2X_2 + \cdots + b_kX_k.$$

This is analogous to the raw-score version of the prediction equation in MR (except that the coefficients in the DA are scaled differently than in the MR analysis):

$$Y' = b_0 + b_1X_1 + b_2X_2 + \cdots + b_kX_k.$$

In the SPSS output, the coefficients for these raw-score versions of the discriminant functions are found in the table headed "Canonical Discriminant Function Coefficients"; this is an optional statistic.

Note that the number of discriminant functions that can be obtained (whether they are stated in terms of z scores or raw X scores) is limited to the minimum of $(k - 1)$, that is, the number of groups on the dependent variable minus 1, and p, the number of X predictor variables that are not completely redundant with each other.

Fisher suggested another approach to group classification that is easier to apply in situations with multiple groups. In this approach, for each of the k groups, we compute a score for each participant on a Fisher classification coefficient that corresponds to each group (i.e., a function for Group 1, Group 2, . . . , Group k). The Fisher discriminant function, C_j, where j = Group 1, 2, . . . , k, can be denoted as follows (Tabachnick & Fidell, 2018):

$$C_j = c_{j0} + c_{j1}X_1 + c_{j2}X_2 + \cdots + c_{jp}X_p.$$

A score for each case is computed for each group, and each participant is predicted to be a member of the group for which that participant has the highest C_j score. Values of C_j can be used to derive other useful information, such as the theoretical probability that the individual participant is a member of each group given his or her scores on the C_j functions. Adjustments that take group size into account, or that otherwise increase the likelihood of classification of cases into any individual group, can easily be made by modifying the size of the c_{j0} term (which is analogous to the b_0 concept or intercept in a regression equation). The values of coefficients for the C_j classification functions, often called Fisher's linear classification coefficients, appear in the SPSS table that has the heading "Classification Function Coefficients" in Figure 10.19. These coefficients are not generally interpreted or used to describe the nature of group differences but they provide a more convenient method for classification of individual cases into groups when the number of groups, k, is greater than 2.

COMPREHENSION QUESTIONS

1. Describe two research situations where DA can be used: one situation where the primary interest is in group classification and another situation where the primary interest is in the description of the nature of group differences.
2. What assumptions about the data should be satisfied for a DA? What types of data screening should be performed to assess whether assumptions are violated?
3. In general, if p is the number of predictor variables and k is the number of groups for the outcome variable in a DA, how many different discriminant functions can be obtained to differentiate among groups? (This question assumes that the X predictor variables have a determinant that is nonzero; that is, no individual X_i predictor variable can be perfectly predicted from scores on one or more other X predictor variables.)
4. Show how the canonical r_c for each discriminant function can be calculated from the eigenvalue (λ) for each function, and verbally interpret these canonical correlations. How is a squared canonical correlation related to an η^2?
5. From the territorial map, how can we describe the nature of group differences on one or more discriminant functions?
6. What information can a data analyst use to evaluate how much information about group membership is contributed by each individual variable? Be sure to distinguish between procedures that examine variables in isolation and procedures that take correlations among predictor variables into account.

DIGITAL RESOURCES

Find **free study tools** to support your learning, including **eFlashcards, data sets, and web resources,** on the accompanying website at **edge.sagepub.com/warner3e.**

CHAPTER 11

MULTIVARIATE ANALYSIS OF VARIANCE

11.1 RESEARCH SITUATIONS AND RESEARCH QUESTIONS

Multivariate analysis of variance (MANOVA) is a generalization of univariate analysis of variance (ANOVA). Recall that in a univariate one-way ANOVA, mean scores on just *one* quantitative Y were compared across groups; for example, in a one-way univariate ANOVA, the null hypothesis of interest is as follows:

$$H_0 : \mu_1 = \mu_2 = \cdots = \mu_k. \tag{11.1}$$

In Equation 11.1, each μ_i term corresponds to the population mean for the score on a single Y outcome variable in one of the k groups.

In a one-way MANOVA, mean scores on *multiple* quantitative outcome variables are compared for participants across two or more groups. When a MANOVA has only one factor or one categorical predictor variable, we call it a one-way MANOVA with k levels. In other words, MANOVA is an extension of univariate ANOVA (ANOVA with just one outcome variable) to situations where there are multiple quantitative outcome variables; we denote the number of quantitative outcome variables in a MANOVA as p. The set of p outcome variables in MANOVA can be represented as a list or vector of Y outcome variables:

$$H_0 = \begin{bmatrix} \mu_{11} \\ \mu_{12} \\ \vdots \\ \mu_{1p} \end{bmatrix} = \begin{bmatrix} \mu_{21} \\ \mu_{22} \\ \vdots \\ \mu_{2p} \end{bmatrix} = \cdots = \begin{bmatrix} \mu_{k1} \\ \mu_{k2} \\ \vdots \\ \mu_{kp} \end{bmatrix}. \tag{11.2}$$

In words, the null hypothesis in Equation 11.2 corresponds to the assumption that when the scores on all p of the Y outcome variables are considered jointly as a set, taking intercorrelations among the Y variables into account, the means for this set of p outcome variables do not differ across any of the populations that correspond to groups in the study. This is a multivariate extension of the univariate null hypothesis that appeared in Equation 11.1.

This multivariate null hypothesis can be written more compactly by using matrix algebra notation. We can use $\boldsymbol{\mu}_i$ to represent the vector of means for the Y outcome variables Y_1 through Y_p in each population ($i = 1, 2, \ldots, k$), which corresponds to each of the k sample groups in the study: $\boldsymbol{\mu}_i = [\mu_{i1}, \mu_{i2}, \mu_{i3}, \ldots, \mu_{ip}]$. When set in a boldface font, $\boldsymbol{\mu}_i$ corresponds to a vector or list of means on p outcome variables; when shown in a regular font, μ_i corresponds to a population mean on a single Y outcome variable. The null hypothesis for a one-way MANOVA can be written out in full, using μ_i to stand for the list of population means for each of the p outcome variables in each of the groups as follows:

$$H_0: \boldsymbol{\mu}_1 = \boldsymbol{\mu}_2 = \cdots = \boldsymbol{\mu}_k. \quad (11.3)$$

In Equation 11.3, each $\boldsymbol{\mu}_i$ term represents the vector or set of means on p different Y outcome variables across k groups. In words, the null hypothesis for a one-way MANOVA is that the population means for the entire *set* of variables $Y_1, Y_2, \ldots, Y_p$ (considered jointly and taking their intercorrelations into account) do not differ across Groups $1, 2, \ldots, k$. If we obtain an omnibus F large enough to reject this null hypothesis, it usually implies at least one significant inequality between groups on at least one Y outcome variable. If the omnibus test is statistically significant, follow-up analyses may be performed to assess which contrasts between groups are significant and which of the individual Y variables (or combinations of Y variables) differ significantly between groups. However, data analysts need to be aware that even when the omnibus or multivariate test for MANOVA is statistically significant, there is an inflated risk for Type I error when numerous univariate follow-up analyses are performed.

11.2 FIRST RESEARCH EXAMPLE: ONE-WAY MANOVA

MANOVA can be used to compare the means of multiple groups in nonexperimental research situations that evaluate differences in patterns of means on several Y outcome variables for naturally occurring groups. The same data are used for discriminant analysis (DA) in the previous chapter and for MANOVA in this chapter; data are in the file talent.sav. The factor, future career choice, had three levels: 1 = business career, 2 = medical career, and 3 = teaching or social work career. Students had scores on six aptitude and achievement tests. The goal of the DA was to see whether membership in one of these three future career choice groups could be predicted from each participant's set of scores on the six tests.

This same set of data can be examined using a MANOVA. The MANOVA examines whether the set of means on the six test scores differs significantly across any of the three career choice groups. Later in this chapter, a one-way MANOVA is reported for the data that were analyzed using DA. This example illustrates (but does not prove) that MANOVA and DA are essentially equivalent for one-way designs.

MANOVA can also be used to test the significance of differences of means in experiments where groups are formed by random assignment of participants to different treatment conditions. Consider the following hypothetical experiment as an example of a one-way MANOVA design. A researcher wants to assess how various types of stressors differ in their impact on physiological responses. Each participant is randomly assigned to one of the following conditions: Group 1 = no stress (baseline), Group 2 = mental stress, Group 3 = cold pressor, and Group 4 = stressful social role play. The name of the factor in this one-way MANOVA is stress, and there are $k = 4$ levels of this factor. The following outcome measures are obtained: Y_1, self-reported anxiety; Y_2, heart rate; Y_3, systolic blood pressure; Y_4, diastolic blood pressure; and Y_5, electrodermal activity (palmar sweating). The research question is whether these four groups differ in the pattern of mean responses on these five outcome measures. For example, it is possible that the three stress inductions (Groups 2 through 4) all had higher scores on self-reported anxiety, Y_1, than the baseline condition (Group 1) and that Group 3, the pain induction by cold pressor, had higher systolic blood pressure and a higher heart rate than the other two stress inductions.

11.3 WHY INCLUDE MULTIPLE OUTCOME MEASURES?

Let us pause to consider two questions. First, what do researchers gain by including multiple outcome measures in a study instead of just one single outcome measure? And second, what are the advantages of using MANOVA to evaluate the nature of group differences on multiple outcome variables instead of doing a univariate ANOVA separately for each of the individual Y outcome variables in isolation?

There are several reasons why it may be advantageous to include multiple outcome variables in a study that makes comparisons across groups. For example, a stress intervention might cause increases in physiological responses such as blood pressure and heart rate, emotional responses such as self-reported anxiety, and observable behaviors such as grimacing and fidgeting. These measures might possibly be interpreted as multiple indicators of the same underlying construct (anxiety) using different measurement methods. In the stress intervention study, measuring anxiety using different types of methods (physiological, self-report, and observation of behavior) provides us with different kinds of information about a single outcome such as anxiety. If we use only a self-report measure of anxiety, it is possible that the outcome of the study will be artifactually influenced by some of the known weaknesses of self-report (such as social desirability response bias). When we include other types of measures that are not as vulnerable to social desirability response bias, we may be able to show that conclusions about the impact of stress on anxiety are generalizable to anxiety assessments that are made using methods other than self-report.

It is also possible that an intervention has an impact on two or more conceptually distinct outcomes; for example, some types of stress interventions might arouse hostility in addition to or instead of anxiety. Whether the Y variables represent multiple measures of one single construct or measures of conceptually distinct outcomes, the inclusion of multiple outcome measures in a study potentially provides richer and more detailed information about the overall response pattern.

When we include more than one quantitative Y outcome variable in a study that involves comparisons of means across groups, one way that we might assess differences across groups would be to conduct a separate one-way ANOVA for each individual outcome measure. For example, in the hypothetical stress experiment described earlier, we could do a one-way ANOVA to see if mean self-reported anxiety differs across the four stress intervention conditions and a separate one-way ANOVA to assess whether mean heart rate differs across the four stress intervention conditions. However, running a series of p one-way ANOVAs (a univariate ANOVA for each of the Y outcome variables) raises a series of problems. The first issue involves the inflated risk for Type I error that arises when multiple significance tests are performed. For example, if a researcher measures 10 different outcome variables in an experiment and then performs 10 separate ANOVAs (one for each outcome measure), the probability of making at least one Type I error in the set of 10 tests is likely to be substantially higher than the conventional $\alpha = .05$ level that is often used as a criterion for significance, unless steps are taken to try to limit the risk for Type I error. (For example, Bonferroni-corrected per comparison alpha levels can be used to assess statistical significance for the set of several univariate ANOVAs.) In contrast, if we report just one omnibus test for the entire set of 10 outcome variables and if the assumptions for MANOVA are reasonably well satisfied, then the risk for Type I error for the single omnibus test should not be inflated.

A second issue concerns the linear intercorrelations among the Y outcome variables. A series of separate univariate ANOVA tests do not take intercorrelations among the Y outcome variables into account, and therefore, the univariate tests do not make it clear whether each of the p outcome variables represents a conceptually distinct and independent outcome or whether the outcome variables are intercorrelated in a way that suggests that they may represent multiple measures of just one or two conceptually distinct outcomes. In the stress experiment mentioned earlier, for example, it seems likely that the measures would be positively intercorrelated and that they might all tap the same response (anxiety). When the outcome measures actually represent different ways of assessing the same underlying construct, reporting a separate univariate ANOVA for each variable may mislead us to think that the intervention had an impact on several unrelated responses when, in fact, the response may be better described as an "anxiety syndrome" that includes elevated heart rate, increased self-report anxiety, and fidgeting behavior. Univariate tests for each individual Y outcome variable also ignore the possibility that statistically controlling for one of the Y variables may change the apparent association between scores on other Y variables and group membership.

A third reason why univariate ANOVAs may not be an adequate analysis is that there are situations in which groups may not differ significantly on any one individual outcome variable but their outcomes are distinguishable when the outcomes on two or more variables are considered jointly. For an example of a hypothetical research situation where this occurs, refer to Figures 10.3 and 10.4; in this hypothetical example, skulls could be accurately classified using information about two different skull measurements, even though each individual measurement showed substantial overlap between the ape and human skull groups.

For all the preceding reasons, the results of a MANOVA may be more informative than the results of a series of univariate ANOVAs. In an ideal situation, running a single MANOVA yields an omnibus test of the overall null hypothesis in Equation 11.2 that does not have an inflated risk for Type I error (however, violations of assumptions for MANOVA can make the obtained p value a poor indication of the true level of risk for Type I error). A MANOVA may be more powerful than a series of univariate ANOVAs; that is, sometimes a MANOVA can detect a significant difference in response pattern on several variables across groups even when none of the individual Y variables differ significantly across groups. However, MANOVA is not always more powerful than a series of univariate ANOVAs. MANOVA provides a summary of group differences that takes intercorrelations among the outcome variables into account, and this may provide a clearer understanding of the nature of group differences, that is, how many dimensions or weighted linear combinations of variables show significant differences across groups. For example, if the set of outcome variables in the stress study included measures of hostility as well as anxiety, the results of a MANOVA might suggest that one stress intervention elicits a pattern of response that is more hostile, while another stress intervention elicits a response pattern that is more anxious, and yet another stress intervention elicits both hostility and anxiety. It is sometimes useful to divide the list of Y outcome variables into different sets and then perform a separate MANOVA for each set of Y outcome variables to detect group differences (Stevens, 2009).

11.4 EQUIVALENCE OF MANOVA AND DA

A one-way MANOVA can be thought of as a DA in reverse. Each of these analyses assesses the association between one categorical or group membership variable and a set of several quantitative variables. DA is usually limited to the evaluation of one categorical variable or factor, while MANOVA can be generalized to include multiple categorical variables or factors, for example, a factorial design. In MANOVA, we tend to speak of the categorical or group membership variable (or factor) as the predictor and the quantitative variables as outcomes; thus, in MANOVA, the outcome variables are typically denoted by $Y_1, Y_2, \ldots, Y_p$. In DA, we tend to speak of prediction of group membership from scores on a set of quantitative predictors; thus, in the discussion of DA, the quantitative variables were denoted by $X_1, X_2, \ldots, X_p$. Apart from this difference in language and notation, however, DA and one-way MANOVA are essentially equivalent.

Let us reconsider the talent.sav data set. Six quantitative test scores (scores on English, reading, mechanical reasoning, abstract reasoning, math, and office skills) were used to predict future career choice; this categorical variable, cargroup, had three levels: 1 = business career, 2 = medical career, and 3 = teaching or social work career. DA was used to evaluate how well membership in these three career choice groups could be predicted from the set of scores on these six tests. We will reexamine this problem as a one-way MANOVA in this chapter; we will use MANOVA to ask whether there are any statistically significant differences across the three career groups for a set of one or more of these test scores. DA and MANOVA of these data yield identical results when we look at the omnibus tests that test the overall null hypothesis in Equation 11.2. However, the two analyses provide different kinds of supplemental information or follow-up analyses that can be used to understand the nature of group differences.

In MANOVA, the researcher tends to be primarily interested in F tests to assess which groups differ significantly; in DA, the researcher may be more interested in the assessment of the relative predictive usefulness of individual discriminating variables. However, the underlying computations for MANOVA and DA are identical, and the information that they provide is complementary (Bray & Maxwell, 1985). Thus, it is common to include some MANOVA results (such as omnibus F tests of group differences) even when a study is framed as a DA and to include DA results as part of a follow-up to a significant MANOVA.

The MANOVA procedures available in SPSS (through the SPSS general linear model [GLM] procedure) provide some tests that cannot be easily obtained from the DA procedure; for example, planned contrasts and/or post hoc tests can be requested as part of MANOVA to evaluate which specific pairs of groups differ significantly. On the other hand, DA provides some information that is not readily available in most MANOVA programs. For example, DA provides coefficients for discriminant functions that provide helpful information about the nature of the association between scores on each quantitative variable and group membership when intercorrelations among the entire set of quantitative variables are taken into account; DA also provides information about the accuracy of classification of individual cases into groups. It is fairly common for a researcher who finds a significant difference among groups using MANOVA to run DA on the same data to obtain more detailed information about the way in which individual quantitative variables differ across groups.

11.5 THE GENERAL LINEAR MODEL

In SPSS, MANOVA is performed using GLM. In its most general form, the SPSS GLM procedure can handle any combination of multiple outcome measures, multiple factors, repeated measures, and multiple covariates. What is the GLM? In GLM, there can be one or multiple predictor variables, and these predictor variables can be either categorical or quantitative. There may also be one or several quantitative outcome variables and one or several quantitative covariates. In practice, the GLM procedure in SPSS permits us to include both multiple predictor and multiple outcome variables in an analysis, although there are some limitations to the ways in which we can combine categorical and quantitative variables. *Linear* indicates that for all pairs of quantitative variables, the nature of the relationship between variables is linear and, thus, can be handled by using correlation and linear regression methods. Many of the statistical techniques that are presented in an introductory statistics course are special cases of the GLM. For example, an independent-samples t test is a GLM with one predictor variable that is dichotomous and one outcome variable that is quantitative. Pearson's r is a GLM with one quantitative predictor and one quantitative outcome variable. Multiple regression is a GLM with one quantitative outcome variable and multiple predictor variables, which may include quantitative and dummy or dichotomous variables. All the GLM analyses encountered so far involve similar assumptions about data structure; for example, scores on the outcome variable must be independent, scores on quantitative variables should be at least approximately normally distributed, relations between pairs of quantitative variables should be linear, there should be no correlation between the error or residual from an analysis and the predicted score generated by the analysis, and so forth. Some specific forms of the GLM involve additional assumptions; for example, analysis of covariance assumes no treatment-by-covariate interaction, and DA assumes homogeneity of variance/covariance matrices across groups. The SPSS GLM procedure allows the user to set up an analysis that combines one or more quantitative dependent variables, one or more factors or categorical predictors, and possibly one or more quantitative covariates.

The questions that can be addressed in a MANOVA include some of the same questions described in the DA chapter. Using MANOVA and DA together, a researcher can assess which contrasts between groups are significant, which variables are the most useful in differentiating among groups, and the pattern of group means on each discriminant function. Issues involving accuracy of prediction of group membership, which were a major focus in the presentation of DA, are usually not as important in MANOVA; the SPSS GLM procedure does not provide information about accuracy of classification into groups.

In addition, MANOVA can address other questions. Do group differences on mean vectors remain statistically significant when one or several covariates are statistically controlled? Do different treatment groups show different patterns or responses across time in repeated-measures studies? Are there interaction effects such that the pattern of cell means on the discriminant functions is much different from what would be predicted on the basis of the additive main effects of the factors? Do the discriminant functions that differentiate among levels on the B factor for the A_1 group have large coefficients for a different subset of the quantitative outcome variables from the subset for the discriminant functions that differentiate among levels of the B factor within the A_2 group?

Like DA and multiple regression, MANOVA takes intercorrelations among variables into account when assessing group differences. MANOVA is appropriate when the multiple outcome measures are moderately intercorrelated. If correlations among the outcome variables are extremely high (e.g., greater than .8 in absolute value), it is possible that the variables are all measures of the same construct; if this is the case, it may be simpler and more appropriate to combine the scores on the variables into a single summary index of that construct. In a typical research situation where correlations among the outcome measures are moderate in size, some outcome variables may appear to be less important when other outcome variables are statistically controlled, or the nature of the relationship between a quantitative variable and group membership may change when other variables are statistically controlled. The magnitude and signs of the correlations among outcome measures influence the statistical power of MANOVA and the interpretability of results in complex ways (see Bray & Maxwell, 1985, for further discussion).

11.6 ASSUMPTIONS AND DATA SCREENING

The assumptions for MANOVA are essentially the same as those for DA and for many other GLM procedures. We will refer to the multiple outcome measures as $Y_1, Y_2, \ldots, Y_p$.

1. Observations on the Y outcome variables should be collected in such a way that the scores of different participants on any one Y_i outcome variable are independent of each other. Systematic exceptions to this rule (such as the correlated observations that are obtained in repeated-measures designs) require repeated measures or other analytic methods that take these correlations among observations into account.

2. Each Y outcome variable should be quantitative and reasonably normally distributed. As discussed in earlier chapters, it is useful to examine univariate histograms to assess whether scores on each Y variable are approximately normally distributed. Univariate outliers should be evaluated; data transformations such as logs may remedy problems with univariate and even some multivariate outliers.

3. Associations between pairs of Y variables should be linear; this can be assessed by obtaining a matrix of scatterplots between all possible pairs of Y variables. More formally, the joint distribution of the entire set of Y variables should be multivariate normal within each group. In practice, it is difficult to assess multivariate normality, particularly when the overall number of scores in each group is small. It may be easier to evaluate univariate and bivariate normality. Some types of deviation from multivariate normality probably have a relatively small effect on Type I error (Stevens, 2009). However, MANOVA is not very robust to outliers.

4. The variance/covariance matrices for the Y outcome variables ($\mathbf{\Sigma}$) should be homogeneous across the populations that correspond to groups in the study. Recall that the variances and covariances that correspond to the elements of this population $\mathbf{\Sigma}$ matrix are as follows:

$$\Sigma = \begin{bmatrix} \text{Var}(Y_1) & \text{Cov}(Y_1, Y_2) & \cdots & \text{Cov}(Y_1, Y_p) \\ \text{Cov}(Y_2, Y_1) & \text{Var}(Y_2) & \cdots & \text{Cov}(Y_2, Y_p) \\ \vdots & \vdots & & \vdots \\ \text{Cov}(Y_p, Y_1) & \text{Cov}(Y_p, Y_2) & \cdots & \text{Var}(Y_p) \end{bmatrix}. \tag{11.4}$$

The null hypothesis for the test of homogeneity of $\mathbf{\Sigma}$ across groups is as follows:

$$H_0\text{: } \mathbf{\Sigma}_1 = \mathbf{\Sigma}_2 = \cdots = \mathbf{\Sigma}_k, \text{ for Groups } 1, 2, \ldots, k. \tag{11.5}$$

In words, this is the assumption that the variances for all the outcome variables ($Y_1, Y_2, \ldots, Y_p$) are equal across populations and the covariances for all possible pairs of outcome variables, such as Y_1 with Y_2, Y_1 with Y_3, and so on, are equal across populations. As in DA, the optional Box's M test provides a test for whether there is a significant violation of this assumption; Box's M is a function of the sample sum of cross products (**SCP**) matrices calculated separately for each of the groups in the design.

What problems arise when Box's M test and/or the Levene test suggests serious violations of the assumption of homogeneity of variances and covariances across groups? Violations of this assumption may be problematic in two ways: They may alter the risk for committing a Type I error (so that the nominal alpha level does not provide accurate information about the actual risk for Type I error), and they may reduce statistical power. Whether the risk for Type I error increases or decreases relative to the nominal alpha level depends on the way in which the homogeneity of variance assumption is violated. When the group that has the largest variances has a small n relative to the other groups in the study, the test of significance has a higher risk for Type I error than is indicated by the nominal alpha level. On the other hand, when the group with the largest variance has a large n relative to the other groups in the study, it tends to result in a more conservative test (Stevens, 2009). Violations of the homogeneity of variance/covariance matrix pose a much more serious problem when the group n's are small and/or extremely unequal. Stevens (2002) states that if Box's M test is significant in a MANOVA where the n's in the groups are nearly equal, it may not have much effect on the risk for Type I error, but it may reduce statistical power.

What remedies may help reduce violations of the assumption of homogeneity of variances and covariances across groups? Transformations (such as logarithms) may be effective in reducing the differences in variances and covariances across groups and, thus, correcting for violations of the homogeneity of variance/covariance matrices. In addition, when Box's

M test is significant, particularly if there are unequal *n*'s in the groups, the researcher may prefer to report **Pillai's trace** instead of Wilks' lambda (Λ) as the overall test statistic; Pillai's trace is more robust to violations of the homogeneity of variances and covariances.

In a factorial ANOVA, unequal or imbalanced *n*'s in the cells imply a nonorthogonal design; that is, some of the variance in scores that can be predicted from the *A* factor is also predictable from the *B* factor. The same problem arises in factorial MANOVA design: Unequal (or imbalanced) *n*'s in the cells imply a nonorthogonal design; that is, there is an overlap in the variance accounted for by the *A* and *B* main effects and their interaction. As in univariate ANOVA, this overlap can be handled in two different ways. SPSS allows the user to choose among four methods of computation for sums of squares (*SS*) in the GLM procedure. Types II and IV methods of computation for sums of squares are rarely used and are not discussed here. *SS* Type I computation of sums of squares partitions the sums of squares in a hierarchical fashion; that is, the effects or factors in the model are entered one at a time, and each effect is assessed statistically controlling for effects that were entered in the earlier steps. For example, in an analysis with *SS* Type I, SS_A could be computed without controlling for *B* or the $A \times B$ interaction, SS_B could be computed controlling only for the *A* factor, and $SS_{A \times B}$ could be computed controlling for the main effects of both *A* and *B*. GLM *SS* Type III, which is similar to the variance partitioning in a standard or simultaneous multiple regression, computes an adjusted sum of squares for each effect in the model while controlling for all other predictors. Thus, for an $A \times B$ factorial design, SS_A would be calculated controlling for the main effect of *B* and the $A \times B$ interaction, SS_B would be calculated controlling for the main effect of *A* and the $A \times B$ interaction, and $SS_{A \times B}$ would be calculated controlling for the main effects of both *A* and *B*. The *SS* Type III method of variance partitioning usually yields a more conservative estimate of the proportion of variance uniquely predictable from each factor. If the cell *n*'s are equal or balanced, the *SS* Types I and III computation methods produce identical results. The factorial MANOVAs presented as examples later in this chapter use *SS* Type III computation methods.

11.7 ISSUES IN PLANNING A STUDY

The issues that were relevant in planning ANOVA designs and in planning regression or DA studies are also relevant for planning MANOVA.

The first design issue involves the number, composition, and size of groups and, also, the treatments that are administered to groups (in experimental research) and the basis for categorization into groups (in nonexperimental studies that compare naturally occurring groups). As in other ANOVA designs, it is desirable to have a reasonably small number of groups and to have a relatively large number of participants in each group. If a factorial design is used, equal (or at least balanced) *n*'s in the cells are desirable, to avoid confounds of the effects of factors and interactions. If treatments are compared, the researcher needs to think about the dosage levels (i.e., are the treatments different enough to produce detectable differences in outcomes?) and the number of levels of groups required (are two groups sufficient, or are multiple groups needed to map out a curvilinear dose-response relationship?). Design issues that are important in one-way and factorial ANOVA are also applicable to MANOVA. MANOVA generally requires a larger within-cell *n* than the corresponding ANOVA, to have sufficient degrees of freedom to estimate the elements of the matrix SCP within each cell. As in other ANOVA designs, the cell *n*'s in MANOVA should not be too small. It is desirable to have sufficient degrees of freedom in each cell to obtain independent estimates for all the elements of the SCP matrix within each group, although in practice, if this requirement is not met, the SPSS GLM procedure sets up an omnibus test on the basis of pooled within-group SCP matrices across all groups. In practice, if there are at least 20 cases in each cell, it may be sufficient to ensure robustness of the univariate *F* tests (Tabachnick & Fidell, 2018). Stevens (2009) recommended a minimum total *N* of at least 20*p*, where *p* is the number of outcome measures. Another suggested minimum requirement for sample size is that there should be more cases than dependent variables in every cell of the MANOVA design. Section 11.12 includes a table that can be used to estimate the per group sample size that is needed to achieve reasonable power given the assumed values of the multivariate effect size; the number of cases, N; the number of outcome variables, p; and the alpha level used to assess statistical significance. Note also that grossly unequal and/or extremely small n's in the cells of a MANOVA will make the impact of violations of the homogeneity of the variance/covariance matrix across groups more serious, and this can result in an inflated risk for Type I error and/or a decrease in statistical power.

When quantitative outcome variables are selected, these should include variables that the researcher believes will be affected by the treatment (in an experiment) or variables that will show differences among naturally occurring groups (in nonexperimental studies). The researcher may have one construct in mind as the outcome of interest; in a study that involves manipulation of stress, for example, the outcome of interest might be anxiety. The researcher might include multiple measures of anxiety, perhaps using entirely different methods of measurement (e.g., a standardized written test, observations of behaviors such as fidgeting or stammering, and physiological measures such as heart rate). Multiple operationalizations of a construct (also called **triangulation of measurement**) can greatly strengthen confidence in the results of a study. Different types of measurement (such as self-report, observations of behavior, and physiological monitoring) have different strengths and weaknesses. If only self-report measures are used, group differences might be due to some of the known weaknesses of self-report, such as social desirability bias, response to perceived experimenter expectancy or demand, or **faking**. However, if the stress manipulation increases scores on three different types of measures (such as self-report, physiological, and behavioral assessments), we can be more confident that we have a result that is not solely attributable to a response artifact associated with one type of measurement (such as the social desirability bias that often occurs in self-report). Thus, MANOVA can be used to analyze data with multiple measures of the same construct.

On the other hand, the multiple outcome measures included in a MANOVA may tap several different constructs. For example, a researcher might administer psychological tests to persons from several different cultures, including assessments of individualism or collectivism, self-esteem, internal or external locus of control, extraversion, openness to experience, conscientiousness, need for achievement, and so forth. It might turn out that the pattern of test scores that differentiates the United States from Japan is different from the pattern of scores that differentiates the United States from Brazil. MANOVA corrects for intercorrelations or overlaps among these test scores whether they are multiple measures of the same construct or measures of several different constructs.

11.8 CONCEPTUAL BASIS OF MANOVA

Recall that in a univariate one-way ANOVA, SS_{total} for the Y outcome variable is partitioned into the following components:

$$SS_{\text{total}} = SS_{\text{between}} + SS_{\text{within}}. \tag{11.6}$$

In univariate ANOVA, the information about the variance of the single outcome variable Y can be summarized in a single SS term, SS_{total}. Recall that the variance of Y equals $SS_{\text{total}}/(N - 1)$. In MANOVA, because we have multiple outcome variables, we need information about the variance of each outcome variable $Y_1, Y_2, \ldots, Y_p$, and we also need information about the covariance of each pair of Y variables. Recall also that the SCP between Y_1 and Y_2 is computed as $\Sigma(Y_1 - M_{Y1})(Y_2 - M_{Y2})$ and that the covariance between Y_1 and Y_2 can be estimated by the ratio SCP/N. The total **SCP** matrix (denoted $\mathbf{SCP}_{\text{total}}$) in a MANOVA has the following elements (all the scores in the entire design are included in the computation of each SS and SCP element):

$$\mathbf{SCP}_{\text{total}} = \begin{bmatrix} SS(Y_1) & \text{SCP}(Y_1, Y_2) & \cdots & \text{SCP}(Y_1, Y_p) \\ \text{SCP}(Y_2, Y_1) & SS(Y_2) & \cdots & \text{SCP}(Y_2, Y_p) \\ \text{SCP}(Y_p, Y_1) & \text{SCP}(Y_p, Y_2) & \cdots & SS(Y_p) \end{bmatrix}. \tag{11.7}$$

If each element of the sample **SCP** matrix is divided by N or df, it yields a sample estimate **S** of the population variance/covariance matrix $\mathbf{\Sigma}$ (as shown in Equation 11.4).

In a one-way MANOVA, the total **SCP** matrix is partitioned into between-group and within-group components:

$$\mathbf{SCP}_{\text{total}} = \mathbf{SCP}_{\text{between}} + \mathbf{SCP}_{\text{within}}. \tag{11.8}$$

Note that the partition of **SCP** matrices in MANOVA (Equation 11.8) is analogous to the partition of sums of squares in univariate ANOVA (Equation 11.6), and in fact, Equation 11.8 would reduce to Equation 11.6 when p, the number of outcome variables, equals 1.

The same analogy holds for factorial ANOVA. In a balanced univariate factorial ANOVA, the SS_{total} is divided into components that represent the main effects and interactions:

$$SS_{\text{total}} = SS_A + SS_B + SS_{A\times B} + SS_{\text{within}}. \tag{11.9}$$

An analogous partition of the **SCP** matrix can be made in a factorial MANOVA:

$$\mathbf{SCP}_{\text{total}} = \mathbf{SCP}_A + \mathbf{SCP}_B + \mathbf{SCP}_{A\times B} + \mathbf{SCP}_{\text{within}}. \tag{11.10}$$

See Tabachnick and Fidell (2018) for a completely worked empirical example that demonstrates the computation of the elements of the total, between-group, and within-group **SCP** matrices.

The approach to significance testing (to assess the null hypothesis of no group differences) is rather different in MANOVA compared with univariate ANOVA. In univariate ANOVA, each SS term is converted to a mean square (MS) by dividing it by the appropriate degrees of freedom (df). In univariate ANOVA, the test statistic for each effect is $F_{\text{effect}} = MS_{\text{effect}}/MS_{\text{within}}$. F values large enough to exceed the critical values for F were interpreted as evidence for significant differences among group means; an eta squared effect size was also reported to indicate what proportion of variance in scores on the individual Y outcome variable was associated with group membership in the sample data.

To summarize information about sums of squares and sums of cross products for a one-way MANOVA (Equation 11.8) with a set of p outcome variables, we need a single number that tells us how much variance for the set of p outcome variables is represented by each of the **SCP** matrix terms in Equation 11.8. The determinant of an **SCP** matrix provides the summary information that is needed. A nontechnical introduction to the idea of a determinant is therefore needed before we can continue to discuss significance testing in MANOVA.

The notation $|\mathbf{SCP}|$ refers to the determinant of the **SCP** matrix. The determinant of a matrix is a way of describing the variance for a set of variables considered jointly, controlling for or removing any duplicated or redundant information contained in the variables (due to linear correlations among the variables). For a 2 × 2 matrix with elements labeled *a*, *b*, *c*, and *d*, the determinant of the matrix is obtained by taking the products of the elements on the major diagonal of the matrix (from upper left to lower right) minus the products of the elements on the minor diagonal (from lower left to upper right), as in the following example:

$$\mathbf{A} = \begin{bmatrix} a & b \\ c & d \end{bmatrix}.$$

The determinant of **A**, $|\mathbf{A}| = (a \times d) - (c \times b)$.

One type of matrix for which a determinant can be computed is a correlation matrix, **R**. For two variables, Y_1 and Y_2, the correlation matrix **R** consists of the following elements:

$$\mathbf{R} = \begin{bmatrix} 1.00 & r_{12} \\ r_{12} & 1.00 \end{bmatrix},$$

and the determinant of **R**, $|\mathbf{R}| = (1 \times 1) - (r_{12} \times r_{12}) = 1 - r^2_{12}$.

Note that $(1 - r^2_{12})$ describes the proportion of nonshared variance between Y_1 and Y_2; it allows us to summarize how much variance a weighted linear composite of Y_1 and Y_2 has after we correct for or remove the overlap or shared variance in this set of two variables that is represented by r^2_{12}. In a sense, the determinant is a single summary index of variance for a set of *p* dependent variables that corrects for multicollinearity or shared variance among the *p* variables. When a matrix is larger than 2 × 2, the computation of a determinant becomes more complex, but the summary information provided by the determinant remains essentially the same: A determinant tells us how much variance linear composites of the variables $Y_1, Y_2, \ldots, Y_p$ have when we correct for correlations or covariances among the *Y* outcome variables.

In MANOVA, our test statistic takes a different form from that in univariate ANOVA; in univariate ANOVA, *F* is the ratio $MS_{\text{effect}}/MS_{\text{error}}$. In MANOVA, the omnibus test statistic is a ratio of determinants of the matrix that represents "error" or within-group sums of cross products in the numerator, compared with the sum of the matrices that provide information about "effect plus error" in the denominator. In contrast to *F*, which is a signal-to-noise ratio that increases in magnitude as the magnitude of between-group differences becomes large relative to error, Wilks' Λ is an estimate of the proportion of variance in outcome variable scores that is *not* predictable from group membership. In a univariate one-way ANOVA, Wilks' Λ corresponds to $SS_{\text{within}}/SS_{\text{total}}$. For a univariate one-way ANOVA, Wilks' Λ is interpreted as the proportion of variance in scores for a single *Y* outcome variable that is *not* predictable from or associated with group membership. One computational formula for multivariate Wilks' Λ, which is essentially a ratio of the overall error or within-group variance to the total variance for the set of *p* intercorrelated outcome variables, is as follows:

$$\Lambda = \frac{|\mathbf{SCP}_{\text{error}}|}{|\mathbf{SCP}_{\text{effect}} + \mathbf{SCP}_{\text{error}}|}. \quad (11.11)$$

$|\mathbf{SCP}_{\text{error}}|$ is a measure of the generalized variance of the set of *p* dependent variables within groups; it is, therefore, an estimate of the within-group variance in scores for a set of one or more discriminant functions. $|\mathbf{SCP}_{\text{effect}} + \mathbf{SCP}_{\text{error}}|$ (which is equivalent to $|\mathbf{SCP}_{\text{total}}|$ in a one-way MANOVA) is a measure of the variance of the set of *p* dependent variables, including both within- and between-group sources of variation and correcting for intercorrelations among the variables. In a one-way MANOVA, Wilks' Λ estimates the ratio of unexplained variance to total variance; it is equivalent to $1 - \eta^2$. In a situation where the number of dependent variables, *p*, is equal to 1, Λ reduces to $SS_{\text{within}}/SS_{\text{total}}$ or $1 - \eta^2$. Usually, a researcher wants the value of Λ to be small (because a small Λ tends to correspond to relatively large between-group differences relative to the amount of variability of scores within groups). Lambda can be converted to an *F* or a chi-square to assess statistical significance.

Thus, while a higher value of F in ANOVA typically indicates a *stronger* association between scores on the Y outcome variable and group membership (other factors being equal), a higher value of Wilks' Λ implies a *weaker* association between scores on the Y outcome variables and group membership. We can calculate an estimate of partial eta squared—that is, the proportion of variance in scores on the entire set of outcome variables that is associated with group differences for one specific effect in our model, statistically controlling for all other effects in the model—as follows:

$$\eta^2_{\text{effect}} = 1 - \Lambda_{\text{effect}}. \tag{11.12}$$

The SPSS GLM procedure converts Wilks' Λ to an F ratio; this conversion is only approximate in many situations. Note that as the value of Λ decreases (other factors being equal), the value of the corresponding F ratio tends to increase. This conversion of Λ to F makes it possible to use critical values of the F distribution to assess the overall significance of the MANOVA, that is, to have an omnibus test of equality that involves all the $Y_1, Y_2, \ldots, Y_p$ variables and all k groups.

Tabachnick and Fidell (2018) provided formulas for this approximate conversion from Λ to F as follows. This information is provided here for the sake of completeness; SPSS GLM provides this conversion. First, it is necessary to calculate intermediate terms, denoted by s and y. In a one-way MANOVA with k groups, a total of N participants, and p outcome variables, $df_{\text{effect}} = k - 1$ and $df_{\text{error}} = N - k - 1$.

$$s = \sqrt{\frac{p^2 \times \left(df_{\text{effect}}\right)^2 - 4}{p^2 + \left(df_{\text{effect}}\right)^2 - 5}}, \tag{11.13}$$

$$y = \Lambda^{1/s}. \tag{11.14}$$

Next, the degrees of freedom for the F ratio, df_1 and df_2, are calculated as follows:

$$df_1 = p \times (df_{\text{effect}}), \tag{11.15}$$

$$df_2 = s \times \left[df_{error} - \frac{p - df_{\text{effect}} + 1}{2}\right] - \left[\frac{p \times df_{\text{effect}} - 2}{2}\right]. \tag{11.16}$$

Finally, the conversion from Λ to an approximate F with df_1, df_2 degrees of freedom is given by

$$\text{Approximate } F\left(df_1, df_2\right) = \left(\frac{1-y}{y}\right)\left(\frac{df_2}{df}\right). \tag{11.17}$$

If the overall F for an effect (such as the main effect for types of stress in the hypothetical research example described earlier) is statistically significant, it implies that there is at least one significant contrast between groups and that this difference can be detected when the entire set of outcome variables is examined. Follow-up analyses are required to assess which of the various contrasts among groups are significant and to evaluate what patterns of scores on outcome variables best describe the nature of the group differences.

11.9 MULTIVARIATE TEST STATISTICS

The SPSS GLM procedure provides a set of four multivariate test statistics for the null hypotheses shown in Equations 11.2 and 11.3, including Wilks' lambda, **Hotelling's trace**, Pillai's trace, and **Roy's largest root**. In some research situations, one of the other three test statistics might be preferred to Wilks' Λ. In situations where a factor has only two levels and the effect has only 1 *df*, decisions about statistical significance based on these tests do not differ. In more complex situations where a larger number of groups are compared, particularly in situations that involve violations of assumptions of multivariate normality, the decision whether to reject the null hypothesis may be different depending on which of the four multivariate test statistics the data analyst decides to report. When assumptions for MANOVA are seriously violated, researchers often choose to report Pillai's trace (which is more robust to violations of assumptions).

All four multivariate test statistics are derived from the eigenvalues that are associated with the discriminant functions that best differentiate among groups. Discriminant functions are created by forming optimal weighted linear composites of scores on the quantitative predictor variables and the information that is provided by the eigenvalue associated with each discriminant function. In MANOVA, as in DA, one or several discriminant functions are created by forming weighted linear composites of the quantitative variables. These weighted linear combinations of variables are constructed in a way that maximizes the values of $SS_{\text{between}}/SS_{\text{within}}$. That is, the first discriminant function is the weighted linear composite of $Y_1, Y_2, \ldots, Y_p$ that has the largest possible $SS_{\text{between}}/SS_{\text{within}}$ ratio. The second discriminant function is uncorrelated with the first discriminant function and has the maximum possible $SS_{\text{between}}/SS_{\text{within}}$. The number of discriminant functions that can be obtained for each effect in a factorial MANOVA is usually just the minimum of these two values: df_{effect} and p. The discriminant functions are obtained by computing eigenvalues and eigenvectors. For each discriminant function D_i, we obtain a corresponding eigenvalue λ_i:

$$\lambda_i = SS_{\text{between}}/SS_{\text{within}}. \tag{11.18}$$

That is, the eigenvalue λ_i for each discriminant function tells us how strongly the scores on that discriminant function are related to group membership. When $SS_{\text{between}}/SS_{\text{within}}$ is large, it implies a stronger association of discriminant function scores with group membership. Thus, large eigenvalues are associated with discriminant functions that show relatively large between-group differences and that have relatively small within-group variances.

Because an eigenvalue contains information about the ratio of within- and between-group sums of squares, it is possible to derive a canonical correlation for each discriminant function from its eigenvalue, that is, a correlation describing the strength of the relation between group membership and scores on that particular weighted linear combination of variables.

$$r_{ci} = \sqrt{\frac{\lambda_i}{1+\lambda_i}}. \tag{11.19}$$

To summarize how well a *set* of discriminant functions collectively explains variance, we need to combine the variance that is (or is not) explained by each of the individual discriminant functions. The summary can be created by multiplying terms that involve the eigenvalues, as in Wilks' Λ, or by adding terms that involve the eigenvalues, as in Pillai's trace or Hotelling's trace. The three multivariate test statistics—Wilks' Λ, Pillai's trace, and Hotelling's trace—represent three different ways of combining the information about between- versus within-group differences (contained in the eigenvalues) across multiple discriminant functions. Note that the value of Wilks' Λ obtained by multiplication in Equation 11.21 is equivalent to the Wilks' Λ obtained by taking a ratio of determinants of SCP matrices as shown earlier in Equation 11.11. Note that the symbol ∏ in Equation 11.20 indicates that the terms that follow should be multiplied:

$$\text{Wilks' } \Lambda = \prod[1/(1+\lambda_i)], \tag{11.20}$$

$$\text{Pillai's trace} = \Sigma[\lambda_i/(1 + \lambda_i)], \tag{11.21}$$

$$\text{Hotelling's trace } T = \Sigma\lambda_i. \tag{11.22}$$

The application of simple algebra helps clarify just what information each of these multivariate tests includes (as described in Bray & Maxwell, 1985). First, consider Wilks' Λ. We know that $\lambda = SS_{\text{between}}/SS_{\text{within}}$. Thus, $1/(1 + \lambda) = 1/[1 + (SS_{\text{between}}/SS_{\text{within}})]$.

We can simplify this expression if we multiply both the numerator and the denominator by SS_{within} to obtain $SS_{\text{within}}/(SS_{\text{within}} + SS_{\text{between}})$; this ratio is equivalent to $(1 - \eta^2)$. Thus, the $1/(1 + \lambda)$ terms that are multiplicatively combined when we form Wilks' Λ each correspond to the proportion of error or within-group variance for the scores on one discriminant function. A larger value of Wilks' Λ therefore corresponds to a larger collective error (within-group variance) across the set of discriminant functions. A researcher usually hopes that the overall Wilks' Λ will be small.

Now, let's examine the terms included in the computation of Pillai's trace. Recall from Equation 11.19 that $\sqrt{\lambda/(1+\lambda)} = r_{ci}$. Thus, Pillai's trace is just the sum of r^2_{ci} across all the discriminant functions. A larger value of Pillai's trace indicates a higher proportion of between-group or explained variance.

Hotelling's trace is the sum of the eigenvalues across discriminant functions; thus, a larger value of Hotelling's trace indicates a higher proportion of between-group or explained variance. Note that **Hotelling's T^2**, discussed in some multivariate statistics textbooks, is different from Hotelling's trace, although they are related, and Hotelling's trace value can be converted to Hotelling's T^2. A researcher usually hopes that Wilks' Λ will be relatively small and that Pillai's trace and Hotelling's trace will be large, because these outcomes indicate that the proportion of between-group (or explained) variance is relatively high.

The fourth multivariate test statistic reported by the SPSS GLM procedure is called Roy's largest root (in some textbooks, such as Harris, 2001, it is called **Roy's greatest characteristic root**, or *gcr*). Roy's largest root is unique among these summary statistics; unlike the first three statistics, which summarize explained variance across all the discriminant functions, Roy's largest root includes information only for the *first* discriminant function:

$$\text{Roy's largest root} = \lambda_1/(1 + \lambda_1). \tag{11.23}$$

Thus, Roy's largest root is equivalent to the squared canonical correlation for the first discriminant function alone. SPSS provides significance tests (either exact or approximate F's) for each of these multivariate test statistics. Among these four multivariate test statistics, Pillai's trace is believed to be the most robust to violations of assumptions; Wilks' Λ is the most widely used and therefore more likely to be familiar to readers. Roy's largest root is not very robust to violations of assumptions, and under some conditions, it may also be less powerful than Wilks' Λ; however, Roy's largest root may be preferred in situations where the researcher wants a test of significance for only the first discriminant function.

To summarize, Wilks' Λ, Pillai's trace, and Hotelling's trace all test the same hypothesis: that a set of discriminant functions considered jointly provides significant discrimination across groups. When significant differences are found, it is typical to follow up this omnibus test with additional analyses to identify (a) which discriminating variables are the most useful, (b) which particular groups are most clearly differentiated, or (c) whether the weighted linear combinations that discriminate between some groups differ from the weighted linear combinations that best discriminate among other groups. All these issues may be considered in the context of more complex designs that may include repeated measures and covariates.

11.10 FACTORS THAT INFLUENCE THE MAGNITUDE OF WILKS' Λ

As in DA, MANOVA is likely to yield significant differences between groups when there are one or several significant univariate differences on *Y* means, although the inclusion of several nonsignificant *Y* variables may lead to a nonsignificant overall MANOVA even when there are one or two *Y* variables that significantly differ between groups. MANOVA sometimes also reveals significant differences between groups that can be detected only by examining two or more of the outcome variables jointly. In MANOVA, as in multiple regression and many other multivariate analyses, adding "garbage" variables that do not provide predictive information may reduce statistical power. MANOVA, like DA, sometimes yields significant group differences for sets of variables even when no individual variable differs significantly across groups. The magnitude and signs of correlations among *Y* variables affect the power of MANOVA and the confidence of interpretations in experimental studies in complex ways (for further explanation, see Bray & Maxwell, 1985).

11.11 EFFECT SIZE FOR MANOVA

In a univariate ANOVA, Wilks' Λ corresponds to $SS_{\text{within}}/SS_{\text{total}}$; this, in turn, is equivalent to $1 - \eta^2$, the proportion of within-group (or error) variance in *Y* scores. In a univariate ANOVA, an estimate of η^2 may be obtained by taking

$$\eta^2 = 1 - \Lambda. \tag{11.24}$$

This equation may also be used to calculate an estimate of effect size for each effect in a MANOVA. Tabachnick and Fidell (2018) suggest a correction that yields a more conservative estimate of effect size in MANOVA, in situations where $s > 1$:

$$\text{Partial } \eta^2 = 1 - \Lambda^{1/s}, \tag{11.25}$$

where the value of *s* is given by Equation 11.13 above.

11.12 STATISTICAL POWER AND SAMPLE SIZE DECISIONS

Stevens (2009) provided a table of recommended minimum cell *n*'s for MANOVA designs for a reasonably wide range of research situations; his table is reproduced here (Table 11.1). The recommended minimum *n* per group varies as a function of the following:

1. The (unknown) population multivariate effect size the researcher wants to be able to detect
2. The desired level of statistical power (Table 11.1 provides sample sizes that should yield statistical power of approximately .70)
3. The alpha level used for significance tests (often set at $\alpha = .05$)
4. The number of levels or groups (*k*) in the factor (usually on the order of three to six)
5. The number of dependent variables, *p* (the table from Stevens provides values for *p* from 3 to 6)

Table 11.1 Sample Size Required per Cell in a k-Group One-Way MANOVA, for Estimated Power of .70 Using α = .05 With Three to Six Outcome Variables

	Number of Groups (k)			
Effect Size[a]	3	4	5	6
Very large	12–16	14–18	15–19	16–21
Large	25–32	28–36	31–40	33–44
Medium	42–54	48–62	54–70	58–76
Small	92–120	105–140	120–155	130–170

Source: Adapted from Stevens (2009).

[a]Stevens provided formal algebraic definitions for multivariate effect size based on work by Lauter (1978). In practice, researchers usually do not have enough information about the strength of effects to be able to substitute specific numerical values into these formulas to compute numerical estimates of effect size. Researchers who have enough information to guess the approximate magnitude of multivariate effect size (small, medium, or large) may use this table to make reasonable judgments about sample size.

Stevens (2009) provided algebraic indices of multivariate effect size, but it would be difficult to apply these formulas in many research situations. In practice, a researcher can choose one of the verbal effect size labels Stevens suggested (these range from "small" to "very large") without resorting to computations that, in any case, would require information researchers generally do not have prior to undertaking data collection. Given specific values for effect size, α, power, k, and p, Table 11.1 can be used to look up reasonable minimum cell n's for many MANOVA research situations.

For example, suppose that a researcher sets $\alpha = .05$ for a MANOVA with $k = 3$ groups and $p = 6$ outcome variables and believes that the multivariate effect size is likely to be very large. Using Table 11.1, the number of participants per group would need to be about 12 to 16 to have statistical power of .70. If the effect size is only medium, then the number of participants required per group to achieve power of .70 would increase to 42 to 54.

Note that in MANOVA, as in many other tests, we have the same paradox. When N is large, we have greater statistical power for Box's M test and the Levene test, which assess whether the assumptions about equality of variances and covariances across groups are violated. However, it is in studies where the group n's are small and unequal across groups that violations of these assumptions are more problematic. When the total N is small, it is advisable to use a relatively large alpha value for the Box's M and Levene tests, for example, to decide that the homogeneity assumption is violated for $p < .10$ or $p < .20$; these larger alpha levels make it easier to detect violations of the assumptions in the small-N situations where these violations are more problematic. On the other hand, when the total N is large, it may make sense to use a smaller alpha level for these tests of assumptions; when total N is large, we might require Box's M test with $p < .01$ or even $p < .001$ as evidence of significant violation of the assumption of the homogeneity of $\mathbf{\Sigma}$ across groups.

11.13 ONE-WAY MANOVA: CAREER GROUP DATA

This analysis uses the same data used for the DA example in the previous chapter. A DA was performed for a small subset of data for high school seniors from the Project Talent survey. Students chose one of three different types of future careers. The career group factor had these three levels: Group 1, business; Group 2, medical; and Group 3, teaching or social work. Scores on six achievement and aptitude tests were used to predict membership in these future career groups. The DA reported in the previous chapter corresponds to the following one-way MANOVA, which compares the set of means on these six test scores across the three career choice groups.

To run a MANOVA, the following menu selections are made: <Analyze> → <General Linear Model> → <Multivariate> (see Figure 11.1). The choice of the <Multivariate> option makes it possible to include a list of several outcome variables. When the SPSS GLM procedure is used to run a univariate factorial ANOVA (i.e., a factorial ANOVA with only one *Y* outcome variable), the <Univariate> option is selected to limit the analysis to just a single outcome variable.

The main dialog box for the GLM multivariate procedure appears in Figure 11.2. For this one-way MANOVA, the categorical variable cargroup was placed in the box headed "Fixed Factor(s)," and the six test scores (English, reading, etc.) were placed in the box headed "Dependent Variables." To request optional statistics such as univariate means and univariate *F* tests and tests of the homogeneity of variance/covariance matrices across groups, we could click the Options button and select these statistics. This was omitted in the present example because this information is already available in the DA results in the previous chapter. To request post hoc tests to compare all possible pairs of group means (in this instance, business vs. medical, medical vs. teaching or social work, and business vs. teaching or social work career groups), the Post Hoc button was used to open up the dialog box that appears in Figure 11.3. The name of the factor for which post hoc tests are requested (cargroup) was placed in the right-hand pane under the heading "Post Hoc Tests for," and a checkbox was used to select "Tukey," that is, the Tukey honestly significant difference test. Note that these post hoc tests are univariate; that is, they assess which groups differ separately for each of the six outcome variables. The SPSS syntax obtained through these menu selections is given in Figure 11.4.

Figure 11.1 Menu Selections for a MANOVA in the SPSS GLM Procedure

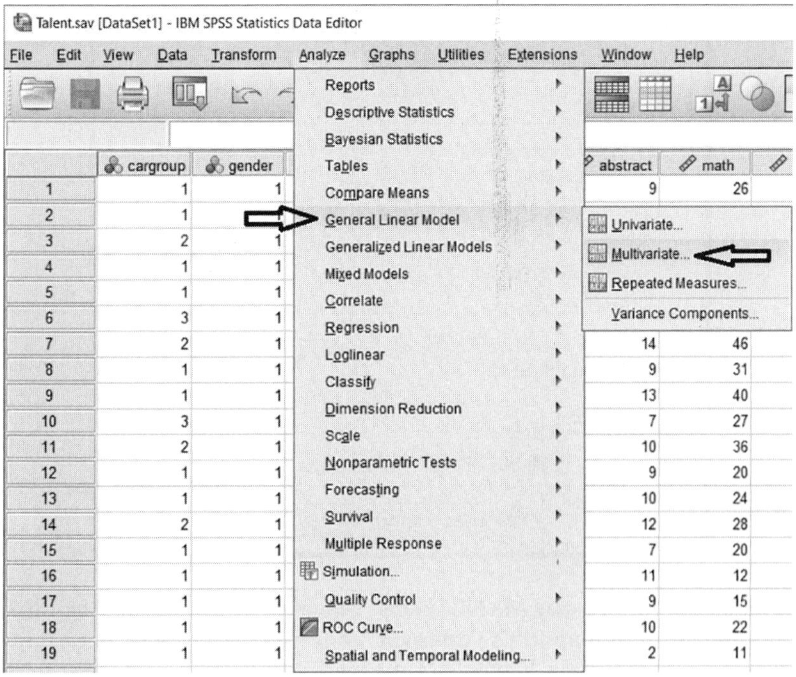

Note: GLM is the SPSS general linear model procedure (not generalized linear model). The data used for this example were previously used in Chapter 10, on discriminant analysis.

Figure 11.2 Main Dialog Box for One-Way MANOVA

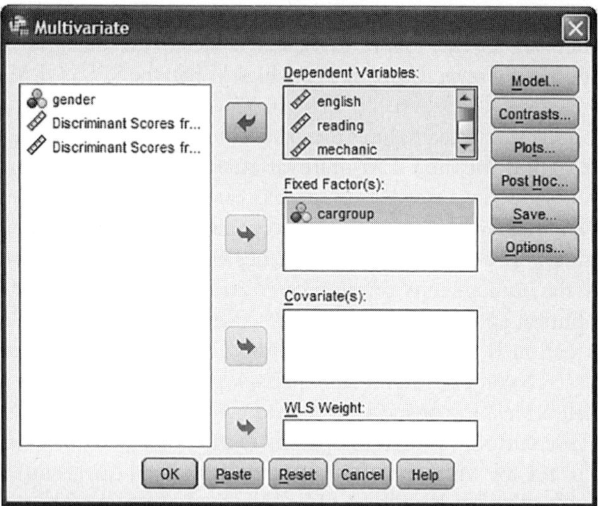

Figure 11.3 Tukey Honestly Significant Difference Post Hoc Tests Requested for One-Way MANOVA

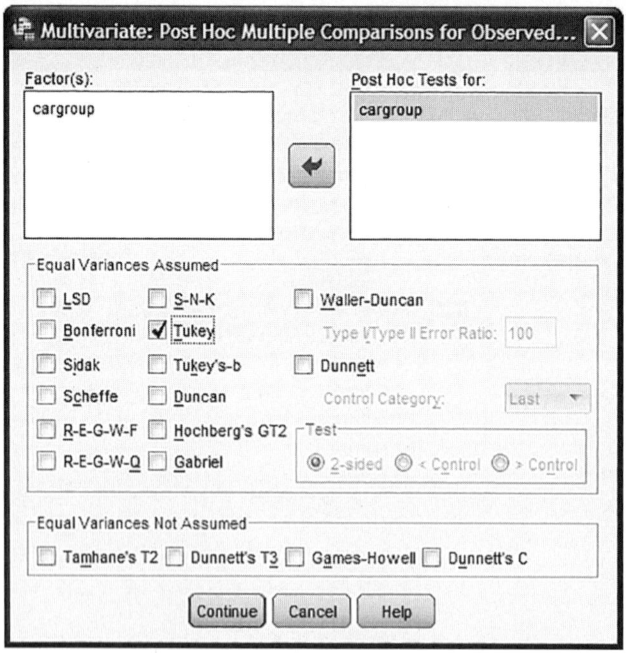

The output for the one-way MANOVA that compares six test scores (English, reading, etc.) across three career groups (that correspond to levels of the factor named cargroup) is given in the next three figures. Figure 11.5 summarizes the multivariate tests for the null hypothesis that appears in Equation 11.2—that is, the null hypothesis that the entire list of means on all six outcome variables is equal across the populations that correspond to the three groups in this study. These results are consistent with the DA results for the same data

Figure 11.4 SPSS Syntax for One-Way MANOVA

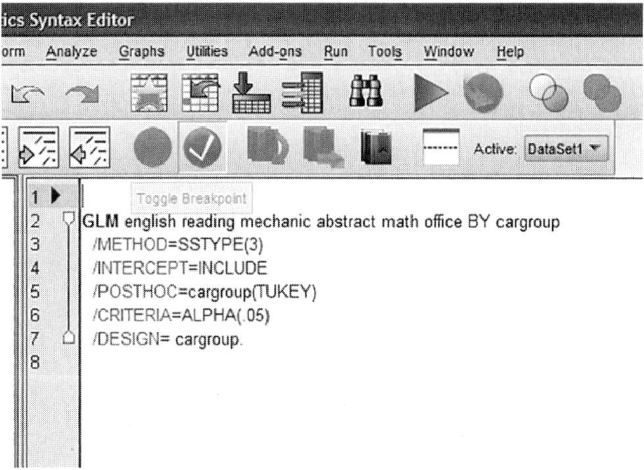

presented earlier. Box's M test was requested; using α = .10 as the criterion for statistical significance, the Box's M did not indicate a significant violation of the assumption of equality of variance/covariance matrices across these three populations (p = .045). If we set α = .05, however, Box's M would be judged statistically significant. The violation of this assumption may decrease statistical power, and it may also make the "exact" p value in the SPSS output an inaccurate estimate of the true risk for committing a Type I error. Because Pillai's trace is somewhat more robust to violations of this assumption than the other multivariate test statistics, this is the test that will be reported. Figure 11.5 includes two sets of multivariate tests. The first part of the figure (effect = intercept) tests whether the vector of means on all p outcome variables is equal to 0 for all variables. This test is rarely of interest and it is not interpreted here. The second part of the figure (effect = cargroup) reports the four multivariate tests of the null hypothesis in Equation 11.2. The main effect for cargroup was statistically significant, Pillai's trace = .317, $F(12, 314)$ = 4.93, partial η^2 = .16. This result suggests that it is likely that at least one pair of groups differs significantly on one outcome variable or on some combination of outcome variables; an effect size of η^2 = .16 could be considered large using the effect size labels suggested in Tables 1.1 and 1.2 in Chapter 1. This omnibus test suggests that at least one group differs significantly from the other two groups (or perhaps all three groups differ significantly) and that this difference may involve one or several of the p outcome variables. Follow-up analyses are required to understand the nature of the cargroup main effect: which groups differed significantly and on which variable(s).

Follow-up questions about the nature of main effects can be examined in two different ways. First, we can examine univariate follow-up tests, for example, Tukey post hoc tests that provide information about whether scores on each individual Y outcome variable (such as scores on the math test) differ between each pair of career groups (such as the business and medicine career groups). As discussed previously, univariate follow-up tests raise many problems because they involve an inflated risk for Type I error and also ignore intercorrelations among the Y outcome variables. Second, we can do multivariate follow-up tests; for example, we can do a DA on the three career groups to assess how the groups differ on weighted linear combinations of the six test scores; these multivariate follow-up analyses provide information about the nature of group differences on Y outcome variables when we control for intercorrelations among the Y outcome variables. It can be useful to perform both univariate and multivariate follow-up analyses; we just need to keep in mind that they provide different kinds of information. In some research situations, the nature of group differences that is detected by using univariate follow-up analyses and detected by multivariate follow-up

Figure 11.5 Output From One-Way MANOVA: Multivariate Tests for Main Effect on Cargroup Factor

Multivariate Tests[c]

Effect		Value	F	Hypothesis df	Error df	Sig.	Partial Eta Squared
Intercept	Pillai's Trace	.987	1917.575[a]	6.000	156.000	.000	.987
	Wilks's Lambda	.013	1917.575[a]	6.000	156.000	.000	.987
	Hotelling's Trace	73.753	1917.575[a]	6.000	156.000	.000	.987
	Roy's Largest Root	73.753	1917.575[a]	6.000	156.000	.000	.987
cargroup	Pillai's Trace	.317	4.926	12.000	314.000	.000	.158
	Wilks's Lambda	.685	5.406[a]	12.000	312.000	.000	.172
	Hotelling's Trace	.456	5.887	12.000	310.000	.000	.186
	Roy's Largest Root	.448	11.734[b]	6.000	157.000	.000	.310

a. Exact statistic

b. The statistic is an upper bound on F that yields a lower bound on the significance level.

c. Design: Intercept+cargroup

analyses may be quite similar. In other research situations, the conclusions about the nature of group differences suggested by multivariate follow-up analyses may be quite different from those suggested by univariate analyses, and in such situations, provided that the assumptions for MANOVA and DA are reasonably well satisfied, it may be preferable to focus primarily on multivariate follow-up tests in the discussion of results.

The univariate Tukey post hoc test results that appear in Figures 11.6 and 11.7 answer the first question: Which career groups differed significantly on which individual variables? Note that these are univariate follow-up tests. While the Tukey procedure provides some protection from the inflated risk for Type I error that arises when we make multiple tests between pairs of groups, this procedure does not control for the inflated risk for Type I error that arises when we run six separate tests, one for each of the outcome variables (such as English, math, etc.). There were no significant differences among the three career groups in scores on English, reading, mechanical reasoning, or abstract reasoning. However, group differences were found for scores on math and office skills. The homogeneous subsets reported in Figure 11.7 indicate that the business group scored significantly lower on math ($M = 20.05$) relative to the other two career groups ($M = 25.59$ for teaching, $M = 26.37$ for medicine); the last two groups did not differ significantly from each other. Also, the business group scored significantly higher on office skills ($M = 28.84$) compared with the other two groups (medicine, $M = 18.37$, and teaching or social work, $M = 20.62$); the last two groups did not differ significantly. The limitation of this follow-up analysis is that it does not correct for intercorrelations among the six test scores.

A multivariate follow-up analysis that takes intercorrelations among the six test scores into account is DA. A DA for this set of data is reported in the previous chapter. The highlights of the results are as follows: Two discriminant functions were formed, but the second discriminant function did not contribute significantly to discrimination between groups, and therefore the coefficients for the second discriminant function were not interpreted. The standardized discriminant function coefficients (from Table 11.1) that form the first discriminant function (D_1) based on the z scores on the six tests are as follows:

$$D_1 = +.42z_{\text{English}} + .02z_{\text{reading}} + .44z_{\text{mechanic}} - .12z_{\text{abstract}} - .79z_{\text{math}} + .75z_{\text{office}}.$$

If we use an arbitrary cutoff value of .50 as the criterion for a "large" discriminant function coefficient, only two variables had discriminant function coefficients that exceeded .5 in absolute value: z_{math}, with a coefficient of −.79, and z_{office}, with a coefficient of +.75. High scores on D_1 were thus associated with low z scores on math and high z scores on office skills.

Figure 11.6 Tukey Post Hoc Comparisons Among Means on Each of Six Outcome Test Scores for All Pairs of Career Groups

Multiple Comparisons

Tukey HSD

Dependent Variable	(I) cargroup	(J) cargroup	Mean Difference (I-J)	Std. Error	Sig.	95% Confidence Interval Lower Bound	95% Confidence Interval Upper Bound
english	business	medicine	1.29	2.189	.826	–3.89	6.47
		teach_social	–.28	1.907	.988	–4.79	4.23
	medicine	business	–1.29	2.189	.826	–6.47	3.89
		teach_social	–1.57	2.522	.808	–7.54	4.40
	teach_social	business	.28	1.907	.988	–4.23	4.79
		medicine	1.57	2.522	.808	–4.40	7.54
reading	business	medicine	–1.71	1.945	.655	–6.31	2.89
		teach_social	–2.46	1.695	.317	–6.47	1.55
	medicine	business	1.71	1.945	.655	–2.89	6.31
		teach_social	–.75	2.241	.940	–6.05	4.55
	teach_social	business	2.46	1.695	.317	–1.55	6.47
		medicine	.75	2.241	.940	–4.55	6.05
abstract	business	medicine	–.62	.586	.539	–2.01	.76
		teach_social	–.61	.510	.457	–1.82	.60
	medicine	business	.62	.586	.539	–.76	2.01
		teach_social	.01	.675	1.000	–1.58	1.61
	teach_social	business	.61	.510	.457	–.60	1.82
		medicine	–.01	.675	1.000	–1.61	1.58
math	business	medicine	–6.32*	1.749	.001	–10.46	–2.18
		teach_social	–5.54*	1.523	.001	–9.14	–1.94
	medicine	business	6.32*	1.749	.001	2.18	10.46
		teach_social	.78	2.014	.921	–3.98	5.55
	teach__social	business	5.54*	1.523	.001	1.94	9.14
		medicine	–.78	2.014	.921	–5.55	3.98
office	business	medicine	10.47*	1.849	.000	6.09	14.84
		teach_social	8.22"	1.611	.000	4.41	12.03
	medicine	business	–10.47*	1.849	.000	–14.84	–6.09
		teach_social	–2.25	2.130	.544	–7.28	2.79
	teach_social	business	–8.22*	1.611	.000	–12.03	–4.41
		medicine	2.25	2.130	.544	–2.79	7.28

The business group had a much higher mean score on D_1 than the other two career groups; the medicine and teaching or social work groups had means on D_1 that were very close together. In this example, the DA leads us essentially to the same conclusion as the univariate ANOVAs: Whether or not we control for the other four test scores, the nature of the group differences was that the office group scored lower on math and higher on office skills than the other two career groups; the other two career groups (medicine and teaching or social work) did not differ significantly on any of the outcome variables. Note that it is possible for the nature of the differences among groups to appear quite different when discriminant functions are examined (as a multivariate follow-up) than when univariate ANOVAs and Tukey post hoc tests are examined (as univariate follow-ups). However, in this example, the multivariate and univariate follow-ups lead to similar conclusions about which groups differed and which variables differed across groups.

Figure 11.7 Selected Homogeneous Subsets: Outcomes From Tukey Tests (for Four of the Six Outcome Test Scores)

english

Tukey HSD[a,b]

cargroup	N	Subset
		1
medicine	27	87.15
business	98	88.44
teach_social	39	88.72
Sig.		.760

Means for groups in homogeneous subsets are displayed.
Based on Type III Sum of Squares
The error term is Mean Square(Error) = 101.462.

a. Uses Harmonic Mean Sample Size = 41.162.

b. Alpha = .05.

mechanic

Tukey HSD[a,b]

cargroup	N	Subset
		1
business	98	9.12
teach_social	39	9.49
medicine	27	10.22
Sig.		.399

Means for groups in homogeneous subsets are displayed.
Based on Type III Sum of Squares
The error term is Mean Square(Error) = 14.813.

a. Uses Harmonic Mean Sample Size = 41.162.

b. Alpha = .05.

math

Tukey HSD[a,b]

cargroup	N	Subset	
		1	2
business	98	20.05	
teach_social	39		25.59
medicine	27		26.37
Sig.		1.000	.899

Means for groups in homogeneous subsets are displayed.
Based on Type III Sum of Squares
The error term is Mean Square(Error) = 64.736.

a. Uses Harmonic Mean Sample Size = 41.162.

b. Alpha = .05.

office

Tukey HSD[a,b]

cargroup	N	Subset	
		1	2
medicine	27	18.37	
teach_social	39	20.62	
business	98		28.84
Sig.		.457	1.000

Means for groups in homogeneous subsets are displayed.
Based on Type III Sum of Squares
The error term is Mean Square(Error) = 72.366.

a. Uses Harmonic Mean Sample Size = 41.162.

b. Alpha = .05.

11.14 2 × 3 FACTORIAL MANOVA: CAREER GROUP DATA

The career group data in talent.sav include information on one additional categorical variable or factor: gender (coded 1 = male and 2 = female). Having this information about gender makes it possible to set up a 2 × 3 factorial MANOVA for the career group data (for which a one-way MANOVA was reported in the previous section and DA was reported in the previous chapter). This 2 × 3 factorial involves gender as the first factor (with two levels, 1 = male and 2 = female) and cargroup as the second factor (with 1 = business, 2 = medicine, and 3 = teaching or social work). The same six quantitative outcome variables that were used in the previous one-way MANOVA and DA examples are used in this 2 × 3 factorial MANOVA.

Prior to performing the factorial MANOVA, the SPSS crosstabs procedure was used to set up a table to report the number of cases in each of the six cells of this design. Figure 11.8 shows the table of cell *n*'s for this 2 × 3 factorial MANOVA design. Note that the cell *n*'s in this factorial MANOVA are unequal (and not balanced). Therefore, this is a nonorthogonal factorial design; there is a confound between student gender and career choice. A higher proportion of women than men chose teaching or social work as a future career. Furthermore, the *n*'s in two cells were below 10. These cell *n*'s are well below any of the recommended minimum group sizes for MANOVA (see Table 11.1). The very small number of cases in some cells in this example illustrates the serious problems that can arise when some cells have small *n*'s (although the overall *N* of 164 persons who had nonmissing scores on all the variables involved in this 2 × 3 MANOVA might have seemed reasonably adequate). This 2 × 3 MANOVA example illustrates some of the problems that can arise when there are insufficient data to obtain a clearly interpretable outcome. For instructional purposes, we will go ahead and examine the results of a 2 × 3 factorial MANOVA for this set of data. However, keep in mind that the very small cell *n*'s in Figure 11.8 suggest that the results of factorial MANOVA are likely to be inconclusive for this set of data.

To run this 2 × 3 factorial MANOVA using the GLM procedure in SPSS, the following menu selections were made: <Analyze> → <General Linear Model> → <Multivariate>.

Figure 11.9 shows the GLM dialog box for this MANOVA. Gender and cargroup were entered as fixed factors. The six dependent variables were entered into the box headed "Dependent Variables." The Options button was clicked to request optional statistics; the Multivariate: Options dialog box appears in Figure 11.10. Descriptive statistics including means were requested; in addition, estimates of effect size and tests for possible violations of the homogeneity of variance/covariance matrices were requested. The SPSS syntax that corresponds to the preceding set of menu selections appears in Figure 11.11.

Because of the small sample sizes in many cells, $\alpha = .20$ was used as the criterion for statistical significance; by this criterion, Box's *M* test (in Figure 11.12) did not indicate a

Figure 11.8 Cell *n*'s for the Gender-by-Cargroup Factorial MANOVA

gender * cargroup Crosstabulation

			cargroup			
			business	medicine	teach_social	Total
gender	male	Count	16	6	4	26
		% within gender	61.5%	23.1%	15.4%	100.0%
	female	Count	84	23	36	143
		% within gender	58.7%	16.1%	25.2%	100.0%
Total		Count	100	29	40	169
		% within gender	59.2%	17.2%	23.7%	100.0%

Note: For two of the male groups (medicine and teaching careers), the cell *n* is less than 10.

Figure 11.9 GLM Dialog Box for 2 × 3 Factorial MANOVA

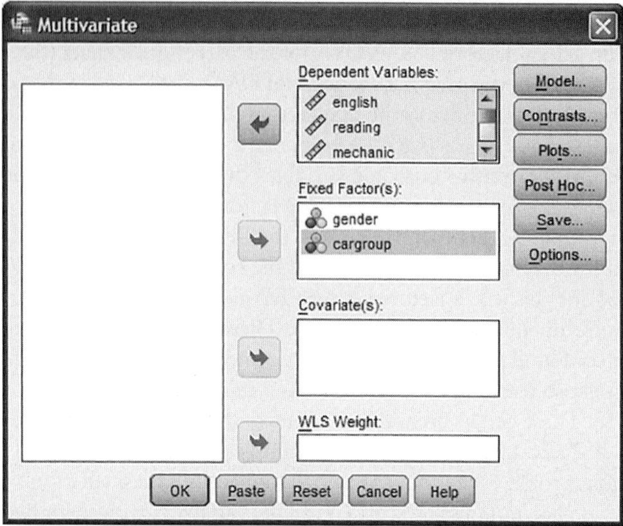

Figure 11.10 Optional Statistics Selected for MANOVA

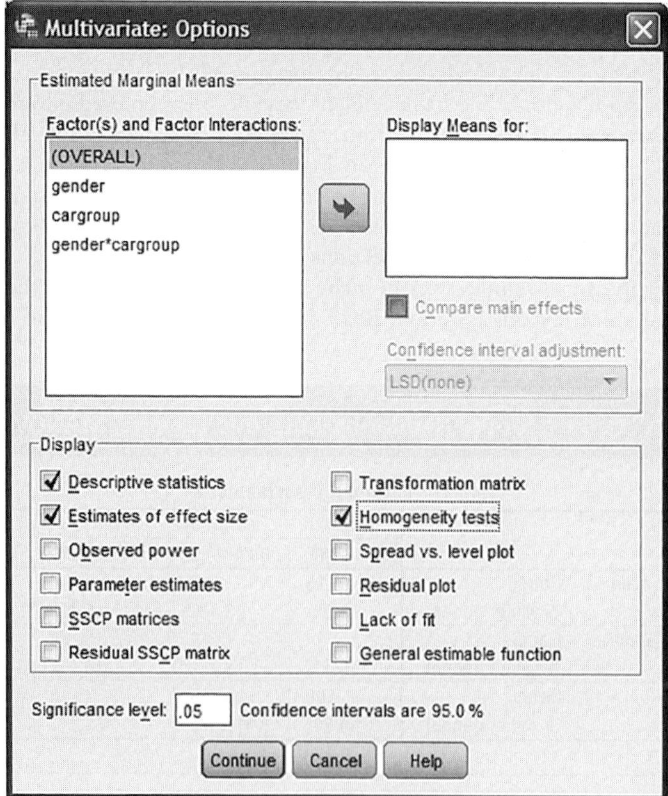

Figure 11.11 SPSS Syntax for GLM 2 × 3 Factorial MANOVA

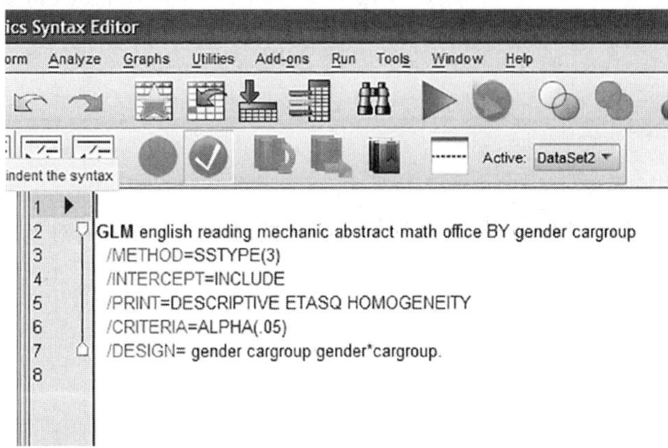
ics Syntax Editor
orm Analyze Graphs Utilities Add-ons Run Tools Window Help
Active: DataSet2
indent the syntax
GLM english reading mechanic abstract math office BY gender cargroup
/METHOD=SSTYPE(3)
/INTERCEPT=INCLUDE
/PRINT=DESCRIPTIVE ETASQ HOMOGENEITY
/CRITERIA=ALPHA(.05)
/DESIGN= gender cargroup gender*cargroup.

Figure 11.12 Selected Output From the 2 × 3 Factorial MANOVA: Box's M Test of Equality of Covariance Matrices

Box's Test of Equality of Covariance Matrices[a]

Box's M	78.722
F	1.097
df1	63
df2	8330.280
Sig.	.279

Tests the null hypothesis that the observed covariance matrices of the dependent variables are equal across groups.

a. Design: Intercept+cargroup+gender+cargroup* gender

statistically significant violation of the assumption of homogeneity of the variance/covariance matrices. In the following example, Wilks' Λ and its associated F values are reported for each effect in the 2 × 3 factorial model.

In a 2 × 3 factorial MANOVA (as in a 2 × 3 factorial ANOVA), we obtain statistical significance tests that address three questions. Is there a significant main effect for Factor A, in this case, gender? In other words, when the entire set of six test scores is considered jointly, do men and women differ significantly in their pattern of test scores? From Figure 11.13, for the main effect of gender, $\Lambda = .67$, $F(6, 153) = 12.53$, $p < .001$, partial $\eta^2 = .33$. Thus, there was a significant gender difference in the overall pattern of test scores on these six aptitude and achievement tests, and the effect size was large (33% of the variance in scores on the single discriminant function that combined the six test scores was associated with gender).

Is there a significant main effect for Factor B, cargroup? From Figure 11.13, for the main effect of cargroup, we obtained $\Lambda = .78$, $F(12, 306) = 3.36$, $p < .001$, partial $\eta^2 = .12$. In other words, it appears that at least one of the three career groups is distinguishable from the other two groups on the basis of some combination of scores on the six tests. Is there a statistically significant interaction between gender and cargroup? From Figure 11.13, the results for this interaction were as follows: $\Lambda = .88$, $F(12, 306) = 1.71$, $p = .063$, partial $\eta^2 = .063$. While a partial η^2 of .063 in the sample suggests a medium effect size for the data in this sample, this interaction was not statistically significant at the conventional $\alpha = .05$ level.

Figure 11.13 Omnibus Multivariate Test Results for the Set of Six Test Scores in the Gender-by-Cargroup MANOVA

Multivariate Tests[c]

Effect		Value	F	Hypothesis df	Error df	Sig.	Partial Eta Squared
Intercept	Pillai's Trace	.974	964.830[a]	6.000	153.000	.000	.974
	Wilks's Lambda	.026	964.830[a]	6.000	153.000	.000	.974
	Hotelling's Trace	37.836	964.830[a]	6.000	153.000	.000	.974
	Roy's Largest Root	37.836	964.830[a]	6.000	153.000	.000	.974
gender	Pillai's Trace	.330	12.534[a]	6.000	153.000	.000	.330
	Wilks's Lambda	.670	12.534[a]	6.000	153.000	.000	.330
	Hotelling's Trace	.492	12.534[a]	6.000	153.000	.000	.330
	Roy's Largest Root	.492	12.534[a]	6.000	153.000	.000	.330
cargroup	Pillai's Trace	.225	3.256	12.000	308.000	.000	.113
	Wilks's Lambda	.781	3.358[a]	12.000	306.000	.000	.116
	Hotelling's Trace	.273	3.459	12.000	304.000	.000	.120
	Roy's Largest Root	.241	6.198[b]	6.000	154.000	.000	.195
gender * cargroup	Pillai's Trace	.125	1.717	12.000	308.000	.062	.063
	Wilks's Lambda	.878	1.713[a]	12.000	306.000	.063	.063
	Hotelling's Trace	.135	1.709	12.000	304.000	.064	.063
	Roy's Largest Root	.092	2.361[b]	6.000	154.000	.033	.084

a. Exact statistic

b. The statistic is an upper bound on F that yields a lower bound on the significance level.

c. Design: Intercept+gender+cargroup+gender * cargroup

This 2 × 3 factorial MANOVA reveals a problem that is not resolvable using the available data. From Figure 11.8, we can see that the cell *n*'s are quite small in several groups; therefore, we do not have good statistical power for detection of an interaction. Even if the gender-by-cargroup interaction had been statistically significant at the .05 level, we simply do not have large enough *n*'s in the groups to make confident statements about the nature of an interaction. In this situation, we cannot reach a clear conclusion on whether there is an interaction, and if there is an interaction, we cannot be confident about understanding the nature of the interaction. There is enough indication of a possible interaction (η^2 = .06 in the sample) so that we cannot completely dismiss the possibility that a properly specified model needs to include a gender-by-cargroup interaction, and we should not proceed to interpret the main effects without taking possible interactions into account. On the other hand, we do not have sufficient information (cell *n*'s are too small) to judge the interaction statistically significant or to provide a good description of the nature of the interaction.

An honest data analyst would decide that, because of the extremely small cell *n*'s, the results of this 2 × 3 factorial MANOVA are inconclusive; these data are not publishable. There is not enough evidence to decide whether a gender-by-cargroup interaction is present and not enough information to describe the nature of the interaction between these factors. It would not be honest to backtrack and present one-way MANOVA (using only cargroup, or only gender, as a factor) because this factorial MANOVA has raised the possibility of an interaction between these factors, and if there is an interaction, a one-way MANOVA (using only gender or only cargroup) would not be a properly specified model. If the data analyst has the chance to recruit larger numbers of cases in the cells with small *n*'s, analysis of a larger data set might provide a clearer description of the data. The follow-up analyses that are presented in subsequent sections serve to illustrate what could be done to further understand the nature of main effects and interactions if a larger data set were available, but these results would not be publishable with such small within-cell *n*'s.

11.14.1 Follow-Up Tests for Significant Main Effects

For instructional purposes only, a few additional follow-up analyses were performed to illustrate what could have been done if there had been enough data to decide whether to focus the interpretation on an interaction or on main effects. If the data had clearly indicated the absence of an interaction between gender and cargroup, then the following additional analyses could have been run to explore the nature of the significant main effects for gender and for cargroup. Multivariate assessment of the nature of group differences could be made by doing a DA separately for each factor. DA results describing the nature of differences across the three career groups represented by the cargroup factor were reported in the previous chapter, and a one-way MANOVA on the cargroup factor was discussed in the preceding section; information from those two analyses could be reported as follow-up information for the MANOVA in this chapter. In the DA, only two out of the six tests (math and office skills) had large weights in the discriminant function that differentiated the three career groups; the same two variables were also statistically significant when univariate post hoc Tukey tests were examined. The Tukey tests reported with the one-way MANOVA (see Figures 11.6 and 11.7) made the nature of the group differences clear: Members of the business career group scored significantly higher on office skills, and also significantly lower on math skills, than members of the medicine and teaching or social work career groups (and the last two groups did not differ from each other on any individual test score variables or on any discriminant functions formed from the test scores). Univariate *F*s were significant for math, $F(2, 158) = 6.83$, $p = .001$, and for office skills, $F(2, 158) = 9.47$, $p = .001$. Table 11.2 shows the univariate means for these variables across the three career groups. The group means shown in Table 11.2 were obtained from Figure 11.15, and the *F* tests and effect sizes were obtained from Figure 11.14.

The direction and magnitude of the differences between groups on these two variables (math and office) were essentially the same whether these variables were examined using DA or by doing two univariate ANOVAs, one on math scores across the three career groups and the other on office scores across the three career groups. In other words, this was not a situation in which controlling for other test scores changed our understanding of how the career groups differed with respect to scores on the math and office tests.

To assess the nature of main effect differences for gender, a DA was performed to see what weighted linear combination of test scores best differentiated men and women, and univariate differences on each test score are reported. Table 11.3 summarizes the male and female group means on the four tests that showed significant gender differences. The nature of the univariate differences was that men tended to score higher on tests of mechanical reasoning and math, whereas women scored higher on tests of English and office skills. In this sample, gender explained more variance in the test scores than was explained by either career choice or the interaction between gender and career choice (partial $\eta^2 = .33$). Four of the six variables showed significant univariate *F*s for gender differences: for English, $F(1, 158) = 5.53$, $p = .02$; for mechanical reasoning, $F(1, 158) = 36.344$, $p < .001$; for math, $F(1, 158) = 9.56$, $p = .002$; and for office, $F(1, 158) = 16.37$, $p < .001$. There were no significant univariate gender differences on the reading or abstract reasoning tests.

A multivariate follow-up analysis to further evaluate the nature of gender differences in test scores is also presented (accompanied again by the warning that, for this data set, we can't be sure whether we should be looking at an interaction of gender and career rather than

Table 11.2 Significant Univariate Differences in Test Scores Across Levels of the Cargroup Factor: Means and Univariate F Ratios for Math and Office Scores

	Business	Medicine	Teaching	F^a	p	Partial η^2
Math	22.5	30.11	24.55	6.84	.001	.080
Office	24.14	16.51	19.13	7.47	.001	.086

[a]Each F ratio in this table had (2, 158) *df*. The *p* values have not been adjusted for inflated risk for Type I error; that is, they correspond to the theoretical risk for Type I error for a single univariate F test. Only outcome variables that had significant univariate main effects for cargroup at the $\alpha = .05$ level are included in this table.

Figure 11.14 Tests of Univariate Differences for Each of the Six Test Scores in the 2 × 3 MANOVA (Gender by Cargroup)

Tests of Between-Subjects Effects

Source	Dependent Variable	Type III Sum of Squares	df	Mean Square	F	Sig.	Partial Eta Squared
Corrected Model	english	948.694[a]	5	189.739	1.943	.090	.058
	reading	283.835[b]	5	56.767	.700	.624	.022
	mechanic	573.758[c]	5	114.752	9.868	.000	.238
	abstract	45.917[d]	5	9.183	1.275	.277	.039
	math	2796.307[e]	5	559.261	9.830	.000	.237
	office	5707.152[f]	5	1141.430	19.353	.000	.380
Intercept	english	459975.928	1	459975.928	4709.674	.000	.968
	reading	72447.144	1	72447.144	893.926	.000	.850
	mechanic	8028.496	1	8028.496	690.429	.000	.814
	abstract	5800.541	1	5800.541	805.134	.000	.836
	math	41331.997	1	41331.997	726.468	.000	.821
	office	24808.601	1	24808.601	420.632	.000	.727
cargroup	english	313.822	2	156.911	1.607	.204	.020
	reading	87.673	2	43.837	.541	.583	.007
	mechanic	25.416	2	12.708	1.093	.338	.014
	abstract	13.202	2	6.601	.916	.402	.011
	math	777.823	2	388.912	6.836	.001	.080
	office	880.704	2	440.352	7.466	.001	.086
gender	english	540.513	1	540.513	5.534	.020	.034
	reading	15.104	1	15.104	.186	.667	.001
	mechanic	422.618	1	422.618	36.344	.000	.187
	abstract	5.519	1	5.519	.766	.383	.005
	math	543.873	1	543.873	9.559	.002	.057
	office	965.260	1	965.260	16.366	.000	.094
cargroup * gender	english	604.573	2	302.287	3.095	.048	.038
	reading	67.485	2	33.743	.416	.660	.005
	mechanic	23.902	2	11.951	1.028	.360	.013
	abstract	20.072	2	10.036	1.393	.251	.017
	math	524.025	2	262.013	4.605	.011	.055
	office	309.867	2	1 54 934	2.627	.075	.032
Error	english	15431.257	158	97.666			
	reading	12804.921	158	81.044			
	mechanic	1837.267	158	11.628			
	abstract	1138.302	158	7204			
	math	8989.321	158	56.894			
	office	9318.726	158	58.979			
Total	english	1294858.000	164				
	reading	195000.000	164				
	mechanic	16872.000	164				
	abstract	15458.000	164				
	math	94137.000	164				
	office	118830.000	164				
Corrected Total	english	16379.951	163				
	reading	13088.756	163				
	mechanic	2411.024	163				
	abstract	1184.220	163				
	math	11785.628	163				
	office	15025.878	163				

at main effects for gender). DA using gender as the categorical variable provides a potential way to summarize the multivariate differences between men and women for the entire set of outcome measures, controlling for intercorrelations among the tests. The nature of the multivariate gender differences can be described by scores on one standardized discriminant function D with the following coefficients (see Figure 11.16):

$$D = +.50z_{\text{English}} + .20z_{\text{reading}} - .75z_{\text{mechanic}} - .02z_{\text{abstract}} - .31z_{\text{math}} + .44z_{\text{office}}.$$

Figure 11.15 Estimated Marginal Means for Univariate Outcomes

1. Grand Mean

Dependent Variable	Mean	Std. Error	95% Confidence Interval	
			Lower Bound	Upper Bound
english	85.802	1.250	83.333	88.272
reading	34.052	1.139	31.803	36.301
mechanic	11.336	.431	10.484	12.188
abstract	9.635	.340	8.965	10.306
math	25.720	.954	23.835	27.605
office	19.927	.972	18.008	21.846

2. cargroup

Dependent Variable	cargroup	Mean	Std. Error	95% Confidence Interval	
				Lower Bound	Upper Bound
english	business	87.839	1.426	85.022	90.657
	medicine	87.036	2.287	82.518	91.554
	teach_social	82.532	2.608	77.381	87.683
reading	business	32.762	1.299	30.195	35.328
	medicine	35.226	2.084	31.111	39.342
	teach_social	34.168	2.376	29.476	38.860
mechanic	business	10.798	.492	9.825	11.770
	medicine	12.167	.789	10.608	13.726
	teach_social	11.043	.900	9.265	12.820
abstract	business	9.345	.387	8.580	10.110
	medicine	10.286	.621	9.059	11.513
	teach_social	9.275	.708	7.876	10.674
math	business	22.500	1.089	20.350	24.650
	medicine	30.107	1.746	26.659	33.555
	teach_social	24.554	1.991	20.622	28.485
office	business	24.143	1.108	21.954	26.332
	medicine	16.512	1.778	13.001	20.023
	teach_social	19.125	2.027	15.122	23.128

3. Gender

Dependent Variable	gender	Mean	Std. Error	95% Confidence Interval	
				Lower Bound	Upper Bound
english	male	82.861	2.301	78.316	87.407
	female	88.744	.978	86.813	90.675
reading	male	34.544	2.096	30.403	38.684
	female	33.560	.891	31.801	35.319
mechanic	male	13.937	.794	12.368	15.505
	female	8.735	.337	8.069	9.401
abstract	male	9.933	.625	8.698	11.167
	female	9.338	.266	8.814	9.863
math	male	28.671	1.757	25.201	32.140
	female	22.770	.746	21.296	24.244
office	male	15.996	1.788	12.464	19.528
	female	23.857	.760	22.356	25.358

(Continued)

Figure 11.15 (Continued)

4. Cargroup × Gender

Descriptive Statistics

	cargroup	gender	Mean	Std. Deviation	N
english	business	male	87.00	8.557	14
		female	88.68	8.737	84
		Total	88.44	8.688	98
	medicine	male	86.83	11.839	6
		female	87.24	8.831	21
		Total	87.15	9.326	27
	teach_social	male	74.75	18.839	4
		female	90.31	11.903	35
		Total	88.72	13.330	39
	Total	male	84.92	11.821	24
		female	88.87	9.613	140
		Total	88.29	10.024	164
reading	business	male	33.21	9.940	14
		female	32.31	8.284	84
		Total	32.44	8.489	98
	medicine	male	37.17	16.437	6
		female	33.29	8.539	21
		Total	34.15	10.524	27
	teach_social	male	33.25	8.539	4
		female	35.09	9.070	35
		Total	34.90	8.926	39
	Total	male	34.21	11.275	24
		female	33.15	8.542	140
		Total	33.30	8.961	164
mechanic	business	male	13.14	3.820	14
		female	8.45	3.220	84
		Total	9.12	3.681	98
	medicine	male	15.67	2.338	6
		female	8.67	4.747	21
		Total	10.22	5.213	27
	teach_social	male	13.00	3.559	4
		female	9.09	2.822	3
		Total	9.49	3.094	39
	Total	male	13.75	3.517	24
		female	8.64	3.384	140
		Total	9.39	3.846	164

Descriptive Statistics

	cargroup	gender	Mean	Std. Deviation	N
abstract	business	male	9.71	2.758	14
		female	8.98	2.430	84
		Total	9.08	2.477	98
	medicine	male	11.33	1.633	6
		female	9.24	2.682	21
		Total	9.70	2.614	27
	teach_social	male	8.75	4.031	4
		female	9.80	3.179	35
		Total	9.69	3.229	39
	Total	male	9.96	2.789	24
		female	9.22	2.674	140
		Total	9.33	2.695	164
math	business	male	25.93	9.458	14
		female	19.07	6.251	84
		Total	20.05	7.159	98
	medicine	male	36.83	7.055	6
		female	23.38	8.617	21
		Total	26.37	9.958	27
	teach_social	male	23.25	12.447	4
		female	25.86	8.374	35
		Total	25.59	8.696	39
	Total	male	28.21	10.413	24
		female	21.41	7.749	140
		Total	22.41	8.503	164
office	business	male	17.57	7.133	14
		female	30.71	6.870	84
		Total	28.84	8.281	98
	medicine	male	13.17	6.735	6
		female	19.86	9.393	21
		Total	18.37	9.199	27
	teach_social	male	17.25	8.808	4
		female	21.00	8.602	35
		Total	20.62	8.583	39
	Total	male	16.42	7.241	24
		female	26.66	9.163	140
		Total	25.16	9.601	164

If we require a discriminant function coefficient to be greater than or equal to .5 in absolute value to consider it "large," then only two variables had large discriminant function coefficients: There was a positive coefficient associated with z_{English} and a negative coefficient associated with z_{mechanic}. Thus, a higher score on D tends to be related to higher z scores on English and lower z scores on mechanical reasoning. When we look at the centroids in Figure 11.16 (i.e., the mean D scores for the male and female groups), we see that the female group scored higher on D (and therefore higher on z_{English} and lower on z_{mechanic}) than the male group. In this multivariate follow-up using DA, we might conclude that the gender differences are best described as higher scores for women on English and higher scores for men

Table 11.3 Univariate Gender Differences in Means for Four Achievement and Aptitude Tests

	Group Means					
	Men		Women	F^a	p	Partial η^2
English	82.86	<	88.74	5.53	.020	.034
Office	16.00	<	23.86	16.37	<.001	.094
Mechanic	13.94	>	8.74	36.34	<.001	.187
Math	28.67	>	22.77	9.56	.002	.057

[a]Each F ratio reported in this table has (1, 158) *df*. The *p* values have not been adjusted for inflated risk for Type I error; that is, they correspond to the theoretical risk for Type I error for a single univariate *F* test. Only the outcome variables that had a significant univariate main effect for gender at the $\alpha = .05$ level are included in this table.

Figure 11.16 Selected Output From Multivariate Follow-Up on the Significant Main Effect for Gender: Discriminant Analysis Comparing Scores on Six Achievement and Aptitude Tests Across Gender Groups (Gender Coded 1 = Male, 2 = Female)

Eigenvalues

Function	Eigenvalue	% of Variance	Cumulative %	Canonical Correlation
1	.546[a]	100.0	100.0	.594

a. First 1 canonical discriminant functions were used in the analysis.

Wilks's Lambda

Test of Function(s)	Wilks's Lambda	Chi-square	df	Sig
1	.647	69.227	6	.000

Standardized Canonical Discriminant Function Coefficients

	Function
	1
english	.502
reading	.196
mechanic	-.752
abstract	-.015
math	-.310
office	.444

Functions at Group Centroids

	Function
gender	1
male	-1.773
female	.304

Unstandardized canonical discriminant functions evaluated at group means on mechanical reasoning; two variables for which statistically significant univariate F's were obtained (math and office) had standardized discriminant function coefficients that were less than +.5 in absolute value. A judgment call would be required. If we are willing to set our criterion for a large standardized discriminant function coefficient at a much lower level, such as +.3, we might conclude that, even when they are evaluated in a multivariate context, math and office scores continue to show differences across gender. If we adhere to the admittedly arbitrary standard chosen earlier (that a standardized discriminant function coefficient must equal or exceed .5 in absolute value to be considered large), then we would conclude that when they are evaluated in a multivariate follow-up, math and office scores do not show differences between men and women.

11.14.2 Follow-Up Tests for Nature of Interaction

If a larger data set had yielded clear evidence of a statistically significant gender-by-cargroup interaction in the previous 2 × 3 factorial MANOVA and if the n's in all the cells had been large enough (at least 20) to provide adequate information about group means for all six cells, then follow-up analyses could be performed to evaluate the nature of this interaction. For an interaction in a factorial MANOVA, there are more numerous options for the type of follow-up analysis. First of all, just as in a univariate factorial ANOVA (a factorial ANOVA with just one Y outcome variable), the researcher needs to decide whether to focus the discussion primarily on the interactions between factors or primarily on main effects. Sometimes both are included, but a statistically significant interaction that has a large effect size can be interpreted as evidence that the effects of different levels of Factor B differ substantially across groups on the basis of levels of Factor A, and in that case, a detailed discussion of main effects for the A and B factors may not be necessary or appropriate. If there is a significant $A \times B$ interaction in a MANOVA, the data analyst may choose to focus the discussion on that interaction. Two kinds of information can be used to understand the nature of the interaction in a MANOVA. At the univariate level, the researcher may report the means and F tests for the interaction for each individual Y outcome variable considered in isolation from the other Y outcome variables. In other words, the researcher may ask, If we examine the means on only Y_1 and Y_3, how do these differ across the cells of the MANOVA factorial design, and what is the nature of the interaction for each separate Y outcome variable? In the career group-by-gender factorial MANOVA data, for example, we might ask, Do men show a different pattern of scores on their math test scores across the three career groups from women?

A second approach to understanding the nature of an interaction in MANOVA is multivariate, that is, an examination of discriminant function scores across groups. We can do an "analysis of simple main effects" using DA. In an analysis of simple main effects, the researcher essentially conducts a one-way ANOVA or MANOVA across levels of the B factor, separately within each group on the A factor. In the empirical example presented here, we can do a MANOVA and a DA to compare scores on the entire set of Y outcome variables across the three career groups, separately for the male and female groups. To assess how differences among career groups in discriminant function scores may differ for men versus women, DA can be performed separately for the male and female groups. In this follow-up analysis, the question is whether the weighted linear combination of scores on the Y outcome variables that best differentiates career groups is different for men versus women, both in terms of which career groups can be discriminated and in terms of the signs and magnitudes of the discriminant function coefficients for the Y variables. It is possible, for example, that math scores better predict career choice for men and that abstract reasoning scores better predict career choice for women. It is possible that the three career groups do not differ significantly on Y test scores in the male sample but that at least one of the career groups differs significantly from the other two career groups in Y test scores in the female sample.

A univariate follow-up to explore the nature of the interaction involves examining the pattern of means across cells to describe gender differences in the predictability of career group from achievement test scores. At the univariate level, the individual Y variables that had statistically significant F's for the gender-by-cargroup interaction were English, $F(2, 158) = 3.095$, $p = .048$, and math, $F(2, 158) = 4.605$, $p = .011$. If we use Bonferroni-corrected per comparison alpha levels, neither of these would be judged statistically significant. The cell means for these two variables (English and math) are given in Table 11.4. For example, there were significant univariate F's for the gender-by-career choice interaction for English, $F(2, 158) = 3.10$, $p = .048$, and for math, $F(2, 158) = 4.61$, $p = .011$. The nature of these univariate interactions was as follows. For men, those with low English scores were likely to choose teaching over business or medicine; for women, differences in English scores across career groups were small, but those with high English scores tended to choose teaching. For men, those with high scores on math tended to choose medicine as a career; for women, those with high scores on math were likely to choose teaching as a career. Thus, these data suggest that the pattern of differences in English and math scores across these three career groups might differ for men and women. However, given the very small n's in two of the cells and a nonsignificant p value for the multivariate interaction, this would not be a publishable outcome.

A multivariate follow-up to this finding of an interaction can be obtained by doing DA to assess differences among the three career groups separately for the male and female samples. This provides a way to describe the multivariate pattern of scores that best differentiates among the three career groups for the male and female participants. Selected output from two separate DAs across the three levels of the career group factor (one DA for male participants and a separate DA for female participants) appears in Figures 11.17 and 11.18.

These analyses provide information about simple main effects, that is, the differences in scores across the three levels of the cargroup factor separately for each gender group. In other words, does the pattern of test scores that predicts choice of career differ for women versus men? A DA to compare career group test scores for men indicated that for men, test scores were not significantly related to career group choice; the lack of significant differences among career groups for men could be due to low power (because of very small sample sizes in two of the cells). For women, the business group could be differentiated

Table 11.4 Univariate Cell Means on English and Math Scores for the Gender-by-Career Interaction

	Business	Medicine	Teaching
English scores: $F(2, 158) = 3.095$, $p = .048$, partial $\eta^2 = .038$[a]			
Male	87.00	86.83	74.75
Female	88.68	87.24	90.31
Math scores: $F(2, 158) = 4.605$, $p = .011$, partial $\eta^2 = .055$[a]			
Male	25.93	36.83	23.25
Female	19.07	23.38	25.86

[a]The p values have not been adjusted for inflated risk for Type I error; that is, they correspond to the theoretical risk for Type I error for a single univariate F test. Only outcome variables that had significant univariate interactions between the gender and cargroup factors at the $\alpha = .05$ level are included in this table.

Figure 11.17 Follow-Up Analysis to Assess the Nature of the Nonsignificant Gender-by-Cargroup Interaction: Discriminant Analysis to Assess Differences in Six Test Scores Across Levels of Career Group (for Men Only)

Log Determinants[d]

CARGROUP	Rank	Log Determinant
business	6	17.017
medicine	.[a]	.[b]
teach_social	.[c]	.[b]
Pooled within-groups	6	20.139

The ranks and natural logarithms of determinants printed are those of the group covariance matrices.

a. Rank < 6

b. Too few cases to be non-singular

c. Rank < 4

d. GENDER = male

Test Results[a,b]

Tests null hypothesis of equal population covariance matrices.

a. No test can be performed with fewer than two nonsingular group covariance matrices.

b. GENDER = male

Selected Discriminant Function Results Across Three Levels of Cargroup for Males

Eigenvalues[b]

Function	Eigenvalue	% of Variance	Cumulative %	Canonical Correlation
1	.554[a]	55.1	55.1	.597
2	.451[a]	44.9	100.0	.558

a. First 2 canonical discriminant functions were used in the analysis.

b. GENDER = male

Wilks's Lambda[a]

Test of Function(s)	Wilks's Lambda	Chi -square	df	Sig.
1 through 2	.443	15.050	12	.239
2	.689	6.889	5	.229

a. GENDER = male

Note: The variance/covariance matrices for the medicine and teaching groups were "not of full rank," because the numbers of cases in these groups were too small relative to the number of variances and covariances among variables that needed to be estimated; there were not enough degrees of freedom to obtain independent estimates of these variances and covariances. In actual research, an analysis with such small *n*'s would not be reported. This example was included here to illustrate the error messages that arise when within-group *n*'s are too small.

Figure 11.18 Follow-Up Analysis to Assess the Nature of Nonsignificant Gender-by-Cargroup Interaction: Discriminant Analysis to Assess Differences in Six Test Scores Across Levels of Career Group (for Women Only)

Eigenvalues[b]

Function	Eigenvalue	% of Variance	Cumulative %	Canonical Correlation
1	.616[a]	97.3	97.3	.618
2	.017[a]	2.7	100.0	.130

a. First 2 canonical discriminant functions were used in the analysis.

b. GENDER = female

Wilks's Lambda[a]

Test of Function(s)	Wilks's Lambda	Chi-square	df	Sig.
1 through 2	.608	66.871	12	.000
2	.983	2.290	5	.808

a. GENDER = female

Standardized Canonical Discriminant Function Coefficients[a]

	Function	
	1	2
english	.347	.430
reading	.191	-.071
mechanic	.187	-.092
abstract	-.116	-.006
math	-.742	.723
office	.798	.440

a. GENDER = female

Functions at Group Centroids[a]

	Function	
CARGROUP	1	-5.18E-04
business	.634	2
medicine	-.959	-.264
teach_social	-.946	.160

Unstandardized canonical discriminant functions evaluated at group means

a. GENDER = female

from the other two career groups. Women who chose business as a career tended to score higher on office skills and lower on math than women who chose the other two careers. It would be interesting to see whether a different pattern of career group differences would emerge if we could obtain data for a much larger number of male participants; the sample of male participants in this study was too small to provide an adequate assessment of career group differences for men.

11.14.3 Further Discussion of Problems With This 2 × 3 Factorial Example

In the "real world," analysis of these 2 × 3 factorial data should be halted when the data analyst notices the small numbers of cases in several cells (in Figure 11.8) or when the data analyst obtains a p value of .063 for the interaction. Together, these outcomes make these data inconclusive; we cannot be sure whether an interaction is present. For instructional purposes, these data were used to demonstrate some possible ways to examine the nature of main effects and interactions, but these results would not be publishable because of small cell sizes.

Note that the "Discussion" section following this "Results" section would need to address additional issues. Gender differences in contemporary data might be entirely different from any differences seen in those data collected prior to 1970. As Gergen (1973) pointed out in his seminal paper "Social Psychology as History," some findings may be specific to the time period when the data were collected; gender differences in achievement tests have tended to become smaller in recent years. Furthermore, causal inferences about either career group or gender differences would be inappropriate because neither of these were manipulated factors.

11.15 SIGNIFICANT INTERACTION IN A 3 × 6 MANOVA

A second empirical example is presented as an illustration of analysis of an interaction in MANOVA. Warner and Sugarman (1986) conducted an experiment to assess how person perceptions vary as a function of two factors. The first factor was type of information; the second factor was individual target persons. Six individual target persons were randomly selected from a set of 40 participants; for these target individuals, three different types of information about expressive behavior and physical appearance were obtained. A still photograph (head and shoulders) was taken to provide information about physical appearance and attractiveness. A standard selection of text from a textbook was given to all participants. A speech sample was obtained for each participant by tape-recording the participant reading the standard text selection out loud. A handwriting sample was obtained for each participant by having each participant copy the standard text sample in cursive handwriting. Thus, the basic experimental design was as follows: For each of the six target persons (with target person treated as a random factor), three different types of information were obtained: a facial photograph, a speech sample, and a handwriting sample. Thus, there were 18 different conditions in the study, such as a handwriting sample for Person 2 and a speech sample for Person 4.

A between-subjects experiment was performed using these 18 stimuli as targets for person perception. A total of 404 undergraduate students were recruited as participants; each participant made judgments about the personality of one target person on the basis of one type of information about that target person (such as the handwriting sample of Person 3). Multiple-item measures that included items that assessed perceptions of friendliness, warmth, dominance, intelligence, and other individual differences were administered to each participant. For each participant rating each target person on one type of information (e.g., photograph, speech sample, or handwriting), scores on the following six dependent variables were obtained: ratings of social evaluation, intellectual evaluation, potency or dominance, activity, sociability, and emotionality.

Because the interaction between type of information and target was statistically significant, an analysis of simple main effects was performed to evaluate how the targets were differentiated within each type of information. The <Data> → <Split File> command in SPSS was used to divide the file into separate groups on the basis of type of information (facial appearance, handwriting, speech); then, a DA was done separately for each type of information to assess the nature of differences in attributions of personality made to the six randomly selected target persons. As discussed in the following "Results" section, each of these three follow-up DAs told a different story about the nature of differences in personality attributions. When presented with information about physical appearance, people rated targets differently on social evaluation; when presented with information about handwriting, people differentiated targets on potency or dominance; and when given only a speech sample, people distinguished targets on perceived activity level.

Results

A 6 × 3 MANOVA was performed on the person perception data (Warner & Sugarman, 1986) using six personality scales (social evaluation, intellectual evaluation, potency, activity, sociability, and emotionality) as dependent variables. The factors were target person (a random factor with six levels) and type of information (a fixed factor with Level 1 = a facial photograph, Level 2 = a handwriting sample, and Level 3 = a speech sample for each target person). Each of the 18 cells in this design corresponded to one combination of target person and type of information, for example, the speech sample for Person 3. Although the 404 participants who rated these stimuli were each randomly assigned to one of the 18 cells in this 3 × 6 design (each participant rated only one type of information for one individual target person), the cell *n*'s were slightly unequal. Type III sums of squares were used to correct for the minor confounding between factors that occurred because of the slightly unequal *n*'s in the cells; that is, the *SS* term for each effect in this design was calculated controlling for all other effects.

Preliminary data screening did not indicate any serious violations of the assumption of multivariate normality or of the assumption of linearity of associations between quantitative outcome variables. Box's *M* test (using $\alpha = .10$ as the criterion for significance) did not indicate a significant violation of the assumption of homogeneity of variance/covariance matrices across conditions. Table 11.5 shows the pooled or averaged within-cell correlations among the six outcome variables. Intercorrelations between measures ranged from –.274 to +.541. None of the correlations among outcome variables was sufficiently large to raise concerns about multicollinearity.

For the overall MANOVA, all three multivariate tests were statistically significant (using $\alpha = .05$ as the criterion). For the target person-by-type of information interaction, Wilks' $\Lambda = .634$, approximate $F(60, 1886) = 2.86$, $p < .001$. The corresponding η^2 effect size of .37 indicated a strong effect for this interaction. The main effect for target person was also statistically significant, with $\Lambda = .833$, approximate $F(30, 1438) = 2.25$, $p < .001$; this also corresponded to a fairly strong effect size ($\eta^2 = .17$). The main effect for type of information was statistically significant, with $\Lambda = .463$, approximate $F(12, 718) = 28.12$, $p < .001$. This suggested that there was a difference in mean ratings of personality on the basis of the type

Table 11.5 Pooled Within-Cell Correlation Matrix for Six Person Perception Outcome Scales in the 3 × 6 MANOVA Example

Scale	Scale				
	1	2	3	4	5
1. Social evaluation					
2. Intellectual evaluation	+.24				
3. Potency/dominance	+.14	+.26			
4. Activity	+.37	+.20	+.38		
5. Sociability/extraversion	+.50	+.01	+.22	+.54	
6. Emotionality	−.16	−.11	−.33	−.06	−.27

Source: Warner and Sugarman (1986).

of information provided. For example, mean ratings of level of activity were higher for tape-recorded speech samples than for still photographs. This effect is not of interest in the present study, and therefore it is not discussed further.

Because the interaction was statistically significant and accounted for a relatively large proportion of variance, the follow-up analyses focused primarily on the nature of this interaction. An analysis of simple main effects was performed; that is, a one-way DA was performed to assess differences among the six target persons, separately for each level of the factor type of information. The goal of this analysis was to evaluate how the perceptions of the target persons differed depending on which type of information was provided. In other words, we wanted to evaluate whether people make different types of evaluations of individual differences among targets when they base their perceptions on facial appearance than when they base their perceptions on handwriting or speech samples.

Table 11.6 reports the standardized discriminant function coefficients that were obtained by performing a separate DA (across six target persons) for each type of information condition; a separate DA was performed on the ratings on the basis of photographs of facial appearance (top panel), handwriting samples (middle panel), and tape-recorded speech sample (bottom panel). For each of these three DAs, five discriminant functions were obtained; coefficients for only the first discriminant function in each analysis are reported in Table 11.6. For all three DAs, the multivariate significance test for the entire set of five discriminant functions was statistically significant using α = .05 as the criterion. Tests of sets of discriminant functions were examined to decide how many discriminant functions to retain for interpretation; because the set of discriminant functions from 2 through 5 was not statistically significant for two of the three types of information, only the first discriminant function was retained for interpretation.

An arbitrary standard was used to evaluate which of the discriminant function coefficients were relatively large; an outcome variable was interpreted as an indication of meaningful differences across groups only if the standardized discriminant

Table 11.6 Simple Main Effects Follow-Up Analysis for Warner and Sugarman (1986)

Scale	Standardized Discriminant Function Coefficient for D_1	Univariate F	Univariate p
Type of information, Level 1: Facial apperance			
Social evaluation	**−.82**	**8.89**	**<.001**
Intellectual evaluation	+.35	3.03	.001
Potency/dominance	−.26	2.10	.065
Activity	−.18	**4.66**	**<.001**
Sociability/extraversion	−.11	**4.87**	**<.001**
Emotionality	−.03	0.45	.81
Type of information, Level 2: Handwriting			
Social evaluation	+.31	1.19	.31
Intellectual evaluation	−.22	2.04	.07
Potency/dominance	**−.89**	**3.71**	**.003**
Activity	−.16	1.81	.11
Sociability/extraversion	+.20	1.50	.20
Emotionality	−.36	.27	.93
Type of information, Level 3: Speech sample			
Social evaluation	+.46	**3.94**	**.002**
Intellectual evaluation	−.18	3.13	.009
Potency/dominance	−.43	.53	.78
Activity	**+1.07**	**7.92**	**<.001**
Sociability/extraversion	−.50	1.99	.08
Emotionality	+.02	0.75	.58

Source: Warner and Sugarman (1986).

Note: Mean personality ratings received by targets were compared separately within each of three conditions: facial appearance, handwriting, and voice recording. Discriminant function coefficients are set in bold if they are greater than .50 in absolute value; F ratios and p values are set in bold if $p < .05$. These represent arbitrary standards for which variables are "important" predictors of group membership.

function coefficient had an absolute value >.50. F ratios were obtained for each of the individual outcome variables to see whether scores differed significantly on each type of evaluation in a univariate context as well as in a multivariate analysis that controls for intercorrelations among outcome measures.

11.16 COMPARISON OF UNIVARIATE AND MULTIVARIATE FOLLOW-UP ANALYSES

The preceding examples have presented both univariate follow-up analyses (univariate *F*'s and univariate Tukey post hoc tests) and multivariate follow-up analyses (discriminant functions that differ across groups). There are potential problems with univariate follow-up analyses. First, running multiple significance tests for univariate follow-ups may result in an inflated risk for Type I error (Bonferroni-corrected per comparison alpha levels may be used to limit the risk for Type I error for large sets of significance tests). Second, these univariate analyses do not take intercorrelations among the individual *Y* outcome variables into account. This can be problematic in two ways. First, some of the *Y* variables may be highly correlated or confounded and may thus provide redundant information; univariate analysis may mislead us into thinking that we have detected significant differences on *p* separate *Y* outcome variables, when in fact many of the outcome variables (or subsets of the *Y* outcome variables) are so highly correlated that they may actually be measures of the same underlying construct. As in previous chapters, we see that the nature and strength of the association between a pair of variables can change substantially when we statistically control for one or more additional variables. In the context of MANOVA, this can lead to several different outcomes. Sometimes in MANOVA, the *Y* variables that show large univariate differences across groups when the variables are examined individually are the same variables that have the largest weights in forming discriminant functions that differ across groups. However, it is possible for a variable that shows large differences across groups in a univariate analysis to show smaller differences across groups when other *Y* outcome variables are statistically controlled. It is possible for suppression to occur; that is, a *Y* variable that was not significant in a univariate analysis may be given substantial weight in forming one of the discriminant functions, or the rank order of means on a Y_i outcome variable may be different across levels of an *A* factor when Y_i is examined in isolation in a univariate analysis than when Y_i is evaluated in the context of a multivariate analysis such as DA. Discussion of the nature of group differences is more complex when multivariate outcomes for some of the Y_i variables differ substantially from the univariate outcomes for these Y_i variables.

The dilemma that we face in trying to assess which individual quantitative outcome variables differ across groups in MANOVA is similar to the problem that we encounter in other analyses such as multiple regression. We can only obtain an accurate assessment of the nature of differences across groups on an individual *Y* outcome variable in the context of a correctly specified model, that is, an analysis that includes all the variables that need to be statistically controlled and that does not include variables that should not be controlled. We can never be certain that the model is correctly specified, that is, that the MANOVA includes just the right set of factors and outcome variables. However, we can sometimes obtain a better understanding of the nature of differences across groups on an individual Y_i outcome variable (such as math scores or ratings of dominance) when we control for other appropriate outcome variables. The results of any univariate follow-up analyses that are reported for a MANOVA need to be evaluated relative to the multivariate outcomes.

It is prudent to keep the number of outcome variables in a MANOVA relatively small for two reasons. First, there should be some theoretical rationale for inclusion of each outcome measure; each variable should represent some outcome that the researcher reasonably expects will be influenced by the treatment in the experiment. If some of the outcome variables are very highly correlated with one another, it may be preferable to consider whether those two or three variables are in fact all measures of the same construct; it may be better to average the raw scores or *z* scores on those highly correlated measures into a single, possibly more reliable outcome measure before doing the MANOVA.

Whatever method of follow-up analysis you choose, keep this basic question in mind. For each contrast that is examined, you want to know whether the groups differ significantly (using an omnibus multivariate test) and what variable or combination of variables (if any) significantly differentiates the groups that are being compared. When we do multiple univariate significance tests in follow-up analyses to MANOVA, there is an inflated risk for Type I error associated with these tests. Finally, our understanding of the pattern of differences in values of an individual Y_i variable may be different when we examine a Y_i variable by itself in a univariate ANOVA than when we examine the Y_i variable in the context of a DA that takes correlations with other Y outcome variables into account. See Bray and Maxwell (1985) and Stevens (2009) for more detailed discussions of follow-up analyses that can clarify the nature of group differences in MANOVA.

11.17 SUMMARY

Essentially, MANOVA is DA in reverse. Both MANOVA and DA assess how scores on multiple intercorrelated quantitative variables are associated with group membership. In MANOVA, the group membership variable is generally treated as the predictor, whereas in DA, the group membership variable is treated as the outcome. DA is generally limited to one-way designs, whereas MANOVA can involve multiple factors, repeated measures, and/or covariates. In addition, when one or more quantitative covariates are added to the analysis, MANOVA becomes MANCOVA: multivariate analysis of covariance.

MANOVA and DA provide complementary information, and they are often used together. In reporting MANOVA results, the focus tends to be on which groups differed significantly. In reporting DA, the focus tends to be on which discriminating variables were useful in differentiating groups and sometimes on the accuracy of classification of cases into groups. Thus, a DA is often reported as one of the follow-up analyses to describe the nature of significant group differences detected in MANOVA.

The research questions involved in a two-way factorial MANOVA include the following:

1. Which levels of the A factor differ significantly on the set of means for the outcome variables $Y_1, Y_2, \ldots, Y_p$?
2. Which levels of the B factor differ significantly on the set of means for $Y_1, Y_2, \ldots, Y_p$?
3. What do differences in the pattern of group means for the outcome variables Y_1, $Y_2, \ldots, Y_p$ tell us about the nature of $A \times B$ interaction?

In reporting a factorial ANOVA—for example, an $A \times B$ factorial—the following results are typically included. For each of these effects—main effects for A and B and the $A \times B$ interaction—researchers typically report the following:

1. An overall significance test (usually Wilks' Λ and its associated F, sometimes Pillai's trace, or Roy's largest root).
2. An effect size, that is, a measure of the proportion of explained variance (1 – Wilks' Λ); this is comparable with $1 - \eta^2$ in an ANOVA.
3. A description of the nature of the differences among groups; this can include the pattern of univariate means and/or DA results.

Thus, when we look at the main effect for the A factor (cargroup), we are asking, What two discriminant functions can we form to discriminate among these three groups? (In general, if a factor has k levels or groups, then you can form $k - 1$ discriminant functions; however, the number of discriminant functions cannot exceed the number of dependent variables, so this can be another factor that limits the number of discriminant functions.) Can we form composites of the z scores on these six variables that are significantly different across these three career groups?

When we look at the main effect for the B factor (gender), we are asking, Can we form one weighted linear combination of these variables that is significantly different for men versus women? Is there a significant difference overall in the "test profile" for men versus women?

When we look at the $A \times B$ interaction effect, we are asking a somewhat more complex question: Is the pattern of differences in these test scores among the three career groups different for women than it is for men? If we have a significant $A \times B$ interaction, then several interpretations are possible. It could be that we can discriminate among men's career choices—but not among women's choices—using this set of six variables (or vice versa; perhaps we can discriminate career choice for women but not for men). Perhaps the pattern of test scores that predicts choice of a medical career for men gives a lot of weight to different variables compared with the pattern of test scores that predicts choice of a medical career for women. A significant interaction suggests that the nature of the differences in these six test scores across these three career groups is not the same for women as for men; there are many ways the patterns can be different.

MANOVA may provide substantial advantages over a series of univariate ANOVAs (i.e., a separate ANOVA for each Y outcome variable). Reporting a single p value for one omnibus test in MANOVA, instead of p values for multiple univariate significance tests for each Y variable, can limit the risk for Type I error. In addition, MANOVA may be more powerful than a series of univariate ANOVAs, particularly when groups differ in their patterns of response on several outcome variables.

COMPREHENSION QUESTIONS

1. Design a factorial MANOVA study. Your design should clearly describe the nature of the study (experimental vs. nonexperimental), the factors (what they are and how many levels each factor has), and the outcome measures (whether they are multiple measures of the same construct or measures of different but correlated outcomes). What pattern of results would you predict, and what follow-up analyses do you think you would need?

2. In words, what null hypothesis is tested by Box's M value in the SPSS GLM results for a MANOVA? Do you want this Box's M test to be statistically significant? Give reasons for your answer. If Box's M is statistically significant, what consequences might this have for the risk for Type I error and for statistical power? What might you do to remedy this problem?

3. Complete each of the following sentences with the correct term (*increases*, *decreases*, or *stays the same*):

    a. As Wilks' Λ increases (and assuming all other factors remain constant), the multivariate F associated with Wilks' Λ _______________.
    b. As Pillai's trace increases, all other factors remaining constant, the associated value of F _______________.
    c. As Wilks' Λ increases, the associated value of partial η^2 _______________.
    d. As Wilks' Λ increases, the corresponding value of Pillai's trace _______________.

4. Suppose that your first discriminant function has an eigenvalue of $\lambda_1 = .5$. What is the value of r_c, the canonical correlation, for this function? What is the value of Roy's largest root for this function?

5. Suppose that you plan to do a one-way MANOVA with five groups, using $\alpha = .05$. There are four outcome variables, and you want to have power of .70. For each of the following effect sizes, what minimum number of cases per cell or group would be needed? (Use Table 11.1 to look up recommended sample sizes.)

    a. For a "very large" effect: _____
    b. For a "moderate" effect: _____
    c. For a "small" effect: _____

6. Find a journal article that reports a MANOVA, and write a brief (about three pages) critique, including the following:

    a. A brief description of the design: the nature of the study (experimental vs. nonexperimental), the factors and number of levels, a description of outcome variables.
    b. Assessment: Was any information included about data screening and detection of any violations of assumptions?
    c. Results: What results were reported? Do the interpretations seem reasonable?
    d. Is there other information that the author could have reported that would have been useful?
    e. Do you see any problems with this application of MANOVA?

7. Describe a hypothetical experimental research situation in which MANOVA could be used.

8. Describe a hypothetical nonexperimental research situation in which MANOVA could be used.

9. In addition to the usual assumptions for parametric statistics (e.g., normally distributed quantitative variables, homogeneity of variance across groups), what additional assumptions should be tested when doing a MANOVA?
10. What information is provided by the elements of a within-group or within-cell **SCP** matrix?
11. How is the Wilks' Λ statistic obtained from the within-group and total **SCP** matrices?
12. In a one-way ANOVA, what does Wilks' Λ correspond to? How is Wilks' Λ interpreted in a one-way ANOVA?
13. Explain how MANOVA is similar to and different from ANOVA.
14. Explain how MANOVA is related to DA.
15. Using your own data or data provided by your instructor, conduct your own factorial MANOVA; use at least four outcome variables. If you obtain a statistically significant interaction, do appropriate follow-up analyses. If you do not obtain a statistically significant interaction, describe in detail what follow-up analyses you would need to do to describe the nature of an interaction in this type of design. Write up your results in the form of a journal article "Results" section.

DIGITAL RESOURCES

Find **free study tools** to support your learning, including **eFlashcards, data sets, and web resources,** on the accompanying website at **edge.sagepub.com/warner3e**.

CHAPTER 12

EXPLORATORY FACTOR ANALYSIS

12.1 RESEARCH SITUATIONS

The term *factor analysis* actually refers to a group of related analytic methods. The methods in this chapter are described as exploratory factor analysis (EFA) because the data analyst does not decide which variables will be associated with each factor prior to the analysis (although the analyst may decide to limit the number of retained factors during analysis). In contrast, confirmatory factor analysis (CFA) involves setting up a model ahead of time that specifies the number of factors or latent variables and which measured indicator variables will be associated with each factor or latent variable; CFA will be discussed in Chapter 15.

Programs such as SPSS offer numerous options for the setup of factor analysis using different computational procedures and decision rules. This chapter describes the basic concepts involved in only two of the analyses that are available through the SPSS factor procedure. The two analyses discussed here are called principal-components (PC) analysis and principal-axis factoring (PAF). PC is somewhat simpler and was developed earlier. PAF is one of the methods that is most widely reported in published journal articles. Section 12.11 explains the difference in computations for PC and PAF. Although the underlying model and the computational procedures differ for PC and PAF, these two analytic methods are frequently used in similar research situations, that is, situations in which we want to evaluate whether the scores on a set of p individual measured X variables can be explained by a small number of **latent variables** called components (when the analysis is done using PC) or **factors** (when the analysis is done using PAF).

The version of factor analysis in this chapter is called exploratory because the analyst usually does not specify ahead of time either the number of factors or the variables that are expected to be strongly associated with each factor. When data analysts have hypotheses about those aspects of factor structure, they can use structural equation modeling (SEM) programs such as Amos to test specific models. That approach is called CFA, although in practice, data analysts sometimes evaluate so many different models that their results should be viewed as exploratory.

Up to this point, the analyses considered in this book have typically involved prediction of scores on one or more measured dependent variables from one or more measured independent variables. For example, in multiple regression, the score on a single quantitative Y dependent variable is predicted from a weighted linear combination of scores on several quantitative or dummy X_i predictor variables. For example, we might predict each individual faculty member's salary in dollars (Y) from number of years in the job (X_1), number of published journal articles (X_2), and a dummy-coded variable that provides information about gender.

In PC and PAF, we will not typically identify some of the measured variables as X predictors and other measured variables as Y outcomes. Instead, we will examine a set of z scores on p measured X variables (p represents the number of variables) to see whether it would make sense to interpret the set of p measured variables as measures of some smaller number of underlying constructs or latent variables. Most of the path models presented here involve **standard scores** or **z scores** for a set of measured variables. For example, $z_{X1}, z_{X2}, z_{X3}, \ldots, z_{Xp}$ represent z scores for a set of p different tests or measurements. The information that we have to work with is the set of correlations among all these p measured variables. The inference we typically want to make involves the question whether we can reasonably interpret some or all the measured variables as measures of the same latent variable or construct. For example, if $X_1, X_2, X_3, \ldots, X_p$ are scores on tests of vocabulary, reading comprehension, understanding analogies, geometry, and solving algebraic problems, we can examine the pattern of correlations among these variables to decide whether all these X variables can be interpreted as measures of a single underlying latent variable (such as "mental ability") or whether we can obtain a more accurate understanding of the pattern of correlations among variables by interpreting some of the variables as a measure of one latent variable (such as "verbal ability") and other variables as measures of some different latent variable (such as "mathematical ability").

A latent variable can be defined as "an underlying characteristic that cannot be observed or measured directly; it is hypothesized to exist so as to explain [manifest] variables, such as behavior, that can be observed" (Vogt, 1999, pp. 154–155). A latent variable can also be called a component (in PC) or a factor (in factor analysis), and it is sometimes also described as a "dimension."

An example of a widely invoked latent variable is the concept of "intelligence." We cannot observe or measure intelligence directly; however, we can imagine a quality that we call "general intelligence" to make sense of patterns in scores on tests that are believed to measure specific mental abilities. A person who obtains high scores on numerous tests that are believed to be good indicators of intelligence is thought to be high on the dimension of intelligence; a person who obtains low scores on these tests is thought to occupy a low position on the dimension of intelligence. The concept of intelligence (or any latent variable, for that matter) is problematic in many ways. Researchers disagree about how many types or dimensions of intelligence there are and what they should be called; they disagree about the selection of indicator variables. They also disagree about which variety of factor analysis yields the best description of patterns in intelligence.

Once a latent variable such as intelligence has been given a name, it tends to become "reified"; that is, people think of it as a "real thing" and forget that it is a theoretical or imaginary construct. This reification can have social and political consequences. An individual person's score on an intelligence quotient (IQ) test may be used to make decisions about that person (e.g., whether to admit Kim to a program for intellectually gifted students). When we reduce information about a person to a single number (such as an IQ score of 145), we obviously choose to ignore many other qualities the person has. If our measurement methods to obtain IQ scores are biased, the use of IQ scores to describe individual differences among persons and to make decisions about them can lead to systematic discrimination among different ethnic or cultural groups. See Gould (1996) for a history of IQ measurement that focuses on the problems that can arise when we overinterpret IQ scores, that is, when we take them to be more accurate, valid, and complete information about some theoretical construct (such as "intelligence") than they actually are.

In PC and PAF, the X_i scores are treated as multiple indicators or multiple measures of one or more underlying constructs or *latent variables* that are not directly observable. A latent variable is an "imaginary" variable, a variable that we create as part of a model that represents the structure of the relationships among the X_i measured variables; information about structure is contained in the **R** matrix of correlations among the X measured variables. When we use the PC method of **extraction**, each latent variable is called a component; when we use the PAF method of extraction, each latent variable is called a factor.

Many familiar applications of factor analysis involve X variables that represent responses to self-report measures of attitude or scores on mental ability tests. However, factor analysis can be used to understand a pattern in many different types of data. For instance, the X variables could be physiological measures; electrodermal activity, heart rate, salivation, and pupil dilation might be interpreted as indications of sympathetic nervous system arousal.

12.2 PATH MODEL FOR FACTOR ANALYSIS

In a factor-analytic study, we examine a set of scores on p measured variables ($X_1, X_2, \ldots, X_p$); the goal of the analysis is not typically to predict scores on one measured variable from scores on other measured variables but rather to use the pattern of correlations among the X_i measured variables to make inferences about the structure of the data, that is, the number of latent variables that we need to account for the correlations among the variables. Part of a path model that corresponds to a factor analysis of z scores on $X_1, X_2, \ldots, X_p$ is shown in Figure 12.1.

Figure 12.1 corresponds to a theoretical model in which the outcome variables ($z_{X1}, z_{X2}, \ldots, z_{Xp}$) are standardized scores on a set of quantitative X measured variables, for example,

Figure 12.1 Path Model for an Orthogonal Factor Analysis With p Measured Variables ($z_{X1}, z_{X2}, \ldots, z_{Xp}$) and p Factors ($F_1, F_2, \ldots, F_p$)

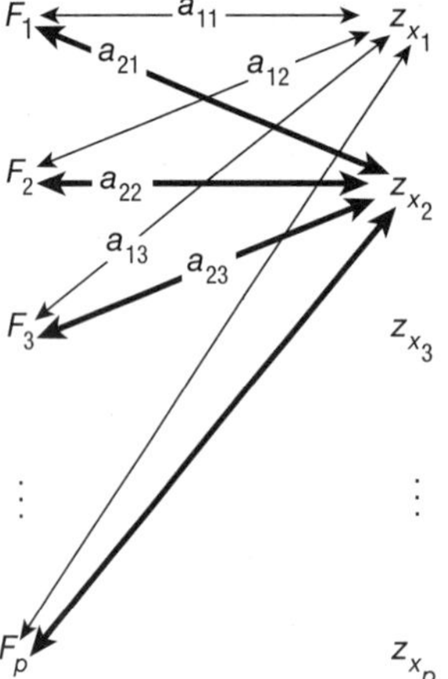

a_{ij} corresponds to the loading or correlation between variable x_{X_i} and Factor F_j

This diagram shows only the paths from z_{X_1} and z_{X_2} to each of the factors. The paths that connect $z_{X_3}, z_{X_4}, \ldots, z_{X_p}$ to the factors are not shown.

To reproduce part of the correlation between X_1 and X_2 by tracing a path via F_1, we multiply $a_{11} \times a_{21}$. To reproduce the part of the correlation between X_1 and X_2 that is explained by F_2, we multiply $a_{12} \times a_{22}$, and in general, to use the path via Factor j, to reproduce part of the X_1, X_2 correlation, we multiply $a_{1j} \times a_{2j}$. The total correlation 1_{12} between X_1 and X_2 is the sum of all the a_{ij} products that correspond to all possible paths via Factors 1 to j.

Note: This example shows orthogonal or uncorrelated factors. If the F factors were correlated (or nonorthogonal), the presence of correlations between factors would be represented by double-headed arrows between each pair of factors such as F_1 and F_2.

z scores on p different tasks that assess various aspects of mental ability. The predictor variables are "latent variables" called factors (or sometimes components); factors are denoted $F_1, F_2, \ldots, F_p$ in Figure 12.1. Each a_{ij} path in Figure 12.1 represents a correlation between a measured z_{Xi} variable and a latent F_j factor. In factor analysis, we estimate correlations between all possible pairs of variables (z_{X1} through z_{Xp}) and all pairs of possible latent variables or factors (F_1 through F_p). These correlations between measured variables and latent variables are also called "loadings."

Coefficients for paths in a "causal" model are estimated in such a way that they can be used to reconstruct the correlations among the measured X variables. For example, the beta coefficients for a standardized multiple regression with just two predictor variables were calculated from the correlations among the three variables (r_{1Y}, r_{2Y}, and r_{12}). By using the tracing rule to identify all possible paths from z_{X1} to z_Y and from z_{X2} to z_Y and by multiplying path coefficients along each path and summing contributions across different paths, we could perfectly reconstruct the observed correlations r_{1Y} and r_{2Y} from the regression path model, including the beta path coefficients.

We can understand the factor analysis model that appears in Figure 12.1 as a representation of a set of multiple regressions. In the first multiple regression, the score on z_{X1} is predicted from scores on all p of the factors ($F_1, F_2, \ldots, F_p$); in the next multiple regression, the score on z_{X2} is predicted from scores on the same set of p factors; and so forth. The path model shown in Figure 12.1 differs from path models that appear in earlier chapters in one important way; that is, in path models in the earlier chapters, all the variables corresponded to things we could actually measure. In Figure 12.1, some of the variables (e.g., $F_1, F_2, \ldots, F_p$) represent theoretical or imaginary constructs or latent variables; the F latent variables are not directly measured or observed. We imagine a set of factors $F_1, F_2, \ldots, F_p$ to set up a theoretical model that will account for the underlying structure in the set of measured X variables, that is, reproduce the correlations among all pairs of measured X variables.

12.3 FACTOR ANALYSIS AS A METHOD OF DATA REDUCTION

We shall see that if we retain all p of the F factors in our model (where p represents the number of measured X variables), we can reproduce all the correlations among the measured X variables perfectly by tracing the paths in models similar to the models that appear in Figure 12.1. However, the goal in factor analysis is usually "data reduction" or "dimension reduction"; that is, we often hope that the scores on p measured X variables can be understood by thinking of them as measurements that are correlated with *fewer* than p latent variables or factors. For example, we may want to know how well we can reconstruct the correlations among the measured X variables using only the paths that involve the first two latent variables F_1 and F_2. In other words, if we drop latent variables F_3 through F_p from the model, can we use this simpler reduced model (with a smaller number of latent variables) to account for the correlation structure of the data? We usually hope that a model that includes a relatively small number of factors (on the order of one to five factors) will account for the pattern of correlations among the X variables. For example, if we can identify a set of two latent variables that do a reasonably good job of reconstructing the correlations among the p measured variables, we may be able to interpret the p variables as measures of just two latent variables, factors, or dimensions. For example, consider a set of mental ability tests: X_1, vocabulary; X_2, analogies; X_3, synonyms and antonyms; X_4, reading comprehension; X_5, algebra; X_6, trigonometry; and X_7, geometry. We would expect to see high positive correlations among variables X_1 through X_4 (because they all measure the same kind of ability, i.e., verbal skills) and high positive correlations among X_5 through X_7 (because they all assess related abilities, i.e., mathematical skills). On the other hand, correlations between the first set of measures (X_1 through X_4) and the second set of measures (X_5 through X_7) might be fairly low because these two sets of tests measure different abilities. If we perform a factor analysis on measurements on X_1 through X_7, we may find that we can reconstruct the correlations among these seven measured X variables through their correlations or relationships with just two latent constructs or factors. The first factor or latent variable, which we might label verbal ability, would have high correlations with variables X_1 through X_4 and low correlations with variables X_5 through X_7. The second latent variable or factor, which we might label mathematical skills, would have high correlations with variables X_5 through X_7 and low correlations with variables X_1 through X_4. If we can show that the correlations of these X predictors with just two factors (verbal and math abilities) are sufficient information to do a reasonably good job of reproducing the observed correlations among these X measured variables, this might lead us to conclude that the $p = 7$ test scores included in the set of measured X variables can be understood as information about just *two* latent variables that cannot be directly observed (verbal and math abilities).

Although we cannot directly measure or observe a latent variable, we can understand the nature of a latent variable by examining how it is correlated with measured X variables, and we can use the z scores for measured X variables to construct a score for each participant for each latent variable, factor, or dimension. The goal of factor analysis and related analytic methods such as PC is to assess the extent to which the various X variables in our data set can be interpreted as measures of one or more underlying constructs or latent variables. In factor analysis, a large Pearson's r between measures X_i and X_j is taken as an indication that X_i and X_j may both measure the same thing. For instance, if verbal SAT score (X_1) and grade in an English course (X_2) have a high positive correlation, we might assume that they are different measures of the same underlying ability (language skill).

A factor-analytic study usually begins with the collection of measurements on many X variables that we think may all measure the "same thing" (e.g., a personality construct such as "extraversion") or, perhaps, a set of related constructs (such as different types of mental ability, e.g., verbal and mathematical). We have two basic questions in mind when we do a factor analysis on a set of X variables (p is the number of variables).

The first question is, How many different theoretical constructs (or components, factors, dimensions, or latent variables) do these p variables seem to measure? In the preceding hypothetical example, we saw that we could represent the correlation structure among a set of $p = 7$ measured variables by using a model that included just two factors or latent variables. The second question is, What (if anything) do these factors mean? Can we name the factors and interpret them as latent variables? In the preceding example, the two factors were named verbal ability and math ability; these labels were based on the type of tests each factor correlated with most highly.

We will answer these two questions (how many latent variables we need and what interpretation we can suggest for each latent variable) by using PC or PAF to convert the correlation matrix of correlations among our p measured variables into a matrix of loadings or correlations that tell us how strongly linearly related each measured X variable is with each component or factor. We can obtain scores for each participant on each latent variable or factor by computing specific weighted linear combinations of each person's scores on z_1 through z_X. In the examples presented in this chapter, the factors are constrained to be orthogonal to each other (i.e., scores on F_1 are almost perfectly uncorrelated with scores on F_2 for all possible pairs of factors).

Data reduction through factor analysis has potential advantages and disadvantages. First consider some potential advantages. Factor analysis can be helpful during the process of theory development and theory testing. Suppose that we analyze correlations among scores on 10 mental ability tests and decide that we can account for the pattern of correlations among scores by interpreting some of the variables as indicators of a latent verbal ability factor and other variables as indicators of a latent math ability factor. A factor analysis may help us decide how many types of mental ability we need to measure (two types in this hypothetical example). Factor analysis can be an aid in theory development; however, it is a good tool for making inferences about the "state of the world" only if we are confident that the battery of tests included in the factor analysis comprehensively covers the entire domain of mental ability tests. Researchers need to understand that factor analysis and related methods such as PC tell us only how many factors (or components) are needed to account for the correlations among the variables that were included in the test battery in the study; this set of variables may not include all the mental abilities or traits that exist out in the real world.

One example of theory development that made extensive use of factor analysis was the development of the "Big Five" model of personality (Costa & McCrae, 1995, 1997). Factor analyses of large sets of personality measures suggested that much of the pattern in self-report personality data may be accounted for by a model that includes five latent variables, factors, or traits: openness to experience, conscientiousness, extraversion, agreeableness, and neuroticism.

In addition, factor analysis may have a practical application in some data analysis situations where a researcher wants to reduce a large number of predictors (in a multiple regression analysis, for example) to a smaller number of predictors. Suppose that the researcher wants to predict starting salary (Y) from the 10 mental ability test scores used in the previous examples. It might not be a good idea to use each of the 10 test scores as a separate predictor in one multiple regression for two reasons. First, when we include several predictor variables that are highly correlated with each other, to a great extent, they compete to predict or explain the same variance in the Y outcome variable. If we include all three of the verbal ability indicator variables (i.e., scores on vocabulary, reading comprehension, and understanding analogies tests: X_1, X_2, and X_3) as predictors in a regression, we may find that none of these three predictors uniquely predicts a significant amount of variance in the Y outcome variable when its correlations with the other two verbal ability measures are statistically controlled. Instead, if we compute a single "verbal ability" **factor score** that summarizes the information contained in variables X_1, X_2, and X_3 and a single "math ability" factor score that summarizes the information contained in the math tests, we can then set up a regression to predict salary (Y) from just two variables: F_1, a factor score that summarizes information about each person's verbal ability, and F_2, a second factor score that summarizes information about each person's math ability. If the factors are orthogonal or uncorrelated, then the new predictor variables F_1 and F_2 will not compete with each other to explain the same variance; F_1 will have a correlation with F_2 that is near 0 if these factor scores are derived from an orthogonal factor analysis. There are potentially several reasons why it may be preferable to report a regression to predict Y from F_1 and F_2 instead of a regression to predict Y from a set of 10 predictor variables, X_1 through X_{10}. This two-predictor regression model may be easier to report and interpret, the predictor variables F_1 and F_2 will not compete with each other to explain the same variance, and this model has only two t tests for the significance of the individual predictors F_1 and F_2. In contrast, a regression with 10 predictors (X_1 through X_{10}) is more difficult to report and interpret; the analysis may involve high levels of multicollinearity among predictors; the degrees of freedom for the error term will be smaller, and therefore, significance tests will have less power; and finally, conducting 10 t tests to assess the predictive contributions for each variable (X_1 through X_{10}) could lead to an inflated risk for Type I error. Thus, one practical use for factor analysis is in data reduction; in some research situations, a data analyst may be able to use a preliminary factor analysis to assess whether the information contained in the scores on a large set of individual measures (X_1 through X_p) can be summarized and replaced by a smaller set of factor scores, for example, scores on just two factors.

Factor analysis is often used during the process of developing multiple-item scales to measure personality traits such as extraversion, political attitudes such as conservatism, or mental abilities such as intelligence. Factor analysis can help a researcher to decide how many different latent variables are needed to understand the responses to test items and to decide which items are good indicators for each latent variable. Factor analysis is also useful for reexamination of existing measures. For example, the Type A personality was originally thought to be one single construct, but factor analysis of self-report items from Type A questionnaires suggested that time urgency, competitiveness/job involvement, and hostility were three separate components of Type A, and further research suggested that hostility was the one component of Type A that was most predictive of coronary heart disease. This resulted in a change in focus in this research domain, with greater interest in measures that are specific to hostility (Linden, 1987).

Factor analysis is also used for theory development and theory testing, although researchers need to be cautious when they try to interpret factors they see in their data as evidence of traits or abilities that exist in the real world. The history of mental ability or IQ testing is closely associated with the development of factor analysis as a statistical method, and the theories about the number of types of intelligence have changed over time as the test batteries became larger and more varied tests were added. There is often a kind of feedback process involved, such that researchers select the variables to include in a factor analysis on the basis of a theory; they may then use the outcome of the factor analysis to help rethink the selection of variables in the next study.

However, there are also some potential disadvantages or problems with the use of factor analysis for data reduction. Factor analysis is sometimes used as a desperation tactic to look for a way to summarize the information in a messy data set (where the selection of variables was not carefully planned). In research situations where long lists of questions were generated without much theoretical basis, factor analysis is sometimes used to try to make sense of the information contained in a poorly selected collection of variables and to reduce the number of variables that must be handled in later stages of analysis. In such situations, the outcome of the study is "data driven." As in other data-driven approaches to analysis, such as statistical selection of predictor variables for multiple regression using forward methods of selection, we tend to obtain empirical results that represent peculiarities of the batch of data due to sampling error; these results may not make sense and may not replicate in different samples.

There are two reasons why some people are skeptical or critical of factor analysis. First, the use of factor analysis as a desperation tactic can lead to outcomes that are quite idiosyncratic to the selected variables and the sample. Second, researchers or readers sometimes mistakenly view the results of **exploratory factor analysis** as "proof" of the existence of latent variables; however, as noted earlier, the set of latent variables that are obtained in a factor analysis is highly dependent on the selection of variables that were measured. A researcher who includes only tests that measure verbal ability in the set of X measured variables will not be able to find evidence of mathematical ability or other potential abilities. The latent variables that emerge from a factor analysis are limited to a great extent by the set of measured variables that are input to the factor analysis procedure. Unless the set of X measured variables are selected in a manner that ensures that they cover the entire domain of interest (such as all known types of mental abilities), the outcome of the study tells us only how many types of mental ability were represented by the measurements included in the study and not how many different types of mental ability exist in the real world. Factor analysis and PC can be useful data analysis tools provided that their limitations are understood and acknowledged.

12.4 INTRODUCTION OF EMPIRICAL EXAMPLE

The empirical example included here uses a subset of data collected by Robert Belli using a modified version of the Bem Sex Role Inventory (BSRI; Bem, 1974). The BSRI is a measure of sex-role orientation. Participants were asked to rate how well each of 60 adjectives described them using Likert-type scales (0 = *not at all* to 4 = *extremely well*). The items in the original BSRI were divided into masculine, feminine, and neutral filler items. Prior measures of masculinity, such as the Minnesota Multiphasic Personality Inventory, implicitly assumed that "masculine" and "feminine" were opposite end points of a single continuum—that is, to be "more masculine," a person had to be "less feminine." If this one-dimensional theory is correct, a factor analysis of items from a test such as the BSRI should yield one factor (which is positively correlated with masculine items and negatively correlated with feminine items). Bem (1974) argued that masculinity and femininity are two separate dimensions and that an individual may be high on one or high on both of these sex-role orientations. Her theory suggested that a factor analysis of BSRI items should yield a two-factor solution; one factor should be correlated only with masculine items and the other factor should be correlated only with feminine items.

For the empirical example presented here, a subset of $p = 9$ items and $N = 100$ participants was taken from the original larger data set collected by Belli (unpublished data). The 9 items included self-ratings of the following characteristics: warm, loving, affectionate, nurturant, assertive, forceful, strong person, dominant, and aggressive. The complete data set for the empirical examples presented is in the file bsri.sav. The correlations among the nine variables in this data set are summarized in Figure 12.2.

To explain how PC and PAF work, we will examine a series of four analyses of this data set. Each of the following four analyses of the BSRI data introduces one or two important concepts. In Analysis 1, we will examine PC for a set of just three BSRI items. In this example, we shall see that when we retain three components, we can reproduce all the correlations among the three X measured items perfectly. In Analysis 2, we will examine what happens when we retain only one component and use only one component to try to reproduce the correlations among the X variables. When the number of components is smaller than the number of variables (p), we cannot reproduce the correlations among the measured variables perfectly; we need to have a model that retains p components to reconstruct the correlations among p variables perfectly. In Analysis 3, a larger set of BSRI items is included ($p = 9$ items), and the PAF method of extraction is used. Finally, Analysis 4 shows how rotation of retained factors sometimes makes the outcome of the analysis more interpretable. In practice, data analysts are most likely to report an analysis similar to Analysis 4. The first three examples are presented for instructional purposes; they represent intermediate stages in the process of factor analysis. The last analysis, Analysis 4, can be used as a model for data analysis for research reports.

12.5 SCREENING FOR VIOLATIONS OF ASSUMPTIONS

PC and factor analysis both begin with a correlation matrix **R**, which includes Pearson correlations between all possible pairs of the X variables included in the test battery (where p is the number of X variables that correspond to actual measurements):

$$\mathbf{R} = \begin{bmatrix} 1 & r_{12} & \cdots & r_{1p} \\ r_{21} & 1 & \cdots & r_{2p} \\ \vdots & \vdots & \vdots & \vdots \\ r_{p2} & r_{p2} & \cdots & 1 \end{bmatrix}. \qquad (12.1)$$

All the assumptions that are necessary for other uses of Pearson's r should be satisfied. Scores on each X_i variable should be quantitative and reasonably normally distributed. Any association between a pair of X variables should be linear. Ideally, the joint distribution of this set of variables should be multivariate normal, but in practice, it is difficult to assess multivariate normality. It is a good idea to examine histograms for each X_i variable (to make sure that the distribution of scores on each X variable is approximately normal, with no extreme outliers) and scatterplot for all X_i, X_j pairs of variables to make sure that all relations between pairs of X variables are linear with no bivariate outliers.

We also assume that **R** is not a **diagonal matrix**; that is, at least some of the r_{ij} correlations that are included in this matrix differ significantly from 0. If all the off-diagonal elements of

Figure 12.2 Correlations Among Nine Items in the Modified BSRI Self-Rated Personality Data

Correlations

		nurturant	affectionate	warm	compas sionate	strong personality	assertive	forceful	dominant	aggressive
nurturant	Pearson Correlation	1	.368 **	.303 **	.331 **	.062	.025	-.149	-.234 *	-.061
	Sig. (2-tailed)		.000	.002	.001	.543	.802	.139	.019	.544
	N	100	100	100	100	100	100	100	100	100
affectionate	Pearson Correlation	.368 **	1	.651* *	.432* *	.266 **	.266 **	.050	.106	.215 *
	Sig. (2-tailed)	.000		.000	.000	.007	.008	.622	.295	.032
	N	100	100	100	100	100	100	100	100	100
warm	Pearson Correlation	.303 **	.651 **	1	.603* *	.261 **	.209 *	.018	.106	.196
	Sig. (2-tailed)	.002	.000		.000	.009	.037	.856	.295	.050
	N	100	100	100	100	100	100	100	100	100
compassionate	Pearson Correlation	.331 **	.432 **	.603 **	1	.171	.053	-.175	-.059	.101
	Sig. (2-tailed)	.001	.000	.000		.088	.603	.081	.557	.315
	N	100	100	100	100	100	100	100	100	100
strong personality	Pearson Correlation	.062	.266 **	.261 **	.171	1	.453 **	.274 **	.312 **	.321 **
	Sig. (2-tailed)	.543	.007	.009	.088		.000	.006	.002	.001
	N	100	100	100	100	100	100	100	100	100
assertive	Pearson Correlation	.025	.266 **	.209 *	.053	.453 **	1	.350 **	.356 **	.322 **
	Sig. (2-tailed)	.802	.008	.037	.603	.000		.000	.000	.001
	N	100	100	100	100	100	100	100	100	100
forceful	Pearson Correlation	-.149	.050	.018	-.175	.274 **	.350 **	1	.467 **	.421 **
	Sig. (2-tailed)	.139	.622	.856	.081	.006	.000		.000	.000
	N	100	100	100	100	100	100	100	100	100
dominant	Pearson Correlation	-.234 *	.106	.106	-.059	.312 **	.356 **	.467 **	1	.381 **
	Sig. (2-tailed)	.019	.295	.295	.557	.002	.000	.000		.000
	N	100	100	100	100	100	100	100	100	100
aggressive	Pearson Correlation	-.061	.215 *	.196	.101	.321 **	.322 **	.421 **	.381 **	1
	Sig. (2-tailed)	.544	.032	.050	.315	.001	.001	.000	.000	
	N	100	100	100	100	100	100	100	100	100

**. Correlation is significant at the 0.01 level (2-tailed).

*. Correlation is significant at the 0.05 level (2-tailed).

R are 0, it does not make sense to try to represent the p variables using a smaller number of factors, because none of the variables correlate highly enough with each other to indicate that they measure the same thing. A matrix is generally factorable when it includes a fairly large number of correlations that are at least moderate (>.3) in absolute magnitude.

It may be instructive to consider two extreme cases. If all the off-diagonal elements of **R** are 0, the set of p variables is already equivalent to p uncorrelated factors; no reduction in the number of factors is possible, because each X variable measures something completely unrelated to all the other variables. On the other hand, if all off-diagonal elements of **R** were 1.0, then the information contained in the X variables could be perfectly represented by just one factor, because all the variables provide perfectly equivalent information. Usually, researchers are interested in situations where there may be blocks, groups, or subsets of variables that are highly intercorrelated with the sets and have low correlations between sets; each set of highly intercorrelated variables will correspond to one factor.

Many things can make Pearson's r misleading as a measure of association (e.g., nonlinearity of relationships, restricted range, attenuation of r due to unreliability). Because the input to factor analysis is a matrix of correlations, any problems that make Pearson's r misleading as a description of the strength of the relationship between pairs of X variables will also lead to problems in factor analysis.

12.6 ISSUES IN PLANNING A FACTOR-ANALYTIC STUDY

It should be apparent that the number of factors (and the nature of the factors) depends entirely on the set of X measurements that are selected for inclusion in the study. If a researcher sets out to study mental ability and sets up a battery of 10 mental ability tests, all of which are strongly correlated with verbal ability (and none of which are correlated with math ability), factor analysis is likely to yield a single factor that is interpretable as a verbal ability factor. On the other hand, if the researcher includes five verbal ability tests and five quantitative reasoning tests, factor analysis is likely to yield two factors: one that represents verbal ability and one that represents math ability. If other mental abilities exist "out in the world" that are not represented by the selection of tests included in the study, these other abilities are not likely to show up in the factor analysis. With factor analysis, it is quite clear that the nature of the outcome depends entirely on what variables you put in. (That limitation applies to all statistical analyses, of course.)

Thus, a key issue in planning a factor-analytic study is the selection of variables or measures. In areas with strong theories, the theories provide a map of the domain that the items must cover. For example, Goldberg's (1999) International Personality Item Pool used Costa and McCrae's (1995, 1997) theoretical work on the Big Five personality traits to develop alternative measures of the Big Five traits. In a research area that lacks a clear theory, a researcher might begin by brainstorming items and sorting them into groups or categories that might correspond to factors, asking experts for their judgment about completeness of the content coverage, and asking research participants for open-ended responses that might provide content for additional items. For example, suppose a researcher is called on to develop a measure of patient satisfaction with medical care. Reading past research might suggest that satisfaction is not unidimensional, but rather there may be three potentially separate components of satisfaction: satisfaction with practitioner competence, satisfaction with the interpersonal communication style or "bedside manner" of the practitioner, and satisfaction with cost and convenience of care. The researcher might first generate 10 or 20 items or questions to cover each of these three components of satisfaction, perhaps drawing on open-ended interviews with patients, research literature, or theoretical models to come up with specific wording. This pool of "candidate" questions would then be administered to a large number of patients, and factor analysis could be used to assess whether three separate components emerged clearly from the pattern of responses.

In practice, the development of test batteries or multiple-item tests using factor analysis is usually a multiple-pass process. Factor analysis results sometimes suggest that the number of factors (or the nature of the factors) may be different from what is anticipated. Some items may not correlate with other measures as expected; these may have to be dropped. Some factors that emerge may not have a large enough number of questions that tap that construct, so additional items may need to be written to cover some components.

Although there is no absolute requirement about sample size, most analysts agree that the number of subjects (N) should be large relative to the number of variables included in the factor analysis (p). In general, N should never be less than 100; it is desirable to have $N > 10p$. Correlations are unstable with small N's, and this, in turn, makes the factor analysis results difficult to replicate. Furthermore, when you have p variables, you need to estimate $[p(p - 1)]/2$ correlations among pairs of variables. You must have more degrees of freedom in your data than the number of correlations you are trying to estimate, or you can run into problems with the structure of this correlation matrix. Error messages such as "determinant = 0," "R is not positive definite," and so forth, suggest that you do not have enough degrees of freedom (a large enough N) to estimate all the correlations needed independently. In general, for factor-analytic studies, it is desirable to have the N's as large as possible.

It is also important when designing the study to include X measures that have reasonably large variances. Restricted range on X's tends to reduce the size of correlations, which will in turn lead to a less clear factor structure. Thus, for factor-analytic studies, it is desirable to recruit participants who vary substantially on the measures; for instance, if your factor analysis deals with a set of mental ability tests, you need to have subjects who are heterogeneous on mental ability.

12.7 COMPUTATION OF FACTOR LOADINGS

A problem that has been raised implicitly in the preceding sections is the following: How can we obtain estimates of correlations between measured variables (X_1 through X_p) and latent variables or factors (F_1 through F_p)? We begin with a set of actual measured variables ($X_1, X_2, X_3, \ldots, X_p$), for instance, a set of personality or ability test items or a set of physiological measures. We next calculate a matrix of correlations (**R**) among all possible pairs of items, as shown in Equation 12.1.

The matrix of loadings or correlations between each measured X variable and each factor is called a factor loading matrix; this matrix of correlations is usually denoted **A**. Each element of this **A** matrix, a_{ij}, corresponds to the correlation of variable X_i with factor or component j, as shown in the path model in Figure 12.1:

$$\mathbf{A} = \begin{array}{c} \\ X_1 \\ X_2 \\ \vdots \\ X_p \end{array} \begin{array}{c} \begin{array}{cccc} F_1 & F_2 & \ldots & F_p \end{array} \\ \left[\begin{array}{cccc} a_{11} & a_{12} & & a_{1p} \\ a_{21} & a_{22} & & a_{2p} \\ & & & \\ a_{p1} & a_{p2} & & a_{pp} \end{array} \right] \end{array}. \qquad (12.2)$$

The a_{ij} terms in Figure 12.1 and Equation 12.2 represent estimated correlations (or "loadings") between each measured X variable (X_i) and each latent variable or factor (F_j). How can we obtain estimates of these a_{ij} correlations between actual measured variables and latent or purely imaginary variables? We will address this problem in an intuitive manner, using logic very similar to the logic that was used in Chapter 4, on partitioning variance in regression with two predictors, to show how estimates of beta coefficients can be derived from correlations for a standardized regression that has just two predictor variables. We will apply the tracing rule to the path model in Figure 12.1. When we have traced all possible paths between each pair of measured z_X variables in Figure 12.1, the correlations or loadings between each z_X and each factor (F)—that is, the values of all the a_{ij} terms—must be consistent with the correlations among the X's. That is, we must be able to reconstruct all the r_{ij} correlations among the X's (the **R** matrix) from the factor loadings (the a_{ij} terms that label the paths in Figure 12.1).

In the examples provided in this chapter (in which factors are orthogonal), a_{ij} is the correlation between variable i and factor j. For example, a_{25} is the correlation between X_2 and F_5. If we retain the entire set of correlations between all p factors and all p of the X variables as shown in the **A** matrix in Equation 12.2, we can use these correlations in the **A** matrix to reproduce the correlation between all possible pairs of measured X variables in the **R** matrix (such as X_1 and X_2) perfectly by tracing all possible paths from X_1 to X_2 via each of the p factors (see the path model in Figure 12.1).

For example, to reproduce the r_{12} correlation between X_1 and X_2, we multiply together the two path coefficients or loadings of X_1 and X_2 with each factor; we sum the paths that connect X_1 and X_2 across all p of the factors, and the resulting sum will exactly equal the overall correlation, r_{12} (see Figure 12.1 for the paths that correspond to each pair of a_{ij} terms):

$$r_{12} = (a_{11}\, a_{21}) + (a_{12}\, a_{22}) + (a_{13}\, a_{23}) + \cdots + (a_{1p}\, a_{2p}). \tag{12.3}$$

In practice, we try to reconstruct as much of the r_{12} correlation as we can with the loadings of the variables on the first factor F_1; we try to explain as much of the residual $(r_{12} - a_{11}a_{21})$ with the loadings of the variables on the second factor F_2, and so forth. We can perfectly reproduce the correlations among p variables with p factors. However, we hope that we can reproduce the observed correlations among X variables reasonably well using the correlations of variables with just a few of the factors. In other words, when we estimate the correlations of loadings of all the X variables with Factor 1, we want to obtain values for these loadings that make the product of the path coefficients that lead from X_1 to X_2 via F_1 $(a_{11}a_{21})$ as close to r_{12} as possible, make the product $a_{21}a_{31}$ as close to r_{23} as possible, and so on for all possible pairs of the X's.

If we look at this set of relations, we will see that our path diagram implies proportional relations among the a's. This is only an approximation and not an exact solution for the computation of values of the a loadings. The actual computations are done using the algebra of eigenvectors and eigenvalues, and they include correlations with all p factors, not only the loadings on Factor 1, as in the following conceptual description. If we consider only the loadings on the first factor, we need to estimate loadings on Factor 1 that come as close as possible to reproducing the correlations between X_1, X_2 and X_1, X_3:

$$r_{12} = a_{11}a_{21}, \tag{12.4}$$

$$r_{13} = a_{11}a_{31}. \tag{12.5}$$

With the application of some simple algebra, we can rearrange each of the two previous equations so that each equation gives a value of a_{11} in terms of a ratio of the other terms in Equations 12.4 and 12.5:

$$a_{11} = r_{12}/a_{21}, \tag{12.6}$$

$$a_{11} = r_{13}/a_{31}. \tag{12.7}$$

In the two previous equations, both r_{12}/a_{21} and r_{13}/a_{31} are equal to a_{11}. Thus, these ratios must also be equal to each other, which implies that the following proportional relationship should hold:

$$r_{12}/a_{21} = r_{13}/a_{31}. \quad (12.8)$$

We can rearrange the terms in Equation 12.8 so that we have (initially unknown) values of the a_{ij} loadings on one side of the equation and (known) correlations between observed X variables on the other side of the equation:

$$a_{31}/a_{21} = r_{13}/r_{12}. \quad (12.9)$$

For the a's to be able to reconstruct the correlations among all pairs of variables, the values of the a coefficients must satisfy ratio relationships such as the one indicated in the previous equation. We can set up similar ratios for all pairs of variables in the model and solve these equations to obtain values of the factor loadings (the a's) in terms of the values of the computed correlations (the r's). In other words, any values of a_{31} and a_{21} that have the same ratio as r_{13}/r_{12} could be an acceptable approximate solution for loadings on the first factor. However, this approach to computation does not provide a unique solution for the factor loadings; in fact, there are an infinite number of pairs of values of a_{31} and a_{21} that would satisfy the ratio requirement given in Equation 12.9.

How do we obtain a unique solution for the values of the a_{ij} factor loadings? We do this by placing additional constraints or requirements on the values of the a_{ij} loadings. The value of each a_{ij} coefficient must be scaled so that it can be interpreted as a correlation and so that squared loadings can be interpreted as proportions of explained variance. We will see later in the chapter that the **sum of squared loadings (SSL)** for each X variable in a PC analysis, summed across all p components, must be equal to 1; this value of 1 is interpreted as the total variance of the standardized X variable and is called a **communality**.

Another constraint may be involved when loadings are calculated. The loadings that define the components or factors are initially estimated in a manner that makes the components or factors uncorrelated with one another (or orthogonal to one another). Finally, we will estimate the factor (or component) loadings subject to the constraint that cross-multiplications of loadings for the first factor yield a **reproduced correlation matrix, R***, that is as close as possible to the observed **R** matrix based on the original data. Then take the residual matrix (the observed **R** matrix minus the **R*** matrix reproduced from loadings on the first factor) and estimate a second factor to reproduce that residual matrix as closely as possible. Then take the residuals from that step, fit a third factor, and so on. By the time we have extracted p factors, we have a $p \times p$ matrix (denoted **A**), and the elements of this **A** matrix consist of the correlations of each of the p measured variables with each of the p factors. This **A** matrix of loadings can be used to exactly reproduce the observed **R** correlation matrix among the X variables.

When we place all the preceding constraints on the factor solution, we can obtain unique estimates for the factor loadings in the **A** matrix in Equation 12.2. However, the initial estimate of factor loadings is arbitrary; it is just one of an infinite set of solutions that would do equally well at reproducing the correlation matrix. Appendix 12A briefly describes how solving the eigenvalue/eigenvector problem for the **R** correlation matrix provides a means of obtaining estimated factor loadings.

12.8 STEPS IN THE COMPUTATION OF PC AND FACTOR ANALYSIS

The term *factor analysis* refers to a family of related analytic techniques, not just a single form of specific analysis. When we do a PC or PAF analysis, there are a series of intermediate steps in the analysis. Some of these intermediate results are reported in SPSS output. At each point in the analysis, the data analyst has choices about how to proceed. We are now ready to take on the problem of describing how a typical package program (such as SPSS) computes a PC or factor analysis and to identify the choices that are available to the data analyst to specify exactly how the analysis is to be performed.

12.8.1 Computation of the Correlation Matrix R

Most PC and PAF analyses begin with the computation of correlations among all possible pairs of the measured *X* variables; this **R** matrix (see Equation 12.1) summarizes these correlations. Subsequent analyses for PC and PAF are based on this **R** matrix. There are other types of factor analysis that are based on different kinds of information about data structure (see Harman, 1976, for further discussion).

It is useful to examine the correlation matrix before you perform a factor analysis and ask, "Is there any indication that there may be groups or sets of variables that are highly intercorrelated with each other and not correlated with variables in other groups?" In the empirical example used in this chapter, nine items from a modified version of the BSRI were factor analyzed; the first result shown is the matrix of correlations among these nine items. These variables were intentionally listed so that the four items that all appeared to assess aspects of being caring (ratings on affectionate, compassionate, warm, and loving) appeared at the beginning of the list, and five items that appeared to assess different types of strength (assertive, forceful, dominant, aggressive, and strong personality) were placed second on the list. Thus, in the correlation matrix that appears in Figure 12.2, it is possible to see that the first four items form a group with high intercorrelations, the last five items form a second group with high intercorrelations, and the intercorrelations between the first and second groups of items were quite low. This preliminary look at the correlation matrix allows us to guess that the factor analysis will probably end up with a two-factor solution: one factor to represent what is measured by each of these two sets of items. In general, of course, data analysts may not know a priori which items are intercorrelated, or even how many groups of items there might be, but it is useful to try to sort the items into groups before setting up a correlation matrix to see if a pattern can be detected at an early stage in the analysis.

12.8.2 Computation of the Initial Factor Loading Matrix A

One important decision involves the choice of a method of extraction. SPSS provides numerous options. The discussion that follows considers only two of these methods of extraction: PC versus PAF. These two extraction methods differ both computationally and conceptually. The computational difference involves the values of the diagonal elements in the **R** correlation matrix. In PC, these diagonal elements have values of 1, as shown in Equation 12.1. These values of 1 correspond conceptually to the "total" variance of each measured variable (when each variable is expressed as a *z* score, it has a variance of 1). PC may be preferred by researchers who are primarily interested in reducing the information in a large set of variables down to scores on a smaller number of components (Tabachnick & Fidell, 2018). In PAF, the diagonal elements in **R** are replaced with estimates of the proportion of variance in each of the measured *X* variables that is predictable from or shared with other *X* variables; PAF is mathematically somewhat more complex than PC. PAF is more widely used because it makes it possible for researchers to ignore the unique or error variance associated with each measurement and to obtain factor loading estimates that are based on the variance that is shared among the measured variables.

When PC is specified as the method of extraction, the analysis is based on the **R** matrix as it appears in Equation 12.1, with values of 1 in the diagonal that represent the total variance of each z_X measured variable. When PAF is specified as the method of extraction, the ones in the diagonal of the **R** matrix are replaced with "communality estimates" that provide information about the proportion of variance in each z_X measured variable that is shared with or predictable from other variables in the data set. One possible way to obtain initial communality estimates is to run a regression to predict each X_i from all the other measured X variables. For example, the initial communality estimate for X_1 can be the R^2 for the prediction of X_1 from $X_2, X_3, \ldots, X_p$. These communality estimates can range from 0 to 1.00, but in general, they tend to be less than 1.00. When a PAF extraction is requested from SPSS, the ones in the diagonal of the **R** correlation matrix are replaced with initial communality estimates, and the loadings (i.e., the correlations between factors and measured X variables) are estimated so that they reproduce this modified **R** matrix that has communality estimates that represent only shared variance in its diagonal. Unlike PC (where the loadings could be obtained in one step), the process of estimating factor loadings using PAF may involve multiple steps or **iterations**. An initial set of factor loadings (correlations between p measured variables and p factors) is estimated and this set of factor loadings is used to try to reproduce **R**, the matrix of correlations with initial communality estimates in the diagonal. However, the communalities implied by (or constructed from) this first set of factor loadings generally differ from the initial communality estimates that were obtained using the R^2 values. The initial communality estimates in the diagonal of **R** are replaced with the new communality estimates that are based on the factor loadings. Then the factor loadings are reestimated from this new version of the correlation matrix **R**. This process is repeated ("iterated") until the communality estimates "converge"; that is, they do not change substantially from one iteration to the next. Occasionally, the values do not converge after a reasonable number of iterations (such as 25 iterations). In this case, SPSS reports an error message ("failure to converge").

In practice, the key difference between PC and PAF is the way they estimate communality for each variable. In PC, the initial set of loadings for p variables on p factors accounts for *all* the variance in each of the p variables. On the other hand, PAF attempts to reproduce only the variance in each measured variable that is shared with or predictable from other variables in the data set.

Whether you use PC or PAF, the program initially calculates a set of estimated loadings that describe how all p of the X variables are correlated with all p of the components or factors. This corresponds to the complete set of estimated a_{ij} loadings in the **A** matrix that appears in Equation 12.2. If you have p variables, this complete loading matrix will have p rows (one for each X variable) and p columns (one for each component or factor). Most programs (such as SPSS) do not output this intermediate result. In Analysis 1 reported below, the complete set of loadings for three variables on three components was requested to demonstrate that when we retain the loadings of p variables on p components, this complete set of loadings can perfectly reproduce the original correlation matrix.

12.8.3 Limiting the Number of Components or Factors

A key issue in both PC and PAF is that we usually hope that we can describe the correlation structure of the data using a number of components or factors that is smaller than p, the number of measured variables. The decision about the number of components or factors to retain can be made using an arbitrary criterion. For each component or factor, we have an eigenvalue that corresponds to the sum of the squared loadings for that component or factor. The eigenvalue provides information about how much of the variance in the set of p standardized measured variables can be reproduced by each component or factor. An eigenvalue of 1 corresponds to the variance of a single z score standardized variable. The default decision rule in SPSS is to retain only the components or factors that have eigenvalues greater than 1, that is, only latent variables that have a variance that is greater than the variance of a single standardized variable.

The eigenvalues can be graphed for each factor ($F_1, F_2, \ldots, F_p$). The factors are rank ordered by the magnitudes of eigenvalues—that is, F_1 has the largest eigenvalue, F_2 the next largest eigenvalue, and so forth. A graph of the eigenvalues (on the Y axis) across factor numbers (on the X axis) is called a **scree plot**. This term comes from geology; *scree* refers to the distribution of the rubble and debris at the foot of a hill. The shape of the curve in a plot of eigenvalues tends to decline rapidly for the first few factors and more slowly for the remaining factors, such that it resembles a graph of a side view of the scree at the foot of a hill. Sometimes data analysts visually examine the scree plot and look for a point of inflection; that is, they try to decide where the scree plot "flattens out," and they decide to retain only factors whose eigenvalues are large enough that they are distinguishable from the "flat" portion of the scree plot. An example of a scree plot appears in Figure 12.27.

Alternatively, it is possible to decide what number of components or factors to retain on the basis of conceptual or theoretical issues. For example, if a researcher is working with a large set of personality trait items and has adopted the Big Five theoretical model of personality, the researcher would be likely to retain five factors or components (to correspond with the five factors in the theoretical model). In some cases, the decision about the number of components or factors to retain may take both empirical information (such as the magnitudes of eigenvalues) and conceptual background (the number of latent variables specified by a theory or by past research) into account. Usually, researchers hope that the number of retained factors will be relatively small. In practice, it is rather rare to see factor analyses reported that retain more than 5 or 10 factors; many reported analyses have as few as 2 or 3 retained factors.

Another consideration involves the number of items or measurements that have high correlations with (or high loadings on) each factor. In practice, data analysts typically want a minimum of three indicator variables for each factor, and more than three indicator variables is often considered preferable. Thus, a factor that has high correlations with only one or two measured variables might not be retained because there are not enough indicator variables in the data set to provide adequate information about any latent variable that might correspond to that factor.

12.8.4 Rotation of Factors

If more than one component or factor is retained, researchers often find it useful to request a "rotated" solution. The goal of factor rotation is to make the pattern of correlations between variables and factors more interpretable. When correlations are close to 0, +1, or –1, these correlation values make it easy for the researcher to make binary decisions. For example, if the measured variable X_1 has a loading of .02 with F_1, the researcher would probably conclude that X_1 is not an indicator variable for F_1; whatever the F_1 factor or latent variable may represent conceptually, it is not related to the measured X_1 variable. On the other hand, if the measured variable X_2 has a loading of +.87 with F_1, the researcher would probably conclude that X_2 is a good indicator variable for the construct that is represented by F_1 and that we can make some inferences about what is measured by the latent variable F_1 by noting that F_1 is highly positively correlated with X_2. From the point of view of a data analyst, solutions that have loadings that are close to 0, +1, or –1 are usually the easiest to interpret. When many of the loadings have intermediate values (e.g., most loadings are on the order of +.30), the data analyst faces a more difficult interpretation problem. Usually, the data analyst would like to be able to make a binary decision: For each *X* variable, we want the loading to be large enough in absolute value to say that the *X* variable is related to Factor 1, or we want the loading to be close enough to 0 to say that the *X* variable is not related to Factor 1.

The initial set of component or factor loadings that are obtained is sometimes not easy to interpret. Often, many variables have moderate to large positive loadings on Factor 1, and many times the second factor has a mixture of both positive and negative loadings that are moderate in size. Numerous methods of "factor rotation" are available through SPSS; the most widely used is **varimax rotation**. Factor rotation is discussed more extensively in Section 12.13. The goal of factor rotation is to obtain a pattern of factor loadings that is easier to interpret.

If only one factor or component is retained, rotation is not performed. If two or more components or factors are retained, sometimes (but not always) the pattern of loadings becomes more interpretable after some type of rotation (such as varimax) is applied to the loadings.

12.8.5 Naming or Labeling Components or Factors

The last decision involves interpretation of the retained (and possibly also rotated) components or factors. How can we name a latent variable? We can make inferences about the nature of a latent variable by examining its pattern of correlations with measured X variables. If Factor 1 has high correlations with measures of vocabulary, reading comprehension, and comprehension of analogies, we would try to decide what (if anything) these three X variables have in common conceptually. If we decide that these three measurements all represent indicators of "verbal ability," then we might give Factor 1 a name such as verbal ability. Note, however, that different data analysts do not always see the same common denominator in the set of X variables. Also, it is possible for a set of variables to be highly correlated within a sample because of sampling error; when a set of X variables are all highly correlated with each other in a sample, they will also tend to be highly correlated with the same latent variable. It is possible for a latent variable to be uninterpretable. If a factor is highly correlated with three or four measured X variables that seem to have nothing in common with each other, it may be preferable to treat the solution as uninterpretable, rather than to concoct a wild post hoc explanation about why this oddly assorted set of variables are all correlated with the same factor in your factor analysis.

12.9 ANALYSIS 1: PC ANALYSIS OF THREE ITEMS RETAINING ALL THREE COMPONENTS

A series of four analyses will be presented for the data in bsri.sav. Each analysis is used to explain one or two important concepts. In practice, a data analyst is most likely to run an analysis similar to Analysis 4, the last example that is presented in this chapter. Each analysis takes the reasoning one or two steps further. The first analysis that is presented is the simplest one. In this analysis, a set of just three personality rating items is included; three components are extracted using the PC method, and all three are retained. Examination of the results from the first analysis makes it possible to see how the component loadings can be used to obtain summary information about the proportion of variance that is explained for each measured variable (i.e., the communality for each variable) and by each component as well as to demonstrate that, when we have p components that correspond to p measured variables, we can use the loadings on these components to reconstruct perfectly all the correlations among the measured variables (contained in the **R** correlation matrix in Equation 12.1).

In SPSS, the menu selections that are used to run a factor analysis are as follows. From the top-level menu in Data View, select the menu options for <Analyze> → <Dimension Reduction> → <Factor>, as shown in Figure 12.3. The main dialog box for the factor analysis procedure is shown in Figure 12.4. The names of the three items that were selected for inclusion in Analysis 1 were moved into the right-hand side pane under the heading "Variables." The Descriptive Statistics button was used to open the Factor Analysis: Descriptives dialog box that appears in Figure 12.5. The following information was requested: **initial solution**, correlation coefficients, and the reproduced correlation coefficients. Next, the Extraction button in the main Factor Analysis dialog box was used to open up the Factor Analysis: Extraction dialog box that appears in Figure 12.6. The pull-down menu at the top was used to select principal components as the method of extraction. Near the bottom of this window, the radio button next to "Fixed number of factors" was clicked to make it possible for the data analyst to specify the number of components to retain; for this first example, we want to see

Figure 12.3 SPSS Menu Selections for Analysis 1: PC Analysis of Three Items Without Rotation; All Three Components Retained

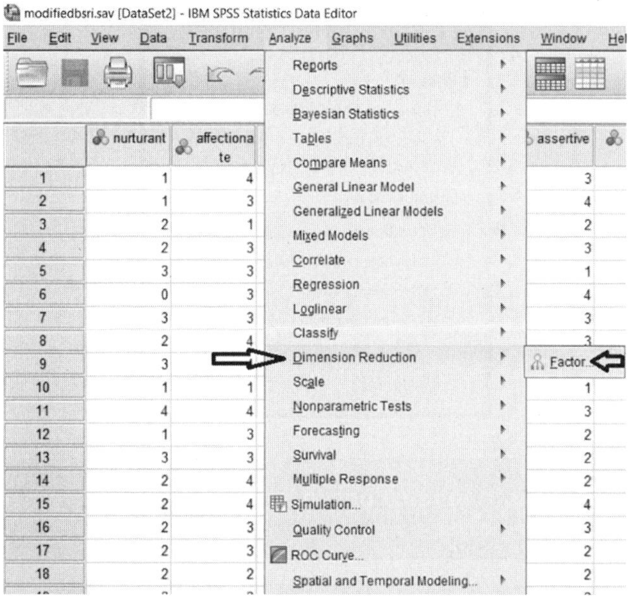

Figure 12.4 Analysis 1: Specification of Three Items to Include in PC Analysis

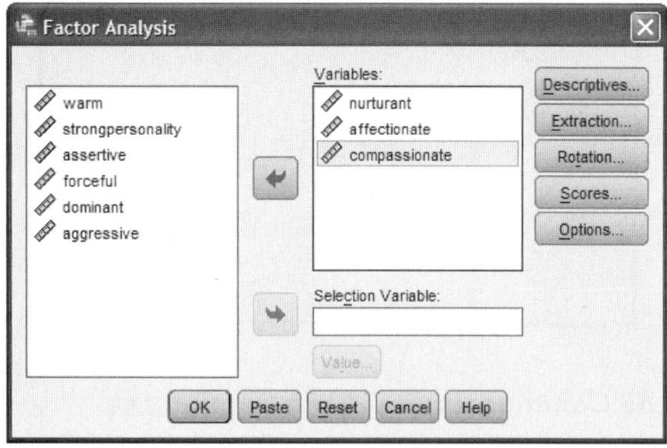

the complete set of loadings (i.e., the correlations of all three measured variables with all three components), and therefore, the number of factors to be retained was specified to be 3. The SPSS syntax that was obtained through these menu selections appears in Figure 12.7.

The output obtained using this syntax appears in Figures 12.8 through 12.12. Figure 12.8 shows the Pearson correlations among the set of three items that were included in Analysis 1, that is, the correlations among self-ratings on nurturant, affectionate, and compassionate; all these correlations were positive, and the r values ranged from .33 to .43. These moderate positive correlations suggest that it is reasonable to interpret the ratings on these three characteristics as measures of the "same thing," perhaps a self-perception of being a caring person.

Figure 12.5 Analysis 1: Descriptive Statistics for PC Analysis Including Correlations Reproduced From the Loading Matrix

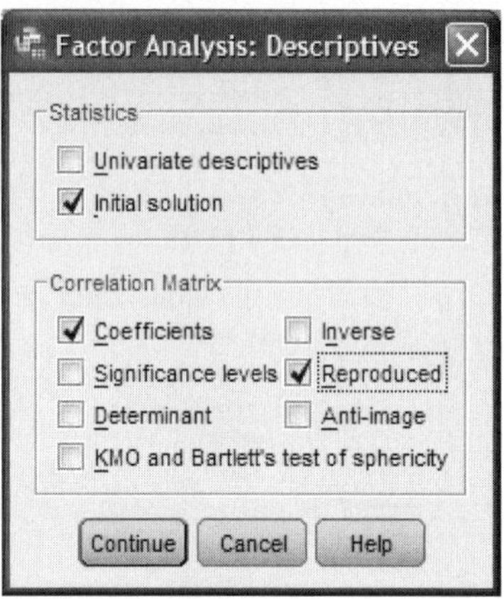

Figure 12.6 Analysis 1: Instructions to Retain All Three Components in the PC Analysis

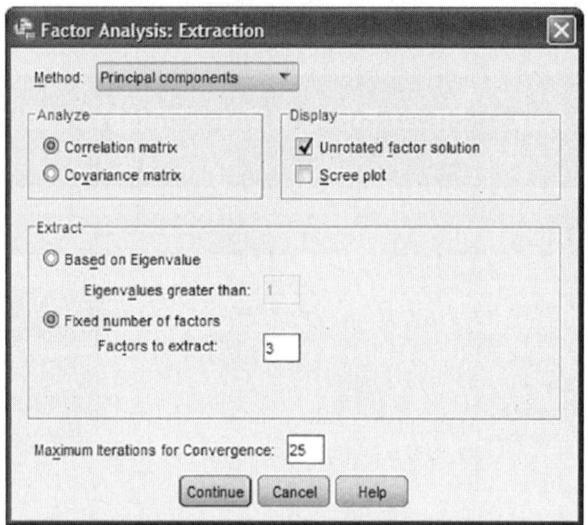

12.9.1 Finding the Communality for Each Item on the Basis of All Three Components

The communalities for each of the three variables in this PC analysis are summarized in Figure 12.9. Each communality is interpreted as "the proportion of variance in one of the measured variables that can be reproduced from the set of three uncorrelated components." The communality for each variable is obtained by squaring and summing the loadings (correlations) of that variable across the retained components. The component matrix appears in Figure 12.11. From Figure 12.11, we can see that the correlations of the variable nurturant with Components 1, 2, and 3, respectively, are .726, .677, and .123. Because each loading is a correlation, a squared loading corresponds to a proportion of explained or predicted variance. Because the three components are uncorrelated with one another, they do not compete to explain the same variance, and we can sum the squared correlations across components to

Figure 12.7 SPSS Syntax for Analysis 1: PC Analysis of Three Items

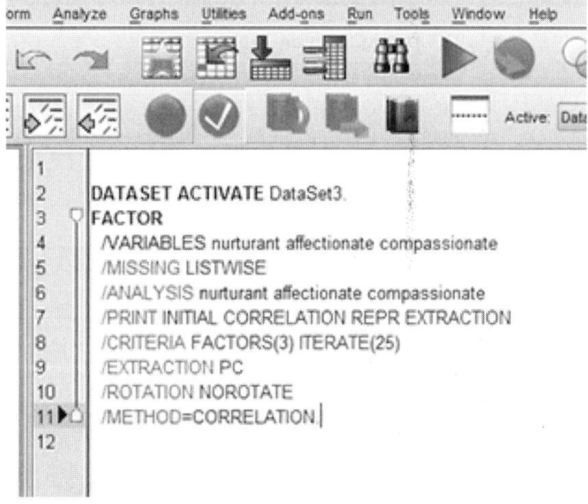

Figure 12.8 Output From Analysis 1: Correlations Among Three Items in PC Analysis

Correlation Matrix

		nurturant	affectionate	compassionate
Correlation	nurturant	1.000	.368	.331
	affectionate	.368	1.000	.432
	compassionate	.331	.432	1.000

Figure 12.9 Analysis 1: Communalities for Each of the Three Variables From the PC Analysis With All Three Components Retained

Communalities

	Initial	Extraction
nurturant	1.000	1.000
affectionate	1.000	1.000
compassionate	1.000	1.000

Extraction Method: Principal Component Analysis.

summarize how much of the variance in self-rated "nurturance" is predictable from the set of Components 1, 2, and 3. In this example, the communality for nurturance is found by squaring its loadings on the three components and then summing these squared loadings: $.726^2 + .677^2 + .123^2 = .5271 + .4583 + .0151 = 1.00$. In other words, for self-ratings on nurturance, about 53% of the variance can be predicted on the basis of the correlations with Component 1, about 46% of the variance can be predicted on the basis of the correlation with Component 2, and about 1% of the variance can be predicted from the correlation with Component 3; when all three components are retained, together they account for 100% of the variance in scores on self-rated nurturance. When the PC method of extraction is used, the communalities

Figure 12.10 Analysis 1: Summary of Variance Explained or Reproduced by Each of the Three Components in Analysis 1

Total Variance Explained

Component	Initial Eigenvalues			Extraction Sums of Squared Loadings		
	Total	% of Variance	Cumulative %	Total	% of Variance	Cumulative %
1	1.756	58.539	58.539	1.756	58.539	58.539
2	.681	22.684	81.223	.681	22.684	81.223
3	.563	18.777	100.000	.563	18.777	100.000

Extraction Method: Principal Component Analysis.

Note: The residuals (differences between actual r_{ij} and predicted r_{ij}') are all 0 to many decimal places.

Figure 12.11 Analysis 1: Loading of (or Correlation of) Each of the Three Measured Variables (Nurturant, Affectionate, and Compassionate) With Each of the Orthogonal Extracted Components (Components 1, 2, and 3)

Component Matrix[a]

	Component		
	1	2	3
nurturant	.726	.677	.123
affectionate	.795	-.205	-.571
compassionate	.773	-.425	.472

Extraction Method: Principal Component Analysis.
a. 3 components extracted.

Figure 12.12 Analysis 1: Reproduced Correlations From Three-Component PC Analysis Solution

Reproduced Correlations

		nurturant	affectionate	compassionate
Reproduced Correlation	nurturant	1.000[b]	.368	.331
	affectionate	.368	1.000[b]	.432
	compassionate	.331	.432	1.000[b]
Residual[a]	nurturant		-3.89E-016	5.0E-016
	affectionate	-4E-016		3.3E-016
	compassionate	5.0E-016	3.33E-016	

Extraction Method: Principal Component Analysis.

a. Residuals are computed between observed and reproduced correlations. There are 0 (.0%) nonredundant residuals with absolute values greater than 0.05.

b. Reproduced communalities

for all variables will always equal 1 when all p of the components are retained. We shall see in the next analysis that the communalities associated with individual measured variables will decrease when we decide to retain only a few of the components.

12.9.2 Variance Reproduced by Each of the Three Components

Another way we can summarize information about the variance explained in Analysis 1 is to sum the loadings and square them for each of the components. To answer the question, How much of the total variance (for a set of three z-score variables, each with a variance of 1.0) is represented by Component 1? we take the list of loadings for Component 1 from Figure 12.11, square them, and sum them as follows. For Component 1, the SSL = $.726^2 + .795^2 + .773^2 = .5271 + .6320 + .5975 \approx 1.756$. Note that this numerical value of 1.756 corresponds to the initial eigenvalue and also to the SSL for Component 1 that appears in Figure 12.10. The entire set of p = 3 z-score variables has a variance of 3. The proportion of this total variance (because each of the three z scores has a variance of 1.0, the total variance for the set of three variables equals 3), represented by Component 1, is found by dividing the eigenvalue or SSL for Component 1 by the total variance. Therefore, in this example, we can see that 58.5% (1.756/3) of the variance in the data can be reproduced by Component 1. The sum of the SSL values for Components 1, 2, and 3 is 3; thus, for a set of three measured variables, a set of three components obtained by PC explains or reproduces 100% of the variance in the correlation matrix **R**.

12.9.3 Reproduction of Correlations From Loadings on All Three Components

The set of loadings in Figure 12.11 has to satisfy the requirement that they can reproduce the matrix of correlations, **R**, among the measured variables. Consider the path model in Figure 12.1; in this path model, the latent variables were labeled as "factors." A similar path model (see Figure 12.13) can be used to illustrate the pattern of relationships between components and z scores on measured variables.

The loadings in Figure 12.11 are the correlations or path coefficients for the PC model for three measured variables and the three components shown in Figure 12.13. To reproduce the correlation between one pair of variables—for example, the correlation between self-rated nurturance and self-rating on affection—we can apply the tracing rule explained in an earlier chapter. There is one path from nurturance to affection via C_1, a second path via C_2, and a third path via C_3. We can reproduce the observed correlation between nurturance and affectionate ($r = .368$) by multiplying the correlations along each path and then summing the contributions for the three paths; for example, the part of the correlation that is reproduced by Component 1 ($.726 \times .795$), the part of the correlation that is reproduced by Component 2 ($.677 \times [-.205]$), and the part of the correlation that is reproduced by Component 3 ($.123 \times [-.571]$). When these products are summed, the overall reproduced correlation is .368. When all three components are retained in the model, there is perfect agreement between the original correlation between each pair of variables (in Figure 12.8) and the reproduced correlation between each pair of variables (in Figure 12.12). Note that the residuals (difference between observed and reproduced correlations for each pair of variables) in Figure 12.12 are all 0 to several decimal places.

At this point, we have demonstrated that we can reproduce all the variances of the z scores on the three ratings, as well as the correlations among the three ratings, perfectly by using a three-component PC solution. However, we do not achieve any data reduction or simplification when the number of retained components (in this example, three components) is the same as the number of variables. The next step, in Analysis 2, will involve assessing how well we can reproduce the correlations when we drop two components from the PC model and retain only the first component.

Figure 12.13 Path Model Showing Correlations Between the Three Variables and the Three Components Obtained in Analysis 1

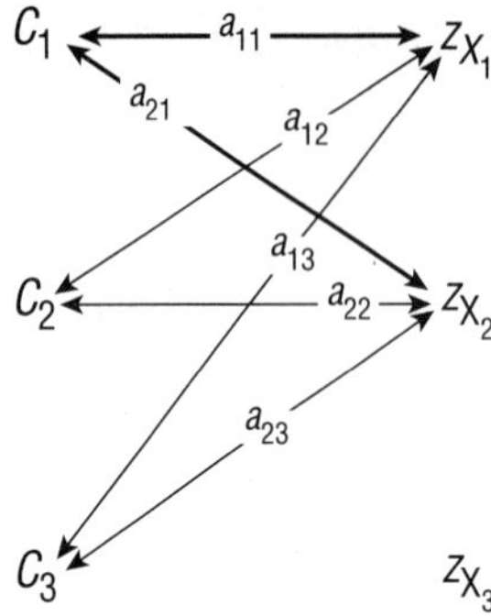

a_{ij} is the correlation between variable X_i and Component C_j

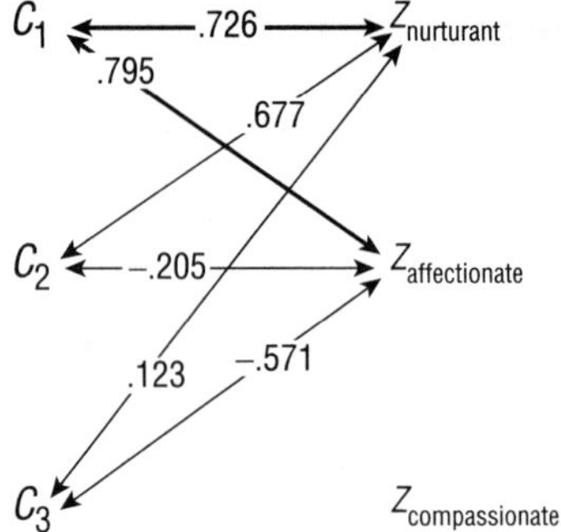

Reconstruction of correlation between nurturant and affectionate based on correlations with C_1, C_2, and C_3:

$r = .726 \times .795 + .677 \times (-.205) + .123 \times (-.571)$

$= .5772 \quad - .1388 \quad - .0702$

$= .368$

Note that this is only a demonstration, not a formal proof, that the correlations among p variables can be reconstructed from a PC model that contains p components.

12.10 ANALYSIS 2: PC ANALYSIS OF THREE ITEMS RETAINING ONLY THE FIRST COMPONENT

In the second analysis, we shall see that—if we decide to retain only one of the three components—we can still reproduce the correlations among the measured variables to some extent, but we can no longer reproduce them perfectly as we could when we had the same number of components as variables (in Analysis 1). One typical goal of PC or factor analysis is to decide how *few* components or factors we can retain and still have enough information to do a reasonably good job of reproducing the observed correlations among measured variables in the **R** correlation matrix. The second analysis reported here is identical to Analysis 1 except for one change in the procedure. Instead of instructing the program to retain all three components (as in Analysis 1), in Analysis 2, the decision rule that was used to decide how many components to retain was based on the magnitudes of the eigenvalues. In Figure 12.14, the option to extract (retain components with) eigenvalues greater than 1 was selected. Apart from this, all the commands were identical to those used in Analysis 1. The SPSS syntax for Analysis 2 appears in Figure 12.15. The output for Analysis 2 appears in Figures 12.16 through 12.19.

12.10.1 Communality for Each Item on the Basis of One Component

The table of communalities in Figure 12.16 reports the communalities for the entire set of three components in the first column. Recall that p, the number of measured variables in this PC analysis, is equal to 3. As in Analysis 1, when we retain all three components, the communality (proportion of variance that can be predicted for each measured

Figure 12.14 Analysis 2: Specifying PC Method of Extraction and Retention of Only Components With Eigenvalues > 1

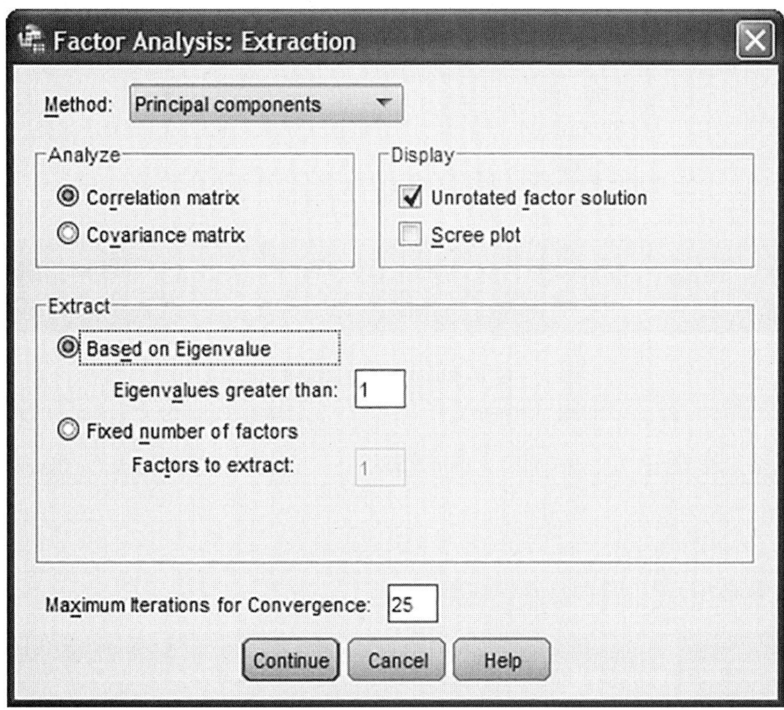

Figure 12.15 Analysis 2: SPSS Syntax

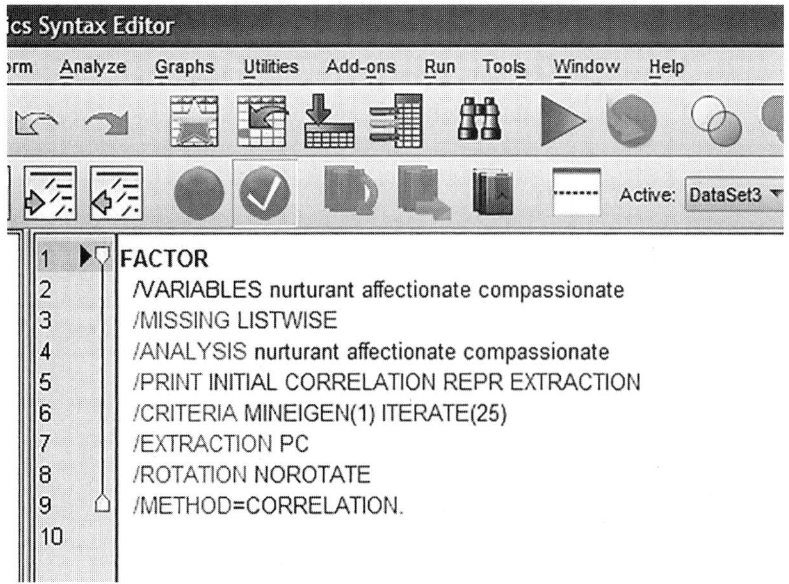

Figure 12.16 Analysis 2: Communalities After Extraction When Only One Component Is Retained

Communalities

	Initial	Extraction
nurturant	1.000	.527
affectionate	1.000	.632
compassionate	1.000	.597

Extraction Method: Principal Component Analysis.

Figure 12.17 Analysis 2: Total Variance Explained by One Retained Component

Total Variance Explained

	Initial Eigenvalues			Extraction Sums of Squared Loadings		
Component	Total	% of Variance	Cumulative %	Total	% of Variance	Cumulative %
1	1.756	58.539	58.539	1.756	58.539	58.539
2	.681	22.684	81.223			
3	.563	18.777	100.000			

Extraction Method: Principal Component Analysis.

Figure 12.18 Analysis 2: Component Matrix for One Retained Component

Component Matrix[a]

	Component
	1
nurturant	.726
affectionate	.795
compassionate	.773

Extraction Method: Principal Component Analysis.

a. 1 components extracted.

Figure 12.19 Analysis 2: Correlations Reproduced From the One-Component Model

Reproduced Correlations

		nurturant	affectionate	compassionate
Reproduced Correlation	nurturant	.527[b]	.577	.561
	affectionate	.577	.632[b]	.614
	compassionate	.561	.614	.597[b]
Residual[a]	nurturant		-.209	-.229
	affectionate	-.209		-.182
	compassionate	-.229	-.182	

Extraction Method: Principal Component Analysis.

a. Residuals are computed between observed and reproduced correlations. There are 3 (100.0%) nonredundant residuals with absolute values greater than 0.05.

b. Reproduced communalities

variable) equals 1. The second column in this table, under the heading "Extraction," tells us the values of the communalities after we make the decision to retain only Component 1. For example, the "after extraction" communality for nurturant in Figure 12.16, $h^2 = .527$, was obtained by squaring the loading or correlation of the variable nurturant with the one retained component. The loadings for three variables on the one retained component appear in Figure 12.18. Because nurturant has a correlation of .726 with Component 1, its communality (the proportion of variance in nurturance ratings that can be reproduced from a one-component model) is $.726^2 = .527$. Almost 53% of the variance in self-ratings on nurturance can be explained when we represent the variables using a one-component model. We usually hope that communalities will be reasonably high for most of the measured variables included in the analysis, although there is no agreed-on standard for the minimum acceptable size of communality. In practice, a communality < .10 suggests that the variable is not very well predicted from the component or factor model; variables with such low communalities apparently do not measure the same constructs as the other variables included in the analysis, and it might make sense to drop variables with extremely low communalities and run the analysis without them to see whether a clearer pattern emerges when the "unrelated" variable is omitted.

12.10.2 Variance Reproduced by the First Component

How much of the variance in the original set of three variables (each with a standard score variance equal to 1) can be reproduced when we retain only one of the three components? To find out how much variance is explained by correlations of the variables with Component 1, all we need to do is square and sum the loadings for Component 1 only (in Figure 12.18). When we do this, we obtain the same numerical result as in Analysis 1; Component 1 has an SSL of 1.756 (see Figure 12.17); compared with the total variance for a set of $p = 3$ variables, this represents 1.756/3 = .585 or almost 59% of the variance. There is no agreed-on minimum value for the proportion or percentage of variance that should be reproduced by the retained component or components for a model to be judged adequate, but a higher proportion of variance tells us that the component does a relatively better job of reproducing variance. In practice, if the first component cannot explain much more than $1/p$ proportion of the variance (where p is the number of measured variables), then transforming the scores into components does not provide us with a useful way of summarizing the information in the data. In this example, with the number of variables $p = 3$, we want the first component to explain more than 33% of the variance.

12.10.3 Cannot Reproduce Correlations Perfectly From Loadings on Only One Component

As in Analysis 1, we can try to reproduce the correlation between any pair of measured variables by tracing the paths that connect that pair of variables in the reduced model that has a limited number of components. In this example, we have limited the number of components to 1. Therefore, when we attempt to reproduce the correlation between one pair of variables (such as the correlation between nurturant and affectionate), the only information we can use is the correlation of each variable with Component 1. Refer to the path model in Figure 12.13. If we use only Component 1, and trace the path from nurturant to affectionate via C_1, the predicted correlation between nurturant and affectionate is equal to the product of the correlations of these variables with Component 1; that is, the estimated correlation between nurturant and affectionate using a one-component model is $.726 \times .795 = .577$. The actual observed correlation between nurturant and affectionate (from Figure 12.8) is $r = .368$. The residual, or error of prediction, is the difference between the actual correlation and the predicted correlation based on the one-component model—in this case, the residual or prediction error for this correlation = .368 − .577 = −.209. The complete set of reproduced correlations based on this model, as well as differences between actual and reproduced correlations, appears in Figure 12.19. We would like these residuals or prediction errors to be reasonably small for most of the correlations in the **R** matrix. SPSS flags residuals that are greater than .05 in absolute value as "large," but this is an arbitrary criterion.

To summarize, for the PC model to be judged adequate after we drop one or more of the components from the model, we would want to see the following evidence:

1. Reasonably large communalities for all the measured variables (e.g., $h^2 > .10$ for all variables; however, this is an arbitrary standard).

2. A reasonably high proportion of the total variance p (where p corresponds to the number of z scores on measured variables) should be explained by the retained components.

3. The retained component or components should reproduce all the correlations between measured variables reasonably well; for example, we might require the residuals for each predicted correlation to be less than .05 in absolute value (however, this is an arbitrary criterion).

4. For all the preceding points to be true, we will also expect to see that many of the measured variables have reasonably large loadings (e.g., loadings greater than .30 in absolute magnitude) on at least one of the retained components.

Note that SPSS does not provide statistical significance tests for any of the estimated parameters (such as loadings), nor does it provide confidence intervals. Judgments about the adequacy of a one- or two-component model are not made on the basis of statistical significance tests but by making arbitrary judgments whether the model that is limited to just one or two components does an adequate job of reproducing the communalities (the variance in each individual measured X variable) and the correlations among variables (in the **R** correlation matrix).

12.11 PC VERSUS PAF

As noted earlier, the most widely used method in factor analysis is the PAF method. In practice, PC and PAF are based on slightly different versions of the **R** correlation matrix (which includes the entire set of correlations among measured X variables). PC analyzes and reproduces a version of the **R** matrix that has ones in the diagonal. Each value of 1.00 corresponds to the total variance of one standardized measured variable, and the initial set of p components must have sums of squared correlations for each variable across all components that sum to 1.00. This is interpreted as evidence that a p-component PC model can reproduce all the variances of each standardized measured variable. In contrast, in PAF, we replace the ones in the diagonal of the correlation matrix **R** with estimates of communality that represent the proportion of variance in each measured X variable that is predictable from or shared with other X variables in the data set. Many programs use multiple regression to obtain an initial communality estimate for each variable; for example, an initial estimate of the communality of X_1 could be the R^2 for a regression that predicts X_1 from $X_2, X_3, \ldots, X_p$. However, after the first step in the analysis, communalities are defined as sums of squared factor loadings, and the estimation of communalities with a set of factor loadings that can do a reasonably good job of reproducing the entire **R** correlation matrix typically requires multiple iterations in PAF.

For some data sets, PC and PAF may yield similar results about the number and nature of components or factors. The conceptual approach involved in PAF treats each X variable as a measurement that, to some extent, may provide information about the same small set of factors or latent variables as other measured X variables, but at the same time, each X variable may also be influenced by unique sources of error. In PAF, the analysis of data structure focused on shared variance and not on sources of error that are unique to individual measurements. For many applications of factor analysis in the behavioral and social sciences, the conceptual approach involved in PAF (i.e., trying to understand the shared variance in a set of X measurements through a small set of latent variables called factors) may be more convenient than the mathematically simpler PC approach (which sets out to represent all of the variance in the X variables through a small set of components). Partly because of the conceptual basis (PAF models only the shared variance in a set of X measurements) and partly because it is more familiar to most readers, PAF is more commonly reported in social and behavioral science research reports than PC. The next two empirical examples illustrate application of PAF to nine items for the data in bsri.sav.

12.12 ANALYSIS 3: PAF OF NINE ITEMS, TWO FACTORS RETAINED, NO ROTATION

Analysis 3 differs from the first two analyses in several ways. First, Analysis 3 includes nine variables (rather than the set of three variables used in earlier analyses). Second, PAF is used as the method of extraction in Analysis 3. Finally, in Analysis 3, two factors were retained on the basis of the sizes of their eigenvalues.

Figure 12.20 shows the initial Factor Analysis dialog box for Analysis 3, with nine self-rated characteristics included as variables (e.g., nurturant, affectionate, . . . , aggressive). None of the additional descriptive statistics (such as reproduced correlations) that were requested in Analyses 1 and 2 were also requested for Analysis 3. To perform the extraction as a PAF, the Extraction button was used to open the Factor Analysis: Extraction dialog box that appears in Figure 12.21. From the pull-down menu near the top of this window, "Principal axis factoring" was selected as the method of extraction. The decision about the number of factors to retain was indicated by clicking the radio button for "Based on Eigenvalue: Eigenvalues

Figure 12.20 Analysis 3: Selection of All Nine Variables for Inclusion in PAF Analysis

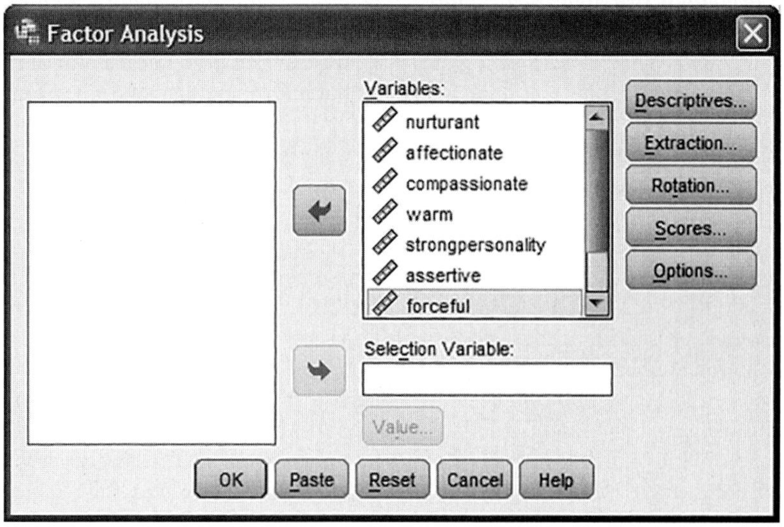

Figure 12.21 Analysis 3: Method of Extraction: PAF Using Default Criterion (Retain Factors With Eigenvalues > 1); Request Scree Plot

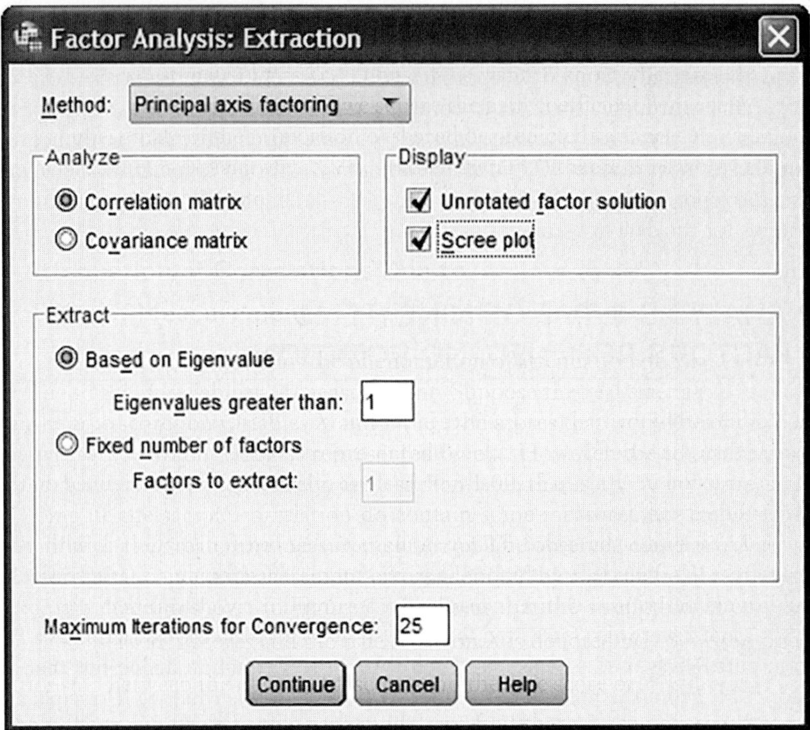

greater than"; the default minimum size generally used to decide which factors to retain is 1, and that was not changed. Under the "Display" heading, a box was checked to request a scree plot; a scree plot summarizes information about the magnitudes of the eigenvalues across all the factors, and sometimes the scree plot is examined when making decisions about the number of factors to retain. The Rotation button was used to open the Factor Analysis: Rotation dialog box that appears in Figure 12.22. The default (indicated by a radio button) is no rotation ("None," under the heading for method), and that was not changed. The box for "Loading plot(s)" under the heading "Display" was checked to request a plot of the factor loadings for all nine variables on the two retained (but not rotated) factors. The Options button opened up the dialog box for Factor Analysis: Options in Figure 12.23; in this box, under the heading "Coefficient Display Format," a check was placed in the checkbox for the "Sorted by size" option. This does not change any computed results, but it arranges the summary table of factor loadings so that variables that have large loadings on the same factor are grouped together, and this improves the readability of the output, particularly when the number of variables included in the analysis is large. The syntax for Analysis 3 that resulted from the menu selections just discussed appears in Figure 12.24. The results of this PAF analysis of nine variables appear in Figures 12.25 through 12.31.

The information that is reported in the summary tables for Analysis 3 includes loadings and sums of squared loadings. The complete 9 × 9 matrix that contains the correlations of all nine measured variables with all nine factors does not appear as part of the SPSS output; the tables that do appear in the printout summarize information that is contained in this larger 9 × 9 loading matrix. For example, each communality is the SSL for one variable across all retained factors. Each SSL for a retained factor corresponds to the SSL for all nine variables on that factor.

Figure 12.22 Analysis 3: No Rotation Requested; Requested Plot of Unrotated Loadings on Factors 1 and 2

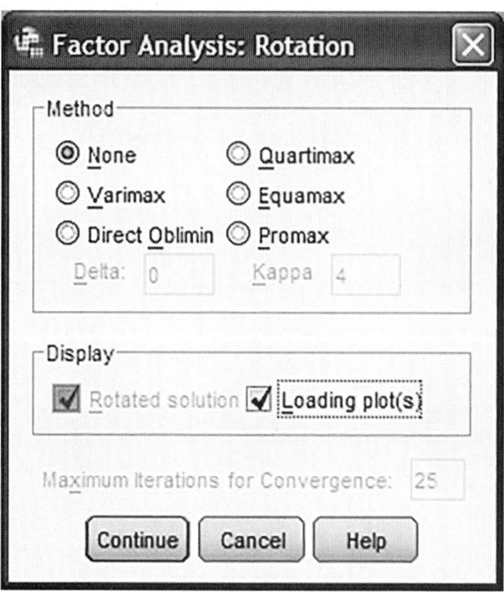

Figure 12.23 Analysis 3: Request for Factor Loadings Sorted by Size

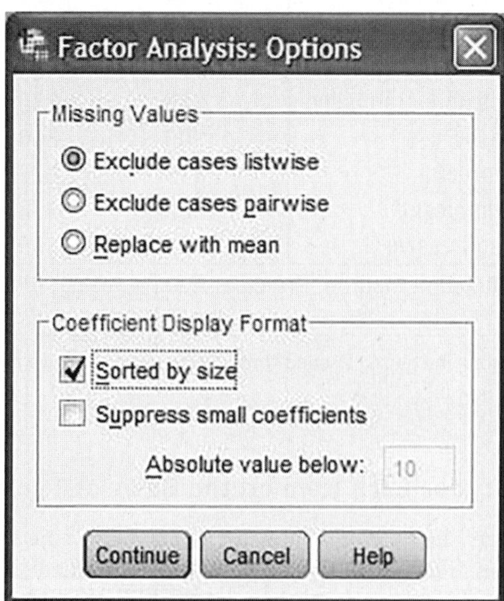

Figure 12.24 Analysis 3: SPSS Syntax for PAF With Nine Items; Retain Factors With Eigenvalues > 1; No Rotation

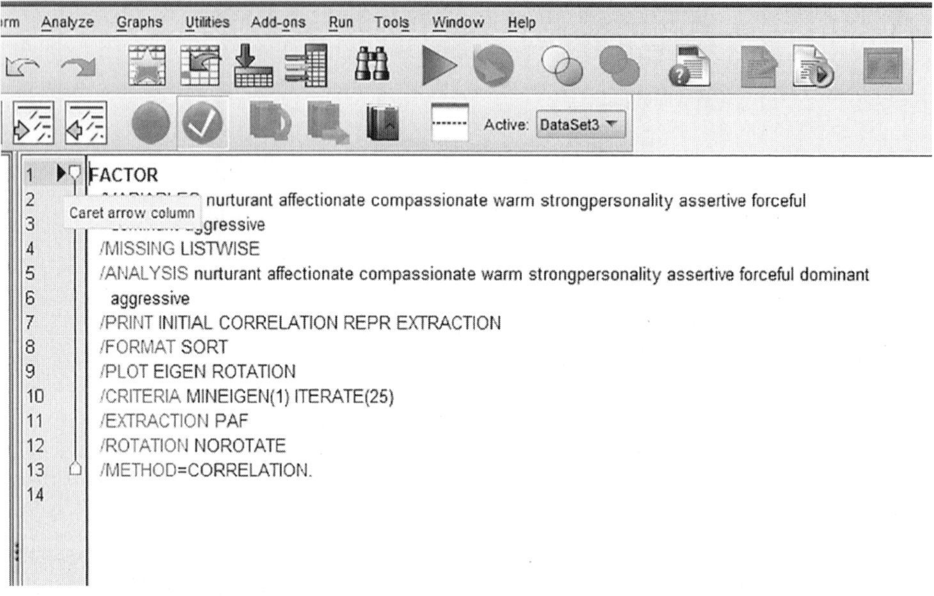

Figure 12.25 Analysis 3: Communalities for PAF on Nine Items

Communalities

	Initial	Extraction
nurturant	.242	.262
affectionate	.484	.570
compassionate	.427	.484
warm	.561	.694
strongpersonality	.285	.330
assertive	.309	.366
forceful	.349	.476
dominant	.339	.458
aggressive	.285	.355

Extraction Method: Principal Axis Factoring.

Note: Based on all nine factors, in column headed "Initial"; based on only two retained factors, in column headed "Extraction."

12.12.1 Communality for Each Item on the Basis of Two Retained Factors

Figure 12.25 reports the communalities for each of the nine variables at two different stages in the analysis. The entries in the first column, headed "Initial," tell us the proportion of variance in each measured variable that could be predicted from the other eight

Figure 12.26 Analysis 3: Total Variance Explained by All Nine Factors (Under Heading Initial Eigenvalues) and by Only the First Two Retained Factors (Under Heading Extraction Sums of Squared Loadings)

Total Variance Explained

	Initial Eigenvalues			Extraction Sums of Squared Loadings		
Factor	Total	% of Variance	Cumulative %	Total	% of Variance	Cumulative %
1	2.863	31.807	31.807	2.346	26.070	26.070
2	2.194	24.380	56.187	1.649	18.327	44.397
3	.823	9.143	65.330			
4	.723	8.032	73.362			
5	.626	6.961	80.323			
6	.535	5.950	86.272			
7	.480	5.331	91.603			
8	.469	5.214	96.817			
9	.286	3.183	100.000			

Extraction Method: Principal Axis Factoring.

Figure 12.27 Analysis 3: Scree Plot of Eigenvalues for Factors 1 Through 9

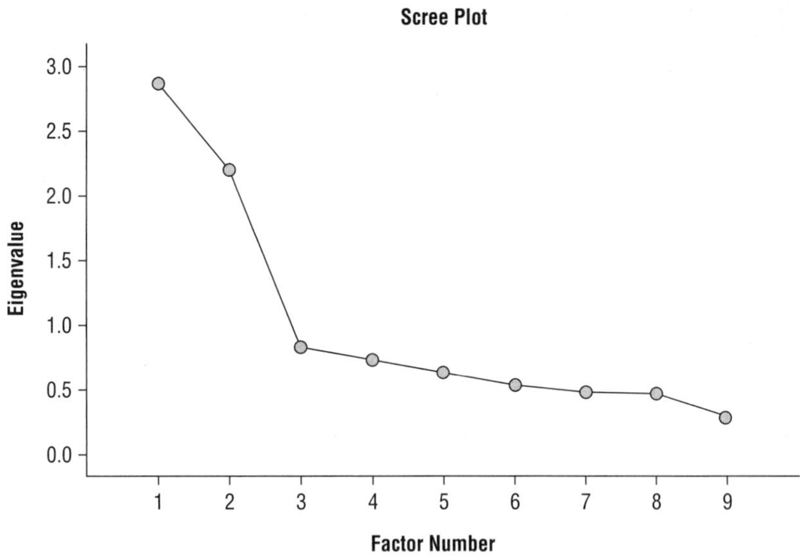

predictor variables in a preliminary multiple regression. The R^2 from this initial regression is used as the initial communality estimate. For example, a regression was performed to predict scores on nurturant from the scores on the other eight measured variables ($R^2 = .242$ for this regression); therefore, the initial estimate of communality was .242; that is, 24.2% of the variance in nurturant could be predicted from the set of scores on the other eight variables. Note that in the PC analysis, the values of the communalities for the initial model were set to 1.00 for all variables. In the PAF analysis, however, these initial

Figure 12.28 Analysis 3: Unrotated Factor Loadings on Two Retained Factors

Factor Matrix[a]

	Factor	
	1	2
warm	.725	-.411
affectionate	.682	-.323
strongpersonality	.534	.212
assertive	.518	.313
aggressive	.482	.351
forceful	.345	.597
dominant	.402	.544
compassionate	.481	-.503
nurturant	.240	-.452

Extraction Method: Principal Axis Factoring.

a. 2 factors extracted. 8 iterations required.

Figure 12.29 Analysis 3: Plot of the Factor Loadings for the Two Retained Unrotated Factors

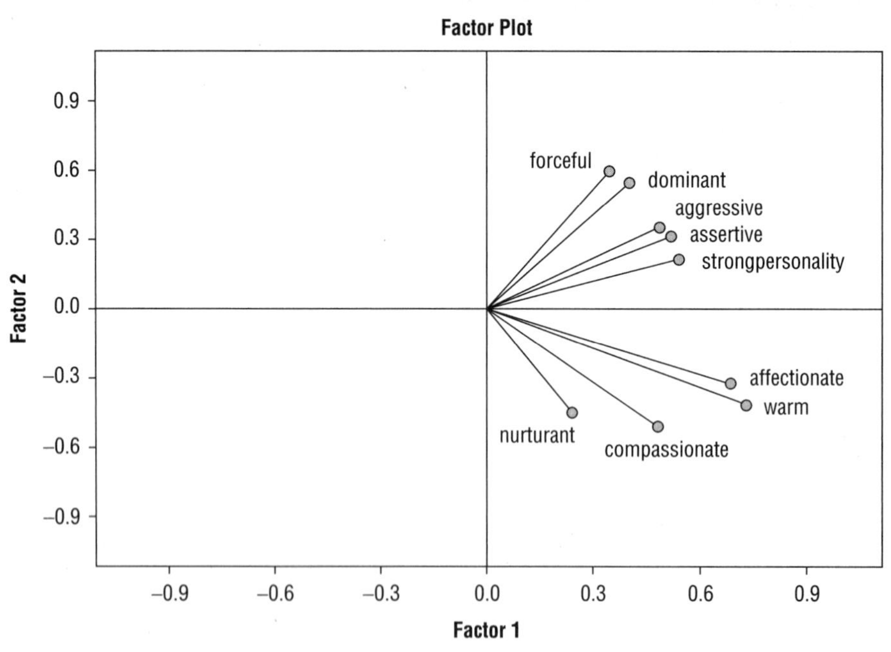

Note: For example, "affectionate" had a loading of +.68 on Factor 1 and a loading of −.32 on Factor 2. The vectors (the lines that connect the dot labeled "affectionate" to the 0,0 origin in the graph) were added in a graphics program; they do not appear in the SPSS plot.

Figure 12.30 Analysis 3: Reproduced Correlations on the Basis of Two Retained (Unrotated) Factors

Reproduced Correlations

		nurturant	affectionate	compas sionate	warm	strong personality	assertive	forceful	dominant	aggressive
Reproduced Correlation	nurturant	.262[b]	.310	.343	.360	.032	-.017	-.187	-.149	-.043
	affectionate	.310	.570[b]	.491	.628	.296	.252	.042	.099	.215
	compassionate	.343	.491	.484[b]	.555	.150	.091	-.134	-.080	.055
	warm	.360	.628	.555	.694[b]	.300	.247	.005	.068	.205
	strongpersonality	.032	.296	.150	.300	.330[b]	.343	.311	.330	.332
	assertive	-.017	.252	.091	.247	.343	.366[b]	.366	.379	.359
	forceful	-.187	.042	-.134	.005	.311	.366	.476[b]	.464	.376
	dominant	-.149	.099	-.080	.068	.330	.379	.464	.458[b]	.385
	aggressive	-.043	.215	.055	.205	.332	.359	.376	.385	.355[b]
Residual[a]	nurturant		.058	-.011	-.057	.029	.043	.038	-.085	-.019
	affectionate	.058		-.058	.024	-.029	.014	.008	.007	.000
	compassionate	-.011	-.058		.048	.021	-.039	-.041	.021	.046
	warm	-.057	.024	.048		-.039	-.038	.014	.038	-.009
	strongpersonality	.029	-.029	.021	-.039		.110	-.037	-.019	-.011
	assertive	.043	.014	-.039	-.038	.110		-.016	-.022	-.038
	forceful	.038	.008	-.041	.014	-.037	-.016		.003	.045
	dominant	-.085	.007	.021	.038	-.019	-.022	.003		-.003
	aggressive	-.019	.000	.046	-.009	-.011	-.038	.045	-.003	

Extraction Method: Principal Axis Factoring.

a. Residuals are computed between observed and reproduced correlations. There are 5 (13.0%) nonredundant residuals with absolute values greater than 0.05.

b. Reproduced communalities

Figure 12.31 Geometric Representation of Correlation Between X and X

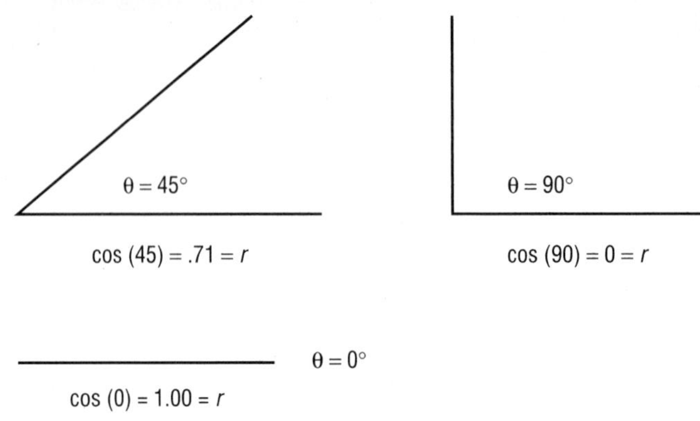

Note: $r = \cos(\theta)$, where θ is the angle between the vectors that represent the X_1 and X_2 vectors (given in degrees).

communalities are based on multiple regressions that tell us what proportion of the variance in each measured variable is predictable from, or shared with, other measured variables. The second column, headed "Extraction," tells us what proportion of variance in each variable is explained after we extract nine factors and decide to retain only two factors. (Information about the number of factors retained appears in later parts of the SPSS output.) For example, in Figure 12.28, we can see that the variable affectionate had loadings or correlations with the two unrotated retained factors that were equal to .682 (for Factor 1) and –.323 (for Factor 2). The communality for affectionate—that is, the proportion of variance in scores on self-ratings of affectionate that can be predicted from the two retained factors—is obtained by summing these squared correlations: $(.682^2) + (-.323)^2 = .569$. This agrees with the communality value of .570 for the variable affectionate in the summary table in Figure 12.25. Thus, the communalities in the "Extraction" column in Figure 12.25 provide information about how well a PAF factor solution that retains only Factors 1 and 2 can reproduce the variance in the nine measured variables. The variable affectionate had a relatively high proportion of predicted variance (.570), while the variable nurturant had a much lower proportion of predicted variance (.262).

12.12.2 Variance Reproduced by Two Retained Factors

We also need summary information that tells us the extent to which the variances and correlations in the set of $p = 9$ variables could be reconstructed, Factor 1, Factor 2, . . . , Factor 9. This information appears in Figure 12.26. In the left-hand panel under the heading "Initial Eigenvalues," for each of the nine factors, there is an eigenvalue (which is equivalent to an SSL). The factors are rank ordered by the sizes of their eigenvalues; thus, Factor 1 has the largest eigenvalue, Factor 2 the second largest eigenvalue, and so forth. The sum of the eigenvalues in the column "Total" under the banner heading "Initial Eigenvalues" will always be p, the number of measured variables. In this example, the sum of the eigenvalues (2.863 + 2.194 + .823 + . . . + .286) = 9, because there were nine measured variables. We can convert each eigenvalue into a percentage of explained variance by dividing each eigenvalue by the sum of the eigenvalues (which is equivalent to p, the number of measured variables) and multiplying this by 100. Thus, for example, in the initial set of nine factors, Factor 1 explained 100 × (2.863/9) = 31.8% of the variance, and Factor 2 explained 100 × (2.194/9) = 24.4% of the variance.

A data analyst needs to decide how many of these nine initial factors to retain for interpretation. The number of retained factors can be based on a conceptual model (e.g., a personality theorist may wish to retain five factors to correspond to the five dimensions in the Big Five personality theory). However, in this example, as in many analyses, the decision was made on the basis of the sizes of the eigenvalues. The radio button that was selected earlier requested retention of factors with eigenvalues > 1.00. When we examine the list of eigenvalues for the initial set of nine factors in the left-hand column of Figure 12.26, we can see that only Factors 1 and 2 had eigenvalues > 1; Factor 3 and all subsequent factors had eigenvalues <1. Using a cutoff value of 1.00 (which corresponds to the amount of variance in one z score or standardized variable) results in a decision to retain only Factors 1 and 2. The right-hand side of the table in Figure 12.26 shows the SSLs for only Factors 1 and 2. After the other seven factors are dropped, SPSS reestimates the factor loadings for the retained factors, and therefore the SSLs for Factors 1 and 2 changed when other factors are dropped from the model. After limiting the model to two factors, Factor 1 predicted or accounted for about 26% of the variance and Factor 2 accounted for about 18% of the variance; together, Factors 1 and 2 accounted for about 44% of the variance in the data (i.e., the variances in scores on measured variables and the pattern of correlations among measured variables).

The eigenvalues for the initial set of nine factors listed in the left-hand column of Figure 12.26 are also shown in a scree plot in Figure 12.27. Sometimes, data analysts use the scree plot as an aid in deciding how many factors to drop from the model. Essentially, we look for a point of inflection; the proportion of variance explained by the first few factors often tends to be large, and then, after two or three factors, the amount of variance accounted for by the remaining factors often levels off. Examination of the scree plot in Figure 12.27 would lead to the same decision as the decision that was made on the basis of the sizes of the eigenvalues. Compared with Factors 1 and 2, Factors 3 through 9 all had quite small eigenvalues that corresponded to very small proportions of explained variance.

The unrotated factor loadings for the two retained factors appear in Figure 12.28 under the heading "Factor Matrix." Each entry in this table corresponds to a factor loading or, in other words, a correlation between one of the measured variables (warm, affectionate, etc.) and one of the retained factors (Factor 1, Factor 2). The goal in factor analysis is often to identify sets or groups of variables, each of which have high correlations with only one factor. When this type of pattern is obtained, it is possible to interpret each factor and perhaps give it a name or label, on the basis of the nature of the measured variables with which it has high correlations. The pattern of unrotated loadings that appears in Figure 12.28 is not easy to interpret. Most of the variables had rather large positive loadings on Factor 1; loadings on Factor 2 tended to be a mixture of positive and negative correlations. The plot in Figure 12.29 shows the factor loadings in a two-dimensional space. This pattern does not make it easy to differentiate Factor 1 and Factor 2 from each other and to identify them with different latent constructs. When more than one factor is retained, in fact, the unrotated loadings are often not very easy to interpret. The last empirical example, Analysis 4, will show how rotation of the factors can yield a more interpretable solution.

12.12.3 Partial Reproduction of Correlations From Loadings on Only Two Factors

A factor analysis, like a PC analysis, should be able to do a reasonably good job of reproducing the observed correlations between all pairs of measured variables. The matrix in Figure 12.30 summarizes the actual versus reproduced correlations for the two-factor model in Analysis 3. Each correlation (e.g., the reproduced correlation between nurturant and affectionate) is reproduced from the factor loadings by tracing the paths that connect these two variables via the latent variables Factor 1 (F_1) and Factor 2 (F_2). We multiply the path coefficients along each path and sum the contributions across the paths. The correlations of nurturant and affectionate with Factor 1, from the factor loading matrix in Figure 12.28, were .240 and .682, respectively; the product of these two correlations is (.24 × .682) = .1637. The correlations of nurturant and affectionate with Factor 2 were –.452 and –.323; the product of these two correlations was .1460. If we sum the predicted correlations on the basis of F_1 (.1637) and F_2 (.1460), we obtain an overall predicted correlation between nurturant and affectionate that equals .1637 + .1460 = .31. This agrees with the reproduced correlation between nurturant and affectionate reported in Figure 12.30. A factor analysis, like a PC analysis, is supposed to do a reasonably good job of reproducing the correlations among observed variables (preferably using a relatively small number of factors).

The factor analysis reported as Analysis 3 has done a reasonably good job of reproducing the variance on each of the measured variables (with communalities that ranged from .262 to .570); it also did a reasonably good job of reproducing the correlations in the correlation matrix **R** (according to the footnote at the bottom of Figure 12.30, only 13% of the prediction errors or residuals for reproduced correlations were larger than .05 in absolute value). Analysis 3 suggests that we can do a reasonably good job of representing the information about the pattern in the **R** data matrix by using just two factors. However, the pattern of (unrotated) factor loadings for the two retained factors in Analysis 3 was not easy to interpret. Rotated factors are often, although not always, more interpretable. Before presenting a final example in which the factors are rotated, we will use a geometric representation of correlations between variables. This geometric representation will make it clear what it means to "rotate" factor axes.

12.13 GEOMETRIC REPRESENTATION OF FACTOR ROTATION

A correlation between two variables X_1 and X_2 can be represented geometrically by setting up two vectors (or arrows). Each vector represents one variable; the cosine of the angle between the vectors is equivalent to the correlation between the variables.

In other words, we can represent the correlation between any two variables by representing them as vectors separated by an angle θ (the angle is given in degrees). The angle between the two vectors that represent variables X_1 and X_2 is related to the correlation between variables X_1 and X_2 through the cosine function. Two variables that have a correlation of 1.00 can be represented by a pair of vectors separated by a 0° angle (as in the bottom example in Figure 12.31). Two variables that have a correlation of 0 (i.e., two variables that are uncorrelated or orthogonal to each other) are represented by a pair of vectors separated by a 90° angle, as in the upper right-hand drawing in Figure 12.31. Angles that are between 0° and 90° represent intermediate amounts of linear correlation; for example, if the vectors that represent variables X_1 and X_2 are at a 45° angle to each other, the correlation between variables X_1 and X_2 is equivalent to the cosine of 45°, or $r = .71$.

A factor loading is also a correlation, so we can represent the loadings (or correlations) between an individual X_1 variable and two factors, F_1 and F_2. So far, in all the examples we have considered, we have constrained the factors or components to be orthogonal to each other or uncorrelated with each other. Therefore, we can diagram Factor 1 and Factor 2 by using vectors that are separated by a 90° angle, as shown in Figure 12.32. We can represent the loadings or correlations of the X_1 variable with these two factors by plotting the X_1 variable as a point. For example, if X_1 has a correlation of .8 with F_1 and a correlation of .1 with F_2, we can plot a point that corresponds to +.8 along the horizontal (F_1) axis and +.1 along the vertical (F_2) axis. When we connect this point to the origin (the 0,0) point in the graph, we obtain a vector that represents the "location" of the X_1 variable in a two-dimensional space relative to Factors 1 and 2. From this graph, we can see that X_1 is much more highly

Figure 12.32 Graphic Representation of Correlation Between Variable X_1 and Factors F_1 and F_2

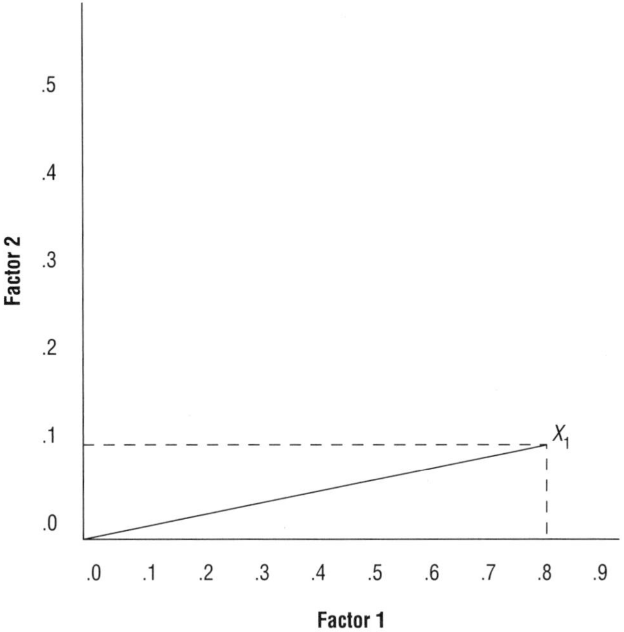

Note: Correlation of X_1 with F_1 = +.80; correlation of X_1 with F_2 = +.1.

correlated with F_1 than it is with F_2; that is, the angle that separates the X_1 vector from the F_1 axis is much smaller than the angle that separates the X_1 vector from the F_2 axis.

Now let's think about the problem of showing the correlations among *p* variables geometrically. We can draw our first two vectors (representing X_1 and X_2) on a flat plane. When we add a third variable X_3, we may or may not need to use a third dimension to set up angles between X_3 and X_1 and X_3 and X_2 that accurately represent the correlation of X_2 with the other variables. As we add more variables, we typically need a higher dimensional space if we want our "bundle" of vectors to be separated by angles that perfectly correspond to the correlations among all possible pairs of variables. In general, we can perfectly represent the correlations among *p* variables in a *p*-dimensional space. We can think of factor analysis as a reduction in the number of dimensions that we use to represent the correlations among variables. Imagine our set of *p* variables as a bundle of vectors in a *p*-dimensional space, like an imaginary set of *p*-dimensional umbrella spokes. We want to know, If we squeeze this bundle of vectors into a lower dimensional space (such as a two- or three-dimensional space), can we still fairly accurately represent the correlations among all pairs of variables by the angles among vectors? Or do we lose this information as we "flatten" the bundle of vectors into a lower dimensional space?

When we decide that a two-factor solution is adequate, our factor solution (a set of loadings, or of correlations between measured variables and imaginary factors) can be understood geometrically. We can represent each X_i variable in terms of its correlations with the axes that represent Factors 1 and 2. The vector that corresponds to X_1 will have a specific location in the two-dimensional space that is defined by the factor axes F_1 and F_2, as shown in Figure 12.32. If a two-dimensional representation is reasonably adequate, we will find that we can reconstruct the correlations among all pairs of measured variables (e.g., X_1 and X_2, X_1 and X_3, X_2 and X_3) reasonably well from the correlations of each of these variables with each of the two retained factors.

When we plotted the correlations or factor loadings for variables X_1 through X_9 (e.g., warm, nurturant, aggressive) with the two unrotated factors in Analysis 3, as shown in Figure 12.29, we could not see a clear pattern. We would like to be able to "name" each latent variable or factor by identifying a set of measured variables with which it has high correlations, and we would like to be able to differentiate two or more latent variables or factors by noticing that they correlated with different groups or sets of measured variables. Figure 12.33 illustrates the pattern that often arises in an unrotated factor solution (in the left-hand side graph). In an unrotated two-factor solution, we often find that all the measured X variables have at least moderate positive correlations with Factor 1 and that there is a mixture of smaller positive and negative correlations of variables with Factor 2, as seen in the hypothetical unrotated factor loadings on the left-hand side of Figure 12.33.

However, if our two-dimensional space is a "plane," and we want to be able to identify the location of vectors in that plane, our choice of reference axes to use is arbitrary. In the original unrotated solution, the factor loadings were calculated so that Factor 1 would explain as much of the variance in the measured X variables as possible; Factor 2 was constrained to be orthogonal to or uncorrelated with Factor 1, and to explain as much of the remaining variance as possible; and so forth. This strategy often yields a set of factor loadings that do a reasonably good job of reproducing all the correlations among measured X variables, but it often does not optimize the interpretability of the pattern.

We can think of factors as the axes that are used as points of reference to define locations of variables in our two-dimensional space. In the present example, F_1 and F_2 correspond to the unrotated factor axes that define a two-dimensional space. The fact that these factors are orthogonal or uncorrelated is represented by the 90° angle between the F_1 and F_2 factor axes. However, the location of the factor axes is arbitrary. We could define locations within this same plane relative to any two orthogonal F_1 and F_2 vectors in this plane.

When we "rotate" a factor solution, we simply move the factor axes to a different location in the imaginary plane and then recalculate the correlation of each X measured variable relative to these relocated or "rotated" axes. On the right-hand side of Figure 12.33, the rotated

Figure 12.33 Hypothetical Graphs of Factor Loadings for Variables X_1 Through X_6 on Factors 1 and 2

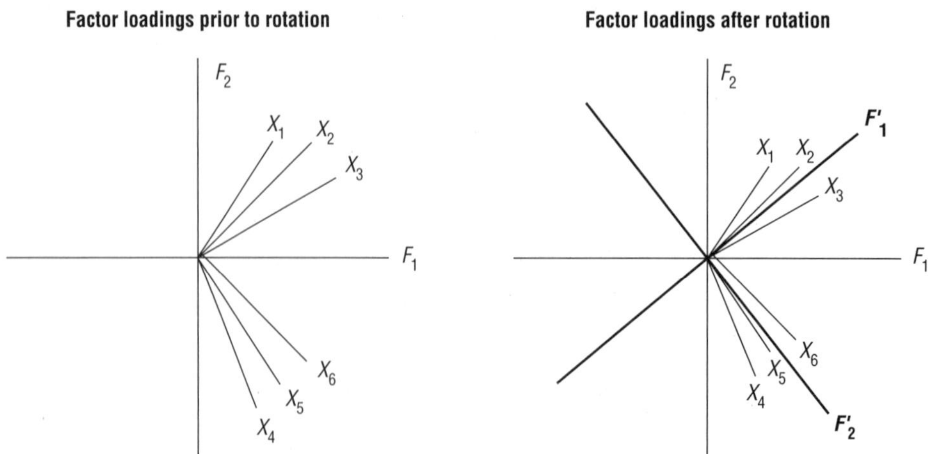

factor axes are labeled F'_1 and F'_2. A factor is most easily interpretable if variables tend to have either very large loadings or loadings near 0; that is, you can say clearly that the factor either *is* or *is not* related to each variable. In this hypothetical example, variables have relatively high correlations with the new rotated F'_1 axis (i.e., the vectors for these first three variables are separated from the F'_1 axis by small angles). X_1, X_2, and X_3 have fairly low correlations with the F'_2 axis. On the other hand, variables X_4, X_5, and X_6 are fairly highly correlated with F'_2 and have very low correlations with F'_1. If we set out to interpret these rotated factors, then, we might say that the F'_1 rotated factor is highly correlated with variables X_1, X_2, and X_3; we can try to name or label the latent variable that corresponds to this rotated factor F'_1 by trying to summarize what common ability of trait the set of variables X_1, X_2, and X_3 might all measure.

There are many different approaches to factor rotation; the most popular rotation method is called varimax. According to the online help in SPSS 25, the varimax rotation method "minimizes the number of variables that have high loadings on each factor . . . this method simplifies the interpretation of factors." Varimax is one of several types of **orthogonal rotation**. In an orthogonal rotation, factor axes are relocated, but the factor axes retain 90° angles to one another after rotation. That is, when we use orthogonal rotation methods, the rotated factors remain orthogonal to or uncorrelated with one another.

Many other rotation methods are available, and in some methods of rotation, we allow the rotated factor axes to become nonorthogonal; that is, we may decide we want to obtain factors that are correlated with one another to some degree. Varimax is the most popular and widely reported rotation method; see Harman (1976) for a discussion of other possible methods of rotation. When a nonorthogonal rotation method is used, the interpretation of factor loadings becomes more complex, and additional matrices are involved in the description of the complete factor solution; see Tabachnick and Fidell (2018) for a discussion of these issues.

A rotated solution is often (although not always) easier to interpret than the unrotated solution. *Rotation* simply means that the factor loadings or correlations of the X variables are recalculated relative to the new rotated factors. Varimax-rotated loadings will typically explain or reproduce the observed correlations as well as the unrotated loadings and so there is no mathematical basis to prefer one solution over the other. The choice between various possible sets of factor loadings (with or without various kinds of rotation) is based on theoretical considerations: Which solution gives you the most meaningful and interpretable pattern? One of the objections to factor analysis is the arbitrariness of the choice of one solution out of many solutions that are all equally good in the mathematical sense (i.e., they all reproduce the observed correlations equally well). Analysis 4 (in a subsequent section) includes varimax **rotated factor loadings**.

Sometimes, even after rotation, one or more items still have relatively large loadings on two or more factors. An item that has a strong correlation with more than one factor can be called "factorially complex." For example, consider what might happen in a factor analysis of scores on the following mental ability tests: X_1 = reading comprehension, X_2 = vocabulary, X_3 = algebra "story problems," X_4 = linear algebra, and X_5 = geometry. We might obtain a two-factor model such that X_1 and X_2 have high correlations with the first factor and X_4 and X_5 have high correlations with the second factor; we might label Factor 1 "verbal ability" and Factor 2 "math ability." However, the X_3 variable (scores on a test that involves algebra story problems) might have moderately high correlations or loadings for both Factors 1 and 2. The X_3 variable could be interpreted as factorially complex; that is, scores on this test are moderately predictable from both verbal ability and math ability; reasonably good reading comprehension is required to set up the problem in the form of equations, and reasonably good math skills are required to solve the equations. Thus, the X_3 test, story problems in algebra, might not be interpretable purely as a measure of verbal ability or purely as a measure of math ability; it might tap both these two different kinds of ability. If the goal of the data analyst is to identify items that each clearly measure only one of the two factors, then he or she might drop this item from future analyses.

12.14 FACTOR ANALYSIS AS TWO SETS OF MULTIPLE REGRESSIONS

One more issue needs to be considered before examining the final factor analysis example. It is possible to calculate and save a "factor score" on each factor for each participant. To understand how this is done, we need to think about factor analysis as two sets of multiple regressions. One set of regression models is used to construct factor scores for individual subjects from their *z* scores on the measured variables.

There is also a second set of regressions implicit in factor analysis; in principle, we could predict the *z* scores for each individual participant on each measured *X* variable from that participant's factor scores. This second set of regression does not have much practical application.

12.14.1 Construction of Factor Scores for Each Individual (F_1, F_2, etc.) From Individual Item *z* Scores

Earlier, factor analysis was described as a method of "data reduction." If we can show that we can reproduce or explain the pattern of correlations among a large number of measured variables using a factor model that includes a small number of factors (e.g., a two-factor model), then it may be reasonable to develop a measure that corresponds to each latent variable or factor. In later data analyses, we can then use this smaller set of measures (perhaps one factor score per retained factor) instead of including the scores on all the *p* measured variables in our analyses.

A path diagram for the regression to construct a factor score on Factor 1 (F_1) from *z* scores on the measured *X* variables appears in Figure 12.34. If you request the "regression" method of computing saved factor scores, the **factor score coefficients** (denoted beta in Figure 12.34) are used as weights. Factor score coefficients can be requested from SPSS as

Figure 12.34 An Example of the First Set of Regressions: The Construction of a Factor Score on Factor 1 From Measured Items $z_{X1}, z_{X2}, \ldots, z_{Xp}$

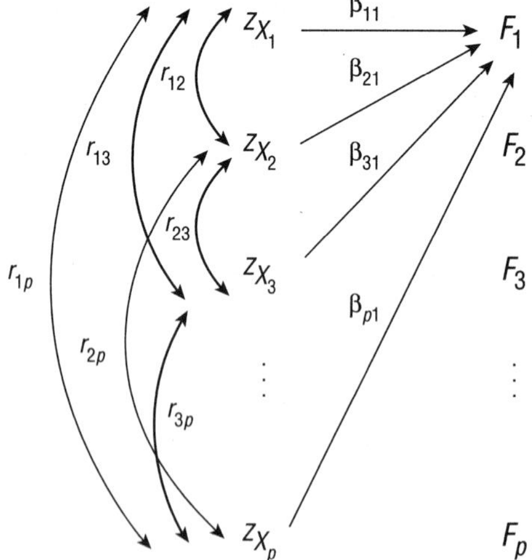

Note: Each β_{ij} coefficient in the diagram represents a factor score coefficient associated with a z_{Xj} predictor variable. To construct a score on F_1 for each participant in the study, we apply the β_{i1} factor score coefficients to that participant's scores on the standardized measured variables, z_{X1} through z_{Xp}, as follows: constructed factor score $F_1 = \beta_{11}z_{X1} + \beta_{21}z_{X2} + \beta_{31}z_{X3} + \cdots + \beta_{p1}z_{Xp}$. Typically, factor scores are constructed only for a small number of retained factors.

an optional output, and the factor scores for each participant can be generated and saved as new variables in the SPSS data file. The factor score coefficients are similar to beta weights in a multiple regression. The z_X predictor variables are usually intercorrelated, and these intercorrelations must be taken into account when factor score coefficients are calculated. Factor scores on each of the rotated final factors can be saved for each subject and then used as variables in subsequent SPSS analyses. This can be quite useful in reducing the number of variables and/or obtaining orthogonal variables for use in later analyses.

Another possible follow-up to a factor analysis is to form a simple unit-weighted composite that corresponds to each factor. For example, if Factor 1 has high positive correlations with X_1, X_2, and X_3, we might combine the information in these three variables by summing their z scores (or possibly the raw scores).

The last analysis that will be reported, Analysis 4, provides factor score coefficients that can be used to construct factor scores on the two retained factors. Figure 12.42 summarizes factor score coefficients from Analysis 4; these factor score coefficients correspond to the beta values in the path model in Figure 12.34. To construct a score for each person on Factor 1, the factor score coefficients in Figure 12.42 are applied to the z scores on the measured variables. A factor score for each participant on Factor 1 can be constructed as follows: $F_1 = .135 \times z_{\text{nurturant}} + .273 \times z_{\text{affectionate}} + .233 \times z_{\text{compassionate}} + .442 \times z_{\text{warm}} + .046 \times z_{\text{strong personality}} + .02 \times z_{\text{assertive}} - .109 \times z_{\text{forceful}} - .096 \times z_{\text{dominant}} - .004 \times z_{\text{aggressive}}$.

12.14.2 Prediction of z Scores for Individual Participant (z_{Xi}) From Participant Scores on Factors (F_1, F_2, etc.)

We can view the factor loadings for an orthogonal solution as coefficients that we could use to predict an individual participant's scores on each individual measured variable from his or her scores on Factor 1, Factor 2, and any other retained factors. That is, $z_{X1} = a_{11}F_1 + a_{12}F_2 + \cdots + a_{1p}F_p$. Examples of path models for regressions that could predict z scores on X_1 and X_2 measured variables from constructed factor scores F_1 and F_2 appear in Figure 12.35. Note that for these regressions, we can use the factor loadings (denoted a's in Figure 12.35) as

Figure 12.35 Two Examples From the Second Set of Multiple Regressions That Correspond to Factor Analysis

Prediction of score on z_{X_1} from scores on two retained orthogonal factors, F_1 and F_2

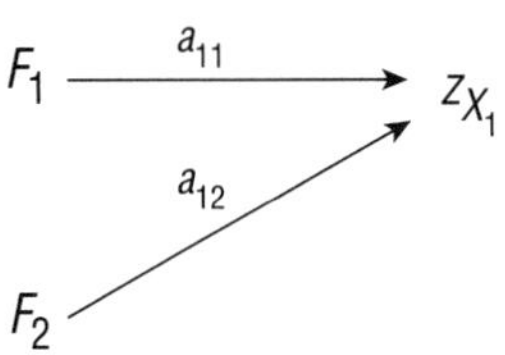

Prediction of score on z_{X_2} from scores on two retained orthogonal factors, F_1 and F_2

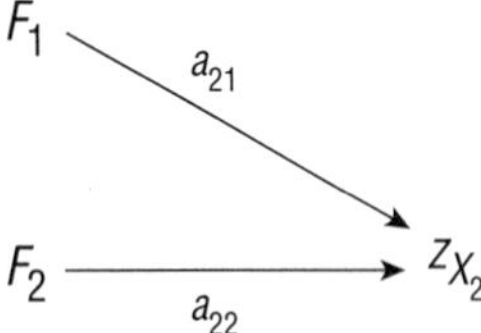

Note: The z score on each measured X variable can be perfectly predicted from the entire set of factor scores ($F_1, F_2, \ldots, F_p$) in the initial factor analysis (as shown in Figure 12.1). In the final stages of a factor analysis, we usually want to see how well we can predict the scores on the z_X items from scores on a small number of retained factors (in this example, two retained factors). This example shows uncorrelated or orthogonal factors. The communality for z_{X1} is obtained by summing the squared correlations of z_{X1} with F_1 and F_2. For example, the communality for z_{X1} when we use only F_1 and F_2 as predictors = $a^2_{11} + a^2_{12}$. To be judged adequate, one thing a factor analysis needs to do is reproduce or predict a reasonably high proportion of variance on the z_X variables. We can set up similar path models to show that scores on $z_{X3}, z_{X4}, \ldots, z_{Xp}$ can be predicted from F_1 and F_2.

regression coefficients, provided that the factors F_1 and F_2 are orthogonal to each other. See Appendix 12A for additional information about the matrix algebra involved in factor analysis; additional matrices are required when a nonorthogonal rotation is applied to factors.

12.15 ANALYSIS 4: PAF WITH VARIMAX ROTATION

The preceding analyses demonstrated the logic involved in PC and PAF one step at a time. The final analysis presented in this chapter is a replication of Analysis 3 (a factor analysis that uses the size of eigenvalues as the criterion for factor retention) with two additional specifications: Varimax rotation is requested (to make the pattern of factor loadings more interpretable), and factor scores are saved for the two retained factors. This final analysis provides a model for a version of factor analysis that is often presented in research reports.

Figure 12.36 shows the radio button selection that was used to request varimax rotation. In addition, a loading plot was requested. In the main Factor Analysis dialog box, the Save button was used to open the Factor Analysis: Factor Scores window that appears in Figure 12.37. Checkbox selections were made to save the factor scores as variables and to display the factor score coefficient matrix; a radio button selection was used to request the "regression" method of computation for the saved factor scores. The SPSS syntax for Analysis 4 appears in Figure 12.38. Selected output for Analysis 4 appears in Figures 12.39 through 12.43. Some parts of the results for Analysis 4 are identical to results from Analysis 3 (e.g., the communalities for individual items), and that information is not repeated here.

Figure 12.36 Analysis 4: PAF With All Specifications the Same as in Analysis 3, With Varimax Rotation Requested

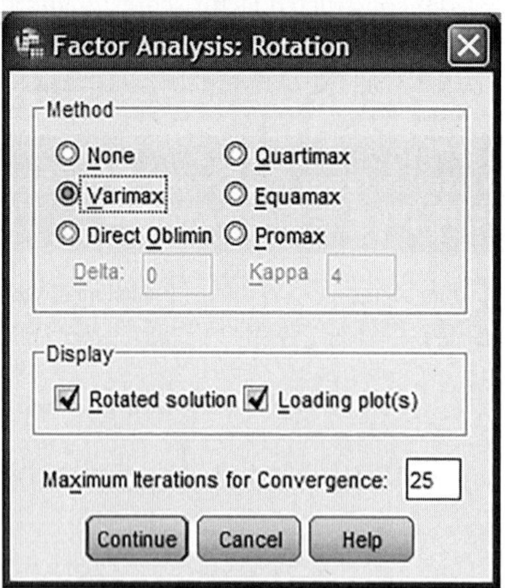

Figure 12.37 Request to Compute and Save Factor Scores From Analysis 4

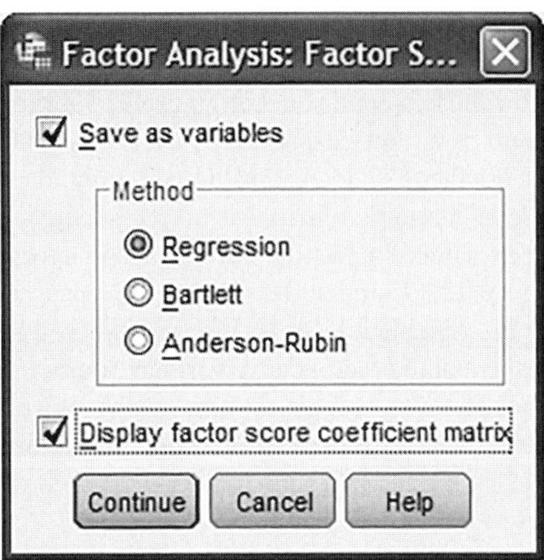

Figure 12.38 Analysis 4 SPSS Syntax: Nine Items; Method of Extraction, PAF; Criterion for Retention of Factors Is Eigenvalue >1; Rotation = Varimax; Factor Scores Are Calculated and Saved

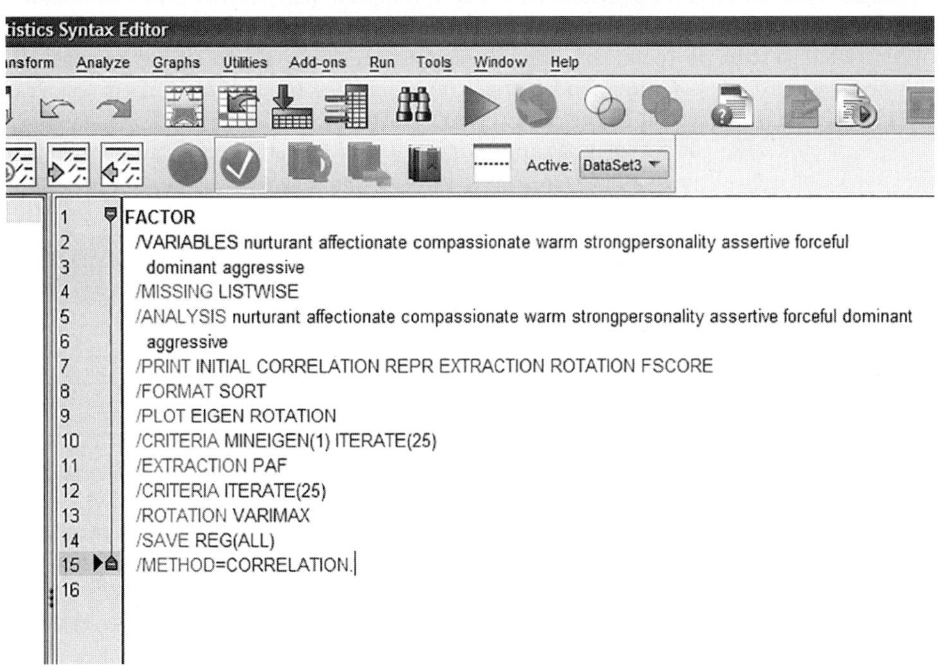

12.15.1 Variance Reproduced by Each Factor at Three Stages in the Analysis

Figure 12.39 replicates the information in Figure 12.26 and adds a third panel to report the results from the varimax rotation. The total variance explained by factors is now reported at three stages in the analysis. In the left-hand panel of Figure 12.39, the initial eigenvalues provide information about the amount of variance explained by the entire set of nine factors extracted by PAF. Because only two of these factors had eigenvalues > 1, only Factors 1 and 2 were retained. After the loadings for these two retained factors were reestimated, the SSLs that summarize the amount of information explained by just the two retained factors (prior to rotation) appear in the middle panel of Figure 12.39, under the heading "Extraction Sums of Squared Loadings." Finally, the panel on the right-hand side of Figure 12.39 reports the amount of variance explained by each of the two retained factors after varimax rotation. Note that the sum of the variance explained by the two factors as a set is the same for the unrotated and the rotated solutions, but the way the variance is apportioned between the two factors is slightly different. After varimax rotation, rotated Factor 2 explained approximately as much variance as rotated Factor 1.

12.15.2 Rotated Factor Loadings

The varimax-rotated factor loadings appear in table form in Figure 12.40 and as a plot in Figure 12.41. It is now clear, when we examine the correlations or loadings in Figure 12.40, that the latent variable represented by Factor 1 is highly positively correlated with self-ratings on warm, affectionate, compassionate, and nurturant. The latent variable represented by Factor 2 is highly positively correlated with self-ratings on dominant, forceful, aggressive, assertive, and strong personality. (An arbitrary cutoff value of .40 was used to decide which factor loadings were "large" in absolute value.)

The plot in Figure 12.41 makes it clear how rotation has improved the interpretability of the factors. After rotation, Factor 1 corresponds to a bundle of intercorrelated variables that all seem to be related to caring (affectionate, warm, compassionate, nurturant). After rotation, Factor 2 corresponds to a bundle of intercorrelated vectors that represent variables that seem to be related to strength (dominant, forceful, aggressive, assertive, and strong personality). It is now easier to label or name the factors, because for each rotated factor, there is a clear group of variables that has high correlations with that factor.

The factor score coefficients that appear in Figure 12.42 can be used to compute and save a score on each of the retained factors. Because this type of information is often reported

Figure 12.39 Analysis 4: PAF With Varimax Rotation: Summary of Total Variance Explained

Total Variance Explained

Factor	Initial Eigenvalues			Extraction Sums of Squared Loadings			Rotation Sums of Squared Loadings		
	Total	% of Variance	Cumulative %	Total	% of Variance	Cumulative %	Total	% of Variance	Cumulative %
1	2.863	31.807	31.807	2.346	26.070	26.070	2.002	22.243	22.243
2	2.194	24.380	56.187	1.649	18.327	44.397	1.994	22.154	44.397
3	.823	9.143	65.330						
4	.723	8.032	73.362						
5	.626	6.961	80.323						
6	.535	5.950	86.272						
7	.480	5.331	91.603						
8	.469	5.214	96.817						
9	.286	3.183	100.000						

Extraction Method: Principal Axis Factoring.

Figure 12.40 Analysis 4: Rotated Factor Loadings on Factors 1 and 2 (Sorted by Size)

Rotated Factor Matrix[a]

	Factor	
	1	2
warm	.804	.217
affectionate	.713	.250
compassionate	.696	-.020
nurturant	.489	-.153
dominant	-.096	.670
forceful	-.174	.667
aggressive	.096	.588
assertive	.148	.587
strongpersonality	.230	.526

Extraction Method: Principal Axis Factoring.
Rotation Method: Varimax with Kaiser Normalization.
a. Rotation converged in 3 iterations.

Figure 12.41 Analysis 4: Plots of Factor Loadings in Varimax-Rotated Factor Space

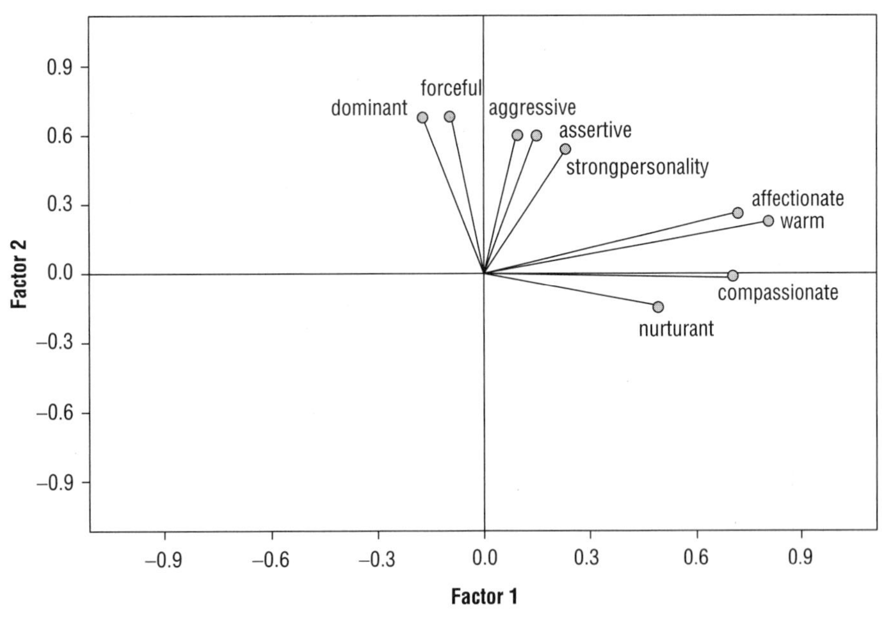

Note: The plot in SPSS did not include the vectors, that is, the lines from the 0,0 origin that represent the correlation of each variable with the two factors; these vectors were added to the SPSS plot in a graphics program.

Figure 12.42 Analysis 4: Factor Score Coefficients Used to Compute Saved Factor Scores

Factor Score Coefficient Matrix

	Factor	
	1	2
nurturant	.135	-.069
affectionate	.273	.067
compassionate	.233	-.055
warm	.442	.063
strongpersonality	.046	.174
assertive	.020	.207
forceful	-.109	.302
dominant	-.096	.290
aggressive	-.004	.203

Extraction Method: Principal Axis Factoring.
Rotation Method: Varimax with Kaiser Normalization.
Factor Scores Method: Regression.

Note: These values correspond to the β_{ij} coefficients that appear in Figure 12.35. β_{ij} corresponds to the factor score coefficient applied to z_{Xi} to construct a score on F_j. To construct a score for each person on Factor 1, we would do the following computation for each participant: Constructed factor score on $F_1 = .135 \times z_{nurturant} + .273 \times z_{affectionate} + .233 \times z_{compassionate} + .442 \times z_{warm} + .046 \times z_{strong\ personality} + .02 \times z_{assertive} - .109 \times z_{forceful} - .096 \times z_{dominant} - .004 \times z_{aggressive}$.

as a follow-up to a factor analysis, discussion of the factor scores appears in Section 12.18, after the results section for Analysis 4.

In words, to obtain high scores on F_1, an individual would need to have relatively high scores on nurturant, affectionate, compassionate, and warm; the scores on the remaining five variables have such small coefficients that they do not make much contribution toward the value of F_1.

12.15.3 Example of a Reverse-Scored Item

Up to this point, we have considered only examples in which all the large factor loadings have been positive. However, researchers sometimes intentionally include **reverse-worded questions** or reverse-scored items. What would the rotated factor loadings look like if people had been asked to rate themselves on "coldness" instead of on warmth? Changing the wording of the item might alter people's responses. Here we will just consider what happens if we reverse the direction of the 0-to-4 rating scale scores on the variable warm. If we compute a new variable named cold by subtracting scores on warmth from the constant value 5, then scores on cold would be perfectly negatively correlated with scores on warmth. If we then rerun Analysis 4 including the new variable cold instead of warmth, we would obtain the rotated factor loadings that appear in Figure 12.43. Note that the rotated factor loading for cold is –.804; from Figure 12.40, the rotated factor loading for warmth was +.804. Reversing the direction of scoring did not change the magnitude of the loading, but it reversed the sign. We could interpret the rotated factor loadings in Figure 12.43 by saying that people who are high on the caring dimension that corresponds to Factor 1 tend to have high scores on nurturant, affectionate, and compassionate and low scores on coldness.

Figure 12.43 Final Rotated Factor Loadings in Analysis 4 if "Warm" Is Replaced With "Cold"

Rotated Factor Matrix[a]

	Factor	
	1	2
cold	-.804	-.217
affectionate	.713	.250
compassionate	.696	-.020
nurturant	.489	-.153
dominant	-.096	.670
forceful	-.174	.667
aggressive	.096	.588
assertive	.148	.587
strongpersonality	.230	.526

Extraction Method: Principal Axis Factoring.
Rotation Method: Varimax with Kaiser Normalization.
a. Rotation converged in 3 iterations.

Note: The variable cold was computed as follows: cold = 5 − warm. Thus, cold is just the reverse-scored version of warm. Note that when Analysis 4 was repeated using the variable cold instead of warm, the magnitudes of all factor loadings remained the same; however, now the variable cold has a negative sign on its factor loading. Thus, higher scores on rotated Factor 1 now correspond to high scores on affectionate, compassionate, and nurturant and to a low score on cold.

12.16 QUESTIONS TO ADDRESS IN THE INTERPRETATION OF FACTOR ANALYSIS

Whether the analysis is performed using PC or PAF, the data analyst typically hopes to answer the following questions about the structure of the data.

1. How many factors (or components or latent variables) are needed to account for (or reconstruct) the pattern of correlations among the measured variables?

   To say this in another way, How many different constructs are measured by the set of *X* variables? In some cases, a single latent variable (such as general intelligence) may be sufficient to reconstruct correlations among scores on a battery of mental tests. In other research situations (such as the Big Five model of personality), we may need larger numbers of latent variables to account for data structure. The Big Five model of personality traits developed by Costa and McCrae (1995) suggests that people differ on five personality dimensions (openness, conscientiousness, extraversion, agreeableness, and neuroticism) and that a model that includes latent variables to represent each of these theoretical constructs can do a good job of reconstructing the correlations among responses to individual items in personality tests. The decision of how many factors or components to retain in the model can be made on the basis of theoretical or conceptual issues. In exploratory analysis, in situations where the data analyst does not have a well-developed theory, empirical criteria (such as the number of factors with eigenvalues > 1 or the scree plot) may be used to make decisions about the number of factors or components to retain. Sometimes decisions about number of factors to retain take both theoretical and empirical issues into account.

2. How "important" are the factors or components? How much variance does each factor or component explain?

   In the initial solution, each factor or component has an eigenvalue associated with it; the eigenvalue represents the amount of variance that can be reconstructed from the correlations of measured variables with that factor or component. In later stages of the analysis, we summarize information about the amount of variance accounted for by each component or factor by computing the SSL; that is, to find out how much variance is accounted for by Factor 1, we square the loadings of each of the p measured variables on Factor 1 and sum the squared loadings across all p variables. We can interpret an eigenvalue as a proportion or percentage of explained variance. If a set of p measured variables are all transformed into z scores or standard scores, and each standard score has a variance of 1, then the total variance for the set of p variables is represented by p. For example, if we do a factor analysis of nine measured variables, the total variance for that data set is equal to 9. The SSL for the entire set of nine variables on nine factors is scaled so that it equals 9. We can interpret the proportion of variance explained by Factor 1 by dividing the eigenvalue or the SSL for Factor 1 by p, the number of measured variables. If we retain two or three factors in the model, we can report the proportion or percentage of variance that is accounted for by each factor and, also, the proportion of variance that is accounted for by the set of two or three factors. There is no standard criterion for how much variance a solution must explain to be considered adequate, but in general, you want the percentage of variance explained by your retained factors to be reasonably high, perhaps on the order of 40% to 70%. If the percentage of explained variance for a small set of retained factors is very low, this suggests that the effort to understand the set of measured variables as multiple indicators of a small number of latent variables has not been very successful and that the measured variables may be measuring a larger number of different constructs.

3. Can we label or name the factors (or components)?

   Usually, we answer this question by looking at the pattern of the rotated factor loadings (although occasionally, it is easier to make sense of unrotated loadings). A factor can be named by considering which measured variables it does (and does not) correlate with highly and trying to identify a common denominator, that is, something all the variables with high loadings on that factor have in common. For instance, a factor with high loadings for tests of trigonometry, calculus, algebra, and geometry could be interpreted as a "math ability" factor. Note that a large negative loading can be just as informative as a large positive one, but we must take the sign into account when making interpretations. If respondents to the BSRI had rated themselves on the adjective cold, instead of warm, there would be a large negative loading for the variable cold on the first factor; this first factor had high positive correlations with variables such as affectionate and compassionate. We would say that high scores on Factor 1 were associated with high ratings on compassionate, affectionate, and loving and low ratings on coldness.

   Before we try to interpret a component or factor as evidence that suggests the existence of a latent variable, each factor should have high correlations with at least three or four of the measured variables. If only two variables are correlated with a factor, the entire factor really represents just one correlation, and a single large correlation might arise just from sampling error. Generally, researchers want to have at least three to five measured variables that correlate with each retained factor; a larger number of items may be desirable in some cases. When factor analysis is used to help develop multiple-item measures for personality tests, the analyst may hope to identify 10 or 15 items that all correlate with a latent variable (e.g., "extraversion"). If this result is obtained, the data analyst is then in a position to construct a 10- or 15-item multiple-item scale to measure extraversion.

You should not try to force an interpretation if there is no clear common meaning among items that load on a factor. Sometimes results are simply due to sampling error and do not reflect any meaningful underlying pattern. Also avoid the error of reification; that is, do not assume that just because you see a pattern that looks like a factor in your data, there must be some "real thing" (an ability or personality trait) out in the world that corresponds to the factor you obtained in your data analysis.

4. How adequately do the retained components or factors reproduce the structure in the original data, that is, the correlation matrix?

   To evaluate how well the factor model reproduces the original correlations between measured variables, we can examine the discrepancies between the reproduced **R** and the actual **R** correlation matrices. This information can be requested from SPSS. We want these discrepancies between observed and reconstructed correlations to be fairly small for most pairs of variables. You can also look at the final communality estimates for individual variables to see how well or poorly their variance is being reproduced. If the estimated communality for an individual variable (which is similar to an R^2 or a proportion of explained variance) is close to 1, then most of its variance is being reproduced; if variance is close to 0, little of the variance of that measured variable is reproduced by the retained components or factors. It may be helpful to drop variables with very low communality and rerun the factor analysis. This may result in a solution that is more interpretable and a solution that does a better job of reconstructing the observed correlations in the **R** matrix.

12.17 RESULTS SECTION FOR ANALYSIS 4: PAF WITH VARIMAX ROTATION

The following "Results" section is based on Analysis 4. In this analysis, nine self-rated personality items were included, PAF was specified as the method of extraction, the rule used to decide how many factors to retain was to retain only factors with eigenvalues > 1, and varimax rotation was requested.

Results

To assess the **dimensionality** of a set of nine items selected from a modified version of the BSRI, factor analysis was performed using PAF, the default criterion to retain only factors with eigenvalues greater than 1, and varimax rotation was requested. The items included consisted of self-reported ratings on the following adjectives: nurturant, affectionate, compassionate, warm, strong personality, assertive, forceful, dominant, and aggressive. The first four items were those identified as associated with feminine sex role stereotyped behavior, and the last five items were associated with masculine sex role stereotyped behavior. Each item was rated on a 5-point scale that ranged from 0 (*does not apply to me at all*) to 4 (*describes me extremely well*).

The correlation matrix that appears in Figure 12.2 indicated that these nine items seemed to form two separate groups. There were moderately high positive correlations among the ratings on nurturant, affectionate, warm, and compassionate; these items were items Bem (1974) identified as sex role stereotyped feminine. There were moderately high positive correlations among the items strong personality, assertive, forceful, dominant, and aggressive; these items were all identified as sex role stereotyped masculine. The correlations between items in these two separate groups of items (e.g., the correlation between nurturant and assertive) tended to be small.

An EFA was performed to evaluate whether a two-factor model made sense for these data. The SPSS factor analysis procedure was used, the method of extraction was PAF, only factors with eigenvalues > 1 were retained, and varimax rotation was requested. Two factors were retained and rotated. Results from this analysis, including rotated factor loadings, communalities, and SSL for the retained factors, are summarized in Table 12.1.

In the initial factor solution that consisted of nine factors, only two factors had eigenvalues greater than 1; therefore, only Factors 1 and 2 were retained and rotated. After varimax rotation, Factor 1 accounted for 22.24% of the variance and Factor 2 accounted for 22.15% of the variance; together, the first two factors accounted for 44.4% of the variance in this data set. [Note that we can find these percentages of variance in the SPSS output in Figure 12.39; we can also compute the SSL for each column in Table 12.1 and then divide the SSL by p, the number of variables, to

Table 12.1 Rotated Factor Loadings From Analysis 4: PAF With Varimax Rotation

	Nine Self-Rated Personality Items		
	Factor 1: "Caring"	Factor 2: "Strong"	Communality
Warm	.80	.22	.69
Affectionate	.71	.25	.57
Compassionate	.70	−.02	.48
Nurturant	.49	−.15	.26
Dominant	−.10	.67	.46
Forceful	−.17	.67	.48
Aggressive	.10	.59	.46
Assertive	.15	.59	.37
Strong personality	.23	.43	.33
Sum of squared loadings	2.00	1.99	
% explained variance	22.24%	22.15%	

obtain the same information about proportion or percentage of explained variance for each factor. In Table 12.1, the SSL for Factor 1 can be reproduced as follows: SSL = $.80^2 + .71^2 + .70^2 + .49^2 + (-.10)^2 + (-.17)^2 + .10^2 + .15^2 + .23^2 = 2.00$.]

Communalities for variables were generally reasonably high, ranging from a low of .26 for nurturant to a high of .69 for warm. [These communalities were reported by SPSS in Figure 12.40 as part of Analysis 3. They can also be reproduced by squaring and summing the loadings in each row of Table 12.1; for example, warm had loadings of .80 and .22, so its communality is $.80^2 + .22^2 = .69$. The communality for each measured variable changes when we go from a nine-factor model to a two-factor model, but the communality for each measured variable does not change when the two-factor model is rotated.]

Rotated factor loadings (see Table 12.1) were examined to assess the nature of these two retained varimax-rotated factors. An arbitrary criterion was used to decide which factor loadings were large; a loading was interpreted as large if it exceeded .40 in absolute magnitude. The four items that had high loadings on the first factor (warm, affectionate, compassionate, and nurturant) were consistent with the feminine sex role stereotype. This first factor could be labeled "femininity," but because this factor was based on such a limited number of BSRI femininity items, a more specific label was applied; for this analysis, Factor 1 was identified as "caring." The five items that had high loadings on the second factor (dominant, forceful, aggressive, assertive, and strong personality) were all related to the masculine sex role stereotype, so this second factor could have been labeled "masculinity." However, because only five items were included in Factor 2, a more specific label was chosen; Factor 2 was labeled "strong." None of the items were factorially complex; that is, none of the items had large loadings on both factors.

12.18 FACTOR SCORES VERSUS UNIT-WEIGHTED COMPOSITES

As a follow-up to this factor analysis, scores for caring and strength were calculated in two different ways. First, as part of Analysis 4, the regression-based factor scores for the two factors were saved as variables. For example, as noted previously, the factor score coefficients that were requested as part of the output in Analysis 4 (that appear in Figure 12.42) can be applied to the z scores on the measured variables to construct a factor score on each factor for each participant: $F_1 = .135 \times z_{\text{nurturant}} + .273 \times z_{\text{affectionate}} + .233 \times z_{\text{compassionate}} + .442 \times z_{\text{warm}} + .046 \times z_{\text{strong personality}} + .02 \times z_{\text{assertive}} - .109 \times z_{\text{forceful}} - .096 \times z_{\text{dominant}} - .004 \times z_{\text{aggressive}}$. We can see from this equation that people who have relatively high positive scores on nurturant, affectionate, compassionate, and warm will also tend to have high scores on the newly created score on Factor 1. We could say that a person who has a high score on Factor 1 is high on the caring dimension; we know that this person has high scores on at least some of the measured variables that have high correlations with the caring factor. (People's scores on strong personality, assertive, forceful, dominant, and aggressive are essentially unrelated to their factor scores on caring.)

A problem with factor score coefficients is that they are optimized to capture the most variance and create factor scores that are most nearly orthogonal only for the sample of participants whose data were used in the factor analysis. If we want to obtain scores for new participants on these nine self-ratings and form scores to summarize the locations of these new participants on the caring and strong dimensions, there are two potential disadvantages to creating these scores by using the factor score coefficients. First, it can be somewhat inconvenient to use factor score coefficients (reported to two or three decimal places) as weights when we combine scores across items. Second, when we apply these factor score coefficients to scores in new batches of data, the advantages that these factor score coefficients had in the data used in the factor analysis (that they generated factors that explained the largest possible amount of variance and that they generated scores on Factor 2 that had extremely small correlations with Factor 1) tend not to hold up when the factor scores are applied to new batches of data.

A different approach to forming scores that might represent caring and strong is much simpler. Instead of applying the factor score coefficients to z scores, we might form simple unit-weighted composites, using the factor analysis results to guide us in deciding whether to give each measured item a weight of 0, +1, or −1. For example, in Analysis 4, we decided that Factor 1 had "high" positive loadings with the variables warm, affectionate, compassionate, and nurturant and that it had loadings close to 0 with the remaining five measured variables. We could form a score that summarizes the information about the caring dimension contained in our nine variables by assigning weights of +1 to variables that have high correlations with Factor 1 and weights of 0 to variables that have low correlations with Factor 1. (If a variable has a large negative loading on Factor 1, we would assign it a weight of −1.) The unit-weighted score for caring based on raw scores on the measured items would be as follows:

$$\text{Caring} = (+1) \times \text{Warm} + (+1) \times \text{Affectionate} + (+1) \times \text{Compassionate} + (+1) \times \text{Nurturant} + 0 \times \text{Dominant} + 0 \times \text{Forceful} + 0 \times \text{Aggressive} + 0 \times \text{Strong personality}.$$

This can be reduced to

$$\text{Caring} = \text{Warm} + \text{Affectionate} + \text{Compassionate} + \text{Nurturant}.$$

Unit-weighted scores for new variables called caring and strong can be calculated in SPSS using the <Transform> → <Compute Variable> menu selections to open up the computation dialog box that appears in Figure 12.44. A unit-weighted composite called strong was computed by forming this sum: dominant + forceful + aggressive + assertive + strong personality. The SPSS data worksheet that appears in Figure 12.45 shows the new variables that were created as a follow-up to Factor Analysis 4. The columns headed "Fac1_1" and "Fac2_1" contain the values of the saved factor scores that were obtained by requesting saved factor scores in Figure 12.37; Fac1_1 corresponds to the following sum of z scores on the measured items: Fac1_1 = $.135 \times z_{\text{nurturant}} + .273 \times z_{\text{affectionate}} + .233 \times z_{\text{compassionate}} + .442 \times z_{\text{warm}} + .046 \times z_{\text{strong personality}} + .02 \times z_{\text{assertive}} - .109 \times z_{\text{forceful}} - .096 \times z_{\text{dominant}} - .004 \times z_{\text{aggressive}}$. (The factor score coefficients appear in column one of Figure 12.42.) The last two columns in the SPSS data view worksheet contain the unit-weighted scores that were computed for the group of variables with high loadings on Factor 1 (the caring factor) and the group of variables with high loadings on Factor 2 (the strong factor):

$$\text{Caring} = \text{Warm} + \text{Loving} + \text{Affectionate} + \text{Compassionate},$$
$$\text{Strong} = \text{Assertive} + \text{Aggressive} + \text{Dominant} + \text{Forceful} + \text{Strong personality}.$$

How much do the results differ for the saved factor scores (Fac1_1, Fac2_1) and the unit-weighted composites of raw scores (caring, strong)? We can assess this by obtaining Pearson correlations among all these variables; the Pearson correlations appear in Figure 12.46. In this empirical example, the saved factor score on Factor 1 (Fac1_1) correlated .966 with the unit-weighted composite score on caring; the saved factor score on Factor 2 (Fac2_1) correlated .988 with the unit-weighted score on strong. The correlations between factor scores and unit-weighted composite scores are not always as high as in this empirical example. In many research situations, scores that are obtained from unit-weighted composites provide

Figure 12.44 Computation of a Unit-Weighted Composite Score Consisting of Items With High Correlations With Factor 1

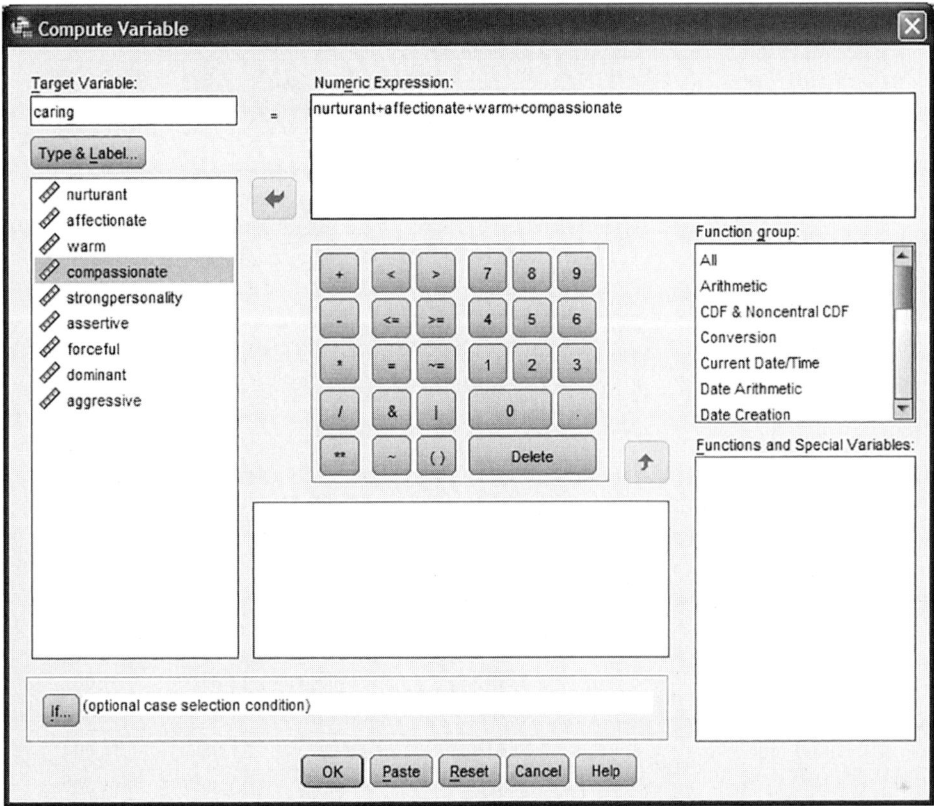

information very similar to scores that are based on weights derived from multivariate models, for example, saved factor scores (Fava & Velicer, 1992; Wainer, 1976).

Note that when all the items that are summed to form a weighted linear composite are measured in the same units, it may be acceptable to form a unit-weighted composite by summing the raw scores. In this example, all the adjectives were rated using the same 0-to-4 rating scale. However, if the variables that are summed have been measured in different units and have quite different variances, better results may be obtained by forming a unit-weighted composite of standard scores or z scores (rather than raw scores). For example, we could form a score on caring by summing unit-weighted z scores on warm, affectionate, compassionate, and nurturant:

$$\text{Caring} = z_{\text{warm}} + z_{\text{affectionate}} + z_{\text{compassionate}} + z_{\text{nurturant}}.$$

12.19 SUMMARY OF ISSUES IN FACTOR ANALYSIS

Factor analysis is an enormously complex topic; this brief introduction does not include many important issues. For example, there are many different methods of factor extraction; only PC and PAF have been covered here. There are many different methods of factor rotation; only one type of orthogonal rotation, varimax rotation, was presented here. If you plan to

Figure 12.45 New Variables in SPSS Data Worksheet

	aggressive	FAC1_1	FAC2_1	caring	strong
1	1	.29713	.58785	11.00	12.00
2	2	-1.46739	1.42055	8.00	16.00
3	2	-1.39082	-.99929	8.00	8.00
4	2	-.03313	-.31450	11.00	10.00
5	2	.06896	-.87869	12.00	8.00
6	3	-.70329	1.75080	9.00	17.00
7	1	1.05039	-.60198	14.00	9.00
8	2	.34961	-.72553	12.00	8.00
9	3	.71092	.69925	14.00	13.00
10	2	-.94452	-1.07962	8.00	7.00
11	1	1.53194	-.59610	16.00	9.00
12	2	-.58120	.09685	9.00	11.00
13	2	-.12203	-.01409	12.00	11.00
14	3	.62385	1.11814	14.00	14.00
15	3	-.35847	1.04601	11.00	14.00
16	3	-.66426	.90945	10.00	14.00
17	2	-1.25180	-.14226	9.00	10.00
18	3	-1.28304	.16525	8.00	11.00
19	3	.38677	-.46811	13.00	10.00
20	2	-.33372	.41234	12.00	13.00
21	3	-1.00704	-.29457	10.00	10.00
22	1	-1.24766	-.36872	9.00	9.00
23	2	1.15733	.02951	14.00	11.00

Note: Fac1_1 and Fac2_1 are the saved factor scores (for rotated Factors 1 and 2, respectively) from the PAF analysis previously reported as Analysis 4. Caring and strong are the unit-weighted composites formed by summing raw scores on the items that had high loadings on F_1 and F_2, as shown in the two previous figures. The saved factor scores and the unit-weighted composite scores provide two different ways of summarizing the information contained in the original nine personality items into scores on just two latent variables: Both caring and Fac1_1 summarize the information in the items nurturant, affectionate, warm, and compassionate. Both strong and Fac2_1 summarize the information in the items strong personality, assertive, forceful, dominant, and aggressive.

Figure 12.46 Correlations Between Saved Factor Scores on Fac1_1 and Fac2_1 (From Analysis 4) and Unit-Weighted Composites

Correlations

		FAC1_1	FAC2_1	caring	strong
FAC1_1	Pearson Correlation	1	.046	.966(**)	.057
	Sig. (2-tailed)		.652	.000	.575
	N	100	100	100	100
FAC2_1	Pearson Correlation	.046	1	.097	.988(**)
	Sig. (2-tailed)	.652		.335	.000
	N	100	100	100	100
caring	Pearson Correlation	.966(**)	.097	1	.117
	Sig. (2-tailed)	.000	.335		.244
	N	100	100	100	100
strong	Pearson Correlation	.057	.988(**)	.117	1
	Sig. (2-tailed)	.575	.000	.244	
	N	100	100	100	100

**Correlation is significant at the 0.01 level (2-tailed).

Note: Fac1_1 and caring are highly correlated; Fac2_1 and strong are highly correlated. Fac1_1 and Fac2_1 have a correlation that is near 0. The correlation between caring and strong is higher than the correlation between Fac1_1 and Fac2_1 but by a trivial amount.

use factor analysis extensively in your research, particularly if you plan to use nonorthogonal rotation methods, you should consult more advanced source books for information (e.g., Harman, 1976).

Factor analysis is an exploratory analysis that is sometimes used as a form of data fishing. That is, when a data analyst has a messy set of too many variables, he or she may run a factor analysis to see if the variables can be reduced to a smaller set of variables. The analyst may also run a series of different factor analyses, varying the set of variables that are included and the choices of method of extraction and rotation until a "meaningful" result is obtained. When factor analysis is used in this way, the results may be due to Type I error; the researcher is likely to discover, and focus on, correlations that are large in this particular batch of data just because of sampling error. Results of factor analysis often replicate very poorly for new sets of data when the analysis has been done in this undisciplined manner. In general, it is preferable to plan thoughtfully, to select variables in a way that is theory driven, and to run a small number of planned analyses.

Results of factor analyses are sometimes "overinterpreted": The researcher may conclude that he or she has determined "the structure of intelligence," or how many types of mental ability there are, when in fact all we can ever do using factor analysis is assess how many factors we need to describe the correlations among the set of measures that we included in our research. The set of measures included in our research may or may not adequately represent the domain of all the measures that should be included in a comprehensive study (of mental ability, for instance). If we do a factor analysis of a test battery that includes only tests of verbal and quantitative ability, for instance, we cannot expect to obtain factors that represent musical ability or emotional intelligence. To a great extent, we get out of factor analysis exactly what we put in when we selected our variables. Conclusions should not be overstated: Factor analysis tells us about structure in our sample, our data, our selection of measures—and this is not necessarily a clear reflection of structure in "the world," particularly if our selection of participants and measures is in any way biased or incomplete.

We can do somewhat better if we map out the domain of interest thoroughly at the beginning of our research and think carefully about what measures should be included or excluded. Keep in mind that for any factor you want to describe, you should include a minimum of three to four measured variables, and ideally, a larger number of variables would be desirable. If we think we might be interested in a musical ability factor, for example, we need to include several measures that should reflect this ability. In research areas where theory is well developed, it may be possible to identify the factors you expect ahead of time and then systematically list the items or measures that you expect to load on each factor. When factor analysis is done this way, it is somewhat more "confirmatory." However, keep in mind that if the theory used to select measurements to include in the study is incomplete, it will lead us to omit constructs that might have emerged as additional factors.

Test developers often go through a process where they factor-analyze huge lists of items (in one study, for example, developers collected all the items from self-report measures of love and did a factor analysis to see how many factors would emerge). On the basis of these initial results, the researchers may clarify their thinking about what factors are important. Items that do not load on any important factor may be dropped. New items may be added to provide additional indicator variables to enlarge factors that did not have enough items with high loadings in the initial exploratory analyses. It may be even more obvious in the case of factor analysis than for other analytic methods covered earlier in this book: The results you obtain are determined to a very great extent by the variables that you choose to include in (and exclude from) the analysis.

APPENDIX 12A

The Matrix Algebra of Factor Analysis

For a review of basic matrix algebra notation and simple matrix operations such as matrix multiplication, see Appendix 5A at the end of Chapter 5. To perform a factor analysis, we begin with a correlation matrix **R** that contains the correlations between all pairs of measured *X* variables. For PC, we leave ones in the diagonal. For PAF, we place initial estimates of communality (usually the squared multiple correlations to predict each *X* from all the other *X*'s) in the diagonal.

Next we solve the "eigenvalue/eigenvector" problem for this matrix **R**.

For the correlation matrix **R**, we want to solve for a list of eigenvalues ($\boldsymbol{\lambda}$) and a set of corresponding eigenvectors (**V**) such that $(\mathbf{R} - \boldsymbol{\lambda}\mathbf{I})\mathbf{V} = 0$.

This equation is formally called the "eigenproblem." This eigenvalue/eigenvector analysis in effect rearranges or "repackages" the information in **R**. When we create new weighted linear composites (factor scores) using the eigenvectors as weights, these new variates preserve the information in **R**. The variances of each of these new weighted linear composites are given by the eigenvalue that corresponds to each eigenvector. Our solution for a $p \times p$ correlation matrix **R** will consist of p eigenvectors, each one with a corresponding eigenvalue. The new weighted linear composites that can be created by summing the scores on the z_X standard scores on measured variables using weights that are proportional to the elements of each eigenvector are orthogonal to or uncorrelated with each other. Thus, a matrix of variances and covariances among them will have all zeros for the covariances.

$\lambda\mathbf{I}$ stands for the matrix that includes the eigenvalues; its diagonal entries, the λ's, correspond to the variances of these "new" variables (the factors) and the off-diagonal zeros stand for the 0 covariances among these new variates or factors.

When we find a set of eigenvectors and eigenvalues that are based on a correlation matrix **R** and that can reproduce **R**, we can say that this eigenproblem solution "diagonalizes" the **R** matrix. We "repackage" the information in the **R** matrix as a set of eigenvalues and eigenvectors; the new variables represented by the eigenvector/eigenvalue pairs have a "diagonal" variance/covariance matrix; that is, in the **L** matrix, all off-diagonal entries are 0. We convert a set of (intercorrelated) *X* variables to a new set of variables (factors) that are uncorrelated with each other, but we do this in a way that preserves the information; that is, we can reconstruct the **R** matrix from its corresponding eigenvalues and eigenvectors. We can rewrite the eigenvalue/eigenvector problem as follows:

$$\mathbf{L} = \mathbf{V'RV},$$

where **L** is the eigenvalue matrix (denoted $\lambda\mathbf{I}$ earlier). The **L** matrix for a 4 × 4 case would be as follows:

$$\mathbf{L} = \begin{bmatrix} \lambda_1 & 0 & 0 & 0 \\ 0 & \lambda_2 & 0 & 0 \\ 0 & 0 & \lambda_3 & 0 \\ 0 & 0 & 0 & \lambda_4 \end{bmatrix}.$$

R is the correlation matrix. **V** is the matrix in which each column corresponds to one of the eigenvectors, and each eigenvector in the **V** matrix has a corresponding eigenvalue in the **L** matrix; for example, column 1 of **V** and λ_1 together constitute one of the solutions to the eigenvalue/eigenvector analysis of the **R** matrix.

As discussed earlier, the eigenvalue that corresponds to a factor is also equivalent to the SSL for that factor; the eigenvalue for each factor provides information about the amount of variance associated with each factor.

Diagonalization of **R** is accomplished by pre- and postmultiplying **R** by the matrix **V** and its transpose. A matrix is called "diagonal" when all its off-diagonal elements equal 0. Once we have "repackaged" the information about *p* intercorrelated variables by extracting *p* orthogonal factors, the correlations among these factors are 0, so we speak of this factor solution as diagonal.

We rescale the values of the eigenvectors such that they can be interpreted as correlations; this requires that $\mathbf{V'V} = \mathbf{I}$. Thus, pre- and postmultiplying **R** by **V′** and **V** does not so much change it (multiplying by the identity matrix **I** is like multiplying by 1) as repackage it. Instead of *p* intercorrelated measured *X* variables, we now have *p* uncorrelated factors or components. We can rewrite the $\mathbf{L} = \mathbf{V'RV}$ equation given earlier in the following form:

$$\mathbf{R} = \mathbf{V'LV},$$

which is just a way of saying that we can reproduce **R**, the original observed correlation matrix, by multiplying together the eigenvectors and their corresponding eigenvalues.

The factor loadings for each factor are obtained by rescaling the values of each eigenvector such that they can be interpreted as correlations. This can be done by multiplying each eigenvector by the square root of its corresponding eigenvalue (the eigenvalue is equivalent to a variance). So we obtain a matrix of factor loadings, **A**, such that each column of **A** corresponds to the loadings of all *p* variables on one of the factors, from the following:

$$\mathbf{A} = \mathbf{V}\sqrt{\mathbf{L}}\,.$$

It follows that the product **AA′** = **R**; that is, we can reproduce the observed correlation matrix **R** from the factor loading matrix **A** and its transpose. **R** = **AA′** is the "fundamental equation" for factor analysis. In words, **R** = **AA′** tells us that the correlation matrix **R** can be perfectly reproduced from the loading matrix **A** (when we retain all p factors that were extracted from the set of p measured variables).

Returning to the path diagram shown in Figure 12.1, **AA′** implies that, for each pair of variables, when we trace all possible paths that connect all pairs of variables via their correlations with the factors, multiply these loadings together for each path, and sum the paths across all the p factors, the resulting values will exactly reproduce each of the bivariate correlations between pairs of measured X variables (provided that the model retains the same number of factors as variables).

Thus, to obtain the elements of the **A** matrix (which are the factor loadings), we do the following:

1. Calculate the sample correlation matrix **R** (and make whatever change in the diagonal entry of **R** that we wish to make, depending on our choice of PC, PAF, or other methods).
2. Obtain the p eigenvalue/eigenvector pairs that are the solutions for the eigenvalue problem for the **R** matrix.
3. Rescale the values in the eigenvectors by multiplying them by the square roots of the corresponding eigenvalues; this scales the values so that they can be interpreted as correlations or factor loadings.
4. In iterative solutions, such as PAF, we then reconstruct **R** from **A** by finding **R** = **AA′**. We compare the diagonal elements in our reconstructed **R** matrix (the communalities) with the communality estimates from the previous step. If they are the same (to some large number of decimal places), we say the solution has "converged" and we stop. If they differ, we put in our most recent communality estimates and redo the entire eigenvalue/eigenvector algebra to generate new loading estimates on the basis of our most recent communalities. Occasionally, the solution does not converge; this leads to an error message.

Additional matrices may also be obtained as part of factor analysis. For instance, the calculation of factor scores (**F**) for each subject on each factor from each subject's z scores on the variables (**Z**) and the factor score coefficient matrix (**B**) can be written compactly as **F** = **ZB**. This equation says that we can construct a factor score for each participant on each factor by multiplying the z scores on each measured variable for each subject by the corresponding factor score coefficients, creating a weighted linear composite for each subject on each factor.

We can write an equation for the **B**'s (the factor score coefficients, which are like beta coefficients in a multiple regression) in terms of the values of the factor loadings (correlations between the X's and the factors) and the intercorrelations among the X's, the **R** matrix:

$$\mathbf{B} = \mathbf{R}^{-1}\mathbf{A}.$$

This equation is analogous to our equation for the betas in multiple regression:

$$\mathbf{B}_i = \mathbf{R}^{-1}{}_{ii}\mathbf{R}_{iy}.$$

The multiple regression in the other direction, in which we predict (standardized) scores on the X's from factor scores, can be compactly written as **Z** = **FA′**. That is, we can multiply each individual subject's factor scores by the factor loadings of the variables and generate predicted scores on the individual variables. For instance, if we know that a subject has a high score on the math factor and a low score on the verbal ability factor, we would end up predicting high scores for that subject on tests that had high loadings on (or high correlations with) the math factor.

To convert the original factor loading matrix **A** to a rotated factor loading matrix, we multiply it by the "transformation matrix" **Λ**:

$$\mathbf{A}_{\text{unrotated}}\mathbf{\Lambda} = \mathbf{A}_{\text{rotated}}.$$

The elements of this transformation matrix **Λ** are the cosines and sines of the angles through which the reference axes or factors are being rotated to give a more interpretable solution; usually, this is not of great intrinsic interest.

If you choose to perform an **oblique rotation** (i.e., one that results in rotated factors that are correlated to some degree), there are some additional matrices that you have to keep track of. You need a ϕ matrix to provide information about the correlations among the oblique factors (for an orthogonal solution, these correlations among factors are constrained to be 0). When you have an oblique factor solution, you also have to distinguish between the factor loading or pattern matrix **A**, which tells you the regression-like weights that predict variables from factors, and the structure matrix **C**, which gives the correlations between the variables and the (correlated or oblique) factors. For an orthogonal solution, **A** = **C**. For an oblique solution, the **A** and **C** matrices differ. For further discussion of the matrix algebra involved in factor analysis, see Tabachnick and Fidell (2018).

APPENDIX 12B

A Brief Introduction to Latent Variables in SEM

A factor can be called a latent variable. A factor is not directly measured; it is defined by its association with measured variables, and it is a way of making sense of associations among measured variables.

SEM provides a broader framework that includes latent variables in both measurement models and path ("causal") models; see Chapter 15. Up until about 1970, most factor-analytic studies used the exploratory or data-driven approach to factor analysis described in this chapter. Work by Jöreskog (1969) on the general analysis of covariance structures led to the development of CFA and other types of structural equation models. In the exploratory factor analyses described in this chapter, the number and nature of the retained factors are determined to a great extent by the sample data (e.g., often we retain only components or factors with eigenvalues greater than 1). In CFA, a researcher typically begins with a theoretically based model that specifies the number of latent variables and identifies which specific measured variables are believed to be indicators of each latent variable in the model. This model usually includes many additional constraints; for example, each time we assume a 0 correlation between one of the measured variables and one of the latent variables in the model, we reduce the number of free parameters in the model. In EFA, the initial p component or p factor solution always reproduced the correlation matrix **R** perfectly; in typical applications of EFA, there are more than enough free parameters (a $p \times p$ matrix of the correlations between p measured variables and p latent variables or factors) to perfectly reproduce the $p(p - 1)/2$ correlations among the measured variables.

CFA and structural equation models, on the other hand, are usually set up to include so many constraints (and thus so few free parameters) that they cannot perfectly reproduce the correlations (or covariances) among the measured variables. Another difference between structural equation models and the EFA described here is that most structural equation programs start with a matrix of variances and covariances among measured variables (rather than the correlation matrix **R**), and the adequacy of the model is judged by how well it reproduces the observed variance/covariance matrix (instead of a correlation matrix). In CFA, the data analyst may compare several different models or may evaluate one theory-based model to assess whether the limited number of relationships among latent and measured variables that are included in the specification of the model is sufficient to reproduce the observed variances and covariances reasonably well. For CFA and structural equation models, statistical significance tests can be performed, both to assess whether the reproduced variance/covariance matrix deviates significantly from the observed variances and covariances calculated for the measured variables and to assess whether the coefficients associated with specific paths in the model differ significantly from 0.

At later stages in research, when more is known about the latent variables of interest, researchers may prefer to test the fit of CFA or structural equation models that are theory based instead of continuing to run exploratory analyses. While a complete description of structural equation models is beyond the scope of this book, the following example provides a brief intuitive conceptual introduction to a structural equation model.

A structural equation model is often represented as a path model; the path model often represents two different kinds of assumptions. First, some parts of the structural equation model correspond to a "measurement model." A measurement model, like a factor analysis, shows how a latent variable is correlated with scores on two or more measured variables that are believed to be indicators of that latent variable or construct. Consider a study by Warner and Vroman (2011); we examined extraversion as a predictor of happiness and also examined the degree to which the effect of extraversion on happiness was mediated by behaviors such as

Figure 12.47 A Measurement Model for the Latent Variable Extraversion

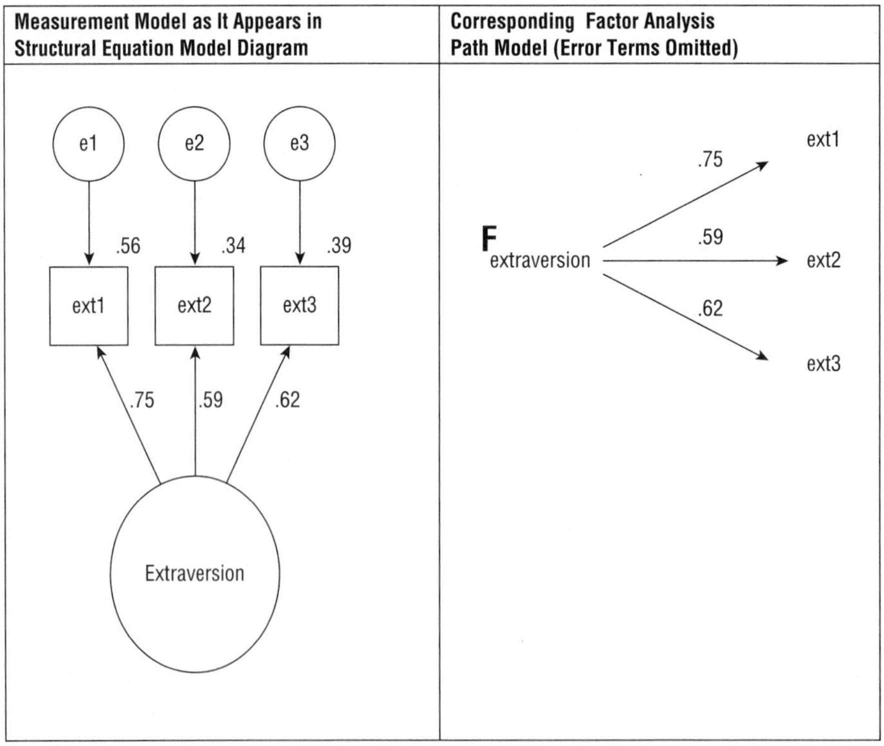

Note: Standardized path coefficients are shown; all were statistically significant using $\alpha = .05$, two tailed, as the criterion.

nurturing social relationships. The following example uses data collected as part of this study that were not reported in Warner and Vroman. For this example, extraversion, social support, and happiness are treated as latent variables. Each latent variable has multiple indicators; these indicators are three items from self-report scales that measure each of these constructs. Indicators can also be scores on multiple-item scales or scores for subsets of items from a scale.

The measurement model for the latent variable named extraversion appears in Figure 12.47. On the left-hand side of this figure, the measurement model is shown graphically using the conventions of SEM programs such as Amos. Each square corresponds to a measured variable. Each circle or oval corresponds to a latent variable. The arrows indicate hypothesized causal connections or, in the case of the measurement model, assumptions that scores on specific indicator variables can be predicted from a latent variable. The right-hand side of Figure 12.47 shows the same relationship among variables in the path model notation that has been used earlier in this chapter. The standardized path coefficients in this model (.75, .59, .62) correspond to the correlations of each indicator variable with the latent variable extraversion, or the loadings of each indicator variable on an extraversion factor. Structural equation models include additional notation to represent sources of error that were not explicitly included in the path models in this chapter. In the full structural equation model that appears in Figure 12.49, you can see that we also assume that the other six measured variables (hap1 through hap3 and sup1 through sup3) are *not* direct indicators of extraversion. The absence of a direct path between, for example, the measured variable hap1 and the latent variable extraversion is as informative about the nature of the theory about relations among variables as the presence of a direct path between ext1 and extraversion.

Path models were introduced in earlier chapters to show different possible kinds of causal and noncausal associations in three-variable research situations. One possible model to describe how X_1 and X_2 are related to a Y outcome variable was a mediated model, that is, a model in which the X_1 variable has part or all of its influence on the Y outcome variable through a path that involves a sequence of hypothesized causal connections; first X_1 "causes" X_2, and then X_2 "causes" Y. (The word *causes* is in quotation marks to remind us that we

Figure 12.48 Hypothesized Causal Paths Among (Measured) Variables

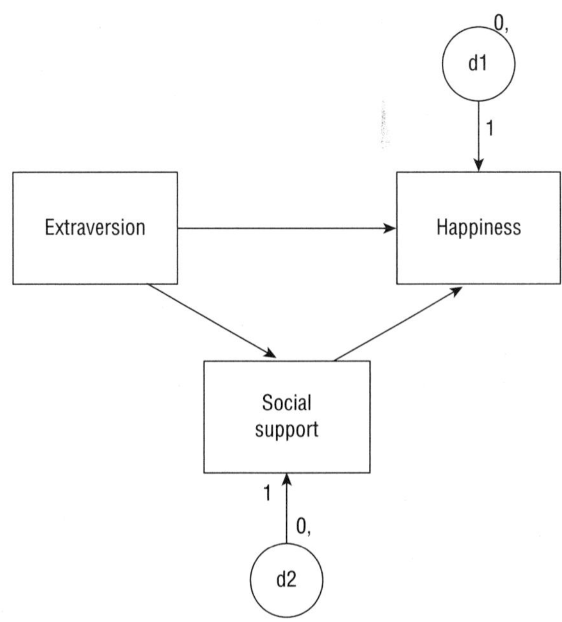

Figure 12.49 A Full Structural Equation Model That Includes Multiple Indicator/Measurement Models for the Latent Variables Extraversion, Social Support, and Happiness and Hypothesized Causal Paths

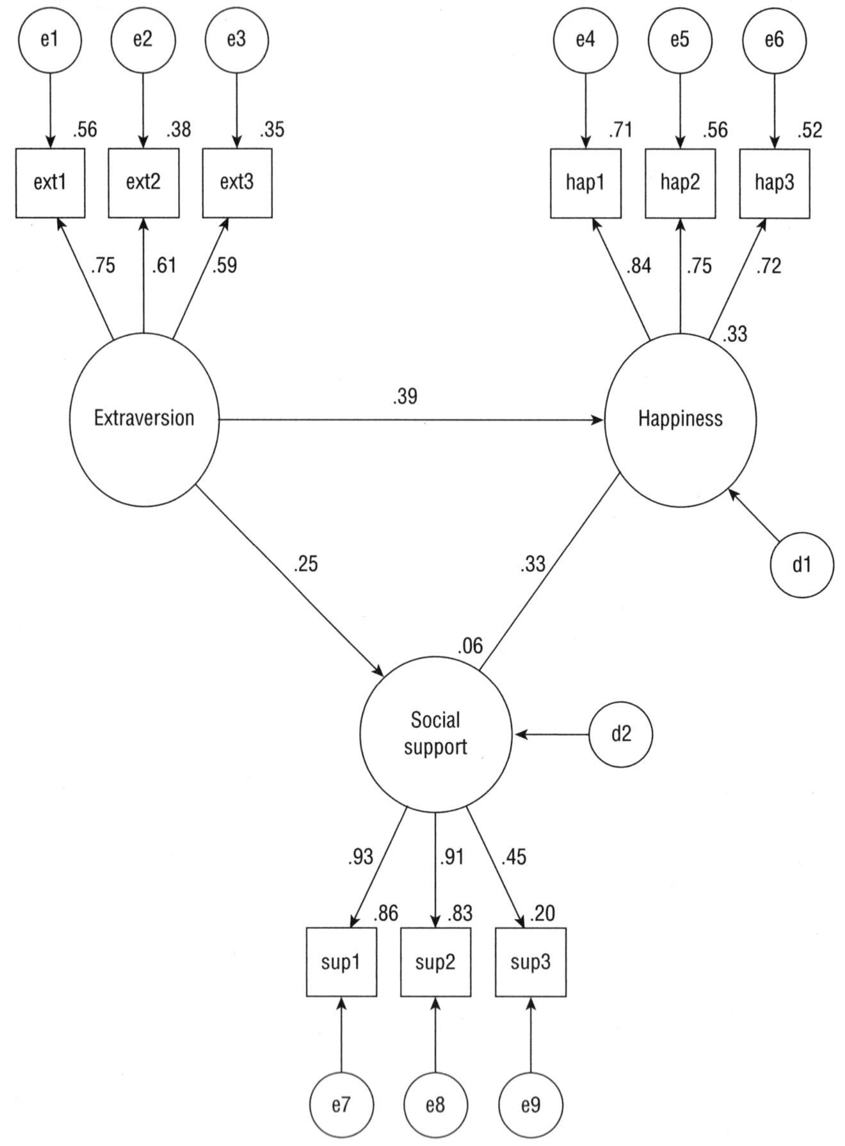

Source: Data from Warner and Vroman (2011).

Note: Standardized path coefficients are shown; all coefficients were statistically significant.

are talking about *hypothesized* causal connections. Analysis of nonexperimental data can yield results that are consistent or inconsistent with various "causal" models, but analysis of nonexperimental data cannot prove causality.)

Many structural equation models involve similar types of paths to show theorized "causal" associations that involve latent variables. Figure 12.48 shows a simple mediated causal model in which extraversion is a predictor of happiness. There is a path that corresponds to a direct

Figure 12.50 Examples of CFA Models

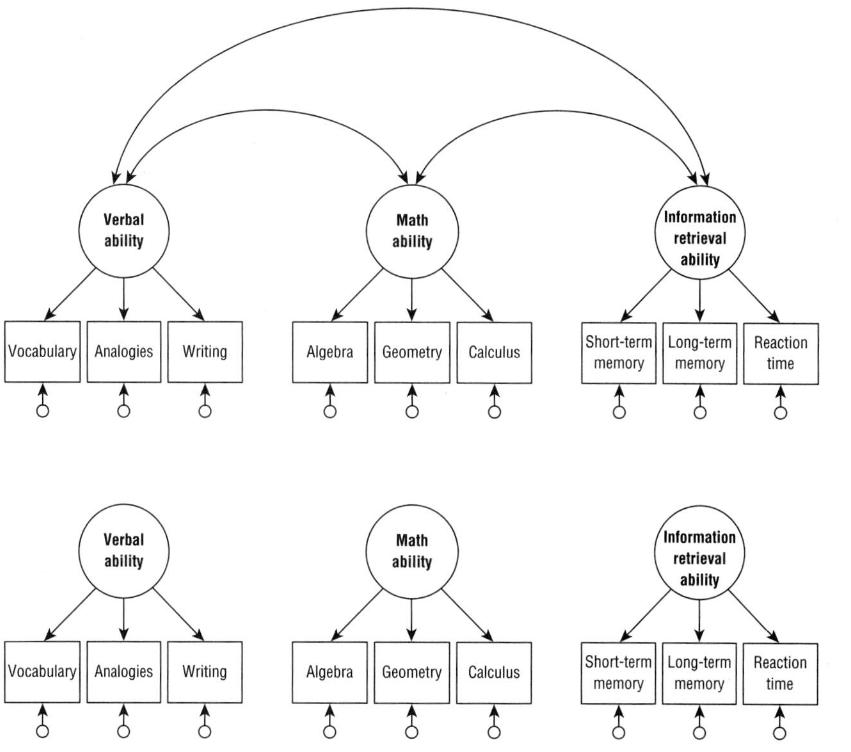

Note: Correlated (nonorthogonal) factors are depicted at the top and uncorrelated (orthogonal) factors at the bottom.

effect of extraversion on happiness and also a path that shows an indirect path from extraversion to happiness through social support.

The model that appears in Figure 12.49 is a full structural equation model that includes three measurement models: one for each of the latent variables (extraversion, social support, and happiness). It also includes the "causal" model that represents a theory that extraversion may have both direct and mediated or indirect effects on happiness. SEM thus combines the idea of latent variables (for which our measured variables serve as multiple indicators), an idea that is central to factor analysis, and causal or path models that describe associations among variables.

SEM programs can also be used to do CFA. See Figure 12.50 for examples of two different CFA models that could be evaluated using a hypothetical set of nine measurements of mental ability. The top panel shows hypothesized factor structure that assumes that three latent variables are needed to model the relationships among the nine measurements; each of the three latent variables is related to three of the indicator variables (as indicated in the path model), and the three latent variables are correlated or nonorthogonal. The bottom model is the same, except that in this more constrained model, the three factors or latent variables are assumed to be orthogonal or uncorrelated (the path coefficients between latent variables are constrained to be 0).

For a CFA or a structural equation model to be judged acceptable, it must be possible to reconstruct the observed variances and covariances of the measured variables reasonably well using the (unstandardized) path coefficients that are estimated for the model. The algebra used to reconstruct variances and covariances from structural equation model coefficients is analogous to the algebra that was used in the chapter about regression with two predictor variables to reconstruct correlations from the beta coefficients in a path model (as discussed in Chapter 4).

COMPREHENSION QUESTIONS

1. For a set of three mental ability tests in the Project Talent data set used in Chapter 10, on discriminant analysis, and Chapter 11, on multivariate analysis of variance, a PC analysis was obtained and all three components were retained. Selected results from this analysis appear in Figure 12.51.

   a. Show that you can reconstruct the communality for the variable mechanic from the loadings of mechanic on all three components in the component matrix above.
   b. Show that you can reconstruct the eigenvalue or SSL for the first component by using the loadings on Component 1 in the component matrix above.
   c. If you had allowed SPSS to use the default method (retain only components with eigenvalues > 1), how many components would have been retained?
   d. For a one-component solution, what proportion of the variance in the data would be accounted for?
   e. If you retained only one component, what proportion of the variance in the variable mechanic would be accounted for by just the first component?
   f. On the basis of this analysis, do you think it is reasonable to infer that these three mental tests tap three distinctly different types of mental ability?

Figure 12.51 Excerpt From PC Analysis for Comprehension Question 1

Communalities

	Initial	Extraction
mechanic	1.000	1.000
abstract	1.000	1.000
math	1.000	1.000

Extraction Method: Principal Component Analysis.

Total Variance Explained

	Initial Eigen values			Extraction Sums of Squared Loadings		
Component	Total	% of Variance	Cumulative %	Total	% of Variance	Cumulative %
1	1.975	65.830	65.830	1.975	65.830	65.830
2	.622	20.740	86.570	.622	20.740	86.570
3	.403	13.430	100.000	.403	13.430	100.000

Extraction Method: Principal Component Analysis.

Component Matrix

	Component		
	1	2	3
mechanic	.790	-.534	.301
abstract	.776	.579	.251
math	.866	-.032	-.499

Extraction Method: Principal Component Analysis.

a. 3 components extracted.

2. Using the data in talent.sav, the Project Talent test score data:

   a. What kinds of preliminary data screening would you need to run to evaluate whether assumptions for factor analysis are reasonably well met? Run these analyses.
   b. Using PAF as the method of extraction, do a factor analysis of this set of test scores: English, reading, mechanic, abstract, and math; the default criterion to retain factors with eigenvalues > 1; varimax rotation; and the option to sort the factor loadings by size; also request that factor scores be computed using the regression method and saved. Report and discuss your results. Do your results suggest more than one kind of mental ability? Do your results convince you that mental ability is "one-dimensional" and that there is no need to have theories about separate math and verbal dimensions of ability?
   c. Compute unit-weighted scores to summarize scores on the variables that have high loadings on your first factor in the preceding analysis. Do this in two different ways: using raw scores on the measured variables and using z scores. (Recall that you can save the z scores for quantitative variables by checking a box to save standardized scores in the descriptive statistics procedure.) Run a Pearson correlation to assess how closely these three scores agree (the saved factor score from the factor analysis you ran in 2b, the sum of the raw scores, and the sum of the z scores). In future analyses, do you think it will make much difference which of these three different scores you use as a summary variable (the saved factor score, sum of raw scores, or sum of z scores)?

3. Describe a hypothetical study in which principal components or factor analysis would be useful either for theory development (e.g., trying to decide how many different dimensions are needed to describe personality traits) or for data reduction (e.g., trying to decide how many different latent variables might be needed to understand what is measured in a multiple-item survey of "patient satisfaction with medical care"). List the variables, describe the analyses that you would run, and describe what pattern of results you expect to see.

4. Using your own data or data provided by your instructor, run a factor analysis using a set of variables that you expect to yield at least two factors, with at least four items that have high loadings on each factor. Write up a "Results" section for this analysis.

5. Locate a journal article that reports either a PC or a PAF analysis (avoid solutions that include nonorthogonal or oblique rotations or that use CFA methods, because this chapter does not provide sufficient background for you to evaluate these). Evaluate the reported results and show that you can reproduce the communalities and SSLs from the table of factor loadings. Can you think of different verbal labels for the factors or components than the labels suggested by the authors? Identify one or more variables that—if they were omitted from the analysis—would greatly change the conclusions from the study.

DIGITAL RESOURCES

Find **free study tools** to support your learning, including **eFlashcards, data sets, and web resources,** on the accompanying website at **edge.sagepub.com/warner3e**.

CHAPTER 13

RELIABILITY, VALIDITY, AND MULTIPLE-ITEM SCALES

13.1 ASSESSMENT OF MEASUREMENT QUALITY

Whether the variables in a study are direct measurements of physical characteristics such as height and weight, observer counts of behavior frequencies, or self-report responses to questions about attitudes and behaviors, researchers need to consider several measurement issues. The analyses described in previous chapters yield meaningful results only if the researcher has a well-specified model and if all the measurements for the variables are reliable and valid. This chapter outlines several important issues that should be taken into account when evaluating the quality of measurements. Self-report measures are used as examples throughout most of this chapter; however, the issues that are discussed (such as reliability and validity) are relevant for all types of measurements. When researchers use well-calibrated, widely used, well-established methods of measurement (whether these are physical instruments, such as scales, or self-report instruments, such as an intelligence test), they rely on an existing body of past knowledge about the quality of the measurement as evidence that the measurement method meets the quality standards that are outlined in this chapter. When researchers develop a new measurement method or modify an existing method, they need to obtain evidence that the new (or modified) measurement method meets these quality standards. The development of a new measurement method (whether it is a physical device, an observer coding system, or a set of self-report questions) can be the focus of a study.

13.1.1 Reliability

A good measure should be reasonably reliable; that is, it should yield consistent results. Low reliability implies that the scores contain a great deal of measurement error. Other factors being equal, variables with low reliability tend to have low correlations with other variables. This chapter reviews basic methods for reliability assessment. To assess reliability, a researcher needs to obtain at least two sets of measurements. For example, to assess the reliability of temperature readings from an oral thermometer, the researcher needs to take two temperature readings for each individual in a sample. Reliability can then be assessed using relatively simple statistical methods; for example, the researcher can compute Pearson's r between the first and second temperature readings to assess the stability of temperature across the two times of measurement.

13.1.2 Validity

A measure is valid if it measures what it purports to measure (Kelley, 1927); in other words, a measure is valid if the scores provide information about the underlying construct or theoretical variable that it is intended to measure. Validity is generally more difficult to assess than reliability. For now, consider the following example: If a researcher wants to evaluate whether a new measure of depression really does assess depression, he or she would need to collect data to see if scores on the new depression measure are related to behaviors, group memberships, and test scores in a manner that would make sense if the test really measures depression. For example, scores on the new depression test should predict behaviors that are associated with depression, such as frequency of crying or suicide attempts; mean scores on the new depression test should be higher for a group of patients clinically diagnosed with depression than for a group of randomly selected college students; also, scores on the new depression test should be reasonably highly correlated with scores on existing, widely accepted measures of depression, such as the Beck Depression Inventory (BDI), the Center for Epidemiologic Studies Depression Scale (CES-D), the Hamilton Depression Rating Scale, and the Zung Self-Rating Depression Scale. Conversely, if scores on the new depression test are not systematically related to depressed behaviors, to clinical diagnoses, or to scores on other tests that are generally believed to measure depression, the lack of relationship with these criteria should lead us to question whether the new test really measures depression.

13.1.3 Sensitivity

We would like scores to distinguish among people who have different characteristics. Some common problems that reduce measurement sensitivity include a limited number of response alternatives and ceiling or floor effects. For example, consider a question about depression that has just two response alternatives ("Are you depressed?" 1 = no, 2 = yes) compared with a rating scale for depression that provides a larger number of response alternatives ("How depressed are you?" on a scale from 0 = *not at all depressed* to 10 = *extremely depressed*). A question that provides a larger number of response alternatives potentially provides a measure of depression that is more sensitive to individual differences; the researcher may be able to distinguish among persons who vary in degree of depression, from mild to moderate to severe. A question that provides only two response alternatives (e.g., "Are you depressed?" no vs. yes) is less sensitive to individual differences in depression.

Some research has systematically examined differences in data obtained using different numbers of response alternatives; in some situations, a 5-point or 7-point rating scale may provide adequate sensitivity. Five-point degree of agreement rating scales (often called Likert scales) are widely used. A Likert scale item for an attitude scale consists of a "stem" statement that clearly represents either a positive or negative attitude about the object of the survey. Here is an example of an attitude statement with Likert response alternatives for a survey about attitudes toward medical care. A high score on this item (a choice of str*ongly agree*, or 5) would indicate a higher level of satisfaction with medical care.

My physician provides excellent health care.

1	2	3	4	5
Strongly disagree	Disagree	Neutral/don't know	Agree	Strongly agree

Other rating scales use response alternatives to report the frequency of a feeling or behavior, instead of the degree of agreement, as in the CES-D (see Appendix 13A; Radloff, 1977). The test taker is instructed to report how often in the past week he or she behaved or felt a specific way, such as "I felt that I could not shake off the blues even with help from my family and friends." The response alternatives for this item are given in terms of frequency:

0 = Rarely or none of the time (less than 1 day)

1 = Some or a little of the time (1–2 days)

2 = Occasionally or for a moderate amount of the time (3–4 days)

3 = Most of the time or all the time (5–7 days)

The use of too few response alternatives on rating scales can lead to lack of sensitivity. Ceiling or floor effects are another possible source of lack of sensitivity. A ceiling effect occurs when scores pile up at the top end of a distribution; for instance, if scores on an examination have a possible range of 0 to 50 points, and more than 90% of the students in a class achieve scores between 45 and 50 points, the distribution of scores shows a ceiling effect. The examination was too easy, and thus, it does not make it possible for the teacher to distinguish differences in ability among the better students in the class. On the other hand, a floor effect occurs when most scores are near the minimum possible value; if the majority of students have scores between 0 and 10 points, then the examination was too difficult.

Note that the term *sensitivity* has a different meaning in the context of medical research. For example, consider the problem of making a diagnosis about the presence of breast cancer (0 = breast cancer absent, 1 = breast cancer present) on the basis of visual inspection of a mammogram. Two different types of error are possible: The pathologist may report that breast cancer is present when the patient really does not have cancer (a false positive) or that breast cancer is absent when the patient really does have cancer (a false negative). For a medical diagnostic test to be considered sensitive, it should have low risks for both these possible types of errors.

13.1.4 Bias

A measurement is biased if the observed X scores are systematically too high or too low relative to the true value. For example, if a scale is calibrated incorrectly so that body weight readings are consistently 5 lb (2.27 kg) lighter than actual body weight, the scale yields biased weight measurements. When a researcher is primarily interested in correlations between variables, bias is not a major problem; correlations do not change when scores on one or both variables are biased. If measurements are used to make clinical diagnoses by comparing scores to cutoff values (e.g., diagnose high blood pressure if systolic blood pressure > 130 mm Hg), then bias in measurement can lead to misdiagnosis.

13.2 COST AND INVASIVENESS OF MEASUREMENTS

13.2.1 Cost

Researchers generally prefer measures that are relatively low in cost (in terms of time requirements as well as money). For example, other factors being equal, a researcher would prefer a test with 20 questions to a test with 500 questions because of the amount of time required for the longer test. Including a much larger number of questions in a survey than really necessary increases the length of time required from participants. When a survey is very long, this can reduce return rate (i.e., a lower proportion of respondents tend to complete surveys when they are very time-consuming) and create problems with data quality (because respondents hurry through the questions or become bored and fatigued as they work on the survey). In addition, including many more questions in a survey than necessary can ultimately create problems in data analysis; if the data analyst conducts significance tests on large numbers of correlations, this will lead to an inflated risk for Type I error.

13.2.2 Invasiveness

A measurement procedure can be psychologically invasive (in the sense that it invades the individual's privacy by requesting private or potentially embarrassing information) or physically invasive (e.g., when a blood sample is taken by venipuncture). In some research situations, the only way to obtain the information the researcher needs is through invasive procedures; however, professional ethical standards require researchers to minimize risk and discomfort to research participants. Sometimes, less direct or less invasive alternative methods are available. For example, a researcher might assess levels of the stress hormone cortisol by taking a sample of an infant's saliva instead of taking a blood sample. However, researchers need to be aware that less invasive measurements usually do not provide the same information that could be obtained from more invasive procedures; for example, analysis of infant saliva does not provide precise information about the levels of cortisol in the bloodstream. On the other hand, performing an invasive measure (such as taking a blood sample from a vein) may trigger changes in blood chemistry because of stress or anxiety.

13.2.3 Reactivity of Measurement

A measure is reactive when the act of measurement changes the behavior or mood or attribute that is being measured. For example, suppose a researcher asks a person to fill out an attitude survey on a public policy issue, but the person has not thought about the public policy issue. The act of filling out the survey may prompt the respondent to think about this issue for the first time, and the person may either form an opinion for the first time or report an opinion even if he or she really does not have an opinion. That is, the attitude that the researcher is measuring may be produced, in part, by the survey questions. This problem is not unique to psychology; Heisenberg's uncertainty principle pointed out that it was not possible to measure both the position and momentum of an electron simultaneously (because the researcher has to disturb, or interfere with, the system to make a measurement).

Reactivity has both methodological and ethical consequences. When a measure is highly reactive, the researcher may see behaviors or hear verbal responses that would not have occurred in the researcher's absence. At worst, when our methods are highly reactive, we may influence or even create the responses of participants. When people are aware that an observer is watching their behavior, they are likely to behave differently than if they believed themselves to be alone. One solution to this problem is not to tell participants that they are being observed, but of course, this raises ethical problems (invasion of privacy).

Answering questions can have unintended consequences for participants' everyday lives. When Rubin (1970) developed self-report measures for liking and loving, he administered his surveys to dating couples at the University of Michigan. One question he asked was, How likely is it that your dating relationship will lead to marriage? Many of the dating couples discussed this question after participating in the survey; some couples discovered that they were in agreement, while others learned that they did not agree. Rubin (1976) later reported that his participants told him that the discovery of disagreements in their opinions about the likelihood of future marriage led to the breakup of some dating relationships.

13.3 EMPIRICAL EXAMPLES OF RELIABILITY ASSESSMENT

13.3.1 Definition of Reliability

Reliability is defined as consistency of measurement results. To assess reliability, a researcher needs to make at least two measurements for each participant and calculate an appropriate statistic to assess the stability or consistency of the scores. The two measurements that are compared can be obtained in many different ways; some of these are given below:

1. A quantitative X variable, such as body temperature, may be measured for the same set of participants at two points in time (test-retest reliability). The retest may be done almost immediately (within a few minutes), or it may be done days or weeks

later. Over longer periods of time, such as months or years, it becomes more likely that the characteristic that is being measured may actually change. One simple form of reliability is test-retest reliability. In this section, we will examine a Pearson correlation between body temperature measurements made at two points in time for the same set of 10 persons.

2. Scores on a categorical X variable (such as judgments about attachment style) may be reported by two different observers or raters. We can assess interobserver reliability about classifications of persons or behaviors into categories by reporting the percentage agreement between a pair of judges. A reliability index presented later in this section (**Cohen's kappa [κ]**) provides an assessment of the level of agreement between observers that is corrected for levels of agreement that would be expected to occur by chance.

3. Scores on X can be obtained from questions in a multiple-item test, such as the CES-D (Radloff, 1977). The 20-item CES-D appears in Appendix 13A. In a later section of this chapter, factor analysis (FA) and **Cronbach's alpha (α)** internal consistency reliability coefficient are discussed as two methods of assessing agreement about the level of depression measured by multiple test questions or items.

13.3.2 Test-Retest Reliability Assessment for a Quantitative Variable

In the first example, we will consider how to assess test-retest reliability for scores on a quantitative variable (temperature). The researcher obtains a sample of $N = 10$ participants and measures body temperature at two points in time (e.g., 8 a.m. and 9 a.m.). In general, Time 1 scores are denoted by X_1, and Time 2 scores are denoted by X_2. To assess the stability or consistency of the scores across the two times, Pearson's r can be computed. Figure 13.1 shows hypothetical data for this example. The variables fahrenheit1 and fahrenheit2 represent body temperatures in degrees Fahrenheit at Time 1 (8 a.m.) and Time 2 (9 a.m.) for the 10 persons included in this hypothetical study.

When Pearson's r is used as a reliability coefficient, it often appears with a double subscript; for example, r_{XX} can be used to denote the test-retest reliability of X. An r_{XX} reliability coefficient can be obtained by calculating Pearson's r between measures of X at Time 1 (X_1) and measures on the same variable X for the same set of persons at Time 2 (X_2). For the data in this example, Pearson's r was obtained by running the correlation procedure on the X_1 and X_2 variables (see Figure 13.2); the obtained r_{XX} value was $r = +.924$, which indicates very strong consistency or reliability. This high value of r tells us that there is a high degree of consistency between temperature readings at Times 1 and 2; for example, Fran had the highest temperature at both times.

There is no universally agreed on minimum standard for acceptable measurement reliability. Nunnally and Bernstein (1994, pp. 264–265) stated that the requirement for reliability varies depending on how scores are used. They distinguished between two different situations: first, basic research, where the focus is on the size of correlations between X and Y variables or on the magnitude of differences in Y means across groups with different scores on X, and second, clinical applications, where an individual person's score on an X measure is used to make treatment or placement decisions. In preliminary or exploratory basic research (where the focus is on the correlations between variables, rather than on the scores of individuals), modest measurement reliability (about .70) may be sufficient. If preliminary studies yield evidence that variables may be related, further work can be done to develop more reliable measures. Nunnally and Bernstein argued that in basic research, increasing measurement reliabilities much beyond $r = .80$ may yield diminishing returns. However, when scores for an individual are used to make decisions that have important consequences (such as medical

Figure 13.1 Hypothetical Data for Test-Retest Reliability Study of Body Temperature

*ch21ReliabilityExample1.sav [DataSet0] - IBM SPSS Statis

File Edit View Data Transform Analyze Graphs Utilities

3 :

	Name	fahrenheit1	fahrenheit2	var
1	ann	99.2	99.1	
2	bob	95.5	96.5	
3	caren	98.6	96.7	
4	dan	98.9	98.4	
5	ed	97.6	97.9	
6	fran	103.2	102.9	
7	george	98.6	98.5	
8	hilary	98.7	98.4	
9	irene	98.7	98.5	
10	jane	98.6	98.3	
11				

Figure 13.2 Pearson Correlation to Assess Test-Retest Reliability of Body Temperature in Degrees Fahrenheit

Correlations

		fahrenheit1	fahrenheit2
fahrenheit1	Pearson Correlation	1	.924**
	Sig. (2-tailed)		.000
	N	10	10
fahrenheit2	Pearson Correlation	.924**	1
	Sig. (2-tailed)	.000	
	N	10	10

**. Correlation is significant at the 0.01 level (2-tailed).

Note: Based on data in Figure 13.1.

diagnosis or placement in a special class for children with low academic ability), reliability of $r = .90$ is the bare minimum, and $r = .95$ would be a desirable standard for reliability (because even very small measurement errors could result in mistakes that could be quite harmful to the individuals concerned).

Note that whether the correlation between X_1 and X_2 is low or high, there can be a difference in the means of X_1 and X_2. If we want to assess whether the mean level of scores (as well as the position of each participant's score within the list of scores) remains the same from Time 1 to Time 2, we need to examine a paired-samples t test to see whether the mean of

X_2 differs from the mean of X_1, in addition to obtaining Pearson's r as an index of consistency of measurements across situations. In this example, there was a nonsignificant difference between mean temperatures at Time 1 (M_1 = 98.76) and Time 2 (M_2 = 98.52), $t(9) = 1.05$, $p = .321$.

In some research situations, the only kind of reliability or consistency that researchers are concerned about is the consistency represented by the value of Pearson's r, but there are some real-life situations where a shift in the mean of measurements across times would be a matter of interest or concern. A difference in mean body temperature between Time 1 and Time 2 could represent a real change in the body temperature; the temperature might be higher at Time 2 if this second temperature reading is taken immediately after intense physical exercise. There can also be measurement artifacts that result in systematic changes in scores across times; for example, there is some evidence that when research participants fill out a physical symptom checklist every day for a week, they become more sensitized to symptoms and begin to report a larger number of symptoms over time. This kind of change in symptom reporting across time might be due to artifacts such as reactivity (frequent reporting about symptoms may make the participant focus more on physical feelings), social desirability, or demand (the participant may respond to the researcher's apparent interest in physical symptoms by helpfully reporting more symptoms), rather than to an actual change in the level of physical symptoms.

13.3.3 Interobserver Reliability Assessment for Scores on a Categorical Variable

If a researcher wants to assess the reliability or consistency of scores on a categorical variable, he or she can initially set up a contingency table and examine the percentage of agreement. Suppose that attachment style (coded 1 = anxious, 2 = avoidant, and 3 = secure) is rated independently by two different observers (Observer A and Observer B) for each of N = 100 children. Attachment style is a categorical variable; on the basis of observations of every child's behavior, each observer classifies each child as anxious, avoidant, or secure. The hypothetical data in Table 13.1 show a possible pattern of results for a study in which each of these two observers makes an independent judgment about the attachment style for each of the 100 children in the study. Each row corresponds to the attachment score assigned by Observer A; each column corresponds to the attachment score assigned by Observer B. Thus, the value 14 in the cell in the upper left-hand corner of the table tells us that 14 children were classified as "anxiously attached" by both observers. The adjacent cell, with an observed frequency of 1, represents one child who was classified as "anxiously attached" by Observer A and "avoidantly attached" by Observer B.

A simple way to assess agreement or consistency in the assignment of children to attachment style categories at these two points in time is to add up the number of times the two judges were in agreement: We have 14 + 13 + 53 = 80; that is, 80 children were placed in the same attachment style category by both observers. This is divided by the total number of children (100) to yield 80/100 = .80 or 80% agreement in classification. Proportion or percentage of agreement is fairly widely used to assess consistency or reliability of scores on categorical variables.

However, there is a problem with proportion or percentage of agreement as a reliability index; fairly high levels of agreement can arise purely by chance. Given that Observer A coded 58/100 (a proportion of .58) of the cases as secure and Observer B coded 65/100 (.65) of the cases as secure, we would expect them both to give the score "secure" to $.58 \times .65 = .377$ of the cases in the sample just by chance. By similar reasoning, they should both code $.17 \times .23 = .0391$ of the cases as anxious and $.18 \times .19 = .0342$ of the cases as avoidant. When we sum these chance levels of agreement across the three categories, the overall level of agreement predicted just by chance in this case would be .377 + .0391 + .0342 = .4503.

Table 13.1 Data for Interobserver Reliability for Attachment Style Category

	Observer B			
Observer A	**1 = Anxious**	**2 = Avoidant**	**3 = Secure**	**Row Total (Proportion)**
1 = Anxious	14	1	8	23
				(.23)
2 = Avoidant	2	13	4	19
				(.19)
3 = Secure	1	4	53	58
				(.58)
Column total (proportion)	17	18	65	100
	(.17)	(.18)	(.65)	

Note: Number of agreements between Observer A and Observer B: 14 + 13 + 53 = 80. Total number of judgments = 100 (sum of all cells in the table, all row totals, or all column totals). P_o = observed level of agreement = number of agreements/total number of judgments = 80/100 = .80. P_c = chance level of agreement for each category, summed across all three categories = (.17 × .23) + (.18 × .19) + (.65 × .58) = .0391 + .0342 + .377 = .4503. Thus Cohen's κ is calculated as follows:

$$\kappa = \frac{(P_o - P_c)}{(1 - P_c)} = \frac{(.80 - .4503)}{(1 - .4503)} = \frac{.3499}{.5499} = .636 \cdot$$

Cohen's kappa (κ) takes the observed proportion of agreement (P_o, in this case, .80) and corrects for the chance level of agreement (P_c, in this case, .4503).

The formula for Cohen's κ is

$$\kappa = \frac{(P_o - P_c)}{(1 - P_c)} = \frac{(.80 - .4503)}{(1 - .4503)} = .636 . \quad (13.1)$$

Cohen's kappa index of interrater reliability in this case (corrected for chance levels of agreement) is .64. Landis and Koch (1977) suggested the following guidelines for evaluating the size of Cohen's κ: .21 to .40, fair reliability; .41 to .60, moderate reliability; .61 to .80, substantial reliability; and .81 to 1.00, almost perfect reliability. By these standards, the obtained κ value of .64 in this example represents a substantial level of reliability. Cohen's kappa is available as one of the contingency table statistics in the SPSS crosstabs procedure, and it can be used to assess the reliability for a pair of raters on categorical codes with any number of categories; it could also be used to assess test-retest reliability for a categorical measurement at two points in time.

13.4 CONCEPTS FROM CLASSICAL MEASUREMENT THEORY

The version of **measurement theory** presented here is a simplified version of classical measurement theory (presented in more detail by Carmines & Zeller, 1979). Contemporary measurement theory has moved on to more complex ideas that are beyond the scope of this book. The following is a brief introduction to some of the ideas in classical measurement theory. We can represent an individual observed score, X_i, as the sum of two theoretical components:

$$X_{ij} = T_i + e_{ij}, \tag{13.2}$$

where X_i represents the observed score on a test for Person i at Time j, T_i is the theoretical "true" score of Person i, and e_{ij} is the error of measurement for Person i at Time j.

The first component, T, usually called the true score, denotes the part of the X score that is stable or consistent across measurements. In theory, this value is constant for each participant or test taker and should represent that person's true score on the measurement. We hope that the T component really is a valid measure of the construct of interest, but the stable components of observed scores can contain consistent sources of bias (i.e., systematic error). For example, if an intelligence quotient (IQ) test is culturally biased, then the high reliability of IQ scores might be due in part to the consistent effect of an individual's cultural background on his or her test performance. It is generally more difficult to assess validity—that is, to assess whether a scale really measures the construct it is supposed to measure—without any "contaminants," such as social desirability or cultural background, than to assess reliability (which is simply an assessment of score consistency or stability).

To see what the T and e components might look like for an individual person, suppose that Jim is Participant 6 in a test-retest reliability study of the verbal SAT; suppose Jim's score is X_{61} = 570 at Time 1 and X_{62} = 590 at Time 2. To estimate T_6, Jim's "true" score, we would find his mean score across measurements; in this case, M = (570 + 590)/2 = 580.

We can compute an error (e_{ij}) term for each of Jim's observed scores by subtracting T from each observed score; in this case, the error term for Jim at Time 1 is 570 – 580 = –10, and the error at Time 2 is 590 – 580 = +10. We can then use the model in Equation 13.2 to represent Jim's scores at Time 1 and Time 2 as follows:

$$X_{61} = T_6 + e_{61} = 580 - 10 = 570,$$

$$X_{62} = T_6 + e_{62} = 580 + 10 = 590.$$

Using this simplified form of classical measurement theory as a framework, we would then say that at Time 1, Jim's observed score is made up of his true score minus a 10-point error; at Time 2, Jim's observed score is made up of his true score plus a 10-point error. The measurement theory assumes that the mean of the measurement errors is 0 across all occasions of measurement in the study and that the true score for each person corresponds to the part of the score that is stable or consistent for each person across measurements.

Errors may be even more likely to arise when we try to measure psychological characteristics (such as attitude or ability) than when we measure physical quantities (such as body temperature). Standardized tests, such as the SAT, are supposed to provide information about students' academic aptitude. What factors could lead to measurement errors in Jim's SAT scores? There are numerous sources of measurement errors (the e component) in exam scores; for example, a student might score lower on an exam than would be expected given his or her ability and knowledge because of illness, anxiety, or lack of motivation. A student might score higher on an exam than his or her ability and general knowledge of the material would predict because of luck (the student just happened to study exactly the same things the teacher decided to put on the exam), cheating (the student copied answers from a better prepared student), randomly guessing some answers correctly, and so forth.

We can generalize this logic as follows. If we have measurements at two points in time, and X_{ij} is the observed score for Person i at Time j, we can compute the estimated true score for Person i, T_i, as $(X_1 + X_2)/2$. We can then find the estimated error terms (e_{i1} and e_{i2}) for

Person i by subtracting this T_i term from each of the observed scores. When we partition each observed score (X_{ij}) into two parts (T_i and e_{ij}), we are using logic similar to analysis of variance. The T and e terms are uncorrelated with each other, and in theory, the error terms e have a mean of 0.

13.4.1 Reliability as Partition of Variance

Reliability analysis (using test-retest correlation, for example) partitions the variance in the test scores into two components: Variance that is stable across times or occasions of measurement, and variance that is not stable across time. When we have divided each observed score into two components (T and e), we can summarize information about the T and e terms across persons by computing a variance for the values of T and a variance for the values of e. A reliability coefficient (e.g., a test-retest Pearson's r) is the proportion of variance in the test scores that is due to T components that are consistent or stable across occasions of measurement. Reliability can be defined as the ratio of the variance of the "true score" components relative to the total variance of the observed X measurements as shown in Equation 13.3:

$$r_{XX} = \frac{s_T^2}{s_T^2 + s_e^2} = \frac{s_T^2}{s_X^2}, \tag{13.3}$$

where s^2_X is the variance of the observed X scores, s^2_T is the variance of the T true scores, s^2_e is the variance of the e or error terms, and $s^2_X = s^2_T + s^2_e$; that is, the total variance of the observed X scores is the sum of the variances of the T true score components and the e errors.

Equation 13.3 defines reliability as the proportion of variance in the observed X scores that is due to T, a component of scores that is stable across the two times of measurement (and different across persons). For a measure with a test-retest reliability of $r_{XX} = .80$, in theory, 80% of the variance in the observed X scores is due to T; the remaining 20% of the variance in the observed scores is due to error.

Note that it is r_{XX}, the test-retest correlation (rather than r^2, as in other situations where the Pearson correlation is used to predict scores on Y from X), that is interpreted as a proportion of variance in this case. To see why this is so, consider a theoretical path model that shows how the observed measurements (X_1 and X_2) are related to the stable characteristic T. Notation and tracing rules for path diagrams were introduced in earlier chapters.

Figure 13.3 corresponds to Equation 13.2; that is, the observed score on X_1 is predictable from T and e_1; the observed score on X_2 is predictable from T and e_2. The path coefficient that represents the strength of the realationship between T and X_1 is denoted by β; we assume that the strength of the relationship between T and X_1 is the same as the strength of the relationship between T and X_2, so the beta coefficients for the paths from $T \rightarrow X_1$ and $T \rightarrow X_2$ are equal.

Figure 13.3 Classic Measurement Theory in Path Model Form

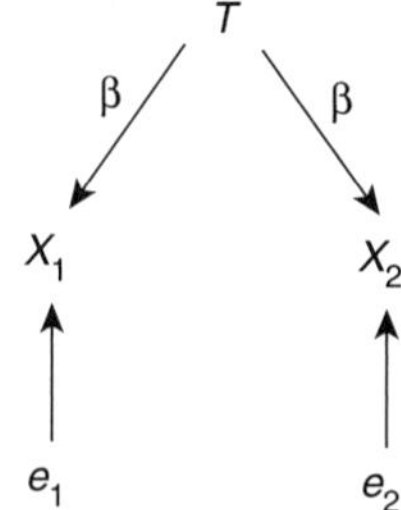

We would like to estimate beta, the strength of the relation between the observed scores X_1 and X_2 and the true score T. We can compute a test-retest r between X_1 and X_2 (we can denote this by r_{XX} or r_{12}). On the basis of the logic of path diagrams and the application of the tracing rule, we know that we can reproduce the observed value of r_{12} by tracing the path from X_1 to X_2 (by way of T) and multiplying the coefficients for each section of the path together. In this case, we see that (by the application of the tracing rule) $\beta \times \beta$ should equal r_{12}. This implies that the best estimate of β is $\sqrt{r_{12}}$. Thus, if we want to know what proportion of variance in X_1 (or in X_2) is associated with T, we square the path coefficient between X_1 and T; this yields r_{12} as our estimate of the proportion of shared or predicted variance. An examination of this path diagram helps explain why it is r_{12} (rather than r^2_{12}) that tells us what proportion of variance in X is predictable from T.

13.4.2 Attenuation of Correlations Due to Unreliability of Measurement

A related path diagram explains the phenomenon of attenuation of correlation between X and Y due to the unreliability of measurement. Suppose the real strength of relationship between the true scores on variables X and Y (i.e., the scores that would be obtained if they could be measured without error) is represented by ρ_{XY}. However, the reliabilities of X and Y (denoted by r_{XX} and r_{YY}, respectively) are less than 1.00; that is, the X and Y measurements used to compute the sample correlation do contain some measurement error; the smaller these reliabilities, the greater the measurement error. The observed correlation between the X and Y scores that contain measurement errors (denoted by r_{XY}) can be obtained from the data. The path diagram shown in Figure 13.4 tells us that the strength of the relationship between the observed X scores and the true score on X is denoted by $\sqrt{r_{XX}}$; the strength of the relationship between the observed Y scores and the true score on Y is denoted by $\sqrt{r_{YY}}$. If we apply the tracing rule to this path model, we find that r_{XY} (the observed correlation between X and Y) should be reproducible by tracing the path from the observed X to the true X, from the true X to the true Y, and from the true Y to the observed Y (and multiplying these path coefficients). Thus, we would set up the following equation on the basis of Figure 13.4:

$$r_{XY} = \rho_{XY}\sqrt{r_{XX}} \times \sqrt{r_{YY}} = \rho_{XY}\sqrt{r_{XX} \times r_{YY}}, \tag{13.4}$$

where r_{XY} is the observed correlation between the observed scores on X and Y, ρ_{XY} is the correlation between the true scores on X and Y (i.e., the correlation between X and Y if they could be measured without error), and r_{XX} and r_{YY} are the reliabilities of the variables. In situations where there is measurement error, the reliability coefficients r_{XX} and r_{YY} will be less than 1. It follows that the observed correlation r_{XY} will be less than ρ_{XY} whenever there is measurement error, and in fact, the smaller the reliabilities of X and Y, the greater the reduction in the size of the observed r_{XY} correlation. This is called attenuation of correlation due to unreliability. In theory, the lower the reliabilities of X and Y, the smaller the observed r_{XY} correlation will tend to be.

It is theoretically possible to correct for attenuation, that is, to calculate an estimate ($\hat{\rho}_{XY}$) for the "true" correlation that would exist between X and Y, if they could be measured without error. An estimate of the correlation between X and Y that is adjusted for attenuation due to unreliability of measurement in X and Y is given by the following equation:

$$\hat{\rho}_{XY} = \frac{r_{XY}}{\sqrt{r_{XX} \times r_{YY}}}, \tag{13.5}$$

Figure 13.4 Path Model for Attenuation of Correlation Between Observed *X* and Observed *Y* as a Function of Reliability of Measurement of *X* and *Y*

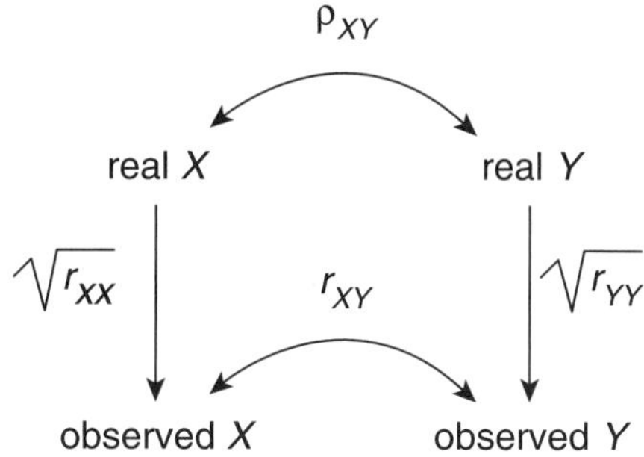

Note: ρ_{XY} is the "real" correlation between X and Y (i.e., the true strength of relationship between variables, if they were measured without error), r_{XY} the observed correlation between X and Y scores that contain measurement error, r_{XX} the estimated reliability of X, and r_{YY} the estimated reliability of Y.

where $\hat{\rho}_{XY}$ is the "attenuation-corrected" estimate of the correlation between X and Y, that is, an estimate of the strength of the relationship between X and Y if they could be measured with perfect reliability; r_{XY} is the observed correlation between the X and Y scores; r_{XX} is the estimated reliability coefficient for the X variable; and r_{YY} is the estimated reliability coefficient for the Y variable.

However, this attenuation-corrected estimate should be interpreted with great caution. If the reliability estimates r_{XX} and r_{YY} are inaccurate, this correction will yield misleading results. In particular, if the reliabilities of X and Y are underestimated, the attenuation-corrected estimate of the true strength of the association between X and Y will be inflated.

Poor reliability of measurement has two major negative consequences. First of all, reliability is a necessary (although not sufficient) condition for validity; thus, if measurements are not reliable, they cannot be valid. Second, when measurements of variables are unreliable, it tends to result in low correlations among variables. In traditional approaches to statistical analysis, using primarily the pre-1970 methods of analysis, we need to make an effort to obtain measurements for each variable in the analysis that are as reliable as possible; to the extent that we use unreliable measures, the outcomes from these traditional data analyses may be misleading. Another way in which we can handle the problems caused by unreliability of measurements is to use structural equation models. In these models, the observed X variables are treated as multiple indicators of latent variables, and the analysis incorporates assumptions about errors of measurement for the observed X variables.

We can typically improve the reliability of measurement for an X variable in a sample in two different ways. One approach is to increase the differences in T across participants by working with samples that are more diverse on the characteristic that is being measured (i.e., we increase the magnitude of s^2_T). Another approach is to reduce the magnitude of errors of measurement by controlling extraneous variables that might influence measurements (i.e., we decrease the magnitude of s^2_e). Combining scores from multiple questions or measurements is another method that may result in more reliable scores; the use of scales that are formed by summing multiple measures is described in the next section.

13.5 USE OF MULTIPLE-ITEM MEASURES TO IMPROVE MEASUREMENT RELIABILITY

For some variables, a single measurement or a single question is sufficient. For example, a single question is generally sufficient to obtain information about age or height. However, to assess personality traits, abilities, attitudes, or knowledge, it is usually necessary to include multiple questions. Using a score that corresponds to a sum (or an average) across multiple measurements provides a measure that is more reliable than a single measure. The inclusion of multiple measures also makes it possible to assess the reliability or consistency of responses. In many situations, we may also be more confident about the validity of a score that summarizes information across multiple measurements than regarding a score based on a single measurement; however, validity of measurement is more difficult to evaluate than reliability.

However, for variables that have complex definitions or that must be assessed indirectly, such as depression, the use of single-item measures can be problematic. A one-item assessment of depression would have serious limitations. First of all, a measure with only two possible score values would provide very limited information about individual differences in the degree of depression; also, the scores could not be normally distributed (and normally distributed scores are desirable for many statistical analyses). Even a question with multiple response options, such as "Rate your degree of depression on a scale from 1 to 4," would produce a limited number of possible values and a non-normal distribution. Second, when only one response is obtained, there is no way of assessing the reliability or consistency of responses. Third, a single question does not capture the complex nature of depression; depression includes not only feelings of sadness and hopelessness but also vegetative symptoms such as fatigue and sleep disturbance. A comprehensive measure of depression needs to include all relevant symptoms. In some surveys, the inclusion of multiple questions helps respondents better understand what kind of information the researcher is looking for. Responses to multiple, specific questions may be far more informative than responses to a single, global, nonspecific question. For instance, when people are asked a global, nonspecific question (such as "Are you satisfied with your medical care?"), they tend to report high levels of satisfaction; but when they are asked about specific potential problems (such as "Did the physician use language you did not understand?" "Did you have enough time to explain the nature of your problems?" or "Was the practitioner sympathetic to your concerns?"), their responses typically become more informative.

To avoid the limitations of single-item measures, researchers often use multiple items or questions to obtain information about variables such as depression. For example, the CES-D (Radloff, 1977) has 20 questions about depression (see the complete CES-D in Appendix 13A). The following example uses scores for the first 5 items from the CES-D for a set of 98 participants in the data set depressionitems.sav. Later analysis of CES-D reliability is based on the full 20-item scale; however, initial examples are limited to the assessment of Items 1 through 5 to keep the empirical examples brief and simple. Multiple-item measures often provide many advantages compared with single-item measures. They can potentially provide finer discriminations of the amount of depression, scores that are more nearly normally distributed, and scores that are generally more reliable than scores based on single-item assessments.

In many research situations, we can obtain more reliable measurements by measuring the variable of interest on multiple occasions or using multiple methods or multiple items and, then, combining the information across multiple measures into a single summary score. Ideas derived from the measurement theory model presented in a previous section suggest that combining two or more scores should result in more reliable measures. As we sum a set of X_i measurements for a particular participant (which could correspond to a series of p different questions on a self-report survey, a series of behaviors, or a series of judgments by observers or raters), in theory, we obtain the following result:

$$\begin{aligned} X_1 &= T + e_1 \\ X_2 &= T + e_2 \\ X_3 &= T + e_3 \\ &\vdots \\ X_p &= T + e_p \\ \hline \sum X &= p \times T + \sum e. \end{aligned}$$

The expected value of Σe, particularly when errors are summed across many items, theoretically approaches 0. That is, we expect the errors to "cancel out" as we sum more and more items. When we use ΣX, the sum of scores on several questions, as a measure, it is likely that a large part of the score obtained from ΣX will correspond to the true score for the participant (because we have $p \times T$ in the expression for ΣX). It is also likely that a small part of the score obtained by taking ΣX will correspond to errors of measurement, because we expect that this sum approaches 0 when many errors are summed. Thus, the reliability of a score based on ΣX (the percentage of variance in the value of ΣX that is due to T) should be higher than the reliability of a score based on a single question.

When we use a sum (or a mean) of scores across multiple items, questions, or measures, there are several potential advantages:

1. Composite measures (scales) are generally more reliable than single scores, and the inclusion of multiple measures makes it possible to assess **internal consistency reliability**.
2. Scores on composite measures generally have greater variance than scores based on single items.
3. The distribution shape for scores on individual items using common methods response alternatives, such as five-point Likert rating scales, are typically non-normal. The distribution shape for scores formed by summing multiple measures that are positively correlated with each other tend to resemble a somewhat flattened normal distribution; the higher the correlation among items, the flatter the distribution for the sum of the scores (Nunnally & Bernstein, 1994). Many statistical tests work better with scores that are approximately normally distributed, and composite scores that are obtained by summing multiple-item measures sometimes approximate a normal distribution better than scores for single-item measures.
4. Scores on composite or multiple-item measures may be more sensitive to individual differences (because a larger number of different score values are possible than on single items that use five-point rating scales, for example).

To summarize information across many items or questions, scores are often summed across items. We will define a summated scale as a sum (or mean) of scores across multiple occasions of measurement. For example, an attitude scale to measure patient satisfaction with medical treatment might include a set of 20 statements about various aspects of quality of medical care; patients may be asked to indicate the degree to which their care providers meet each of the quality standards included in the survey. It may make sense to compute a single score to summarize overall patient satisfaction with the quality of medical care by summing responses across all 20 items. However, it is also possible that, on closer examination, we might decide that the set of 20 items may provide information about more than one type of satisfaction. For example, some items might assess satisfaction with rapport, empathy, or bedside manner; some items might assess perceived technical competence of the practitioner; and other questions might assess issues of accessibility, for example, length of waiting time and difficulty in scheduling appointments.

Note that the use of multiple-item scales is not limited to self-report data. We can also combine other types of multiple measurements to form scales (such as physiological measures or ratings of behavior made by multiple judges or observers).

13.6 COMPUTATION OF SUMMATED SCALES

This section describes assumptions that are required when scores are summed across items to create a total score for a scale. We assume that all items measure the same construct (such as depression) and are scored in the same direction (such that a higher number of points correspond to more depression for all items). Several methods for combining scores across items are described, including different ways of handling missing data. An empirical demonstration suggests that, in some research situations, it may not make much difference which method is used to combine scores. Cronbach's alpha reliability is used to describe the internal consistency reliability for the total score on the basis of a sum of items.

13.6.1 Assumption: All Items Measure the Same Construct and Are Scored in the Same Direction

When we add together scores on a list of items or questions, we implicitly assume that these scores all measure the same underlying construct and that all questions or items are scored in the same direction. Items included in psychological tests such as the CES-D are typically written so that they assess slightly different aspects of a complex variable such as depression (e.g., low self-esteem, fatigue, sadness). To evaluate empirically whether these items can reasonably be interpreted as measures of the same underlying latent variable or construct, we usually look for at least moderate correlations among the scores on the items.[1] If scores on a set of measurements or test items are at least moderately correlated with one another, this is consistent with a belief that the items may all be measures of the same underlying construct. However, high correlations among items can arise for other reasons and are not necessarily proof that the items measure the same underlying construct; high correlations among items may occur due to sampling error or because items share some measurement such as social desirability bias. Later in this chapter, we will see that the most widely reported method of evaluating reliability for summated scales, Cronbach's alpha, is based on the mean of the interitem correlations.

13.6.2 Initial (Raw) Scores Assigned to Individual Responses

Suppose that a web-based program (such as SurveyMonkey or Qualtrics) is used to collect people's responses to the 20-item CES-D (Radloff, 1977; see Appendix 13A for the complete list of questions). Programs such as SurveyMonkey display multiple-choice questions that require one response as radio buttons, as in Figure 13.5.

When data are initially downloaded from SurveyMonkey, scores for responses are numbered (1 = first response option, 2 = second response, 3 = third response, 4 = fourth response). The raw-score numerical codes may be acceptable for many measures. However, CES-D responses are conventionally assigned codes of 0 = rarely, 1 = some, 2 = occasionally, and

Figure 13.5 Radio-Button Format for Typical Multiple-Choice Item

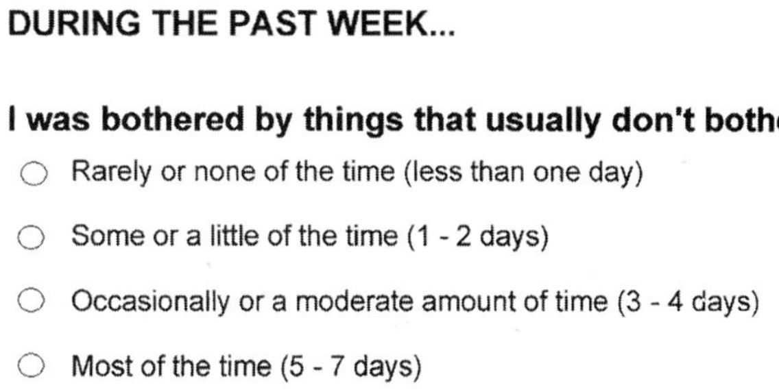

DURING THE PAST WEEK...

I was bothered by things that usually don't bother me.

- ○ Rarely or none of the time (less than one day)
- ○ Some or a little of the time (1 - 2 days)
- ○ Occasionally or a moderate amount of time (3 - 4 days)
- ○ Most of the time (5 - 7 days)

3 = most of the time. If we want to compare scores from our sample to scores obtained in past research and to test norms, the scores assigned to responses must be consistent with scores used in past research. To make the scoring of these items consistent with the usual scores for the CES-D, prior to other analyses, SPSS compute statements were used to subtract 1 from all scores shown in originaldepressionitems.sav. Data for the first five items of the CES-D, with corrected raw scores that range from 0 to 3, appear in depressionitems.sav. For some purposes (correlation, FA), subtracting 1 from each score makes no difference. However, comparison of sample means to published test norms is not valid if the points assigned to responses in the sample are not the same as those used in the development of test norms.

13.6.3 Variable Naming, Particularly for Reverse-Worded Questions

Notice the names of the variables in depressionitems.sav (dep1, dep2, dep3, revdep4, dep5). Most items were given names that remind us what they measure ("dep" is short for depression) and numbers that correspond to the item numbers on the CES-D (in Appendix 13A). Including the item number in the variable name makes it easier to look up the exact item wording on the questionnaire. The fourth item was named "revdep4" because it is a reverse-worded item. Most items on the CES-D are worded so that reporting a higher frequency or higher numerical score indicates a greater degree of depression. For Question 3, "I felt that I could not shake off the blues even with help from my family or friends," the response that corresponds to 3 points ("I felt this way most of the time, 5–7 days per week") indicates a higher level of depression than the response that corresponds to 0 points ("I felt this way rarely or none of the time"). It is desirable to have a high score on a scale correspond to "more of" the characteristic in the name of the scale. Readers typically expect that a high score on a scale that is called a depression scale corresponds to a high level of depression. (If a high score on a depression scale indicates a low level of depression, this is likely to cause confusion.)

Four of the items on the CES-D (Items 4, 8, 12, and 16) are reverse worded. For example, Question 4 asks how frequently the respondent "felt that I was just as good as other people." The response to this question that would indicate the *highest* level of depression corresponds to the *lowest* frequency of occurrence (0 = *rarely or none of the time*). Reverse-worded items are often included in scales to try to minimize yea-saying bias. Not all scales include reverse-worded items. Note that respondents sometimes find reverse-worded items confusing, particularly if they include words such as *not*; double negatives tend to be difficult for readers. When reverse-worded items are included in a multiple-item measure, the scoring on these items must be recoded (so that a high score on each item corresponds to the same thing, i.e., a higher level of depression) before we sum scores across items or do a reliability analysis.

It is helpful to assign variable names that provide information about the initial direction of scoring. In the sample data set for the CES-D scores, items were initially named dep1 to dep20; however, reverse-worded items were initially named revdep4, revdep8, revdep12, and revdep16; the inclusion of "rev" in these variable names reminds us that initially these items were scored in a direction reverse from the other 16 items and that these items must be recoded so that the direction of scoring is consistent with other items before summing scores across items to create a total depression score. Methods for recoding are discussed in Section 13.6.5.

13.6.4 Factor Analysis to Assess Dimensionality of a Set of Items

In Chapter 12, exploratory factor analysis was presented as a way to examine the pattern of correlations among a set of measured variables to assess whether the variables can reasonably be interpreted as measures of one (or several) latent variables or constructs. FA is often used in development of multiple-item measures; FA can help us to decide how many different latent variables or dimensions may be needed to account for the pattern of correlations among test items and to characterize these dimensions by examining the groups of items that have high correlations with each factor. The following analyses use the data that appear in depressionitems.sav; these are scores on Items 1 through 5 from the CES-D (revdep4 has not yet been recoded to be scored in the same direction as the other four items in this set).

If all five items have large loadings or correlations with the same factor, it may make sense to interpret them all as measures of the same latent variable or underlying construct. On the other hand, if FA suggests that more than one factor is required to describe the pattern of correlations among items, the researcher should consider the possibility that the items may measure more than one construct or that some items should not be included in the scale. Principal-axis factoring and the user-determined decision to retain one factor were used. Only one factor was retained because, in theory, all five of these items are supposed to be indicators of the same latent variable, depression. (In addition, factor scores were saved for use later in this chapter; the variable name for the saved factor score is "REGR factor score 1 for analysis 1.") The Factor Analysis dialog box for this analysis appears in Figure 13.6. Selected output from this FA appears in Figure 13.7.

Figure 13.6 Factor Analysis of Items 1 to 5 on the CES-D

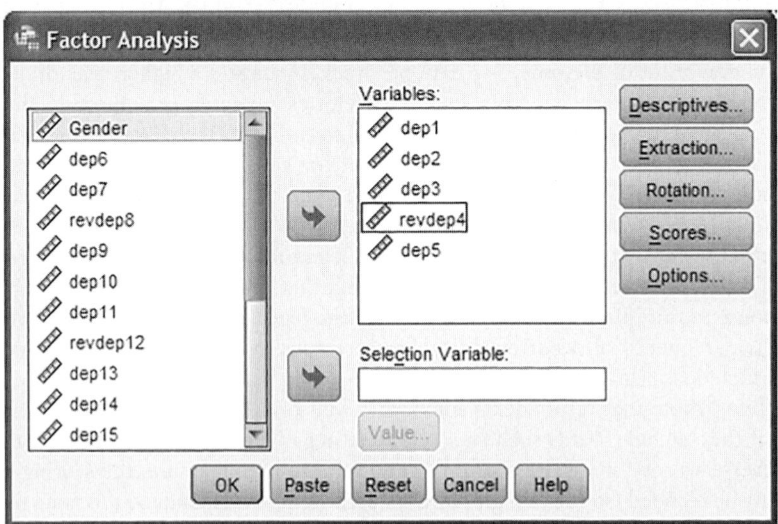

Note: This analysis uses revdep4, the original reverse-worded item. See Appendix 13A for a complete list of all 20 CES-D items.

Figure 13.7 SPSS Output From Principal-Axis Factor Analysis of Items 1 to 5 on the CES-D

Communalities

	Initial	Extraction
dep1	.341	.456
dep2	.195	.247
dep3	.434	.714
revdep4	.237	.227
dep5	.055	.005

Extraction Method: Principal Axis Factoring.

Total Variance Explained

	Initial Eigenvalues			Extraction Sums of Squared Loadings		
Factor	Total	% of Variance	Cumulative %	Total	% of Variance	Cumulative %
1	2.170	43.400	43.400	1.649	32.982	32.982
2	1.089	21.775	65.175			
3	.737	14.743	79.918			
4	.605	12.109	92.026			
5	.399	7.974	100.000			

Extraction Method: Principal Axis Factoring.

Factor Matrix[a]

	Factor
	1
dep1	.676
dep2	.497
dep3	.845
revdep4	−.476
dep5	.070

Extraction Method: Principal Axis Factoring.
a. 1 factors extracted. 14 iterations required.

Factor Score Coefficient Matrix

	Factor
	1
dep1	.258
dep2	.140
dep3	.605
revdep4	−.097
dep5	−.014

Extraction Method: Principal Axis Factoring.
Factors Scores Method: Regression.

The first factor accounted for approximately 43% of the variance (in the initial solution that included all five factors); after the model was reduced to one factor, the one retained factor accounted for approximately 33% of the variance. If we adopt an arbitrary cutoff of .40 in absolute value as the criterion for a "large" factor loading, then items dep1 to revdep4 all had large loadings on this single retained factor. For the items dep1 to dep3—"couldn't shake the blues," "bothered by things," and "poor appetite"—the loadings or correlations were large and positive. The reverse-worded item revdep4 ("I felt that I was just as good as other people") had a correlation of –.48 on this factor; this is as expected because a high score on this item corresponds to a low level of depression. The item dep5, "I had trouble concentrating," had a correlation with this retained factor that was close to 0. This suggests that dep5 may not measure the same latent variable as the other four depression items. In later analyses, we will see additional evidence that this item dep5, trouble concentrating, was not a "good" measure of depression; that is, it did not correlate highly with the other items in this set of questions about depression.

13.6.5 Recoding Scores for Reverse-Worded Items

Prior to recoding, the item named revdep4 is scored in the opposite direction from the other four CES-D items (i.e., high scores on dep1, dep2, dep3, and dep5 indicate more depression, while a high score on revdep4 indicates less depression). Before doing reliability analysis and summing scores for this set of items, the scores on the fourth depression item (revdep4) must be recoded so that they are in the same direction as the other items. It is helpful to create a new variable when making this change in direction of scoring. In the following example, revdep4 is the name of the SPSS variable in the original data set for the reverse-worded item "I felt that I was just as good as other people" (Question 4 on the CES-D). After changing the direction of scoring so that it is consistent with other items, the new name for this item is dep4.

One simple method to reverse the scoring on this item (so that a higher score corresponds to more depression) is as follows: Create a new variable (dep4) that corresponds to 3 – revdep4. In this case, the variable revdep4 has scores of 0, 1, 2, and 3; to reverse the direction of scoring, we need to apply a transformation that changes 0 to 3, 1 to 2, 2 to 1, and 3 to 0. (We could also use a recode command that explicitly assigns these new values.) In this example, it is easier to subtract each score from 3 (which is the highest possible score value for this item). To do this, use the SPSS menu selections <Transform> → <Compute Variable> to open the Compute Variable dialog box (see Figure 13.8). The new or "target" variable (dep4, scored in the same direction as the other items on the depression scale) is placed in the box labeled "Target Variable." The equation to compute a score for this new variable as a function of the score on an existing variable is placed in the box labeled "Numeric Expression" (in this case, the expression is 3 – revdep4).[2]

For a person with a score of 3 on revdep4, the score on the recoded version of the variable (dep4) will be 3 – 3, or 0. (There was one participant with a system missing code on revdep4; that person also was assigned a system missing code on dep4.) It is a good idea to paste the syntax for all recoding operations into an SPSS Syntax window and save the syntax so that you have a record of all the recoding operations that have been applied to the data. Figure 13.9 summarizes the SPSS syntax that reverses the direction of scoring on all of the reverse-worded CES-D items (for Questions 4, 8, 12, and 16).

13.6.6 Summing Scores Across Items to Compute a Total Score: Handling Missing Data

Assuming that FA indicates that all of the items seem to measure the same construct, and after any reverse-worded items have been recoded so that all items are scored in the same direction, it is possible to combine scores across items to form a total score (on depression,

Figure 13.8 Computing a Recoded Variable (dep4) From Reverse-Scored Item revdep4

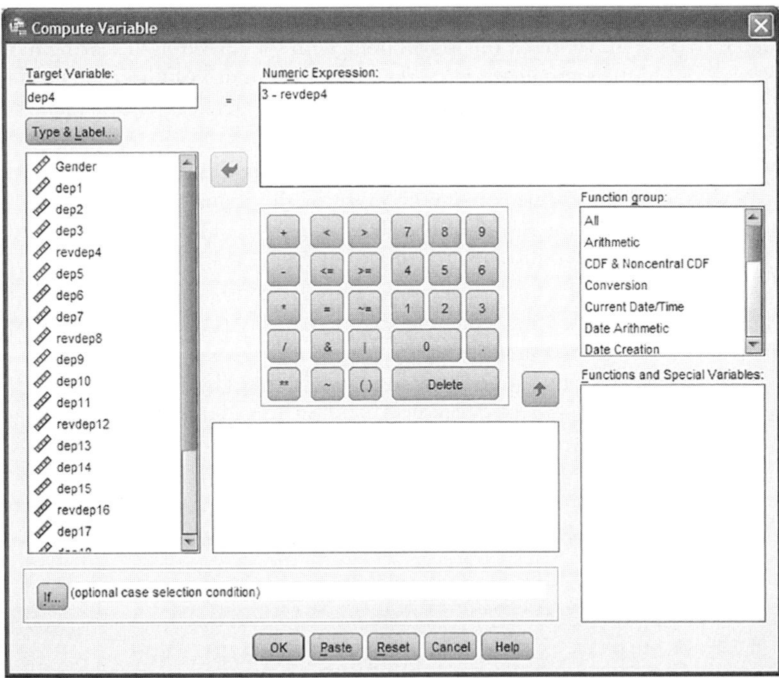

Figure 13.9 SPSS Syntax for Reverse-Scored Items

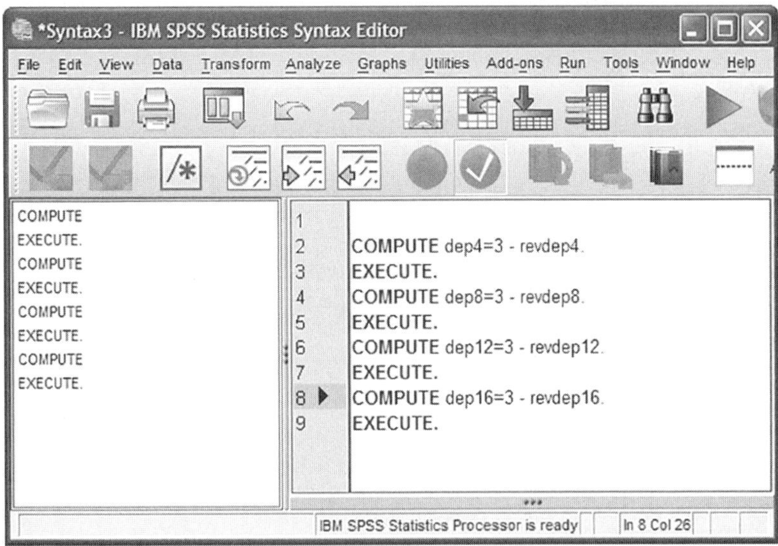

Note: These commands compute new items (dep4, dep8, dep12, dep16) that are recoded versions of the raw scores for reverse-worded items revdep4, revdep8, revdep12, and revdep16.

for example). Within SPSS, there are several ways to do this, and these methods differ in the way they handle missing data. If there are no missing scores, the three methods described in this section will yield identical results. When there are missing scores on any items, the data analyst needs to decide which of these three methods is most appropriate.

In the examples that follow, BriefCESD is the sum of the first five items on the CES-D, dep1 to dep5. Dep4 is the score on the fourth item, with the direction of scoring modified so that dep4 is scored in the same direction as the other four items. All three of the methods of combining scores use the SPSS compute variable procedure, accessed by the menu selections <Transform> → <Compute Variable>.

We can combine scores across items using the plus sign as shown in Figure 13.10. If a participant has a missing value for one or more of the items included in this sum, the participant's score on the sum BriefCESD will be a system missing code. In other words, a participant only obtains a valid score for the total BriefCESD if he or she has provided valid responses to all items included in the sum. This is a conservative approach to handling missing values, but if there are many missing scores, this may result in a situation where few participants have scores for the sum BriefCESD.

We can combine scores using the SUM function followed by a list of items in parentheses (items in the list are separated by commas), as shown in Figure 13.11. The value that is returned for each participant is the sum of scores across all nonmissing or valid items. This is not usually what data analysts want; the magnitude of this sum depends on the number of questions that are answered. There may be unusual situations in which this type of sum is what the data analyst wants, but usually this is not the preferred method.

Figure 13.10 Computation of Brief Five-Item Version of Depression Scale: Adding Scores Across Items Using Plus Signs

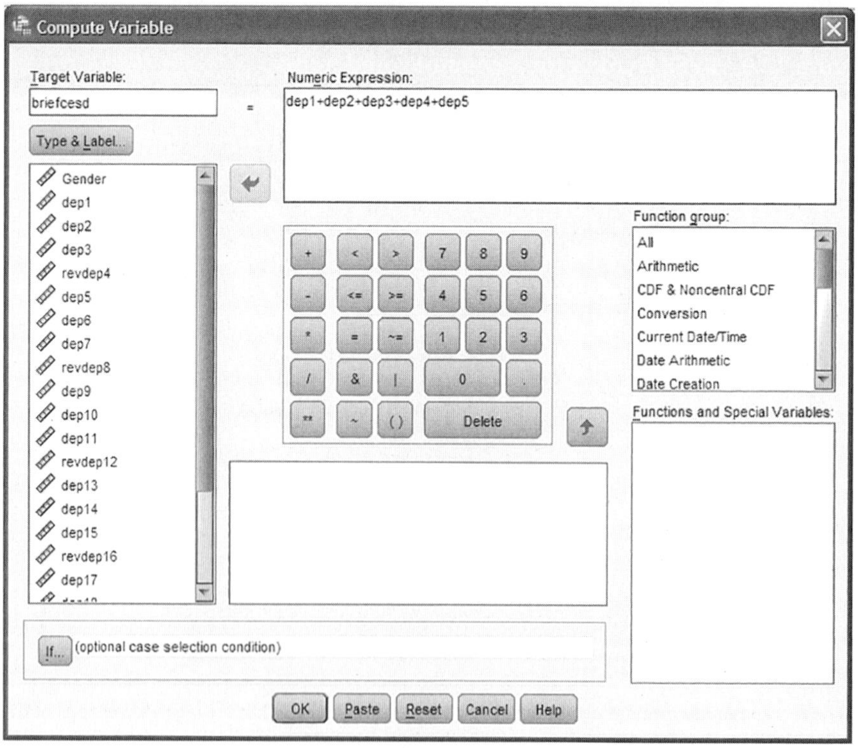

Figure 13.11 Computing the Sum of Scores Across Five Items Using the SPSS SUM Function

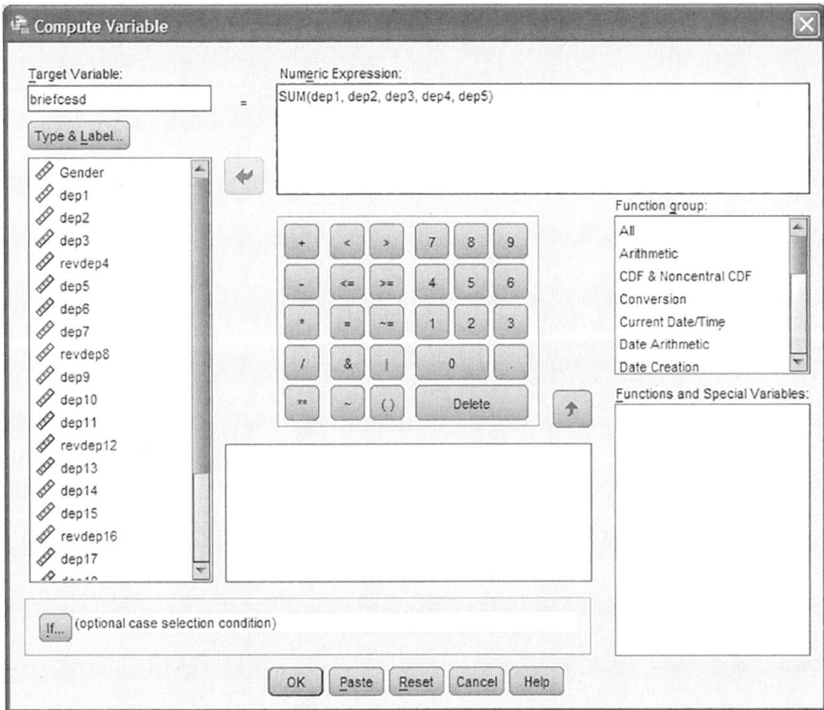

A third approach is to combine scores across items using the MEAN function followed by a list of items in parentheses; we may also want to multiply this mean by the number of items in the scale, as shown in Figure 13.12. The MEAN function returns the average across items for which a participant has provided valid responses; thus, a participant who has not answered some items still receives a score for the BriefCESD scale on the basis of all the items that were answered. Sometimes a mean across items is the form in which data analysts want to report scales (if items are all scored on a 1-to-5 degree of agreement scale, then the mean of responses across items tells us the average degree of agreement). In other situations, as with the CES-D, we may want to compare scores for people in our sample with normative data where the total score is defined as the sum across items; in this case, multiplying the mean response times the number of items gives us a score that is comparable with the test norms in terms of minimum and maximum possible scores.

Whatever method is used to add items, it is useful to provide information to readers that helps them interpret the value of the total score. Given the scoring procedures, what are the possible minimum and maximum scores? For the full 20-item CES-D, with each item scored on a 0-to-3 scale, the minimum possible score is 0 and the maximum possible score is 60 (i.e., 20 × 3); a higher score indicates more depression.

13.6.7 Comparison of Unit-Weighted Summed Scores Versus Saved Factor Scores

The previous section assumed that we do not attach different weights to items and that all items are given equal weight when combining scores across items. There are several ways to

Figure 13.12 Combining Scores From Five Items Using the SPSS MEAN Function (Multiplied by Number of Items)

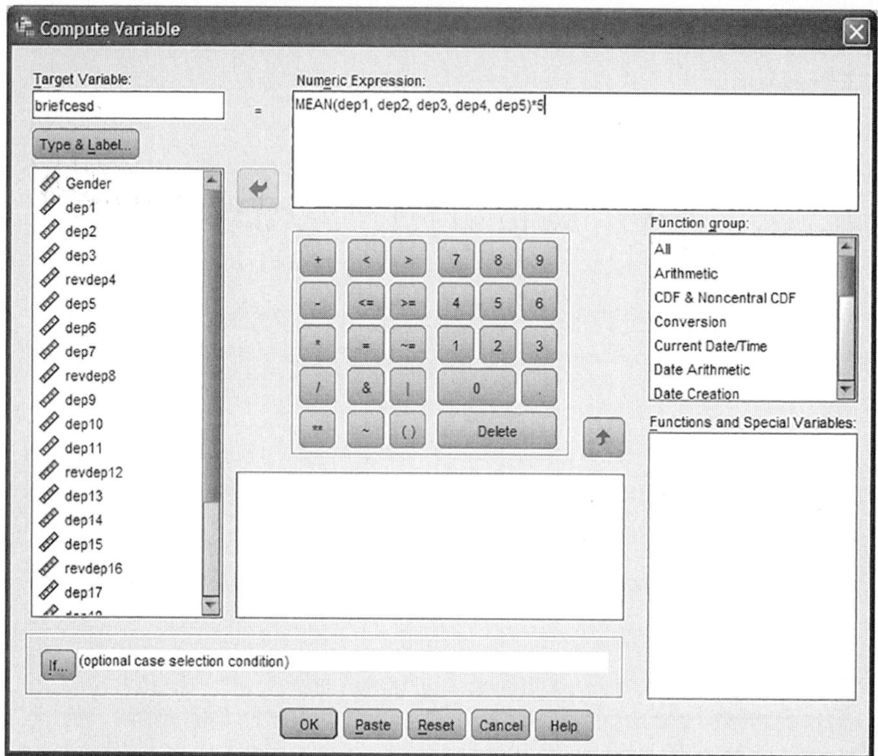

weight scores for individual items; three of these are discussed in this section. First, a total scale score can be created just by summing the raw scores across items, as in Section 13.6.5. Second, if the items that form a scale have different means and variances, then it may make more sense to sum z scores across items. Third, other analyses discussed earlier in this book, such as FA and discriminant analysis (DA), use empirically derived weights for each item or variable to form an "optimal" composite (e.g., saved factor scores from a FA or discriminant function scores).

13.6.7.1 Simple Unit-Weighted Sum of Raw Scores

We refer to a scale as "unit weighted" if we give each variable a weight of 1 (to indicate that it is included in a scale), 0 (to indicate that it is not included), and sometimes –1 (to indicate that it is included but scored in the opposite direction from other items in the sum). A simple unit-weighted combination of scores for dep1 to dep5 (dep4 is coded in the same direction as the other four items) is given by this expression:

$$\text{BriefCESD} = \text{dep1} + \text{dep2} + \text{dep3} + \text{dep4} + \text{dep5}. \tag{13.6}$$

13.6.7.2 Unit-Weighted Sum of *z* Scores

Summing raw scores may be reasonable when the items are all scored using the same response alternatives or measured in the same units. However, there are occasions when researchers want to combine information across variables that are measured in quite different units. Suppose a sociologist wants to create an overall index of socioeconomic status (SES) by combining information about the following measures: annual income in dollars, years of education, and occupational prestige rated on a scale from 0 to 100. If raw scores (in dollars, years, and points) were summed, the value of the total score would be dominated by the value of annual income. If we want to give these three factors (income, education, and occupational prestige) equal weight when we combine them, we can convert each variable to a z score or standard score and then form a unit-weighted composite of these z scores:

$$z_{\text{total}} = z_{X1} + z_{X2} + \cdots + z_{Xp}. \quad (13.7)$$

To create a composite of z scores on income, education, and occupational prestige to summarize information about SES, you could compute SES = $z_{\text{income}} + z_{\text{education}} + z_{\text{occupationalprestige}}$.

13.6.7.3 Saved Factor Scores or Other Optimally Weighted Linear Composites

You may have noticed by now that whenever we encounter the problem of combining information from multiple predictor variables, as in multiple regression (MR) or DA or from multiple outcome variables (as in multivariate analysis of variance [MANOVA]), we have handled it in the same way: by forming a "weighted linear composite" of scores on the X's. This composite may be a sum of weighted raw scores or of weighted z scores. For example, the goal in MR is to find the weighted linear composite $Y' = a + b_1X_1 + b_2X_2 + \cdots + b_kX_k$, such that the Y' values are as close as possible to the actual Y scores. The goal in DA is to find one (or several) discriminant function(s), $D_i = d_{i1}z_{x1} + d_{i2}z_{x2} + \cdots + d_{ip}z_{xp}$, such that the new variable D_i has the largest possible $SS_{\text{between groups}}$ and the smallest possible $SS_{\text{within groups}}$. The b coefficients in MR are often referred to as slopes because they describe slopes when the data are represented in the form of scatterplots; at this point, however, it is more helpful to think of coefficients such as b and d as "weights" that describe how much importance is given to scores on each predictor variable. Note also that the sign of the coefficient is important; some variables may be given positive weights and others negative weights when forming the optimally weighted linear composite.

In general, a multivariate analysis such as regression yields weights or coefficients that optimize prediction of some outcome (such as scores for a quantitative Y outcome variable in an MR or group membership in a DA). We might apply this predictive equation to new cases, but in general, we cannot expect that predictions made using the regression or discriminant function coefficients will continue to be optimal when we apply these coefficients to data in new studies.

In FA we create weighted linear composites of variables, called factor scores; these are created by summing z scores on all the variables using weights called factor score coefficients. The goal of FA is somewhat different from the goal of MR or DA. In MR and DA, the goal is to combine the scores on a set of X predictors in a manner that optimizes the prediction of some outcome: a continuous Y score in MR or group membership in DA. In FA, the goal is to obtain a small set of factor scores such that the scores on Factors 1, 2, and so forth are orthogonal to one another and in such a way that the X variables can be grouped into sets of variables that each has high correlations with one factor and are understood as measures of a small number of underlying factors or constructs. In other words, in FA, the correlations among X's are taken as a possible indication that the X's might measure the same construct, whereas in MR, correlations between X's and Y are usually interpreted to mean that X is useful as a predictor of Y, and we do not necessarily assume that the X's are measures of the same construct (although in some MR analyses, this might be the case).

Each of these multivariate methods (MR, DA, and FA) can calculate and save scores on the basis of the "optimal weighted linear combination" of variables in that analysis; for example, factor scores can be saved as part of the results of an FA. A saved factor score for the first, Factor F_1 (for the analysis in Figure 13.7), corresponds to a weighted sum of the z scores on the variables included in that FA (dep1, dep2, dep3, revdep4, dep5), where the weights are the factor score coefficients (f_{ij} is the factor score coefficient to create a score on Factor i using z scores on variable j):

$$F_1 = f_{11} \times z_{\text{dep1}} + f_{12} \times z_{\text{dep2}} + f_{13} \times z_{\text{dep3}} + f_{14} \times z_{\text{revdep4}} + f_{15} \times z_{\text{dep5}}. \quad (13.8)$$

Thus, another way to create a total score for a list of items is to save factor scores from an FA of the items (or to use weights derived from some other analysis, such as discriminant function analysis).

13.6.7.4 Correlation Between Unit-Weighted Sums Versus Factor Scores

How do we decide whether to use unit-weighted composites (sum of dep1 to dep5) or optimally weighted composites (such as saved factor scores for dep1 to dep5) to combine the information across multiple items?

A demonstration (not a formal proof) appears in Figure 13.13. Summary scores were computed for the first five items in the CES-D in three ways. BriefCESD is just the sum of dep1 to dep5. ZbriefCESD is the sum of the z scores on dep1 to dep5. Finally, "REGR factor score 1 for analysis 1" is the saved factor score from the FA of the first five CES-D items (reported in Figure 13.7). Figure 13.13 shows the sample correlations among these three versions of the scale; all correlations among these three different combinations of the first five items were above .90. In this example, it did not make much difference whether the total score was computed using raw scores or z scores or whether it was formed using unit weights or factor score coefficients. (However, if items have vastly different means and standard deviations, as in the previous example of an SES scale, use of z scores might yield different and more interpretable results than use of raw scores.)

In many research situations, it makes surprisingly little difference in the outcome whether the researcher uses weights derived from a multivariate analysis (such as MR, DA, MANOVA, or FA) or simply forms a unit-weighted composite of scores (Cohen, 1990; Fava & Velicer, 1992; Nunnally, 1978). As the title of a classic paper by Wainer (1976) suggests, in many situations, "It don't make no nevermind" whether researchers combine scores by just summing them or by forming weighted composites. Of course, a researcher should not assume that

Figure 13.13 Correlations Among Three Versions of the CES-D

Correlations

		zbriefcesd	briefcesd	REGR factor score 1 for analysis 1
zbriefcesd	Pearson Correlation	1	.943**	.915**
	Sig. (2-tailed)		.000	.000
	N	97	97	97
briefcesd	Pearson Correlation	.943**	1	.947**
	Sig. (2-tailed)	.000		.000
	N	97	97	97
REGR factor score 1 for analysis 1	Pearson Correlation	.915**	.947**	1
	Sig. (2-tailed)	.000	.000	
	N	97	97	97

**. Correlation is significant at the 0.01 level (2-tailed).

Note: zbriefcesd = $z_{\text{dep1}} + z_{\text{dep2}} + z_{\text{dep3}} + z_{\text{dep4}} + z_{\text{dep5}}$. briefcesd = dep1 + dep2 + dep3 + dep4 + dep5. REGR factor score 1 for analysis 1: saved score on Factor 1 from FA of five CES-D items in Figure 13.7.

these three methods of scoring always yield highly similar results. It is easy to obtain correlations among scores derived from different scoring methods to evaluate this, as shown in Figure 13.13. When all three methods of scoring yield essentially equivalent information, then it may be sufficient to provide other users of the multiple-item scale with the simplest possible scoring instructions (i.e., simply sum the raw scores across items, making sure that scores on any reverse-worded items are appropriately recoded).

13.6.7.5 Choice Among Methods of Combining Items

An advantage of using optimal weights (such as factor score coefficients) when combining scores is that the performance of the weighted linear composite will be optimized for the sample of participants in the study. For example, discriminant function scores will show the largest possible mean differences across groups in a study. However, a disadvantage of using optimal weights (given by discriminant function coefficients or factor score coefficients) is that they are "optimal" only for the sample of participants whose data were used to estimate those coefficients. Another disadvantage of using optimal weights derived from multivariate analyses such as FA is that this requires relatively cumbersome computations. If you want to create a new multiple-item scale that other people will use in their research, other researchers may prefer a scale that has simple scoring instructions (such as "sum the scores across all 20 items") to a scale that has complex scoring instructions (such as "convert each score into a *z* or standard score and use factor score coefficients").

The advantage of using unit-weighted composites as the scoring method is that unit-weighted scoring procedures are simpler to describe and apply in new situations. In many research situations, the scores that are obtained by applying optimal weights from a multivariate analysis are quite highly correlated with scores that are obtained using unit-weighted composites; when this is the case, it makes sense to prefer the unit-weighted composites because they provide essentially equivalent information and are easier to compute.

13.7 ASSESSMENT OF INTERNAL HOMOGENEITY FOR MULTIPLE-ITEM MEASURES: CRONBACH'S ALPHA RELIABILITY COEFFICIENT

The internal consistency reliability of a multiple-item scale tells us the degree to which the items on the scale measure the same thing. If the items on a test all measure the same underlying construct or variable, and if all items are scored in the same direction, then the correlations among all the items should be positive. If we perform an FA on a set of five items that are all supposed to measure the same latent construct, we would expect the solution to consist of one factor that has large correlations with all five items on the test.

13.7.1 Conceptual Basis of Cronbach's Alpha

We can summarize information about positive intercorrelations between the items on a multiple-item test by calculating Cronbach's alpha reliability. Cronbach's alpha has become the most popular form of reliability assessment for multiple-item scales. As seen in an earlier section, as we sum a larger number of items for each participant, the expected value of Σe_i approaches 0, while the value of $p \times T$ increases. In theory, as the number of items (p) included in a scale increases, assuming other characteristics of the data remain the same, the reliability of the measure (the size of the $p \times T$ component compared with the size of the Σe component) also increases. Cronbach's alpha provides a reliability coefficient that tells us, in theory, how reliable our estimate of the "stable" entity that we are trying to measure is, when we combine scores from p test items (or behaviors or ratings by judges). Cronbach's alpha (in effect) uses the mean of all the interitem correlations (for all pairs of items or measures) to assess the stability or consistency of measurement.

Cronbach's alpha can be understood as a generalization of the **Spearman-Brown prophecy formula**; we calculate the mean interitem correlation to assess the degree of agreement among individual test items, and then, we predict the reliability coefficient for a p-item test from the correlations among all these single-item measures. Another possible interpretation of Cronbach's alpha is that it is, essentially, the average of all possible split-half reliabilities. Here is one formula for Cronbach's α from Carmines and Zeller (1979, p. 44):

$$\alpha = \frac{p\bar{r}}{[1+\bar{r}(p-1)]}, \tag{13.9}$$

where p is the number of items on the test, and $\bar{r}$ is the mean of the interitem correlations. The size of Cronbach's alpha depends on the following two factors:

As p (the number of items included in the composite scale) increases, and assuming that stays the same, the value of Cronbach's alpha increases.

As (the mean of the correlations among items or measures) increases, assuming that the number of items p remains the same, the Cronbach's alpha increases.

It follows that we can increase the reliability of a scale by adding more items (but only if doing so does not decrease the mean interitem correlation) or by modifying items to increase (either by dropping items with low item-total correlations or by writing new items that correlate highly with existing items). There is a trade-off: If the interitem correlation is high, we may be able to construct a reasonably reliable scale with few items, and of course, a brief scale is less costly to use and less cumbersome to administer than a long scale. Note that all items must be scored in the same direction prior to summing. Items that are scored in the opposite direction relative to other items on the scale would have negative correlations with other items, and this would reduce the magnitude of the mean interitem correlation.

Researchers usually hope to be able to construct a reasonably reliable scale that does not have an excessively large number of items. Many published measures of attitudes or personality traits include between 4 and 20 items for each trait. Ability or achievement tests (such as IQ) may require much larger numbers of measurements to produce reliable results.

Note that when the items are all dichotomous (such as true/false), Cronbach's alpha may still be used to assess the homogeneity of response across items. In this situation, it is sometimes called a **Kuder-Richardson 20 (KR-20)** reliability coefficient. Cronbach's alpha is not appropriate for use with items that have categorical responses with more than two categories.

13.7.2 Empirical Example: Cronbach's Alpha for Five Selected CES-D Items

Ninety-eight students filled out the 20-item CES-D (items shown in Appendix 13A) as part of a survey. The names given to these 20 items in the SPSS data file that appears in depressionitems.sav were dep1 to dep20. As a reminder that some items were reverse worded, initial variable names for reverse-worded items were revdep4, revdep8, revdep12, and revdep16; it was necessary to recode the scores on these reverse-worded items. The recoded values were placed in variables with the names dep4, dep8, dep12, and dep16. The SPSS reliability procedure was used to assess the internal consistency reliability of their responses. The value of Cronbach's alpha is an index of the internal consistency reliability of the depression score formed by summing the first five items. In this first example, only the first five items (dep1,

Figure 13.14 SPSS Menu Selections for the Reliability Procedure

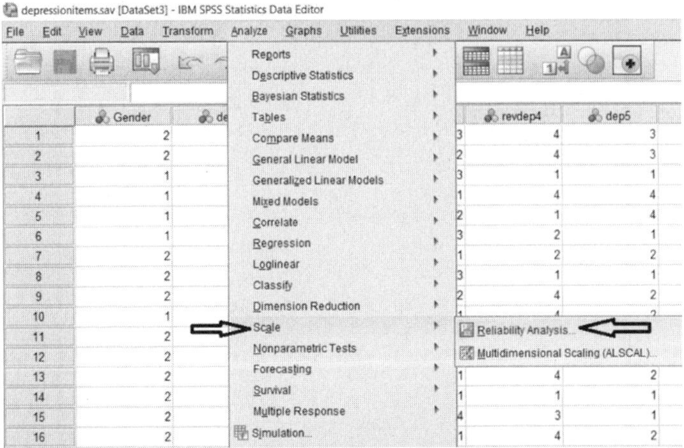

Figure 13.15 SPSS Reliability Analysis for Five CES-D Items: dep1, dep2, dep3, dep4, and dep5

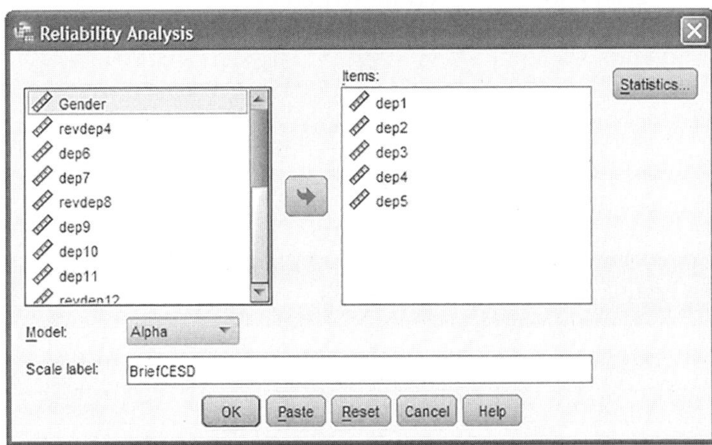

dep2, dep3, dep4, and dep5) were included. To run SPSS reliability, the following menu selections were made, starting from the top-level menu for the SPSS data worksheet (see Figure 13.14): <Analyze> → <Scale> → <Reliability Analysis>.

The Reliability Analysis dialog box appears in Figure 13.15. The names of the five items on the CES-D were moved into the variable list for this procedure. The Statistics button was clicked to request additional output; the Reliability Analysis: Statistics window appears in Figure 13.16. In this example, "Scale if item deleted" in the "Descriptives for" box and "Correlations" in the "Inter-Item" box were checked. The syntax for this procedure appears in Figure 13.17, and the output appears in Figure 13.18.

The "Reliability Statistics" panel in Figure 13.18 reports two versions of Cronbach's alpha for the scale that includes five items. For the sum dep1 + dep2 + dep3 + dep4 + dep5, Cronbach's alpha estimates the proportion of the variance in this total that is due to $p \times T$, the part of the score that is stable or consistent for each participant across all five items. A score can be formed by summing raw scores (the sum of dep1, dep2, dep3, dep4, and dep5), z scores,

Figure 13.16 Statistics Selected for SPSS Reliability Analysis

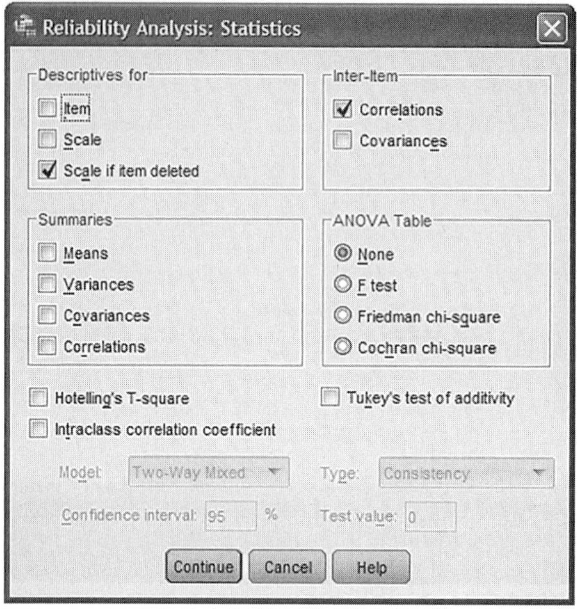

Figure 13.17 SPSS Syntax for Reliability Analysis

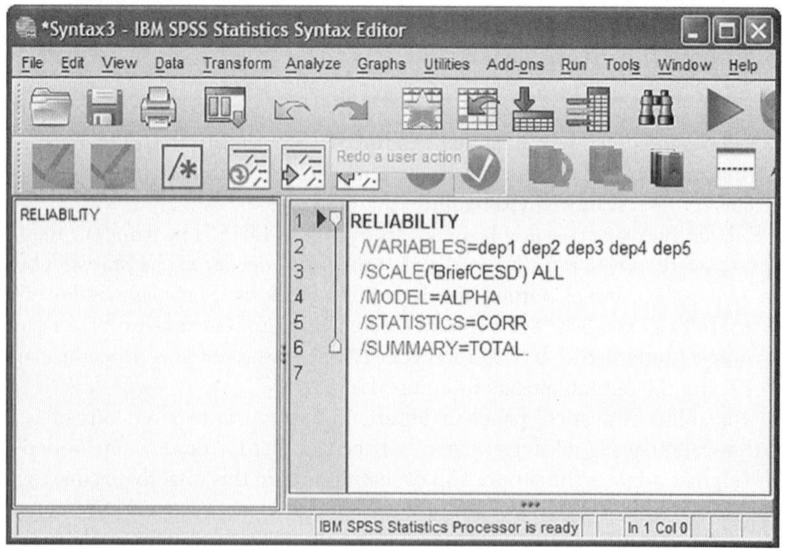

Figure 13.18 SPSS Output From the First Reliability Procedure for Scale: BriefCESD

Scale: CESD items 1 through 5

Reliability Statistics

Cronbach's Alpha	Cronbach's Alpha Based on Standardized Items	N of Items
.585	.614	5

Inter-Item Correlation Matrix

	bothered by things	did not feel like eating	could not shake off blues	trouble keeping mind on what doing	not as good as others
bothered by things	1.000	.380	.555	.062	.302
did not feel like eating	.380	1.000	.394	.074	.213
could not shake off blues	.555	.394	1.000	.115	.446
trouble keeping mind on what doing	.062	.074	.115	1.000	−.129
not as good as others	.302	.213	.446	−.129	1.000

Item-Total Statistics

	Scale Mean if Item Deleted	Scale Variance if Item Deleted	Corrected Item-Total Correlation	Squared Multiple Correlation	Cronbach's Alpha if Item Deleted
bothered by things	3.67	5.786	.511	.341	.455
did not feel like eating	3.70	5.941	.398	.195	.504
could not shake off blues	3.61	4.845	.615	.434	.365
trouble keeping mind on what doing	2.82	7.042	.032	.055	.703
not as good as others	3.47	5.710	.294	.237	.562

or standardized scores ($z_{dep1} + z_{dep2} + \cdots + z_{dep5}$). The first value, $\alpha = .59$, is the reliability for the scale formed by summing raw scores; the second value, $\alpha = .61$, is the reliability for the scale formed by summing z scores across items. In this example, these two versions of Cronbach's alpha (raw score and standardized score) are nearly identical. They generally differ from each other more when the items that are included in the sum are measured using different scales with different variances (as in the earlier example of an SES scale based on a sum of income, occupational prestige, and years of education).

Recall that Cronbach's alpha, like other reliability coefficients, can be interpreted as a proportion of variance. Approximately 60% of the variance in the total score for depression, which is obtained by summing the z scores on Items 1 through 5 from the CES-D, is shared across these five items. A Cronbach's α reliability coefficient of .61 would be considered unacceptably poor reliability in most research situations. Subsequent sections describe two different things researchers can do that may improve Cronbach's alpha reliability: deleting poor items or increasing the number of items.

A correlation matrix appears under the heading "Inter-Item Correlation Matrix." This reports the correlations between all possible pairs of items. If all items measure the same underlying construct, and if all items are scored in the same direction, then all the correlations in this matrix should be positive and reasonably large. Note that the same item that had a small loading on the depression factor in the preceding FA (trouble concentrating) also tended to have low or even negative correlations with the other four items. The "Item-Total Statistics" table shows how the statistics associated with the scale formed by summing all five items would change if each individual item were deleted from the scale. The corrected item-total correlation for each item is its correlation with the sum of the other four items in the scale; for example, for dep1, the correlation of dep1 with the "corrected total" (dep2 + dep3 + dep4 + dep5) is shown. This total is called "corrected" because the score for dep1 is not included when we assess how dep1 is related to the total. If an individual item is a "good" measure, then it should be strongly related to the sum of all other items in the scale; conversely, a low item-total correlation is evidence that an individual item does not seem to measure the same construct as other items in the scale. The item that has the lowest item-total correlation with the other items is, once again, the question about trouble concentrating. This low item-total correlation is yet another piece of evidence that this item does not seem to measure the "same thing" as the other four items in this scale.

The last column in the "Item-Total Statistics" table reports "Cronbach's alpha if item deleted"; that is, what is Cronbach's alpha for the scale if each individual item is deleted? For the item that corresponded to the trouble concentrating question, deletion of this item from the scale would increase Cronbach's α to .70. Sometimes the deletion of an item that has low correlations with other items on the scale results in an increase in α reliability. In this example, we can obtain slightly better reliability for the scale if we drop the item trouble concentrating, which tends to have small correlations with other items on this depression scale; the sum of the remaining four items has a Cronbach's α of .70, which represents slightly better reliability.

13.7.3 Improving Cronbach's Alpha by Dropping a "Poor" Item

The SPSS reliability procedure was performed on the reduced set of four items: dep1, dep2, dep3, and dep4. The output from this second reliability analysis (in Figure 13.19) shows that the reduced four-item scale had Cronbach's α reliabilities of .703 (for the sum of raw scores) and .712 (for the sum of z scores). A review of the column headed "Cronbach's Alpha if Item Deleted" in the new "Item-Total Statistics" table indicates that the reliability of the scale would become lower if any additional items were deleted from the scale. Thus, we have obtained slightly better reliability from the four-item version of the scale (Figure 13.19) than for a five-item version of the scale (Figure 13.18). The four-item scale had better reliability because the mean interitem correlation was higher after the item trouble concentrating was deleted.

13.7.4 Improving Cronbach's Alpha by Increasing the Number of Items

Other factors being equal, Cronbach's alpha reliability tends to increase as p, the number of items in the scale, increases. For example, we obtain a higher Cronbach's alpha when we use all 20 items in the full-length CES-D than when we examine just the first five items. The output from the SPSS reliability procedure for the full 20-item CES-D appears in Figure 13.20. For the full scale formed by summing scores across all 20 items, Cronbach's α was .88.

Figure 13.19 Output for the Second Reliability Analysis: Scale Reduced to Four Items

Scale: CESD Items 1 Through 4 (Item Dep5 Omitted)

Reliability Statistics

Cronbach's Alpha	Cronbach's Alpha Based on Standardized Items	N of Items
.703	.712	4

Inter-Item Covariance Matrix

	bothered by things	did not feel like eating	could not shake off blues	not as good as others
bothered by things	.605	.250	.408	.247
did not feel like eating	.250	.718	.316	.190
could not shake off blues	.408	.316	.895	.444
not as good as others	.247	.190	.444	1.111

Item-Total Statistics

	Scale Mean if Item Deleted	Scale Variance if Item Deleted	Corrected Item-Total Correlation	Squared Multiple Correlation	Cronbach's Alpha if Item Deleted
bothered by things	2.18	4.625	.541	.341	.617
did not feel like eating	2.21	4.811	.407	.194	.686
could not shake off blues	2.11	3.810	.633	.421	.542
not as good as others	1.98	4.166	.410	.204	.702

Note: dep5, trouble concentrating, has been dropped.

Figure 13.20 SPSS Output: Cronbach's Alpha Reliability for the 20-Item CES-D

Case Processing Summary

		N	%
Cases	Valid	94	95.9
	Excluded[a]	4	4.1
	Total	98	100.0

a. Listwise deletion based on all variables in the procedure.

Reliability Statistics

Cronbach's Alpha	N of Items
.880	20

13.7.5 Other Methods of Reliability Assessment for Multiple-Item Measures

13.7.5.1 Split-Half Reliability

A **split-half reliability** for a scale with p items is obtained by dividing the items into two sets (each with $p/2$ items). This can be done randomly or systematically; for example, the first set might consist of odd-numbered items and the second set might consist of even-numbered items. Separate scores are obtained for the sum of the Set 1 items (X_1) and the sum of the Set 2 items (X_2), and Pearson's r (r_{12}) is calculated between X_1 and X_2. However, this r_{12} correlation between X_1 and X_2 is the reliability for a test with only $p/2$ items; if we want to know the reliability for the full test that consists of twice as many items (all p items, in this example), we can "predict" the reliability of the longer test using the Spearman-Brown prophecy formula (Carmines & Zeller, 1979):

$$r_{XX} = \frac{2 \times r_{12}}{1 + r_{12}}, \tag{13.10}$$

where r_{12} is the correlation between the scores on the basis of split-half versions of the test (each with $p/2$ items), and r_{XX} is the reliability for a score on the basis of all p items. Depending on the way in which items are divided into sets, the value of the split-half reliability can vary.

13.7.5.2 Parallel-Forms Reliability

Sometimes it is desirable to have two versions of a test that include different questions but that yield comparable information; these are called parallel forms. Parallel forms of a test, such as the Eysenck Personality Inventory, are often designated Form A and Form B. Parallel forms are particularly useful in repeated-measures studies where we would like to test some ability or attitude on two occasions but want to avoid repeating the same questions. **Parallel-forms reliability** is similar to split-half reliability, except that when parallel forms are developed, more attention is paid to matching items so that the two forms contain similar types of questions. For example, consider Eysenck's Extraversion scale. Both Form A and Form B include similar numbers of items that assess each aspect of extraversion—for instance, enjoyment of social gatherings, comfort in talking with strangers, sensation seeking, and so forth. Computing Pearson's r between scores on Form A and Form B is a typical way of assessing reliability; in addition, however, a researcher wants scores on Form A and Form B to yield the same means, variances, and so forth, so these should also be assessed.

13.8 VALIDITY ASSESSMENT

The validity of a measurement essentially refers to whether the measurement really measures what it purports to measure. In psychological and educational measurement, the degree to which scores on a measure correspond to the underlying construct that the measure is supposed to assess is called **construct validity**. (Some textbooks used to list construct validity as one of several types of measurement validity; in recent years, many authors have used the term *construct validity* to subsume all the forms of validity assessment described below.)

For some types of measurement (such as direct measurements of simple physical characteristics), validity is reasonably self-evident. If a researcher uses a tape measure to obtain information about people's heights (whether the measurements are reported in centimeters, inches, feet, or other units), he or she does not need to go to great lengths to persuade readers that this type of measurement is valid. However, there are many situations where the characteristic of interest is not directly observable, and researchers can only obtain indirect information about it. For example, we cannot directly observe intelligence (or depression), but we may infer that a person is intelligent (or depressed) if he or she gives certain types of responses to large numbers of questions that researchers agree are diagnostic of intelligence (or depression). A similar problem arises in medicine, for example, in the assessment of blood pressure. Arterial blood pressure could be measured directly by shunting the blood flow out of the person's artery through a pressure measurement system, but this procedure is invasive (and generally, less invasive measures are preferred). The commonly used method of blood pressure assessment uses an arm cuff; the cuff is inflated until the pressure in the cuff is high enough to occlude the blood flow; a human listener (or a microphone attached to a computerized system) listens for sounds in the brachial artery while the cuff is deflated. At the point when the sounds of blood flow are detectable (the Korotkoff sounds), the pressure on the arm cuff is read, and this number is used as the index of systolic blood pressure, that is, the blood pressure at the point in the cardiac cycle when the heart is pumping blood into the artery. The point of this example is that this common blood pressure measurement method is quite indirect; research had to be done to establish that measurements taken in this manner were highly correlated with measurements obtained more directly by shunting blood from a major artery into a pressure detection system. Similarly, it is possible to take satellite photographs and use the colors in these images to make inferences about the type of vegetation on the ground, but it is necessary to do validity studies to demonstrate that the type of vegetation that is identified using satellite images corresponds to the type of vegetation that is seen when direct observations are made at ground level.

As these examples illustrate, it is quite common in many fields (such as psychology, medicine, and natural resources) for researchers to use rather indirect assessment methods, either because the variable in question cannot be directly observed or because direct observation would be too invasive or too costly.

In cases such as these, whether the measurements are made through self-report questionnaires, by human observers, or by automated systems, validity cannot be assumed; we need to obtain evidence to show that measurements are valid.

For self-report questionnaire measurements, two types of evidence are used to assess validity. One type of evidence concerns the content of the questionnaire (**content validity** or **face validity**); the other type of evidence involves correlations of scores on the questionnaire with other variables (criterion-oriented validity).

13.8.1 Content and Face Validity

Both content and face validity are concerned with the content of the test or survey items. Content validity involves the question whether test items represent all theoretical dimensions or content areas. For example, if depression is theoretically defined to include low self-esteem, feelings of hopelessness, thoughts of suicide, lack of pleasure, and physical symptoms of fatigue, then a content-valid test of depression should include items that assess all these symptoms. Content validity may be assessed by mapping out the test contents in a systematic way and matching them to elements of a theory or by having expert judges decide whether the content coverage is complete.

A related issue is whether the instrument has *face validity*; that is, does it appear to measure what it says it measures? Face validity is sometimes desirable, when it is helpful for test takers to be able to see the relevance of the measurements to their concerns, as in some evaluation research studies where participants need to feel that their concerns are being taken into account.

If a test is an assessment of knowledge (e.g., knowledge about dietary guidelines for blood glucose management for diabetic patients), then content validity is crucial. Test questions should be systematically chosen so that they provide reasonably complete coverage of the information (e.g., What are the desirable goals for the proportions and amounts of carbohydrate, protein, and fat in each meal? When blood sugar is tested before and after meals, what ranges of values would be considered normal?).

When a psychological test is intended for use as a clinical diagnosis (of depression, for instance), clinical source books such as the fourth edition of the *Diagnostic and Statistical Manual of Mental Disorders* (American Psychiatric Association, 2013) might be used to guide item selection, to ensure that all relevant facets of depression are covered. More generally, a well-developed theory (about ability, personality, mood, or whatever else is being measured) can help a researcher map out the domain of behaviors, beliefs, or feelings that questions should cover to have a content-valid and comprehensive measure.

However, sometimes, it is important that test takers should *not* be able to guess the purpose of the assessment, particularly in situations where participants might be motivated to "fake good," "fake bad," lie, or give deceptive responses. There are two types of psychological tests that (intentionally) do not have high face validity: **projective tests** and empirically keyed objective tests. One well-known example of a projective test is the Rorschach test, in which people are asked to say what they see when they look at ink blots; a diagnosis of psychopathology is made if responses are bizarre. Another is the thematic apperception test, in which people are asked to tell stories in response to ambiguous pictures; these stories are scored for themes such as need for achievement and need for affiliation. In projective tests, it is usually not obvious to participants what motives are being assessed, and because of this, test takers should not be able to engage in impression management or faking. Thus, projective tests intentionally have low face validity.

Some widely used psychological tests were constructed using **empirical keying** methods; that is, test items were chosen because the responses to those questions were empirically related to a psychiatric diagnosis (such as depression), even though the question did not appear to have anything to do with depression. For example, persons diagnosed with depression tend to respond "false" to the Minnesota Multiphasic Personality Inventory (MMPI) item "I sometimes tease animals"; this item was included in the MMPI depression scale because the response was (weakly) empirically related to a diagnosis of depression, although the item does not appear face valid as a question about depression (Wiggins, 1973).

Face validity can be problematic; people do not always agree about what underlying characteristic(s) a test question measures. Gergen, Hepburn, and Fisher (1986) demonstrated that when items taken from one psychological test (the Rotter Internal/External Locus of Control scale) were presented to people out of context and people were asked to say what trait they thought the questions assessed, they generated a wide variety of responses.

13.8.2 Criterion-Oriented Validity

Content validity and face validity are assessed by looking at the items or questions on a test to see what material it contains and what the questions appear to measure. Criterion-oriented validity is assessed by examining correlations of scores on the test with scores on other variables that should be related to it if the test really measures what it purports to measure. If the CES-D really is a valid measure of depression, for example, scores on this scale should be correlated with scores on other existing measures of depression that are thought to be valid, and they should predict behaviors that are known or theorized to be associated with depression.

13.8.2.1 Convergent Validity

Convergent validity is assessed by checking to see if scores on a new test of some characteristic X correlate highly with scores on existing tests that are believed to be valid measures of that same characteristic. For example, do scores on a new brief IQ test correlate highly with scores on well-established IQ tests such as the Wechsler Adult Intelligence Scale or the Stanford-Binet? Are scores on the CES-D closely related to scores on other depression measures such as the BDI? If a new measure of a construct has reasonably high correlations with existing measures that are generally viewed as valid, this is evidence of *convergent validity*.

13.8.2.2 Discriminant Validity

Equally important, scores on X should *not* correlate with things the test is not supposed to measure (**discriminant validity**). For instance, researchers sometimes try to demonstrate that scores on a new test are *not* contaminated by social desirability bias by showing that these scores are not significantly correlated with scores on the Crown-Marlowe Social Desirability scale or other measures of social desirability bias.

13.8.2.3 Concurrent Validity

As the name suggests, concurrent validity is evaluated by obtaining correlations between scores on the test with current behaviors or current group memberships. For example, if persons who are currently clinically diagnosed with depression have higher mean scores on the CES-D than persons who are not currently diagnosed with depression, this would be one type of evidence for concurrent validity.

13.8.2.4 Predictive Validity

Another way of assessing validity is to ask whether scores on the test predict future behaviors or group membership. For example, are scores on the CES-D higher for persons who later commit suicide than for people who do not commit suicide?

13.8.3 Construct Validity: Summary

Many types of evidence (including content, convergent, discriminant, concurrent, and predictive validity) may be required to establish that a measure has strong construct validity, that is, that it really measures what the test developer says it measures, and it predicts the behaviors and group memberships that it should be able to predict. Westen and Rosenthal (2003) suggested that researchers should compare a matrix of obtained validity coefficients or correlations with a target matrix of predicted correlations and compute a summary statistic to describe how well the observed pattern of correlations matches the predicted pattern. This provides a way of quantifying information about construct validity on the basis of many different kinds of evidence.

Although the preceding examples have used psychological tests, validity questions certainly arise in other domains of measurement. For example, referring to the example discussed earlier, when the colors in satellite images are used to make inferences about the types and amounts of vegetation on the ground, are those inferences correct? Indirect assessments are sometimes used because they are less invasive (e.g., as discussed earlier, it is less invasive to use an inflatable arm cuff to measure blood pressure) and sometimes because they are less expensive (broad geographical regions can be surveyed more quickly by taking satellite photographs than by having observers on the ground). Whenever indirect methods of assessment are used, validity assessment is required.

Multiple-item assessments of some variables (such as depression) may be useful or even necessary to achieve validity as well as reliability. How can we best combine information from multiple measures? This brings us back to a theme that has arisen repeatedly throughout the book; that is, we can often summarize the information in a set of p variables or items by creating a weighted linear composite or, sometimes, just a unit weight sum of scores for the set of p variables.

13.9 TYPICAL SCALE DEVELOPMENT PROCESS

If an existing multiple-item measure is available for the variable of interest, such as depression, it is usually preferable to use an existing measure for which we have good evidence about reliability and validity. However, occasionally, a researcher would like to develop a measure for some construct that has not been measured before or develop a different way of measuring a construct for which the existing tests are flawed. An outline of a typical research process for scale development appears in Figure 13.21. In this section, the steps included in this diagram are discussed briefly. Although the examples provided involve self-report questionnaire data, comparable issues are involved in combining physiological measures or observational data.

13.9.1 Generating and Modifying the Pool of Items or Measures

When a researcher sets out to develop a measure for a new construct (for which there are no existing measures) or a different measure in a research domain where other measures have been developed, the first step is the generation of a pool of "candidate" items. There are many ways in which this can be done. For example, to develop a set of self-report items to measure "Machiavellianism" (a cynical, manipulative attitude toward people), Christie and Geis (1970) drew on the writings of Machiavelli for some items (and also on statements by P. T. Barnum,

Figure 13.21 Possible Steps in the Development of a Multiple-Item Scale

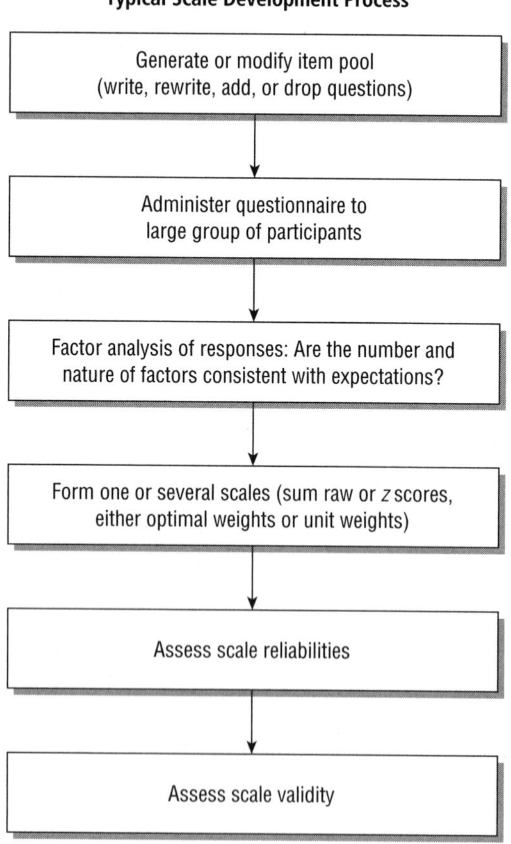

another notable cynic). To develop measures of love, Rubin (1970) drew on writings about love that ranged from the works of classic poets to the lyrics of popular songs. In some cases, items are borrowed from existing measures; for example, a number of research scales have used items that are part of the MMPI. However, there are copyright restrictions on the use of items that are part of **published tests**.

Brainstorming by experts, as well as interviews, focus groups, or open-ended questions with members of the population of interest, can also provide useful ideas about items. For example, to develop a measure of college student life space, including numbers and types of material possessions, Brackett (2004) interviewed student informants, visited dormitory rooms, and examined merchandise catalogs popular in that age group.

A theory can be extremely helpful as guidance in initial item development. The early interview and self-report measures of the global Type A behavior pattern drew on a developing theory that suggested that persons prone to cardiovascular disease tend to be competitive, time urgent, job-involved, and hostile. The behaviors that were identified for coding in the interview thus included interrupting the interviewer and loud or explosive speech. The self-report items on the Jenkins Activity Survey, a self-report measure of global Type A behavior, included questions about eating fast, never having time to get a haircut, and being unwilling to lose in games even when playing checkers with a child (Jenkins, Zyzanski, & Rosenman, 1979).

It is useful for the researcher to try to anticipate the factors that will emerge when these items are pretested and FA is performed. If a researcher wants to measure satisfaction with health care and believes that there are three separate components to satisfaction (evaluation of practitioner competence, satisfaction with rapport or "bedside manner," and issues of cost and convenience), then he or she should pause and evaluate whether the survey includes sufficient items to measure each of these three components. Keeping in mind that a minimum of 4 to 5 items are generally desired for each factor or scale and that not all candidate items may turn out to be good measures, it may be helpful to have something like 8 or 10 candidate items that correspond to each construct or factor that the researcher wants to measure.

13.9.2 Administer the Survey to Participants

The survey containing all the candidate items should be pilot tested on a relatively small sample of participants; it may be desirable to interview or debrief participants to find out whether items seemed clear and plausible and whether response alternatives covered all the options people might want to report. A pilot test can also help the researcher judge how long it will take for participants to complete the survey. After making any changes judged necessary on the basis of the initial pilot tests, the survey should be administered to a sample that is large enough to be used for FA. Ideally, these participants should vary substantially on the characteristics that the scales are supposed to measure (because a restricted range of scores on T, the component of the X measures that taps stable individual differences among participants, will lead to lower interitem correlations and lower scale reliabilities).

13.9.3 Factor-Analyze Items to Assess the Number and Nature of Latent Variables or Constructs

FA can be performed on the scores for all items. If the number of factors that are obtained and the nature of the factors (i.e., the groups of variables that have high loadings on each factor) are consistent with the researcher's expectations, then the researcher may want to go ahead and form one scale that corresponds to each factor. If the FA does not turn out as expected—for example, if the number of factors is different from what was anticipated or if the pattern of variables that load on each factor is not as expected—the researcher needs to make a decision. If the researcher wants to make the FA more consistent with a priori theoretical constructs, he or she may need to go back to Step 1 to revise, add, and drop items. If the researcher sees patterns in the data that were not anticipated from theoretical evaluations (but the patterns make sense), he or she may want to use the empirical factor solution (instead of the original conceptual model) as a basis for grouping items into scales. Also, if a factor that was not anticipated emerges in the FA, but there are only a few items to represent that factor, the researcher may want to add or revise items to obtain a better set of questions for the new factor.

In practice, a researcher may have to go through these first three sets several times; that is, he or she may run FA, modify items, gather additional data, and run a new FA several times until the results of the FA are clear, and the factors correspond to meaningful groups of items that can be summed to form scales.

Note that some scales are developed on the basis of the predictive utility of items rather than on the factor structure; for these, DA (rather than FA) might be the data reduction method of choice. For example, items included in the Jenkins Activity Survey (Jenkins et al., 1979) were selected because they were useful predictors of a person having a future heart attack.

13.9.4 Development of Summated Scales

After FA (or DA), the researcher may want to form scales by combining scores on multiple measures or items. There are numerous options at this point.

1. One or several scales may be created (depending on whether the survey or test measures just one construct or several separate constructs).
2. Composition of scales (i.e., selection of items) may be dictated by conceptual grouping of items or by empirical groups of items that emerge from FA. In most scale development research, researchers hope that the items that are grouped to form scales can be justified both conceptually and empirically.
3. Scales may involve combining raw scores or standardized scores (z scores) on multiple items. Usually, if the variables use drastically different measurement units (as in the example above where an SES index was formed by combining income, years of education, and occupational prestige rating), z scores are used to ensure that each variable has equal importance.
4. Scales may be based on sums or means of scores across items.
5. Weights used to combine scores may be optimal weights (e.g., the factor score coefficients obtained through an FA) or unit weights (+1, 0, −1).

13.9.5 Assess Scale Reliability

At a minimum, the internal consistency of each scale is assessed, usually by obtaining Cronbach's alpha. Test-retest reliability should also be assessed if the construct is something that is expected to remain reasonably stable across time (such as a personality trait), but high test-retest reliability is not a requirement for measures of things that are expected to be unstable across time (such as moods).

13.9.6 Assess Scale Validity

If there are existing measures of the same theoretical construct, the researcher assesses convergent validity by checking to see whether scores on the new measure are reasonably highly correlated with scores on existing measures. If the researcher has defined the construct as something that should be independent of verbal ability or not influenced by social desirability, he or she should assess discriminant validity by making sure that correlations with measures of verbal ability and social desirability are close to 0. To assess concurrent and predictive validity, scores on the scale can be used to predict current or future group membership and current or future behaviors, which it should be able to predict. For example, scores on Zick Rubin's Love Scale (Rubin, 1970) were evaluated to see if they predicted self-rated likelihood that the relationship would lead to marriage and whether scores predicted which dating couples would split up and which ones would stay together within the year or two following the initial survey.

13.9.7 Iterative Process

At any point in this process, if results are not satisfactory, the researcher may "cycle back" to an earlier point in the process; for example, if the factors that emerge from FA are not clear or if internal consistency reliability of scales is low, the researcher may want to generate new items and collect more data. In addition, particularly for scales that will be used in clinical diagnosis or selection decisions, normative data are required; that is, the mean, variance, and distribution shape of scores must be evaluated on the basis of a large number of people (at least several thousand). This provides test users with a basis for evaluation. For example, for the BDI (Beck, Ward, Mendelson, Mock, & Erbaugh, 1961), the following interpretations for scores have been suggested on the basis of normative data for thousands of test takers: scores from 5 to 9, normal mood variations; 10 to 18, mild to moderate depression; 19 to 29, moderate to severe depression; and 30 to 63, severe depression. Scores of 4 or below on the BDI may be interpreted as possible denial of depression or faking good; it is very unusual for people to have scores that are this low on the BDI.

13.9.8 Create the Final Scale

When all the criteria for good quality measurement appear to be satisfied (i.e., the data analyst has obtained a reasonably brief list of items or measurements that appears to provide reliable and valid information about the construct of interest), a final version of the scale may be created. Often such scales are first published as tables or appendixes in journal articles. A complete report for a newly developed scale should include the instructions for the test respondents (e.g., what period of time should the test taker think about when reporting frequency of behaviors or feelings?); a complete list of items, statements, or questions; the specific response alternatives; indication whether any items need to be reverse coded; and scoring instructions. Usually, the scoring procedure consists of reversing the direction of scores for any reverse-worded items and then summing the raw scores across all items for each scale. If subsequent research provides additional evidence that the scale is reliable and valid, and if the scale measures something that has a reasonably wide application, at some point, the test author may copyright the test and perhaps have it distributed on a fee-per-use basis by a test publishing company. Of course, as years go by, the contents of some test items may become dated. Therefore, periodic revisions may be required to keep test item wording current.

13.10 A BRIEF NOTE ABOUT MODERN MEASUREMENT THEORIES

The preceding material in this chapter discusses scale construction and reliability from the perspective of classical test theory. A more modern psychometric approach is item response theory (IRT; Embretson & Reise, 2000). Although some of the basic concepts of IRT are relatively simple, the techniques are complex. IRT is now extensively used in commercial testing of ability and achievement, such as the SAT. The basic idea is that scores on each individual item can be assessed to evaluate their difficulty (on the SAT, what proportion of test takers choose the correct answer?) and the degree to which each item provides information about individual differences on the construct being measured. For example, items that are higher in difficulty may provide more information to differentiate among high-ability test takers, while items lower in difficulty may provide more useful information about differences in lower ability ranges. An item response function and item information function are mapped for each question or item on a test (see Morizot, Ainsworth, & Reise, 2007). IRT can be used to make better decisions about item inclusion and to make inferences about people's location on an underlying ability or trait dimension that are less dependent on the "difficulty" of items that happen to be included in a particular multiple-item measure. An application of IRT that is familiar to many students is the tailored testing approach in computer versions of the SAT (Wainer, 2000). When a student answers a question correctly, more difficult items are presented; if a student answers a question incorrectly, less difficult items are presented. While the use of IRT in the development of personality and attitude measures has been relatively limited, this method has great potential. For an introduction, see Morizot et al. (2007).

13.11 REPORTING RELIABILITY

When a journal article reports the development of a new scale, the process of scale development should be thoroughly documented. This type of report usually includes FA to assess dimensionality of items; test-retest reliability, in addition to internal homogeneity reliability; and assessment of validity. Often the research report includes a complete list of items, response alternatives, and scoring instructions, including information about reverse-worded items.

When researchers use a previously developed multiple-item measure such as the CES-D, they usually assess Cronbach's alpha reliability within their own sample. Reliability information is often reported in the "Methods" section of the journal article under "Measures" or "Materials." Sometimes Cronbach's alpha for each multiple-item measure is reported in a table along with other sample descriptive statistics (such as mean, standard deviation, minimum, and maximum).

13.12 SUMMARY

To summarize, measurements need to be reliable. When measurements are unreliable, it leads to two problems. Low reliability may imply that the measure is not valid (if a measure does not detect *anything* consistently, it does not make much sense to ask *what* it is measuring). In addition, when researchers conduct statistical analyses, such as correlations, to assess how scores on an X variable are related to scores on other variables, the relationship of X to other variables becomes weaker as the reliability of X becomes smaller. To put it more plainly, when a researcher has unreliable measures, relationships between variables usually appear to be weaker. It is also essential for measures to be valid: If a measure is not valid, then the study does not provide information about the theoretical constructs that are of real interest. It is also desirable for measures to be sensitive to individual differences, unbiased, relatively inexpensive, not very invasive, and not highly reactive.

Research methods textbooks point out that each type of measurement method (such as direct observation of behavior, self-report, physiological or physical measurements, and archival data) has strengths and weaknesses. For example, self-report is generally low cost, but such reports may be biased by social desirability (i.e., people report attitudes and behaviors that they believe are socially desirable, instead of honestly reporting their actual attitudes and behaviors). When it is possible to do so, a study can be made much stronger by including multiple types of measurements (this is called "triangulation" of measurement). For example, if a researcher wants to measure anxiety, it would be desirable to include direct observation of behavior (e.g., "um"s and "ah"s in speech and rapid blinking), self-report (answers to questions that ask about subjective anxiety), and physiological measures (such as heart rates and cortisol levels). If an experimental manipulation has similar effects on anxiety when it is assessed using behavioral, self-report, and physiological outcomes, the researcher can be more confident that the outcome of the study is not attributable to a methodological weakness associated with one form of measurement, such as self-report.

The development of a new measure can require a substantial amount of time and effort. It is relatively easy to demonstrate reliability for a new measurement, but the evaluation of validity is far more difficult and the validity of a measure can be a matter of controversy. When possible, researchers may prefer to use existing measures for which data on reliability and validity are already available.

For psychological testing, a useful online resource is the American Psychological Association FAQ on testing: www.apa.org/science/programs/testing/index.

Another useful resource is a directory of published research tests on the Educational Testing Service Test Link site, www.ets.org/test_link/about, which has information on about 20,000 published psychological tests.

Although most of the variables used as examples in this chapter were self-report measures, the issues discussed in this chapter (concerning reliability, validity, sensitivity, bias, cost-effectiveness, invasiveness, and reactivity) are relevant for other types of data, including physical measurements, medical tests, and observations of behavior.

APPENDIX 13A

The CES-D

INSTRUCTIONS: Using the scale below, please circle the number before each statement which best describes how often you felt or behaved this way DURING THE PAST WEEK.

0 = Rarely or none of the time (less than 1 day)

1 = Some or a little of the time (1–2 days)

2 = Occasionally or a moderate amount of time (3–4 days)

3 = Most of the time (5–7 days)

The total CES-D depression score is the sum of the scores on the following 20 questions, with Items 4, 8, 12, and 16 reverse scored.

1. I was bothered by things that usually don't bother me.
2. I did not feel like eating; my appetite was poor.
3. I felt that I could not shake off the blues even with help from my family or friends.
4. I felt that I was just as good as other people. [Reverse worded]
5. I had trouble keeping my mind on what I was doing.
6. I felt depressed.
7. I felt that everything I did was an effort.

8. I felt hopeful about the future. [Reverse worded]
9. I thought my life had been a failure.
10. I felt fearful.
11. My sleep was restless.
12. I was happy. [Reverse worded]
13. I talked less than usual.
14. I felt lonely.
15. People were unfriendly.
16. I enjoyed life. [Reverse worded]
17. I had crying spells.
18. I felt sad.
19. I felt that people dislike me.
20. I could not get "going."

A total score on CES-D is obtained by reversing the direction of scoring on the four reverse-worded items (4, 8, 12, and 16), so that a higher score on all items corresponds to a higher level of depression, and then summing the scores across all 20 items.

Source: Radloff (1977).

APPENDIX 13B

Web Resources on Psychological Measurement

American Psychological Association	www.apa.org/science/programs/testing/index
ETS Test Collection	www.ets.org/test_link/about
Goldberg's International Personality Item Pool: royalty-free versions of scales that measure "Big Five" personality traits	ipip.ori.org
Health and Psychosocial Instruments	Available through online access to databases at some universities
Mental Measurements Yearbook Test Reviews online	marketplace.unl.edu/buros/
PsychWeb information on psychological tests	https://www.psywww.com/resource/bytopic/testing.html
Robinson et al. (1991): a full list of scales is available in their book *Measures of Personality and Social Psychological Attitudes*	library.csun.edu/mstover/db/robinson.html

COMPREHENSION QUESTIONS

1. List and describe the most common methods of reliability assessment.
2. One way to interpret a reliability coefficient (often an r value) is as the proportion of variance in X scores that is stable across occasions of measurement. Why do we interpret r (rather than r^2) as the proportion of variance in X scores that is due to T, a component of the score that is stable across occasions of measurement? (See Figure 13.3.)
3. Explain the terms in this equation: $X = T + e$. You might use a specific example; for instance, suppose that X represents your score on a measure of life satisfaction and T represents your true level of life satisfaction.

   The next set of questions focuses on Cronbach's alpha, an index of internal consistency reliability.
4. How is α related to p, the number of items, when the mean interitem correlation is held constant?
5. How is α related to the mean interitem correlation when p, the number of items in the scale, is held constant?
6. Why is it important to make certain that all the items in a scale are scored in the same direction before you enter your variables into the reliability program to compute Cronbach's alpha?
7. Why are reverse-scored questions often included in surveys and tests? For example, in a hostility scale, answering yes to some questions might indicate greater hostility, while on other questions, a "no" answer might indicate hostility.
8. What is the Kuder-Richardson 20 (KR-20) statistic?
9. How can you use the item-total statistics from the reliability program to decide whether reliability could be improved by dropping some items?
10. If a scale has 0 reliability, can it be valid? If a scale has high reliability, such as $\alpha = .9$, does that necessarily mean that it is valid? Explain.
11. What is the difference between α and standardized α? Why does it make sense to sum z scores rather than raw scores across items in some situations?
12. What is an "optimally weighted linear composite"? What is a "unit-weighted linear composite"? What are the relative advantages and disadvantages of these two ways of forming scores?
13. Three different methods were presented for the computation of scale scores: summing raw scores on the X variables, summing z scores on the X variables, and saving factor scores from an FA of a set of X variables. (Assume that all the X variables are measures of a single latent construct.) How are the scores that are obtained using these three different methods of computation typically related?
14. What is meant by attenuation of correlation due to unreliability?
15. How can we correct for attenuation of correlation due to unreliability? Under what circumstances does correction for attenuation due to unreliability yield inaccurate results?

16. Discuss the kinds of information that you need to assess reliability of measurements.
17. Discuss the kinds of information that you need to assess the validity of a measure.
18. Why is it generally more difficult to establish the validity of a measure than its reliability?
19. In addition to reliability and validity, what other characteristics should good quality measurements have?

DATA ANALYSIS EXERCISE

Scale Construction and Internal Consistency Reliability Assessment

Select a set of at least four items that you think could form an internally consistent, unidimensional scale. Make sure that all items are scored in the same direction by creating reverse-scored versions of some items, if necessary. Create unit-weighted scores on your scale using the SPSS compute statement. Keep in mind that in some cases, it may make more sense to sum *z* scores than raw scores. Run the SPSS reliability procedure to assess the internal reliability or internal consistency of your scale. Also, run correlations between factor scores (created using the <Save> command) and a summated scale (created using the compute statement to add raw scores, *z* scores, or both).

On the basis of your findings, answer the following questions:

1. In terms of Cronbach's alpha, how reliable is your scale? Could the reliability of your scale be improved by dropping one or more items, and if so, which items?
2. How closely related are scores on the different versions of your scale, that is, factor scores saved from the FA versus scores created by summing raw scores and/or *z* scores?

NOTES

[1] There are a few multiple-item scales for which items are not expected to be correlated. For example, Holmes and Rahe (1967) created a scale to summarize the amount of adaptation that a person was required to make because of recent major life events such as death of a spouse, loss of a job, or having a child. Each event (such as death of a spouse) was assigned points on the basis of subjective evaluation of the amount of readjustment that event would typically require. A person's score on Holmes and Rahe's Social Readjustment Rating Scale is just the sum of the points for all of the events that person has experienced during the previous 6 months. Occurrences of these events (e.g., loss of a job, having a child) are probably relatively independent of one another; for this scale, we would not expect to find high correlations across items.

[2] You could use the same variable name for the item before and after recoding as in the example: dep4 = 3 – dep4; however, if you use the same variable name for both before and after reverse scoring, it is easy to lose track of whether reverse scoring has been applied. I recommend that you retain the raw score (in this example, revdep4) and create a new variable (dep4); this minimizes the possibility of confusion.

DIGITAL RESOURCES

Find **free study tools** to support your learning, including **eFlashcards, data sets, and web resources,** on the accompanying website at **edge.sagepub.com/warner3e**.

CHAPTER 14

MORE ABOUT REPEATED MEASURES

14.1 INTRODUCTION

This chapter assumes familiarity with simple analyses for repeated-measures or paired-samples data: the paired-samples *t* test and one-way repeated-measures analysis of variance (ANOVA). See Chapters 14 and 15 in Volume I (Warner, 2020). Techniques used in factorial ANOVA and multivariate ANOVA (MANOVA) provide additional ways to evaluate repeated-measures data, as discussed in this chapter. Multilevel modeling (Grimm & Ram, 2016) provides flexible ways to handle repeated-measures data that are not included in this discussion.

14.2 REVIEW OF ASSUMPTIONS FOR REPEATED-MEASURES ANOVA

One-way repeated-measures ANOVA requires the following conditions and assumptions.

1. The same variable (using the same measurement scale) is measured at each point in time. For this example, the repeated-measures variables are five measures of heart rate (hr), hr1 through hr5, so this assumption is satisfied.

2. Scores on these variables are quantitative. This assumption is satisfied; heart rate is a quantitative variable.

3. Scores are independent of one another (across participants). If participants had done the task in pairs, each person might have influenced the other's heart rate (through shared anxiety, for example). Participants did the study individually, so observations should be independent across participants. (Of course, the repeated measures hr1 through hr5 are dependent or correlated, i.e., hr1 will have a high positive correlation with hr2, hr3, and so forth; this form of nonindependence occurs because people tend to have consistently high or low heart rates across occasions of measurement; repeated-measures ANOVA takes this kind of nonindependence into account.)

4. Scores are at least approximately normally distributed (with no extreme outliers). This was assessed by obtaining histograms and boxplots for each of the variables hr1 through hr5 and could be assessed using other data-screening methods. SPSS commands and output for these were discussed earlier and are not included here. All variables had distribution shapes that were reasonably close to normal.

5. There were no extreme outliers. Using the sizes of z scores as a criterion, no cases were identified or removed as outliers.

6. Associations among variables hr1 through hr5 should be linear. Scatterplots for pairs of heart rate measures appeared reasonably linear with no bivariate outliers.

7. There should be no participant-by-time (or participant-by-treatment) interaction.

8. The variances for the contrasts1 (in this case, M2 – M1, M3 – M1, etc.) should not differ significantly; in other words, the sphericity assumption must not be violated. This chapter discusses MANOVA as a possible alternative analysis for situations where this assumption is violated.

14.3 FIRST EXAMPLE: HEART RATE AND SOCIAL STRESS

Data for examples in this chapter, in the file hrbpsocialstress.sav, are a subset of data collected in dissertation research conducted by Mooney (1990). She conducted a repeated-measures study to evaluate the effects of a stressful social role play on heart rate and blood pressure. This example focuses on heart rate measured at five points in time.

1. Participants ($N = 63$) came to the lab one at a time. A finger cuff was fitted to measure heart rate and systolic and diastolic blood pressure. A baseline measure of heart rate at Time 1 (hr1) was obtained.

2. Participants read and signed an informed consent form explaining the upcoming stressful social role play; hr2 was taken.

3. Participants mentally rehearsed what they would say (hr3).

4. Participants were video-recorded during the role play (hr4).

5. Ten minutes after the end of the role play, a final measure (hr5) was taken to evaluate whether heart rate had returned to baseline.

The analysis will be a one-way repeated-measures ANOVA; the within-*S* or repeated-measures factor is time (Times 1 through 5). The dependent variable is hr. (As an exercise, students can conduct the same one-way repeated-measures ANOVA using one of the other dependent variables, either systolic or diastolic blood pressure.)

14.4 TEST FOR PARTICIPANT-BY-TIME OR PARTICIPANT-BY-TREATMENT INTERACTION

Individual participants may differ in pattern of change in heart rate across levels of time or treatment situations. For example, some individuals may return to their baseline heart rates at Time 5, and others may not. If we graph lines to represent hr scores for each individual participant, and if there is a participant-by-time interaction, at least some of these lines would not be parallel. Another way to say this is that the lines for at least some individual participants would show a different pattern than the line that illustrates the average hr response pattern for the entire sample (look ahead to Figure 14.6).

Interaction is sometimes called "nonadditivity." Additivity exists when individual scores can be predicted from the main effects of two or more factors. Nonadditivity occurs when an additional term (an interaction term) is required to represent effects for specific combinations of levels of factors. A test for nonadditivity (interaction of participant with item or time or treatment) is available from the SPSS reliability procedure (this procedure is more often used to obtain Cronbach's α reliabilities for multiple item scales, but its repeated-measures ANOVA features are also useful). In this example, the items are measures of heart rate on five separate occasions.

To run the SPSS reliability, make the following menu selections: <Analyze> → <Scale> → <Reliability Analysis>. Within the main Reliability Analysis dialog box (Figure 14.1), enter the names of the repeated measures in the "Items" pane.

In the Reliability Analysis: Statistics dialog box, select Tukey's test of additivity. I also selected "Correlations" under "Inter-Item."

Figure 14.1 Main Dialog Box for SPSS Reliability Procedure

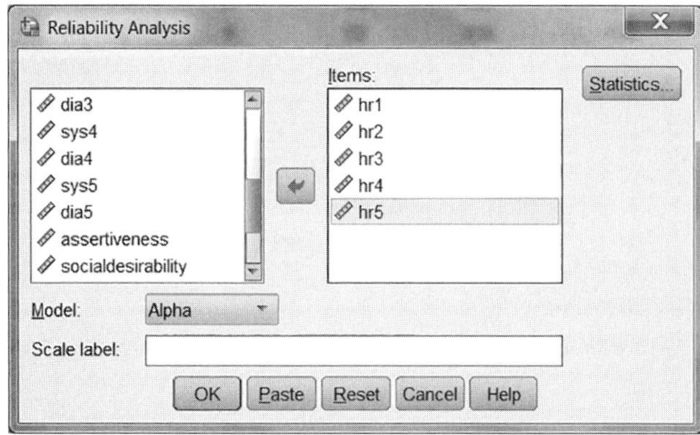

Figure 14.2 Reliability Analysis: Statistics Dialog Box for SPSS Reliability

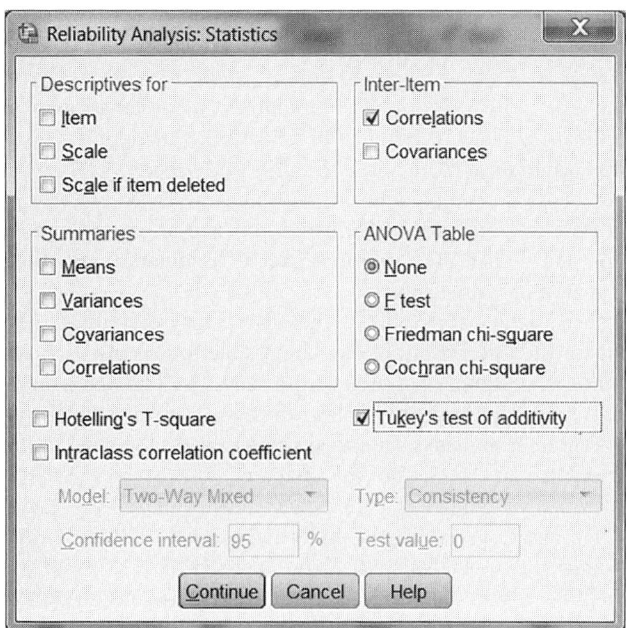

Figure 14.3 Correlations Among Variables hr1 to hr5

Inter-Item Correlation Matrix

	hr1	hr2	hr3	hr4	hr5
hr1	1.000	.717	.833	.688	.831
hr2	.717	1.000	.660	.617	.624
hr3	.833	.660	1.000	.794	.812
hr4	.688	.617	.794	1.000	.717
hr5	.831	.624	.812	.717	1.000

Figure 14.4 Friedman and Tukey Tests for Nonadditivity

ANOVA with Friedman's Test and Tukey's Test for Nonadditivity

			Sum of Squares	df	Mean Square	Friedman's Chi-Square	Sig
Between People			23337.987	62	376.419		
Within People	Between Items		6868.825	4	1717.206	61.375	.000
	Residual	Nonadditivity	274.849[a]	1	274.849	10.187	.002
		Balance	6663.926	247	26.979		
		Total	6938.775	248	27.979		
	Total		13807.600	252	54.792		
Total			37145.587	314	118.298		

Grand Mean = 80.65

a. Tukey's estimate of power to which observations must be raised to achieve additivity = -.874.

Correlations among hr1 to hr5 appear in Figure 14.3. As one might expect, there are high positive correlations among measures; participants whose heart rates were high on one occasion of measurement also tended to have relatively high heart rates on other occasions. In other words, participant heart rate was consistent across time (or across occasions of measurement).

This is not a trivial observation, by the way. For physiological measures such as heart rate, we cannot automatically assume that measurement results are consistent across time. It is important to assess reliability and validity for physiological or physical measures (not just self-report measures).

Nonadditivity (participant-by-time interaction) was statistically significant (see Figure 14.4), Friedman's $\chi^2(1) = 10.19$, $p = .002$. Tukey's estimate of the power to which the heart rate measures would need to be raised to achieve additivity was –.874.

Recall that the model for simple one-way repeated-measures ANOVA presented earlier did not include a term for this interaction; it did not take into account an interaction that we now know is present. When the "no Participant × Treatment interaction" assumption is violated, the data analyst has three choices.

First, if relevant variables are available, the analyst may examine these to see if any of them may account for the interaction. It may be possible to identify participant characteristics that predict different response patterns across the time. The pattern or profile of heart rate might vary as a function of cardiovascular fitness, or body mass index, or sex, for example. These kinds of interactions can be assessed using mixed-model between-*S* and within-*S* ANOVA, as discussed in a later section of this chapter.

Second, the data analyst might apply the power transformation suggested in the footnote to the SPSS output in Figure 14.4 (but only if this makes sense). In this example, the exponent was –.874. Let's round this to –1. If we take X to the power of –1, $X^{-1} = 1/X$. In this example, taking the inverse of heart rate (hr^{-1} or 1/hr) is a meaningful thing to do. Interbeat interval (in seconds) is the average time from one beat to the next. It is obtained by dividing 60 (the number of seconds per minute) by hr (the number or beats per minute). Interbeat interval is a common way to report cardiovascular data. (When hr scores were converted to interbeat interval, the nonadditivity test was no longer statistically significant; these results are not presented here.) However, for many dependent variables, the suggested power transformation might not yield interpretable results.

The least desirable option is to just mention that an interaction appears to be present and note that this should be investigated in further research.

Subsequent examples continue to use heart rate (rather than interbeat interval) because heart rate is more familiar.

14.5 ONE-WAY REPEATED-MEASURES RESULTS FOR HEART RATE AND SOCIAL STRESS DATA

This section briefly reviews one-way repeated measures applied to Mooney's (1990) data. This reviews ideas from the introduction to one-way repeated measures and adds discussion of the MANOVA results from the general linear model (GLM). A one-way repeated-measures ANOVA was performed to assess differences in mean heart rate from Times 1 through 5. "Repeated" contrasts were chosen for the time factor (to replace the default choice, polynomial contrasts). As always with the GLM, descriptive statistics including means were requested (these do not appear by default). Effect sizes and plots were also requested. The SPSS syntax is:

```
GLM hr1 hr2 hr3 hr4 hr5
/WSFACTOR=Time 5 Repeated
/METHOD=SSTYPE(3)
/PLOT=PROFILE(Time)
/EMMEANS=TABLES(OVERALL)
/EMMEANS=TABLES(Time)
/PRINT=ETASQ
/CRITERIA=ALPHA(.05)
/WSDESIGN=Time.
```

First look at the information about means for hr1 through hr5 across time (Figures 14.5 and 14.6).

Mean heart rate increased from baseline to consent, consent to rehearsal, and rehearsal to role play. Mean heart rate decreased after the role play. ANOVA will test whether the differences among means are statistically significant. To choose the most appropriate test, we need to know whether the data violated the sphericity assumption.

The sphericity assumption (Figure 14.7) was significantly violated for these data, $W = .496$, $\chi^2(9) = 42.34$, $p < .001$. The nature of this assumption is explained in the next section of this chapter. If the sphericity assumption is violated, downwardly adjusted degrees of freedom can be used to assess the statistical significance of F ratios in univariate repeated-measures ANOVA. (*Univariate repeated measures* refers to the computations that include *SS*, *MS*, and F ratios.)

The ε (epsilon) values on the right-hand side of the sphericity test results (Figure 14.7) index the degree to which data violate the sphericity assumption. The value of ε can be used as a multiplier to reduce the degrees of freedom for the univariate repeated-measures F ratio.

Figure 14.5 Means and Standard Deviations for Heart Rate Measured at Five Points in Time

Descriptive Statistics

	Mean	Std. Deviation	N
hr1	74.71	8.611	63
hr2	80.49	9.707	63
hr3	82.65	9.966	63
hr4	88.24	11.871	63
hr5	77.16	8.928	63

Figure 14.6 Graph of Mean Heart Rate During Social Stress Study

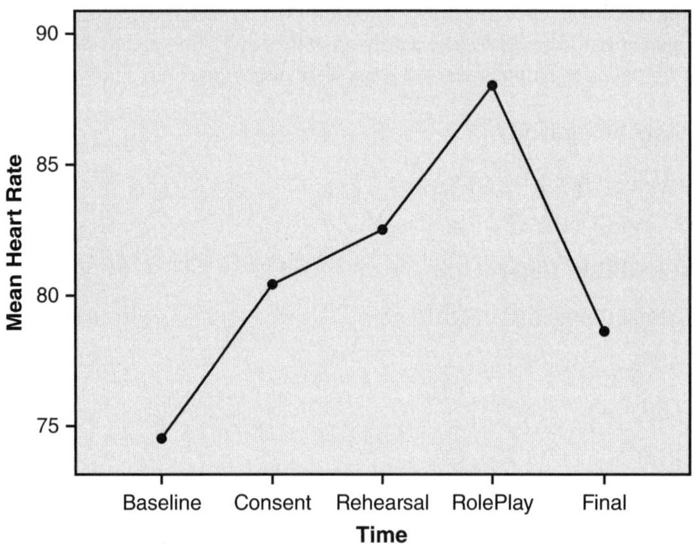

Figure 14.7 Test of the Sphericity Assumption for the Heart Rate and Social Stress Study

Mauchly's Test of Sphericity[a]

Measure: MEASURE_1

					Epsilon[b]		
Within Subjects Effect	Mauchly's W	Approx. Chi-Square	df	Sig.	Greenhouse-Geisser	Huynh-Feldt	Lower-bound
Time	.496	42.343	9	.000	.764	.808	.250

Tests the null hypothesis that the error covariance matrix of the orthonormalized transformed dependent variables is proportional to an identity matrix.

a. Design: Intercept
Within Subjects Design: Time

b. May be used to adjust the degrees of freedom for the averaged tests of significance. Corrected tests are displayed in the Tests of Within-Subjects Effects table.

Values of ε can range from 1 (no violation of assumption) down to a lower limit of $(k - 1)$, where k is the number of levels for the repeated-measures factor. The lower limit is not generally used. In this example, $k = 5$ levels of time, so the minimum possible value of ε is 1/4 = .25. Two different versions of ε have been suggested (Greenhouse-Geisser and Huynh-Feldt). The Greenhouse-Geisser version is more conservative.

Note that the *df* for the Greenhouse-Geisser test (in Figure 14.8) are obtained by taking the *df* for the sphericity-assumed test ($df = 4$ for time) and multiplying it by the ε of .764 to obtain a downwardly adjusted *df* of 2.203 for time. The same adjustment is made to *df* error. The value of the *F* ratio does not change; however, the distribution of *F* that is used to evaluate whether *F* is significant has fewer *df* and requires a slightly larger obtained value of *F* to judge the outcome statistically significant than is used in the sphericity assumed situation. It is possible to find that the *F* test in the sphericity assumed situation is statistically significant and then find that *F* is not significant when one of these two versions of downwardly corrected *df* are used. In this example, even after downward adjustment of *df*, the *F* ratio for main effect of time is still statistically significant.

Because the sphericity assumption was significantly violated, and in order to use the more conservative correction, the Greenhouse-Geisser-corrected *df* values are reported here.

Using the Greenhouse-Geisser correction for violation of the sphericity assumption, the main effect for time was statistically significant, $F(2.203, 138.774) = 26.384$, $p < .001$, partial $\eta^2 = .295$.[2] Mean heart rate differed significantly across these five points in time.

Repeated-measures contrasts (in Figure 14.9) showed that each increase in heart rate (from Times 1 to 2, 2 to 3, 3 to 4, and 4 to 5) was statistically significant, and that the decrease in heart rate from Times 4 to 5 was also statistically significant.

Figure 14.8 Univariate *F* Tests of Within-Subjects Effects in Heart Rate and Social Stress Study

Tests of Within-Subjects Effects

Measure: MEASURE_1

Source		Type III Sum of Squares	df	Mean Square	F	Sig.	Partial Eta Squared
Time	Sphericity Assumed	6377.206	4	1594.302	26.384	.000	.295
	Greenhouse-Geisser	6377.206	2.203	2895.099	26.384	.000	.295
	Huynh-Feldt	6377.206	2.286	2789.797	26.384	.000	.295
	Lower-bound	6377.206	1.000	6377.206	26.384	.000	.295
Error(Time)	Sphericity Assumed	15227.594	252	60.427			
	Greenhouse-Geisser	15227.594	138.774	109.730			
	Huynh-Feldt	15227.594	144.012	105.738			
	Lower-bound	15227.594	63.000	241.708			

Figure 14.9 "Repeated" Contrasts to Compare Means at Adjacent Levels of Time Factor for Heart Rate and Social Stress Study

Tests of Within-Subjects Contrasts

Measure: MEASURE_1

Source	time	Type III Sum of Squares	df	Mean Square	F	Sig.
time	Level 1 vs. Level 2	2103.111	1	2103.111	43.365	.000
	Level 2 vs. Level 3	293.587	1	293.587	4.454	.039
	Level 3 vs. Level 4	1966.730	1	1966.730	37.574	.000
	Level 4 vs. Level 5	7733.397	1	7733.397	112.695	.000
Error(time)	Level 1 vs. Level 2	3006.889	62	48.498		
	Level 2 vs. Level 3	4086.413	62	65.910		
	Level 3 vs. Level 4	3245.270	62	52.343		
	Level 4 vs. Level 5	4254.603	62	68.623		

A limited number of additional tests can be done to evaluate comparisons of means not included in the set of contrasts. One comparison of interest is whether hr5 differed from hr1; that is, did participant heart rate return to the original baseline value? Additional comparisons among levels of a repeated-measures factor can be made using paired-samples *t* tests. If large numbers of these *t* tests are examined, the analyst should consider possible inflated risk for Type I error; Bonferroni-corrected α levels can be used to reduce this problem.

Procedures to obtain a paired-samples *t* test appeared in Volume I, Chapter 14 (Warner, 2020), and are not repeated here. A paired-samples *t* test indicated that Time 5 mean heart rate was significantly higher than Time 1 mean heart rate, $t(62) = 3.803$, $p < .001$, two tailed. Although this was a statistically significant difference, it was not a large difference in clinical or practical terms. For hr5, $M = 77.16$, $SD = 8.93$; for hr1, $M = 74.71$, $SD = 8.61$; the difference between means was $M_5 - M_1 = 2.44$. From a clinical point of view, a heart rate difference of less than 2.5 beats per minute is not very important. This outcome suggests that by Time 5, mean heart rate had returned to the Time 1 baseline level. The difference in mean heart rate from Time 1 baseline ($M_1 = 74.71$) to Time 4 role play ($M_4 = 88.24$), an increase of 13.53 beats per minute, was larger, possibly large enough to be of interest to clinicians.

14.6 TESTING THE SPHERICITY ASSUMPTION

Contrasts between group means (denoted *C*) are an important part of repeated measures. To understand this aspect of repeated measures, we need to review contrasts. An individual contrast can be the same as the *d* difference score examined in the paired-samples *t* test. For example, Contrast 1 (C_1) could correspond to the difference in heart rate at baseline (Time 1) and after informed consent (Time 2); $C_1 = (M_2 - M_1)$. C_2 could be the difference between baseline (Time 1) and rehearsal (Time 3); $C_2 = (M_3 - M_1)$. The number of contrasts in a set is $(k - 1)$, where *k* is the number of levels for the within-*S* factor. For this study with five levels, there could be four contrasts. Just as you used SPSS to compute a *d* or difference score for the paired-samples *t* test, you could create scores for each of these *C* variables. The null hypothesis for each contrast is:

$$H_0\text{: } C = 0. \tag{14.1}$$

Another way to state the null hypothesis for the omnibus *F* test in repeated-measures ANOVA is that, as a set, the population means for the $(k - 1)$ contrasts across levels are zero. Repeated-measures ANOVA assumes homogeneity of population variances for scores of these *C* variables; this assumption is called sphericity. It is formally stated as follows (Field, 2018):

$$H_0\text{: } \sigma^2_{C1} = \sigma^2_{C2} = \cdots = \sigma^2_{Ck}. \tag{14.2}$$

If the variances of the contrasts do not differ from each other significantly, then the *F* ratio and its corresponding *p* value should provide accurate information about the risk for Type I error. If the variances of contrasts are unequal the *p* value for the *F* test in repeated-measures ANOVA may seriously underestimate the true risk for Type I error.

Adjusted degrees of freedom to evaluate significance of the *F* ratio can be used in situations where the sphericity assumption is violated. However, a data analyst has another option: Repeated-measures data can be analyzed using MANOVA. (Each contrast is one of the multiple dependent variables in the MANOVA, and in MANOVA, the dependent variables do not need to have equal variances. MANOVA makes a different assumption: that the matrix of variances and covariances for the multiple dependent variables is equal across groups.)

14.7 MANOVA FOR REPEATED MEASURES

A MANOVA approach handles analysis of repeated-measures data in a different manner. Scores on the repeated-measures hr1 to hr5 are transformed into a set of $k - 1$ contrasts. SPSS provides several types of contrasts (including polynomial, difference, simple, repeated, etc.). The omnibus test (e.g., Wilks' Λ) for MANOVA summarizes information across these contrasts (and has the same value no matter how contrasts are coded by the GLM). Selection and interpretation of specific contrasts in repeated-measures ANOVA were covered earlier and are not repeated here. MANOVA creates a vector of new variables (C_1, C_2, C_3, C_4); each C term corresponds to one of the k - 1 contrasts:

$$\begin{bmatrix} C_1 = X_2 - X_1 \\ C_2 = X_3 - X_2 \\ C_3 = X_4 - X_3 \\ C_4 = X_5 - X_4 \end{bmatrix} \tag{14.3}$$

The null hypothesis for the MANOVA is that the corresponding population vectors for each of these contrasts equals a vector with all elements equal to 0:

$$H_0: \begin{bmatrix} \mu_2 - \mu_1 \\ \mu_3 - \mu_2 \\ \mu_4 - \mu_3 \\ \mu_5 - \mu_4 \end{bmatrix} = \begin{bmatrix} 0 \\ 0 \\ 0 \\ 0 \end{bmatrix} \tag{14.4}$$

If we cannot reject this null hypothesis, we do not have evidence for differences among these five means (μ_1, μ_2, μ_3, μ_4, μ_5). These contrasts do not have to be orthogonal because MANOVA corrects for correlations among dependent variables, and they do not have to have equal variances, because MANOVA can handle multiple outcome variables with unequal variances (or even measured using different scales). Unlike the univariate version of repeated-measures ANOVA (which uses *SS*, *MS*, and *F*), MANOVA does not require sphericity.

Instead of computing sums of squares and mean squares for a single outcome variable, and then using these to set up an *F* ratio (as in the univariate repeated-measures ANOVA reported earlier in this chapter), MANOVA obtains multivariate test statistics such as Wilks' Λ that summarize information across the contrasts. A data analyst may choose to use MANOVA as a way to avoid the problem of violation of the sphericity assumption or may prefer it for other reasons; under some circumstances, the MANOVA test has greater statistical power than the univariate *F* test. Algina and Kesselman (1997) suggested the use of MANOVA under these conditions:

1. If the number of levels (k) is less than or equal to 4, use MANOVA if the number of participants is greater than $k + 15$.
2. If the number of levels (k) is between 5 and 8, use MANOVA if the number of participants is greater than $k + 30$.

Under some circumstances, MANOVA may have greater statistical power than a univariate approach to repeated measures. When sample sizes are small, MANOVA may be less powerful than univariate repeated measures. Power depends on many factors in this situation including sizes of correlations among repeated measures.

Figure 14.10 MANOVA for Heart Rate and Social Stress Study

Multivariate Tests[a]

Effect		Value	F	Hypothesis df	Error df	Sig.	Partial Eta Squared
time	Pillai's Trace	.764	47.795[b]	4.000	59.000	.000	.764
	Wilks' Lambda	.236	47.795[b]	4.000	59.000	.000	.764
	Hotelling's Trace	3.240	47.795[b]	4.000	59.000	.000	.764
	Roy's Largest Root	3.240	47.795[b]	4.000	59.000	.000	.764

a. Design: Intercept
Within Subjects Design: time

b. Exact statistic

14.8 RESULTS FOR HEART RATE AND SOCIAL STRESS ANALYSIS USING MANOVA

Let's return to the empirical example (heart rate measured at five points in time in Mooney's [1990] social stress study). The MANOVA results appear in Figure 14.10 (these were obtained as part of the earlier GLM results).

The most common multivariate test statistics include Pillai's trace, Wilks' Λ, Hotelling's trace, and Roy's largest root. Usually only one of these is reported; Wilks' Λ is most widely used. The null hypothesis tested by Wilks' Λ is that the vector of contrasts equals a vector of zeros (Equation 14.8). For these data, the omnibus test (Figure 14.10) is statistically significant, Wilks' $\Lambda = .236$, $F(4,59) = 47.79$, $p < .001$, partial $\eta^2 = .764$. The null hypothesis that mean heart rate was the same across all five points in time can be rejected. Tests of the individual contrasts can be examined to evaluate which pairs of means showed statistically significant differences (i.e., which of the four contrasts were significant). The type of contrast selected for this analysis was "repeated" (i.e., compare means of adjacent times).

Recall that Wilks' Λ can be interpreted as an effect size (somewhat comparable to $1 - \eta^2$, with a range of 0 to 1). A smaller value of Wilks' Λ indicates a smaller proportion of unexplained or error variance (and thus a larger effect). An η^2 effect size can be obtained by subtracting Wilks' Λ from 1, in this instance, $1 - .236 = .764$. That is a very large effect size.

The contrasts between different levels of time were reported earlier (Figure 14.9). For these data, all of the repeated contrasts were statistically significant. (Keep in mind that a statistically significant Wilks' Λ does not necessarily imply that all contrasts are significant; it suggests that one or more contrast may be significant.)

14.9 DOUBLY MULTIVARIATE REPEATED MEASURES

A repeated-measures study can be called doubly multivariate if multiple measures are included at multiple points in time. This study included measures of systolic and diastolic blood pressure, as well as heart rate, at each of the five times. Multiple measures are specified in the first GLM repeated-measures dialog box. Systolic blood pressure (sys) and diastolic blood pressure (dia) are added to the analysis.

The data analyst types "hr" into the "Measure Name" box, then clicks Add, then does the same for the names of the other two measures. After Define is clicked, the next dialog box looks like what is shown in Figure 14.11.

There is no requirement that the name of each variable include what is being measured (such as hr or sys) or the time when it is measured (1, 2, . . . , 5), but it should be obvious that systematic naming of variables makes it much easier to assign variables to levels of the within-*S* factor. Also note that the three different measures (in this example, heart rate and systolic and diastolic blood

Figure 14.11 Repeated Measures Define Factors Dialog Box for Doubly Multivariate GLM Repeated Measures

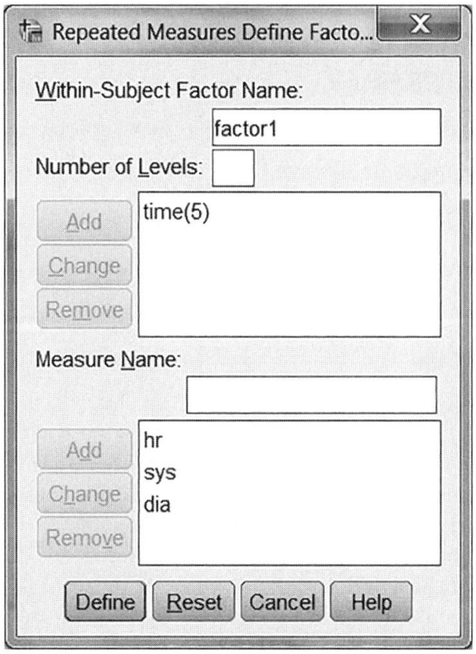

Repeated Measures Define Facto...

Within-Subject Factor Name:

factor1

Number of Levels:

Add

Change

Remove

time(5)

Measure Name:

Add

Change

Remove

hr

sys

dia

Define Reset Cancel Help

Figure 14.12 Repeated Measures Dialog Box for Specification of Doubly Multivariate Analysis

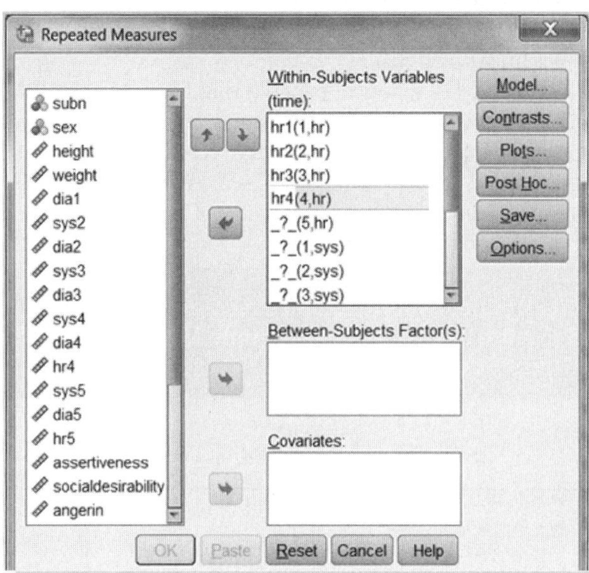

pressure) do not have to be in the same units, although in this instance, systolic and diastolic blood pressure are both given in millimeters of mercury. As in the previous example, means for the levels of the time factor were requested by clicking the Options button, and repeated contrasts were selected. Here is the SPSS syntax for this doubly multivariate repeated measures:

```
GLM hr1 hr2 hr3 hr4 hr5 sys1 sys2 sys3 sys4 sys5 dia1 dia2 dia3 dia4 dia5
/WSFACTOR=Time 5 Repeated
/MEASURE=hr sys dia
/METHOD=SSTYPE(3)
/PLOT=PROFILE(Time)
/EMMEANS=TABLES(Time)
/PRINT=ETASQ
/CRITERIA=ALPHA(.05)
/WSDESIGN=Time.
```

Results appear in the following figures. The sphericity assumption (output not shown) was significantly violated for each of the three measures. Under these circumstances, the analyst might report adjusted degrees of freedom (such as Greenhouse-Geisser); in this example, the choice among tests with different *df* terms for the *F* ratios does not make a difference in the decisions about statistical significance.

MANOVA handles multiple outcome variables by forming the best possible weighted linear combinations of scores on the variables, that is, the combinations that are maximally different across groups. In this situation, MANOVA finds the weighted linear combinations of scores (for the variables hr, sys, and dia) that differ maximally across the five levels of time. The null hypothesis is that, even for the best possible ways of combining variables, there is no significance differences in means for this set of variables across groups.

This omnibus test (Figure 14.13) was statistically significant, Wilks' $\Lambda = .259$, $F(12, 661.729) = 36.79$, $p < .001$, partial $\eta^2 = .363$. This suggests that two (or possible more than two) of the five levels of time differed significantly on one (or more) weighted linear combinations of the variables hr, sys, and dia.

Univariate follow-up analyses appear in the following figures. (The GLM does not provide truly multivariate follow-up tests.) Essentially, what you see below is what you would have obtained if you had done a separate one-way repeated-measures ANOVA for each one of the three variables.

The tests in Figure 14.14 are univariate in two senses of the word. First, they are univariate in that each variable (such as hr) is assessed without taking any correlations with other

Figure 14.13 Heart Rate, Systolic Blood Pressure, and Diastolic Blood Pressure Across Time in Doubly Multivariate Social Stress Study

Multivariate[a,b]

Within Subjects Effect		Value	F	Hypothesis df	Error df	Sig.	Partial Eta Squared
Time	Pillai's Trace	.813	23.430	12.000	756.000	.000	.271
	Wilks' Lambda	.259	36.795	12.000	661.729	.000	.363
	Hotelling's Trace	2.595	53.765	12.000	746.000	.000	.464
	Roy's Largest Root	2.488	156.758[c]	4.000	252.000	.000	.713

a. Design: Intercept
Within Subjects Design: Time

b. Tests are based on averaged variables.

c. The statistic is an upper bound on F that yields a lower bound on the significance level.

outcome variables (such as sys and dia) into account. A separate one-way repeated-measures ANOVA is reported for each of the three outcome variables. Second, for each of these one-way repeated-measures ANOVAs, the univariate version of the test (i.e., the one that uses *SS* and *F* ratios) is reported rather than MANOVA results.

All three of the univariate ANOVAs above were statistically significant; that is, the null hypothesis that hr was equal across all five levels of time could be rejected (and the same conclusion for sys and dia). Follow-up tests (contrasts of means on each variable for Time 1 vs. 2, Time 2 vs. 3, and so forth) appear in Figure 14.15.

Figure 14.14 Univariate Repeated Measures for Each Dependent Variable Measure in Social Stress Study

Univariate Tests

Source	Measure		Type III Sum of Squares	df	Mean Square	F	Sig.	Partial Eta Squared
time	hr	Sphericity Assumed	6868.825	4	1717.206	61.375	.000	.497
		Greenhouse-Geisser	6868.825	3.057	2246.770	61.375	.000	.497
		Huynh-Feldt	6868.825	3.234	2124.136	61.375	.000	.497
		Lower-bound	6868.825	1.000	6868.825	61.375	.000	.497
	sys	Sphericity Assumed	48571.225	4	12142.806	138.983	.000	.692
		Greenhouse-Geisser	48571.225	3.005	16165.957	138.983	.000	.692
		Huynh-Feldt	48571.225	3.175	15300.043	138.983	.000	.692
		Lower-bound	48571.225	1.000	48571.225	138.983	.000	.692
	dia	Sphericity Assumed	12719.949	4	3179.987	87.265	.000	.585
		Greenhouse-Geisser	12719.949	2.476	5136.893	87.265	.000	.585
		Huynh-Feldt	12719.949	2.587	4916.482	87.265	.000	.585
		Lower-bound	12719.949	1.000	12719.949	87.265	.000	.585
Error(time)	hr	Sphericity Assumed	6938.775	248	27.979			
		Greenhouse-Geisser	6938.775	189.546	36.607			
		Huynh-Feldt	6938.775	200.490	34.609			
		Lower-bound	6938.775	62.000	111.916			
	sys	Sphericity Assumed	21667.575	248	87.369			
		Greenhouse-Geisser	21667.575	186.281	116.316			
		Huynh-Feldt	21667.575	196.824	110.086			
		Lower-bound	21667.575	62.000	349.477			
	dia	Sphericity Assumed	9037.251	248	36.441			
		Greenhouse-Geisser	9037.251	153.524	58.865			
		Huynh-Feldt	9037.251	160.407	56.340			
		Lower-bound	9037.251	62.000	145.762			

Figure 14.15 Repeated-Measures Contrasts for Heart Rate, Diastolic Blood Pressure, and Systolic Blood Pressure

Tests of Within-Subjects Contrasts

Source	Measure	time	Type III Sum of Squares	df	Mean Square	F	Sig.	Partial Eta Squared
time	hr	Level 1 vs. Level 2	2103.111	1	2103.111	43.365	.000	.412
		Level 2 vs. Level 3	293.587	1	293.587	4.454	.039	.067
		Level 3 vs. Level 4	1966.730	1	1966.730	37.574	.000	.377
		Level 4 vs. Level 5	7733.397	1	7733.397	112.695	.000	.645
	sys	Level 1 vs. Level 2	10934.921	1	10934.921	81.163	.000	.567
		Level 2 vs. Level 3	2746.921	1	2746.921	20.507	.000	.249
		Level 3 vs. Level 4	20664.778	1	20664.778	181.367	.000	.745
		Level 4 vs. Level 5	13347.444	1	13347.444	83.831	.000	.575
	dia	Level 1 vs. Level 2	1515.571	1	1515.571	44.972	.000	.420
		Level 2 vs. Level 3	706.683	1	706.683	12.788	.001	.171
		Level 3 vs. Level 4	7296.571	1	7296.571	160.340	.000	.721
		Level 4 vs. Level 5	4064.063	1	4064.063	84.613	.000	.577
Error(time)	hr	Level 1 vs. Level 2	3006.889	62	48.498			
		Level 2 vs. Level 3	4086.413	62	65.910			
		Level 3 vs. Level 4	3245.270	62	52.343			
		Level 4 vs. Level 5	4254.603	62	68.623			
	sys	Level 1 vs. Level 2	8353.079	62	134.727			
		Level 2 vs. Level 3	8305.079	62	133.953			
		Level 3 vs. Level 4	7064.222	62	113.939			
		Level 4 vs. Level 5	9871.556	62	159.219			
	dia	Level 1 vs. Level 2	2089.429	62	33.700			
		Level 2 vs. Level 3	3426.317	62	55.263			
		Level 3 vs. Level 4	2821.429	62	45.507			
		Level 4 vs. Level 5	2977.937	62	48.031			

The contrast results (Figure 14.15) tell us that changes in each variable (hr, sys, and dia) were significant for all points in time (Time 2 vs. 1, Time 3 vs. 2, etc.). We do not have a test of whether Time 5 (final baseline mean) differs significantly from Time 1 (initial baseline) for each of these three variables; we could obtain these using paired-samples *t* tests.

As in earlier GLM examples, output (not shown here) also includes a test whether the grand mean for each variable equals 0 (test of the "intercept"). That test is of no interest in this situation.

It is unusual to find (in actual data) that every test turns out to be statistically significant along with such large effect sizes. The large number of statistical significance tests (*p* values) that appear in these results should raise a warning flag. The risk that a set of *k* tests includes at least one instance of Type I error increases dramatically as *k*, the number of tests, increases. It is not possible to calculate the increase in risk in this situation, because observations are correlated and thus, these tests are not independent. In applications of doubly multivariate repeated measures, the analyst would probably examine the omnibus multivariate tests (such as Wilks' Λ) first and do follow-up tests only if the omnibus test is statistically significant. Even after a significant omnibus test, results of large numbers of follow-up tests should be interpreted with caution. The *p* values that appear in SPSS results have not been corrected for inflated risk for Type I error. The analyst might use Bonferroni-corrected α levels as the criterion for statistical significance or some other method of correction for inflated risk for Type I error for follow-up tests.

14.10 MIXED-MODEL ANOVA: BETWEEN-*S* AND WITHIN-*S* FACTORS

We now consider adding one between-*S* factor along with one within-*S* factor. This between-*S* factor can represent participant characteristics. The example in this section uses sex as the between-*S* factor. Another possible example of a between-*S* participant characteristics factor would be groups based on body mass index scores (e.g., below healthy weight, healthy weight, slightly overweight, more overweight). A completely different type of between-*S* factor can be used; in some studies we want to compare groups that have received different types of treatment of intervention. Section 14.11 considers a between-*S* factor that represents different types of treatments.

In this example, the between-*S* factor represents a participant characteristic (sex). The question becomes whether men and women show a different response across time (or under a series of different treatment conditions). The earlier test of participant-by-time interaction indicated significant nonadditivity. Adding one or more between-*S* factors to the analysis can be used to try to identify specific participant characteristics that might account for that interaction.

14.10.1 Mixed-Model ANOVA for Heart Rate and Stress Study

A mixed-model ANOVA was performed using the heart rate and social stress data in hrbpsocialstress.sav, with sex as the between-*S* factor and time as the within-*S* factor. The between-*S* factor is added in the second GLM Repeated Measures dialog box in Figure 14.16.

Plots were requested with a separate line for each sex; the Add button must be clicked to move the Time × Sex plot specification to the list of requested plots.

The SPSS GLM syntax for this analysis is as follows:

```
GLM hr1 hr2 hr3 hr4 hr5 BY sex
/WSFACTOR=Time 5 Repeated
/METHOD=SSTYPE(3)
```

Figure 14.16 Specification of Between-S Factor in GLM Repeated Measures

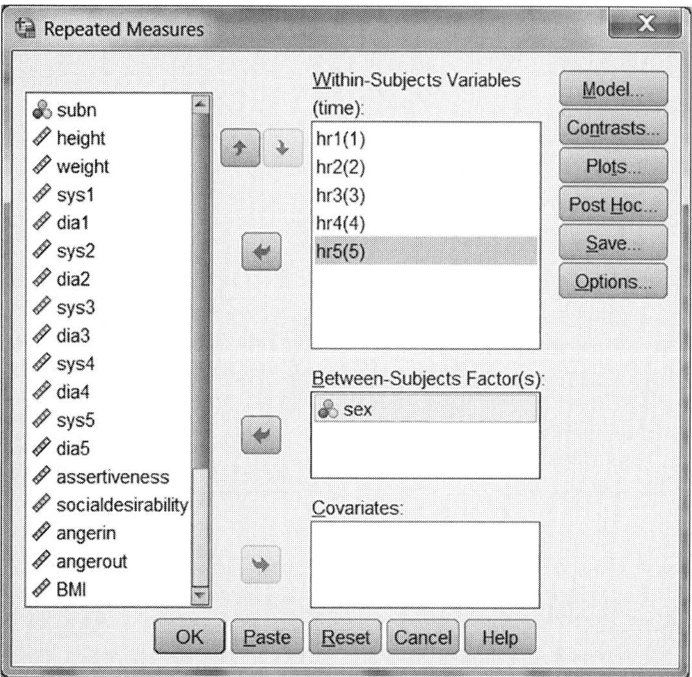

Figure 14.17 Request Plot of Mean Heart Rate Across Levels of Time Separated by Sex

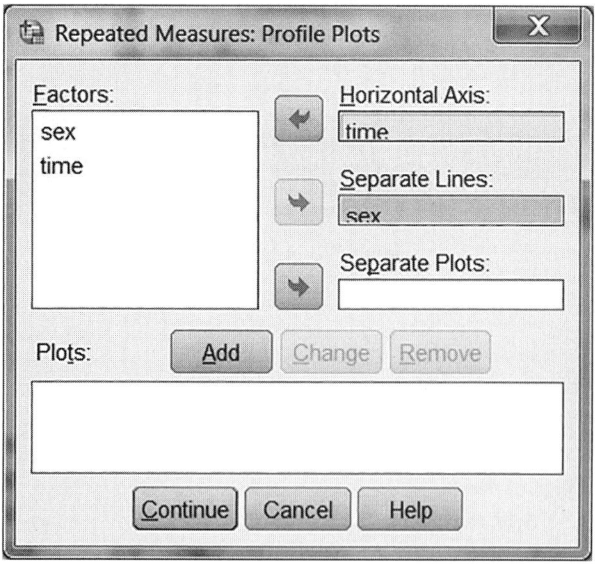

```
/PLOT=PROFILE(Time Time*sex)
/EMMEANS=TABLES(Time)
/PRINT=ETASQ
/CRITERIA=ALPHA(.05)
/WSDESIGN=Time
/DESIGN=sex.
```

The MANOVA results (Figure 14.18) indicated that there was not a significant Time × Sex interaction.

The lack of significant interaction indicates that plots of group means for men and women across levels of time should yield roughly parallel means, and that is what we see in Figure 14.19.

Both men and women showed increasing heart rate (with similar amounts of increase) across Times 1, 2, 3, and 4 and a similar decrease in heart rate from Time 4 to 5. The absence of a significant sex-by-time interaction tells us that sex differences do not account for the participant-by-time interaction reported in Section 14.4.

Even though mean heart rate for women was higher than mean heart rate for men in the graph in Figure 14.19, the main effect for sex was not statistically significant (Figure 14.20).

14.10.2 Interaction of Intervention Type and Times of Assessment in Hypothetical Experiment With Follow-Up

Now let's consider a different type of between-*S* factor: a factor that represents different types of treatments. For this hypothetical study, groups of participants receive one of three treatments in a positive psychology study: no treatment, relaxation, or meditation. This is a between-*S* factor; each participant is assigned to only one treatment condition. The within-*S* factor is three times of assessment (before a positive psychology intervention, 6 weeks after, and 12 weeks after). The outcome variable is a score on a well-being measure. The complete (hypothetical) data are in the file wellbeinginterventions.sav. Type of positive psychology intervention was coded 1 = control, 2 = relaxation, 3 = meditation. The dependent variable, well-being (WB), was rated on a scale from 1 to 30 at Times 1, 2, and 3 (WB1, WB2, WB3). The GLM repeated-measures menu selections are similar to those in the previous example; here is the GLM syntax:

Figure 14.18 MANOVA Results for Sex-by-Time Analysis in Social Stress Study

Multivariate Tests[a]

Effect		Value	F	Hypothesis df	Error df	Sig.
time	Pillai's Trace	.759	45.572[b]	4.000	58.000	.000
	Wilks' Lambda	.241	45.572[b]	4.000	58.000	.000
	Hotelling's Trace	3.143	45.572[b]	4.000	58.000	.000
	Roy's Largest Root	3.143	45.572[b]	4.000	58.000	.000
time * sex	Pillai's Trace	.045	.690[b]	4.000	58.000	.602
	Wilks' Lambda	.955	.690[b]	4.000	58.000	.602
	Hotelling's Trace	.048	.690[b]	4.000	58.000	.602
	Roy's Largest Root	.048	.690[b]	4.000	58.000	.602

a. Design: Intercept + sex
Within Subjects Design: time

b. Exact statistic

```
GLM WB1 WB2 WB3 BY Intervention
/WSFACTOR=time 3 Polynomial
/METHOD=SSTYPE(3)
/POSTHOC=Intervention(TUKEY)
/PLOT=PROFILE(time*Intervention)
/EMMEANS=TABLES(OVERALL)
/EMMEANS=TABLES(Intervention)
/EMMEANS=TABLES(time)
/EMMEANS=TABLES(Intervention*time)
```

Figure 14.19 Plot of Heart Rate Across Five Times for Men and Women

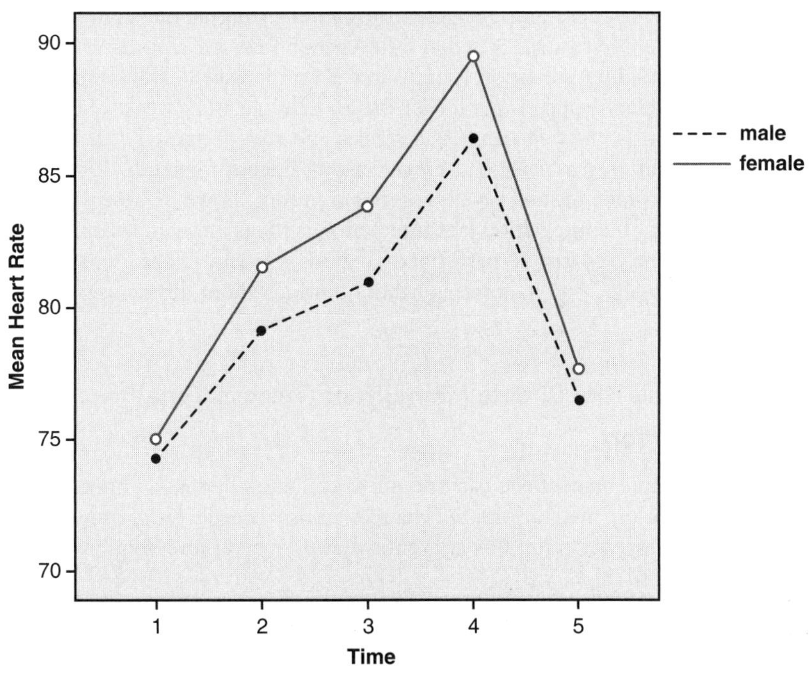

Figure 14.20 Test of Sex Main Effect for Heart Rate in Social Stress Study

Tests of Between-Subjects Effects

Measure: MEASURE_1

Transformed Variable: Average

Source	Type III Sum of Squares	df	Mean Square	F	Sig.
Intercept	1977657.800	1	1977657.800	5241.270	.000
sex	321.216	1	321.216	.851	.360
Error	23016.772	61	377.324		

```
/PRINT=ETASQ
/CRITERIA=ALPHA(.05)
/WSDESIGN=time
/DESIGN=Intervention.
```

The Mauchly test for the sphericity assumption was not statistically significant (results not shown). Selected results appear in the next figures.

When the between-*S* factor is type of intervention, and the within-*S* factor is time, researchers usually hope for a statistically significant interaction between type of treatment and time as evidence that different treatment conditions had different outcomes. Figure 14.21 shows a statistically significant Time × Intervention interaction, Wilks' $\Lambda = .187$, $F(4, 52) = 17.09$, $p < .001$, partial $\eta^2 = .568$. A plot of cell means (separate line for each intervention, and time on the *X* axis) indicates the nature of the interaction.

Figure 14.22 suggests that the three groups had similar well-being scores at baseline (a researcher would hope for this as evidence that groups are approximately equivalent prior to treatment). It appears that people in the no-intervention (control) group showed no improvement in well-being at 6 weeks and a slight drop in well-being at the 12-week follow-up. (Follow-up tests are needed to assess whether differences noted in examination of the plot are statistically significant.) The relaxation group showed an increase in well-being at 6 weeks, and then their well-being dropped to a level close to baseline at 12 weeks. The meditation group showed an increase in well-being at 6 weeks, and this decreased only slightly at 12 weeks (so the intervention may have had longer lasting benefit for them). There were also significant main effects for time and type of intervention (not shown). These would be interpreted in the context of the interaction (the interaction is the primary focus of interest).

Two follow-up analyses can be performed (analysis of main effects, separately for the within-*S* and between-*S* factors) to answer further questions about differences among group means.

14.10.3 First Follow-Up: Simple Main Effect (Across Time) for Each Intervention

An analysis of simple main effects can be done to evaluate whether, within each treatment group (control, relaxation, meditation), the changes across three levels of time were statistically significant. An easy way to do this is to split files by type of intervention and then run

Figure 14.21 MANOVA Results for Type of Positive Psychology Intervention

Multivariate Tests[a]

Effect		Value	F	Hypothesis df	Error df	Sig.	Partial Eta Squared
TIME	Pillai's Trace	.776	45.067[b]	2.000	26.000	.000	.776
	Wilks' Lambda	.224	45.067[b]	2.000	26.000	.000	.776
	Hotelling's Trace	3.467	45.067[b]	2.000	26.000	.000	.776
	Roy's Largest Root	3.467	45.067[b]	2.000	26.000	.000	.776
TIME * Intervention	Pillai's Trace	.933	11.814	4.000	54.000	.000	.467
	Wilks' Lambda	.187	17.089[b]	4.000	52.000	.000	.568
	Hotelling's Trace	3.714	23.212	4.000	50.000	.000	.650
	Roy's Largest Root	3.532	47.678[c]	2.000	27.000	.000	.779

a. Design: Intercept + Intervention
Within Subjects Design: TIME

b. Exact statistic

c. The statistic is an upper bound on F that yields a lower bound on the significance level.

Figure 14.22 Mean Well-Being for Positive Psychology Interventions

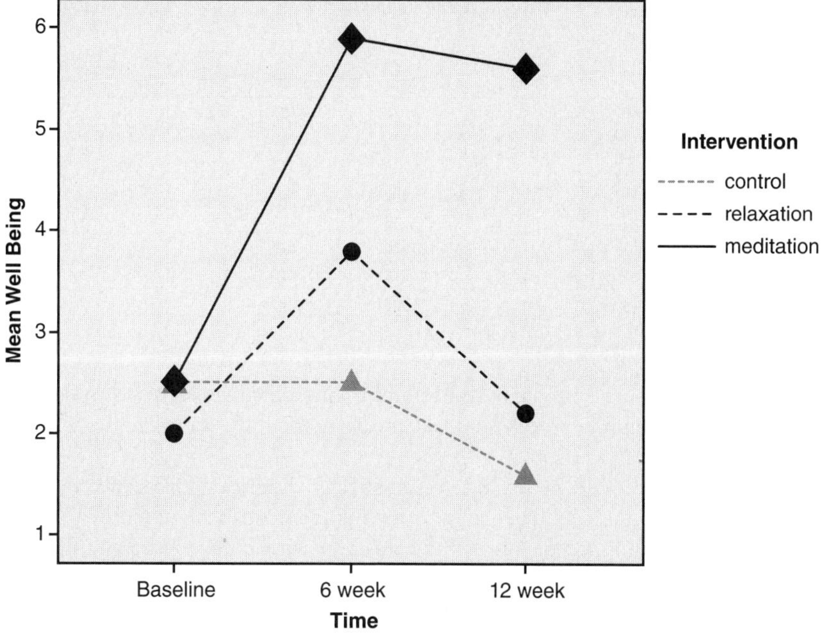

Figure 14.23 Split File Dialog Box

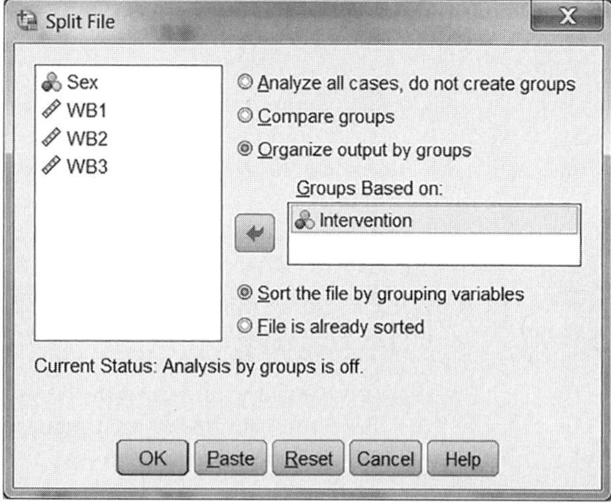

a repeated-measures one-way ANOVA with time as the within-*S* factor. To do this, choose <Data> from the top-level SPSS menu, scroll down, and choose <Split File> (not <Split into Files>). In the Split File dialog box, choose the radio button for "Organize output by groups" and move Intervention into the "Groups Based on" panel as shown in Figure 14.23.

Then run a one-way within-*S* repeated-measures analysis using the GLM (be sure to omit the between-*S* factor for intervention type). Three one-way repeated-measures ANOVAs will be obtained, one for each type of intervention. The means compared in these analyses are the means that fall on the same lines in the plot in Figure 14.22 and have the same markers.

Figure 14.24 Analysis of Simple Main Effects for the Control Group

Tests of Within-Subjects Effects[a]

Measure: MEASURE_1

Source		Type III Sum of Squares	df	Mean Square	F	Sig.	Partial Eta Squared
TIME	Sphericity Assumed	5.400	2	2.700	9.228	.002	.506
	Greenhouse-Geisser	5.400	1.175	4.595	9.228	.010	.506
	Huynh-Feldt	5.400	1.246	4.333	9.228	.008	.506
	Lower-bound	5.400	1.000	5.400	9.228	.014	.506
Error(TIME)	Sphericity Assumed	5.267	18	.293			
	Greenhouse-Geisser	5.267	10.576	.498			
	Huynh-Feldt	5.267	11.215	.470			
	Lower-bound	5.267	9.000	.585			

a. Intervention = control

The first repeated-measures analysis of simple main effects result is shown below (for the control group). This examined the set of three means marked by gray triangles in the interaction plot in Figure 14.22.

The results in Figure 14.24 indicate a statistically significant difference in mean well-being across time within the control group. Why might a no-treatment (control) group show change across time? Suppose that the study was run with college students during spring semester. The control-group measures might show the effect of being at different points in the semester (beginning, after midterm, and just before final examinations). In the absence of intervention, well-being might tend to go down during a semester. Analyses within the relaxation and meditation groups (not shown here) also showed statistically significant differences across the three levels of time.

14.10.4 Second Follow-Up: Comparisons of Intervention Groups at the Same Points in Time

Means can also be compared for each point in time, to answer the question, At each point in time, did means differ across treatment conditions? Note that you need to "turn off" the <Split File> command used earlier before doing the next analysis (refer back to Figure 14.23; you would select the radio button for "Analyze all cases, do not create groups"). Because information about well-being at Time 1 is contained in the variable WB1, the analysis we need is a between-*S* comparison of means on WB1 across the three intervention groups (this is between-*S* because different participants were assigned to each of the interventions). Differences across interventions will also be examined for WB2 and WB3. This can be obtained using the one-way between-*S* ANOVA procedure. In the plot in Figure 14.25, the means compared by each one-way between-*S* ANOVA fall within the same ellipse.

Did well-being differ among the three interventions at each point in time? In studies where the first time point represents a baseline prior to any intervention, one would expect that the three groups would have similar means. Immediately after intervention, and later after intervention, means would be expected to differ for groups that received different interventions.

At Time 1, before intervention, the three groups did not have significantly different means in well-being (this refers to the *F* test for WB1 in Figure 14.26). Researchers usually hope that groups do not differ prior to intervention; preexisting differences create problems when trying to interpret later outcomes. At Times 2 and 3, there were significant differences among group means. Tukey honestly significant difference post hoc tests are not shown; these would provide further information about which pairs of group means differ significantly at each point in time.

Figure 14.25 Between-*S* Comparison of Means Across Interventions Separately for Each Time

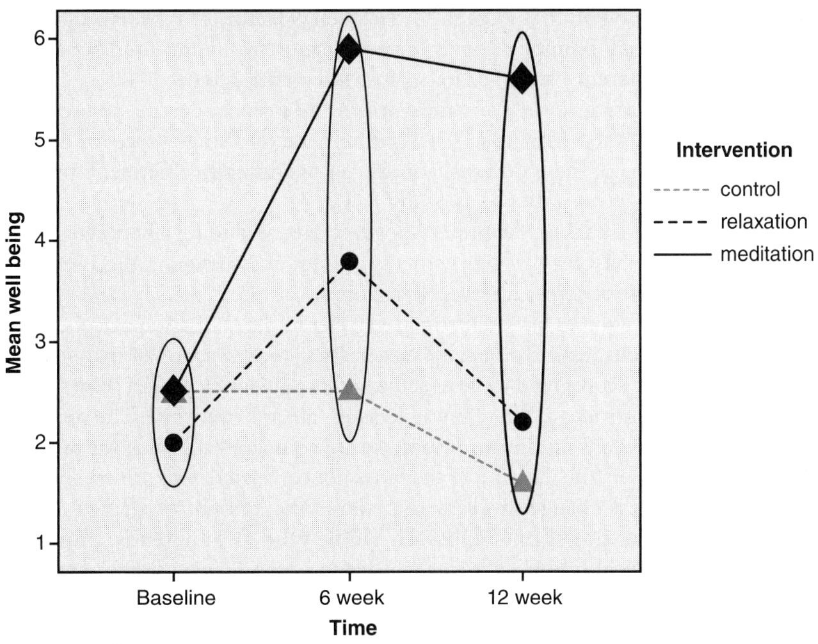

Figure 14.26 Simple Main-Effects Analysis of Well-Being Within Each Time Period

ANOVA

		Sum of Squares	df	Mean Square	F	Sig.
WB1	Between Groups	1.667	2	.833	.776	.470
	Within Groups	29.000	27	1.074		
	Total	30.667	29			
WB2	Between Groups	58.867	2	29.433	19.383	.000
	Within Groups	41.000	27	1.519		
	Total	99.867	29			
WB3	Between Groups	93.067	2	46.533	29.632	.000
	Within Groups	42.400	27	1.570		
	Total	135.467	29			

Note: Output is from the SPSS one-way ANOVA procedure.

14.10.5 Comparison of Repeated-Measures ANOVA With Difference-Score and ANCOVA Approaches

The simple pretest–posttest design was introduced in Chapter 14 in Volume I (Warner, 2020), on the paired-samples *t* test. The paired-samples *t* test can be used to evaluate whether the mean for an outcome variable of interest differs before versus after an intervention. Suppose that X_2 is score at posttest (after intervention) and X_1 is score at pretest (before intervention). The paired-samples *t* test is based on *d*, the difference score: $d = X_2 - X_1$. For example, consider evaluation of a drug education program. Suppose that all students in a school are given a self-report questionnaire about drug use the first week of school (X_1), then they go through the drug education program, then they do the same self-report of drug use after the intervention (X_2). If the mean difference score *d* does not differ significantly from 0, there is no evidence that the intervention is effective. A between-*S* factor for type of intervention can be added, and then we can examine *d* scores for two or more different interventions, and ask whether the mean *d* (difference score) differs across intervention type.

In an analysis of change scores (or gain scores or difference scores), we use the X_2 and X_1 scores for each subject to compute a $X_2 - X_1$ difference or change score for each subject. Then we compare means of these difference scores across different treatment groups (using an independent-samples *t* test or between-*S* ANOVA).

If we use the GLM to analyze the pretest–posttest data, including a between-*S* treatment type factor and a within-*S* factor that represents two times of assessment, this is equivalent to comparing difference scores across intervention groups.

The use of difference scores has been criticized (refer to the discussion in Appendix 14A to the paired-samples *t* test chapter in Volume I [Warner, 2020] for discussion of analysis of pretest–posttest data). Difference scores tend to be negatively correlated with pretest scores. Analysis of covariance (ANCOVA) provides a different way to assess change over time. The pretest score is used as a covariate, and means on the posttest scores are compared adjusting for pretest scores. When this approach is used, the measure of change is not correlated with pretest score.

Do we want to assess change in a way that allows the amount of change to be correlated with pretest level or not? Either approach can be valid. The potential problem is that conclusions about statistical significance of changes (and even which groups have better and worse outcomes) can turn out different when using ANCOVA versus difference scores or repeated measures (Lord's paradox). While this does not happen very often, it is a concern. Data analysts who report a treatment-by-time interaction in a repeated-measures ANOVA are sometimes asked whether the results would be the same using ANCOVA to control for pretest scores or other analyses that assess change across time.

There are more sophisticated ways to assess change over time, such as growth curve analysis or longitudinal structural equation modeling (SEM). Those topics are beyond the scope of this textbook. For further information see Grimm and Ram (2016); Heck, Thomas, and Tabata (2013); Robson and Pevalin (2015); Singer and Willett (2003); or Snijders and Bosker (2011).

14.11 ORDER AND SEQUENCE EFFECTS

In previous examples, the within-*S* factor was time or trials. However, in some repeated-measures studies, the within-*S* factor represents different treatments; that is, all participants receive the same set of treatments. Order effects can be a problem when each participant receives two or more treatments. Here is a hypothetical example of a study in which every subject receives all four treatments (Prozac, Zoloft, a new drug, or a placebo).

The advantage of this type of design is that the researcher can obtain a large amount of data from each participant. This is particularly helpful when patients or nonhuman animal subjects are costly to recruit and care for. However, order effects (the order in which treatments are received) can be a serious problem. The apparent impact of each drug may differ depending upon whether it is administered first, second, third, or fourth. Also, depression may be gradually decreasing for most patients over time, independent of whether any treatments are received. There can also be carryover effects. For instance, Prozac has a relatively long "washout" time; traces of the drug may be present 6 weeks after it is discontinued. If the no-drug interval between drug treatments is only 2 weeks, the effect of any drug that is given following Prozac will be confounded with effects of Prozac. One drug might produce lasting changes that alter the effects of later drugs, or one drug might not be completely gone from the patient's system when the next drug is introduced.

Figure 14.27 Hypothetical Design: Each Participant Tested in All Four Drug Conditions

	Name	Prozac	Zoloft	NewDrug	Placebo
1	Ann	21.0	17.00	12.00	34.00
2	Bob	18.0	17.00	11.00	42.00
3	Chris	.	.	.	.
4	Zack	.	.	.	.
5	Will	.	.	.	.
6	Betty	.	.	.	.
7	John	.	.	.	.
8	Amy	.	.	.	.
9	Cara	.	.	.	.
10	Kim	.	.	.	.
11	Brook	.	.	.	.
12	Lee	.	.	.	.

In general, it is not a good idea to use repeated measures if any of the treatments produce irreversible or long-lasting changes, or if the effect of a treatment is likely to differ depending on what other treatment preceded it. Leaving long time intervals between different treatments can help reduce carryover effects but is not always effective.

When each participant receives every treatment, it is important to control for (or balance) order effects. Order effects can be balanced with treatment effects by varying the order of presentation. Let Z = Zoloft, N = new drug, and P = placebo (I have omitted Prozac). Depression is measured after each treatment. These three treatments could be administered in six different orders:

Order 1: Z N P

Order 2: Z P N

Order 3: P Z N

Order 4: P N Z

Order 5: N P Z

Order 6: N Z P

In general, if the number of treatments in a study = k, there are $k!$ possible orders. For $k = 3$, $k! = 3 \times 2 \times 1 = 6$. If at least 1 (or better yet, several) participants are randomly assigned to each treatment order, the data can be examined to see whether ordinal position matters. For example, does everyone feel better at Time 3, no matter which drug they receive at that point in time? In addition, possible sequence effects can be examined: Does Zoloft have different effects when it is given after placebo, after the new drug, or after no other drug?

Because the number of possible orders is $k!$, it is clear that the number of all possible orders increases very rapidly for $k = 3, 4$, and so forth. It is possible to use a subset of k out of the $k!$ possible orders if the k orders are chosen such that each treatment occurs once in first place, once in second place, and so forth. This is called a Latin square design. Here is an example:

Order 1: Z N P

Order 2: P Z N

Order 3: N P Z

If equal n's of participants are assigned to Orders 1, 2, and 3, there is no confound between order and treatment effects. However, note that Z appears twice after P, and never after N, so this design does not provide information about some possible sequence effects. The next section discusses analysis of possible order effects in a hypothetical study.

14.12 FIRST EXAMPLE: ORDER EFFECT AS A NUISANCE

Suppose an alcohol seller wants to evaluate taste quality ratings for three types of alcoholic beverages: vodka (V), gin (G), and rum (R). All participants in the study will taste and rate all three beverages. Because each participant tastes all three drinks, order effects might occur. For example, depending on how much they drink in each trial, participants could be inebriated by the third drink; it's possible that no matter which drink is tasted third, that drink is most favorably evaluated. Carryover effects can occur; possibly the taste of gin lingers for a long time, and people may evaluate the taste of rum differently when it follows gin (which has a strong and distinctive taste) than when it does not. Order and sequence effects are examined by looking at the main effect for order and the interaction between order and treatment type. Generally, researchers do not want statistically significant order or sequence effects.

In the following hypothetical study, every participant tastes and rates three types of drink: vodka, gin, and rum. To avoid problems with nonindependence of observation, taste tests are run individually. Only three orders of presentation are included in this hypothetical study; these form a Latin square, so that order effects are not confounded with drink effects.

If the investigator needs to know what happens when people taste rum after vodka, note that this Latin square design does not provide that information. Additional orders would need to be included to evaluate that possibility.

Each participant is randomly assigned to taste the drinks in one of the three orders. Order is a between-*S* factor with three levels. Drink is a within-*S* factor with three levels. The hypothetical data for this example are in a file named drinks.sav. Here is the GLM syntax:

```
GLM gin vodka rum BY order
/WSFACTOR=Drink 3 Polynomial
/METHOD=SSTYPE(3)
/PRINT=DESCRIPTIVE ETASQ
/CRITERIA=ALPHA(.05)
/WSDESIGN=Drink
/DESIGN=order.
```

Table 14.1 Latin Square for Order of Drinks in Hypothetical Tasting Study

	Time 1	Time 2	Time 3
Order 1	Gin	Rum	Vodka
Order 2	Vodka	Gin	Rum
Order 3	Rum	Vodka	Gin

Figure 14.28 Analysis of Within-*S* Effects: Drink Type and Interaction of Drink × Order

Tests of Within-Subjects Effects

Measure: MEASURE_1

Source		Type III Sum of Squares	df	Mean Square	F	Sig.	Partial Eta Squared
Drink	Sphericity Assumed	36.578	2	18.289	42.753	.000	.781
	Greenhouse-Geisser	36.578	1.936	18.893	42.753	.000	.781
	Huynh-Feldt	36.578	2.000	18.289	42.753	.000	.781
	Lower-bound	36.578	1.000	36.578	42.753	.000	.781
Drink * order	Sphericity Assumed	16.489	4	4.122	9.636	.000	.616
	Greenhouse-Geisser	16.489	3.872	4.258	9.636	.000	.616
	Huynh-Feldt	16.489	4.000	4.122	9.636	.000	.616
	Lower-bound	16.489	2.000	8.244	9.636	.003	.616
Error(Drink)	Sphericity Assumed	10.267	24	.428			
	Greenhouse-Geisser	10.267	23.232	.442			
	Huynh-Feldt	10.267	24.000	.428			
	Lower-bound	10.267	12.000	.856			

Figure 14.29 Analysis Comparing Order of Presentation of Drinks

Tests of Between-Subjects Effects

Measure: MEASURE_1
Transformed Variable: Average

Source	Type III Sum of Squares	df	Mean Square	F	Sig.
Intercept	836.356	1	836.356	1636.348	.000
order	88.178	2	44.089	86.261	.000
Error	6.133	12	.511		

Partial results appear in Figure 14.28 and 14.29. (The sphericity test is not shown.) Researchers usually want significant differences among the "treatments" of interest (in this example, significant differences in ratings given to different types of drinks: rum, gin, and vodka). There was a statistically significant difference in the ratings given to different drinks, $F(2, 24) = 42.75$, $p < .001$, $\eta^2 = .781$ (Figure 14.28). GLM does not provide drink names in the summary table of means; to identify the drinks, look back at the GLM menu selections or syntax used to define the design.

Of course, a table presented in a journal article would include drink names instead of uninformative drink numbers. In this imaginary study, rum had the highest mean taste rating; gin had the lowest. Contrasts can be used to make pairwise comparison of means across drinks (not shown here).

There were also statistically significant effects for order, and the order-by-drink interaction (Figures 14.28 and 14.29). To evaluate what is happening, examine the mean drink ratings separately for all orders (Figure 14.31).

The one rating that stands out as unusually high is the rating for rum in Order 3, the order in which rum was tasted first, $M = 8.600$ (Figure 14.31). Rum was rated best tasting overall, but that may be partly because it received an unusually high rating in Order 3 (in which it was tasted first). The taste rating of rum depends on context (other drinks).

Figure 14.30 Mean Taste Ratings of Drinks

3. drink

Measure: MEASURE_1

drink	Mean	Std. Error	95% Confidence Interval Lower Bound	Upper Bound
1	3.267	.189	2.856	3.678
2	4.200	.149	3.875	4.525
3	5.467	.183	5.069	5.864

Note: 1 = gin, 2 = vodka, 3 = rum.

Figure 14.31 Mean Taste Rating for Each Drink Separately by Order

order	drink	Mean	Std. Error	Lower Bound	Upper Bound
1 **GRV**	1 **G**	2.200	.327	1.488	2.912
	2 **V**	3.200	.258	2.637	3.763
	3 **R**	3.800	.316	3.111	4.489
2 **VGR**	1 **G**	2.800	.327	2.088	3.512
	2 **V**	4.000	.258	3.437	4.563
	3 **R**	4.000	.316	3.311	4.689
3 **RVG**	1 **G**	4.800	.327	4.088	5.512
	2 **V**	5.400	.258	4.837	5.963
	3 **R**	8.600	.316	7.911	9.289

In some research situations where the analysis indicates significant order effects or interactions, the data analyst might conclude that having each participant experience all of the treatments leads to unwanted complications. In such situations, leaving a longer time interval between treatments might help reduce possible carryover. Doing the imaginary drink rating study as completely between-*S* would be a more effective way to get rid of order effects.

14.13 SECOND EXAMPLE: ORDER EFFECT IS OF INTEREST

In some research situations, order effects provide information of interest (for example, contrast effects). Kenrick and Gutierres (1980) set up a study to assess whether men rated the physical attractiveness of a female friend lower if they made the rating just after rating the physical attractiveness of a female television actor. In their design, each male student rated two photos: a female actor and a female friend. Students were randomly assigned to one of two orders: Order 1, female friend first, and Order 2, female actor first. They hypothesized that female friends would be rated lower in attractiveness when they were rated after seeing the female actor photo.

Data for a hypothetical version of this study are in the file actorvsfriend.sav. The analysis was GLM repeated measures with person being rated (female actor vs. friend) as the within-*S* factor and order (female friend first, female actor first) as the between-*S* factor. In this example, researchers hope for evidence of order effects, and a statistically significant Person Rated × Order effect was found. The SPSS syntax is as follows:

```
GLM friendrating actressrating BY order
/WSFACTOR=Person 2 Polynomial
/METHOD=SSTYPE(3)
/PRINT=DESCRIPTIVE ETASQ
/CRITERIA=ALPHA(.05)
/WSDESIGN=Person
/DESIGN=order.
```

There was a statistically significant interaction for person rated by order, Wilks' Λ = .766, $F(1, 18) = 5.486$, $p = .031$, $\eta^2 = .234$ (Figure 14.32). To understand the results, consider a plot of the cell means (Figure 14.33). Post hoc analysis can be done to compare the means that are contained in the ellipses. Consider the lower left-hand ellipse. When friend is rated second

Figure 14.32 MANOVA for Interaction Between Person Rated (Female Friend, Female Actor) by Order

Multivariate Tests[a]

Effect		Value	F	Hypothesis df	Error df	Sig.	Partial Eta Squared
person_rated	Pillai's Trace	.763	57.943[b]	1.000	18.000	.000	.763
	Wilks' Lambda	.237	57.943[b]	1.000	18.000	.000	.763
	Hotelling's Trace	3.219	57.943[b]	1.000	18.000	.000	.763
	Roy's Largest Root	3.219	57.943[b]	1.000	18.000	.000	.763
person_rated * order	Pillai's Trace	.234	5.486[b]	1.000	18.000	.031	.234
	Wilks' Lambda	.766	5.486[b]	1.000	18.000	.031	.234
	Hotelling's Trace	.305	5.486[b]	1.000	18.000	.031	.234
	Roy's Largest Root	.305	5.486[b]	1.000	18.000	.031	.234

a. Design: Intercept + order
Within Subjects Design: person_rated

b. Exact statistic

Figure 14.33 Attractiveness Ratings of Female Friends Versus Female Actors

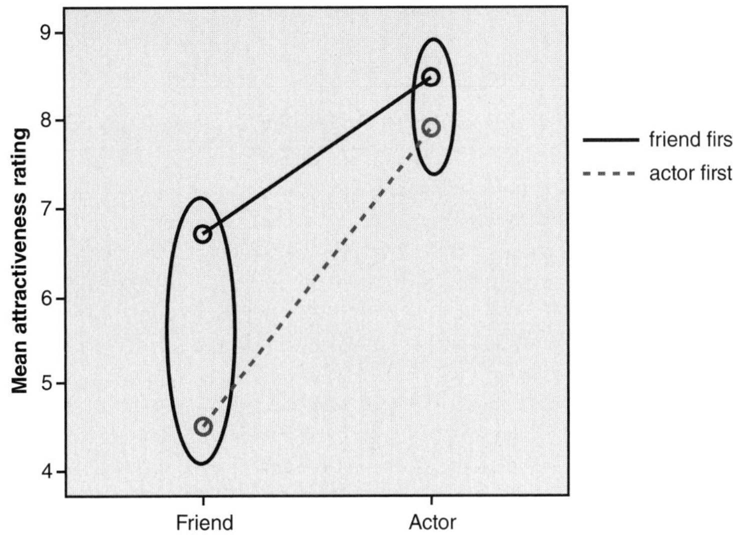

(the lower of the two means), she receives a much lower rating than the female actor. This may be due to a contrast effect. (That is, after a person looks at a highly attractive individual, other people appear less attractive.)

Post hoc comparisons can be done to assess which of these cell means differ significantly. The result that appears clearly distinct from the others: Female friends are rated lowest when they are rated after female actors (this refers to the mean in the lower left of the plot above).

Subsequent research suggested that people rate their own attractiveness, as well as the attractiveness of their friends, lower after seeing attractive people in mass media. Maybe we would all benefit from watching less television and paying less attention to Instagram and Snapchat.

14.14 SUMMARY AND OTHER COMPLEX DESIGNS

The examples in this chapter show several ways in which concepts, designs, and designs and analyses covered in many previous chapters can be combined. It is possible to test participant-by-treatment interaction. The sphericity assumption was explained in more detail. When the sphericity assumption is violated, researchers may choose to use adjusted *df* tests such as the Greenhouse-Geisser; another option is the MANOVA approach to repeated measures, which does not require the assumption of sphericity. A repeated-measures ANOVA design can include one or more within-*S* and one or more between-*S* factors. A repeated-measures doubly multivariate study can include measures of multiple outcome variables. Tukey post hoc tests can be used to make comparisons across levels of between-*S* factors; contrasts or paired-samples *t* tests can be used to compare means across levels of within-*S* factors. Order effects can be evaluated.

Repeated measures is one of several possible ways to assess change over time (for example, in pretest vs. posttest studies). ANCOVA is another possible approach. Unfortunately, these two analyses applied to the same data do not always yield the same results.

More sophisticated methods for analysis of change over time have been developed (such as growth curve and longitudinal SEM).

One or more covariates can also be added to a mixed-model ANOVA design; no examples were presented.

The one-between-*S* and one-within-*S* factor ANOVA examples presented here can be viewed as a special case of profile analysis. Profile analysis can be used to make other kinds of comparisons not discussed here. For further discussion, see Tabachnick and Fidell (2018).

COMPREHENSION QUESTIONS

1. Repeated-measures ANOVA is used only when repeated measures are made on the same people or cases: true or false? Why?
2. In a repeated-measures study to assess effect of crowding on hostility at four points in time, some participants (such as Jack) show increasing hostility across times of assessment, while other participants, such as Jill, show decreasing hostility across time. (They are in the same treatment group.) What term would you use to describe this pattern of response? Does the basic one-way repeated-measures ANOVA introduced in Volume I, Chapter 15 (Warner, 2020), tell us anything about this?
3. What analysis can you run to assess whether there may be a participant-by-time (or participant-by-treatment) interaction?
4. Suppose that a group of students all take the same test of math anxiety (MA) at four points in time during a semester. The test results are in the repeated-measures variables MA_1 to MA_4. Would you expect MA_1, MA_2, MA_3, and MA_4 to be correlated with one another? If so, why?
5. If there are high correlations among these measures, does this violate assumptions needed for one-way repeated-measures ANOVA?
6. True or false:
   a. A Latin square design is needed only when the within-*S* factor levels correspond to different treatments.
   b. A Latin square design includes different orders of presentation of treatments such that each treatment is given once in each ordinal position.
   c. A Latin square design allows the researcher to detect possible sequence effects.
   d. In a design that controls for order effects using a Latin square, the researcher always hopes that the main effect for order and interactions involving order will not be statistically significant.
7. What test statistic is used to assess whether the sphericity assumption is violated?
8. What range of possible values is there for the Greenhouse-Geisser ε?
9. What does the magnitude of ε tell us? How is the ε value used when doing *F* tests?
10. If the sphericity assumption is violated, what problem is likely to occur?
11. Although both the sphericity assumption in repeated-measures ANOVA and the homogeneity of $\mathbf{\Sigma}$ test in MANOVA examine variance/covariance matrices, they have different null hypotheses. Explain briefly how these tests differ.
12. Does MANOVA require the sphericity assumption?
13. What kinds of test statistics appear in the univariate part of repeated-measures ANOVA output?
14. What kinds of test statistics appear in the multivariate (MANOVA) part of repeated-measures output?
15. What effect size information is provided for the univariate version of repeated-measures ANOVA? For the multivariate (MANOVA) version?
16. What are two possible reasons to report MANOVA instead of the univariate approach to repeated-measures ANOVA?

17. What are the sample size guidelines for the use of MANOVA in repeated measures?

18. When the omnibus test for differences among means on the within-*S* factor is statistically significant, what additional information is needed to evaluate the nature of these differences?

19. What is a doubly multivariate repeated-measures design? Give a specific example using variables different from those used in this chapter.

20. In the output from a doubly multivariate repeated-measures analysis, are the follow-up tests multivariate or univariate? How do you know?

21. When one or more between-*S* factors are combined with one or more within-*S* factors, what is this type of factorial design called? What additional information does this analysis provide (compared with the one-way repeated-measures ANOVA)?

22. If there is a significant between-*S* by within-*S* factor interaction, what additional analyses provide information about the nature of this interaction?

23. In a mixed-model ANOVA with a between-*S* factor with levels that represent different treatment groups and a between-*S* factor that represents times of assessment before intervention and at one or more follow-up times after intervention:

    a. Do we want to see statistically significant differences among Time 1 means across treatment groups? Why or why not?
    b. Do you think the interaction term or the main effect for the between-*S* variable would be more informative about the impact of the intervention? Why?

NOTES

[1]Options for types of contrasts within levels of a repeated-measures factor are discussed in Chapter 15 in Volume I (Warner, 2020), on one-way repeated measures.

[2]The difference between the partial eta squared (η^2) effect size reported by SPSS and the simple η^2 that was discussed in earlier situations involving *t* tests and ANOVA was discussed earlier in Chapter 16 in Volume I, on factorial ANOVA. A simple η^2 answers the question, What proportion of the total variance in the outcome variable (such as heart rate) is predictable from the factor (time, in this example)? A partial η^2 answers the question, After you partial out or remove variance due to other factors (in this example, differences among participants) from the total variance, what proportion of that remaining variance can be predicted from the factor of interest? In this case, the partial η^2 tells us what part of the remaining variance, after we remove variance in heart rate scores that is predictable from individual differences among participants, is predictable from the time factor. Generally, partial η^2 is smaller than simple η^2 (unless something unusual is happening, such as suppression).

DIGITAL RESOURCES

Find **free study tools** to support your learning, including **eFlashcards, data sets, and web resources**, on the accompanying website at **edge.sagepub.com/warner3e**.

CHAPTER 15

STRUCTURAL EQUATION MODELING WITH AMOS

A Brief Introduction

15.1 WHAT IS STRUCTURAL EQUATION MODELING?

Structural equation modeling (SEM) is a general analytic method that evaluates relations among variables on the basis of the information in the **variance/covariance matrix**. Analyses discussed in previous chapters were also based on the variance/covariance matrix or the corresponding correlation matrix. Many analyses discussed in previous chapters can be viewed as special or limited cases of SEM or, alternatively, as components that can be included and combined in an SEM analysis. SEM can include path models, multiple regression, moderation, mediation, and factor analysis. SEM is now widely used in many fields (MacCallum & Austin, 2000).

Why use SEM? Structural equation models that include latent variables provide a way to avoid an assumption that is commonly violated when regression and path models are used. Regression analyses implicitly assume that variables are assessed perfectly reliably (i.e., without measurement error). SEM provides a way to handle measurement error through the inclusion of **measurement models** that use multiple observed variables for each construct and shows their relation to latent variables. A measurement model in SEM represents assumptions about the way an unobserved or latent variable is related to **measured variables**, similar to factor analysis, where latent variables are called factors. Later in the chapter, Figure 15.25 is a graphic representation of a measurement model for selected items from the Bem Sex Role Inventory previously examined in the factor analysis chapter. Each oval in Figure 15.25 corresponds to a factor or latent variable, each rectangle corresponds to a measured variable, and the standardized path coefficients for paths that lead from each latent variable to a measured variable correspond to factor loadings. Measured variables are also called observed variables or indicators. The use of multiple indicators for a factor such as femininity provides information about measurement reliability for this factor; internal consistency reliability is higher when factor loadings are large in absolute value. Use of latent variables as predictors and outcomes in SEM provides a way of dealing with measurement error and reliability issues instead of ignoring them.

A second part of many SEM analyses is a **structural model** (also called a path or "causal" model[1]) based on hypotheses about direct and indirect relations among variables. Look ahead to the model in Figure 15.4, which corresponds to the example used in the mediation analysis chapter. This model shows two paths from age to blood pressure: a direct path c and an indirect path from age to blood pressure via weight, the $a \times b$ path.

Look ahead to Figure 15.36 for a diagram of a structural equation model that includes measurement models for both predictor variables along with a structural model. A structural equation model can simultaneously examine measurement issues (how closely indicator variables are related to latent variables) and assess causal hypotheses that involve both latent and measured variables.

Even for simple analyses such as regression or mediation analysis, the use of an SEM program provides advantages. When normality assumptions are violated, some of the estimation methods available in SEM programs provide better information about confidence intervals (CIs) and statistical significance than **ordinary least squares (OLS)** methods. Analysis of mediation models can be done all at one time, instead of running separate regressions and piecing results together, and it is easy to include multiple step mediated paths. Amos provides graphs of path models. These graphs make the pattern of hypothesized relationships easier to understand, particularly for complex models. Finally, learning SEM is a step toward understanding other advanced analytic techniques that include latent variables.

Requirements and assumptions for SEM data are discussed, including power and sample size issues. The chapter introduces structural equation models gradually using empirical examples discussed in earlier chapters. The first example uses SEM to analyze the simple path model in the mediation chapter. The use of the Amos Graphics interface to set up and test this model is explained step by step for this mediation model. The second empirical example is a **confirmatory factor analysis (CFA)**[2] that includes measurement models to reexamine the masculinity and femininity factors in the Bem Sex Role Inventory data; these data were used earlier in exploratory factor analysis (EFA) in Chapter 12. CFA allows users to test a specific hypothesized factor structure model (number of factors and specific variables that load on each factor), while EFA searches for structure in the data with few a priori hypotheses. The third empirical example involves prediction of a latent well-being variable (with indicator variables that include positive affect, life satisfaction, and happiness) from stress and social support using new unpublished data. This third model includes both structural and measurement components.

Introductions to SEM generally assume familiarity with path models, factor analysis, and the idea that a model should be able to reproduce the structure in the original data (the correlation or variance/covariance matrix). This chapter reviews these fundamental concepts and makes explicit connections with earlier topics. Coverage of topics in earlier chapters provides the information needed to understand SEM. For example, the discussion of exploratory factor analysis explains the ability of a model to reproduce the correlation matrix.

IBM SPSS Amos Version 24 is used for the SEM analyses in this chapter. Some analysts prefer other programs, such as Mplus, LISREL, or EQS. Kline (2016) provided an evaluation and comparison of numerous SEM programs. Byrne described SEM using Amos (Byrne, 2016), EQS (Byrne, 2013), Mplus (Byrne, 2011), and LISREL and similar programs (Byrne, 1998).

15.2 REVIEW OF PATH MODELS

When path or causal models were introduced, a path model was used to represent a regression analysis for standardized quantitative variables (with two predictors, z_{X1} and z_{X2}, and one outcome variable, z_Y). Assume that other assumptions for regression are met (normally distributed data, linearly related variables). It is convenient to use z scores for all variables; this makes it easy to demonstrate that the total effect of each z_{Xi} predictor on z_Y, the r_{X1Y} correlation, can be divided into one direct effect and one indirect effect. The path model that represents regression with two correlated predictors appears in Figure 15.1.

Recall that a doubled headed arrow (**noncausal path**) represents the hypothesis that the connected variables are correlated, but not causally associated. In this example z_{X1} and z_{X2} are correlated, but neither is thought of as a hypothesized cause of the other. A unidirectional arrow (**causal path**) represents the hypothesis that the variable at the origin of the arrow causes the variable at the end of the arrow. Chapter 4 introduced an implicit model for regression with correlated predictors. The model in Figure 15.1 corresponds to the hypotheses that z_{X1} causes z_Y, z_{X2} causes z_Y, and z_{X1} is correlated with z_{X2}.

Figure 15.1 Path Model for Prediction of z_Y From Correlated Independent Variables z_{X1} and z_{X2}

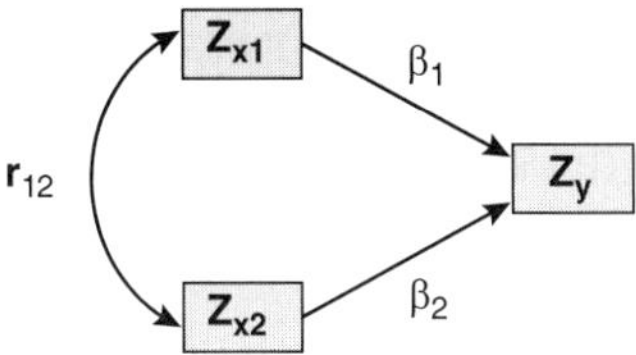

The path diagram in Figure 15.1 can be translated into the following regression equation. To translate a path model diagram into one (or more) equations, first, identify the dependent variable(s). A dependent variable has one or more unidirectional causal arrows pointing toward it. Each dependent variable will be the predicted variable in its own regression equation. There is only one dependent variable, z_Y, in this first example. Structural equation models often have more than one dependent variable. Second, notice which variables have unidirectional arrows that point toward each dependent variable. There are two independent variables that have arrows that point to z_Y: z_{X1} and z_{X2}. These are the predictor variables in the regression equation. The standardized version of the regression equation that corresponds to Figure 15.1 is:

$$z_Y' = \beta_1 z_{X1} + \beta_2 z_{X2}. \quad (15.1)$$

The use of the tracing rule to identify all the paths in the model that lead from each predictor to the dependent variable was introduced earlier. There are two paths from z_{X1} to z_Y. First, there is the direct path from z_{X1} to z_Y (with the coefficient β_1). There is also an indirect path from z_{X1} to z_Y via z_{X2}; this path has two steps. On the basis of the tracing rule, to combine coefficients along a path, multiply them. For the two-step indirect path from z_{X1} to z_Y, the first step is from z_{X1} to z_{X2} (this path has a coefficient of r_{12}). The second step is from z_{X2} to z_Y; this path has a coefficient of β_2. The strength of association of z_{X1} to z_Y by way of the indirect path is $r_{12} \times \beta_2$. The strength of the direct path is given by β_1. On the basis of the tracing rule, to combine information from different paths, their contributions are summed. The total correlation of z_{X1} with z_Y can be reproduced by summing the direct and indirect paths, as in the following equation:

$$r_{1Y} = \beta_1 + r_{12} \times \beta_2. \quad (15.2)$$

Total effect = Direct effect + Indirect effect

The same reasoning can be used to show that the total correlation between z_{X2} and z_Y can be reproduced from a direct and an indirect path:

$$r_{2Y} = \beta_2 + r_{12} \times \beta_1. \quad (15.3)$$

Equations 15.2 and 15.3 are called the normal equations for regression. The correlations among the variables (r_{12}, r_{1Y}, r_{2Y}) are obtained from the data. The unknowns are the path coefficients β_1 and β_2. The two equations in two unknowns can be solved to give values of the unknowns β_1 and β_2 in terms of the three correlations (as shown in the chapter that used two predictors in regression as an introduction to statistical control). In other words, in a regression with two predictors, the standardized path coefficients (β values) are a function of the correlations among all the variables. (The raw score *b* coefficients are computed from sums of squares and sums of cross products that contain information in terms of original units of measurement.)

This example used standardized variables (*z* scores and β coefficients). The model fit of a multiple regression was assessed by showing that the path model and β values can perfectly reproduce the correlation matrix **R** (provided that all predictor variables have direct paths to z_Y and are allowed to be correlated with each other). Similarly, fit of a factor analysis model was assessed by evaluating how well the correlation matrix **R** could be reproduced by factor loadings for retained factors.

In SEM, computations are based on the variance/covariance matrix that summarizes information about raw scores. Structural equation model fit is assessed by evaluating how well the path model and its raw-score coefficients can reproduce the elements of the variance/covariance matrix. There is an algebra of variances and covariances (Ullman, 2007), analogous to the algebra described above based on correlations.

15.3 MORE COMPLEX PATH MODELS

Multiple regression with more than two predictors can also be diagrammed as a simple path model. Consider the example in Figure 15.2.

In this multiple regression model, each z_{Xi} predictor has one direct path to z_Y (via β_i), and three indirect paths (via each of the other three predictor variables). Regression divides the total relationship of each z_{Xi} predictor into one direct or unique part, and the sum of multiple indirect paths to z_Y that are due to correlations with other predictors. The β coefficients for this model can be obtained from OLS multiple linear regression (they can also be obtained using SEM, if desired).

Structural equation models can include more specific assumptions about how variables may be related by direct and indirect paths, instead of just treating them as correlated predictors. The model in Figure 15.3 includes hypothesized mediated causal paths. As in regression, path coefficients for SEM can be given in standardized or unstandardized units. To be consistent with the prior discussion, standardized units are used here.

Figure 15.2 Path Model for Standardized Regression With Four Correlated Predictors z_{X1}, z_{X2}, z_{X3}, and z_{X4}

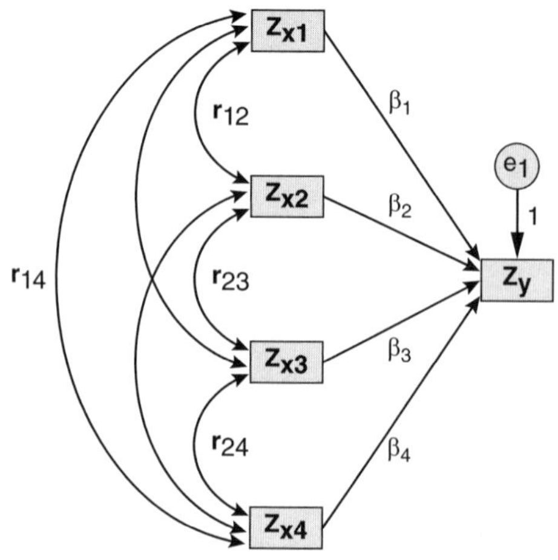

Figure 15.3 A Path Model Representing Two Mediated Paths for Prediction of z_Y From Four Predictors

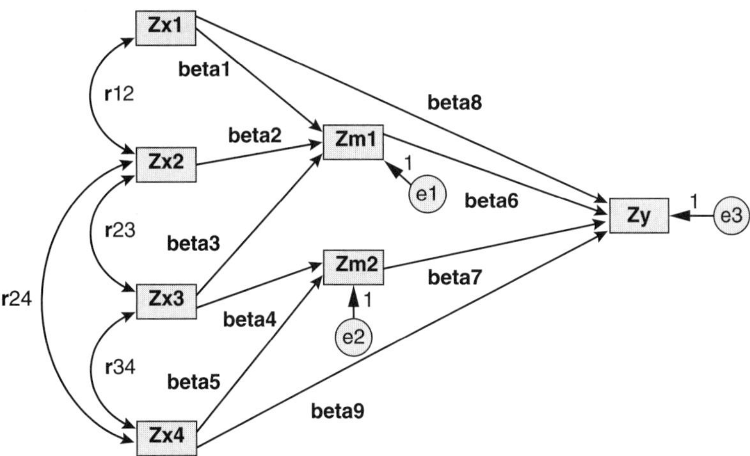

The path model in Figure 15.3 represents more complex hypotheses about which variables are related to one another and how they are related (e.g., causally or noncausally, directly or indirectly or both). On the basis of the path model in Figure 15.3, it is possible to answer questions about hypothesized relations among variables such as the following:

- Which variable (or variables) are "final" dependent variable? z_Y.
- Which variables are predictors? z_{X1}, z_{X2}, z_{X3}, and z_{X4}.
- Which variables are hypothesized to be mediators? z_{M1} and z_{M2}.
- Which pairs of z_X predictor variables are correlated with each other? The absence of a path is just as important as the presence. This model includes a correlation path between z_{X1} and z_{X2}. The absence of paths between z_{X1} and the other predictors z_{X3} and z_{X4} tells us that the model assumes that these are uncorrelated. (Amos produces a warning message if a model specifies uncorrelated predictors; however, it allows models with uncorrelated predictors to run.)
- What path or paths exist between z_{X1} and z_Y? Z_{X1} has a direct path to z_Y (the path marked β_8) and an indirect path mediated by z_{M1} ($\beta_1 \times \beta_6$). z_{X1} has an additional indirect path to z_Y because z_{X1} is correlated with z_{X2}, and z_{X2} has a direct path to z_Y. The product of coefficients along this path is $r_{12} \times \beta_2 \times \beta_6$.

Therefore, the total association of z_{X1} with z_Y represented by the assumptions in this model is the sum of these three paths: $\beta_8 + (\beta_1 \times \beta_6) + (r_{12} \times \beta_2 \times \beta_6)$. As we shall see, a model that omits some of the possible paths usually cannot perfectly reproduce total correlations and covariances between variables. The raw-score coefficients for the model in Figure 15.3 probably will not perfectly reproduce the variance/covariance matrix; the standardized coefficients probably will not perfectly reproduce the correlation matrix **R**. The magnitude of differences between the reconstructed variance/covariance matrix on the basis of the structural equation model and its parameter estimates, and the variance/covariance matrix for the original data, provides information about model fit.

Descriptions of path models are often called causal models. That is because causal *hypotheses* are represented by unidirectional arrows. Discussion of associations among variables is often in terms of "effects" (e.g., the total effect of z_{X1} on z_Y can be divided into direct and indirect effects). Be careful not to let these terms lead you astray. Usually the data used in path and SEM analyses come from nonexperimental research. Analysis of correlational data may yield results that are consistent with, or inconsistent with, a preferred causal model. Analysis of purely correlational data cannot prove causality, although sophisticated models may be able to rule out some rival hypotheses.

If a structural equation model fits well, and thus is consistent with the pattern in the data, that result does not prove that the causal hypotheses included in the model are correct. In many situations, replacing a unidirectional arrow that corresponds to a causal hypothesis with a bidirectional noncausal arrow (or sometimes with a unidirectional arrow that reverses the direction of causation) does not change model fit. Sometimes minor differences in assumptions about causal versus correlational paths lead to equivalent models (i.e., models based on different causal or mediation hypotheses that fit the data equally well). Results of SEM analysis do not provide a basis to prefer one equivalent model to others. If a structural equation model does not fit well, that does not prove that the causal hypotheses in the model are incorrect, although it does fail to show results consistent with the model. MacKinnon, Krull, and Lockwood (2000) explained that we cannot distinguish among mediation, confounding, and suppression models. If the path from X_1 to Y becomes nonsignificant after controlling for X_2, that outcome could be explained by saying that X_2 mediates the effects of X_1 on Y, that X_1 is confounded with X_2 as a competing predictor, or that X_2 is a suppressor variable. Because all three models predict the same empirical outcome, if we obtain that outcome, we cannot know which model is correct.

15.4 FIRST EXAMPLE: MEDIATION STRUCTURAL MODEL

The first example demonstrates that SEM can be used to estimate coefficients for the mediation model discussed in Chapter 9, on mediation. The data in ageweightbp.sav ($N = 30$) are used. A sample of 30 is too small for a publishable result. The variable for systolic blood pressure is labeled SBP in all output.

The model in Figure 15.4 represents the hypothesis that the effect of age on blood pressure is mediated (at least partially) by body weight. Age has two paths to blood pressure: the direct path (c) and an indirect path ($a \times b$). This model corresponds to two regression equations. The first equation predicts weight from age to obtain the a coefficient. The second equation predicts blood pressure from both age and weight to obtain the b and c coefficients. Note that in Amos, error variables (e_1, e_2) must be explicitly included for each dependent variable. Amos and SEM can be used to simultaneously obtain results similar to those from the two regressions that were run separately using OLS earlier. Amos can also provide bootstrapped CIs for model parameters, including the ($a \times b$) indirect path. The first empirical example uses the same hypothetical data (ageweightbp.sav) as in the mediation chapter.

These are the questions we can ask about the model in Figure 15.4:

- Is this model a good fit? Can it reproduce the correlations (and variances/covariances) among variables? Path models that include direct paths between every pair of variables, as in this example, can reproduce correlations and covariances perfectly. This model will have perfect fit to the data (that does not tell us anything about the strength of prediction among variables). Models introduced in Examples 2 and 3 will not fit perfectly, and that is the usual situation in applications of SEM. For these later models, information about goodness of fit will be needed.

Figure 15.4 Hypothesized Model for Partial Mediation of Effects of Age on Blood Pressure by Weight

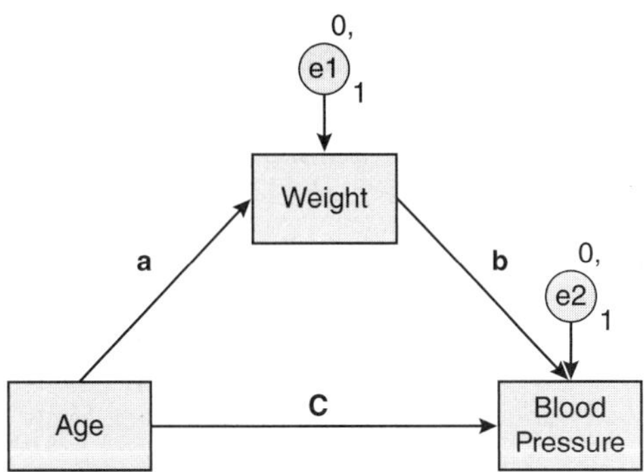

- Which path coefficients in this model are statistically significant, how large are they, and what signs do they have? In this situation, we would expect all three variables to be positively correlated with each other. As people age, weight increases, and systolic blood pressure increases; and to some extent, as weight increases, blood pressure increases.
- What are the R^2 values for prediction of weight and blood pressure? If these variables predict only a trivial amount of variance in blood pressure (e.g., $R^2 < .01$), this is not a very informative model.
- What are the contributions of direct versus indirect paths? This is assessed by looking at the statistical significance (and the relative magnitude) of path *c*, versus $a \times b$.

The indirect path or mediation hypothesis path ($a \times b$) is of particular interest. If $a \times b$ is statistically significant, we can say that a statistically significant part of the total effect of age on blood pressure can be accounted for by the indirect path from age to blood pressure via weight. A data analyst may want to argue that a significant $a \times b$ path is evidence for (or even "proves") mediation. That is too strong a conclusion to reach when data are correlational. Mediation is a two-step causal hypothesis, including the hypotheses that first age causes weight and then weight causes blood pressure. As noted earlier, when we have only correlational data, we cannot make causal inferences. A statistically significant result for $a \times b$ would be consistent with the mediation hypothesis; but it could equally well be interpreted in other ways (for example, as evidence of correlated predictors or spuriousness; MacKinnon et al., 2000). Statistical significance for the *c* path coefficient for the direct path is also evaluated. If $a \times b$ is significant and *c* is not significant, this would be consistent with a hypothesis that effects of age on blood pressure are completely mediated by weight. If $a \times b$ is significant and *c* is also significant, a possible interpretation is partial mediation of effects of age on blood pressure through weight.

15.5 INTRODUCTION TO AMOS

IBM Amos Version 24, used for all analyses in this chapter, is an add-on program for SPSS. A separate license is required. It does not have the full range of capabilities some other SEM programs offer (such as EQS, Mplus, and LISREL). However, Amos provides a simple way to specify path models. Running analyses in Amos Graphics involves the following steps, each discussed in more detail in subsequent sections.

1. **Prior to using Amos, conduct preliminary data screening** using SPSS Base. Variables should have a multivariate normal distribution (or at least univariate normal distributions); all relations between pairs of variables must be linear; and extreme outliers should not be present. Amount and pattern of missing values should be evaluated, and replacements for missing values should be imputed. (Those steps are not included in the examples in this chapter.) Amos will impute missing values if the input data set has missing data, but this will not provide you with assessment of amount of pattern of missing data, and some useful statistics, such as modification indexes (MIs), are not available when Amos does missing value imputation.
2. **Open the Amos Graphics program.** The shortcut icon is in pale green. Other Amos files are represented by the same icon in different colors, for example, text output files are represented by the same icon in purple. Note that the output file tends to be docked (hidden; it may seem to disappear). Many other "junk" files generated by Amos can be deleted after analysis.
3. **Use Amos drawing tools to create a path diagram that represents the hypothesized model.** This first example is the same as the mediation model examined earlier (weight as a possible mediator of the effects of age on blood pressure).
4. **Open the SPSS data file.**
5. **Give each measured variable a name that corresponds to a variable in the SPSS file.**
6. **View and edit analysis properties** to specify how the analysis will be performed and what output you want.
7. **Run the analysis** and check whether it ran successfully.
8. **If there were model specification errors, make corrections** to variable names and/or changes in the path model and run the analysis again.
9. **View and interpret the text and graphic output.** Some optional Amos output includes information about violations of normality assumptions and the presence of multivariate outliers. If problems are found, the analyst needs to reexamine the data and make corrections.

Each of the three SEM examples that follow introduces different concepts and includes different types of output.

Instructions for the use of Amos involve many menu selections, and navigation among screens can be confusing. There are many helpful videos on YouTube that walk you through the sequence of choices step by step. If you lose track of where you are in the following discussion of Amos screenshots, a YouTube video will help you understand how to navigate among menu selections and screen views.

15.6 SCREENING AND PREPARING DATA FOR SEM

15.6.1 SEM Requires Large Sample Sizes

Two kinds of sample size guidelines have been suggested. Some authors simply suggest total minimums for N for SEM in general or an N of cases for each variable in the model. Wolf, Harrington, Clark, and Miller (2013) reviewed widely used general guidelines:

> Various rules-of-thumb have been advanced, including (a) a minimum sample size of 100 or 200 (Boomsma, 1982, 1985), (b) 5 or 10 observations per estimated parameter (Bentler & Chou, 1987; see also Bollen, 1989), and (c) 10 cases per variable (Nunnally, 1967).

They noted that these general guidelines do not take specific features of models into account and may over- or underestimate sample sizes needed for adequate statistical power. Some specific features of CFA models that affect sample size requirements include number of factors, number of indicator variables, magnitude of loadings of indicator variables on latent variables, correlations among latent variables, and amount of missing data.

Taking features of specific models into account, statistical power analysis can be used to find required sample sizes; this can be power for significance test of individual path coefficients or power for the entire model (including model fit indexes such as the root mean square error of approximation [RMSEA], introduced in a later section). Kline (2016) and Preacher and Coffman (2006) provided detailed discussion of statistical power in SEM along with references to sources of tables and software to evaluate sample size requirements; also see Lee, Cai, and McCallum (2012).

15.6.2 Evaluating Assumptions for SEM

Types of variables: Consider the types of variables. SEM can handle both quantitative and dummy-coded variables as predictors. Amos SEM cannot handle categorical variables as outcomes; this requires some form of logistic regression. Amos SEM is also not the most appropriate analysis for dependent variables that represent counts of events or behaviors (better analyses for counted outcome variables are based on models such as zero-inflated Poisson regression).

Distribution and linearity assumptions: Ideally, variables for SEM would have a multivariate normal distribution, and relations among variables would be linear. Linearity is not assessed within Amos and must be examined in preliminary SPSS data screening. SEM analysis is based on covariances; covariances, like correlations, are not suited to capture relations between variables that are non-normally distributed, have extreme outliers, or are not linearly related. Amos can provide some information about violations of normality assumptions and the presence of multivariate outliers, but more detailed information can be obtained by doing preliminary data screening in SPSS. The most common SEM estimation method, maximum likelihood estimation assumes that data are multivariate normal; it is most important that dependent variables be normally distributed.[3]

No extreme outliers: Covariance estimates, like correlations, may be disproportionately influenced by outliers. Chapter 2 discussed identification and handling of outliers. Amos can provide information about the existence of outliers; however, these cannot be modified within Amos. Outliers can be identified and handled within SPSS.

Missing values: Amos replaces missing values by default. However, it does not provide information about amount or pattern of missing data; that information can be obtained from the SPSS Missing Values add-on. When Amos replaces missing values, some useful information (such as **modification indexes**) is not available. Like other SEM programs, Amos is a bit of a black box. When it makes decisions about issues such as handling missing values, the data analyst has less control, and less information about what was done, than when these are managed in SPSS.

15.7 SPECIFYING THE STRUCTURAL EQUATION MODEL (VARIABLE NAMES AND PATHS)

Amos provides a user-friendly graphical interface so that the data analyst can specify the model by drawing a path diagram. (Other SEM programs may require input in the form of matrices of parameters to be estimated.) When Amos is opened, the initial screen view appears as in Figure 15.5.

Across the top of the Amos Graphics main page there is a menu bar used to open and save files, to specify how the analysis will be performed, and to view files such as text output. The left-hand panel has Amos tools for drawing, copying, deleting, and so forth (Figure 15.6).

Just to the right of the tools (in the middle) there is a column that consists of small windows with headings: "Group number 1," "XX: Default model," and so forth. (The analyses in this chapter all use just one group; comparisons of models across groups are possible.) The large blank space on the right-hand side is the drawing area where the path model is created; it is also one of the places where path coefficients can be viewed after analysis.

15.7.1 Drawing the Model Diagram

It is a good idea to sketch the model on a sheet of paper before you draw the model diagram. It is conventional to show a "flow" of causality either from left to right or from top to bottom. To enlarge the drawing area, you can select <View> → <Interface Properties> to resize it. The drawing does not have to be neat for the analysis to run. However, if you want to

Figure 15.5 Initial Screen View in Amos Graphics

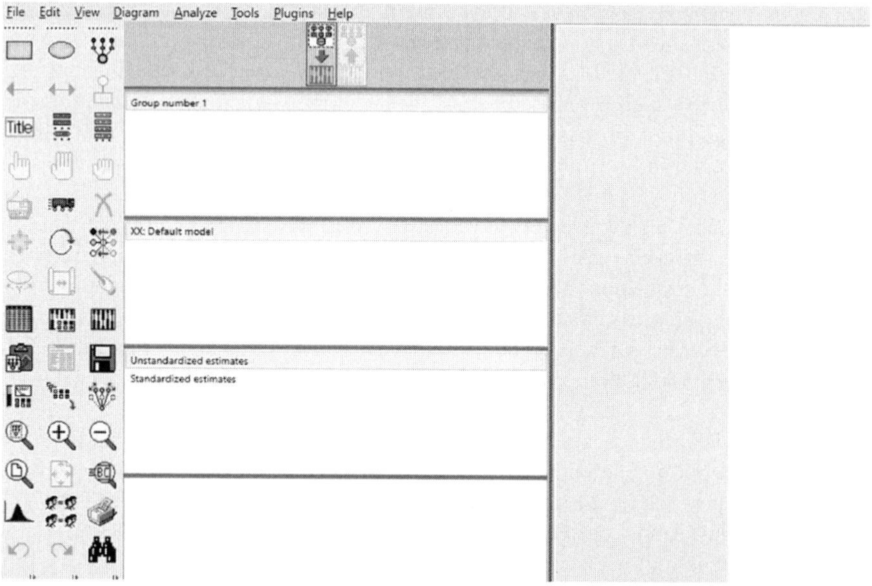

Figure 15.6 Amos Drawing Tools

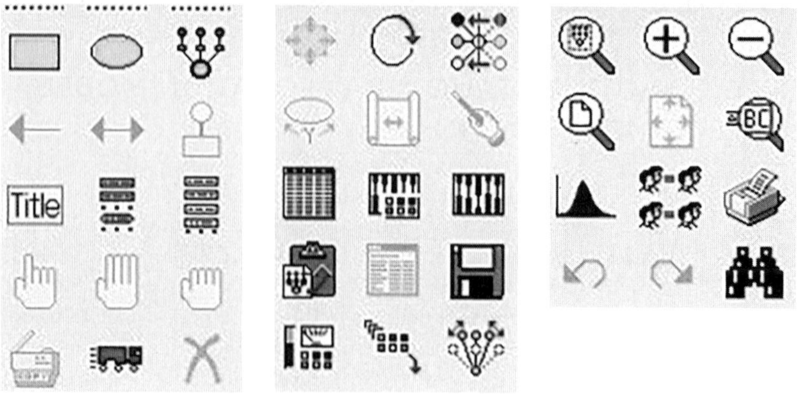

use the diagram as the basis for publishable graphs, it helps to make objects similar in size, to align them, and to avoid lines that cross over other elements in the diagram if possible. *Warning: As you edit your diagram, save it frequently. It is easy to accidentally exit from a project without saving your work.* From the top-level menu, select <File> → <Save as> to save your work.

- The rectangle tool creates a rectangle or square that corresponds to an observed or indicator variable.
- If you want to create multiple objects of the same size, use the duplicate tool  .
- The single-headed and double-headed arrow tools  are used to draw paths.
- The error tool  is used to create an error term for each dependent variable in the path model (error terms must be explicitly included in SEM path models).
- The moving truck tool  moves objects in the path model.
- If you want to "grab" several pieces of the graph and move them as a unit, the hand icons  can be used to select and deselect one, or several, objects in the graph:
- The delete tool 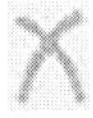 is used to delete objects from the graph.
- The clipboard  is used to copy an Amos path model into the Windows clipboard so that it can be pasted into other applications (such as Word or PowerPoint).

The first example does not include latent variables with multiple indicators; those will be needed in the second and third examples. A latent variable appears as an ellipse. To draw a latent variable, use the ellipse tool , then use the Amos "lollipop" tool to add one or several indicator variables to the latent variable.

Note that error terms need a path coefficient of 1 to link them to variables and that one of the indicator variables for each latent variable must have a path coefficient of 1. (These are scaling constraints.) Amos supplies these automatically. If you delete an indicator variable later, and it was the one that had this fixed coefficient, note that you need to add a coefficient value of 1 for one of the remaining indicator variables.

To specify the model you want in the form of a path diagram, similar to the diagram in Figure 15.4, you need to draw the path model in the blank area on the right-hand side of the main window in Amos (which appeared in Figure 15.5). To draw the path model that appears in Figure 15.4, begin with rectangles to represent the three observed variables. Left-click on the rectangle tool, move the cursor over to the blank worksheet on the right, then hold the left mouse button down and move the cursor until you have the dimensions you want, then release the mouse button. Your worksheet should now contain a rectangle.

For this analysis, the path model needs to include the following additional elements: Rectangles must be added for the other observed variables (weight, blood pressure), three causal paths must be added, and each of the two dependent variables (weight and blood pressure) need explicit error terms. To add paths to the model, left-click on the unidirectional arrow tool, left-click on the initial causal variable (rectangle) in the path model and continue to hold the mouse button down, and drag the mouse until the cursor points at the outcome or dependent variable, then release the mouse button. An arrow will appear in the diagram. For this model, you need three unidirectional arrows: from age to weight, from weight to blood pressure, and from age to blood pressure, as shown in Figure 15.4.

Each dependent variable (a variable is dependent if it has a unidirectional arrow pointing toward it) must have an explicit error term. To create the error terms shown in Figure 15.4,

Figure 15.7 Object Properties Dialog Box

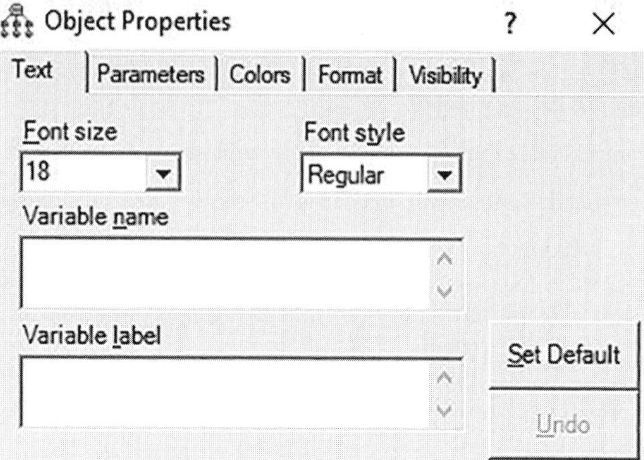

left-click on the error term tool, move the mouse to position the cursor over a dependent variable such as weight, and left-click again. An error term (a circle with an arrow that points toward the observed variable) will appear in the path model. Note that this arrow has a coefficient of 1 associated with it; this predetermined value for this path is required so that Amos can scale the error term to be consistent with the variance of the observed variable. You must give each error term a name; names for error terms must not correspond to the names of any SPSS observed variables. It is conventional to give error terms brief names such as e1 and e2, as shown in Figure 15.4. Names for error variables must be typed in. To do this, right-click on the circle that corresponds to an error term; in the menu that appears, click on Object Properties (Figure 15.7).

Under the "Text" tab of the Object Properties dialog box in Figure 15.7, type in a variable name in the "Variable name" pane. Names are required for all variables, both measured and latent.

15.7.2 Open SPSS Data File and Assign Names to Measured Variables

Each rectangle also needs a variable name. It is convenient to copy these names directly from the SPSS input data file. To open an SPSS data file, go to the top-level menu and make the following selections: <File> → <Data Files>. This opens the Data Files dialog box that appears in Figure 15.8. Click the File Name button and navigate to the location of the data file. (The first time you open this window, the default directory contains Amos examples. You need to navigate to one of your own directories to locate your own data files and save files.) When you have located the SPSS data file (for this example, it is the file named ageweightbp.sav), highlight it, then click the Open button and then the OK button. This returns you to the main Amos window.

Each rectangle must have a variable name. To see a list of variable names in the SPSS file, click on <View> → <Variables in Dataset>. Variable names in this list (Figure 15.9) can be highlighted, then dragged and dropped into the corresponding rectangles.

During this process, if you make a mistake or want to redraw some element of the model, you can use the delete tool to remove any variable or path from the model. Amos has other tools that can be used to make the elements of these path model diagrams look nicer, such as

Figure 15.8 Dialog Box to Open SPSS File

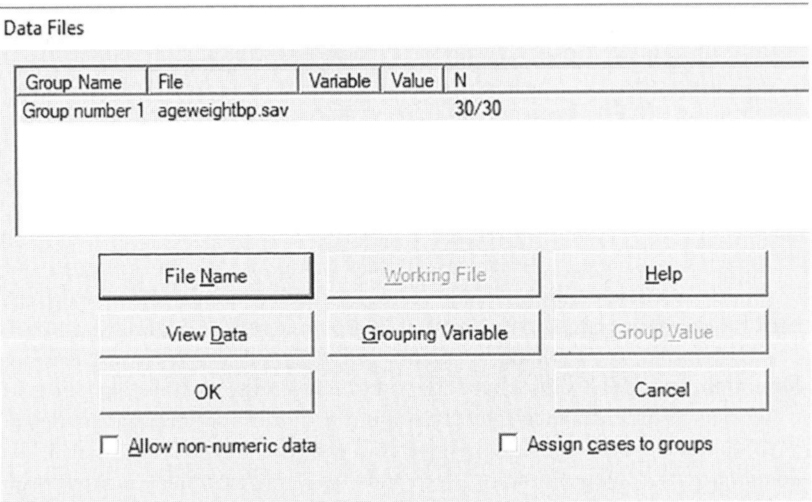

Figure 15.9 List of Variables in SPSS Data File

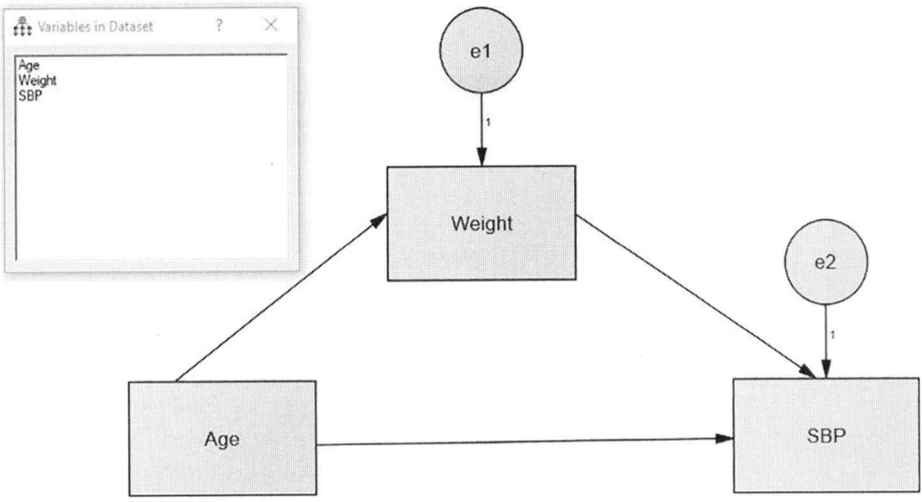

the moving truck. The magic wand tool "fixes" the locations of arrowheads, for example. Be sure to save the model before you go on to run analyses.

One of the choices represented near the top and center of the Amos main page is a toggle between an edit mode and a view results on diagram mode: . The choice on the left (highlighted, with a down arrow shown) makes it possible to edit the model diagram. To view output (specifically, path coefficients within the diagram), highlight the right-hand image (with the up arrow).

15.8 SPECIFY THE ANALYSIS PROPERTIES

One more step is needed before you can run the analysis: specification of the way analysis is done and the results to display. In the main Amos window that appears in Figure 15.5, go to <View> → <Analysis Properties>; you will find a dialog box with multiple tabs across the top, as shown in Figure 15.10.

Note the series of tabs across the top of the window from left to right in Figure 15.10. Only a few of these are used in this example. Click on the "Estimation" tab to specify the estimation method. "Maximum likelihood" is already selected by default. Alternative estimation methods may be preferable for data that violate multivariate normality assumptions (University of Texas at Austin, Department of Statistics and Data Science, n.d.). Check the box for "Estimate means and intercepts." The radio button for "Fit the saturated and independence models" is also chosen by default. (The **saturated model** is one in which there are direct paths for all pairs of measured variables. The saturated model must have a perfect fit; that is, it can reproduce the variance/covariance matrix for all measured variables perfectly. The **independence model** does not have any paths between any variables, and it can't reproduce the variance/covariance model well. The fit indices for the saturated and independence models provide a context for evaluation of the fit of the user-specified [also called "default"] model, but, information about fit for saturated and independence models is rarely reported.)

Next, still in the Analysis Properties dialog box, click on the "Output" tab. In the checklist that appears for the "Output tab" (Figure 15.11), check the boxes for "Standardized estimates," "Squared multiple correlations," "Sample moments," "Implied moments," "Indirect, direct & total effects," and "Tests for normality and outliers." Unstandardized estimates of path coefficients (like *b* coefficients in regression) appear by default. Standardized estimates are equivalent

Figure 15.10 Analysis Properties Dialog Box, "Estimation" Tab

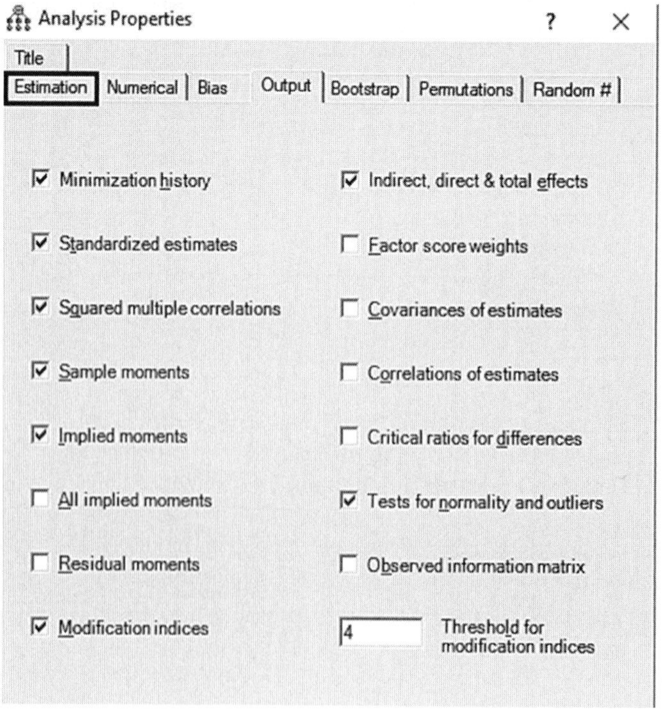

Figure 15.11 Analysis Properties Dialog Box, "Output" Tab

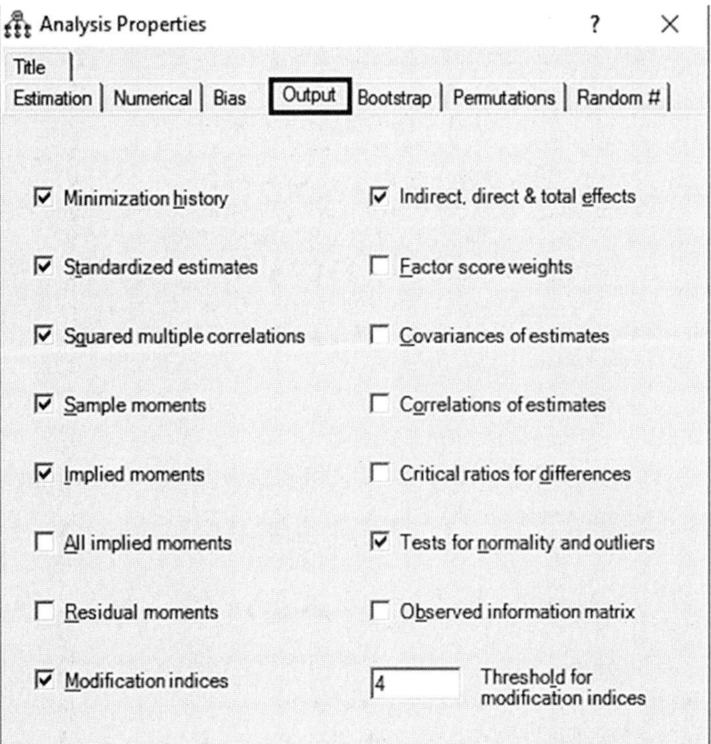

to β coefficients in regression, and squared multiple correlations are the same as R^2 values. *Sample moments* refers to the information about variables from the sample data, including the variance/covariance matrix and the means for all variables; this command also provides the correlation matrix. Implied sample moments are corresponding values of variances and covariances that can be reproduced by using the algebra of covariances to summarize the information for unstandardized path coefficients for the paths included in the model (or applying the tracing rule to reproduce the correlations between all pairs of variables). Indirect, direct, and total effects provide information about the strength of association for each direct path and any indirect paths (in the first example, this is $a \times b$). Tests for normality and outliers provide information about violations of assumptions of normality and presence of multivariate outliers. If problems are detected, these cannot be corrected within Amos; the analyst can return to the original data file in SPSS to make further evaluations and apply suitable remedies.

There is an increasing expectation that authors should report CIs. SEM programs provide CIs based on **bootstrapping** (Carpenter & Bithell, 2000). Bootstrapping is an empirical resampling method. Thousands of different samples are drawn from the data set, and the structural equation model estimates are calculated for every model. For each parameter (such as the $a \times b$ indirect path), the sampling distribution is set up. This empirical sampling distribution is used to estimate the *SE* for the statistic and to generate CIs. To request bootstrapped CIs, go to the "Bootstrap" tab and check the box for "Perform bootstrap." (Note that Amos, unlike Mplus, cannot do bootstrapping if the data set contains missing values. Missing value replacement can be done outside Amos.) The default is 2,000; for this example, only 200 bootstrap samples were requested. In actual research, 2,000 to 5,000 bootstrapped samples are often requested. Also check "Bias-corrected confidence intervals" and specify the level of

Figure 15.12 Analysis Properties Dialog Box, "Bootstrap" Tab

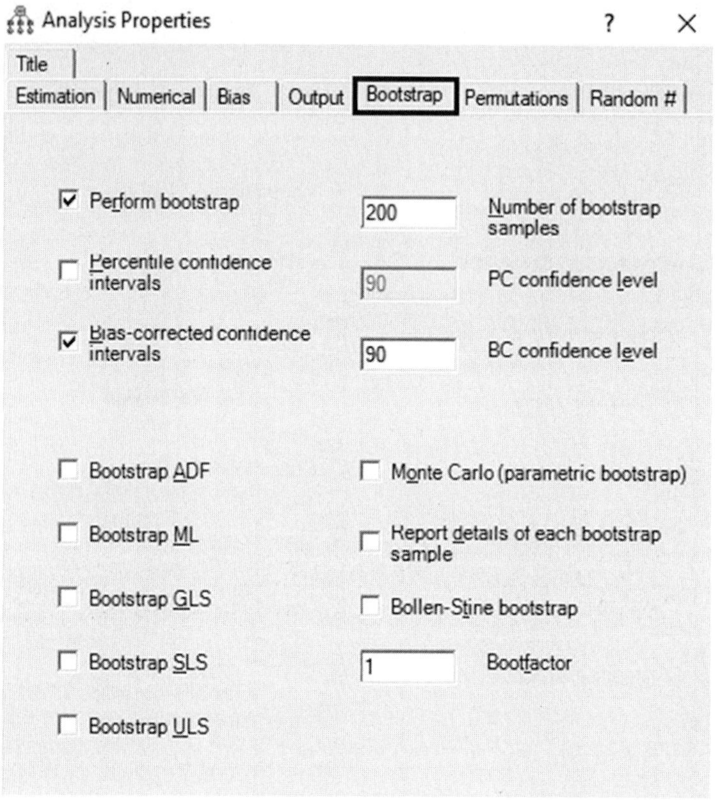

Analysis Properties

Title | Estimation | Numerical | Bias | Output | Bootstrap | Permutations | Random #

☑ Perform bootstrap — 200 — Number of bootstrap samples

☐ Percentile confidence intervals — 90 — PC confidence level

☑ Bias-corrected confidence intervals — 90 — BC confidence level

☐ Bootstrap ADF

☐ Monte Carlo (parametric bootstrap)

☐ Bootstrap ML

☐ Report details of each bootstrap sample

☐ Bootstrap GLS

☐ Bollen-Stine bootstrap

☐ Bootstrap SLS

1 — Bootfactor

☐ Bootstrap ULS

confidence (90% is the default that appears in Figure 15.12; this was changed to 95% for the first example). For the test of H_0: $a \times b = 0$, the bootstrapped CI provides a generally preferred alternative to the Sobel test discussed in Chapter 9, on mediation.

To finish work in the Analysis Properties dialog box, click the X in the upper right-hand corner to close it. This returns you to the main Amos screen.

15.9 RUNNING THE ANALYSIS AND EXAMINING RESULTS

The next step is to run the requested analysis. From the top-level menu, make the following menu selections: <Analyze> → <Calculate Estimates>. After you do this, preliminary results information appears in the center column of the worksheet in Figure 15.13.

Three elements of this page are circled. The bottom ellipse should be examined first. If the analysis ran successfully, a chi-square result appears here; if not, there is an error message. In this first example, χ^2 and its *df* both equal 0. If the analysis did not run, you need to correct mistakes in the model. Many Amos error messages are easy to understand. "Model specification error" usually means that explicit error terms, or constraints such as values of 1 for path coefficients for error terms, are missing.

If the model ran successfully, results appear in two places: in the labeled diagram in the drawing area and in the <View> → <Text Output> windows. The middle ellipsis in Figure 15.13 provides a choice to view either raw or standardized path coefficients superimposed on the path diagram. Generally, these coefficients should not be interpreted until after assessment of overall model fit. However, for this analysis, because all three variables were

Figure 15.13 Amos Main Page View After Analysis

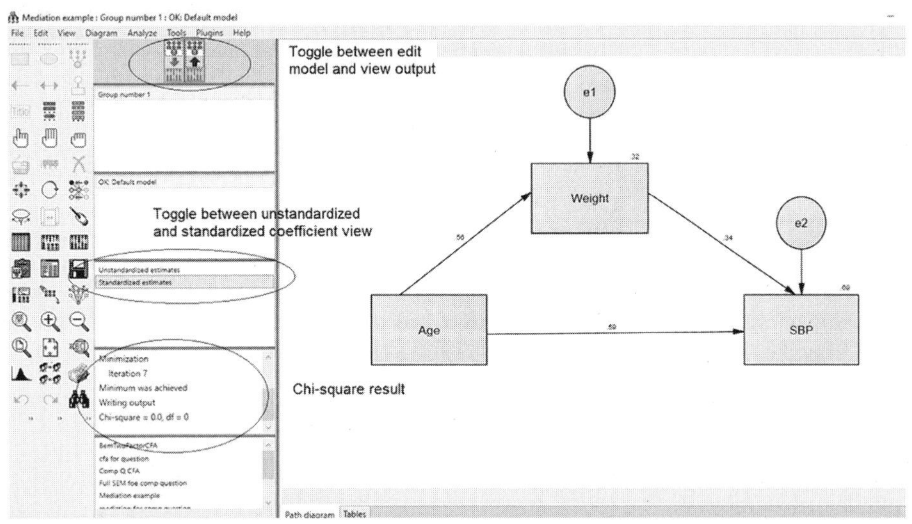

Figure 15.14 Navigation Pane for Mediation Model Text Output

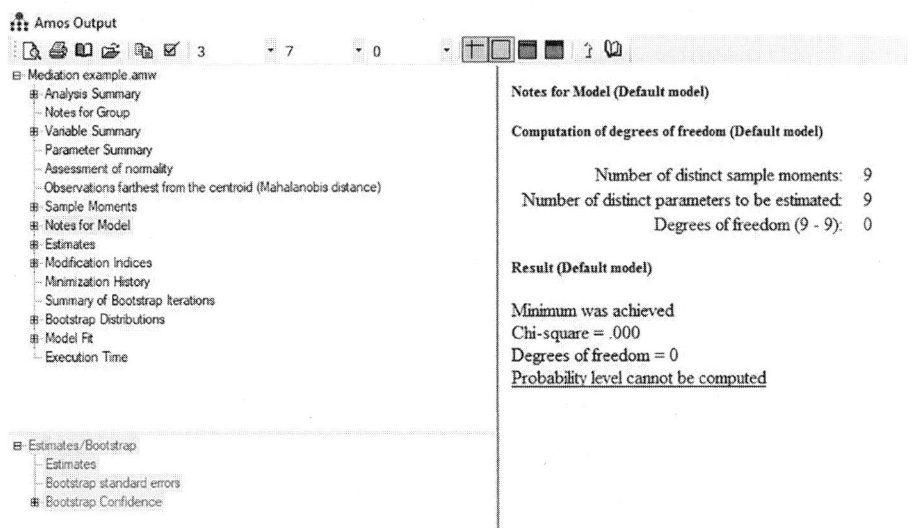

connected by direct paths, the model has perfect fit. This is an example of a **fully identified model** (a model in which the number of pieces of information available from the data equals the number of parameters estimated). It is not necessary to look at other model fit indexes for this model before examining estimates of coefficients. It will be necessary to examine model fit indexes before looking at coefficients in the upcoming examples.

Substantial additional information appears in the Amos text output file. From the top-level menu in the Amos main window that appears in Figure 15.5, choose <View> → <Text Output>. A navigation pane appears as shown in Figure 15.14. Each part of the results can be selected by highlighting an entry in this navigation pane.

"Default model" in the "Notes for Model" results that appear on the right-hand side of Figure 15.14 refers to the model represented by the input path diagram. "Number of distinct sample moments" refers to the number of means, variances, and covariances that represent the information from the data (these statistics are examples of "moments"; Amos is short for "analysis of moment structures"). For $k = 3$ variables, there are 3 means, 3 variances, and $k \times (k - 1)/2 = 3$ distinct covariances; this gives a total of 9 distinct sample moments. The number of free parameters to be estimated can be worked out by examining the diagram to ask how many means, intercepts, path coefficients, variances, and error variances Amos needs to estimate. This can be difficult for even moderately complex models; fortunately, Amos provides this information. In the "Notes for Model" section, Amos reports that 9 parameters were estimated. The *df* for the chi-square is the number of distinct sample moments (pieces of information in the data) minus the number of parameters estimated; in this example, 9 – 9 = 0 *df*. As noted earlier, this is an example of a fully identified model. A fully identified structural equation model perfectly reproduces the variance/covariance matrix (just as a regression analysis can perfectly reproduce the matrix of correlations among variables). The model in this first example has a χ^2 of 0 and fits perfectly; that is what happens when every variable has a direct path to every other variable. The models discussed in later examples in this chapter will not fit perfectly and for those examples, further evaluation of model fit will be needed.

Returning to the navigation pane in Figure 15.14, to view the path coefficient estimates and statistical significance tests, click on "Estimates" in the navigation tree. Results appear in Figure 15.15. Age is given in years, body weight in pounds, and systolic blood pressure (SBP) in millimeters of mercury.

Figure 15.15 Estimates of Path Coefficients for Direct Paths

Estimates (Group number 1 - Default model)

Scalar Estimates (Group number 1 - Default model)

Maximum Likelihood Estimates

Regression Weights: (Group number 1 - Default model)

			Estimate	S.E.	C.R.	P	Label
Weight	<---	Age	1.432	.390	3.669	***	
SBP	<---	Weight	.490	.180	2.718	.007	
SBP	<---	Age	2.161	.458	4.716	***	

Standardized Regression Weights: (Group number 1 - Default model)

			Estimate
Weight	<---	Age	.563
SBP	<---	Weight	.340
SBP	<---	Age	.590

Means: (Group number 1 - Default model)

	Estimate	S.E.	C.R.	P	Label
Age	58.300	3.170	18.391	***	

Intercepts: (Group number 1 - Default model)

	Estimate	S.E.	C.R.	P	Label
Weight	78.508	23.710	3.311	***	
SBP	-28.046	27.002	-1.039	.299	

Figure 15.16 R^2 Values for the Two Dependent Variables in the Mediation Model Example

Squared Multiple Correlations: (Group number 1 - Default model)

	Estimate
Weight	.317
SBP	.690

The first part of results in Figure 15.15 shows the unstandardized path coefficients, along with a statistical significance test called a C.R. C.R. stands for "critical ratio." The C.R. is obtained by dividing each unstandardized regression coefficient or path by its *SE* (as in a *t* or *z* ratio). Comparing these values with the results obtained using OLS regression, the coefficients are the same; however, *SE* estimates differ. The *p* value (shown as a capital *P* in Amos output) appears as *** when it is zero to three or more decimal places. Clicking on any element in the output (such as a specific *p* value) opens a text box that provides an explanation of each term. Although the C.R. values are not identical to the *t* values obtained when the same analysis was performed using linear regression earlier, the path coefficient estimates from Amos and judgments about their statistical significance are the same as for the linear regression results. The standardized estimates (equivalent to β coefficients obtained from multiple regression) also appear in Figure 15.15.

Scroll down the page for the "Estimates" part of the text output to find the squared multiple correlations (R^2 values). These appear in Figure 15.16; they indicate the proportion of variance predicted for each of the two dependent variables.

As shown in Figure 15.16, approximately 32% of the variance in weight and 69% of the variance in SBP were predicted by the model. To see how the total effect for the initial predictor variable age is divided between direct and indirect effect ($a \times b$) estimates, click on the plus sign next to "Estimates" in the navigation pane (the navigation pane appears in Figure 15.14); then click on the plus sign next to "Matrices," then select "Indirect." Results appear in Figure 15.17. Estimates are in terms of raw score (unstandardized) coefficients unless otherwise noted.

15.10 LOCATING BOOTSTRAPPED CI INFORMATION

In Chapter 9, on mediation, the null hypothesis H_0: $a \times b = 0$ was evaluated using the Sobel test in an online calculator. Using Amos, bootstrapped CIs (with associated *p* values) can be obtained for all coefficients in the model, both for indirect (and direct) effects. Bootstrapping is generally viewed as a better way to evaluate statistical significance and set up CIs. The navigation selections required to view these CIs are somewhat complicated.

To obtain information about statistical significance and CI for the indirect effect, examine output for the bootstrapped CIs. To locate this information, click the left mouse button on "Estimates" in the navigation pane list (left side in Figure 15.14) of available output. Select "Matrices" and click to open it. From the choices within the "Matrices" output, select "Indirect" effects (move the cursor to highlight this list entry and then left-click on it). Now you will see the "Estimates/Bootstrap" menu as in the lower left-hand side of Figure 15.18. Left-click on "Bias corrected percentile method."

The right-hand panel in Figure 15.18 shows the 95% CI results for the estimate of the unstandardized $a \times b$ indirect effect (the effect of age on SBP, mediated by weight). Note that the layout is unusual: the uppermost table in Figure 15.18 lists the lower bounds for each CI, while the middle table in Figure 15.18 lists the upper bounds for each CI. The lower and upper limits of the 95% CI are .122 and 2.338. The result of a statistical significance test for H_0: $a \times b = 0$, based on sampling error information from bootstrapping, is p =.011. Note that a bias-corrected percentile method CI is not symmetrical around the sample statistic.

Figure 15.17 Total, Direct, and Indirect Estimates

Standardized Total Effects (Group number 1 - Default model)

	Age	Weight
Weight	.563	.000
SBP	.782	.340

Direct Effects (Group number 1 - Default model)

	Age	Weight
Weight	1.432	.000
SBP	2.161	.490

Standardized Direct Effects (Group number 1 - Default model)

	Age	Weight
Weight	.563	.000
SBP	.590	.340

Indirect Effects (Group number 1 - Default model)

	Age	Weight
Weight	.000	.000
SBP	.701	.000

Standardized Indirect Effects (Group number 1 - Default model)

	Age	Weight
Weight	.000	.000
SBP	.192	.000

Note: The only indirect effect in this model is from age (column variable) to SBP (row variable). Thus, the only nonzero entries in the indirect effect tables, $a \times b = .701$ unstandardized and .192 standardized, are for age (column variable, predictor) as a predictor of SBP (row variable, dependent). More complex models will have larger numbers of indirect paths.

The navigation pane in Figure 15.18 can also be used to locate the bias-corrected percentile method CI for the direct effects and for standardized versions of the estimates. Details are not shown here and are not included in later examples to limit length. Past research reports have not usually reported all these CIs; however, with increasing focus on the importance of CIs, it may be important to include this information in the future.

The unstandardized path coefficients, or the standardized path coefficients and R^2 values, can be displayed superimposed on the graph of the model. To see these, return to the Amos main window and use the terms in the center ellipse (as marked in Figure 15.13) to toggle between the view of standardized versus unstandardized coefficients. The properties of these coefficients (size, font) can be edited using the Object Properties dialog box to improve readability and appearance.

Two additional kinds of information are discussed here. Amos can provide the observed sample variance/covariance matrix, the correlation matrix, and the means for all variables, as shown in Figure 15.19.

Figure 15.18 Navigation Selections to Locate Bootstrapped CI Results

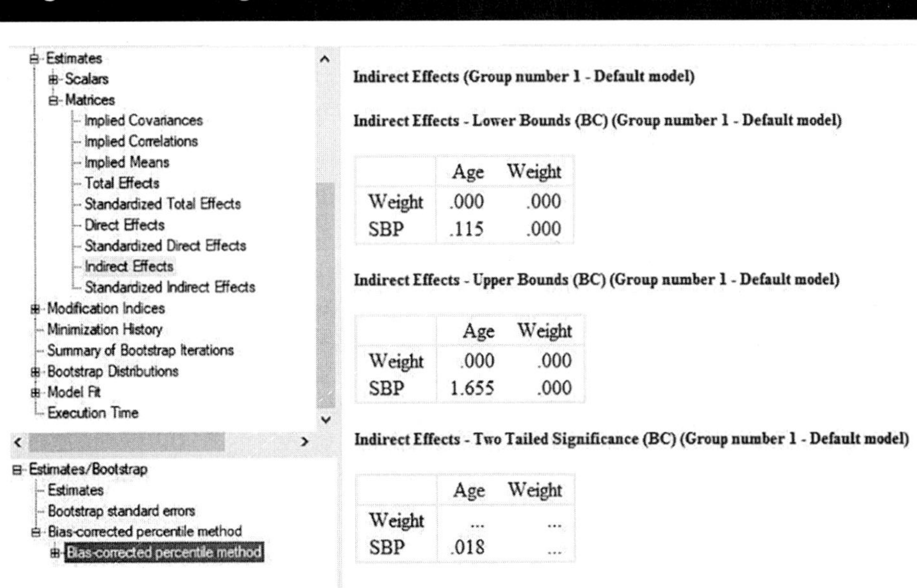

Figure 15.19 Observed Sample Moments: Means, Variances, Covariances (and Correlations)

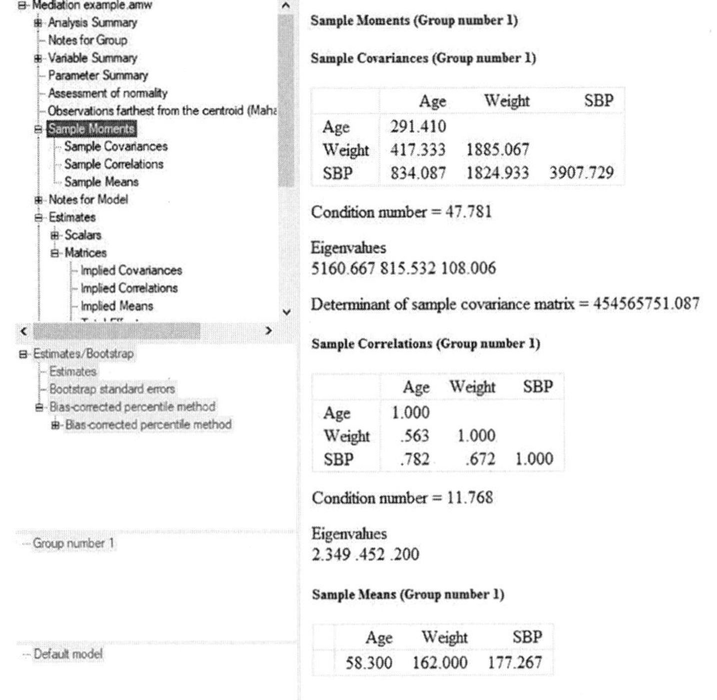

Figure 15.20 Implied or Reproduced Correlations for Mediation Model

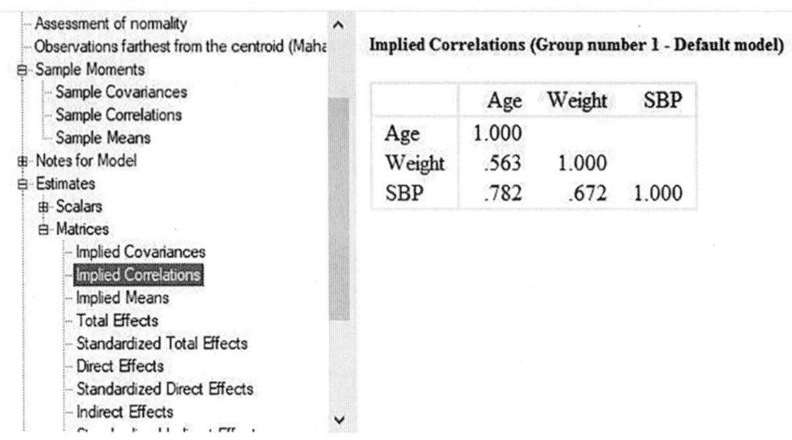

Implied Correlations (Group number 1 - Default model)

	Age	Weight	SBP
Age	1.000		
Weight	.563	1.000	
SBP	.782	.672	1.000

Figure 15.21 Assessment of Normality for Each Variable in Mediation Model

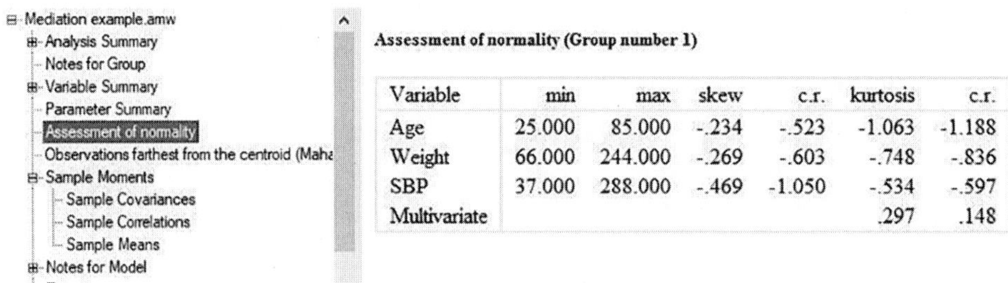

Assessment of normality (Group number 1)

Variable	min	max	skew	c.r.	kurtosis	c.r.
Age	25.000	85.000	-.234	-.523	-1.063	-1.188
Weight	66.000	244.000	-.269	-.603	-.748	-.836
SBP	37.000	288.000	-.469	-1.050	-.534	-.597
Multivariate					.297	.148

Amos can also show the values of correlation (or covariance matrix) that are implied by (can be reproduced by) the path model coefficients. The implied correlations (i.e., correlations reproduced from the standardized path coefficients in the model) appear in Figure 15.20.

In this example, if the tracing rule is applied, the implied (also called reproduced) correlations in Figure 15.20 exactly match the observed correlations in Figure 15.19. This happens because this model is fully identified (all variables have direct paths between them). In later examples, implied correlations and covariances will not perfectly match the observed correlations and covariances, and the question will arise: How well or poorly can the model reproduce observed correlations and covariances?

Two kinds of information are available to assess departures from normality and multivariate outliers. Figure 15.21 shows assessment of skewness and kurtosis for each individual variable.

There are formal tests for skewness and kurtosis. A normal distribution has skewness and kurtosis of 0 (for the way SPSS calculates these indexes), and therefore a statistically significant value for the skewness and kurtosis tests in Figure 15.21 indicates problems with univariate normality. In this example, there were no significant departures from univariate normality.

A second type of information is provided to identify multivariate outliers (Figure 15.22). Cases with large values of Mahalanobis d^2 are candidates for identification as multivariate outliers, as discussed earlier. (Output is shown only for the first seven cases.) If there are problems with normality assumptions or outliers, the analyst might return to SPSS to try to remedy these problems, or choose an estimation method that works well when data violate these assumptions (University of Texas at Austin, Department of Statistics and Data

Figure 15.22 Mahalanobis Distances for Individual Cases in SPSS Data File

⊟ Mediation example.amw
⊞ Analysis Summary
Notes for Group
⊞ Variable Summary
Parameter Summary
Assessment of normality
Observations farthest from the centroid (Maha
⊟ Sample Moments
Sample Covariances
Sample Correlations
Sample Means
⊞ Notes for Model
⊟ Estimates
⊞ Scalars

Observations farthest from the centroid (Mahalanobis distance) (Group number 1)

Observation number	Mahalanobis d-squared	p1	p2
29	9.812	.020	.458
30	6.940	.074	.660
28	6.615	.085	.477
3	6.200	.102	.369
22	5.548	.136	.386
1	5.048	.168	.393
2	4.723	.193	.357

Figure 15.23 Standardized Path Coefficients and R^2 for Hypothesized Mediation Model

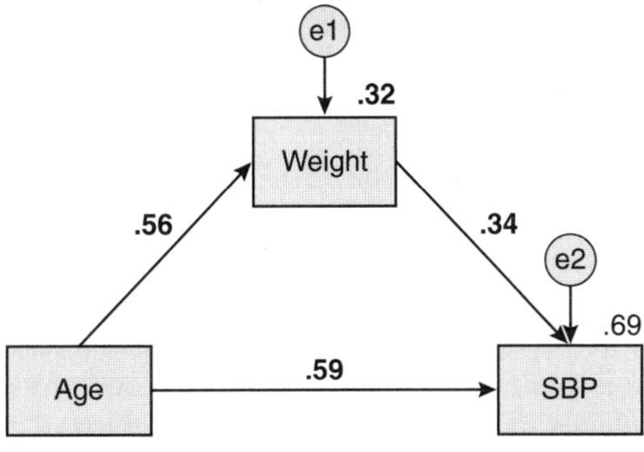

Note: While it is not customary to show this, the variables are actually z_{age}, z_{weight}, and z_{SBP} in this standardized version of the model.

Science, n.d.). In this example, no significant problems with normality were found, and no cases were deleted as outliers.

15.11 SAMPLE RESULTS SECTION FOR MEDIATION EXAMPLE

Following is an example of a "Results" section for the mediation example above.

Results

A path model was set up to evaluate the way age predicts SBP; it included a direct path from age to SBP and an indirect path from age to SBP via weight, to represent the hypothesis that effects of age may be partially mediated by weight. Standardized path coefficients and R^2 values appear in the model diagram in Figure 15.23. This model was fully identified and had perfect fit. Both dependent variables were well predicted ($R^2 = .317$ for prediction of SBP from weight, and $R^2 = .69$ for prediction of SBP from age) [from Figure 15.16]. All three unstandardized direct path coefficients were statistically significant at the $\alpha = .01$ level. The unstandardized coefficient for prediction of weight from age, Path *a*, was 1.432, C.R. = 3.67, $p < .001$; the coefficient for prediction of SBP from weight, Path *b*, was .49, C.R. = 2.72, $p = .007$; and the direct path from age to SBP, Path *c*, was 2.161, C.R. = 4.72, $p < .001$. [Values are from Figure 15.15.] Figure 15.17 displays the corresponding standardized path coefficients and the R^2 values for weight and blood pressure. From Figure 15.17, the $a \times b$ indirect effect was .701 (unstandardized), with a 95% bias corrected percentile method CI on the basis of 200 bootstrap samples [.115, 1.655] [from Figure 15.18]. On the basis of bootstrapping, the $a \times b$ unstandardized effect was statistically significant, $p = .018$, two tailed. Both the direct path from age to SBP and the indirect path from age to SBP via weight were statistically significant. These results are consistent with the hypothesis that effects of age on blood pressure may be partially mediated by weight. However, alternative causal and noncausal models of associations among these variables cannot be ruled out.

Using Cohen's (1988) effect size guidelines, the standardized direct path coefficients can all be labeled as large or close to large effects. Cohen's suggested cutoff for a large effect is $\beta > .371$ (in absolute value). The $a \times b$ standardized indirect path estimate (.192) [from Figure 15.17] would be judged a medium to small effect by Cohen's standards. The R^2 values for prediction of weight and systolic blood pressure were also large by Cohen's standards ($R^2 > .138$). These effect sizes are large enough to be theoretically meaningful.

Assessment of practical or clinical significance requires different information. Given that the three variables were measured in meaningful units, it is instructive to examine the unstandardized path coefficients. [As noted earlier, body weight was given in pounds.] The unstandardized direct effect for a 1-year increase in age was a 2.161 mm Hg increase in SBP (Figure 15.17). The unstandardized indirect effect for effects of age on SBP through weight was approximately .701 mm Hg (Figure 15.17). These values may not seem large. However, in practice, this model might be used to think about outcomes for patients with ages 10 or 20 years apart. The predicted direct effect on blood pressure for a 10-year increase in age would be 21.61 mm Hg. That is a sufficiently large increase in blood pressure to be clinically important. The predicted indirect effect involving weight would be a 7.01 mm Hg increase.

It is possible that other hypothesized mediating variables that were not assessed in this hypothetical study (such as age-related accumulation of arterial plaque) may be more important than body weight in explaining age-related increases in blood pressure.

Arguably, it might not be necessary to take measurement error into account in this situation by including latent variables. Age and weight can be reliably measured. However, blood pressure can vary substantially across occasions of measurement. This problem could be handled by using an average blood pressure across numerous occasions of measurement as the dependent variable or by treating blood pressure as a latent variable with multiple indicators that include numerous blood pressure observations.

15.12 SELECTED STRUCTURAL EQUATION MODEL TERMINOLOGY

Example 1 did not require evaluation of model fit. The next two examples include latent variables and do not include all possible paths among variables, and so they do require assessment of model fit; thus, additional SEM terminology is needed.

Model identification: Assessment of model identification can be complex (see Kenny & Milan, 2012). One part of identification is evaluated by comparing the number of unknowns (free parameters) to be estimated with the amount of information (number of distinct sample moments) in the data. The amount of information from the data that is used in SEM is formally defined as the number of distinct moments. (Moments are distribution characteristics that include mean, variance, and covariance.) For k variables, number of distinct moments includes the k means, the k variances, and the $[k \times (k - 1)]/2$ covariances. Free parameters to be estimated include all the path coefficients and variances of error terms and latent variables. These can be "counted up" by hand, but this can be difficult for complex models. Amos "Notes for Model" output provides this information. The *df* for the overall model χ^2 goodness of fit index is based upon the number of pieces of information in the data minus the number of parameters to be estimated. Analysts need to know whether their models are identified.

Model fit: In the discussion of regression analysis earlier, the standardized regression analysis coefficients perfectly reproduced the matrix of observed correlations between X_1, X_{12}, and Y. The total association of each X_i with Y was divided into a direct path and one or more indirect paths. Similar logic is used to assess structural equation model fit, with one important difference. SEM assesses model fit by taking the variance/covariance matrix among the measured X variables as the starting point for analysis and evaluating how well the unstandardized model coefficients and path model can reproduce the original variance/covariance matrix. As the number of paths between variables is decreased, model fit typically becomes poorer. As in factor analysis, the data analyst is interested in how restricted the structural equation model can be and still do a reasonably good job of reproducing or accounting for relationships among observed variables.

Just-identified model: In a just-identified model, the number of pieces of information in the data equal the number of parameters to be estimated. The mediation path model in example 1 was a just-identified model. The *df* for χ^2, and the value of χ^2, were both 0. The model fits the data perfectly (it can reproduce the variance/covariance or correlation matrix perfectly). Multiple regression is an example of a just-identified model because it includes a direct path from each variable to every other variable. Note that a model can fit perfectly and yet have low R^2 values that indicate that the Y variable is not well predicted by the X independent variables. Whether dependent variables can be well predicted is a separate question from model fit. SEM can be used to obtain path coefficients for a just-identified model, but in most situations, people who use SEM are interested in overidentified models.

Overidentified model: In an overidentified model, the number of pieces of information in the data is greater than the number of model parameters to be estimated. The *df* term is positive, and χ^2 will usually be greater than 0. In other words, the parameters and paths in the model will generally not be sufficient to perfectly reproduce the variance/covariance matrix. These models are usually of greater interest to SEM analysts. One way to create an overidentified model is to take a just-identified model and drop one or more paths. In the mediation analysis example, the original model that included a direct path from X to Y, X to M, and M to Y is just identified. If the analyst wants to test the model that says the effects of X and Y are completely mediated by M, then the analyst can drop the direct path from X to Y. This reduced model will probably not fit quite as well as the fully identified model; however, the reduced model is more parsimonious, and it represents a theory that can account for some relationships in other ways than simply saying everything is correlated with everything else.

Underidentified model: In underidentified models, the number of pieces of information in the data is less than the number of parameters to be estimated; the *df* for the overall model would be negative. An example of a model that is (initially) underidentified is exploratory factor analysis. The number of parameters to be estimated, including factor loadings, variances of factors, and correlations between factors, is larger than the number of unique elements in the correlation matrix. Many different values for factor loadings (in the initial solution using k factors to represent k variables) would provide solutions that reproduce the correlation matrix (as shown in the factor analysis chapter). To resolve this problem, numerous constraints on parameters are imposed; for example, in an initial orthogonal solution, factors are constrained to be uncorrelated, and factor loadings have scaling constraints that make them interpretable as correlations. That reduces the number of free parameters so that a unique solution can be obtained. SEM is not used with underidentified models; these are models with too many paths.

Equivalent models: When causal path models are evaluated to see how well they can reproduce correlations (or covariances), some causal models that make different assumptions about which paths are causal (or direction of causality) may have identical fit. For example, the following three models will have identical fit: $X \rightarrow M \rightarrow Y$, $X \leftrightarrow M \leftrightarrow Y$, and $Y \rightarrow M \rightarrow X$. What these models have in common is a direct path from X to M, and a direct path from M to Y, but no direct path from X to Y. These models differ in that they represent different "causal" hypotheses. If these three path models are each tested in Amos, they will have the same fit. How can analysts decide which causal model is correct? The answer is, they can't. If $X \rightarrow M \rightarrow Y$ is a good fit, that result is consistent with the causal hypothesis that the effect of X on Y is completely mediated by M, but it is equally consistent with any other hypothesized model that has a path from X to M (causal or noncausal) and a path from M to Y (causal or noncausal) but no direct path from X to Y. Structural equation models are more complex; however, reversing the direction of one or more of the causal paths in an SEM or changing a causal to a noncausal path sometimes yield the same fit. It is important to keep in mind that when the paths and coefficients in a structural equation model do a good job of reproducing the variances and covariances in the data, that structural equation model is consistent with the data; however, the result is not proof that the specific causal hypotheses that correspond to directional arrows in the diagram are correct.

Recursive models: A structural equation model is nonrecursive if there are causal loops; for example, X causes M, M causes Y, and then Y causes X. In a recursive model, causation flows in only one direction; a variable that is "caused by" X cannot in turn be represented as a cause of X. Amos can analyze only recursive models. In the "Notes for Groups" section of the text output, Amos indicates if the model is recursive (and it also reports the sample size N).

Nested models: Structural Equation Model A is nested in another Structural Equation Model B if Model A can be obtained by taking Model B and dropping one or more paths. Model A, with fewer paths, will generally have worse fit than Model B, with more paths. Another way to create nesting is to fix the value of one or more causal path coefficients. Chi squared can be used to compare the fit of nested models; in this case, the change of fit is obtained by subtracting the smaller χ^2 from the larger χ^2; df is the difference in number of free parameters between the models. Often, analysts hope that this χ^2 difference will be statistically significant so that it is clear that one model fits significantly better than the other. Examples of nested model comparison can be found in Byrne (2016). When models are not nested, other criteria (such as the Akaike information criterion [AIC]) can be used to compare fit (Huang, 2017).

Inadmissible solution: Occasionally attempts to estimate coefficients for models yield implied correlations > 1 and negative error variances. Such solutions are unacceptable or inadmissible.

15.13 SEM GOODNESS-OF-FIT INDEXES

Amos provides numerous global fit indexes (to assess how well the model paths and parameters can reconstruct the original variance/covariance matrix). Kline (2016) recommended reporting the following four indexes as a minimum set of fit statistics.

1. **Chi-square goodness-of-fit index** with its degrees of freedom and *p* value. The χ^2 goodness-of-fit index is labeled CMIN in Amos output. In SEM, a small and nonsignificant value for χ^2 is preferred. The magnitude of χ^2 depends on the magnitude of the squared discrepancies between actual and reconstructed variances and covariances; a large χ^2 means there are large discrepancies (poor model fit). Chi-square values also increase as a function of *N*, sample size. For models with *N* of about 75 to 200 cases, χ^2 can be a reasonable measure of fit; however, when *N* becomes large, χ^2 is almost always statistically significant. Kline (2016) cautions that with sample sizes typical in published research (about 200 to 300 cases), a statistically significant χ^2 is evidence of poor model fit. In practice, models are often judged adequate even with statistically significant χ^2 when there are large samples and satisfactory results for other fit indexes.

2. **Root mean square error of approximation (RMSEA)** and its 90% or 95% CI: Smaller values of RMSEA indicate better fit. Many authorities suggest that a model can be judged a good fit if RMSEA is ≤ .06 or ≤ .08 (Schreiber, Stage, King, Nora, & Barlow, 2006).

3. **Bentler comparative fit index (CFI):** Larger values of CFI indicate good fit. CFI has a range from 0 to 1. A model may be judged a good fit for CFI > .95 (Schreiber et al., 2006).

4. **Standardized root mean square residual (SRMR):** Kline (2006) recommends that this should also be reported; SRMR > .10 may indicate poor fit. Amos does not provide SRMR.

15.14 SECOND EXAMPLE: CONFIRMATORY FACTOR ANALYSIS

The second example examines the measurement model part of SEM. Measurement models involve one or more latent variables; each latent variable is associated with multiple indicators (variables that are observed or measured). A latent variable should have at least three indicators; more than five indicators can become difficult to manage in drawing a graph of the model. When a structural equation model includes only measurement models for latent variables, and does not include causal paths, it is called CFA.

15.14.1 General Characteristics of CFA

A CFA is a structural equation model that has only measurement model components, without any causal paths. The measurement model part of SEM is similar to EFA, except that the model in CFA is usually much more constrained. In EFA with *k* variables, factor analysis evaluates how well a smaller number of factors (fewer factors than variables) can reproduce the observed correlation matrix. The results of primary interest are the number of retained factors, and the loadings of observed variables on these factors. On the basis of the loadings, each factor (or latent variable) can be given a name. A factor is a latent variable; it is not actually measured, but we can make inferences about it by examining its correlation with variables that are measured.

Recall that factor analysis models can be thought of as two sets of regressions. In one set of regressions, each factor is defined as a weighted linear combination of the measured *X* variables (the coefficients in these regressions are called factor score coefficients). In another set of regressions, the arrows are reversed, and the measured *X* variables are predicted from the factors (the coefficients in these regressions correspond to factor loadings). In an orthogonal factor analysis, the factors are constrained to be uncorrelated with one another. The factor loadings for Factor 1 are correlations between each variable and Factor 1. Chapter 12, on factor analysis, explained that in an initial factor analysis solution of k variables, there are k factors. The matrix of correlations among the k variables can be perfectly reproduced from information about factor loadings for k factors. Usually a goal of factor analysis is to reduce the number of factors from k to some smaller number such as two or three or five factors. As the number of factors is reduced, the ability to reproduce the original matrix of correlations among all variables (**R**) becomes weaker. The goal is to reduce the number of factors to a smaller, interpretable, meaningful set of factors that still do a reasonably good job of reproducing **R**.

In a structural equation model diagram, a latent variable (similar to a factor) is represented by an ellipse. Measured or observed variables (also called indicators) are represented by rectangles. Figure 15.24 represents a CFA; each indicator is predicted by one of the factors.

The regression to predict each measured variable from a latent variable or factor can be represented as a path model; this is called a measurement model. In Figure 15.24, the measurement model for Factor 1 has paths that link it to measured variables (or indicators) X_1, X_2,

Figure 15.24 Schematic Diagram for a CFA With Three Correlated Factors

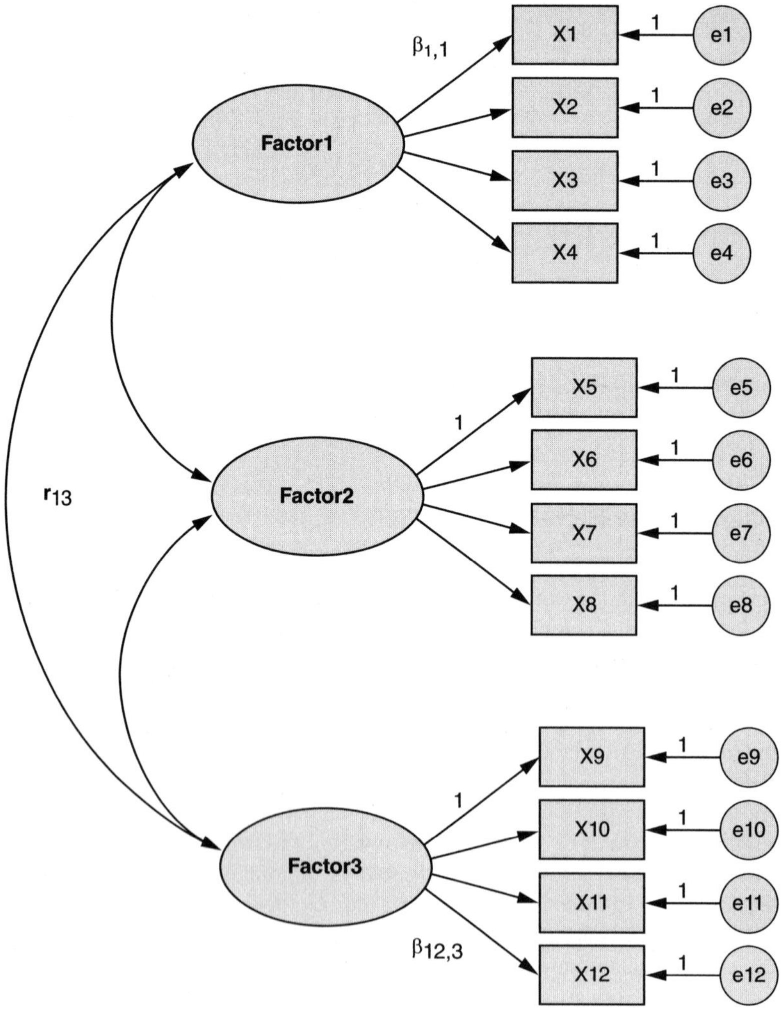

X_3, and X_4. The standardized path coefficients represent factor loadings. Figure 15.24 represents a CFA that has three factors or latent variables and four indicator variables or measured variables per factor. The "meaning" of Factor 1 is inferred from the kinds of variables that have high loadings. For example, if X_1 through X_4 represent scores on vocabulary, reading, grammar, and analogies, Factor 1 might be called a verbal ability factor. In this example, the double-headed arrows among the three latent variables indicate that the factors may be correlated (as in a nonorthogonal or oblique rotation of a factor solution).

When coefficients are given in standardized terms, then the double-headed arrows between factors correspond to correlations between factors, and the single-headed arrows from each factor to one indicator variable correspond to factor loadings. Notice that in this model, each indicator or observed variable is associated with only one latent variable or factor. The absence of a path from X_1 to Factor 2 represents an explicit assumption or constraint that variable X_1 does not load on, is not related to, Factor 2.

It should be noted that the common designation of output from SEM as confirmatory may be misleading, depending on decisions of the data analyst. If the data analyst uses SEM to run large numbers of different CFAs and then selects one best fitting model, this should be characterized as exploratory.

In EFA, the number of factors is usually not specified a priori (although it can be), and each variable can have a nonzero loading on every factor. In CFA, the number of factors is specified a priori along with the set of variables that load on each factor. (It is possible for an X variable to load on more than one factor in CFA.)

Chapter 12 presented EFA analyses of trait ratings for sex role orientation using items from the Bem Sex Role Inventory (data in the file bsri.sav, with an N of 496). The same data set is used here for CFA. Bem argued that femininity was not the opposite end of a single dimension from masculinity (i.e., these were not the opposite ends of a single factor). Instead, she argued that masculinity and femininity represented two separate and orthogonal (uncorrelated) factors or dimensions. EFA results for selected Bem items were consistent with this idea. A CFA can be used to test this hypothesis using a constrained model. In this model, three indicators are allowed to correlate only with BemFem, and the other three variables are allowed to correlate only with BemMasc. A correlation path was included between the two factors. (This path could be omitted; in this example, a path was included, but its coefficient was expected to be not statistically significant.) As in Example 1, the data analyst would draw the diagram for the model using Amos drawing tools, name all latent variables and error terms, and give each rectangle the name of a measured variable in the SPSS data file.

Note that this CFA is constrained in ways that the EFA was not. First, this CFA is limited to only two factors. When the number of factors is not set by the data analyst, EFA retains and rotates factors on the basis of eigenvalues, and that could yield more than two factors. The CFA places **constraints on SEM model parameters**. For example, the variable warm must load on the BemFem factor, and its loading on BemMasc is constrained to zero (because there is no direct path from warm to BemMasc). Note also the only way to try to reproduce the correlation of variable X_1 with X_{12} in this model would be through the following three-step path: loading of X_1 on Factor 1 ($\beta_{1,1}$), correlation of Factor 1 with Factor 3 (r_{13}), and loading of X_{12} on Factor 3 ($\beta_{12,3}$). Unless this model represents the structure in the data very well, it will not be able to reproduce the observed correlations among variables well.

By including a covariance or correlation path between the two factors, this CFA allows factors to be correlated (oblique, in EFA terms). EFA generally extracts orthogonal or uncorrelated factors, and these can be allowed to be correlated by using oblique forms of rotation.

To run this analysis, the model in Figure 15.25 was drawn using Amos drawing tools. Each latent variable was initially drawn using the ellipse tool; then the "lollipop" tool was used to add each indicator variable along with its error term. The variable names warm, sympthtc, and so on, were copied from the variable names in the file bsri.sav. Names for the factors (BemFem and BenMasc) and error terms were entered by editing object properties.

Figure 15.25 CFA for Bem Sex Role Data

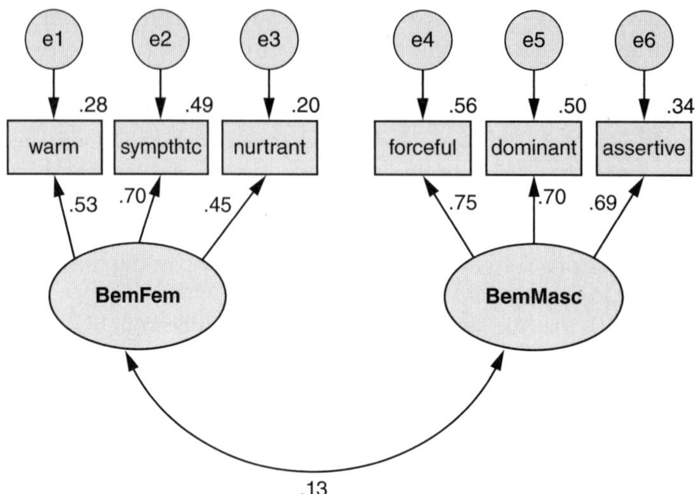

The analysis was specified by going to <View> → <Analysis Properties> and making the following selections: In the "Estimation" tab, "Maximum likelihood," "Fit the saturated and independence models," and "Estimate means and intercepts" were selected. In the "Output" tab, "Standardized estimates" and "Squared multiple correlations" were requested. (Additional selections could be made, as in Example 1, to assess normality and multivariate outliers; however, indirect paths are not of interest in this CFA example.)

At least three indexes of overall model fit should be examined. Model fit information appears in Figure 15.26. First, $\chi^2(8) = 23.127, p = .003$ (in the Amos output, this is called CMIN). This χ^2 is based on the differences between the *estimated* or *implied* variances and covariances and the *observed* variances and covariances. Differences are (essentially) squared and summed to obtain an overall measure of fit; a large χ^2 indicates poor model fit. The value of χ^2 increases with sample size; the total N for this data set was $N = 496$; therefore, the statistically significant χ^2 might be due more to large sample size than to poor model fit. RMSEA = .062 was below the larger of the two cutoff values (RMSEA ≤ .08) suggested by Schreiber et al. (2006), and its 90% CI was reasonably narrow at [.033, .092]. The CFI of .965 was large enough to be judged a good fit. Overall this appears to be evidence of acceptable model fit.

Given that model fit is judged acceptable, the next step is to evaluate the coefficients to see whether they are statistically significant and have signs and magnitudes that make sense.

Results of significance tests for model paths appear in Figure 15.27. Note that the paths that were constrained to be 1 when the model was set up cannot be tested for statistical significance (BemFem to warm and BemMasc to forceful). The other four "factor loading" coefficients were all statistically significant, and all loadings were fairly large and positive, as expected. The covariance (or correlation) between the two factors was also examined (output not shown here). The path between BemFem and BemMasc had C.R. = −1.872, $p = .061$, two tailed. The corresponding standardized coefficient (correlation) between factors was −.13. This essentially 0 correlation is consistent with Bem's hypothesis these two factors should be orthogonal.

Byrne (2016) and Kline (2016) recommended that when a structural equation model includes both measurement and structural paths, the measurement value parts of the model should be evaluated before analyzing the entire model. However, it is worth noting that

Figure 15.26 Selected Model Fit Indexes for Bem Sex Role Inventory CFA

Model Fit Summary

CMIN

Model	NPAR	CMIN	DF	P	CMIN/DF
Default model	19	23.127	8	.003	2.891
Saturated model	27	.000	0		
Independence model	6	457.084	21	.000	21.766

Baseline Comparisons

Model	NFI Delta1	RFI rho1	IFI Delta2	TLI rho2	CFI
Default model	.949	.867	.966	.909	.965
Saturated model	1.000		1.000		1.000
Independence model	.000	.000	.000	.000	.000

Parsimony-Adjusted Measures

Model	PRATIO	PNFI	PCFI
Default model	.381	.362	.368
Saturated model	.000	.000	.000
Independence model	1.000	.000	.000

RMSEA

Model	RMSEA	LO 90	HI 90	PCLOSE
Default model	.062	.033	.092	.222
Independence model	.205	.189	.221	.000

when a measurement model is incorporated into a full structural equation model (as in the next section), the path coefficients for the measurement model are estimated in the context of all the variables in the full model. A measurement model that fits well when examined in isolation may not be the best choice in the context of a full structural model (Warner & Rasco, 2014).

15.15 THIRD EXAMPLE: MODEL WITH BOTH MEASUREMENT AND STRUCTURAL COMPONENTS

The third example includes both measurement components (similar to CFA) and structural components. The following example uses unpublished data from a survey of well-being measures for a sample of $N = 350$ university students (the file is named wbsocial.sav). All outliers and missing values were removed from this file prior to analysis.

Figure 15.28 shows the model that we tested to examine associations between two well-established predictors of well-being (social support, denoted support, and stress) and several widely used measures of well being that included negative affect (NA), positive affect (PA), life satisfaction (LS), and happiness (happy). In most past research, NA has been used as one of the multiple indictors for global well-being; as an indicator of well-being, it has a negative sign for its path. In the model in Figure 15.28, NA is treated as a separate dependent variable rather than as an indicator of well-being. This makes it possible to assess whether NA has different associations with the predictors support and stress than with the other positive well-being measures.

As in Example 1, output for standardized path coefficients, indirect paths, and squared multiple correlations was requested. In the "Output" tab, the box for "Modification indices" was also checked. MIs can help identify changes in a model that may improve model fit. (MIs cannot be obtained if there are missing data in the SPSS file.)

Figure 15.27 Unstandardized Path Coefficients for Bem CFA

Estimates (Group number 1 - Default model)

Scalar Estimates (Group number 1 - Default model)

Maximum Likelihood Estimates

Regression Weights: (Group number 1 - Default model)

			Estimate	S.E.	C.R.	P	Label
warm	<---	BemFem	1.000				
sympthtc	<---	BemFem	1.466	.274	5.357	***	
nurtrant	<---	BemFem	1.043	.171	6.095	***	
forceful	<---	BemMasc	1.000				
dominant	<---	BemMasc	.946	.097	9.792	***	
asertive	<---	BemMasc	.691	.072	9.577	***	

Standardized Regression Weights: (Group number 1 - Default model)

			Estimate
warm	<---	BemFem	.530
sympthtc	<---	BemFem	.701
nurtrant	<---	BemFem	.449
forceful	<---	BemMasc	.751
dominant	<---	BemMasc	.704
asertive	<---	BemMasc	.585

Figure 15.28 Structural Equation Model to Predict Well-Being and Negative Affect From Stress and Social Support

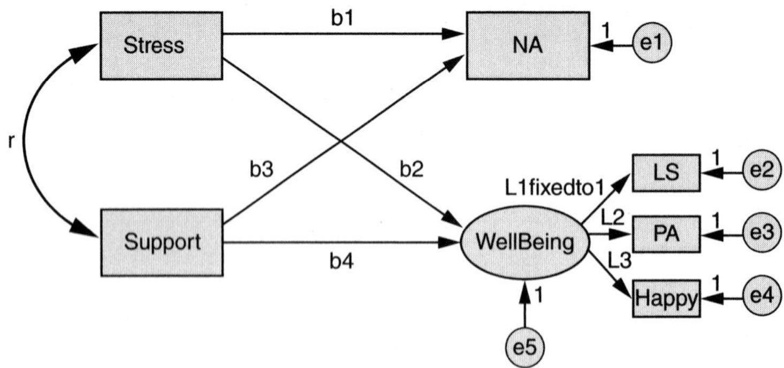

Here is a simple verbal explanation of the model. Stress and support are observed predictor variables that are allowed to be correlated with each other. In the unstandardized model, path *r* represents their covariance; in the standardized version, path *r* represents their correlation. Each variable is assessed as a predictor of two outcomes: negative affect

(NA), represented by one measured variable; and the latent variable well-being, with three indicator variables: life satisfaction (LS), positive affect (PA), and happiness (happy). The use of LS and PA variables as indicators for an overall well-being latent variable is common in positive psychology research. This model departs from common practice in two ways. NA, which is usually represented as a negative indicator of well-being, is treated as a separate dependent variable instead of as a (negative) indicator for the well-being latent variable. The variable happiness was added in order to have a minimum of three indicators for the well-being latent variable. NA was separated from the positive outcome measures because of both theoretical arguments and empirical research suggesting that different variables do well at predicting NA versus PA (Warner & Rasco, 2014). There are no mediation hypotheses. Each of the predictors has a direct path to each *Y* outcome and also an indirect path due to the correlation between predictors. The analysis will evaluate the following:

- How well does this path model fit the data (i.e., how well do the unstandardized path coefficients reproduce the matrix of variances and covariances for all measured variables (stress, support, NA, LS, PA, and happiness).
- If the model has adequate fit, we can go on to ask additional questions. In this example, one modification to the model needs to be made to improve fit.
- Which path coefficients are (or are not) statistically significant? If the measurement model makes sense, we would expect the paths that correspond to factor loadings (L2 and L3) to have significant path coefficients and positive loadings on the well-being latent variable.
- How well do the variables in this model predict well-being and negative affect? For this, we examine the R^2 values. It is possible that a model reproduces the variance/covariance matrix acceptably but, at the same time, does not predict the dependent variables well.

In the "Output" tab in the Analysis Properties dialog box, the box for "Modification indices" was checked; this provides information that is useful when an initial model does not fit adequately. Large MI values indicate where changes to the model may improve fit.

We would usually expect many of the causal path coefficients (b1 through b4) to be statistically significant. However, the analyst may expect that some independent variables are not good predictors of some dependent variables. In this example, on the basis of past well-being research, I would expect paths b1 and b4 (where a positive variable predicts a positive outcome, and a negative variable predicts a negative outcome) to be large, positive, and statistically significant. It is less clear what will happen with paths b2 and b3 (stress predicting well-being and support predicting negative affect). Some past literature suggests that PA and NA are sometimes predictable from different independent variables (Warner & Rasco, 2014). For this reason, NA was kept separate from the other three positive well-being measures. In the extreme case we might find that stress predicts only negative affect (and not well-being) and that support predicts only well-being (and not negative affect).

Because this model is underidentified (which is usually the case in SEM), it is necessary to assess model fit before going on to look at other results. Model fit information appears in Figure 15.29.

For the model in Figure 15.28, $\chi^2(7) = 32.632, p < .001$. The CFI for this model was .955; that is above the value of .95 often recommended as a criterion for good model fit. However, RMSEA was .102, 90% CI [.069, .139]. Usually the criterion for good model fit is RMSEA < .06 or .08. These results suggest that there is room for improvement in model fit.

For suggestions about changes in the model that could improve model fit, look at the output for modification indexes (Figure 15.30). Large values of MIs identify parts of the model that could be changed to improve model fit. In this example, there was only one large value of an MI, corresponding to a potential path between error terms e1 and e3. This tells us that adding an arrow to represent a nonzero covariance or correlation between these two error terms, e1 and e3, could improve model fit.

In this instance, adding correlated errors is reasonable. The measures of PA and NA (the indicators with which these error terms are associated) both came from the same questionnaire, used the same question types and rating scales, and asked about mood during the same time frame (the past 2 weeks). Other measures, such as stress, happiness, and life satisfaction, asked about life in general (rather than a limited time period). We might expect PA and NA to be correlated because of common measurement method (Newsom, 2017). It is not generally

Figure 15.29 Fit Indexes for Initial Model to Predict Well-Being and Negative Affect

CMIN

Model	NPAR	CMIN	DF	P	CMIN/DF
Default model	14	32.632	7	.000	4.662
Saturated model	21	.000	0		
Independence model	6	588.629	15	.000	39.242

Baseline Comparisons

Model	NFI Delta1	RFI rho1	IFI Delta2	TLI rho2	CFI
Default model	.945	.881	.956	.904	.955
Saturated model	1.000		1.000		1.000
Independence model	.000	.000	.000	.000	.000

RMSEA

Model	RMSEA	LO 90	HI 90	PCLOSE
Default model	.102	.069	.139	.007
Independence model	.331	.308	.354	.000

Figure 15.30 Output for Modification Index Results for Prediction of Well-Being and Negative Affect

Covariances: (Group number 1 - Default model)

	M.I.	Par Change
e1 <--> e3	22.270	6.607

Figure 15.31 Standardized Coefficients for Modified Structural Equation Model With Added Correlated Error Path

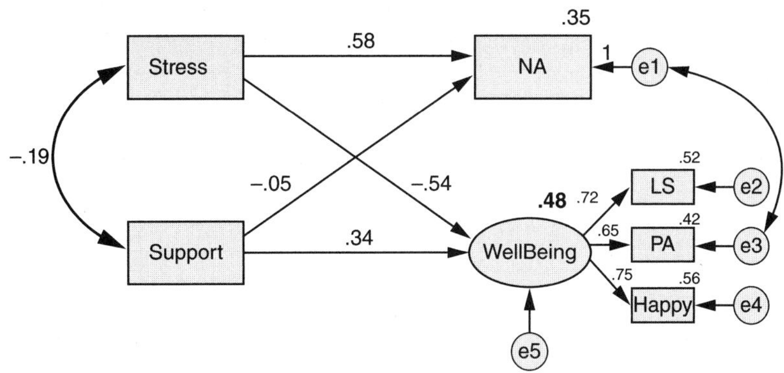

a good idea to add correlated error terms based on MI results unless there are reasons why this makes sense. In this example there were logical reasons to think something else (time frame of questions) could lead to a correlation between PA and NA that was not explained by the existing measurement model. The correlation between error terms e1 and e3 was added; the modified model appears in Figure 15.31.

For the modified model in Figure 15.33, $\chi^2(6) = 9.416$, $p = .151$, CFI = .994, and RMSEA = .04, 90% CI [.000, .087]. These all indicate adequate model fit. It makes sense to go forward and examine path coefficient and R^2 results for this version of the model.

All path coefficients in the modified model were statistically significant, except that social support was not a statistically significant predictor of negative affect. Some data analysts might choose to trim the model by dropping this nonsignificant path. However, it is sometimes preferable to retain the model as originally specified, it if can be made to fit adequately. In this case, the lack of a significant path between social support and NA is potentially interesting and retaining it in the model would highlight that this is not a statistically significant effect. The model suggests that while higher social support is associated with higher well-being, it is not associated with less negative affect. The covariances between the two predictors, and between the correlated error terms, were both statistically significant (Figure 15.33).

Figure 15.32 Path Coefficient Estimates for the Modified Well-Being Model in Figure 15.31

			Estimate	S.E.	C.R.	P	Label
Well Being	←	Overall Social Support Cohen ISEL	.228	.036	6.347	***	
Well Being	←	Perceived Social Stress Cohen	-.366	.039	-9.444	***	
LIFE_SATISFACTION	←	Well Being	1.000				
Positive Affect past 2 weeks	←	Well Being	1.033	.100	10.339	***	
Negative Affect past 2 weeks	←	Perceived Social Stress Cohen	.647	.048	13.491	***	
HAPPINESS	←	Well Being	.190	.017	11.386	***	
Negative Affect past 2 weeks	←	Overall Social Support Cohen ISEL	-.058	.048	-1.219	.223	

15.16 COMPARING STRUCTURAL EQUATION MODELS

One of the advantages of SEM is that it provides ways to compare different models. Three different types of comparisons are common and extremely useful.

15.16.1 Comparison of Nested Models

Model B is nested in Model A if Model B is based on the same cases and variables, but Model B allows at least one parameter that is constrained in Model A to be freely estimated. Model B (the less restricted model) will have a smaller *df* than Model A. Model B is the reduced, or less restricted, model. Model A, which has more constrained parameters and a larger *df*, is called the full model. Model B will fit better than Model A (or at least as well as Model A), and Model B will have smaller χ^2 (better fit) than Model A. The difference in fit can be assessed by finding this difference: $\chi^2_A - \chi^2_B$. This difference in χ^2 values is also χ^2 distributed, with *df* equal to the number of restrictions in Model A that are dropped in Model B. It is not always easy to evaluate whether one model is nested in another model. In Amos, a user runs separate SEM analyses for Models A and B and computes the difference in χ^2 and *df* by hand. As an example, consider the models in Figures 15.28 and 15.31. The first model (Figure 15.28) has more constrained parameters and is the more restricted model. The absence of a path between e1 and e3 constrains the estimated covariance or correlation between these error terms to be 0. The second model (in Figure 15.31) drops that constraint by adding a covariance path between e1 and e3; in other words, it allows this covariance to be a free parameter. The model in Figure 15.31 fits better than the original model in Figure 15.28. We can evaluate the difference in model fit by finding the difference between the χ^2 values (32.632 for the first model – 9.416 for the second model = χ^2 difference of 23.216). This difference is distributed as a χ^2 with 1 *df*. For $\alpha = .001$, the critical value for χ^2 with 1 *df* is 10.8; thus, the difference in model fit is statistically significant ($p < .001$). The model fits significantly better with the added correlated error terms than without.

15.16.2 Comparison of Non-Nested Models

The analyst may want to compare models that use different predictors or have path structures that do not correspond to a nested model situation. This can be done by

Figure 15.33 Statistical Significance of Covariance Paths in Modified Well-Being Model

Covariances: (Group number 1-Default model)

			Estimate	S.E.	C.R.	P	Label
Overall Social Support Cohen ISEL	<-->	Perceived Social Stress Cohen	-6.262	1.794	-3.490	***	
e3	<-->	e1	6.663	1.445	4.610	***	

comparing AIC values for the models; the model with the lower AIC value is judged a better fit (Huang, 2017).

15.16.3 Comparisons of Same Model Across Different Groups

All three examples in this chapter used only one group. However, Amos can be used to evaluate differences between groups. Group comparisons can be used to test CFA, or structural, or combined CFA and structural models across groups. Byrne (2016) provided details about setting up these comparisons. Here are examples of questions that can be answered using group comparisons.

- **Moderation: differences in structural models across groups:** Sex is one example of groups to compare. A possible question could be, Does sex moderate the effects of social support on life satisfaction? For example, is the effect of social support on life satisfaction stronger for women than for men (e.g., is the standardized path coefficient larger for women)? In this example, the two groups consist of men and women. The same path model can be set up within each group. When groups are compared, the analyst has many options. At one extreme, all path coefficients can be constrained to be equal across groups (if there are sex differences, this fully constrained model will not fit well). At the other extreme, none of the path coefficients may be constrained to be equal across groups (this will provide the best fit for each group but will not identify which of the specific paths are the reason for differences). In practice, it is often useful to do something in between these extremes, that is, to constrain some paths to be equal for men and women and to leave other paths free to differ across groups. In this example, one might allow the path from social support to life satisfaction to differ between groups while constraining all other paths to be equal across groups in Model 1. All paths could be constrained to be equal for men and women in Model 2. Significance tests can be used to see if Model 1 fits better than Model 2; if it does, that tells us that the social support/life satisfaction effect differs significantly for men and women or, to put it another way, that sex moderates the effects of social support on life satisfaction.
- **Measurement invariance: comparison of CFA across groups:** Another example is comparison of measurement structural models of CFA models across groups (this can address both reliability issues and potential differences in the meaning of latent variables (i.e., the indicator variables most strongly related to each latent variable). Here are several types of group comparison.

    a) *Different types of people* (for example, men vs. women, children vs. adults, citizens of the United States vs. the United Kingdom). It is possible that the latent variables that describe differentiation of emotions are different for children versus adults, for example.

b) *Different circumstances* (for example, data collected at two different points in time). This assesses whether reliability or measurement structure change across time.

c) *Different subsets of a large data set.* In general, *cross-validation* refers to the following situation. A large data set is divided randomly into two halves. Exploratory analyses are conducted using only the first half of the data. After a model is selected (for example, one regression equation), that analysis is run on the other half of the data to evaluate how well the model works with "new" data. This can be viewed as a type of replication. This can also be done in SEM. Various models can be evaluated using the first half of the data. When a final model has been selected, this model can be compared across the groups that correspond to the cases in the first half of the data and the second half of the data. A successful cross-validation would require that the model fit does not differ substantially between these two groups. (In general, values such as R^2 do tend to shrink when a model developed using one batch of data is applied to a new batch of data.)

15.16.4 Other Uses of SEM

SEM can assess mean structures (e.g., whether means of latent variables differ across groups or across time periods). It can be used in latent growth models and multilevel SEM (Kline, 2016).

15.17 REPORTING SEM

Reports of SEM should include the following. Full details were not included in presentations of results in this chapter to limit length. This following list is based on recommendations made by Kline (2016) and Schreiber et al. (2006).

1. A clear statement: What analysis was done, with what variables, to answer what question or questions? Kline (2016) noted that it is essential to ground the model in theory. The paths that are present (and the paths that are missing) should be justified. If control variables are needed, they should be included. What questions will be answered by comparing models (either nested, or across groups)? Will analysis be confirmatory, in that it examines only models specified ahead of time, or will it be exploratory, because numerous variations or modifications of the model will be examined?
2. Model identification: Is the model fully identified or underidentified? Is the model recursive? How many parameters will be estimated, and how many distinct sample moments are there (this information is available in the "Notes for Model" section of the text output file). As noted earlier, model identification requires more than information about number of parameters and distinct sample moments, and it can be difficult to evaluate model identification; see Kenny and Milan (2012).
3. Specific information about the program(s) and versions of programs used for analyses and the type of estimation used. The default estimation method, maximum likelihood, was used in all examples in this chapter. For bootstrapping, include the number of bootstrap samples.
4. Information about sample size decisions (ideally, based on statistical power analysis).

5. Evaluation of potential violations of assumptions and explanation of remedies that were applied. This includes evaluation of normality, linearity, outliers, and missing values. Some of this information can be obtained in advanced data screening in SPSS; additional information about possible violations of assumptions, including tests for univariate normality and multivariate outliers, can be requested from Amos, as shown in the first example.
6. Descriptive information about all variables included in the structural equation model, including *M*, *SD*, and other information such as minimum and maximum, the variance/covariance matrix, and/or the correlation matrix.
7. If multiple models are examined or compared, model fit information is needed for each of the models.
8. Assessment of overall model fit. Usually this includes χ^2, CFI, and RMSEA. Specify the standards used to judge the adequacy of model fit. Model comparisons should be reported. For nested models, χ^2 can be used to compare model fit. For other kinds of comparisons, AIC values may be examined.
9. Further information about results is typically provided only for those that fit adequately and/or have best comparative fit.
10. Unstandardized path coefficients and their *SE*, C.R., and *p* values; standardized path coefficients; evaluation of coefficients in terms of effect size; R^2 for each outcome variable; evaluation of total, direct, and indirect effects; and bootstrapped CIs (these are sometimes included only for indirect paths, such as $a \times b$ in the first example, but could be included for unstandardized path coefficients as well). Keep in mind that a model with good fit may nevertheless tell you that the outcome variables are not predicted very well.
11. Interpretation of the model relative to initial hypotheses. If competing models were compared, was the best fitting model the one that was expected? Which paths were (or were not) statistically significant? Were signs and magnitudes of path coefficients consistent with predictions? Do R^2 values indicate good or poor prediction of outcome variables? In CFA, are correlations between latent variables or factors as expected? Were modifications to the originally proposed model needed to achieve good fit, and if so, what modifications? (Results from one model should not be reported as if only one confirmatory analysis had been performed, if in fact that model was the result of exploratory analysis of many different models; this is a form of *p*-hacking.)

15.18 SUMMARY

There are many reasons why it is worthwhile to learn how to use Amos (or other SEM programs). These include the following:

1. Regression and path models that include only measured (observed) variables implicitly assume that these have no measurement error. That assumption is usually wrong in practice. The inclusion of latent variables with measurement models in SEM is a way of taking measurement reliability issues into account. (Reliability for a latent variable is higher if its factor loadings are high. The use of multiple measures is a way to improve reliability.)
2. Amos is one of the programs that provides bootstrapping. Bootstrapping (available in SEM and other advanced procedures) is the generally preferred method to test

the statistical significance of indirect effects in mediated models. Bootstrapping is more robust to violations of assumptions of normality.

3. Structural equation models can handle complexity that would be difficult or impossible to manage using OLS regression. In addition to measurement models, SEM can include multiple predictors; multiple outcomes; mediation, including multiple-step mediation; and moderation.
4. Structural equation models can be compared across groups.
5. Knowledge of SEM is a step toward understanding other advanced topics that include latent variables.
6. SEM can be fun.[4]

This chapter is not a complete introduction to SEM. It should be sufficient for those who want to read typical journal article reports of relatively simple SEM analyses. Byrne (2016) and Kline (2016) discuss many important issues not covered here. Those who want to conduct their own SEM analyses should consult them for a much more comprehensive treatment.

COMPREHENSION QUESTIONS

1. Answer the following questions about the diagram in Figure 15.34.
   a. What is this type of SEM analysis called?
   b. What parts of the diagram correspond to each of the following: latent variables, measured or indicator variables, and error terms?
   c. Which variables need to be given the names of variables in the SPSS data file?

Figure 15.34 Schematic Structural Equation Model

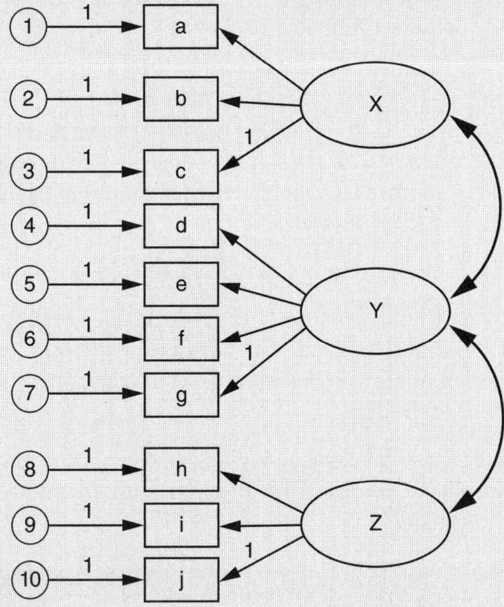

2. Answer the following questions about Figure 15.35.
   a. Does this diagram include a measurement model? Any latent variables? Any error terms?
   b. Trace all possible paths from X_2 to Y (e.g., which values will be multiplied, which values will be added, and which paths are direct versus indirect).
   c. What mediation hypothesis or hypotheses does this model include (identify the paths that are involved by name, e.g., *XX*).
   d. If the *QQ* path coefficient is not statistically significant, how might you interpret that result?
   e. Do you think this model is likely to have perfect fit? Why or why not?
   f. Suppose there is a Model 2, identical to the model in Figure 15.35, except that *QQ* is dropped from Model 2. How does this change any mediation hypotheses? Which is a more restricted model, 1 or 2? Which model is likely to have a larger χ^2 value, Model 2 or Model 1?
   g. Is Model 1 recursive? Explain briefly.
   h. Suppose that Model 1 fits really badly. Think of a path that could be added that might improve model fit (there is more than one possible path).

Figure 15.35 Hypothetical Mediated Path Model

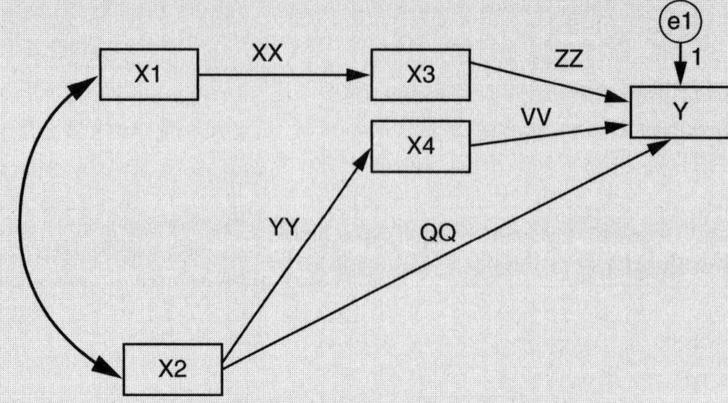

3. How is model fit evaluated in SEM? There are two parts to this question: What information from the data does the model have to reproduce reasonably well to be judged a good fit? What are three fit indexes that are commonly reported, and what arbitrary "cutoff" values are used to decide what these fit index values mean?

4. What does it mean for Model B to be nested in Model A? What information can be used to compare the fit for nested models?

5. If you create a multiple regression model, with all of the predictors correlated with one another, and run it in Amos, is this an overidentified, underidentified, or fully identified model? How well will it fit the data?

6. Answer these questions about the model in Figure 15.36.
   a. In words, how would you describe the associations among variables that are hypothesized in this model?

Figure 15.36 Structural Equation Model With Latent Predictor Variables

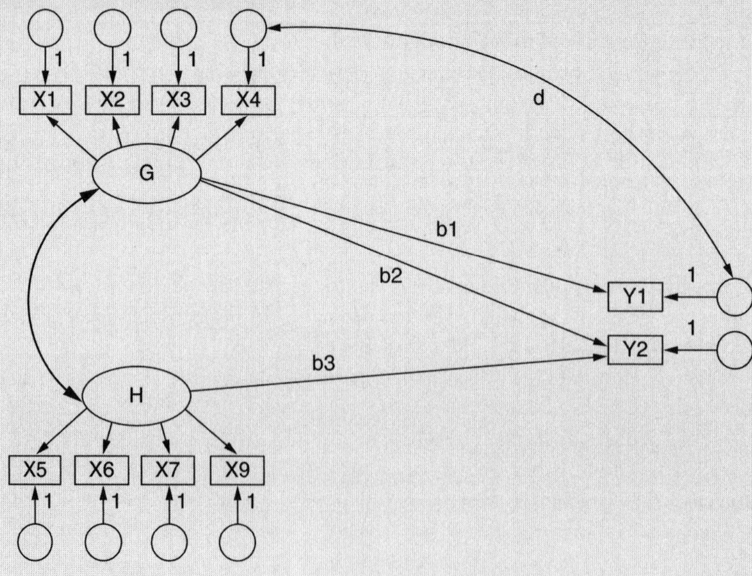

b. Which parts of the diagram represent measurement model(s)? Which parts of the diagram represent structural models?
c. How many measured or observed variables from the SPSS file are included in this model?

7. How does EFA (exploratory factor analysis) differ from CFA (confirmatory factor analysis)?
8. How are the correlation between r_{XY} and the covariance between X and Y related? How are they different?
9. Which type of model is usually of greater interest in SEM: just identified, overidentified, or underidentified? Explain your reasoning.
10. What is the basic idea behind bootstrapping?
11. Paths that are missing in a structural equation model are just as important as paths that are present. Discuss briefly.
12. What do we mean by "constraints" on a model? Give one example.
13. What are the C.R. values in Amos text output, and how are they interpreted?
14. What does the $a \times b$ path in a simple X, M, Y mediation model represent? How is the statistical significance of $a \times b$ assessed using the Sobel test, and what information can be obtained from SEM to assess this?
15. What problem can arise when a data analyst runs a model, changes some parameters, runs the model again, changes the parameters again, runs the model, changes the parameters again, repeats this many times, and stops after finally obtaining good model fit, reporting only the final model as if it were a confirmatory result?
16. What is the difference in the type of information obtained by looking at goodness-of-fit indexes (such as χ^2, CFI, and RMSEA) and the significance of path coefficients and R^2 values within the model? Can you have a model that fits well and yet has very small R^2 values?
17. Discuss the model you see in Figure 15.37. Does it include measurement models? A structural model?

Figure 15.37 Complex Structural Equation Model

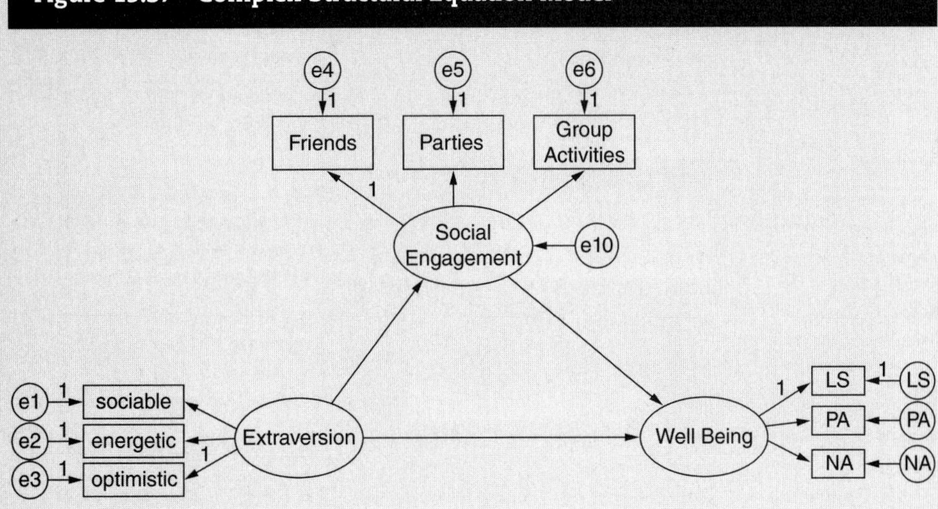

NOTES

[1]The use of annoying quotation marks around the term *causal* is a reminder that although descriptions of structural equation models use causal language, we cannot infer causation from the correlational data generally used in SEM. Results consistent with a model that represents causal hypotheses do not prove that causation is present unless the structural equation model includes manipulated variables in an experiment. SEM can include dummy predictor variables to represent treatment group membership. (However, dummy-coded variables cannot be dependent variables in standard SEM.)

[2]CFA is truly confirmatory only when the analyst runs one model (or a small number of models) that are based on theory. If CFA methods are used to examine large numbers of models, and if repeated model modifications are made to improve fit, results should be described as exploratory.

[3]If significant non-normality is present and cannot be remedied in other ways, the Bollen-Stine p value may be used to assess overall model fit. To obtain this, in the Analysis Properties dialog box in Amos, go to the Bootstrap tab and check the boxes for "Perform bootstrap" and "Bollen-Stine bootstrap." The model is judged adequate when Bollen-Stine p is >.05. In addition, bootstrapped CIs can be obtained for parameter estimates. To obtain these, in the Bootstrap tab, uncheck "Bollen-Stine bootstrap" and check "Perform bootstrap," "Percentile confidence intervals," and "Bias-corrected confidence intervals" (University of Texas at Austin, Department of Statistics and Data Science, n.d.).

[4]For useful information and fun with SEM, see Alan Reifman's SEM course page at http://reifman-sem.blogspot.com/. Each year he and his students compose lyrics for *SEM: The Musical*, with titles such as "RMSEA" (sung to the tune of the Village People's "Y.M.C.A.," for those of you old enough to remember).

DIGITAL RESOURCES

Find **free study tools** to support your learning, including **eFlashcards, data sets, and web resources,** on the accompanying website at **edge.sagepub.com/warner3e**.

CHAPTER 16

BINARY LOGISTIC REGRESSION

16.1 RESEARCH SITUATIONS

16.1.1 Types of Variables

Binary logistic regression can be used to analyze data in studies where the outcome variable is binary or dichotomous. In medical research, a typical binary outcome of interest is whether each patient survives or dies. The outcome variable, "death," can be coded 0 = alive and 1 = dead. The goal of the study is to predict membership in a target group (e.g., membership in the "dead" group) from scores on one or several predictor variables. Like multiple linear regression, a binary logistic regression model may include one or several predictor variables. The predictor variables may be quantitative variables, dummy-coded categorical variables, or both. For example, a researcher might want to predict risk for death from gender, age, drug dosage level, and severity of the patient's medical problem prior to treatment.

16.1.2 Research Questions

The basic research questions that arise in binary logistic regression studies are similar to the research questions that arise in multiple linear regression except that the inclusion of a group membership outcome variable requires some modification to the approach to analysis. Researchers need to assess statistical significance and effect size for an overall model, including an entire set of predictor variables. They may test competing models that include different predictor variables. When an overall model is significant, the researcher examines the contribution of individual predictor variables to assess whether each predictor variable is significantly related to the outcome and the nature and strength of each predictor variable's association with the outcome.

Unlike most analyses presented in earlier chapters, logistic regression is not a special case of the general linear model; it uses different summary statistics to provide information about overall model fit and the nature of the relationship between predictors and group membership. The overall strength of prediction for a multiple linear regression model is indexed by multiple R, and the significance of multiple R is assessed by an F ratio; the nature of the predictive relationship between each predictor variable and the score on the outcome variable is indexed by a raw score or standardized slope coefficient (b or β), and the significance of each individual predictor is assessed by a t test. A logistic regression includes some familiar terms (such as raw score regression coefficients, denoted $\boldsymbol{B}$). In addition, it provides additional information about odds and odds ratios. Odds and odds ratios will be explained before examining an empirical example of binary logistic regression.

16.1.3 Assumptions Required for Linear Regression Versus Binary Logistic Regression

Analyses that are special cases of the general linear model (such as multiple linear regression and discriminant analysis [DA]) involve fairly restrictive assumptions. For a multiple linear regression in which scores on a quantitative Y variable are predicted from scores on quantitative X variables, the assumptions include the following: a multivariate normal distribution for the entire set of variables and, in particular, a univariate normal distribution for scores on Y; linear relations between scores on Y and scores on each X variable, as well as between each pair of X predictor variables; and uniform error variance across score values of the X variable (e.g., variance of Y is homoscedastic across score values of X). For a DA in which group membership is predicted from scores on several quantitative X variables, an additional assumption is required: homogeneity of elements of the variance/covariance matrix for the X variables across all groups. The use of a binary outcome variable in binary logistic regression clearly violates some of these assumptions. For example, scores on Y cannot be normally distributed if Y is dichotomous.

In contrast, logistic regression does not require such restrictive assumptions. The assumptions for logistic regression (from Wright, 1995) are as follows:

1. The outcome variable is dichotomous (usually the scores are coded 1 and 0).
2. Scores on the outcome variable must be statistically independent of each other.
3. The model must be correctly specified; that is, it should include all relevant predictors, and it should not include any irrelevant predictors. (Of course, this is essential for all statistical analyses, although textbooks do not always explicitly state this as a model assumption.)
4. The categories on the outcome variable are assumed to be exhaustive and mutually exclusive; that is, each person in the study is known to be a member of one group or the other but not both.

Note that binary logistic regression does not require normally distributed scores on the Y outcome variable, a linear relation between scores on Y and scores on quantitative X predictor variables, or homogeneous variance of Y across levels of X. Because it requires less restrictive assumptions, binary logistic regression is widely viewed as a more appropriate method of analysis than multiple linear regression or DA in many research situations where the outcome variable corresponds to membership in two groups. Even though fewer assumptions are involved in binary logistic regression, preliminary data screening is still useful and important. For example, it is a good idea to examine the expected frequency in each cell. Like the chi-square test for simple contingency tables, a binary logistic regression does not perform well when many cells have expected frequencies less than five. In addition, extreme outliers on quantitative predictor variables should be identified. Recommendations for data screening are provided in Section 16.8.1.

16.2 FIRST EXAMPLE: DOG OWNERSHIP AND ODDS OF DEATH

Friedmann, Katcher, Lynch, and Thomas (1980) conducted a survey of 92 male patients who initially survived a first heart attack. Table 16.1 summarizes information about a binary predictor variable (Did the patient own a dog?) and the primary outcome of interest (Was the patient alive or dead at the end of the first year after the heart attack?). The data in Table 16.1 are contained in the SPSS file dog.sav, with the following variable names and value labels. The variable dogowner was coded 0 for persons who did not own dogs and 1 for persons who did own

Table 16.1 Data From Friedmann et al.'s (1980) Survey: 92 Men Who Initially Survived a First Heart Attack

	Patient Does Not Own Dog ($Y = 0$)	Patient Owns a Dog ($Y = 1$)	Total N
Patient alive ($X = 0$)	28 (71.8%)	50 (94.3%)	78
Patient dead ($X = 1$)	11 (28.2%)	3 (5.7%)	14
Total N	39	53	92

Note: The X predictor variable was dogowner, coded 0 for a patient who does not own a dog and 1 for a patient who owns a dog. The Y outcome variable was death, coded 0 for a patient who was alive at the end of 1 year and 1 for a patient who was dead at the end of 1 year.

dogs. The variable death was coded 0 for persons who were alive at the end of 1 year and 1 for persons who had died by the end of the 1-year follow-up. The question whether dog ownership is statistically related to survival status was addressed in Chapter 17 in Volume I (Warner, 2020) using simple analyses, including computation of percentages and a chi-square test of association.

Using χ^2, the pattern of results in this table was described by comparing probabilities of death for dog owners and nonowners of dogs. Among the 39 people who did not own dogs, 11 died by the end of the year; thus, for nonowners of dogs, the probability of death in this sample was 11/39 = .282. Among the 53 people who owned dogs, 3 died; therefore, the probability of death for dog owners was 3/53 = .057. On the basis of a **chi-square test** of association, $\chi^2(1) = 8.85$, $p < .01$, this difference in proportions or probabilities was judged to be statistically significant. The researchers concluded that the risk for death was significantly lower for dog owners than for people who did not own dogs.

In this chapter, the same data will be used to illustrate the application of binary logistic regression. In binary logistic regression, the goal of the analysis is to predict the odds of death on the basis of a person's scores on one or more predictor variables. Logistic regression with just one binary predictor variable is a special case that arises fairly often; therefore, it is useful to consider an example. In addition, it is easier to introduce concepts such as odds using the simplest possible data. A binary logistic regression analysis for the dog ownership/death data appears in Section 16.7. A second empirical example, which includes one quantitative and one categorical predictor variable, appears in Section 16.10.

16.3 CONCEPTUAL BASIS FOR BINARY LOGISTIC REGRESSION ANALYSIS

Let's first think about a simple, but inadequate, way to set up an analysis of the dog ownership and death data using a simple multiple linear regression model. Consideration of the problems that arise in this simple model will help us understand why a different approach was developed for binary logistic regression. We could set up an equation to predict each individual's probability of membership in the target group, $\hat{p}$ (such as the dead patient group), from a weighted linear combination of that person's scores on one or several predictor variables as follows:

$$\hat{p}_i = B_0 + B_1 X_1 + B_2 X_2 + \ldots + B_k X_k, \tag{16.1}$$

where $\hat{p}_i$ is the estimated probability that person i is a member of the "target" outcome group that corresponds to a code of 1 (rather than the group that is coded 0). In this example, the target group of interest is the dead group and, therefore, $\hat{p}_i$ corresponds to the estimated probability that person i is dead given his or her scores on the predictor variables. B_0 is the intercept, and the B_i values are the regression coefficients that are applied to raw scores on the predictor variables. For example, we could predict how likely an individual is to be in the "dead" group on the basis of the person's score on dog ownership (0 = does not own a dog, 1 = owns a dog). If we had information about additional variables, such as each individual's age, severity of coronary artery disease, and systolic blood pressure, we could add these variables to the model as predictors.

16.3.1 Why Ordinary Linear Regression Is Inadequate When Outcome Is Categorical

The simple regression model shown in Equation 16.1 is inadequate when the outcome variable is dichotomous for several reasons. One difficulty with this model is that probabilities, by definition, are limited to the range between 0 and 1. However, estimated values of $\hat{p}_i$ obtained by substituting a person's scores into Equation 16.1 would not necessarily be limited to the range 0 to 1. Equation 16.1 could produce estimated probabilities that are negative or greater than 1, and neither of these outcomes makes sense. We need to set up a model in a way that limits the probability estimates to a range between 0 and 1.

If one or more of the predictors are quantitative variables, additional potential problems arise. The relation between scores on a quantitative X predictor variable and the probability of membership in the "dead" group is likely to be nonlinear. For example, suppose that a researcher systematically administers various dosage levels of a toxic drug to animals to assess how risk for death is related to dosage level. Figure 16.1 shows results for hypothetical data. Level of drug dosage appears on the X axis, and probability of death appears on the Y axis.

Figure 16.1 Sigmoidal Curve: Predicted Probability of Death (Y Axis) as a Function of Drug Dose (X Axis)

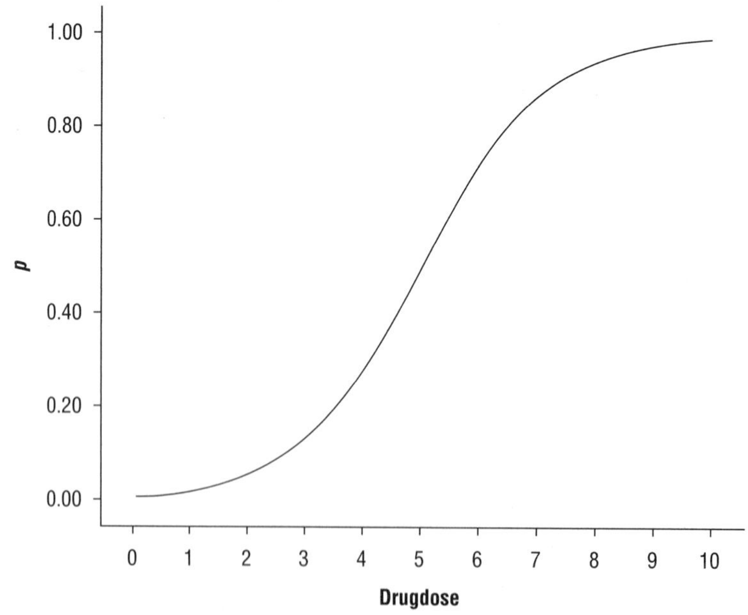

Note: X axis, drug dose in milligrams ranging from 0 to 10 mg; Y axis, predicted probability of death ranging from 0 to 1.

The curve in Figure 16.1 is a **sigmoidal function** or S curve. If the underlying relation between the X predictor variable (such as drug dosage in milligrams) and the Y outcome variable (a binary-coded variable, 0 = alive, 1 = dead) has this form, then ordinary linear regression is not a good model. In this example, a linear model approximately fits the data for values of X between 3 and 7 mg, but a linear model does not capture the way the function flattens out for the lowest and highest scores on X. In this example, the relation between scores on the X variable and probability of membership in the target group is nonlinear; the nonlinearity of relationship violates one of the basic assumptions of linear regression.

A second problem is that when the underlying function has a sigmoidal form (and the Y scores can only take on two values, such as 0 and 1), the magnitudes of prediction errors are generally not uniform across scores on the X predictor variable. The actual outcome scores for Y can only take on two discrete values: $Y = 0$ or $Y = 1$. Figure 16.2 shows the actual Y scores for these hypothetical data superimposed on the curve that represents predicted probability of death at each drug dosage level. The actual Y values are much closer to the prediction line for drug dosage levels $X < 2$ and $X > 7$ than for intermediate drug dosage levels (X between 3 and 7). In other words, the probability of death is predicted much more accurately for animals that receive less than 2 mg or more than 8 mg of the drug. This pattern violates the assumption of homoscedasticity of error variance that is required for ordinary linear regression analysis.

Figure 16.2 Actual Outcome Scores Superimposed on Predicted Probability Curve

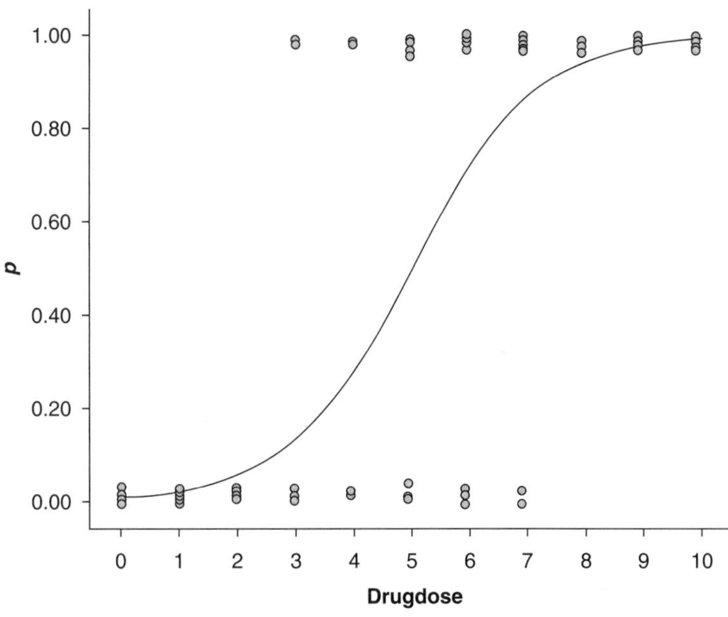

Note: Dots represent actual Y outcomes (Y = 0 = alive, Y = 1 = dead) for each case. A small random error was added to each score to create a jittered plot; this makes it easier to see the number of cases at each drug dose level. This plot illustrates heteroscedasticity of error variance. The sigmoidal function predicts outcome (alive vs. dead) quite well for animals who received very low doses of the drug (0–2 mg) and for animals who received very high doses of the drug (8–10 mg). For animals who received intermediate dosages (between 3 and 7 mg), the prediction errors are much larger. This violates one of the assumptions for ordinary multiple regression (the assumption of homoscedasticity of error variance across levels of X).

16.3.2 Modifying the Method of Analysis to Handle a Binary Categorical Outcome

How can we alter the model in Equation 16.1 in a way that avoids the potential problems just identified? We need to have a dependent variable in a form that avoids the problems just identified and a model that represents the potential nonlinearity of the relationship between scores on X and probability of group membership. At the same time, we want to be able to obtain information about probability of group membership from the scores on this new outcome variable. We need a transformed outcome variable that can give us predicted probabilities that are limited to a range between 0 and 1, that is approximately normally distributed, and that has a linear relationship to scores on quantitative X predictor variables. We can obtain an outcome variable for which the scores satisfy the requirements by using a **logit** as our outcome variable. In binary logistic regression, the outcome variable is called a logit; a logit is denoted by L_i in this chapter (in some textbooks, the logit is denoted by the symbol π). What is a logit (L_i)? How does this new form for the outcome variable solve the problems that have just been identified? How can we use the logit to make inferences about the probability of group membership as a function of scores on predictor variables?

A logit is a log of odds (or sometimes a log of a ratio of odds for two different groups or conditions). To understand the logit, we need to understand each of the terms involved in this expression. What are odds? How does the log transformation typically change the shape of a distribution of scores?

16.4 DEFINITION AND INTERPRETATION OF ODDS

Odds are obtained by dividing the number of times an outcome of interest *does* happen by the number of times when it *does not* happen. For example, consider the data in Table 16.1. The main outcome of interest was whether a patient died (coded 0 = patient survived, 1 = patient died). In the entire sample of $N = 92$ people, 14 people died and 78 people survived. The odds of death for this entire group of 92 persons are found by dividing the number of dead people (14) by the number of survivors (78): 14/78 = .179. Conversely, we can calculate the odds of survival; to find this, we divide the number of survivors by the number of dead persons, 78/14 = 5.57. Note that 5.57 is just the reciprocal of .179. We can say that the odds of survival for patients in this study were greater than 5 to 1 or that the odds of death for patients in this study were less than 1 in 5.

What possible values can odds take on, and how are those values interpreted? The minimum value of odds is 0; this occurs when the frequency of the event in the numerator is 0. The maximum value for odds is $+\infty$; this occurs when the divisor is 0. To interpret odds, we ask whether the value of the odds is greater than 1 or less than 1. When odds are less than 1, the target event is less likely to happen than the alternative outcome. When odds are greater than 1, the target event is more likely to happen than the alternative outcome. When the odds exactly equal 1, the target event has an equal chance of happening versus not happening. To illustrate the interpretation of odds, consider a simple experiment in which a student tosses a coin 100 times and obtains heads 50 times and tails 50 times. In this case, the odds of getting heads is the number of times when heads occur divided by the number of times when heads do not occur: 50/50 = 1.00. This is called "even odds"; an odds of 1 means that the two outcomes (heads or tails) are equally likely to occur. Consider another simple possibility. Suppose that a prize drawing is held and 10 tickets are sold. One ticket will win the prize, while the other nine tickets will not win. The odds of winning are 1/9 (or .1111); the odds of not winning are 9/1 (or 9). These examples show us how odds are interpreted. When odds are close to 1, the two outcomes (such as heads vs. tails for a coin toss) are equally likely to happen. When odds are less than 1, the outcome of interest is unlikely to happen, and the closer the odds get to 0, the less likely the outcome. When odds are greater than 1, the event of interest is more likely to occur than not; the higher the odds, the more likely the event.

Returning to the dog ownership and death data, the odds of death for the entire sample, given above as odds = .179, tells us that a randomly chosen person from this entire sample of 92 people is less likely to be dead than alive. Conversely, the value of the odds of surviving given above, odds = 5.57, tells us that a randomly selected individual from this group of N = 92 is much more likely to be alive than dead. A randomly selected person from this sample of 92 people is between 5 and 6 times as likely to be alive as to be dead.

The language of odds is familiar to gamblers. When the odds are greater than 1, the gambler is betting on an outcome that is likely to happen (and the larger the odds, the more likely the event). When the odds are less than 1, the gambler is betting on an outcome that is unlikely to happen (and the closer to zero the odds become, the less likely the event is).

An odds ratio is a comparison of the odds of some target event across two different groups or conditions. We can compute odds of death separately for each of the two groups in Table 16.1. For the nonowners of dogs, the odds of death are 11/28 = .393. For the owners of dogs, the odds of death are 3/50 = .060. To compare owners and nonowners of dogs, we set up a ratio of these two odds. Throughout this process, we have to be careful to keep track of which group we place in the numerator. Suppose that we use the group of nonowners of dogs as the reference group when we set up a ratio of odds. The ratio of odds that describes the relatively higher odds of death for nonowners of dogs compared with owners of dogs is found by dividing the odds of death for nonowners (.393) by the odds of death for owners (.060). Therefore, the odds ratio in this example is .393/.060 = 6.55. In words, the odds of death are about 6.5 times greater for nonowners of dogs than for dog owners. If we set up the odds ratio the other way, that is, if we divide the odds of death for dog owners by the odds of death for nonowners, .060/.393 = .153, we can say that the odds of death are less than one sixth as high for owners of dogs as for nonowners of dogs in this sample. Whichever group we choose to use in the denominator, the nature of the relationship remains the same, of course: Dog owners in this study were about 6 times less likely to die than people who did not own dogs.

An odds ratio has an advantage over a probability ($\hat{p}_i$). Unlike a probability (which cannot exceed 1), an odds ratio has no fixed upper limit. However, an odds ratio still has a fixed lower limit of 0; values of an odds ratio do not tend to be normally distributed, and values of the odds ratio do not tend to be linearly related to scores on quantitative predictor variables (Pampel, 2000). These are not desirable characteristics for an outcome variable, but these problems can be remedied by applying a simple data transformation: the **natural logarithm**. The base 10 logarithm of X is the power to which 10 must be taken to obtain the value of X. For example, suppose we have a score value X (human body weight in pounds), for instance, X = 100 lb. The base 10 log of X = 100 is the power to which 10 must be raised to equal X = 100. In this case, the log is 2; $10^2 = 100$. Natural logarithms use a different base. Instead of 10, the **natural log** is based on the mathematical constant e; the value of e to five decimal places is 2.71828. The natural log of X, usually denoted ln(X), is the power to which e must be raised to obtain X. Examples of X and the corresponding values of ln(X) appear in Table 16.2.

The operation that is the inverse of natural log is called the **exponential function**; the exponential of X is usually denoted by either e^X or exp(X). By applying the exponential function, we can convert the ln(X) value back into the X score. For example, $e^0 = 1$, $e^{.5} = 1.64872$, $e^1 = 2.71828$, and $e^2 = 7.38905$. Some calculators have the ln(X) function (input X, output the natural logarithm of X) and an exp or e^X function (input X, output e raised to the Xth power). Online calculators are also available to provide these values. Functions that involve the mathematical constant e are often encountered in the biological and social sciences; a familiar function that involves the constant e is the formula for the normal or Gaussian distribution. Functions that involve e often have very useful properties. For example, although the normal curve is defined for values of X that range from $-\infty$ to $+\infty$, the total area under the curve is finite and is scaled to equal 1.00. This property is useful because slices of the (finite) area

Table 16.2 Examples of Values of X, $a = \ln(X)$, and $\exp(a)$		
X	**$a = ln(X)$**	**exp(a) or e^a**
0	Undefined	
1	0	1
1.64872	.5	1.64872
2.71828 (e)	1	2.71828 (e)
7.38905 (e^2)	2	7.38905 (e^2)

under the normal curve can then be interpreted as proportions or probabilities. The mathematical constant e is often included in curves that model growth or that involve sigmoidal or S-shaped curves (similar to Figure 16.1). To summarize, taking the natural log of an odds (or a ratio of odds) to create a new outcome variable called the logit, denoted by L_i, provides an outcome variable that has properties that are much more suitable for use as a dependent variable in a regression model. SPSS and other computer programs provide regression coefficients and the exponential of each regression coefficient as part of the output for binary logistic regression. A data analyst is not likely to need to compute natural logs or exponentials by hand when conducting binary logistic regression because these transformations are usually provided where they are needed as part of the output from computer programs.

16.5 A NEW TYPE OF DEPENDENT VARIABLE: THE LOGIT

Taking the natural log of an odds (or a ratio of odds) converts it to a new form (L_i, called the logit) that has much more desirable properties. It can be shown that L_i scores have no fixed upper or lower limit, L_i scores tend to be normally distributed, and in many research situations, L_i is linearly related to scores on quantitative predictor variables (see Pampel, 2000). Thus, scores on L_i satisfy the assumptions for a linear model, and we can set up a linear equation to predict L_i scores from scores for one or several predictor variables. One disadvantage of using L_i as the dependent variable is that values of L_i do not have a direct intuitive interpretation. However, predicted values of L_i from a binary logistic regression can be used to obtain information about odds and to predict the probability of membership in the target group ($\hat{p}_i$) for persons with specific values of scores on predictor variables. These probability estimates are more readily interpretable.

At this point, we can modify Equation 16.1. On the basis of the reasoning just described, we will use L_i (the logit) rather than $\hat{p}_i$ (a predicted probability) as the dependent variable:

$$L_i = B_0 + B_1 X_1 + \cdots + B_k X_k. \quad (16.2)$$

Equation 16.2 shows that the scores for the logit (L_i) are predicted as a linear function of scores on one or several predictor variables, $X_1, \ldots, X_k$. As in a multiple linear regression, each predictor can be either a quantitative variable or a binary categorical variable. For example, in a medical study where the outcome is patient death versus patient survival, and the predictors include patient gender, age, dosage of drug, and severity of initial symptoms, we would assess the overall fit of the model represented by Equation 16.2; we can use the coefficients associated with the variables in Equation 16.2 to obtain information about the nature and strength of the association of each predictor variable with patient outcome.

Logistic regression is often used to compare two or more different models. The first model examined here, called a **null model** or a **constant-only model**, generates one predicted odds value that is the same for all the people in the entire study; it does not take scores on any *X* predictor variables into account. A comparison model or "full model" includes scores on all the *X* predictor variables and predicts different odds and probabilities for people who have different scores on the predictor variables. A researcher may evaluate several models that include different sets of predictors (King, 2003). The goal is generally to identify a small set of predictor variables that makes sense in terms of our theoretical understanding and that does a good job of predicting group membership for most individuals in the sample.

16.6 TERMS INVOLVED IN BINARY LOGISTIC REGRESSION ANALYSIS

In the study of dog ownership and death (Table 16.1), the null or constant-only model for this research problem can be written as follows:

$$L_i = B_0. \tag{16.3}$$

That is, the null model predicts a logit score (L_i) that is the same for all members of the sample (and that does not differ for owners vs. nonowners of dogs). The full model for this first example in this chapter, where we want to predict odds of death from an *X* predictor variable that represents dog ownership, takes the following form:

$$L_i = B_0 + B_1X. \tag{16.4}$$

Note that the numerical value of the B_0 coefficient in Equation 16.4 will generally not be equal to the numerical value of B_0 in Equation 16.3. Equation 16.4 predicts different values of log odds of death for people whose *X* score is 0 (people who don't own dogs) and people whose *X* score is 1 (people who own dogs). The two different predicted values of L_i are obtained by substituting the two possible values of *X* ($X = 0$ and $X = 1$) into Equation 16.4 and then simplifying the expression. For people with scores of $X = 0$, $L_i = B_0$. For people with scores of $X = 1$, $L_i = B_0 + B_1$. Of course, in research situations where the *X* predictor variable has more than two possible values, the model generates a different L_i value for each possible score value on the *X* variable. In a later section, SPSS will be used to obtain estimates of the coefficients in Equations 16.3 and 16.4. In this simple example (with only one binary predictor variable), it is instructive to do some by-hand computations to see how the logit L_i is related to the odds of death and to the probability of death in each group. The logit (L_i) is related to the predicted probability ($\hat{p}_i$) of membership in the target group (in this empirical example, the probability of death) through the following equation:

$$L_i = \ln\left[\frac{\hat{p}_i}{1-\hat{p}_i}\right]. \tag{16.5}$$

Conversely, given a value of L_i, we can find the corresponding value of $\hat{p}_i$:

$$p_i = \frac{e^{Li}}{1+e^{Li}}. \tag{16.6}$$

For a specific numerical example of a value of L_i, consider the null model in Equation 16.3. The null model predicts the same odds of death for all cases; this model does not include information about dog ownership when making a prediction of odds. For the entire set of

92 people in this study, 14 were dead and 78 were alive. The odds of death for the entire sample is the number of dead persons (14) divided by the number of persons who survived (78) = 14/78 = .179. L_i, the log odds of death for the null model, can be obtained in either of two ways. It can be obtained by taking the natural log of the odds (odds = .179): L_i = ln(odds of death) = ln(.179) = −1.718.

In this example, the probability of death (p) for the overall sample of N = 92 is found by dividing the number of dead persons (14) by the total number of persons: 14/92 = .152. We can use Equation 16.5 to find the value of L_i from this probability:

$$L_i = \ln\left[\frac{\hat{p}_i}{1-\hat{p}_i}\right] = \ln\left[\frac{.152}{1-.152}\right] = \ln\left[\frac{.152}{.848}\right] = \ln(.179) = -1.718$$

Look again at the data in Table 16.1. The L_i value of −1.718 for the null model is consistent with the odds of dying (odds = .179) observed in the overall sample of N = 92. When we use the null model, we can say that (if we ignore information about dog ownership) the odds of death for any individual in this sample is .179, that is, less than 1 in 5. Alternatively, we can report that the probability of death for the entire sample is p = .152.

When we add a predictor variable (dog ownership), we want to see if we can improve the prediction of odds of death by using different predicted odds for the groups owners of dogs and nonowners of dogs. When we examine the SPSS output for this binary logistic regression, we will demonstrate that the coefficients for this model are consistent with the observed odds and probabilities of death for each group (the group of people who own dogs vs. the group of people who do not own dogs).

16.6.1 Estimation of Coefficients for a Binary Logistic Regression Model

Notice that L_i is not a directly measured variable; scores on L_i are constructed in such a way that they maximize the goodness of fit, that is, the best possible match between predicted and actual group memberships. The problem is to find the values of $B_0, B_1, B_2, \ldots, B_k$ that generate values of L_i, which in turn generate probabilities of group membership that correspond as closely as possible to actual group membership. (For example, a person whose predicted probability of being dead is .87 should be a member of the dead group; a person whose predicted probability of being dead is .12 should be a member of the surviving group.) There is no simple ordinary least squares (OLS) analytic solution for this problem. Therefore, estimation procedures such as maximum likelihood estimation (MLE) are used. MLE involves brute-force empirical methods. One simple MLE approach involves grid search. For example, if a person did not know that the formula for the sample mean M was $\Sigma X/M$ but did know that the best estimate of M was the value for which the sum of squared deviations had the minimum possible value, he or she could obtain an MLE estimate of M by systematically computing the sum of squared deviations for all possible values of M within some reasonable range and choosing the value of M that resulted in the smallest possible sum of squared deviations. In situations for which we have simple analytic procedures (such as ordinary linear regression), we can compute the coefficients that provide the optimal model fit by simply carrying out some matrix algebra. For models that do not have simple analytic solutions, such as logistic regression, computer programs search for the combination of coefficient values that produce the best overall model fit.

16.6.2 Assessment of Overall Goodness of Fit for a Binary Logistic Regression Model

The overall "fit" measure that is most often used to assess binary logistic regression models is the **log likelihood (LL)** function. LL is analogous to the sum of squared residuals in multiple linear regression; that is, the larger the (absolute) value of LL, the poorer the agreement between the probabilities of group membership that are generated by the logistic regression model and the actual group membership. Although LL is called a "goodness-of-fit" measure, this description is potentially confusing. Like many other goodness-of-fit measures, a larger absolute value of LL actually indicates worse model fit; it would be more intuitively natural to think of LL as a "badness-of-fit" measure. Each case in the sample has an actual group membership on the outcome variable; in the dog ownership/survival status example, the outcome variable Y had values of 0 for patients who survived and 1 for patients who died. SPSS can be used to find numerical estimates for the coefficients in Equation 16.3, and Equation 16.4 can then be used to obtain an estimate of the probability of being a member of the "dead" group for each individual patient. The LL function essentially compares the actual Y scores (0, 1) with the logs of the estimated probabilities for individual patients:

$$\text{LL} = \Sigma\left[\left(Y_i \times \ln \hat{p}_i\right) + \left(1 - Y_i\right) \times \ln\left(1 - \hat{p}_i\right)\right], \tag{16.7}$$

where Y_i is the actual group membership score for person i (coded $Y_i = 0$ for those who survive and $Y_i = 1$ for those who died) and $\hat{p}_i$ is the predicted probability of membership in the target group—in this example, the "dead" group—for each person. A different predicted probability is obtained for each score value on one or more X predictor variables.

In this example, there is only one predictor variable (dog ownership). For each person, the value of $\hat{p}_i$ was obtained by using Equation 16.4 to find the LL L_i from that individual's score on the predictor variable; then, Equation 16.5 can be used to convert the obtained L_i value into a $\hat{p}_i$ value. Because there is only one predictor variable and it has only two possible values, the equation will generate just two probabilities: the probability of death for owners of dogs and the probability of death for nonowners of dogs. When many predictor variables are included in the analysis, of course, different L_i and $\hat{p}_i$ probability estimates are obtained for all possible combinations of scores on the predictor variables.

Taking the natural log of the $\hat{p}_i$ values, $\ln(\hat{p}_i)$, always results in a negative value (because the value of $\hat{p}_i$ must be less than 1 and the natural log of scores below 1 are all negative numbers). Thus, LL is a sum of negative values, and it is always negative.

When LL values are compared, an LL that is larger in absolute value indicates poorer model fit. That is, LL tends to be closer to zero when actual group memberships are close to the predicted probabilities for most cases; as predicted probabilities get further away from actual group memberships, the negative values of LL become larger in absolute value.

Multiplying LL by –2 converts it to a chi-square distributed variable. SPSS can report **–2 log likelihood (–2LL)** for the null or constant-only model (if the "iteration requested" option is selected) as well as for the full model that includes all predictor variables. The null or constant-only model corresponds to Equation 16.3. This model predicts that all the cases are members of the outcome group that has a larger N (regardless of scores on other variables). The general case of a full model that includes several predictor variables corresponds to Equation 16.2; a model with just one predictor variable, as in the study of dog ownership and death, corresponds to Equation 16.4.

Taking the difference between –2LL for the full model and –2LL for the null model yields a chi-square statistic with k degrees of freedom (df), where k is the number of predictors included in the full model:

$$\chi^2 = -2(\text{LL}_{\text{null model}} - \text{LL}_{\text{full model}}) \text{ or } [-2\text{LL}_{\text{null}} - (-2\text{LL}_{\text{full}})]. \tag{16.8}$$

Researchers usually hope that this chi-square statistic will be large enough to be judged statistically significant as evidence that the full model produces significantly less prediction error than the null model. In words, the null hypothesis that is tested by this chi-square statistic is that the probabilities of group membership derived from the full binary logistic regression model are not significantly closer to actual group membership than probabilities of group membership derived from the null model. When a large chi-square is obtained by taking the difference between LL for the full and null models, and the obtained chi-square exceeds conventional critical values from the table of the chi-square distribution, the researcher can conclude that the full model provides significantly better prediction of group membership than the null model. Obtaining a large chi-square for the improvement in binary logistic regression model fit is analogous to obtaining a large *F* ratio for the test of the significance of the increase in multiple *R* in a multiple linear regression model as predictor variables are added to the model.

16.6.3 Alternative Assessments of Overall Goodness of Fit

Other statistics have been proposed to describe the goodness of fit of logistic regression models. Only widely used statistics that are reported by SPSS are discussed here; for a more thorough discussion of measures of model fit, see Peng, Lee, and Ingersoll (2002). In OLS multiple linear regression, researchers generally report a multiple *R* (the correlation between the Y' predicted scores generated by the regression equation and the actual *Y* scores) to indicate how well the overall regression model predicts scores on the dependent variable; in OLS multiple linear regression, the overall R^2 is interpreted as the percentage of variance in scores on the dependent variable that can be predicted by the model. Binary logistic regression does not actually generate predicted Y' scores; instead, it generates predicted probability values ($\hat{p}_i$) for each case, and so the statistic that assesses overall accuracy of prediction must be based on the $\hat{p}_i$ estimates. Binary logistic regression does not yield a true multiple *R* value, but SPSS provides pseudo-*R* values that are (somewhat) comparable with a multiple *R*. **Cox and Snell's R^2** was developed to provide something similar to the "percentage of explained variance" information available from a true multiple R^2, but its maximum value is often less than 1.0. Similar to a ϕ coefficient for a 2 × 2 table, the Cox and Snell R^2 has a restricted range when the marginal distributions of the predictor and outcome variables differ. **Nagelkerke's R^2** is more widely reported; it is a modified version of the Cox and Snell R^2. Nagelkerke's R^2 is obtained by dividing the obtained Cox and Snell R^2 by the maximum possible value for the Cox and Snell R^2 given the marginal distributions of scores on the predictor and outcome variables; thus, Nagelkerke's R^2 can take on a maximum value of 1.0.

16.6.4 Information About Predictive Usefulness of Individual Predictor Variables

If the overall model is significant, information about the contribution of individual predictor variables can be examined to assess which of them make a statistically significant contribution to the prediction of group membership and the nature of their relationship to group membership (e.g., As age increases, does probability of death increase or decrease?). As in a multiple linear regression, it is possible to estimate coefficients for a raw score predictive equation. However, because binary logistic regression has such a different type of outcome variable (a logit), the interpretation of the *B* slope coefficients for a logistic regression is quite different from the interpretation of *b* raw score slope coefficients in ordinary linear regression. The *B* coefficient in a logistic regression tells us by how many units the log odds ratio increases for a one-unit increase in score on the *X* predictor variable.

For statistical significance tests of individual predictors, the null hypothesis can be stated as follows:

$$H_0: B_i = 0. \tag{16.9}$$

That is, the null hypothesis for each of the X_i predictor variables ($X_1, X_2, \ldots, X_k$) is that the variable is unrelated to the log odds ratio outcome variable L_i. In ordinary multiple linear regression a similar type of null hypothesis was tested by setting up a *t* ratio:

$$t = \frac{b_i}{SE_{bi}}. \tag{16.10}$$

For binary logistic regression, the test that SPSS provides for the null hypothesis in Equation 16.9 is the **Wald χ^2 statistic**:

$$W = \left[\frac{B_i}{SE_{Bi}}\right]^2. \tag{16.11}$$

To test the significance of each raw score slope coefficient, we divide the slope coefficient (B_i) by its standard error (SE_{Bi}) and square the ratio; the resulting value has a chi-square distribution with 1 *df*, and SPSS reports an associated *p* value. The *B* coefficient is not easy to interpret directly, because it tells us how much L_i (the logit or log odds) is predicted to change for each one-unit change in the raw score on the predictor variable. It would be more useful to be able to talk about a change in the odds than a change in the log of the odds. The operation that must be applied to L_i to convert it from a log odds to an odds is the exponential function. That is, e^{L_i} provides information about odds. To evaluate how a one-unit change in X_i is related to change in the odds ratio, we need to look at e^{Bi}, also called $\exp(B_i)$. The value of e^{Bi} can be interpreted more directly. For a one-unit increase in the raw score on X_i, the predicted odds change by e^{Bi} units. If the value of e^{Bi} is less than 1, the odds of membership in the target group go down as scores on X_i increase; if the value of e^{Bi} equals 1, the odds of membership in the target group do not change as X_i increases; and if the value of e^{Bi} is greater than 1, the odds of membership in the target group increase as X_i increases. The distance of e^{Bi} from 1 indicates the size of the effect (Pampel, 2000). The percentage of change (%Δ) in the odds ratio that is associated with a one-unit increase in the raw score on X_i can be obtained as follows:

$$\%\Delta = (e^{Bi} - 1) \times 100. \tag{16.12}$$

For example, if $e^{Bi} = 1.35$, then $\%\Delta = (e^{Bi} - 1) \times 100 = (1.35 - 1) \times 100 = 35\%$; that is, the odds of death increase by 35% for a one-unit increase in the score on X_i. In addition to tests of statistical significance, researchers should also evaluate whether the effect size or strength of predictive relationship for each predictor variable in the model is sufficiently large to be considered practically and/or clinically significant (Kirk, 1995; Thompson, 1999).

16.6.5 Evaluating Accuracy of Group Classification

One additional way to assess the adequacy of the model is to assess the accuracy of classification into groups that is achieved when the model is applied. This involves setting up a contingency table to see how well actual group membership corresponds to predicted group membership. For each case in the sample, once we have coefficients for the equations above, we can compute $\hat{p}_i$, the predicted probability of membership in the "target" group (such as the "dead patient" group), for each case. We can then classify each case into the target group if $\hat{p}_i > .50$ and into the other group if $\hat{p}_i < .50$. Finally, we can set up a contingency table to summarize the correspondence between actual and predicted group memberships. For a binary logistic regression, this is a 2 × 2 table. The percentage of correctly classified cases can be obtained for each group and for the overall sample. However, in the dog owner/death data set, both owners and nonowners of dogs are more likely to be alive than to be dead at the end of 1 year. Thus, when we use the model to predict group membership, the predicted group membership for all $N = 92$ members of the sample is the "alive" group. This is an example of a data set for which accuracy of prediction of group membership is poor even though there is a strong association between odds of death and dog ownership. In this sample, people who do not own dogs have odds of death 6 times as high as people who do own dogs, which seems to be a substantial difference; however, even the group of people who do not own dogs is much more likely to survive than to die. The second empirical example (presented in Section 16.10) is an example of a research situation where group membership can be predicted more accurately.

In addition, residuals can be examined for individual cases (this information was not requested in the following example of binary logistic regression analysis). Examination of individual cases for which the model makes incorrect predictions about group membership can sometimes be helpful in detecting multivariate outliers, or perhaps in identifying additional predictor variables that are needed (Tabachnick & Fidell, 2018).

16.7 LOGISTIC REGRESSION FOR FIRST EXAMPLE: PREDICTION OF DEATH FROM DOG OWNERSHIP

16.7.1 SPSS Menu Selections and Dialog Boxes

To run a binary logistic regression for the data shown in Table 16.1, the following menu selections are made: <Analyze> → <Regression> → <Binary Logistic> (as shown in Figure 16.3).

The main dialog box for the binary logistic regression procedure appears in Figure 16.4. The name of the dependent variable (in this example, death) is placed in the window for the dependent variable. The name for each predictor variable (in this example, dogown) is placed in the "Covariates" pane. If a predictor is a categorical variable, it is necessary to define it as categorical and to specify which group will be treated as the reference group when odds ratios are set up.

The Categorical, Save, and Options buttons were clicked to open additional SPSS dialog boxes. In the Logistic Regression: Define Categorical Variables dialog box that appears in Figure 16.5, the independent variable *X*, the variable dogown, was identified as a categorical predictor variable. Next to "Reference Category," the radio button "Last" was selected, and the Change button was clicked. This instructs SPSS to use the group that has the largest

Figure 16.3 SPSS Menu Selections to Run Binary Logistic Regression Procedure

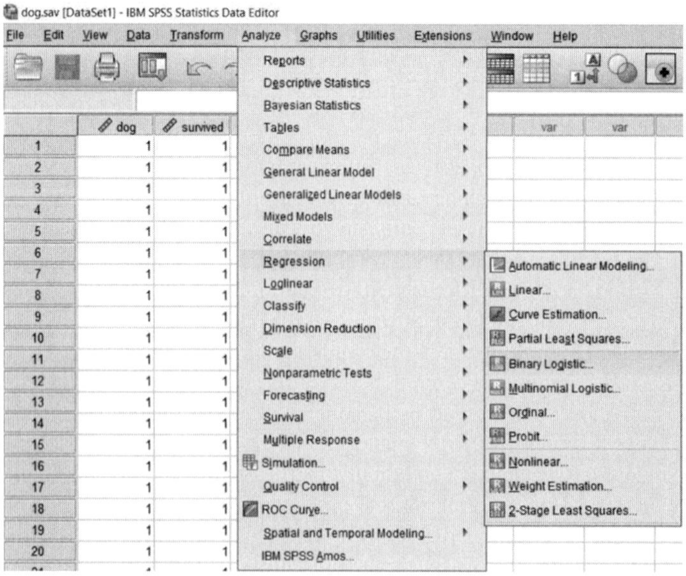

Figure 16.4 Main Dialog Box for SPSS Binary Logistic Regression

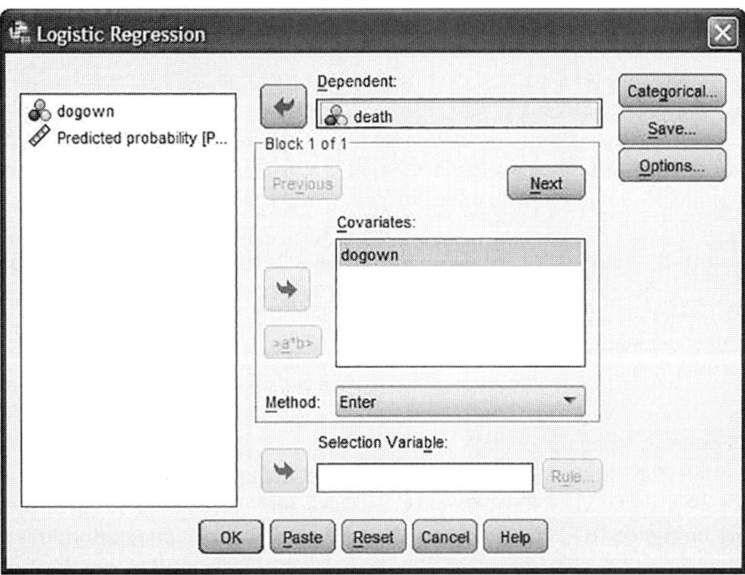

Figure 16.5 Defining Categorical Predictor Variables for Binary Logistic Regression

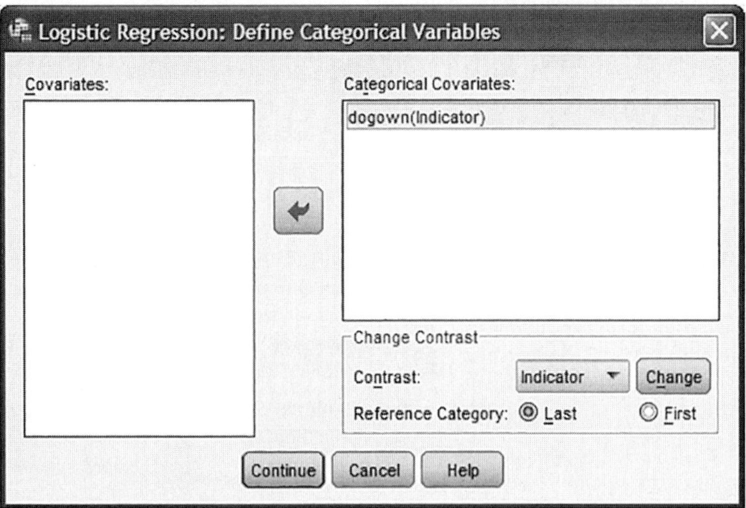

numerical code (in this example, $X = 1$, the person is a dog owner) as the reference group when setting up a ratio of odds. Note that the interpretation of outcomes must be consistent with the way scores are coded on the binary outcome variable (e.g., 1 = dead, 0 = alive) and the choice of which group to use as the reference group when setting up a ratio of odds to compare groups on a categorical variable. In this example, the analysis was set up so that the coefficients in the binary logistic regression can be interpreted as information about the comparative odds of death for people who do *not* own dogs compared with people who *do* own dogs. Different computer programs may have different default rules for the correspondence of the 0 and 1 score codes to the reference group and the comparison group, respectively. When a data analyst is uncertain what comparison the program provides when a particular menu selection is used to identify the reference group, applying the commands to a data set for which the direction of differences in odds between groups is known (such as this dog ownership/death data set) can help clear up possible confusion.

In the Logistic Regression: Save dialog box that appears in Figure 16.6, under "Predicted Values," the "Probabilities" box was checked to request that SPSS save a predicted probability (in this example, a predicted probability of death) for each case. The Logistic Regression: Options dialog box in Figure 16.7 was used to request an iteration history and a confidence interval (CI) for the estimate of e^{Bi}. The value of e^{Bi} tells us, for each one-unit change in the raw score on the corresponding X_i variable, how much change we predict in the odds of membership in the target group. If e^{Bi} is greater than 1, the odds of membership increase as X scores increase; if e^{Bi} is less than 1, the odds of membership in the target group decrease as X scores increase; and if e^{Bi} equals 1, the odds of membership in the target group do not change as X scores increase. The 95% CI provides information about the amount of sampling error associated with this estimated change in odds.

Finally, the Paste button in the Logistic Regression dialog box (Figure 16.4) was used to display the SPSS syntax that was generated by these menu selections in the syntax window that appears in Figure 16.8. It is helpful to save the syntax as documentation; this provides a record of the commands that were used. The syntax can also be used to run the same analysis again later, with or without modifications to the commands.

Figure 16.6 Save Predicted Probability (of Death) for Each Case

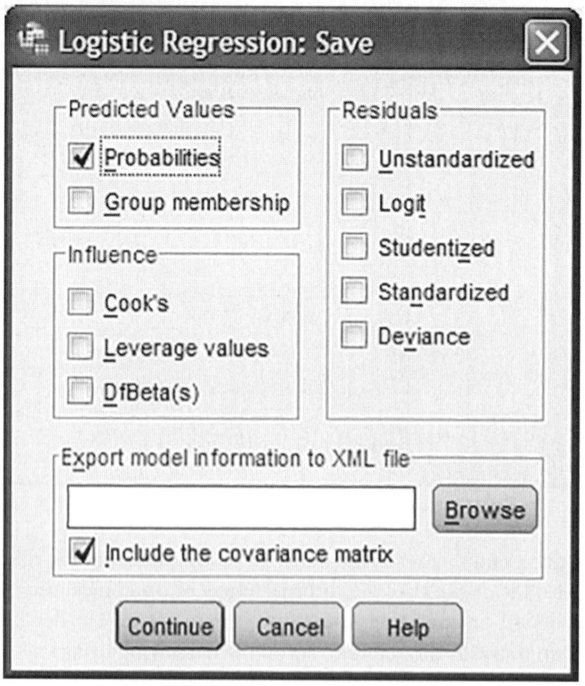

Figure 16.7 Options for Binary Logistic Regression

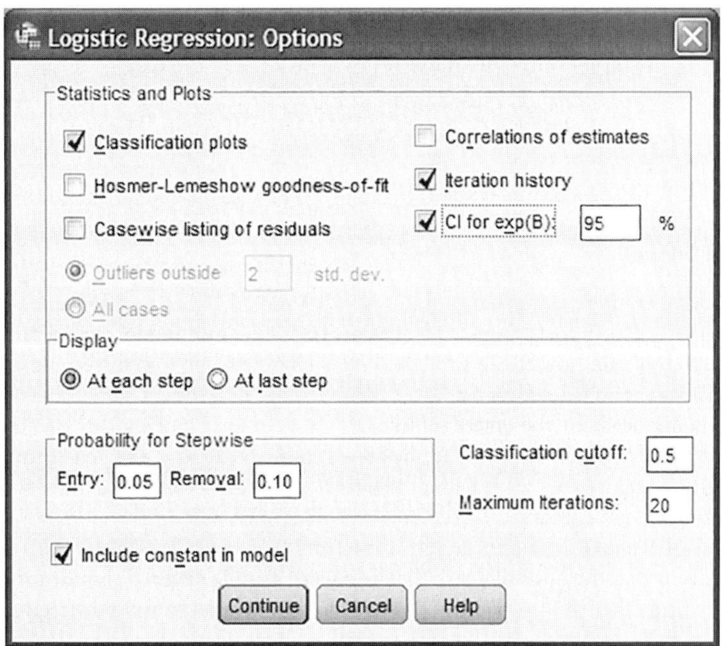

Figure 16.8 SPSS Syntax for Binary Logistic Regression to Predict Death From Dog Ownership

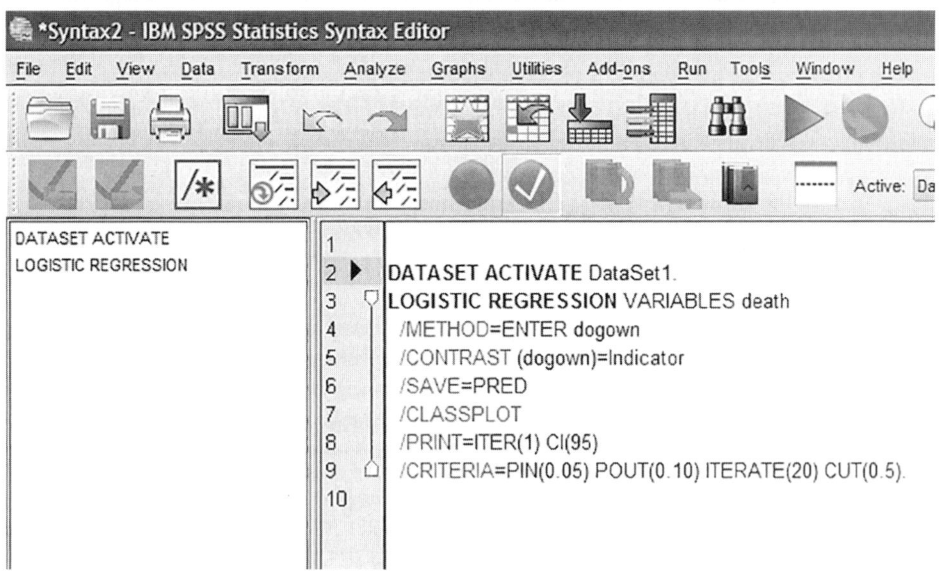

16.7.2 SPSS Output

Selected SPSS output generated by these commands appears in Figures 16.9 through 16.17. Figure 16.9 provides information about the coding for the dependent variable, death; for this variable, a score of 0 corresponded to "alive" and a score of 1 corresponded to "dead." In other words, the target outcome of interest was death, and the results of the model can be interpreted in terms of odds or probabilities of death. In this example, the odds were calculated using the group of nonowners of dogs as the reference group. The coefficients from the binary logistic regression, therefore, are interpreted as information about the comparative odds of death for the dog owner group relative to the group of people who did not own dogs.

16.7.2.1 Null Model

The Block 0 results that appear in Figures 16.10 and 16.11 are results for the null or constant-only model, which correspond to Equation 16.3. The null model is generally used only as a baseline to evaluate how much predicted odds change when one or more predictor variables are added to the model. The null hypothesis for the null model is that the odds of being alive versus being dead for the entire sample of $N = 92$ men, ignoring dog ownership status, equal 1. Thus, if 50% of the sample is in the alive group and 50% in the dead group, the value of B_0 would be 0, and log(B_0) would equal 1. Log(B_0) is interpreted as the odds of death when group membership on categorical predictor variables such as dog ownership is ignored. In this example, the value of B_0 differs significantly from 0, $B_0 = -1.718$, Wald $\chi^2(1) = 35.019$, $p < .001$. This tells us that the odds of death for the overall sample differed significantly from 1.00 or, in other words, that the probability of death for the overall sample differed significantly from .50. Exp(B_0) = .179 tells us that the overall odds of death (for the entire sample of $N = 92$) was .179. In other words, approximately one sixth of the 92 participants in this entire sample died. This null model (that does not include any predictor variables) is used as a baseline to evaluate how much closer the predicted probabilities of group membership become to actual group membership when one or more predictor variables are added to the model.

16.7.2.2 Full Model

The Block 1 model in this example corresponds to a model that uses one categorical predictor variable (does each participant own a dog, yes or no) to predict odds of death. The fit

Figure 16.9 SPSS Output for Binary Logistic Regression to Predict Death From Dog Ownership: Internal Coding for Predictor and Outcome Variables

Dependent Variable Encoding

Original Value	Internal Value
alive	0
dead	1

Categorical Variables Codings

		Frequency	Parameter coding (1)
dogowner	no dog	39	1.000
	own dog	53	.000

Figure 16.10 SPSS Output for Binary Logistic Regression to Predict Death From Dog Ownership: Null Model

Block 0: Beginning Block

Iteration History[a,b,c]

Iteration		-2 Log likelihood	Coefficients
			Constant
Step 0	1	79.831	-1.391
	2	78.481	-1.686
	3	78.469	-1.717
	4	78.469	-1.718

a. Constant is included in the model.

b. Initial -2 Log Likelihood: 78.469

c. Estimation terminated at iteration number 4 because parameter estimates changed by less than .001.

Note: No predictors included.

Figure 16.11 Coefficients for Binary Logistic Regression Null Model

Variables in the Equation

		B	S.E.	Wald	df	Sig.	Exp(B)
Step 0	Constant	-1.718	.290	35.019	1	.000	.179

of this one-predictor model is assessed by evaluating whether the goodness of fit for this one-predictor model is significantly better than the fit for the null model. Selected SPSS output for the Block 1 model appears in Figures 16.12 through 16.15. The first question to consider in the evaluation of the results is, Was the overall model statistically significant? When dog ownership was added to the model, did the LL goodness-of-fit (or "badness"-of-fit) measure that assessed differences between predicted and actual group memberships decrease significantly? The chi-square test for the improvement in fit for the full model compared with the null model reported in Figure 16.13 was obtained by subtracting the LL value for the null model from the LL value for the full model and multiplying this difference by –2 (see Equation 16.8). The *df* for this improvement in model fit chi-square is *k*, where *k* is the number of predictor variables in the full model; in this example, $df = 1$. The omnibus test in Figure 16.13, $\chi^2(1) = 9.011$, $p = .003$, indicated that the full predictive model had a significantly smaller LL "badness-of-fit" measure than the null model. If the conventional $\alpha = .05$ is used as the criterion for statistical significance, this model would be judged statistically significant. In other words, the full predictive model with just one predictor variable, dog ownership, predicts odds of death significantly better than a null model that does not include any predictors. Other goodness-of-fit statistics such as the Nagelkerke R^2 value of .163 appear in Figure 16.14. Figure 16.15 shows the table of actual versus predicted group memberships; in this study, members of both groups (owners and nonowners of dogs) were more likely to survive than to die, and therefore, the best prediction about group membership for both owners and nonowners of dogs was that the person would be alive. When we are trying to predict an outcome, such as death, that has a relatively low rate of occurrence across both groups in a study, an outcome similar to this one is common. Our analysis provides useful information about differences in odds of death for the two groups (people who do not own dogs have a higher odds of death), but it does not make good predictions about death outcomes for individual cases.

The data analyst can go on to ask whether each individual predictor variable makes a statistically significant contribution to the prediction of group membership and how much the odds of membership in the target group (dead) differ as scores on predictor variables change.

Figure 16.12 SPSS Output for Binary Logistic Regression to Predict Death From Dog Ownership Model That Includes "Dog Owner" Variable as Predictor

Iteration History[a,b,c,d]

Iteration		-2 Log likelihood	Coefficients	
			Constant	dogowner(1)
Step 1	1	73.691	-1.774	.902
	2	69.790	-2.487	1.553
	3	69.463	-2.772	1.837
	4	69.458	-2.813	1.878
	5	69.458	-2.813	1.879

a. Method: Enter

b. Constant is included in the model.

c. Initial -2 Log Likelihood: 78.469

d. Estimation terminated at iteration number 5 because parameter estimates changed by less than .001.

Figure 16.13 Chi-Square Test for Improvement in Model Fit Relative to Null Model

Omnibus Tests of Model Coefficients

		Chi-square	df	Sig.
Step 1	Step	9.011	1	.003
	Block	9.011	1	.003
	Model	9.011	1	.003

Figure 16.14 Model Summary: Additional Goodness-of-Fit Measures

Model Summary

Step	-2 Log likelihood	Cox & Snell R Square	Nagelkerke R Square
1	69.458[a]	.093	.163

a. Estimation terminated at iteration number 5 because parameter estimates changed by less than .001.

Figure 16.15 Classification Table for Dog Ownership/Death Study

Classification Table[a,b]

Observed			Predicted: death: alive	Predicted: death: dead	Percentage Correct
Step 0	death	alive	78	0	100.0
		dead	14	0	.0
	Overall Percentage				84.8

a. Constant is included in the model.

b. The cut value is .500

Figure 16.16 SPSS Output for Binary Logistic Regression to Predict Death From Dog Ownership Predictive Equation

Variables in the Equation

		B	S.E.	Wald	df	Sig.	Exp(B)	95.0% C.I.for EXP(B) Lower	Upper
Step 1[a]	dogowner(1)	1.879	.693	7.357	1	.007	6.548	1.684	25.456
	Constant	-2.813	.594	22.402	1	.000	.060		

a. Variable(s) entered on step 1: dogowner.

In this empirical example, the question is whether the odds of death are higher (or lower) for nonowners of dogs than for dog owners. The coefficients for the estimated predictive model appear in Figure 16.16. The row that begins with the label "Constant" provides the estimated value for the B_0 coefficient; the B_0 coefficient is used to obtain information about odds of death for the reference group (in this example, the odds of death for dog owners). The row that begins with the name of the predictor variable, dogowner(1), provides the B_1 coefficient associated with the categorical predictor variable. This B_1 coefficient is used to obtain information about the difference in odds of death for the other group (nonowners of dogs) compared with the reference group (dog owners). The numerical values for the predictive model in Equation 16.5, $L_i = B_0 + B_1X$, were as follows. From Figure 16.16, we find that the coefficients for this model are $L_i = -2.813 + 1.879 \times X$. For each coefficient, a Wald chi-square was calculated using Equation 16.11. Both coefficients were statistically significant; for B_0, Wald $\chi^2 = 7.357$, $p = .007$; for B_1, Wald $\chi^2 = 22.402$, $p < .001$.

It is easier to interpret **exp(*B*)** (also denoted e^{Bi}) than to interpret *B* because the exp(*B*) value is directly interpretable as a change in odds (while *B* itself represents a change in log odds). Because B_0 was significantly different from 0, we know that the odds of death in the "dog owner" group were significantly different from "even odds"; that is, the risk of death was different from 50%. Because $\exp(B_0) = .060$, we know that the odds of death for the dog owner group were .060. Because these odds are much less than 1.00, it tells us that dog owners were less likely to be dead than to be alive.

The B_1 coefficient was also statistically significant, which tells us that the odds of death were significantly different for the group of nonowners of dogs than for the dog owner group. The $\exp(B_1) = 6.548$ value tells us that the odds of death were about 6.5 times higher for nonowners than for owners of dogs. In other words, people who did not own dogs had much higher odds of death (more than 6 times as high) compared with people who did own dogs. A 95% CI for the value of $\exp(B_1)$ was also provided; this had a lower limit of 1.684 and an upper limit of 25.456. The fact that this CI was so wide suggests that because of the rather small sample size, $N = 92$, the estimate of change in odds is not very precise. This CI suggests that the actual difference in odds of death for nonowners of dogs compared with dog owners could be as low as 1.7 or as high as 25.

One issue to keep in mind in interpretation of results is the nature of the original research design. This study was not an experiment; dog ownership was assessed in a survey, and dog ownership was not experimentally manipulated. Therefore, we cannot interpret the results as evidence that dog ownership causally influences death. We would make causal interpretations on the basis of a logistic regression analysis only if the data came from a well-controlled experiment. A second issue involves evaluation whether the difference in odds is large enough to have any practical or clinical significance (Thompson, 1999). In this sample, the odds of death were more than 6 times as high for people who did not own dogs compared with people who did own dogs. That sounds like a substantial difference in odds. However, information about odds ratios should be accompanied by information about probabilities. In this case, it is helpful to include the sample probabilities of death for the two groups. From Table 16.1, approximately 6% of dog owners died in the first year after a heart attack, while about 28% of nonowners of dogs died within the first year after a heart attack. This additional information makes it clear that for both groups, survival was a higher probability outcome than death. In addition, the probability of death was substantially higher for people who did not own dogs than for people who did own dogs.

Recommendations for reporting logistic regression outlined by Peng et al. (2002) state that reports of logistic regression should include predicted probabilities for selected values of each X predictor variable, in addition to information about the statistical significance and overall fit of the model, and the coefficients associated with individual predictor variables. In this example, the SPSS save procedure created a new variable (Pre_1); this is the predicted probability of death for each case. These predicted probabilities should be examined and reported. Figure 16.17 shows the saved scores for this new variable Pre_1. For each member of the dog owner group, the value of Pre_1 corresponds to a predicted probability of death = .057; for each member of the non-dog-owning group, Pre_1 corresponded to a predicted probability of death = .282. The predicted values of death for each score on the categorical predictor variable (dog ownership) are summarized in the table in Figure 16.18. These correspond to the proportions of cases in the original data in Table 16.1.

For these dog ownership/death data, it is relatively easy to show how predicted probabilities are obtained from the binary logistic regression results. First, we substitute the possible score values for X ($X = 0$ and $X = 1$) into the logistic regression equation:

$$L_i = B_0 + B_1X = -2.813 + 1.879X.$$

When we do this, we find that the predicted value of the logit L_0 for the $X = 0$ reference group (dog owners) is –2.813; the predicted value of the logit L_1 for the $X = 1$ group (nonowners) is $-2.813 + 1.879 = -.934$. We can then use Equation 16.6 to find the estimated probability of death for members of each group based on the value of L_i for each group. For $X = 0$, participants who are dog owners ($L_0 = -2.813$),

$$\hat{p}_0 = \frac{e^{L0}}{1+e^{L0}} = \frac{e^{-2.813}}{1+e^{-2.813}} = .0556.$$

For $X = 1$, participants who are not dog owners ($L_1 = -.934$),

$$\hat{p}_1 = \frac{e^{L1}}{1+e^{L1}} = \frac{e^{-.934}}{1+e^{-.934}} = .2821.$$

Note that the predicted probabilities generated from the logistic regression model are the same as the probabilities that are obtained directly from the frequencies in Table 16.1: The proportion of dog owners who died is equal to the number of dead dog owners divided

Figure 16.17 SPSS Data View Worksheet With Saved Estimated Probabilities of Death

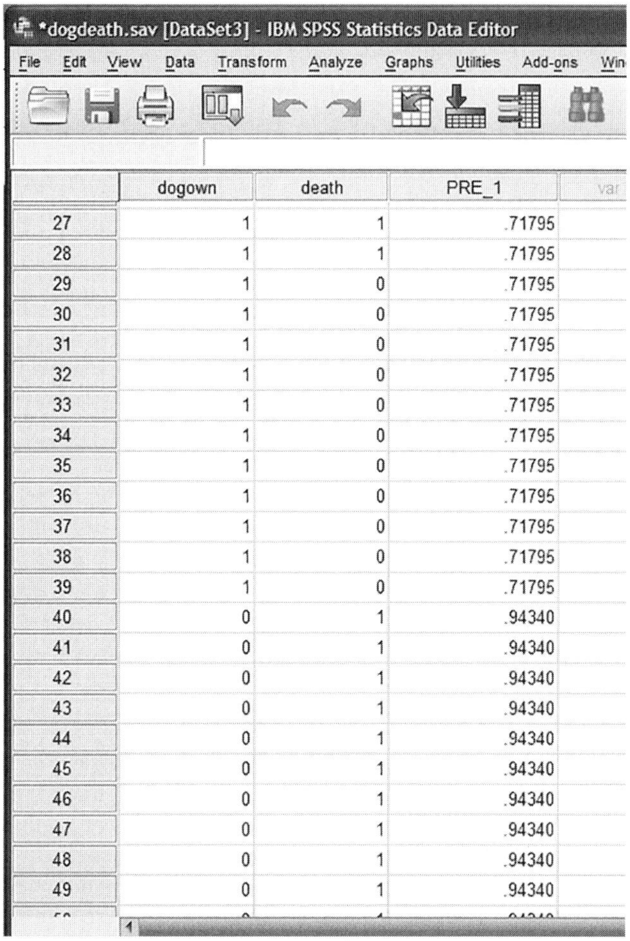

*dogdeath.sav [DataSet3] - IBM SPSS Statistics Data Editor

File Edit View Data Transform Analyze Graphs Utilities Add-ons Win

	dogown	death	PRE_1	var
27	1	1	.71795	
28	1	1	.71795	
29	1	0	.71795	
30	1	0	.71795	
31	1	0	.71795	
32	1	0	.71795	
33	1	0	.71795	
34	1	0	.71795	
35	1	0	.71795	
36	1	0	.71795	
37	1	0	.71795	
38	1	0	.71795	
39	1	0	.71795	
40	0	1	.94340	
41	0	1	.94340	
42	0	1	.94340	
43	0	1	.94340	
44	0	1	.94340	
45	0	1	.94340	
46	0	1	.94340	
47	0	1	.94340	
48	0	1	.94340	
49	0	1	.94340	

Figure 16.18 Predicted Probability of Death for Two Groups in Friedmann et al.'s (1980) Study: Nonowners Versus Owners of Dogs

no dog	Predicted probability	.28205
own dog	Predicted probability	.05660

by the total number of dog owners = 3/53 = .0556. The proportion of nonowners of dogs who died is equal to the number of dead nonowners divided by the total number of nonowners = 11/39 = .2821. The preceding example demonstrates how binary logistic regression can be used to analyze simple data (the dog ownership/death data) that were analyzed earlier in the textbook using simpler methods. A chi-square test of association was reported for the dog ownership data in an earlier chapter. When the data are this simple, we may not gain a great deal by performing the more complicated binary logistic regression. However, this example provides a good starting point for understanding how binary logistic regression works. A second example (reported in Section 16.10) demonstrates a binary logistic regression with one quantitative predictor and one categorical predictor. One major advantage of binary logistic regression over a simple chi-square test is that we can add additional predictor variables to a binary logistic regression and assess their individual predictive contributions while controlling for correlations among predictor variables.

16.7.3 Results for the Study of Dog Ownership and Death

This following model "Results" section was set up in accordance with suggested reporting guidelines for logistic regression in Tabachnick and Fidell (2018).

Results

A binary logistic regression analysis was performed to predict 1-year survival outcomes for male patients who had initially survived a first heart attack. The outcome variable death was coded 0 = alive and 1 = dead. One predictor variable was included in the model; this was a response to a survey question about whether each patient owned a dog. In the SPSS data file, the variable dog ownership was initially coded 0 for nonowners of dogs and 1 for dog owners. The binary logistic regression procedure in SPSS was used to perform the analysis. Data from 92 cases were included in this analysis.

A test of the full model (with dog ownership as the predictor variable) compared with a constant-only or null model was statistically significant, $\chi^2(1) = 9.011, p = .003$. The strength of the association between dog ownership and death was relatively weak with Cox and Snell's $R^2 = .093$ and Nagelkerke's $R^2 = .163$.

Table 16.3 summarizes the raw score binary logistic regression coefficients, Wald statistics, and the estimated change in odds of death for nonowners of dogs compared with dog owners, along with a 95% CI. Table 16.4 reports the odds and the predicted probability of death for each of the two groups (owners and nonowners of dogs). The odds of death for the entire sample was .179. The odds of death for the dog owner group was .06. The odds of death for nonowners of dogs was .393 (number of dead divided by number of surviving persons in the group of nonowners = 11/28 = .393). Thus, the odds of death by the end of the first year were approximately 6.5 times higher for people who did not own dogs than for people who did own dogs (.393/.06 = 6.5). In other words, dog ownership was associated with a lower risk for death. However, both owners and nonowners of dogs were more likely to be alive than to be dead at the end of the 1-year follow-up. For dog owners, the probability

Table 16.3 Binary Logistic Regression Analysis: Prediction of Death 1 Year After Heart Attack From Dog Ownership

					95% CI for exp(*B*)	
Predictor Variable	***B***	**Wald Chi-Square Test**	***p***	**exp(*B*)**	**Lower**	**Upper**
Dog owner	1.879	7.36	.007	6.55	1.68	25.46
Constant	2.813	22.402	<.001	.06		

Table 16.4 Predicted Odds and Probability of Death for Nonowners Versus Owners of Dogs

	Odds of Death	Probability of Death
Nonowners	.393	.282
Owners	.060	.057
Total sample	.179	.152

of death during the 1-year follow-up period was .057; for nonowners of dogs, the probability of death was .282. Note that the probabilities derived from this binary logistic regression analysis correspond to the row percentages (dead vs. alive) in the original contingency table in Table 16.1.

16.8 ISSUES IN PLANNING AND CONDUCTING A STUDY

16.8.1 Preliminary Data Screening

Binary logistic regression involves at least one categorical variable (the Y outcome variable). One of the most important issues that should be addressed in preliminary data screening is the distribution of scores on the binary outcome variable, Y. When the outcome variable Y is binary, it has only two possible values. If the proportion of people in the two groups deviates greatly from a 50/50 split, and if the total N of cases in the study is very small, the number of cases in the smaller outcome group may simply be too small to obtain meaningful results. For example, suppose that the outcome of interest is the occurrence of a relatively rare disease. The researcher has a total N of 100; 97 of the people in this sample do not have the disease and only 3 people in the sample have the disease. No matter what kind of statistical analysis is performed, the outcome of the analysis will be largely determined by the characteristics of the 3 people in the "disease" group. If a researcher is interested in a disease that is so rare that it occurs only three times per 100 people, the researcher needs to obtain a much larger total N, or sample a much larger number of cases in the disease group, to obtain enough data for a binary logistic regression analysis.

Binary logistic regression sometimes includes one or more categorical predictor variables. For any pair of categorical variables, it is useful to set up a table to show the cell frequencies. Like the chi-square test of association for contingency tables, binary logistic regression may not produce valid results when there are one or several cells that have expected cell frequencies < 5. If a preliminary look at tables reveals that more than 20% of the cells have expected values < 5, this situation should be remedied. Groups for which one or more expected frequencies are < 5 can be either combined with other groups or dropped from the analysis, whichever decision seems more reasonable.

Binary logistic regression may also involve one or several quantitative predictor variables. If there are extreme outliers on any of these variables, the data analyst should evaluate whether these extreme scores are incorrect or whether the scores indicate individuals who differ so much from the majority of cases in the sample that they might be thought of as members of a different population. Rules for identification and handling of outliers should be established before looking at the data. In some situations, the analysis will produce more believable results if extreme scores are removed from the sample or made less extreme by changing the scores. For example, suppose that the predictor variable X is systolic blood pressure and the majority of participants have X scores between 100 and 130 (i.e., within the "normal" range). If one participant has a systolic blood pressure of 230, that score could be viewed as an outlier. This large value might be due to measurement error or it could represent a valid case of unusually high blood pressure. If the researcher defines the population of interest in the study as people with normal scores on systolic blood pressure, the blood pressure score of 230 could be dropped from the data set (because that individual is not a member of the population of interest, i.e., persons with normal blood pressure). Another possible way to handle outlier scores is to recode them to the value of the next highest score; if the second highest blood pressure in the sample is 130, the score of $X = 230$ could be recoded to $X = 130$.

16.8.2 Design Decisions

Binary logistic regression uses MLE rather than OLS to estimate model coefficients. If there are few cases for each observed combination of scores on predictor variables, the reliability of estimates tends to be low. Peduzzi, Concato, Kemper, Holford, and Feinstein (1996) suggested a minimum N that is at least 10 times k, where k is the number of independent variables in the model. In addition, as described in the previous section, when contingency tables are set up for each pair of categorical variables in the model (including the binary outcome variable), there should be few cells with expected frequencies below 5. If many cells have expected frequencies < 5, the researcher may need to drop variables, combine scores for some categories, or omit some categories.

Statistical power in multivariate analyses depends on many factors, including the strength of the association between each predictor variable and the outcome, the degree to which assumptions are violated, the size and sign of correlations among predictor variables, and the sample size (not only the total N but also the sizes of N's within cells in the design). Therefore, it is difficult to provide recommendations about the sample size required to have adequate statistical power in binary logistic regression. The sample sizes suggested above (at least 10 times as many cases as predictor variables and few cells with expected cell frequencies < 5) are minimal requirements, and larger N's may be required to have acceptable statistical power.

Another design decision that has an effect on statistical power involves the choice of dosage levels for categorical or interval-level predictor variables in the model. The second empirical example in this chapter examines how increasing dosages of a drug are related to death. In this hypothetical data set, the drug dosage levels range from 0 to 10 mg in increments of 1 mg. For some drugs, a 1-mg increase might be large enough to have a noticeable impact. However, for other drugs, increases of 10 mg or even 1 g might be required to see noticeable changes in outcomes across different dosage levels. The same issue was discussed in connection with the choice of treatment dosage levels for studies that involve comparing group means using *t* tests and analysis of variance.

Other factors being equal, a researcher is more likely to obtain a significant increase in odds as N, the overall sample size, increases; when all the groups formed by combinations of categorical variables have expected cell frequencies > 5; and when the dosage difference that corresponds to each one-point increase in a predictor variable is large enough to produce a noticeable effect on the outcome.

16.8.3 Coding Scores on Binary Variables

It is conventional to code group membership on binary outcome variables using values of 0 and 1 to represent group membership. Using values of 0 versus 1 makes the coefficients of the logistic regression model easier to interpret. It is helpful to think about the nature of the research question and the clearest way to phrase the outcome when assigning these codes to groups. Researchers in social science and medicine often want to report how much a risk factor (such as smoking) increases the odds of some negative outcome (such as death) or how much a treatment or intervention (such as a low-dose aspirin) decreases the odds of some negative outcome (such as a heart attack). The "target" outcome, the one for which we want to report odds of occurrence, is very often a negative outcome. Interpretation of the results from a logistic regression depends on how scores are coded on the binary outcome variable and on which score value the computer program treats as the "target" outcome. In the preceding example, the outcome variable death was coded 0 = alive and 1 = dead; SPSS reports odds for membership in the target group that has a score value of 1 on the dependent variable, so in this empirical example, the odds that were reported were odds of death. Results are usually more interpretable if a code of 1 is assigned to the group with the more negative outcome (death, heart attack, dropping out of school, etc.), and a code of 0 is assigned to the group with a more positive outcome (survival, no heart attack, graduation from school, etc.).

The coding of categorical predictor variables in binary logistic regression is also potentially confusing, and the handling of these codes may differ across programs. When a predictor variable is identified as categorical, users of the SPSS binary logistic regression procedure use a radio button selection to identify whether the "first" or "last" group will be used as the reference group when odds ratios are set up to compare risks across groups. In SPSS, the "first" group is the group with a lower score on the binary predictor variable; the "last" group is the group with a higher score on the binary predictor. In the dog owner/death example presented earlier, we wanted to be able to evaluate how much higher the odds of death were for nonowners of dogs compared with dog owners. That is, we wanted to use the dog owner group as the "reference group" when we set up odds. In the original SPSS data file dog.sav, the variable dogowner was coded 0 = does not own dog and 1 = owns a dog. If we want to use the dog owner group as the reference group, we need to click the radio button in the Logistic Regression: Define Categorical Variables that corresponds to "Reference Group: Last". After this selection was made, SPSS assigned new internal code values to the categorical predictor variable; in this example, the internal codes used by SPSS that appear in Figure 16.9 were 1 for "no dog" and 0 for "owns dog." Given these codes, the value of $\exp(B_0)$ is interpreted as information about odds of death for the reference group (dog owners), and the value of $\exp(B_1)$ is interpreted as the increase (or decrease) in odds of death for the comparison group of nonowners of dogs.

In many applications of binary logistic regression, a risk factor is assessed to see how much it increases the odds of a negative outcome (such as death). When a binary predictor variable represents the presence versus absence of a risk factor, it often makes more sense to set up the internal coding for this variable in SPSS so that the group *without* the risk factor is used as the *reference* group (and has an internal SPSS code of 0) and the group that *has* the risk factor is compared with the reference group. (It is less confusing to code scores initially as yes/no, 0 vs. 1.) For the variable dog owner, a value of 0 was given to persons who did not own a dog and a value of 1 was given to persons who did own a dog. To make sure that SPSS used the dog owner group as the reference group when setting up ratios of odds, a radio button selection was made to identify the last group (the group with the higher score) as the reference group.

If binary logistic regression were used to assess the association between being a smoker versus being a nonsmoker and having lung cancer versus not having lung cancer, data analysts who use SPSS would probably want to set up the scores on the binary categorical variables so that the undesirable medical outcome (lung cancer) corresponds to a score of 1 on the outcome variable and to use nonsmokers (i.e., people who do not have the risk factor of interest) as the reference group. It is important to look carefully at the codes that SPSS uses internally (as shown in Figure 16.9) to make certain that they correspond to the comparisons that you want to make. The group with an internal SPSS code of 0 is used as the reference group; the other group is compared with this reference group. In other common applications of binary logistic regression, the presence of a treatment or intervention is evaluated to see how much it decreases the risk for a negative outcome (such as death). Consider the well-known experiment that assessed the effects of low-dose aspirin on risk for heart attack (Steering Committee of the Physicians' Health Study Research Group, 1989). Results from this study appear in Table 16.5.

Table 16.5 Data From an Experiment to Assess Effect of Low-Dose Aspirin on Risk for Heart Attack in Men

	Placebo ($X = 0$)	Low-Dose Aspirin ($X = 1$)
No heart attack ($Y = 0$)	10,845	10,933
Heart attack ($Y = 1$)	189	104

Source: Steering Committee of the Physicians' Health Study Research Group (1989).

The predictor variable was the treatment group: Patients were given either a placebo (treatment = 0) or low-dose aspirin (treatment = 1). The outcome variable was occurrence of a heart attack (no heart attack = 0, heart attack = 1). If the researcher wants to report results in the form of a statement such as "receiving aspirin reduces the risk for having a heart attack," then it is more convenient to assign a code of 1 to persons who have heart attacks (i.e., to make having a heart attack the target outcome). It is also more convenient in this case to specify the first group on the predictor variable (i.e., the people who receive placebo) as the reference group. These code assignments ensure that the risk for death is the target outcome for which odds and probabilities are calculated and that the odds ratio will indicate how much lower the risk for heart attack is for people who took aspirin compared with the reference group of people who did not take aspirin.

Users of programs other than SPSS need to be aware that the handling of binary-coded predictor and outcome variables differs across programs. Comparing the results from the binary logistic regression (such as predicted probabilities) with simpler statistics based on the original 2 × 2 contingency table can be a helpful way to verify that the odds ratios in the SPSS output correspond to the comparisons that the data analyst had in mind. The data analyst must keep track of the codes on binary variables and understand how the computer program uses these codes to identify the target response and the reference group when setting up odds to interpret odds derived from the analysis correctly.

16.9 MORE COMPLEX MODELS

The empirical example reported above represented the simplest possible type of binary logistic regression with just one binary predictor variable. There are several ways in which binary logistic regression analysis can be made more complex. These can generally be understood by analogy to complex models in multiple linear regression:

1. Binary logistic regression may include a categorical predictor variable that has more than two possible categories. For example, suppose we wanted to report risk for death across four different disease diagnosis categories, such as Stage 1 through Stage 4 lung cancer. In the initial SPSS data file, stages of lung cancer could be coded as 1, 2, 3, and 4. However, the researcher might not want to treat this as a quantitative variable; instead, he or she may want to compare the groups. This can be done in SPSS by identifying lung cancer stage as a categorical variable and then indicating which of the four categories should be used as the reference group. If the first group (Stage 1) is used as the reference category, then the reported odds ratios will tell us how much higher the odds of death are for Stages 2, 3, and 4 compared with Stage 1.

2. Binary logistic regression may use quantitative predictor variables such as age, body mass index, and drug dosage. If age is used to predict risk for death from heart disease, and if age is entered in years, binary logistic regression will tell us how much the odds

of death increase for each 1-year increase in age. (The association between age and odds of death would appear greater, of course, if we used scores that corresponded to age groups, e.g., 1 = 0 to 10, 2 = 20 to 29, . . . , 9 = 90 to 99 years old.)

3. Binary logistic regression may use more than one predictor variable (and these variables may be categorical, quantitative, or a mixture of both). In this type of model, as in a multiple linear regression, we estimate the increase in odds ratio associated with a one-unit increase in score on each individual predictor variable, statistically controlling for all the other predictor variables in the model. Note that if all the predictor variables are categorical, and the outcome variable is also categorical, log-linear analysis might be used rather than logistic regression.

4. Interaction terms may be added to a binary logistic regression model by forming product terms between pairs of predictor variables. The potential problems that arise when interaction terms are included in other analyses are similar to the issues that arise when including interaction terms in multiple linear regression (Jaccard, Turrisi, & Wan, 1990).

5. All predictors may be entered in one step as in a direct or simultaneous or standard multiple linear regression. Alternatively, predictor variables may be entered one at a time or in groups, in a series of steps, in a sequence that is determined by the data analyst; this is analogous to hierarchical multiple linear regression. Alternatively, statistical decision rules may be used to decide on the order of entry of predictor variables and to decide which variables will be excluded from the final model; this is analogous to statistical regression. As described in the chapter about multiple regression, statistical methods of predictor variable selection such as forward, backward, or stepwise regression can substantially increase the risk for Type I error so that the real risk for Type I error is much higher than the nominal *p* values that appear on the SPSS output. SPSS does not adjust the *p* values to correct for the possible inflated risk for Type I error that may arise when a program selects a relatively small number of predictor variables out of a relatively large set of candidate predictor variables. If statistical methods are used to enter predictor variables into a binary logistic regression model, then *p* values on the SPSS printout may substantially underestimate the true risk for Type I error. King (2003) described alternative methods for selecting best subsets of predictors in logistic regression.

6. There is a generalization of the binary logistic regression procedure to situations where the outcome variable has more than two possible outcomes; this is sometimes called polytomous logistic regression (see Hosmer & Lemeshow, 2000).

The list above is not exhaustive, but it includes some of the most widely used variations of binary logistic regression. For further discussion and detailed presentation of empirical examples with larger numbers of predictor variables, see Hosmer and Lemeshow (2000) or Tabachnick and Fidell (2018).

16.10 BINARY LOGISTIC REGRESSION FOR SECOND EXAMPLE: DRUG DOSE AND SEX AS PREDICTORS OF ODDS OF DEATH

The second empirical example includes one quantitative variable and one binary predictor variable. While this is still a very simple case, it illustrates a few additional basic concepts. Consider a hypothetical experiment to assess the increase in risk for death as a function of increased drug dosage of a toxic drug. The researcher assigns 10 rats each (5 male, 5 female) to the following dosage levels of the drug: 0, 1, 2, . . . , 10 mg. The total number of animals was $N = 110$ (10 animals at each of the 11 dosage levels). The outcome variable that is recorded is whether the rat survives (0) or dies (1) after receiving the drug. The predictor variable gender is categorical and is coded 0 = male and 1 = female. The binary logistic regression model has the following form:

$$L_i = B_0 + B_1 \times \text{Drug dose} + B_2 \times \text{Gender},$$

where L_i is the logit, as described earlier in this chapter.

Like the previous example, this equation includes a categorical predictor. Gender was identified as a categorical predictor variable. The last gender group (i.e., the gender group that had a code of 1, female) was used as the reference category. For each animal in the study, the model will be used to calculate an estimated value of L_i (on the basis of the animal's gender and the drug dosage received) and then to use the value of L_i to estimate probability of death for each combination of gender and drug dosage.

Prior to conducting a binary logistic regression, contingency tables were set up to assess whether any of the cells in this design had expected frequencies less than 5 and also to see whether the two predictor variables (gender and drug dosage) were independent of each other. The results appear in Figure 16.19. Visual examination of the frequencies in the drug dose–by–death table indicates that as drug dosage increased from 0 to 10 mg, the proportion of rats that died also increased. The cross-tabulation of gender and drug dosage indicates that the design was balanced; that is, the same proportions of male and female rats were tested at each drug dosage level and, therefore, the predictor variables in this experiment were not confounded or correlated with each other. The pattern of cell frequencies in the gender-by-death table indicates a slightly higher proportion of deaths among male rats than female rats. The logistic regression model will provide information about the increase in odds of death for

Figure 16.19 Preliminary Cross-Tabulations of Variables in Drug Dose/Death Study

drugdose * death Crosstabulation

Count

		death		
		survived	died	Total
drugdose	0	10	0	10
	1	10	0	10
	2	10	0	10
	3	8	2	10
	4	6	4	10
	5	5	5	10
	6	4	6	10
	7	2	8	10
	8	0	10	10
	9	0	10	10
	10	0	10	10
Total		55	55	110

drugdose * gender Crosstabulation

Count

		gender		
		male	female	Total
drugdose	0	5	5	10
	1	5	5	10
	2	5	5	10
	3	5	5	10
	4	5	5	10
	5	5	5	10
	6	5	5	10
	7	5	5	10
	8	5	5	10
	9	5	5	10
	10	5	5	10
Total		55	55	110

gender * death Crosstabulation

Count

		death		
		survived	died	Total
gender	male	26	29	55
	female	29	26	55
Total		55	55	110

each 1-mg increase in drug dosage. It can also be used to generate a predicted probability of death for each animal as a function of two variables: drug dosage and gender.

An additional overall goodness-of-fit measure that is frequently reported when the model includes quantitative predictor variables is the Hosmer and Lemeshow (2000) goodness-of-fit test. This test is done by rank-ordering cases by their scores on predicted probability of membership in the target group ($\hat{p}_i$), dividing them into deciles (e.g., highest 10%, next 10%, . . . , lowest 10%), and doing a chi-square to assess whether the observed frequencies of membership in the target group are closely related to the predicted frequencies on the basis of the model within each decile. A large value of chi-square for the Hosmer and Lemeshow goodness-of-fit test (and a correspondingly small p value, i.e., $p < .05$) would indicate that the predicted group memberships generated by the model deviate significantly from the actual group memberships; in other words, a large chi-square for the Hosmer and Lemeshow goodness-of-fit test indicates poor model fit. Usually, the data analyst hopes that this chi-square will be small and that its corresponding p value will be large (i.e., $p > .05$).

The commands that are required to set up the logistic regression are similar to those used in the previous example except that two variables were included in the list of covariate or predictor variables (drug dose and gender). When more than one predictor variable is included in the model, the data analyst can decide to enter all the predictor variables in one step by leaving the method of entry at the default choice "Enter"; this is similar to standard multiple linear regression, as described in an earlier chapter. In this analysis, both the predictor variables were entered in the same step.

The SPSS commands and syntax for this binary logistic regression appear in Figures 16.20 through 16.24. The commands were similar to the ones for the previous simpler example, with the following exceptions: Two predictor variables, drug dose and gender, were

Figure 16.20 Variables for Second Binary Logistic Regression to Predict Death From Drug Dose and Gender

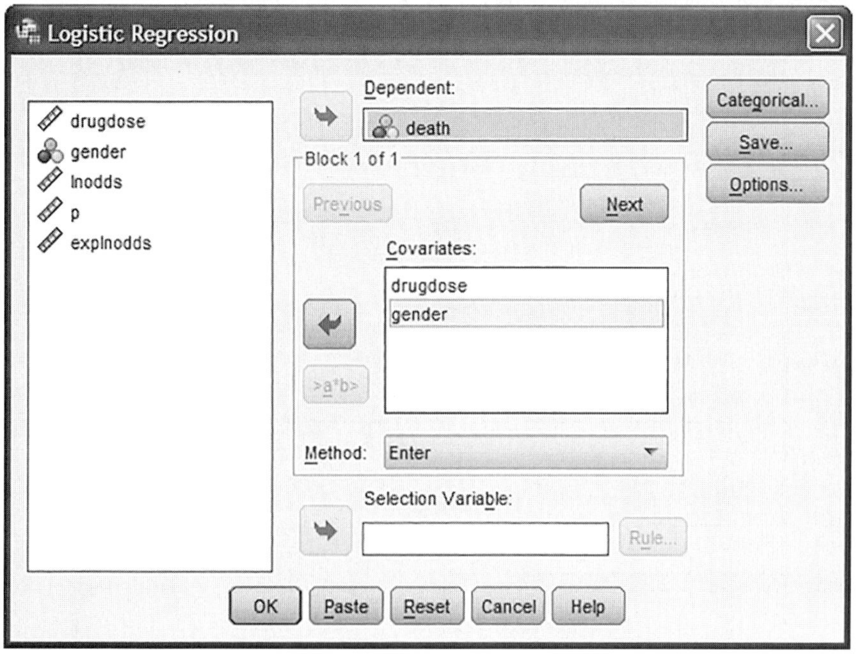

identified by placing them in the list of covariates. Only one of these variables, gender, was further defined as categorical. To open the Logistic Regression: Options dialog box, click the button marked Options in the main Logistic Regression dialog box.

Figure 16.21 Defining Gender as a Categorical Predictor Variable in the Second Binary Logistic Regression Example

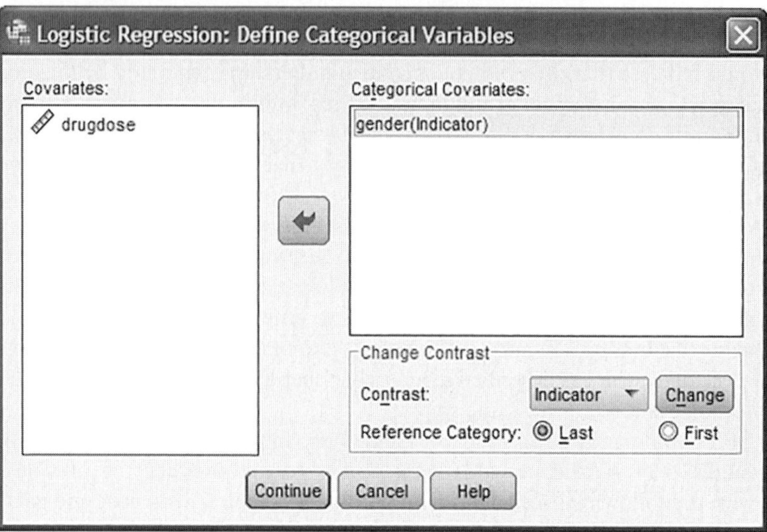

Figure 16.22 Options for Second Binary Logistic Regression Example

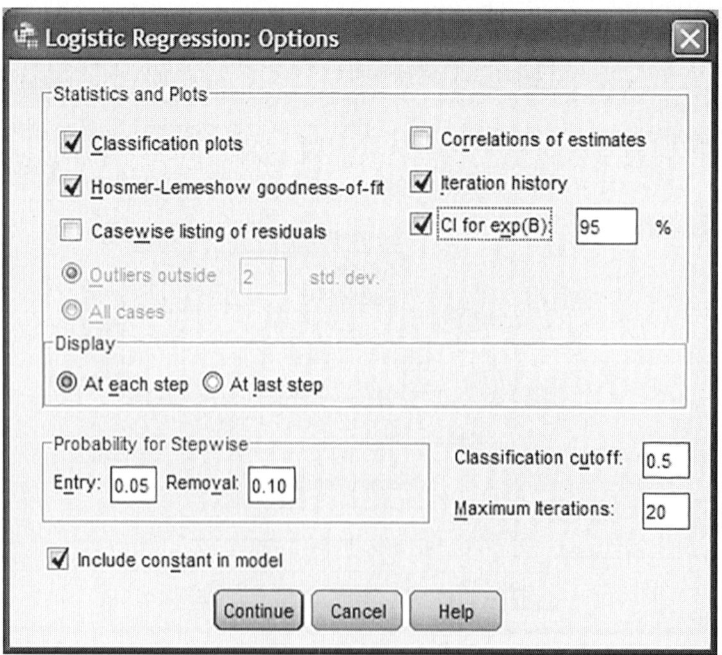

Figure 16.23 Command to Save Predicted Probabilities for Second Binary Logistic Regression Example

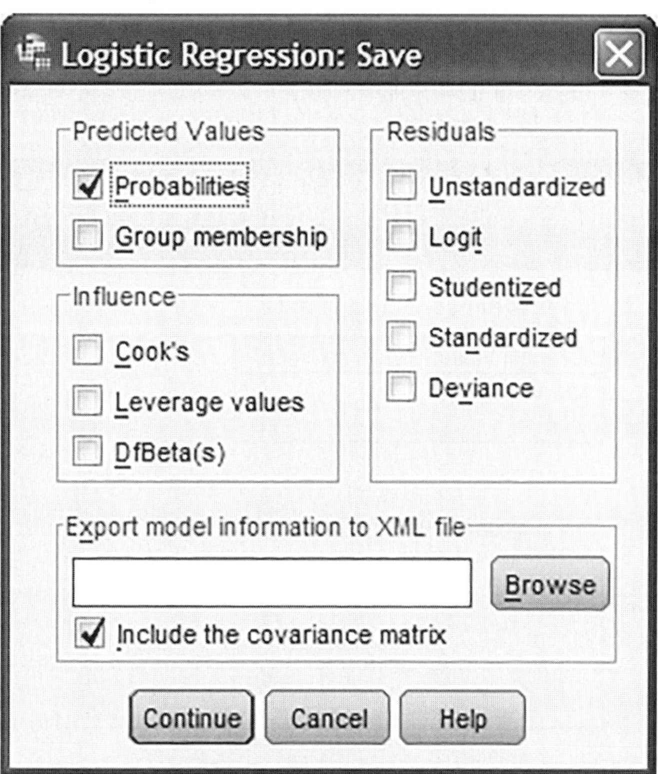

Figure 16.24 SPSS Syntax for Second Binary Logistic Regression Example

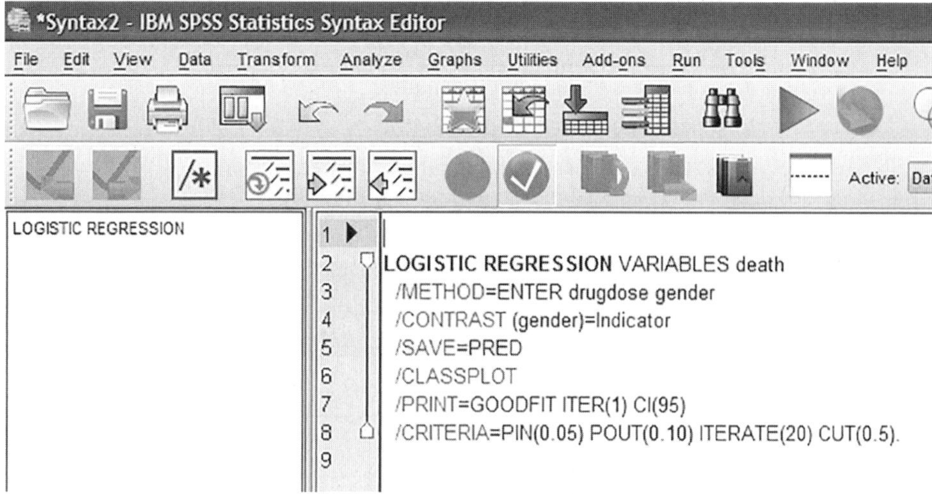

Selected SPSS output from this binary logit regression with two predictors appears in Figures 16.25 through 16.34. Figure 16.25 shows that the internal SPSS coding for the variable gender is male = 1 and female = 0. This means that the female group will be used as the reference group; the odds ratio associated with gender will tell us if males have higher or lower odds of death relative to females. Figure 16.29 provides the omnibus chi-square test for the overall model (i.e., prediction of log odds of death from both drug dose and gender). The overall model, including both drug dosage and gender as predictors, was statistically

Figure 16.25 Coding of Categorical Variables for Second Binary Logistic Regression Example: Prediction of Death From Drug Dose and Gender

Dependent Variable Encoding

Original Value	Internal Value
survived	0
died	1

Categorical Variables Codings

			Parameter coding
		Frequency	(1)
gender	male	55	1.000
	female	55	.000

Figure 16.26 Null Model for Second Binary Logistic Regression Example

Iteration History[a,b,c]

Iteration		-2 Log likelihood	Coefficients
			Constant
Step 0	1	152.492	.000

a. Constant is included in the model.

b. Initial -2 Log Likelihood: 152.492

c. Estimation terminated at iteration number 1 because parameter estimates changed by less than .001.

Figure 16.27 Regression Coefficients for the Null Model for the Second Binary Logistic Regression Example

Variables in the Equation

		B	S.E.	Wald	df	Sig.	Exp(B)
Step 0	Constant	.000	.191	.000	1	1.000	1.000

Figure 16.28 Output for Full Binary Logistic Model to Predict Death From Drug Dose and Gender

Iteration History[a,b,c,d]

Iteration		-2 Log likelihood	Coefficients		
			Constant	drugdose	gender(1)
Step 1	1	77.997	-2.545	.487	.218
	2	67.479	-3.984	.756	.411
	3	65.606	-4.903	.926	.548
	4	65.495	-5.197	.980	.592
	5	65.494	-5.220	.984	.596
	6	65.494	-5.220	.984	.596

a. Method: Enter

b. Constant is included in the model.

c. Initial -2 Log Likelihood: 152.492

d. Estimation terminated at iteration number 6 because parameter estimates changed by less than .001.

Figure 16.29 Chi-Square Test for Improvement in Model Fit: Full Model Compared With Null Model

Omnibus Tests of Model Coefficients

		Chi-square	df	Sig.
Step 1	Step	86.998	2	.000
	Block	86.998	2	.000
	Model	86.998	2	.000

Figure 16.30 Additional Goodness-of-Fit Measures for Second Binary Logistic Regression Example

Model Summary

Step	-2 Log likelihood	Cox & Snell R Square	Nagelkerke R Square
1	65.494[a]	.547	.729

a. Estimation terminated at iteration number 6 because parameter estimates changed by less than .001.

Figure 16.31 Hosmer and Lemeshow Test From Second Binary Logistic Regression Example

Hosmer and Lemeshow Test

Step	Chi-square	df	Sig.
1	4.174	8	.841

Figure 16.32 Contingency Table for Hosmer and Lemeshow Test for Second Binary Logistic Regression Example

Contingency Table for Hosmer and Lemeshow Test

		death = survived		death = died		
		Observed	Expected	Observed	Expected	Total
Step 1	1	10	9.925	0	.075	10
	2	10	9.801	0	.199	10
	3	10	9.485	0	.515	10
	4	8	8.739	2	1.261	10
	5	6	7.240	4	2.760	10
	6	5	5.000	5	5.000	10
	7	4	2.760	6	7.240	10
	8	2	1.261	8	8.739	10
	9	0	.515	10	9.485	10
	10	0	.275	20	19.725	20

Figure 16.33 Classification Table for Second Binary Logistic Regression Model

Classification Table[a]

			Predicted		
			death		Percentage Correct
	Observed		survived	died	
Step 1	death	survived	47	8	85.5
		died	8	47	85.5
	Overall Percentage				85.5

a. The cut value is .500

Figure 16.34 Final Binary Logistic Model to Predict Death From Drug Dose and Gender

Variables in the Equation

		B	S.E.	Wald	df	Sig.	Exp(B)	95.0% C.I. for EXP(B) Lower	Upper
Step 1[a]	drugdose	.984	.181	29.612	1	.000	2.676	1.877	3.815
	gender(1)	.596	.639	.870	1	.351	1.815	.519	6.347
	Constant	-5.220	1.055	24.464	1	.000	.005		

a. Variable(s) entered on step 1: drugdose, gender.

significant, $\chi^2(2) = 86.998, p < .001$. This means the LL "badness of fit" was significantly lower for this full model than for the null or constant-only model. The Cox and Snell R^2 of .547 and the Nagelkerke R^2 of .729 in Figure 16.30 indicate that the association between these two combined predictor variables and odds of death was strong.

The **classification table** in Figure 16.33 indicates that the percentage of animals correctly classified as surviving was 85.5%; the percentage of animals correctly classified as dead was also 85.5% (these two percentages of correct classification are not necessarily always equal in other samples). The numerical estimates of the coefficients for the logit regression model appear in Figure 16.34.

$$\ln(\text{Odds}) = L_i = -5.220 + .984 \times \text{Drug dose} + .596 \times \text{Gender}.$$

The Wald ratio for the coefficient associated with gender was not statistically significant, $\chi^2(df = 1) = .870, p = .351$. Thus, there was no significant difference in odds of death for male versus female animals. Exp(*B*) for gender was 1.815; this indicates that the odds of death for the male group tended to be higher than the odds of death for the female group, but the nonsignificant Wald test tells us that this difference was too small to be judged statistically significant. The Wald chi-square for the coefficient associated with drug dose was statistically significant, $B = +.984$, $\chi^2(1) = 29.612, p < .001$. Exp(*B*) for drug dose was 2.676. This indicates that for each 1-mg increase in drug dose, the predicted odds of death almost tripled. The 95% CI for exp(*B*) ranged from 1.877 to 3.815. We can conclude that increasing the drug dosage significantly increased the odds of death.

It is useful to include predicted probabilities (either in tabular form or as a graph) when the results of logistic regression are reported, because predicted probabilities are much easier to understand than odds. The predicted probabilities of death for each drug dosage level appeared as a graph in Figure 16.1. This graph makes the nature of the relationship represented by the model very clear. At a drug dosage of 0 mg, the predicted probability of death is approximately 0. As drug dosage increases from 0 to 1 to 2 mg, the predicted probability of death increases very slowly. For drug dosages between 3 and 7 mg, the predicted probability of death increases rapidly as a function of drug dose—for example, 50% of the animals that received 5 mg of the drug died—and the curve passes through this point. As the drug dosage continues to increase from 8 to 10 mg, the predicted probability of death increases more slowly (and levels off at .99, within rounding error of 1.00). The predicted probabilities of death for each participant (on the basis of drug dose and gender) can also be summarized in table form as shown in Figure 16.35.

Figure 16.35 Summary Table: Probability of Death Tabled by Drug Dose and Gender

Drug Dose:	Gender:	
	male	female
0	.00972	.00538
1	.02558	.01426
2	.06562	.03726
3	.15816	.09381
4	.33448	.21687
5	.57346	.42556
6	.78245	.66463
7	.90585	.84131
8	.96260	.93413
9	.98568	.97432
10	.99460	.99024

16.11 COMPARISON OF DISCRIMINANT ANALYSIS WITH BINARY LOGISTIC REGRESSION

The use of DA to predict group membership was discussed in Chapter 10. DA is one of the forms of the general linear model. DA requires some rather restrictive assumptions about data structure, and the use of DA is problematic when the sizes of the two groups are drastically unequal (e.g., when there is a 95% vs. 5% split of cases into Group 0 and Group 1) and in situations where the underlying relationship between scores on the predictor and scores on the outcome is nonlinear. Many research situations (such as diagnosis of uncommon diseases) do involve small numbers in one of the groups. Logistic regression and DA can both be used to predict membership in one of two groups; however, logistic regression offers some potential advantages compared with DA. Logistic regression does not require such stringent assumptions about data structure, and it provides a better model for situations where the probabilities of group membership become small (e.g., 10% and less). Logistic regression provides information about odds and differences in odds. In some ways, having the information in this form is an advantage; researchers can readily provide effect size information for individual predictors by reporting how much higher the odds of disease becomes for each one-unit increase in scores on each predictor variable. However, there are two potential disadvantages of logistic regression. The values of coefficients in a logistic regression cannot be obtained using OLS estimation methods; instead, MLE methods are required. Maximum likelihood estimates tend to be unstable unless sample sizes are fairly large. In addition, because the concept of odds is relatively unfamiliar to many readers, reports given in terms of odds may be more difficult for some readers to understand.

In situations where the sizes of the groups are substantially unequal (e.g., a 95% vs. 5% split between Groups 0 and 1), the overall sample size is large, and the multivariate normality assumptions and homogeneity of variance/covariance matrix assumptions that are required for DA are violated, logistic regression provides a better method for prediction of group membership. In practice, many researchers have developed a strong preference for the use of logistic regression rather than DA in clinical prediction studies.

16.12 SUMMARY

As noted in an earlier section, incomplete reports about the results of logistic regression analyses can lead to misleading impressions about the strength of relationships between variables. Researchers need to include probabilities along with odds when they report results. This information can help lay readers (and professional audiences) better understand the true magnitude of increase in risk in studies that examine risk factors as predictors of negative outcomes and the true strength of the benefits in studies that examine how treatments or interventions reduce the likelihood of negative outcomes.

There are two reasons why reporting only information about changes in odds can be misleading. First, most people do not have a good intuitive grasp of the concepts of odds ratios. Second, information about changes in odds can be difficult to interpret without the knowledge of baseline levels of risk, for example, the probability of the negative event when treatment is withheld. For example, suppose that a new cholesterol-reducing drug cuts a person's risk for heart attack in half. The practical or clinical significance of the drug for individual patients is arguably different if the drug reduces a preexisting probability of heart attack of 50% to 25% than if the drug reduces a baseline risk or probability of heart attack of about 1% to 0.5%. Converting log odds to probability estimates for selected cases with representative scores on the predictor variables provides extremely useful information.

The terminology used to report magnitude of risk and effectiveness of treatment (such as relative risk, relative risk reduction, and absolute risk) is confusing, not only to lay readers but even to some researchers. An illustrative example was reported in a comment published in the *New England Journal of Medicine*. Schwartz, Woloshin, and Welch (1999) discussed mass media reports about a study (Schulman et al., 1999) that examined how often physicians recommend cardiac catheterization for hypothetical patients with chest pain; predictor variables included patient ethnicity and gender. The odds ratio for referral for catheterization for Black compared with White patients was reported in the journal as .60. The outcome of Schulman et al.'s study was reported in mass media as evidence that Black patients were "40% less likely" to be referred for cardiac testing than White patients. Schwartz et al. argued that this mass media statement, in isolation, was misleading to many readers; they argued that simpler statistics (84.7% of hypothetical Black patients vs. 90.6% of hypothetical White patients were referred for catheterization) would provide readers with a better sense of the real magnitude of the difference in referral rates. The difficulty involved in translating odds ratios into everyday language in a way that is easy for people to understand is further discussed by Schwartz (2003).

On the basis of recommendations by Tabachnick and Fidell (2018) and Peng et al. (2002), the following information should be included in reports of results for binary logistic regression analysis:

1. Preliminary data screening and a clear statement about the variables and options involved in the analysis.
2. Overall model fit. This should include information about statistical significance, usually in the form of a chi-square test. It should also include some discussion of effect size. For the overall model, statistics such as Nagelkerke's R^2 provide information about effect size.
3. Information about individual predictors should be presented. This includes the model coefficients, which can be used to derive odds and estimated probabilities; statistical significance tests; a discussion of the nature or direction of the association (e.g., as drug dose increases, do the odds of death increase or decrease?). CIs should be reported. Effect size information should be provided, and researchers need to evaluate whether the association between the predictor and outcome variables is strong enough to be of any practical or clinical importance. In the dog ownership/death sample, for example, odds of death were about 6 times higher for nonowners of dogs than for dog owners.
4. Other goodness-of-fit information may be useful if there is at least one quantitative predictor variable, for example, the Hosmer and Lemeshow goodness-of-fit test.
5. Predicted probabilities as a function of reasonable and representative score values can be summarized either as a figure (as in Figure 16.1) or in table form (as in Figure 16.35). For example, a report of the dog ownership/death study should make it clear that death was not a highly likely outcome even for the group that had higher odds of death (nonowners of dogs).
6. Additional information can also be provided; for example, an analysis of residuals can be useful in identifying any multivariate outliers or in detecting violations of assumptions.

To this list I add:

7. Further comments about Item 5: Reports should explain clearly what results mean in terms of both relative risk and absolute risk. Noordzij, van Diepen, Caskey, and Jager (2017) cited an example: Relative risk for death from infection was 82 times higher for patients on dialysis than for a comparison group in the general population. However, the (absolute) risk for occurrence of death from infections was 0.01 per

1,000 patient-years in the general population; 82 times this low base rate tells us that the (absolute) risk for death for the dialysis patients was .82 per 1,000 patient-years, that is, only about one patient on dialysis died in the study for each 1,000 patient-years included. That does not mean infection isn't a problem worthy of concern, of course, but absolute risk provides a less alarming perception than relative risk. When messages are phrased only in terms of relative risk, they lead people to feel that risks from exposures to threats are very high and that treatments can be very effective at reducing risks (Malenka, Baron, Johansen, Wahrenberger, & Ross, 1993).

8. Clear explanation about limits to generalizability. An estimate for risk for death (or any other outcome) in any one study is based on specific types of people (sex, age, and other characteristics), with specific types of risk factors or treatments, observed over a specific time period (and so forth). If you are not similar to people in the study, or if you are concerned about risks over shorter or longer periods of time than included in the study, or if the number of cases in the study and length of follow-up are not sufficient to draw reasonable conclusions, then results of that study may not apply to you. The questions readers often have—What is my lifetime risk for a specific disease? What is my life expectancy given my health situation? How much increase in life expectancy can I expect from a course of treatment?—are often not answered by information in research reports (and certainly not well explained in news posts about the research).

9. Readers should also bear in mind that like all other statistics, estimate of risk and relative risk depend on which variables were (and were not) statistically controlled in the analysis. It is possible that some researchers add and drop covariates in a series of analyses and then report only the "best" outcome (this is *p*-hacking).

A readable discussion of risk reporting for researchers and practitioners is provided by Barratt et al. (2004). Explanations for patients and consumers are given in Irwig, Irwig, Sweet, and Trevena (2007).

This chapter presented two simple examples of binary logistic regression. The first example used one binary predictor (dog ownership) to predict odds of death. The second example used one binary predictor (gender) and one quantitative variable (dosage of a toxic drug) to predict odds of death. Either of these analyses could be extended in several ways. For example, we might want to add a term to test for a possible drug dose-by-gender interaction to the second analysis. We might want to use multiple predictor variables to assess their relative contributions; for example, among a set of risk factors that includes age, gender, smoking, high-density lipoprotein cholesterol, and hostility, how strongly are scores on each of these predictors related to odds of death from heart attack when other correlated predictors are statistically controlled? Other forms of logistic regression provide ways to analyze situations where the categorical outcome variable has more than two possible outcomes; this is sometimes called polytomous logistic regression (Hosmer & Lemeshow, 2000).

COMPREHENSION QUESTIONS

1. Examine the data in Table 16.5, which report the outcomes for the Physicians' Health Study that assessed the effect of a low-dose aspirin regimen on risk for heart attack.

   a. For the group of patients that had the low-dose aspirin regimen, calculate the odds of death.
   b. For the group of patients that received placebo, calculate the odds of death.
   c. Compute the odds ratio that tells us how much more likely a heart attack was for patients in the placebo group (compared with the aspirin group).
   d. Compute the odds ratio that tells us how much less likely a heart attack was for patients in the aspirin group (compared with the placebo group).
   e. In simple language, how did the aspirin regimen change the odds of heart attack?
   f. Now look at the data in a different way. What percentage of the aspirin regimen group had a heart attack? What percentage of the placebo group had a heart attack? Does looking at these two percentages give you a different impression about the impact of aspirin?
   g. Describe two different statistical analyses that you could apply to these data to assess whether heart attack is significantly related to drug regimen. (Hint: For a 2 × 2 table, simpler analyses were described in earlier chapters.)

2. The data summarized in Table 16.6 (from a study by Goodwin, Schylsinger, Hermansen, Guze, & Winokur, 1973) show how "becoming a heavy smoker" is related to "having a biological parent who was diagnosed alcoholic." Either enter these data into SPSS by hand or use the data set goodwin.sav (available on the website for the textbook). Run a binary logistic regression to predict smoking status from parental alcohol diagnosis and write up your results, including tables that summarize all relevant information. In simple language, do people whose biological parent is diagnosed with alcoholism have significantly higher odds of being heavy smokers?

Table 16.6 Diagnosis of Parental Alcoholism as Predictor of Heavy Smoking

	Parent Not Diagnosed With Alcoholism, $X = 0$	Parent Diagnosed With Alcoholism, $X = 1$	
Adult child not a heavy smoker, $Y = 0$	41	6	47
Adult child is a heavy smoker, $Y = 1$	37	49	86
Total	78	55	133

Source: Goodwin et al. (1973).

3. Table 16.7 shows how values of $\hat{p}_i$ are related to values of odds and values of the logit L_i. To do this exercise, you will need a calculator that has the natural log function; you can find a calculator that has this function online (e.g., at https://www.rapidtables.com/calc/math/Ln_Calc.html).

Table 16.7 Selected Corresponding Values of Probabilities (p), Odds, and the Logit (L_i)

$\hat{p}_i$	$(1 - \hat{p}_i)$	Odds = $\hat{p}_i/(1 - \hat{p}_i)$	Logit = $L_i = \ln[\hat{p}_i/(1 - \hat{p}_i)]$
.1		.111	−2.20
.2	.8		−1.39
.3	.7	.429	−.847
.4	.6	.667	
.5	.5		0
.6	.4	1.50	
.7	.3	2.33	.847
.8	.2	4.00	1.39
.9	.1	9.00	2.20

Source: Adapted from Pampel (2000, p. 14).

Note: For one of the comprehension questions, the task is to fill in the blank cells of this table.

a. Fill in the values in the empty cells in Table 16.7.
b. Consider the set of scores on $\hat{p}_i$. Do these scores satisfy the assumptions that are generally required for scores on a Y outcome variable in a multiple linear regression? In what way do these scores violate these assumptions?
c. An odds of 1.00 corresponds to values of p = _____ and $(1 - p)$ = _____.
d. An odds of 1.00 corresponds to a logit L_i value of _____.
e. Look at the column of scores for L_i, the logit. Do these scores satisfy the assumptions that are generally required for scores on a Y outcome variable in a multiple linear regression? Can you see any way that these scores violate these assumptions?
f. On the basis of this table, what advantages are there to using L_i, rather than $\hat{p}_i$, as the dependent variable in a logistic regression? Hint: You might want to graph the values in the last column of Table 16.7 to see what distribution shape they suggest.

DIGITAL RESOURCES

Find **free study tools** to support your learning, including **eFlashcards, data sets, and web resources**, on the accompanying website at **edge.sagepub.com/warner3e**.

CHAPTER 17

ADDITIONAL STATISTICAL TECHNIQUES

It is tempting, if the only tool you have is a hammer, to treat everything as if it were a nail.

—Maslow (1966)

17.1 INTRODUCTION

The research questions people ask can be limited by the statistical tools they know. If all people know is analysis of variance (ANOVA), their studies are likely to be limited to comparisons of group means. By now, you have learned some methods that make it possible to ask additional questions. At this point you should know:

- Analyses for many situations with multiple predictors and multiple outcomes, including both categorical and quantitative variables.
- Some methods for statistical control.
- Some methods for data reduction (e.g., factor analysis, discriminant functions).
- Use of latent variables to model relationships among measured variables (e.g., in factor analysis and structural equation models).
- Use of path models in which paths represent "causal" and noncausal hypotheses and that include direct versus indirect paths.

This list of techniques is, at best, "the end of the beginning." The following timetable for selected developments in statistics may give you a sense how far we have come, and how much farther it is possible to go in understanding analytic techniques.

17.2 A BRIEF HISTORY OF DEVELOPMENTS IN STATISTICS

You may be surprised to learn how long ago some techniques you have learned were developed. The ideas covered in introductory statistics include a small subset of techniques that were developed prior to 1950 and, rarely, later developments. There have been many developments since that time. The adoption of new ideas and techniques by researchers often lags behind the introduction of these techniques, often by decades. Here is a partial timeline of the development of statistics. Where a specific year is given, it corresponds to the date of important publication(s); a range of years is the life span.

1654 Blaise Pascal and Pierre de Fermat create the mathematical theory of probability (on the basis of their correspondence about gambling).

1657 Christiaan Huygens writes the first book about probability theory.

1663 John Graunt estimates the population of London on the basis of parish records.

1702–1761 Thomas Bayes develops the interpretation of probability now known as Bayes' theorem.

1749–1827 Pierre-Simon Laplace is a coinventor of Bayesian statistics.

1777–1855 Carl Friedrich Gauss invents least squares estimation methods (with Adrien-Marie Legendre) and uses maximum likelihood estimation.

1796–1874 Adolphe Quetelet pioneers the use of probability and statistics in social sciences.

1805 Legendre introduces method of least squares for curve fitting.

1807 Jean-Baptiste Fourier develops Fourier series, the basis for later development of spectral analysis.

1808 Gauss (with contributions from Laplace) derives the normal distribution.

1854 John Snow's map of cases in a London neighborhood identifies the source of a cholera epidemic as a contaminated water pump in Broad Street, beginning the modern study of epidemiology.

1820–1910 Florence Nightingale applies statistics and graphics to the analysis of medical data and revolutionized hospital care.

1877 Francis Galton describes regression to the mean and introduces the concept of correlation.

1898 Ladislaus von Bortkiewicz shows that deaths of soldiers from horse kicks follow a predictable pattern: the Poisson distribution.

1857–1936 Karl Pearson develops the Pearson chi squared test and Pearson correlation.

1863–1945 Charles Spearman develops the rank correlation coefficient.

1839–1914 Charles Pierce introduces blinded, controlled, randomized experiments (before Fisher), uses logistic regression, and improves treatment of outliers. He also introduces the terms *confidence* and *likelihood.*

1876–1937 William Gosset (known as Student) develops the Student t distribution and t test.

1890–1962 Ronald A. Fisher develops analysis of variance and numerous other statistical techniques. (The F ratio was given this name in honor of Fisher's work.)

1894 Pearson develops principal-component analysis (later developed into factor analysis); he also introduces the term *standard deviation.*

1918 Sewall Wright develops path analysis.

1934 Alexander Aitken develops generalized least squares.

1935 Fisher's book *The Design of Experiments* explains how to evaluate whether the results of experiments are significant

1937 Jerzy Neyman introduces confidence intervals and the foundations of sampling.

1939–1945 Alan Turing and colleagues break the German Enigma code using advanced Bayesian techniques, saving millions of lives during the war.

1970 George Box and colleagues develop models to describe serial dependence in time-series data (ARIMA models).

1970 Karl Gustav Jöreskog and colleagues develop a general method for analysis of covariance structures, the foundation for structural equation modeling and other latent variable analyses.

1972 David Cox develops the proportional-hazards model (the basis for survival analysis).

1977 John Tukey introduces the boxplot (also called the box and whiskers diagram) to display median, quartiles, and spread of data in one graph.

1979 Bradley Efron introduces bootstrapping.

1980 Richard McCleary, Richard Hay, and colleagues outline methods for analysis of time-series data before and after interventions.

1988 Jacob Cohen introduces effect size, statistical power, and the basis for meta-analysis. Cohen is a critic of statistical significance testing.

1995 Multilevel modeling (also called hierarchical linear modeling) is developed through work by Peter M. Blau, Harvey Goldstein, and others.

1997 The term *big data* first appears in print.

This list is not comprehensive, and developments have usually depended on contributions from others in addition to named individuals. See Stigler (1990) and Tankard (1984) for more information and the poster timeline posted by the Royal Statistical Society (n.d.) at http://www.statslife.org.uk/images/pdf/timeline-of-statistics.pdf.

This chapter briefly describes some questions you can address using some statistical methods not covered in this book. One or two simple examples are presented for each technique; this is not a how-to guide. This discussion does not include all additional advanced techniques.

17.3 SURVIVAL ANALYSIS

Danner, Snowdon, and Friesen's (2001) longitudinal study of a group of nuns provides a compelling example of the use of survival analysis. At entry into the convent, each nun wrote an autobiographical essay; the amount of positive word use was evaluated in each essay. On the basis of positive word use, the nuns were divided into quartiles (four groups): the top 25% in amount of positive word use, the bottom 25% in positive word use, and the two intermediate groups. Ages at death were examined, and a model was developed to describe probability of survival in each of these four groups that differed in amount of positive word use. Results appear in Figure 17.1.

There is a separate survival function for each group of nuns. (These lines are functions estimated from patterns in sample data.) The proportion of nuns alive is plotted as a function of age. By age 90, about 60% of the nuns who used the most positive words in their early autobiographical essays were alive. By contrast, only 30% of the nuns who used the fewest positive words were expected to be alive by age 90. Survival analysis can be applied to cases other than humans and to binary outcomes other than death. For example, we could ask about the probabilities that different kinds of light bulbs are still burning after hundreds of hours of use.

Figure 17.1 Survival Analysis Results for Nuns

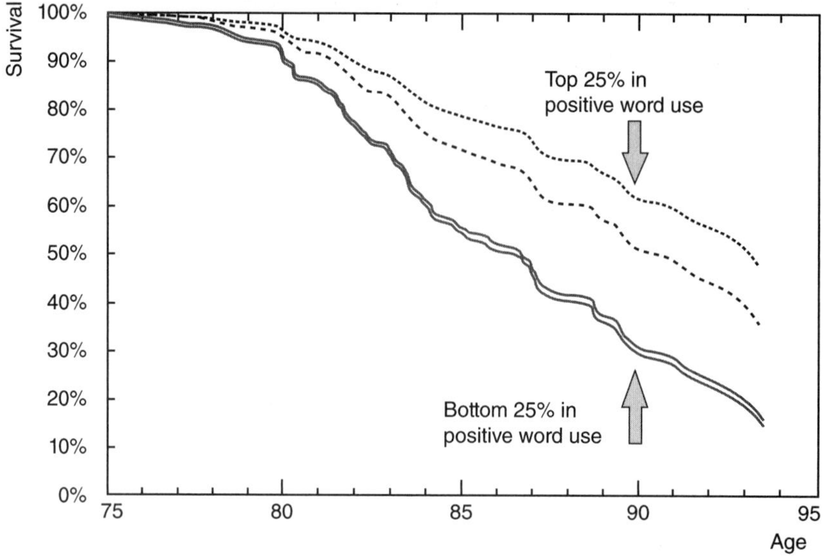

Source: Adapted from Danner et al. (2001).

17.4 CLUSTER ANALYSES

In earlier discussion of discriminant analysis, group memberships were known ahead of time. However, the question can arise whether it makes sense to divide people in a sample into different groups on the basis of their scores on several variables.

Two questions arise:

- First, how many groups should we consider?
- Second, how do these groups differ; how can the groups be described?

As a hypothetical example, suppose we want to find out whether patients who come to a headache clinic can be thought of as different groups. We measure fatigue, depression, anger, and anxiety.

Clusters are obtained subject to a criterion that the cases have small differences (or distances) among scores within groups and large differences (or distances) between groups. There are many ways to define difference or distance (see Everitt, Landau, Leese, & Stahl, 2011).

A dendrogram (or tree diagram) illustrates the way cases can be divided into clusters. In Step 1 (top of diagram in Figure 17.2), all cases are included in one cluster. At each level in the dendrogram, each cluster can be divided into smaller clusters. In the final step (bottom of diagram), each cluster consists of just one case. The first question is, As groups are divided and subdivided (moving from top to bottom of this diagram), at what stage are the clusters optimal? Researchers typically want a reasonably small number of clusters, with small within-group distances among the patterns of scores on cases and with interpretable differences among groups. The diagram below shows a hypothetical example in which 16 cases are divided among clusters. (In actual applications, the number of cases would be larger, and the numbers of members in the clusters would not have to be equal.) In the hypothetical example

Figure 17.2 Dendrogram for Hypothetical Cluster Analysis of 16 Cases

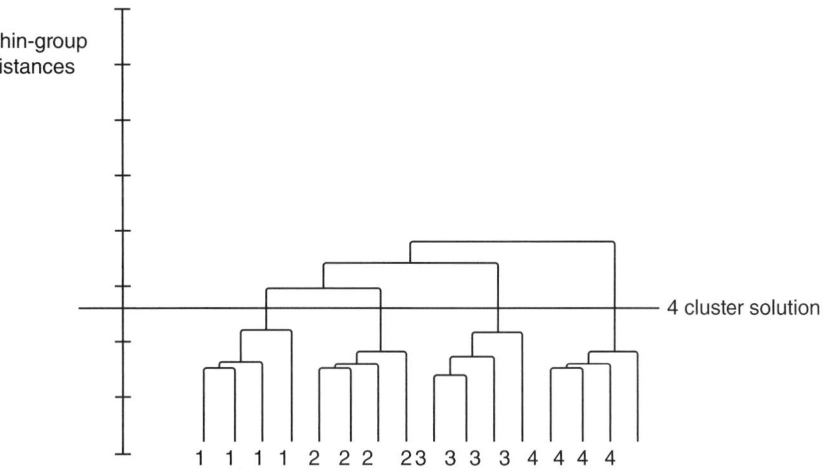

Figure 17.3 Hypothetical Profiles of Means for Variables for Two Clusters of Headache Patients

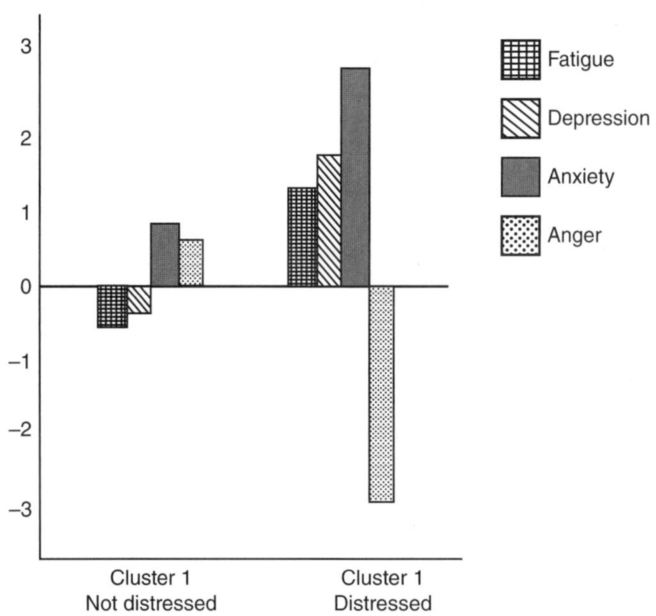

in Figure 17.2, a four-cluster solution is shown. The numbers across the bottom of the diagram indicate to which of the four clusters each case was assigned.

After clusters are obtained, researchers want to describe them. For sake of simplicity, the hypothetical bar chart of group means in Figure 17.3 compares only two hypothetical clusters. We could describe Cluster 1 as not very distressed; their scores are about average on all four measures of psychopathology. We could describe Cluster 2 as distressed, but with the qualifying information that their scores are high only on fatigue, anxiety, and depression (not on anger). Knowledge about these characteristics might be helpful in designing treatment programs specific to different types of patients (i.e., treatments that address the differing needs of subgroups). For more about cluster analysis, see Everitt et al. (2011) or Kaufmann and Rousseeuw (2009).

17.5 TIME-SERIES ANALYSES

Time-series data consist of a series of measurements of the same variable, usually made at equally spaced points in time (ideally, at least 50 observations). Many types of variables can be obtained as time series, for instance, social indicators (such as crime rate assessed once a month), psychological variables (such as once-a-day mood ratings), physiological variables (such as heart rate measured once per minute), stock market indexes (at closing time once a day), and so forth. A common notation for a time-series variable is X_t, where t refers to the time at which each measure is made. Daily measures of mood made on 84 days could be denoted $X_1, X_2, X_3, \ldots,$ X_{84}. It is possible to have concurrent time series for two or more variables; for example, wife's mood on 84 days could be denoted W_1 to W_{84}, and husband's mood on 84 days could be denoted H_1 to H_{84}. Similar to scores in repeated-measures ANOVA designs, time-series observations are usually correlated with one another (or serially dependent). In repeated-measures ANOVA, we assume that the pattern of serial dependence is simple (and that assumption is often violated). In time-series analysis, models (equations) are developed to describe potentially more complex patterns of serial dependence.

17.5.1 Describing a Single Time Series

To evaluate time-series models, we also need a notation for residuals; the prediction error at time t can be denoted e_t. A time-series model must have independent (uncorrelated) residuals to be judged adequate. The concept of lag is also needed. For simplicity, consider a set of mood measures made once a day for a week, starting on Monday, denoted X_t. $X_{(t-1)}$ denotes the observation that is made one time unit earlier than X_t (i.e., the lag 1 value of X).

Table 17.1 Arrangement of Scores for X_t and X_{t-1} in a Data File That Contains Daily Mood Ratings for 7 Days

Row in SPSS Data File	Column 1 Variable X_t (Unlagged)	Column 2 Variable $X_{(t-1)}$ (the Day Before X_t) Lagged Value of X_t
1	X_{Monday}	(No observation available)
2	$X_{Tuesday}$	X_{Monday}
3	$X_{Wednesday}$	$X_{Tuesday}$
4	$X_{Thursday}$	$X_{Wednesday}$
5	X_{Friday}	$X_{Thursday}$
6	$X_{Saturday}$	X_{Friday}
7	X_{Sunday}	$X_{Saturday}$

Note: Column 1 contains values for X_t, mood on the present day. Column 2 contains values for X_{t-1}, mood on the previous day. Column 2 has the same score values as column 1, lagged as shown.

The correlation between X_1 and $X_{(t-1)}$, which can be denoted $r_{(t,\ t-1)}$, is called the lag 1 autocorrelation ("auto" because X is correlated with itself). For many time series, this is a moderate positive correlation. This correlation tells us how well we can predict today's mood from yesterday's mood, across all days in the time series.

SPSS has a lag function that will create a new variable that is the lagged value of X. When we correlate the mood ratings in the X_t column with the (time-shifted or lagged) mood ratings in the X_{t-1} column, we answer the question, How well can values of X at Time t be predicted from values of X at Time $t - 1$? How well can mood ratings for each day be predicted from the mood on the prior day?

Before asking other questions about time-series data, analysts usually examine the time series for trends or cycles. If these are present, they are usually removed before further model fitting.

Model fitting involves finding an equation that describes the pattern in a time series; if the model accounts completely for serial dependence, then residuals from the model should be uncorrelated with each other. Box, Jenkins, Reisel, and Ljung (2015) described ARIMA models for time series. "AR" denotes autoregressive serial dependence; "I" denotes an integrated or summed process; and "MA" denotes a moving average. A model may include any combination of these three sources of serial dependence, and each type of dependence can exist at varying time lags of 1, 2, 3, and so on.

If there is no a priori knowledge of the type of model that is needed, the data analyst usually examines the pattern in a series of lagged autocorrelations to decide what models might fit the data. A series of lagged autocorrelations would be as follows:

Lag 1 X_t with X_{t-1},

Lag 2 X_t with X_{t-2},

Lag 3 X_t with X_{t-3},

and so on. A first-order autoregressive process is easiest to understand. It has a simple signature pattern in the lagged autocorrelation function. Suppose that the lag 1 correlation is .60. If the process that generates the data is first-order autoregressive, we expect to see an exponential decline in the values of the correlations across lags, for example:

Lag 1 autocorrelation: .6;

Lag 2 autocorrelation: $.6^2 = .36$;

Lag 3 autocorrelation: $.6^3 = .216$;

and so on. If this pattern appears in the lagged autocorrelations, the data analyst would fit a first-order **autoregressive model** to the time series. In simple regression notation, the equation that represents a first-order autoregressive model, in which X_t is predicted from X_{t-1}, and where e_t is the residual for the time t scores, can be written:

$$X_t = b \times X_{t-1} + e_t. \tag{17.1}$$

The data analyst then examines the lagged autocorrelations for the values of the e_t residuals. If the model has accounted for all serial dependence in the data, then the autocorrelations among residuals should be (close to) 0 at all time lags. If some autocorrelations among residuals are not close to zero, additional terms may be needed in the model.

A second order autoregressive model predicts X_t from both X_{t-1} and X_{t-2}:

$$X_t = b_1 \times X_{t-1} + b_2 \times X_{t-2} + e_t. \tag{17.2}$$

An **integrated model** is one in which today's measure is just yesterday's measure plus some random error. (In effect, it is a first-order autoregressive model with a b coefficient of 1.) Integrated models tend to show trend or drift.

$$X_t = X_{t-1} + e_t. \tag{17.3}$$

A **moving average model** is a bit more difficult to imagine; it predicts today's observation from one or several lagged residuals from earlier days. This is meaningful in industrial quality control analysis, but in my experience with time-series data in physiology and psychology, MA terms have not been needed to achieve satisfactory model fit.

A complete model may include one or more of these components: AR, I, and MA. If trends or cycles are present, these are usually removed before trying to fit AR or MA models.

If we have a second time series, Y_t, we can include lagged Y terms as predictors of X (and/or lagged X terms as predictors of Y). For example, we might predict time-series measures of infant behavior from time-series measures of parent attempts to stimulate or soothe the infant. This can be useful in modeling social interaction (where X and Y correspond to measures of partner behaviors) and of coordination in physiological systems.

If a good model can be obtained to describe the pattern in a time series, the model may be used to predict or forecast future values for the time series. (Don't get excited. This doesn't work well for the stock market.)

17.5.2 Interrupted Time Series: Evaluating Intervention Impact

One of the simplest possible quasi-experiments is as follows. Make one pretest measure of the variable of interest. Do an intervention. Make one posttest measure of the same variable of interest after the intervention. Compare mean posttest with mean pretest to assess intervention impact. The most obvious analysis for this type of data is paired-samples *t*; other analyses, such as analysis of covariance using pretest X as a covariate, are also possible. This design has many weaknesses (also called threats to internal validity; Campbell & Stanley, 1963). Among these are the following.

With only one preintervention measure, we don't know if we have a stable baseline; we also don't know whether the variable might have been increasing (or decreasing) on its own, even in the absence of intervention. An X variable, such as vocabulary size, may be increasing over time because of maturation. Figure 17.4 illustrates the problem. The possible underlying trend is shown as a line. If we have only two measures (one pre- and one postintervention), we have information about only one mean before the intervention and one mean after, but we don't know whether the difference in means is due to an ongoing trend that was not changed by the intervention or to other temporal processes such as maturation or fatigue.

Figure 17.4 One Mean Before Intervention and One Mean After Intervention and Potential Trend Line (Not Strong Evidence That the Intervention Had an Impact)

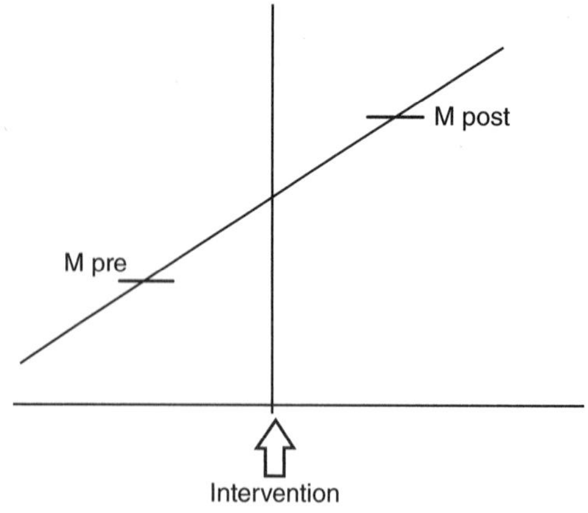

With only one postintervention measure, we also don't know whether any treatment impact lasted very long. If we don't make postintervention assessments at the optimal time, we may miss a temporary intervention effect that occurred at an earlier or later time than our assessment. McDowall, McCleary, and Bartos (2019) identified potential four types of intervention impact across time; see Table 17.2. Variations and combinations of these patterns are possible. The *X* axis represents time; *I* is the point in time when the intervention occurs; and the *Y* axis represents scores on the outcome variable of interest.

Table 17.2 Four Possible Types of Intervention Impact in Time-Series Studies

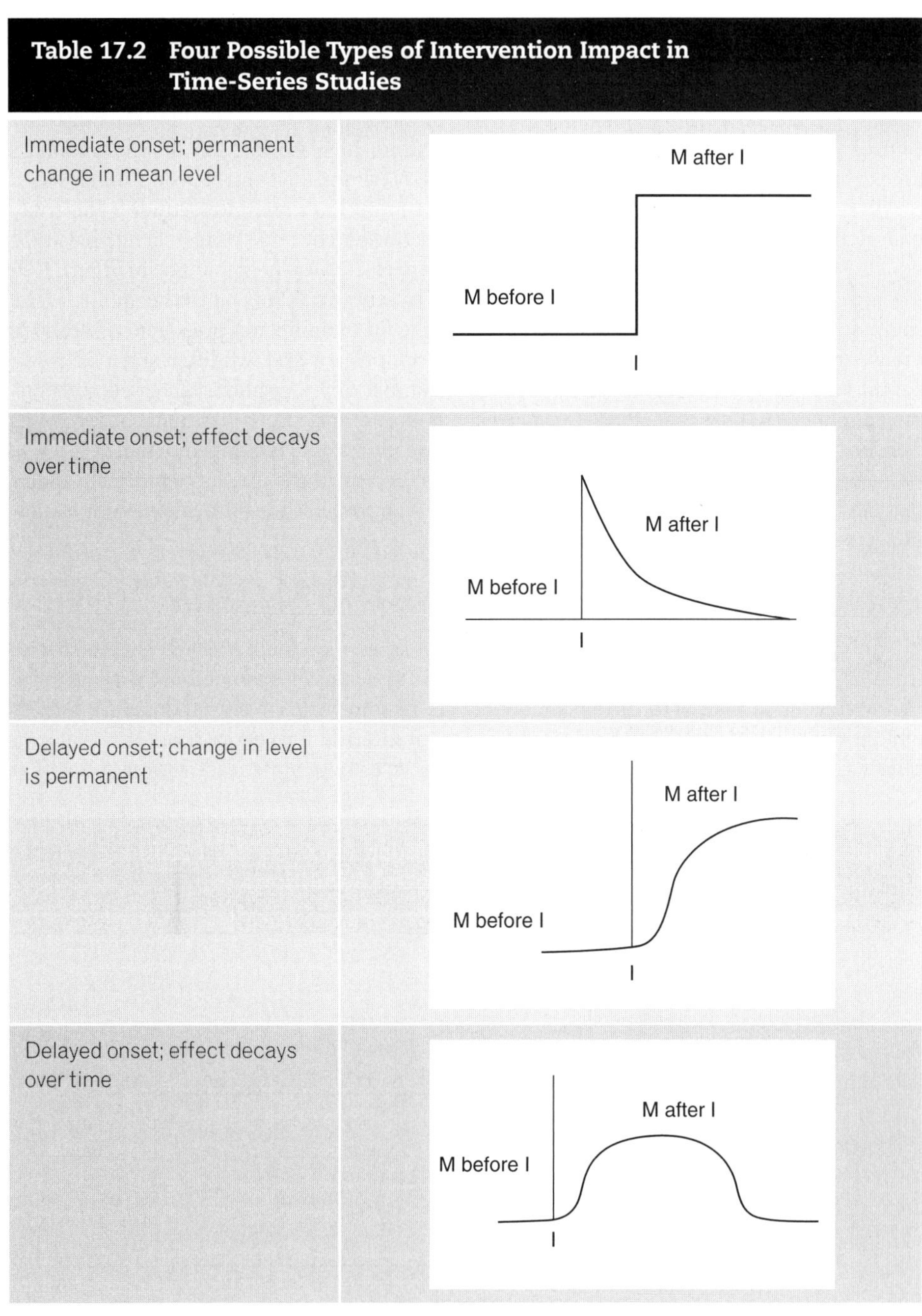

Immediate onset; permanent change in mean level	M before I; I; M after I
Immediate onset; effect decays over time	M before I; I; M after I
Delayed onset; change in level is permanent	M before I; I; M after I
Delayed onset; effect decays over time	M before I; I; M after I

Source: McDowall et al. (2019).

It's useful to have these models of intervention impact in mind in all studies that evaluate intervention impact, even if you never collect time-series data. These patterns make it clear that the timing of assessments after interventions can be crucial. Some treatments require time to begin to take effect. Some interventions may lead to permanent change, while effects of other interventions may wear off over time. It isn't always feasible to do long-term follow-up, but information whether any changes are long lasting is extremely important in practical applications.

Here is an example of an interrupted time-series study that examined the effect of noise exposure on blood pressure in monkeys (Peterson et al., 1984). Each monkey had mean arterial blood pressure (MAP) monitored continuously. After a 3-month baseline period, half of the monkeys were exposed to a daily noise program that mimicked the sounds a human industrial worker would hear during the day. The control group was not exposed to noise. Figure 17.5 (adapted from Peterson et al., 1984) shows daily average MAP for the noise and control groups. There was an immediate (and lasting) increase in MAP for the monkeys exposed to noise. There was not a similar change in MAP for the control group animals.

A second example illustrates the evaluation of more than one intervention. The study evaluated whether modification of policies for prescription coverage would change prescription drug use by clients (Soumerai, Ceccarelli, & Koppel, 2017). The upper line in Figure 17.6 shows that the average number of prescriptions per patient was about 6 at the beginning of the study. After a 3-drug limit was placed on coverage, the mean number of prescriptions dropped to just under 3. Later this cap on number of prescriptions covered was lifted and a $1 copay per prescription was adopted; at that point, mean number of prescriptions increased, and continued to increase over time, until it reached about 4 prescriptions. (If additional observations were obtained, it seems possible that prescription use might have continued to increase.)

Time-series designs for experiments cannot get rid of all the threats to internal validity that are inherent in pretest–posttest studies of intervention, but they provide more information than just one before and after assessment.

17.5.3 Cycles in Time Series

In classic Box/Jenkins and econometric time-series modeling, the existence of cycles (such as annual cycles in sales) is sometimes seen as a nuisance, something that must be removed from data prior to further model fitting. In some other research domains, such as biology and physiology, the existence of cycles is of interest. The proportion of variance in time-series data that is associated with sinusoidal cycles (e.g., up-and-down variations that

Figure 17.5 Daily Average Mean Arterial Blood Pressure (MAP) Among Monkeys in Noise Versus Control Conditions

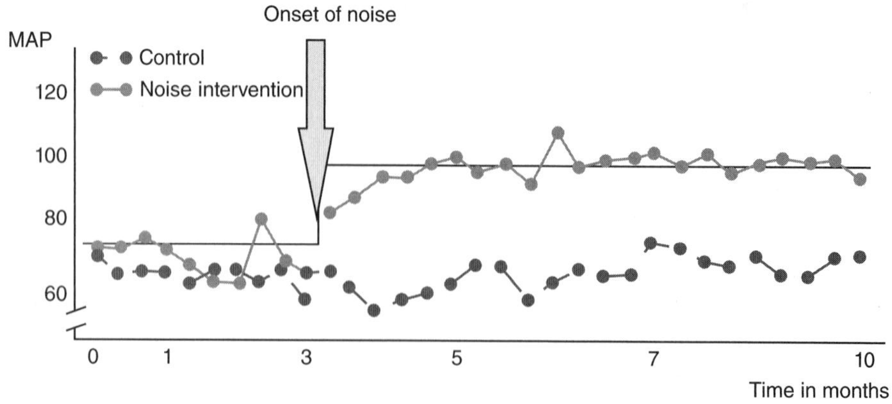

Source: Adapted from Peterson et al. (1984).

Figure 17.6 Mean Number of Prescriptions After Two Changes in Drug Coverage

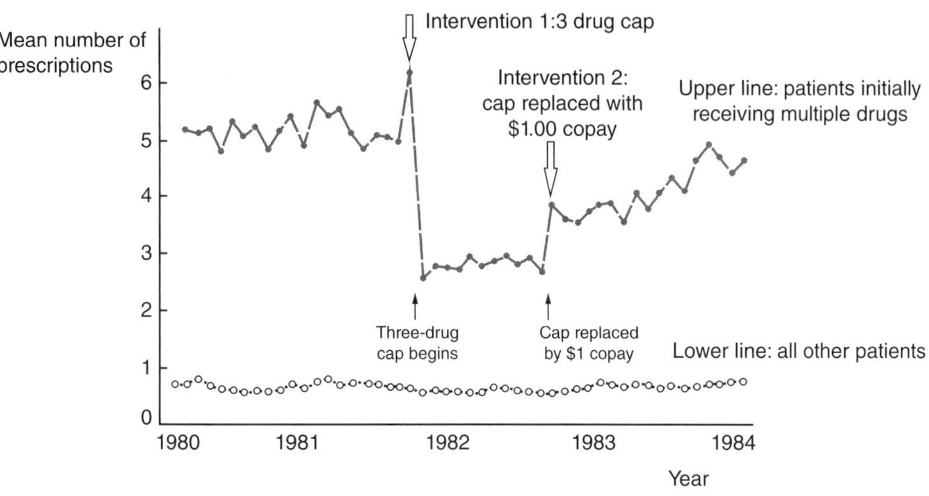

Source: Adapted from Figure 1 in Soumerai, S. B., Ceccarelli, R., & Koppel, R. (2017). Licensed under a Creative Commons Attribution 4.0 International License, http://creativecommons.org/licenses/by/4.0.

Figure 17.7 Weekly Cycles in Mood Over 12 Weeks

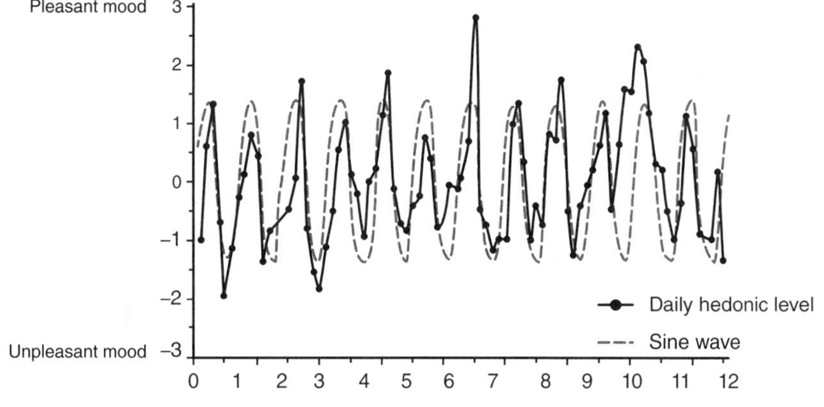

Source: Adapted from Larsen and Kasimatis (1990).

have the smooth shape of a sine or cosine function) can be assessed using a set of interrelated techniques that includes harmonic analysis, periodogram analysis, and spectral analyses. Periodogram and spectral analyses are based on the insight that a time series of length N can be represented by a sum of $N/2$ sinusoidal waveforms (Warner, 1998). If the proportion of variance for a time series (or frequencies) is uniform across all frequencies or cycle lengths, there is not a noticeable cyclic pattern in the time series; the time series is "white noise." However, if one or a few specific cycles fit the data much better than others, we can sometimes see this cyclicity or periodicity by eye, as in the following example.

Larsen and Kasimatis (1990) assessed daily mood for 84 days in a sample of college students. For this analysis, each data point is the average mood across all participants for each day. The solid and somewhat irregular line in Figure 17.7 shows the obtained daily moods.

The dotted line shows a fitted sine wave with a period (cycle length) of 7 days; this dotted line fits the observed time-series data well. What is going on? The students reported negative moods on Mondays, and then their moods tended to improve during the week; the most positive moods were reported on weekends. This shows that a psychological variable (mood) is related to a shared social and work schedule (the 7-day week).

Many biological or physiological variables are strongly cyclic, with the most noticeable variations having a 24-hour cycle length (circadian rhythms). Some variables also exhibit shorter cycles during the day and/or longer cycles over the period of a year. Cycles are involved in regulation of many processes.

17.5.4 Coordination or Interdependence Between Time Series

17.5.4.1 Shared Trends (Often Spurious)

Often researchers want to know whether two time series are related. One common source of spurious (misleading or silly) correlations between time series is a shared trend (often called a secular trend), as in the example in Figure 17.8, provided by Tyler Vigen (n.d.).

There is close correspondence between suicides by hanging and U.S. science spending. (If we look at enough different time series, and cherry-pick examples, it is easy to find many other variables that show similar trends, often upward trends, over time.) Of course, this does not imply that science spending causes suicide, or that suicide causes science spending. The reason for the high correlation is that both time series show an upward trend across time. Shared or similar trend, by itself, is not sufficient evidence for causal influence. To get rid of this problem, an analyst can remove trends from both time series and examine whether the residuals from trend are correlated.

17.5.4.2 Associations Between Cycles in Time Series

Organisms (including humans) wouldn't function or self-regulate well if each physiological process went off on its own and had no connection with other physiological processes.

Figure 17.8 Graph of Two Time Series: U.S. Spending on Science and Suicides

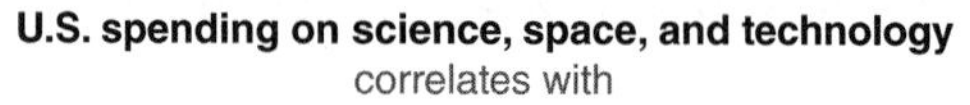

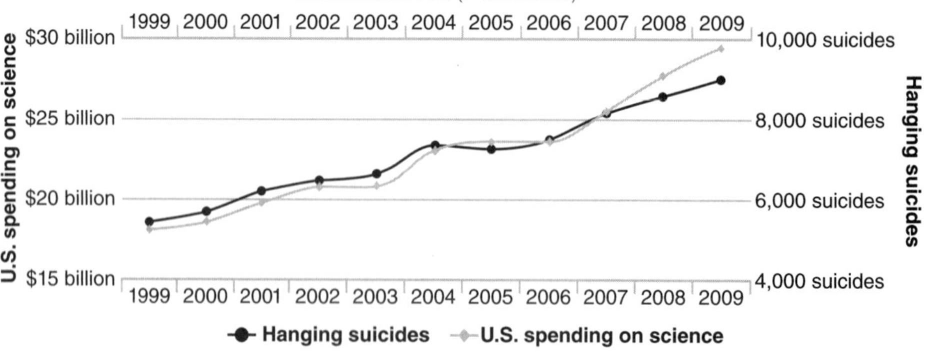

Source: Vigen, T. (n.d.) Spurious correlations. Retrieved from https://www.tylervigen.com/spurious-correlations. Licensed under a Creative Commons Attribution 4.0 International License, http://creativecommons.org/licenses/by/4.0.

Respiratory sinus arrhythmia (RSA) is a form of coupling or coordination between two physiological processes: respiration and heart rate. Figure 17.9 illustrates RSA. The solid smooth line represents the respiratory cycle; at the peak, the lungs are expanded; at the minimum, the lungs relax. This diagram shows about 2½ breaths. The jagged line shows the electrocardiographic trace; each sharp peak corresponds to a heart contraction. Notice that when the lungs are expanding, heartbeats are closer together (the heart beats more rapidly). When the lungs are contracted, the heartbeats are farther apart (the heart beats more slowly). The strength of the association between heart rate and respiration (RSA) varies across people and situations. Porges and his colleagues (Lewis, Furman, McCool, & Porges, 2011; Porges, 2011) suggested that stronger coordination or coupling between the cardiovascular and respiratory systems indicates better function.

Researchers have also examined synchronized cycles in social interaction, for example, mother–infant interaction (Lester, Hoffman, & Brazelton, 1985). Social coordination can take other forms than shared cycles; time-series models and Markov-chain analysis have been used to assess the strength of contingency in dyads. Many different terms are used to describe the association between partner behaviors over time, including *synchrony*, *entrainment*, *coordination*, and *coordinated interpersonal timing*. For further discussion of applications of time-series analysis for social interaction, see Warner, 1979, 1988, 1991, 1992, 1996, 1998, 2002a, 2002b; Warner, Malloy, Schneider, Knoth, & Wilder, 1987; Warner & Mooney, 1988; Warner, Waggener, & Kronauer, 1983. The strength and nature of coordination may be related to relationship quality, evaluation of social interaction, and later life outcomes such as attachment.

17.5.4.3 Correlated Residuals From Trends and Cycles

Shared trends in time series (and shared cycles, such as shared seasonal cycles) usually don't provide strong evidence that two time-series variables are "really" or directly related. Often, data analysts examine how the residuals from time series X (after removing trends, cycles, and any other sources of serial dependence) are correlated with residuals from time series Y (after removing trends, cycles, and any other sources of serial dependence). Y may be predicted from X at some time lag.

Figure 17.9 Respiratory Sinus Arrhythmia: Coupling Between Heart Rate and Respiration

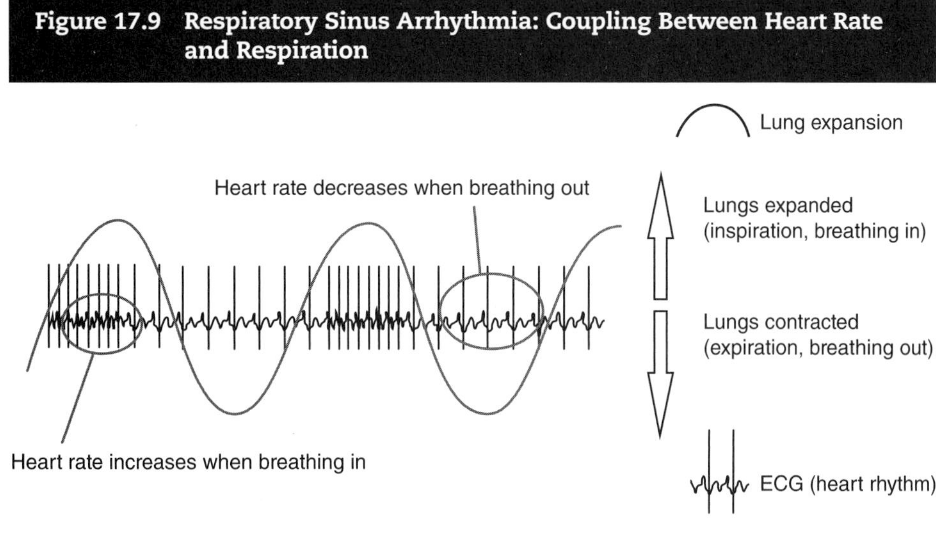

Source: Adapted from https://support.mindwaretech.com/2017/09/all-about-hrv-part-4-respiratory-sinus-arrhythmia.

Note: ECG = electrocardiogram.

17.5.4.4 Sampling Frequency and Selection of Time Unit

The sampling frequency (e.g., whether each observation correspond to a second, an hour, a day, a week, a month, a year, or some other time unit) is a crucial part of time-series design. These are analogous to magnification settings on a microscope: At different sampling frequencies, different structures may appear.

17.6 POISSON AND BINOMIAL REGRESSION FOR ZERO-INFLATED COUNT DATA

Sometimes the outcome variable for a regression analysis represents a count of the number of times a somewhat uncommon behavior has occurred for each case. For example, we might count the number of times each person in a sample has used a drug such as marijuana. For some behavior counts, the most frequent outcome is zero. This can result in a distribution like the one in Figure 17.10.

There are two modes in this distribution; more than 30% of persons in the sample reported that they never used marijuana; among those who did use marijuana, the number of occasions varied, with many persons reporting use nine or more times a month. It is obvious that this distribution is non-normal in shape. There are forms of regression based on zero-inflated Poisson or binomial distributions. *Zero-inflated* just means that there is a large mode at zero; in the hypothetical data in Figure 17.10, more than 30% of persons reported using marijuana zero times.

Johnson et al. (2011) examined a similar question about days of crack use; they asked two separate questions:

- Which variable, or variables (if any) can predict which persons never used crack (a score of 0) and which used crack 1 or more days?

Figure 17.10 Hypothetical Distribution: Number of Days Marijuana Used in Past Month

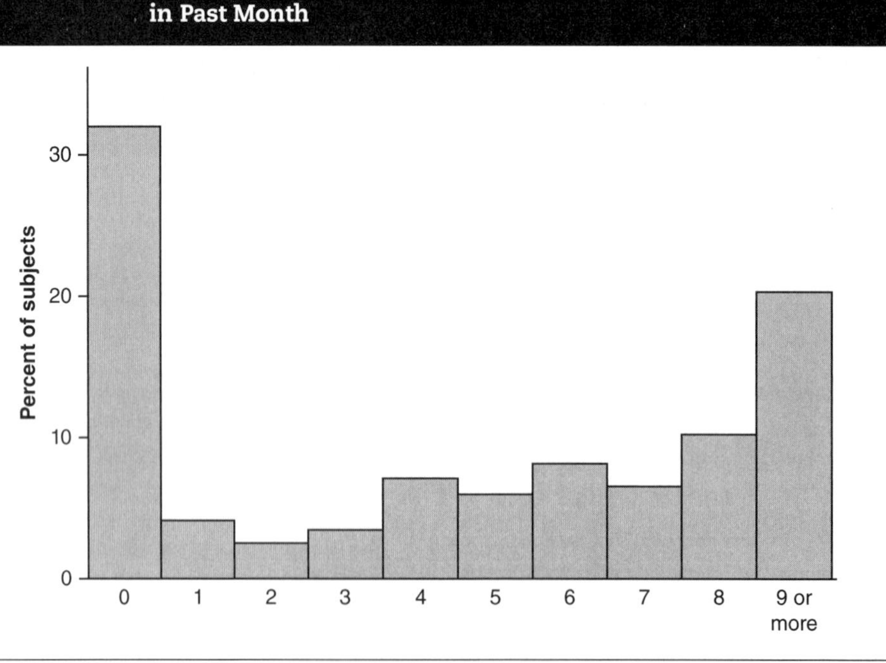

Source: Adapted from Wagner, Riggs, and Mikulich-Gilbertson (2015).

- Which variable or variables (if any) can predict the number of times that the drug takers used crack (excluding those who never used crack)?

They reported that diagnosis of major depressive disorder at the beginning of the study predicted whether women would use crack or not but did not predict the number of days of crack use among users.

Wagner, Riggs, and Mikulich-Gilbertson (2015) explained that the challenge in analysis of this type of data is choosing the correct model for regression (e.g., zero-inflated or not, Poisson vs. binomial). The generalized linear model procedure in SPSS can perform Poisson regression and provide tests whether assumptions for use of these models are satisfied; see tutorials at https://statistics.laerd.com/spss-tutorials/poisson-regression-using-spss-statistics.php.

17.7 BAYES' THEOREM

Bayesian logic begins with the idea that a prior estimate of probability of an event (such as probability that a patient has a disease) can be revised or updated as new evidence becomes available. Conversely, as new evidence becomes available, we should continue to take prior evidence into account. Bayes' theorem provides a simple introduction to this way of thinking.

An example of a question of interest is, How likely is it that a patient has a specific disease? We will call the disease A. The probabilities that follow are hypothetical. Two kinds of information are relevant. It will not surprise you that one piece of relevant information is a lab test result. If a test result indicates the possible presence of Disease A, that evidence will lead us to think the patient may have Disease A. However, lab tests are susceptible to error.

- A false-positive result occurs if a test result indicates Disease A, but the patient does not have Disease A.
- A false-negative result occurs if a test result does not indicate Disease A, but the patient does have Disease A.

It may surprise you to learn that a second piece of relevant information is the base rate. The base rate is the percentage of patients who come in for tests who have Disease A.

Bayes' theorem shows how this information (test result, knowledge of false-positive and false-negative error rates, and base rate) can be combined. The prior probability that a patient has Disease A (base rate) is adjusted after new information (the test result) becomes available. In everyday thinking, people tend to ignore base rate information or, at least, not give it sufficient weight.

Here is a hypothetical numerical example (numerical values are from Lowry, 2019).

- The disease is rare (it occurs in only 0.5% of the general population). We can say that the base rate or prevalence of prior probability of the disease is .005.
- A clinical test is 99% effective in detecting the presence of this disease. When the disease is present, the test yields an accurate positive result in 99% of cases.
- The test yields false-positive results in 5% of the cases where the disease is not present. A false-positive result occurs when the test reports that a person has the disease when the person does not have the disease.

Consider a patient who has a positive test result (e.g., a test result indicating that the patient has Disease A). How likely is it that the patient really has the disease? Pause and try to answer this question.

Formally, what we want to know is, How likely is it that the patient has the disease (A) if that patient has a positive test result for that disease (B)? This conditional probability is written $\Pr(A \mid B)$. To find $\Pr(A \mid B)$ we need to transform the information included in the description of the situation above into a set of conditional and unconditional probabilities. Lowry (2019) provides details.

The outcome of interest is formally written $\Pr(A \mid B)$, that is, the conditional probability that the disease is present if the test result is positive. In simpler language, if the test result indicates disease, how probable is it that the disease is present?

Bayes' theorem is stated formally as follows:

$$\Pr(A \mid B) = [\Pr(B \mid A) \times \Pr(A)]/\Pr(B). \quad (17.4)$$

In words, the terms in this equation are as follows: On the left side of the equation, we have $\Pr(A \mid B)$. This is the conditional probability that the patient has the disease given that the patient has a positive test result (a result indicating that disease may be present). The three pieces of information listed in Table 17.3 are needed to use Equation 17.4 to obtain the conditional probability we need.

Substituting those values into Equation 17.4 for Bayes' theorem, we find:

$$\Pr(A \mid B) = (.99 \times .005)/.0547 = .0905.$$

In words, the probability that the disease is present (A) given that the test result is positive (B) is .0905. Compare this with your earlier guess. If your earlier guess was much higher than 9.05% (and I'm betting that it was), then you did not give prior evidence (the very low base rate of the disease) enough weight when you combined new evidence (the test result) with prior knowledge. When a disease is extremely rare, false-positive and false-negative rates for tests need to be very low to make accurate diagnoses.

The methods you have learned in this book are often called "frequentist" (non-Bayesian) because they focus on information in a current sample to answer questions. Bayes' theorem and other Bayesian analyses update probabilities that a hypothesis is correct as more evidence or information becomes available. For further discussion of Bayesian methods, see Kruschke and Liddell (2018).

17.8 MULTILEVEL MODELING

Multilevel modeling (MLM), sometimes called hierarchical modeling, is useful in situations where units or cases have a nested model structure. Nested structures arise quite often in research in institutional settings. In a hypothetical study of differences between nations in child

Table 17.3 The Probabilities Used to Compute the Posterior Probability That Disease Is Present, Given a Positive Test Result

$\Pr(B \mid A) = .99$	This is the conditional probability of obtaining a positive test result when the disease is present.
$\Pr(A) = .005$	This unconditional probability represents the base rate or probability of the disease. It is the prior probability of the disease (i.e., before test results are known).
$\Pr(B) = .0547$	This is the unconditional probability of a positive test result (a result that says the disease is present, whether the disease is present or not).

achievement, each individual child can be nested in (a member of) one classroom. Each classroom can be nested in (part of) one school. Each school can be nested in (located in) one nation. (If we have a series of measures of achievement for each child at several points in time, these time-point measures are nested within child.)

The basic idea in MLM is that we need to include information at all these levels (e.g., individual, classroom, school, and nation) in the analysis. When models for MLM analysis are written in the form of regression-like equations, it can appear complex. However, a few hypothetical examples (using a few levels of nesting) may help you get a sense how this kind of analysis works.

Suppose you want to know how well cognitive behavioral therapy (CBT) works to reduce patient anxiety, compared to patient-centered therapy (PCT). If the study includes only one therapist to administer each type of treatment, there is a complete confound between abilities and other characteristics of individual therapist and type of therapy. A better study would include several therapists for each type of therapy and multiple clients for each therapist. This would take into account that changes in anxiety also depend on therapist ability and patient responsiveness. This could be set up as a nested design, as shown in Figure 17.11. The numbers of therapists and patients for each therapist could be larger, of course.

The analysis could examine whether CBT yields lower anxiety than PCT (or vice versa), taking into account that therapists differ in ability and that patients differ in response to therapy. Anxiety could be measured for each patient, perhaps at multiple points in time (not shown in Figure 17.11). If each patient is assessed for anxiety at multiple points in time, such as before, during, and after therapy, the design includes a repeated-measures component. The analysis can examine the growth curve or change over time for individual patients to see how that is related to therapist and type of therapy. For example, most of Dr. Q's patients might tend to show steep decreases in anxiety across time, while most of Dr. M's patients might show little change in anxiety across time.

There are many good sources for further discussion of MLM, including Adelson and Owen (2012); Grimm and Ram (2016); Heck, Thomas, and Tabata (2013); Humphrey and LeBreton (2018); Kahn (2011); Robson and Pevalin (2015); and Snijders and Bosker (2011).

17.9 SOME FINAL WORDS

It is tempting to form strong beliefs about relationships between variables when you have done a sophisticated analysis, particularly when sample sizes are extremely large. There are many reasons why this temptation should be resisted.

Figure 17.11 Patients (P1, P2, etc.) Nested Within Therapists (Dr. Q, etc.) and Therapists (Dr. Q, etc.) Nested Within Types of Therapy (CBT, PTC)

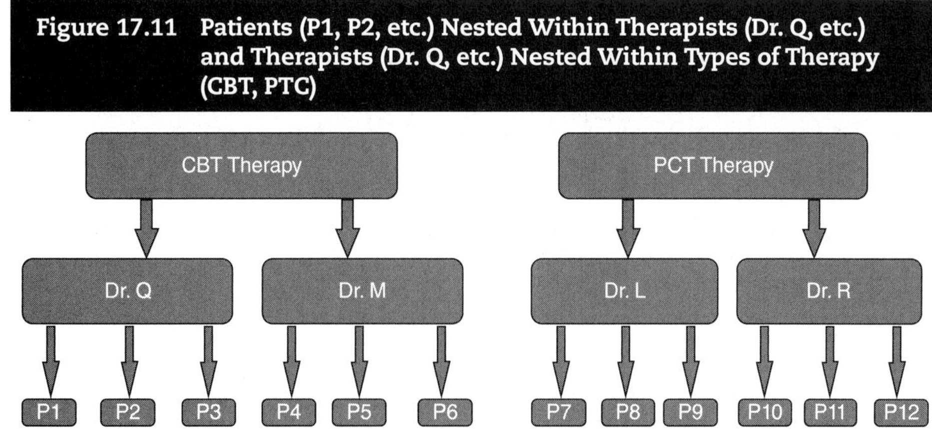

The results of an analysis depend on what you put in (and, equally important, on what you leave out). "What you put in" refers to the selection of variables but also to the selection of participants, interventions, situations, time periods, and all other components of research design; it includes decisions made during data screening and data analysis. We can describe all those things collectively as "context." In many fields, such as psychology and medicine, results can differ depending on context.

During the research process, we make our way through a "garden of forking paths," in the words of Gelman and Loken (2013). We can end up in different places depending on the choices we make. For example, the outcomes for some analyses depend on what covariates are (and are not) included in final analyses. It is important that we document, report, and justify our choices (about everything from omitting outliers to selection of covariates to include in the final analysis).

We need to be mindful that small p values reported in statistical significance tests may not indicate strong treatment effects or strong predictive relationship. We should avoid language such as "highly significant" (when p is less than .001, for example) because this can lead people to assume that statistical significance implies practical or clinical importance. A p value is not information about effect size.

The goal of statistical analysis should not be to obtain the largest possible multiple R or the smallest possible p value. The goal should be to obtain good quality evidence and to report research in ways that makes it possible for readers to understand the limitations of that evidence. For some questions in science, we don't have enough information yet to draw conclusions.

In the beginning of Volume I (Warner, 2020), I suggested that it might be better to think in terms of degrees of belief, instead of a dichotomy between belief and disbelief. The philosopher David Hume said that "a wise [person]. . . proportions his [or her] belief to the evidence" (Schmidt, 2004).

DIGITAL RESOURCES

Find **free study tools** to support your learning, including **eFlashcards, data sets, and web resources,** on the accompanying website at **edge.sagepub.com/warner3e**.

GLOSSARY

–2 log likelihood (–2LL): In binary logistic regression, –2LL is analogous to the sum of squared residuals in a multiple linear regression; the larger the absolute value of –2LL, the worse the agreement between predicted probabilities and actual group membership. LL values are computed to compare fits for different binary logistic regression models. Multiplying an LL value by –2 converts it to a chi-square distributed variable. Critical values of the chi-square distribution can be used to assess the statistical significance of one model or the significance of change in goodness of fit when variables are added to a model. See also *log likelihood (LL)*.

a: In a mediated causal model, an *a* coefficient corresponds to the path from X_1 (the initial cause) to *M* (the hypothesized mediating variable). For purposes of significance tests, the raw-score regression coefficient is used; in diagrams to represent the path model, the standardized version (beta coefficient) usually appears.

Adjusted means: In an analysis of covariance, adjusted *Y* means are created by subtracting the part of the scores that are related to or predictable from the covariates; this adjustment may correct for any confound of the covariates with treatment and/or suppress some of the error variance or noise and increase the statistical power of tests of group differences.

Aroian test: A *z* test for H_0: $ab = 0$ that uses a different method to estimate SE_{ab} from the one proposed by Sobel (1982).

Autoregressive model: A model for serial dependence in a time series such that *X* at Time *t* is predictable from *X* at Time $t - 1$ and possibly *X* at earlier times such as $t - 2$ and $t - 3$. It is called "auto" because values of *X* are predicted from values of *X*.

B: The coefficient for a predictor variable (*X*) in a logistic regression model. For a one-unit increase in *X*, there is a *B*-unit change in the log odds ratio. The value of *B* is not easy to interpret directly; instead, *B* is used to calculate other values that are more directly interpretable, such as the odds ratio and the probability of membership in each group.

b: In a mediated causal model, the *b* coefficient corresponds to the path from *M* (the mediating variable) to *Y* (the final outcome variable). For purposes of significance tests, the raw-score regression coefficient is used; in diagrams to represent the path model, the standardized version (beta coefficient) usually appears.

Backward method of entry: A form of statistical multiple regression that begins with all candidate predictor variables included in the equation; in each step, the predictor variable for which the removal leads to the smallest decrease in the R^2 for the model is dropped. The process of variable deletion ends when dropping a variable would result in a significant decrease in the total R^2.

Bentler comparative fit index (CFI): A goodness-of-fit index based on information that compares model fit for the specified model with model fit for the null model the (which represents an assumption that none of the variables are related). The CFI is downwardly adjusted as the number of free parameters is increased. It has a range from 0 to 1.00; values > .95 are generally viewed as indicating adequate model fit.

Bootstrapping: An empirical method for estimation of standard errors and confidence intervals, used in situations where there is no simple formula to calculate these, or where violations of assumptions (such as multivariate normality) make estimating these problematic. It is a resampling method. In effect, the program draws many samples from the data set (if number of bootstraps = 2,000, then it draws 2,000 samples) and calculates all the model parameters for each sample. An empirical distribution of the 2,000 estimates can then be set up to summarize how each parameter (such as the $a \times b$ product for an indirect path in the mediation analysis example) varies across 2,000 samples. This empirical sampling distribution is used to estimate *SE* and confidence intervals.

***C* (level of confidence):** When setting up a confidence interval, this is usually arbitrarily set at 95% or 90%. If all assumptions for use of confidence intervals are correct, then in the long run, if we set up thousands of confidence intervals using samples from the same population, *C*% of the confidence intervals are expected to contain μ; (1 – *C*%) are expected not to contain μ.

c: In a mediation analysis, *c* represents the "total" relationship between the initial cause X_1 and the outcome *Y*.

c′: In a mediation analysis, *c*′ represents the direct path from X_1 to *Y*, when the indirect path from X_1 to *Y* (via the mediating variable *M*) is controlled for in the analysis.

Canonical correlation (r_c): The correlation between scores on a discriminant function and group membership. This is analogous to an η coefficient in an ANOVA. The term *canonical* implies that there are multiple variables on one side (or both sides) of the model. The canonical correlation, r_c, for each discriminant function is related to the eigenvalue for each function.

Causal model: A causal model is often represented by a path diagram; unidirectional arrows represent hypotheses that variables are causally related, and bidirectional arrows represent hypotheses that variables are not causally associated. These models are sometimes used to help data analysts decide what partial correlation and regression analyses to perform on data obtained from nonexperimental research. The term *causal model* is potentially misleading; we cannot make causal inferences about variables on the basis of nonexperimental data. A causal model is purely theoretical; it represents *hypotheses* about the ways in which two or more variables may be related to one another, and these hypothesized relations may include both causal and noncausal associations.

Causal path: In a causal model, a unidirectional arrow that points from X toward Y denotes the theoretical existence of a causal connection in which X causes or influences Y ($X \rightarrow Y$).

Centering: Scores on a predictor variable are centered by subtracting the sample mean from the scores on each variable. When both predictors are quantitative, scores on both X_1 and X_2 should be centered before creating a product term, $X_1 \times X_2$, to represent an X_1-by-X_2 interaction. (That is, X_1–centered = X_1 – sample mean of X_1.) Centering reduces the correlation between the product term (X_1 and X_2) and its components, X_1 and X_2; centering does not change the statistical significance or effect size for the interaction, but it can make X_1 and X_2 appear to be stronger predictors of Y.

Centroid: A multivariate mean; in discriminant analysis, the centroid of a group corresponds to the mean values of the scores on D_1, D_2, and any other discriminant functions. In a graph, the centroid is the midpoint of the cluster of data points in the space defined by axes that represent the discriminant functions.

Change score: In a simple pretest–posttest design in which the same behavior or attitude is measured twice, a change score (sometimes denoted d for "difference") is computed by subtracting the pretest score from the posttest score: $d = Y_{post} - Y_{pre}$. The null hypothesis of interest is either H_0: $d = 0$ (for studies of a single group) or H_0: $d_1 = d_2 = \cdots = d_k$ (for comparisons of treatment impact across k groups).

Chi-square goodness-of-fit index: One of several indexes of fit that describes how well the overall model paths and coefficients reproduce the variance/covariance information in the data. A large χ^2 indicates poor model fit. For large-N samples, χ^2 tends to be statistically significant even for reasonably good fit.

Chi-square test: For the overall logistic regression model, the chi-square test tests the null hypothesis that the overall model including all predictor variables is not predictive of group membership. Typically, the value of chi-square is obtained by taking the difference between the –2LL value for the model that includes all predictor variables of interest and the –2LL value for a model that includes only a constant. Using binary logistic regression, a data analyst usually hopes for a large χ^2 as evidence that the predictor variables are useful in predicting likelihood of group membership.

Classification errors: When the predicted group membership for a case (on the basis of the values of discriminant function scores or a binary logistic regression) is not the same as the actual group membership for that case, this is referred to as a classification error.

Classification table: To assess how well a logistic regression or a discriminant analysis predicts group membership, a contingency table can be set up to show how well or how poorly the predicted group memberships correspond to the actual group memberships.

Cohen's *d*: A unit-free effect size that is not related to N, sample size. It describes a difference between means in numbers of standard deviations. Cohen's $d = (M - \mu_{hyp})/SD$ for the one-sample t test. Cohen's $d = (M_1 - M_2)/SD_{pooled}$ for the independent-samples t test.

Cohen's kappa: A statistic that assesses the degree of agreement in the assignment of categories made by two judges or observers (correcting for chance levels of agreement).

Communality: A communality, sometimes denoted h^2, is an estimate of the proportion of variance in each of the original p variables that is reproduced by a set of retained components or factors. Communality is obtained for each variable by summing and squaring the correlations (or loadings) for that variable across all the retained factors.

Confirmation bias in publication: The tendency for studies with results that are not statistically significant to remain unpublished.

Confirmatory factor analysis (CFA): A model that represents indicator variables with paths (loadings) that relate them to latent variables (factors). Unlike exploratory factor analysis, a confirmatory factor analysis model is usually highly restricted. For example, the number of factors is fixed, and often each indicator variable is related to only one latent variable.

Consolidated Standards of Reporting Trials (CONSORT): These standards include proposals developed by the CONSORT Group to improve reporting of results from clinical trials, particularly when there is participant attrition across stages of the study (see http://www.consort-statement.org).

Constant-only model: See *null model*.

Constraints on SEM model parameters: There are many ways model parameters can be constrained. One type of constraint is created when there is no direct path between two variables. (The absence of a path is equivalent to constraining the path coefficient between these variables to be 0.) Another constraint occurs when a fixed numerical value is specified ahead of time for a path coefficient. Another occurs when two or more path coefficients are set to equal each other.

Construct validity: The degree to which an X variable really measures the construct that it is supposed to measure. In practice, construct validity can be assessed by examining whether scores on X have the sizes and signs of correlations with other variables (which may include experimental manipulations) that they should have, according to theories about the nature of the variables and their interrelationships.

Content validity: The degree to which the content of questions in a self-report measure covers the entire domain of material that should be included (based on theory or assessments by experts).

Control variable: See *covariate*.

Convergent validity: The degree to which a new measure, X', correlates with an existing measure, X, that is supposed to measure the same construct.

Correctly specified model: A model is correctly specified if it includes all the variables that should be taken into account (e.g., all variables that are confounded with or causally related to X_1 or Y, or all variables that interact with X_1 or Y) and, also, if it does not include any variable that should not be included. A well-developed theory can be helpful in deciding what variables need to be included in a model; however, we can never be certain that the models we use to try to explain relations among variables in nonexperimental studies are correctly specified. Thus, we can never be certain that we have taken all potential rival explanatory variables into account; also, we can never be certain that our statistical analysis provides accurate information about the nature of the relation between X_1 and Y.

Correlated-samples *t* test: See *paired-samples* t *test*.

Covariate: In analysis of covariance or a multiple regression, a covariate is a variable that is included so that its association with the Y outcome and its confound with other predictors can be statistically controlled when assessing the predictive importance of other variables that are of greater interest. Covariates are often quantitative variables (but may be dummy variables).

Cox and Snell's R^2: This is one of the pseudo-R^2 values provided by SPSS as an overall index of the strength of prediction for a binary logistic regression model.

Cronbach's alpha (α): An index of internal consistency reliability that assesses the degree to which responses are consistent across a set of multiple measures of the same construct, usually self-report items.

Data-driven regression: In data-driven regression, variables are entered on the basis of statistical criteria (e.g., in each step, the variable that increases R^2 the most is entered, only if its F ratio exceeds a preset criterion).

Determinant of a matrix: The determinant of a sum of cross products, or correlation matrix, is a single-number summary of the variance for a set of variables when intercorrelations among variables are taken into account.

Diagonal matrix: A matrix in which all the off-diagonal elements are 0.

Difference score: See *change score*.

Dimension reduction analysis: When a discriminant analysis yields several discriminant functions, the data analyst may think of each discriminant analysis as a "dimension" along which groups differ. The discriminant functions are rank ordered in terms of their strength of association with group membership, and sometimes, only the first one or two discriminant functions provide useful information about the prediction of group membership. When a researcher looks at significance tests for sets of discriminant functions and makes the judgment to drop some of the weaker discriminant functions from the interpretation and retain only the first few discriminant functions, it is called a dimension reduction analysis. The descriptions of differences among groups are then described only in terms of their differences on the dimensions that correspond to the retained discriminant functions.

Dimensionality: The dimensionality of a set of X variables involves the following question: How many different constructs do these X variables seem to measure? The most common way of assessing dimensionality is to do factor analysis or principal components analysis and to decide how many factors or components are needed to reproduce the correlation matrix **R** adequately. If these factors or components are meaningful, they may be interpreted as dimensions, latent variables, or measures of different constructs (such as different types of mental ability or different personality traits).

Direct-difference *t* test: See *paired-samples* t *test*.

Direct effect: This is denoted c' in a mediated path model; it represents any additional predictive association between X_1 and Y, after controlling for or taking into account the mediated or indirect path from X_1 to Y via the mediating variable (M or X_2).

Discriminant function: A weighted linear combination of scores on discriminating variables; the weights are calculated so that the discriminant function has the maximum possible between-groups variance and the minimum possible within-groups variance. For convenience, discriminant function scores are generally scaled so that, like z scores, they have variance equal to 1.0.

Discriminant function coefficients: These are the weights given to individual predictor variables when we compute scores on discriminant functions. Like regression coefficients, these discriminant function coefficients can be obtained for either a raw score or standard (z) score version of the model; the coefficients that are applied to standardized scores are generally more interpretable.

Discriminant validity: When our theories tell us that a measure X should be unrelated to other variables such as Y, a correlation of near 0 is taken as evidence of discriminant validity, that is, evidence that X does not measure things it should not be measuring.

Disordinal: In a disordinal interaction, the lines for a graph of group means cross. This tells us that the rank order of A means differ across levels of the B factor.

Dummy-coded dummy variable: Dummy coding involves treating each dummy variable as a yes/no question about group membership; a "yes" response is coded 1 and a "no" response is coded 0. In dummy coding of dummy variables, members of the last group receive codes of 0 on all the dummy variables.

Effect: In analysis of variance, this term has a specific technical meaning. The "effect" of a particular treatment, the treatment given to Group i, is estimated by the mean score on the outcome variable for that group (M_i) minus the grand mean (M_Y). The effect for Group i is usually denoted by α_i; this should not be confused with other uses of alpha (alpha is used elsewhere to refer to the risk for Type I error in significance test procedures and to Cronbach's alpha reliability coefficient). The "effect" α_i represents the "impact" of the treatment or participant characteristics for Group i. When $\alpha_1 = +9.5$, for example, this tells us that the mean for Group 1 was 9.5 points higher than the grand mean. In the context of a well-controlled experiment, this may be interpreted as the magnitude of treatment effect. In studies that involve comparisons of naturally occurring groups, the effect term is not interpreted causally but rather as descriptive information about the difference between each group mean and the grand mean.

Effect-coded dummy variable: Effect coding involves treating each dummy variable as a yes/no question about group membership; a "yes" response is coded 1, and a "no" response is coded 0. However, the last group (the group that does not correspond to a yes answer on any of the effect-coded dummy variables) receives a code of –1 on all the dummy variables.

Eigenvalue (λ): In the context of discriminant analysis, an eigenvalue is a value associated with each discriminant function; for each discriminant function, D_i, its corresponding eigenvalue, λ_i, equals $SS_{between}/SS_{within}$. When discriminant functions are being calculated, the goal is to find functions that have the maximum possible values of λ, that is, discriminant functions that have the maximum possible between-group variance and the minimum possible within-group variance. In the context of factor analysis or principal-components analysis, an eigenvalue is a constant value that is associated with one of the factors in a factor analysis (or one of the components in a principal-components analysis). The eigenvalue is equal to the sum of the squared loadings on a factor (or on a component). When the eigenvalue for a factor or component is divided by k, where k is the number of variables included in the factor analysis, it yields the proportion of the total variance in the data that is accounted for or reproducible from the associated factor or component.

Empirical keying: A method of scale construction in which items are selected for inclusion in the scale because they have high correlations with the criterion of interest (e.g., on the Minnesota Multiphasic Personality Inventory, items for the depression scale were selected because they predicted a clinical diagnosis of depression). Items selected in this way may not have face validity; that is, it may not be apparent to the test taker that the item has anything to do with depression.

Equivalent models: Models are equivalent when they have identical goodness of fit but include paths that correspond to different hypotheses about which variables are causally related, or direction of causality.

Error variance: The proportion of variance in scores for a dependent variable that is not predictable from independent variables included in the analysis.

Error variance suppression: When we obtain a substantially smaller $SS_{residual}$ term by including a covariate in our analysis (as in Equation 8.1), we can say that the covariate X_c acts as a "noise suppressor" or an "error variance suppressor."

exp(B): The exponential of B, which can also be represented as e^B. This is used in binary logistic regression.

Experimental control: Researchers may use a variety of strategies to "control" variables other than the manipulated independent variable, which could influence their results. For example, consider the possible influence of gender as a third variable in the study of the effects of caffeine on anxiety. Women and men may differ in baseline anxiety, and they may differ in the way they respond to caffeine. To get rid of or control for gender effects, a researcher could hold gender constant, that is, conduct the study only on men (or only on women). Limiting the participants to just one gender has a major drawback: It limits the generalizability of the findings (the results of a study run only with male participants may not be generalizable to women). The researcher might also include gender as a variable in the study; for example, the researcher might assign 10 women and 10 men to each group in the study (the caffeine group and the no caffeine group).

Exploratory factor analysis: An exploratory factor analysis initially estimates k loadings for k measured variables on k factors. In later steps, a reduced number of factors are examined to assess how well they can reconstruct the correlation matrix among the variables. In contrast to confirmatory factor analysis, exploratory factor analysis does not set constraints a priori on the number of factors or which variables load on which factors. (SPSS permits the user to specify the number of factors to be retained.)

Exponential function: The exponential function applied to a specific numerical value such as a is simply e^a (i.e., the exponential of a is given by the mathematical constant e raised to the ath power). It is sometimes written as exp(a).

Extraction: This is a computation that involves estimating a set of factor or component loadings; loadings are correlations between actual measured variables and latent factors or components.

Face validity: The degree to which it is obvious what attitudes or abilities a test measures from the content of the questions posed. High face validity may be desirable in some contexts (when respondents need to feel that the survey is relevant to their concerns), but low face validity may be preferable in other situations (when it is desirable to keep respondents from knowing what is being measured to avoid faking, social desirability bias, or deception).

Factor: In the context of factor analysis (Chapter 12), a factor is a latent or an imaginary variable that can be used to reconstruct the observed correlations among measured X variables. By looking at the subset of X variables that have the highest correlations with a factor, the researcher decides what underlying dimension or construct might correspond to that factor. For example, if a set of mental ability test scores are factor analyzed, and the first factor has high positive loadings (positive correlations) with the scores on the variables vocabulary, reading comprehension, analogies, and anagrams, the researcher might decide to label this factor "verbal ability."

Factor: In the context of analysis of variance, a categorical predictor variable is usually called a factor. In an experiment, the levels of a factor typically correspond to different types of treatment, or different dosage levels of the same treatment, administered to participants by the researcher. In nonexperimental studies, the levels of a factor can correspond to different naturally occurring groups, such as political party or religious affiliation.

Factor score: A score for each individual participant that is calculated by applying the factor score coefficients for each factor to z scores on all items.

Factor score coefficients: Coefficients that are used to construct scores on each factor for each participant from z scores on each individual X variable. These are analogous to β coefficients in a multiple regression.

Faking: Sometimes test takers make an effort to try to obtain either a high or low score on an instrument by giving answers that they believe will lead to high (or low) scores; informally, this can be called faking good (or faking bad).

False positive: A study finding is "false positive" if the null hypothesis H_0 is rejected when H_0 is true.

First-order partial correlation: In a first-order partial correlation, just one other variable is statistically controlled when the relation between X_1 and Y is assessed.

Forest plot: A plot that summarizes results of meta-analysis by showing a confidence interval for the effect size of each individual study, along with the confidence interval for averaged effect size.

Forward method of entry: In this method of statistical regression, the analysis begins with none of the candidate predictor variables included; at each step, the predictor that contributes the largest R^2 increment is entered. The addition of predictor variables stops when we arrive at the situation where none of the remaining predictors would cause a significant increase in R^2 (the criterion for significance to enter may be either $p < .05$ or a user-specified F-to-enter).

F-to-enter: In the forward method of statistical regression, the minimum value of F that a variable must have for its R^2 increment before it is entered into the equations.

Fully identified model: Model identification is difficult to assess completely (Kenny & Milan, 2012). One part of this involves comparison of the number of parameters to be estimated versus the number of distinct sample moments. A fully identified structural equation model has $df = 0$; the variance/covariance matrix can be perfectly reconstructed from the data. Multiple regression is an example of a fully identified model.

Gain score: See *change score*.

gcr: See *Roy's largest root*.

Goodman test: A z test for H_0: $ab = 0$ that uses a different method to estimate SE_{ab} from the one proposed by Sobel (1982).

h^2: See *communality*.

Homogeneity of regression assumption: In analysis of covariance, we assume no treatment-by-covariate interactions; this is also called the homogeneity of regression assumption. (The raw score slope b to predict Y from X_c is assumed to be equal or homogeneous across all groups in the analysis of covariance.)

Homogeneity of variance/covariance matrices: Just as univariate analysis of variance assumes equality of the population variances of the dependent variable across groups, discriminant analysis assumes homogeneity of Σ, the population variance/covariance matrix, across groups.

Hotelling's T^2: Just as multivariate analysis of variance is a multivariate generalization of analysis of variance (i.e., multivariate analysis of variance compares vectors of means across multiple groups), Hotelling's T^2 is a multivariate generalization of the t test (Hotelling's T^2 compares vectors of means on p variables across two groups). Note, however, that Hotelling's T^2 is not the same statistic as Hotelling's trace, which is part of the SPSS general linear model output.

Hotelling's trace: One of the multivariate test statistics provided by the SPSS GLM procedure to test the null hypothesis H_0: $\mu_1 = \mu_2 = \cdots = \mu_k$. Hotelling's trace is the sum of the eigenvalues across all the discriminant functions; a larger value of Hotelling's trace indicates larger between-group differences on mean vectors.

Imputation of missing values: This refers to a systematic method of replacing missing scores with reasonable estimates. Estimated scores for missing values are obtained by making use of relationships among variables and a person's scores on nonmissing variables. This can be quite simple (e.g., replace all missing values of blood pressure with the mean blood pressure across all subjects) or complex (e.g., calculate a different estimated blood pressure for each subject based on that subject's scores on other variables that are predictively related to blood pressure).

Inadmissible solution: An analysis that has one or more implied correlations greater than zero in absolute value and/or negative error variance estimates. Even if model fit is otherwise acceptable, this result is not acceptable.

Inconsistent mediation: This occurs when one of the signs for a path in a mediate model is negative; the direct (c') and indirect ($a \times b$) effects of X_1 on Y are opposite in sign.

Increment in R^2: See R^2_{inc}.

Incremental sr^2: In a sequential or statistical regression, the additional proportion of variance explained by each predictor variable at the step when it first enters the analysis. When only one variable enters at a step, sr^2_{inc} is equivalent to the R^2 increment, or R^2_{inc}. SPSS can provide the sr_i value (the semipartial correlation of X_i with Y, controlling for all other predictors in the equation) in regression; however, SPSS calls this a "part" correlation (and you must square it by hand to obtain sr^2).

Independence model: In structural equation modeling, a model that has no paths among any variables; it represents that assumption that none of the variables are related, causally or noncausally. Also called the null or nil model.

Indirect effect: This corresponds to the two-step path from X_1 to Y, via the mediating variable X_2 or M; the strength of this indirect path is found by multiplying the path coefficient for X_1 to X_2 (a) by the path coefficient for X_2 to Y (b). This is often called a "mediated" effect, although statistical results by themselves are not sufficient to prove causation or mediation.

Initial solution: When a set of p variables is factor analyzed, the initial solution consists of p factors. This full solution (same number of factors as variables) is called the initial solution. In later stages, the number of factors is usually reduced, and the solution may be rotated to improve interpretability. The factor loadings for the initial solution are not usually reported.

Integrated model: In time-series modeling, an integrated model is one in which X at Time t is predicted by X at Time $t - 1$ plus a random error.

Interaction: An interaction between predictors can be incorporated into a regression equation by using the product of the two variables that interact as a predictor. When one or both of the variables are dichotomous, the interpretation of the nature of the interaction (in terms of different slopes predicted for different groups) is straightforward. When both variables involved in an interaction are quantitative, product terms for interactions are more difficult to interpret. Note that a widely used term for interaction in regression analyses is *moderation*. We can say that X_1 and X_2 interact to predict Y or that X_2 moderates the predictive relationship between X_1 and Y.

Internal consistency reliability: Consistency or agreement across a number of measures of the same construct, usually multiple items on a self-report test. Cronbach's alpha is the most popular measure of internal consistency reliability when each measure is quantitative. When scores are dichotomous, the internal consistency reliability coefficient is sometimes called the Kuder-Richardson 20, or KR-20, but this is just a special case of Cronbach's alpha.

Iteration: In this process, model parameters are estimated multiple times until some goodness-of-fit criterion is reached. In principal-axis factoring, an iterative process occurs; communality for each variable is initially estimated by the calculation of an R^2 (to predict each X_i from all the other X variables). Then, factor loadings are obtained, and the communality for each X_i is reestimated by squaring and summing the loadings of X_i across all factors. The loadings are then reestimated using these new communality estimates as the diagonal elements of $\mathbf{R}$, the correlation matrix that has to be reproduced by the factor loadings. This iterative process ends when the estimates of communality do not change (by more than .001) at subsequent steps; that is, the estimates converge. By default, SPSS allows 25 iterations.

Just-identified model: A model is just identified if it includes direct paths between all variables and has 0 degrees of freedom. Models for regression analysis, in which each predictor is correlated with every other predictor and each predictor has a direct path to the dependent variable, are just identified. These models can perfectly reproduce the variances and covariances obtained from the original data.

Kuder-Richardson 20 (KR-20): The name given to Cronbach's alpha when all items are dichotomous. See also *internal consistency reliability*.

Latent variable: An "imaginary" variable used to understand why observed X's are correlated. In factor analysis, it is assumed that measured X scores are correlated because they measure the same underlying construct or latent variable. Each factor potentially corresponds to a different latent variable. However, factor solutions do not always have meaningful interpretations in terms of latent variables. Structural equation models can include latent variables with measurement models to show how these are related to measured variables.

Levels of a factor: Each group in an analysis of variance corresponds to a level of the factor. Depending on the nature of the study, levels of a factor may correspond to different amounts of treatment (for instance, if a researcher manipulates the dosage of caffeine, the levels of the caffeine factor could be 0, 100, 200, and 300 mg). In other cases, levels of a factor may correspond to qualitatively different types of treatment (for instance, a study might compare Rogerian, cognitive behavioral, and Freudian therapy as the three levels of a factor called "type of psychotherapy"). In some studies where naturally occurring groups are compared, the levels of a factor correspond to naturally occurring group memberships (e.g., gender, political party).

Leverage: An index that indicates whether a case is disproportionately influential; when a case with a large leverage value is dropped, one or more regression slopes change substantially.

Listwise deletion: A method of handling missing data in SPSS (and many other programs); if a participant has missing data for any of the variables included in an analysis, that participant's data are excluded from all the computations for that analysis. If more than 5% of data are missing, use of listwise deletion may yield a biased sample. See also *pairwise deletion*.

Little's test of MCAR: A test of whether missing values on one variable are related to scores on another variable, for all pairs of variables in the data set. A small p value suggests that data are not missing completely at random.

Log likelihood (LL): In binary logistic regression, the log likelihood is an index of goodness of fit. For each case, the log of the predicted probability of group membership (such as the log of p =.89) is multiplied by the actual group membership code (e.g., 0 or 1). The sum of these products across all cases is a negative value (because probabilities must be less than 1 and the natural logs of values less than 1 are negative numbers). The larger the absolute value of the log likelihood, the worse the agreement between the probabilities of group membership generated by the logistic regression model and the actual group memberships. When coefficients for the logistic regression are calculated using maximum likelihood estimation, the aim is to find the values of coefficients that yield a log likelihood value that is small in absolute value (and that therefore correspond to estimated probabilities that agree relatively well with actual group membership). SPSS reports –2LL for each model and for the change in fit between models.

Logistic regression: A regression analysis for which the outcome variable is categorical; the goal of the analysis is prediction of group membership. Because categorical variables do not conform to ordinary linear regression assumptions, different computational procedures are required.

Logit: A natural logarithm of an odds (in this book, a logit is denoted L_i; in some texts, the logit is denoted π). The dependent variable in a binary logistic regression takes the form of a logit. Also called a log odds ratio.

Measured variables: The X variables in a structural equation model that are measured in the study. In an Amos path model, these are represented as rectangles. Also called observed or indicator variables in structural equation modeling.

Measurement model: A measurement model in a structural equation model consists of a latent variable and its indicator variables. When structural equation modeling includes only measurement models, it is called confirmatory factor analysis. Use of multiple indicators provides a way to evaluate reliability and take unreliability into account when estimating other model parameters.

Measurement theory: The set of mathematical assumptions that underlie the development of reliability indexes is called measurement theory. Only classical measurement theory is discussed in the reliability chapter. The equation $X_i = T + e_i$ represents the assumption that each observed X score consists of a T component that is stable across occasions of measurement and an e_i component that represents random errors that are unique to each occasion of measurement.

Mediated effect: The effect of X_1 on Y is said to be mediated if the effect of X_1 on Y is carried by or transmitted by a mediating variable X_2 or M. Several conditions must be met for this hypothesis to be plausible; it must be reasonable to hypothesize that X_1 causes Y, X_1 causes X_2, and X_2 causes Y. Empirical support needed includes an *ab* product (which represents the strength of this hypothesized mediated path) that is statistically significant and large enough to be of some practical importance.

Missing at random (MAR): One of three possible patterns in missing data; it is ignorable. Type B missingness is not present (although Type A missingness may occur). Type A missingness can be removed using multiple imputation to replace missing values; therefore, this is considered ignorable as a potential source of bias. See *Type A missingness* and *Type B missingness*.

Missing completely at random (MCAR): A pattern in missing data that involves neither Type A nor Type B missingness. It is ignorable. See *Type A missingness* and *Type B missingness*.

Missing not at random (MNAR): A pattern of missing values that has Type B missingness. It is not ignorable. Even using state-of-the-art methods for missing values, final results will be biased. (A few journal articles refer to this as NMAR.) See *Type B missingness*.

Model fit: The degree to which the variance/covariance matrix that is reproduced from the paths and coefficients of a structural model matches the variances and covariances estimated using the original data set. Unless the structural equation model includes all possible paths (and has $df = 0$), the reproduced matrix for the model will usually differ from the matrix obtained using the original data. Model fit indices are different ways to quantify the magnitude of differences between the variances and covariances reproduced by the model, and the variances and covariances calculated from the original data.

Model identification: Assessment of model identification often involves complex considerations and can be quite difficult (Kenny & Milan, 2012). For simple models in Chapter 15, identification was assessed by looking at two things: first, the number of distinct sample moments, and second, the number of free parameters that need to be estimated. Some ways to avoid model identification problems include the following: (a) Each error term, and one of the indicator variables for each latent variable, must have its path coefficient fixed at 1 as a scaling constraint; Amos automatically adds these when models are diagrammed, but sometimes users accidentally edit these out. (b) There should be at least two or, better, at least three indicators for each latent variable. (c) The model must be recursive (it must not have "loops").

Moderation: This term, as defined by Baron and Kenny (1986), is synonymous with interaction. Occasionally writers confuse the terms mediation and moderation, perhaps because they sound similar; however, these are quite different. Mediation includes a hypothesized causal sequence such as X_1 causes X_2 and then X_2 causes Y. Moderation involves a situation in which the slope to predict Y from X_1 differs across scores on the X_2 variable.

Moderator variable: A predictor that has an interaction with another predictor variable.

Modification index: For each parameter, the modification index indicates how much overall model fit can be improved by changing that parameter. For example, if there is no direct path between X_1 and X_9, a large value of the modification index corresponding to that (missing) path may indicate that adding that direct path will improve model fit. Sometimes modification index values suggest that adding correlations between error terms can improve model fit. Newsom (2017) stated that any model changes based on modification index values must be theoretically justified.

Monotonic relationship: Scores on X and Y are monotonically related if an increase in X is always associated with an increase or no change (and not a decrease) in Y. If X and Y are monotonically related, if both are converted to ranks, the ranks are identical.

Moving average model: A time-series model in which X at Time t is predicted not only by an external random variable e_t at Time t but also by earlier values of that random variable such as e_{t-1} and e_{t-2}.

Multicollinearity: The degree of intercorrelation among predictor variables; perfect multicollinearity exists when one variable is completely predictable from one or more other variables.

Multiple imputation (MI): One of numerous methods to calculate replacement scores for missing values.

Nagelkerke's R^2: In binary logistic regression, this is one of the pseudo-R^2 values provided by SPSS as an overall index of the strength of prediction for the entire model.

Natural log: The base e logarithm; that is, the natural log of the value a is the power to which the mathematical constant e must be raised to obtain the number a. Often abbreviated as ln.

Natural logarithm: See *natural log*.

Nested models: In binary logistic regression, Model A (e.g., $Y_i = a + b_1 \times X_1$) is nested in Model B (e.g., $Y_i = a + b_1 \times X_1 + b_2 \times X_2$) if all the variables in Model A are also included in Model B, but Model B has one or more additional variables that are not included in Model A. Researchers often want to test whether the fit of Model B is significantly better than the fit of Model A. In the simplest application of binary logistic regression, two nested models are compared. In some situations, Model A is a null model with no predictor variables, while Model B is a full model with all predictor variables included. For each model, an index of fit is obtained (−2 log likelihood [−2LL]). The difference between −2LLB and −2LLA has a χ^2 distribution with k degrees of freedom, where k is the number of predictors in Model B. If this χ^2 is large enough to be judged statistically significant, the data analyst can conclude that the overall model, including the entire set of independent variables, predicted group membership significantly better than a null model. Nesting has other meanings in structural equation modeling, complex analysis of variance designs, and multilevel modeling.

Nonadditivity: See *interaction*.

Noncausal path: In a causal model, a bidirectional arrow is used to represent a situation where two variables are noncausally associated with each other; they are correlated, confounded, or redundant.

Nonequivalent comparison group: See *nonequivalent control group*.

Nonequivalent control group: When individual participants cannot be randomly assigned to treatment and/or control groups, we often find that these groups are nonequivalent; that is, they are unequal on their scores on many participant characteristics prior to the administration of treatment. Even when a random assignment of participants to groups occurs, sometimes nonequivalence among groups occurs because of "unlucky randomization." If it is not possible to use experimental controls (such as matching) to ensure equivalence, analysis of covariance is often used to try to correct for or remove this type of nonequivalence. However, the statistical control for one or more covariates in analysis of covariance is not guaranteed to correct for all sources of nonequivalence; also, if assumptions of analysis of covariance are violated, the adjustments it makes for covariates may be incorrect.

Null model: In binary logistic regression, the null model represents prediction of group membership that does not use information about any predictor variables. As one or more predictor variables are added to the logistic regression, significance tests are used to evaluate whether these predictors significantly improve prediction of group membership compared with the null model.

Oblique rotation: A rotation in which the factors are allowed to become correlated to some degree. This is useful in situations where the constructs we think we are measuring are correlated to some degree, and oblique rotation may enhance interpretability.

Odds: Odds involve comparison of the number of cases for two possible outcomes. For example, in a study of patients with cancer, if 40 patients die and 10 patients survive, the odds of death in that sample are found by dividing the number of deaths by the number of survivals, 40/10, to obtain odds of 4 to 1 (an individual patient selected at random is 4 times as likely to die as to survive).

OLS: See *ordinary least squares (OLS)*.

Optimal weighted linear composite: In analyses such as regression and discriminant analysis, raw scores or z scores on variables are summed using a set of optimal weights (regression coefficients in the case of multiple regression and discriminant function coefficients in the case of discriminant analysis). In regression, the regression coefficients are calculated so that the multiple correlation R between the weighted linear composite of scores on X predictor variables has the maximum possible value, and the sum of squared prediction errors has the smallest possible value. In a multiple regression of the form $Y' = b_0 + b_1X_1 + b_2X_2 + b_3X_3$, the weights b_1, b_2, and b_3 are calculated so that the multiple correlation R between actual Y scores and predicted Y' scores has the maximum possible value. In discriminant function analysis, the coefficients for the computation of discriminant function scores from raw scores on the X predictor variables are estimated such that the resulting discriminant function scores have the maximum possible sum of squares between groups and the minimum possible sum of squares within groups.

Ordinary least squares (OLS): A statistic is the best ordinary least squares estimate if it minimizes the sum of squared prediction errors; for example, M is the best ordinary least squares estimate of the sample mean because it minimizes the sum of squared prediction errors, $\Sigma(X - M)^2$, that arises if we use M to predict the value of any one score in the sample chosen at random.

Orthogonally coded dummy variable: Dummy variables that are coded so that they represent independent contrasts. Tables of orthogonal polynomials can be used as a source of coefficients that have already been worked out to be orthogonal. Orthogonality of contrasts can also be verified either by finding a zero correlation between the dummy variables that represent the contrasts or by cross-multiplying the corresponding contrast coefficients.

Orthogonal rotation: A factor rotation in which the factors are constrained to remain uncorrelated or orthogonal to one another.

Overidentified model: A model is overidentified if there are more pieces of information in the data (variances and covariances) than free parameter estimates and the model *df* are positive. Structural equation modeling is frequently conducted using over identified models. These models generally cannot perfectly reconstruct the original information from the data (the elements of the variance/covariance matrix).

Paired-samples *t* test: A form of the *t* test that is appropriate when scores come from a repeated-measures study, a pretest–posttest design, matched samples, or other designs where scores are paired in some manner. Also called a correlated-samples *t* test or a direct difference *t* test.

Pairwise deletion: A method of handling missing data in SPSS, such that the program uses all the available information for each computation; for example, when computing many correlations as a preliminary step in regression, pairwise deletion would calculate each correlation using all the participants who had nonmissing values on that pair of variables. See also *listwise deletion*.

Parallel-forms reliability: When a test developer creates two versions of a test (which contain different questions but are constructed to include items that are matched in content), parallel forms reliability is assessed by giving the same group of people both Form A and Form B and correlating scores on these two forms.

Part correlation: Another name for the semipartial correlation. The value SPSS denotes as part correlation in regression output corresponds to the semipartial correlation.

Partition of variance: The variability of scores (as indexed by their sum of squares) can be partitioned or separated into two parts: the variance explained by between-group differences (or treatment) and the variance not predictable from group membership (due to extraneous variables). The ratio of between-group (or explained) variation to total variation, eta squared, is called the proportion of explained variance. Researchers usually hope to explain or predict a reasonably large proportion of variance for the outcome variable. Partition of *SS* was introduced in discussion of one-way analysis of variance, and estimated partitions of variance are also provided by regression and multivariate analyses.

Path model: A diagram in which hypothesized causal associations between variables are represented by unidirectional arrows, noncausal associations between variables are represented by bidirectional arrows, and the lack of either type of arrow between variables represents an assumption that there is no direct relationship between the variables, either causal or noncausal.

Pillai's trace: One of the multivariate test statistics used to test the null hypothesis in MANOVA, H_0: $\mu_1 = \mu_2 = \cdots = \mu k$. It is the sum of the squared canonical correlations (each canonical correlation is a correlation between scores on one discriminant function and group membership) across all discriminant functions. A larger value of Pillai's trace implies that the vectors of means show larger differences across groups. Among the multivariate test statistics, Pillai's trace is thought to be generally more robust to violations of assumptions (although it is not the most robust test under all conditions).

Pretest–posttest design: In this design, the same response is measured for the same participants both before and after an intervention. This may be in the context of a true experiment (in which participants are randomly assigned to different treatments) or a quasi-experiment (one that lacks comparison groups or has nonequivalent comparison groups). If the study has just one group, the data may be analyzed using the correlated-samples or direct-difference *t* test, repeated-measures analysis of variance, or analysis of covariance. If the study has multiple groups, an analysis of variance of gain scores, a repeated-measures analysis of variance, or an analysis of covariance may be performed. If the conclusions about the nature of the effects of treatments differ when different analyses are run, the researcher needs to think carefully about possible violations of assumptions and the choice of statistical method.

Projective tests: Tests that involve the presentation of ambiguous stimuli (such as Rorschach inkblots or thematic apperception test drawings). Participants are asked to interpret or tell stories about the ambiguous stimulus. Because it is not obvious to the test taker what the test is about, the responses to the stimuli should not be influenced by social desirability or self-presentation style. For example, thematic apperception test stories are scored to assess how much achievement imagery a respondent's story includes.

Published tests: Psychological tests that have been copyrighted and published by a test publishing company (such as EDITS or PAR). It is a copyright violation to use these tests unless the user pays royalty fees to the publisher; some published tests are available only to persons who have specific professional credentials such as a PhD in clinical psychology.

R^2: The squared multiple *R*, which can be interpreted as the proportion of variance in *Y* that can be predicted from *X*.

R^2_{inc}: The increase in *R2* from one step in a sequential or statistical regression to the next step; it represents the contribution of one or more variables added at that step.

Recursive model: If a structural model has causal loops (for example, *X* causes *Y*, *Y* causes *Z*, *Z* causes *X*) it is called nonrecursive. Special methods not discussed in Chapter 15 are required to estimate parameters for nonrecursive models.

Regression plane: When two *X* variables are used to predict scores on *Y*, a three-dimensional space is needed to plot the scores on these three variables. The equation to predict *Y* from X_1 and X_2 corresponds to a plane that intersects this three-dimensional space, called the regression plane.

Reproduced correlation matrix: Squaring the factor loading matrix **A** (i.e., finding the matrix product **A′A**) reproduces the correlation matrix **R**. If we retain *p* factors for a set of *p* variables, we can reproduce **R** exactly. What we usually want to know, in factor analysis, is whether a reduced set of factors reproduces **R** reasonably well.

Reverse-worded questions: These are questions that are worded in such a way that a strong level of *disagreement* with the statement indicates more of the trait or attitude that the test is supposed to measure. For example, a reverse-worded item on a depression scale could be "I am happy"; strong disagreement with this statement would indicate a higher level of depression. Reverse-worded questions are included to avoid inflation of scores because of yea-saying bias.

Root mean square error of approximation (RMSEA): This model fit index calculates the size of the standardized residual correlations. It ranges from 0 (perfect fit) to 1 (poor fit). Criteria for adequate model fit vary; usually a model is considered adequate fit if RMSEA < .08 or < .06. Ideally, the upper boundary of the confidence interval for RMSEA should not exceed .10. Some of the statistical power analyses for structural equation modeling focus on RMSEA.

Rotated factor loadings: Factor loadings (or correlations between *X* variables and factors) that have been reestimated relative to rotated or relocated factor axes. The goal of factor rotation is to improve the interpretability of results.

Roy's greatest characteristic root: See *Roy's largest root.*

Roy's largest root: This is a statistic that tests the significance of just the first discriminant function; it is equivalent to the squared canonical correlation of scores on the first discriminant function with group membership. It is generally thought to be less robust to violations of assumptions than the other three multivariate test statistics reported by the SPSS GLM procedure (Hotelling's trace, Pillai's trace, and Wilks's Λ).

Saturated model: A model that includes direct paths from each variable to every other variable. It represents the (usually uninteresting) hypothesis that everything is related to everything.

s_b: The standard error of the raw score regression slope to predict the outcome variable *Y* from the mediating variable (X_2 or *M*).

Scree plot: A plot of the eigenvalues (on the *Y* axis) by factor number 1, 2, . . ., *p* (on the *X* axis). "Scree" refers to the rubble at the foot of a hill, which tends to level off abruptly at some point. Sometimes, visual examination of the scree plot is the basis for the decision regarding the number of factors to retain. Factors that have low eigenvalues (and thus correspond to the point where the scree plot levels off) may be dropped.

SE_{ab}: One notation for the standard error of the *ab* product that represents that strength of the mediated path. Different methods of estimating this standard error have been proposed by Sobel (1982), Aroian (1947), and Goodman (1960).

Second-order partial correlation: A second-order partial correlation between X_1 and *Y* involves statistically controlling for two variables (such as X_2 and X_3) while assessing the relationship between X_1 and *Y*.

Sequential regression: In this form of regression, variables are entered one at a time (or in groups or blocks) in an order determined by the researcher. Sometimes this is called hierarchical regression.

Sigmoidal function: This is a function that has an S-shaped curve. This type of function usually occurs when the variable plotted on the *X* axis is a probability, because probabilities must fall between 0 and 1. For example, consider an experiment to assess drug toxicity. The manipulated independent variable is level of drug dosage, and the outcome variable is whether the animal dies or survives. At very low doses, the probability of death would be near 0; at extremely high doses, the probability of death would be near 1; at intermediate dosage levels, the probability would increase as a function of dosage level.

Simultaneous multiple regression: See *standard multiple regression*.

Sobel test: A widely used significance test proposed by Sobel (1982) to test H_0: $ab = 0$. This tests whether a mediating variable (X_2) transmits the influence of the independent variable X_1 to the Y outcome variable.

Spearman-Brown prophecy formula: When a correlation is obtained to index split-half reliability, that correlation actually indicates the reliability or consistency of a scale with $p/2$ items. If the researcher plans to use a score based on all p items, this score should be more reliable than a score that is based on $p/2$ items. The Spearman-Brown prophecy formula "predicts" the reliability of a scale with p items from the split-half reliability.

Split-half reliability: A type of internal consistency reliability assessment that is used with multiple-item scales. The set of p items in the scale is divided (either randomly or systematically) into two sets of $p/2$ items, a score is computed for each set, and a correlation is calculated between the scores on the two sets to index split-half reliability. This correlation can then be used to "predict" the reliability of the full p-item scale by applying the Spearman-Brown prophecy formula.

Spurious correlation: A correlation between X_1 and Y is said to be spurious if the correlation drops to 0 when you control for an appropriate X_2 variable, and/or if the correlation is "silly" and it does not make sense to propose a causal model in which X_2 mediates a causal connection between X_1 and Y.

sr^2: A common notation for a squared semipartial (or squared part) correlation.

sr^2_{inc}: See *Incremental* sr^2.

Standard multiple regression: In standard multiple regression, no matter how many X_i predictors are included in the equation, each X_i predictor variable's unique contribution to explained variance is assessed controlling for or partialling out all other X predictor variables. In Figure 4.3, for example, the $X1$ predictor variable is given credit for predicting only the variance that corresponds to Area a; the $X2$ predictor variable is given credit for predicting only the variance that corresponds to Area b; and Area c, which could be predicted by either X_1 or X_2, is not attributed to either predictor variable.

Standard score: The distance of an individual score from the mean of a distribution expressed in unit-free terms (i.e., in terms of the number of standard deviations from the mean). If μ and σ are known, the z score is given by $z = (X - \mu)/\sigma$. When μ and σ are not known, a distance from the mean can be computed using the corresponding sample statistics, M and s. If the distribution of scores has a normal shape, a table of the standard normal distribution can be used to assess how distance from the mean (given in z-score units) corresponds to proportions of area or proportions of cases in the sample that correspond to distances of z units above or below the mean.

Standardized root mean square residual (SRMR): The standardized distance between observed correlations (calculated from the original data) and the correlations that are reconstructed from structural equation model parameters. A value of 0 indicates perfect fit.

Statistical control: When information is available about at least one additional variable (Z), it is possible to evaluate the relationship between an X predictor and a Y outcome variable using statistical methods to partial out or remove variation associated with Z. Sometimes controlling for Z makes the association between X and Y appear stronger, but there are many ways that inclusion of a control variable can change our understanding of the way X and Y may be related. In paired-samples designs, the control variable is "persons."

Statistical regression: A method of regression in which the decisions to add or drop predictors from a multiple regression are made on the basis of statistical criteria (such as the increment in R^2 when a predictor is entered). This is sometimes called stepwise regression, but that term should be reserved for a specific method of entry of predictor variables. See also *stepwise method of entry*.

Stepwise method of entry: This method of statistical (data-driven) regression combines forward and backward methods. At each step, the predictor variable in the pool of candidate variables that provides the largest increase in R^2 is added to the model; also, at each step, variables that have entered are reevaluated to assess whether their unique contribution has become nonsignificant after other variables are added to the model (and dropped, if they are no longer significant).

Structural model: The part of a structural equation model that represents causal (and correlational) paths among latent variables and measured variables. The paths that connect multiple indicator variables to latent variables are not considered part of the structural model.

Structure coefficient: The correlation between a discriminant function and an individual discriminating variable (analogous to a factor loading); it tells us how closely related a discriminant function is to each individual predictor variable. These structure coefficients can sometimes be interpreted in a manner similar to factor loading to name the dimension represented by each discriminant function. However, discriminant functions are not always easily interpretable.

Sum of cross products matrix (SCP): A sum of cross products matrix for a multivariate analysis such as discriminant analysis or multivariate analysis of variance includes the sum of squares (*SS*) for each of the quantitative variables and the sum of cross products for each pair of quantitative variables. In discriminant analysis, the quantitative variables are typically denoted X; in multivariate analysis of variance, they are typically denoted Y. *SS* terms appear in the diagonal of the SCP matrix; SCP terms are the off-diagonal elements. In multivariate analysis of variance, evaluation of the SCPs computed separately within each group, compared with the SCP for the total set of data, provides the information needed for multivariate tests to evaluate whether lists of means on the Y variables differ significantly across groups, taking into account the covariances or correlations among the Y variables. Note that dividing each element of the SCP matrix by *df* yields the sample variance/covariance matrix **S**.

Sum of squared loadings (SSL): For each factor, the sum of squared loadings is obtained by squaring and summing the loadings of all variables with that factor. In the initial solution, the SSL for a factor equals the eigenvalue for that factor. The proportion of the total variance in the p variables that is accounted for by one factor is estimated by dividing the SSL for that factor by p.

Suppression: X_2 is said to suppress the X_1, Y relationship if the relationship between X_1 and Y either gets stronger or reverses sign when X_2 is statistically controlled. There are several different types of suppression.

Suppressor variable: An X_2 variable is called a suppressor variable relative to X_1 and Y if the partial correlation between X_1 and Y controlling for X_2 ($r_{1Y \cdot 2}$) is larger in absolute magnitude than $r1_Y$ or if $r1_{Y \cdot 2}$ is significantly different from 0 and opposite in sign to r_{1Y}. In such situations, we might say that the effect of the X_2 variable is to suppress or conceal the true nature of the relationship between X_1 and Y; it is only when we statistically control for X_2 that we can see the true strength or true sign of the X_1, Y association.

t_b: The t test for the raw score regression coefficient to predict the final outcome variable Y from the mediating variable (X_2 or M) in a mediation model.

Territorial map: A graphical representation of the scores of individual cases on discriminant functions (SPSS plots only scores on the first two discriminant functions, D_1 and D_2). This map also shows the boundaries that correspond to the classification rules that use scores on D_1 and D_2 to predict group membership for each case.

Tolerance: For each X_i predictor variable, the proportion of variance in X_i that is not predictable from other predictor variables already in the equation (i.e., tolerance = $1 - R^2$ for a regression to predict X_i from all the other X's already included in the regression equation). A tolerance of zero would indicate perfect multicollinearity, and a variable with zero tolerance cannot add any new predictive information to a regression analysis. Generally, predictor variables with higher values of tolerance may possibly contribute more useful predictive information. The maximum possible tolerance is 1.00, and this would occur when X_i is completely uncorrelated with other predictors.

Total effect: The overall strength of association between the initial predictor variable X_1 and the final outcome variable Y in a mediated model (ignoring or not controlling for the mediating variable M or X_2); denoted c.

Tracing rule: Procedures for tracing all possible paths from each causal variable to each outcome variable in a path model. The (reproduced) total correlation between two variables in a path model using standardized variables is obtained by tracing all routes or paths between those variables, with some limitations.

Treatment-by-covariate interaction: An assumption required for analysis of covariance is that there must not be a treatment (A)–by–covariate (X_c) interaction; in other words, the slope that predicts Y from X_c should be the same within each group or level of A.

Triangulation of measurement: The use of multiple and different types of measurement to tap the same construct, for example, direct observation of behavior, self-report, and physiological measures. Triangulation is undertaken because different methods of measurement have different advantages and disadvantages, and we can be more confident about results that replicate across different types of measures.

Type A missingness: If the existence of missing values is related to scores on one or more other variables, this is Type A missingness. For example, consider missing values for questions about emotional stress. It is possible that missingness is related to other variables, such as sex. Perhaps men are more likely to avoid answering this question than women.

Type B missingness: Missing values for Y are related to what the Y values would have been if the persons with missing values had answered. For example, if people with high scores on depression are more likely not to report a score on Y (depression) than people with low scores on depression, there is Type B missingness. Type B missingness is difficult to detect and to remedy. It is not ignorable (it leads to bias in parameter estimates). Type B missingness is present in missing not at random data, but not in missing at random or missing completely at random data. This is not a widely used term; I introduce it to try to simplify distinctions among patterns of missingness.

Underidentified model: A model that has more free parameters to estimate than pieces of information in the data; it does not have a unique solution. Principal components and factor analyses are examples of analyses that are underidentified unless scaling constraints are placed on estimates (e.g., factor loadings must be interpretable as correlations).

Underpowered: A study is underpowered if the sample size is too small (relative to the effect size) to have a reasonable chance of rejecting H_0 when H_0 is false.

Unweighted mean: This is a mean that combines information across several groups or cells (such as the mean for all the scores in one row of a factorial analysis of variance) and is calculated by averaging cell means together without taking the numbers of scores within each cell or group into account.

User-determined order of entry in regression: In user-determined order of entry, the data analyst decides the order of entry of predictor variables.

Variance/covariance matrix: This is a matrix that summarizes all the variances and all the possible covariances for a list of variables. The population matrix is denoted by $\mathbf{\Sigma}$; the corresponding sample variance/covariance matrix is denoted by **S**. These matrices are used in discriminant analysis, multivariate analysis of variance, and structural equation models (among other analyses).

Varimax rotation: The most popular method of factor rotation. The goal is to maximize the variance of the absolute values of loadings of variables within each factor, that is, to relocate the factors in such a way that, for each variable, the loading on the relocated factors is as close to 0, or as close to +1 or −1, as possible. This makes it easy to determine which variables are related to a factor (those with large loadings) and which variables are not related to each factor (those with loadings close to 0), which improves the interpretability of the solution.

Wald χ^2 statistic: For each B coefficient in a logistic regression model, the corresponding Wald function tests the null hypothesis that $B = 0$. The Wald χ^2 is given as $[B/SE_B]^2$; the sampling distribution for the Wald function is a chi-square distribution with 1 degree of freedom. Note that in an ordinary linear regression, the null hypothesis that $b = 0$ is tested by setting up a t ratio where $t = b/SE_b$; the Wald chi-square used in binary logistic regression is comparable to the square of this t ratio.

Wilks' lambda (Λ): An overall goodness-of-fit measure used in discriminant analysis. It is the most widely used multivariate test statistic for the null hypothesis shown in multivariate analysis of variance: H_0: $\mu_1 = \mu_2 = \cdots = \mu_k$. Unlike Hotelling's trace and Pillai's trace, larger values of Wilks' Λ indicate smaller differences across groups. Wilks' Λ may be calculated in two different ways that yield the same result: as a ratio of determinants of the sum of cross products matrices or as the product of $1/(1 + \lambda_i)$ across all the discriminant functions. It can be converted to an estimate of effect size; $\eta^2 = 1 - \Lambda$. In a univariate analysis of variance, Wilks' Λ is equivalent to $(1 - \eta^2)$, that is, the proportion of unexplained or within-group variance on outcome variable scores. Like Hotelling's trace and Pillai's trace, it summarizes information about the magnitude of between-group differences on scores across all discriminant functions. Unlike these two other statistics, however, larger values of Wilks' Λ indicate smaller between-group differences on vectors of means or on discriminant scores. In a discriminant analysis, the goal is to *minimize* the size of Wilks' Λ.

***z* score:** The formula to calculate a standard score or z score is $z = (X - M)/SD$. A distribution of z scores has $M = 0$ and $SD = 1$.

REFERENCES

Abelson, R. P. (1997). On the surprising longevity of flogged horses: Why there is a case for the significance test. *Psychological Science*, *8*(1), 12–15.

Abelson, R. P., & Rosenberg, M. J. (1958). Symbolic psycho-logic: A model of attitudinal cognition. *Behavioral Science*, *3*, 1–13.

Acock, A. C. (2005). Working with missing values. *Journal of Marriage and Family*, *67*(4), 1012–1028. doi:10.1111/j1741-3737.2005.00191.x

Adelson, J. L., & Owen, J. (2012). Bringing the psychotherapist back: Basic concepts for reading articles examining therapist effects using multilevel modeling. *Psychotherapy*, *49*(2), 152–162.

Aguinas, H., Gottfredson, R. K., & Joo, H. (2013). Best-practice recommendations for defining, identifying, and handling outliers. *Organizational Research Methods*, *16*(2), 270–301. doi:10.1177/1094428112470848

Aiken, L. S., & West, S. G. (1991). *Multiple regression: Testing and interpreting interactions*. Thousand Oaks, CA: Sage.

Algina, J., & Kesselman, H. J. (1997). Detecting repeated measures effects with univariate and multivariate statistics. *Psychological Methods*, *2*, 208–218.

Algina, J., & Olejnik, S. (2003). Sample size tables for correlation analysis with applications in partial correlation and multiple regression analysis. *Multivariate Behavioral Research*, *38*(3), 309–323.

Allison, P. D. (2002). *Missing data*. Sage University Paper: Quantitative Applications in the Social Sciences, No. 36. Thousand Oaks, CA: Sage.

American Psychiatric Association. (2013). *Diagnostic and statistical manual of mental disorders* (5th ed.). Washington, DC: Author.

Aroian, L. A. (1947). The probability function of the product of two normally distributed variables. *Annals of Mathematical Statistics*, *18*, 265–271.

Aschwanden, C. (2015). Science isn't broken. It's just a hell of a lot harder than we gave it credit for. *Slate*. Retrieved from https://fivethirtyeight.com/features/science-isnt-broken/

Asendorpf, J. B., Conner, M., de Fruyt, F., de Houwer, J., Denissen, J.J.A., Fiedler, K., . . . Wicherts, J. M. (2013). Recommendations for increasing replicability in psychology. *European Journal of Personality*, *27*, 108–119. doi:10.1002/per.1919

Atkins, D. C., & Gallop, R. J. (2007). Rethinking how family researchers model infrequent outcomes: A tutorial on count regression and zero-inflated models. *Journal of Family Psychology*, *21*, 726–735.

Baker, M. (2015). First results from psychology's largest reproducibility test: Crowd-sourced effort raises nuanced questions about what counts as replication. *Nature*. Retrieved from https://www.nature.com/news/first-results-from-psychology-s-largest-reproducibility-test-1.17433

Baker, M., & Dolgin, E. (2017). An open-science effort to replicate dozens of cancer-biology studies is off to a confusing start. *Nature*, *541*(7637), 269–270. doi:10.1038/541269a

Baron, R. M., & Kenny, D. A. (1986). The moderator-mediator variable distinction in social psychological research: Conceptual, strategic and statistical considerations. *Journal of Personality and Social Psychology*, *51*, 1173–1182.

Barratt, A., Wyer, P. C., Hatala, R., McGinn, T., Dans, A. L., Keitz, S., . . . For, G. G. (2004). Tips for learners of evidence-based medicine: 1. Relative risk reduction, absolute risk reduction and number needed to treat. *Canadian Medical Association Journal*, *171*(4), 353–358.

Bates, T. (July 23, 2017). *Changing the default* p-*value threshold for statistical significance ought not be done, and is the least of our problems*. Retrieved from https://medium.com/@timothycbates/changing-the-default-p-value-threshold-for-statistical-significance-ought-not-be-done-in-isolation-3a7ab357b5c1

Beck, A. T., Ward, C. H., Mendelson, M., Mock, J., & Erbaugh, J. (1961). An inventory for measuring depression. *Archives of General Psychiatry*, *4*, 561–571.

Begley, C. G., & Ellis, L. M. (2012). Drug development: Raise standards for preclinical cancer research. *Nature*, *483*, 531–533. doi:10.1038/483531a

Bem, S. L. (1974). The measurement of psychological androgyny. *Journal of Consulting and Clinical Psychology*, *42*, 155–162.

Benjamin, D. J., Berger, J., Johannesson, M., Nosek, B. A., Wagenmakers, E.-J., Berk, R., . . . Johnson, V. (2017, July 22). Redefine statistical significance. *PsyArKiv*. Retrieved from http://psyarxiv.com/mky9j

Bentler, M., & Chou, C. H. (1987). Practical issues in structural modeling. *Sociological Methods & Research*, *16*, 78–117.

Berry, W. D. (1993). *Understanding regression assumptions* (Quantitative Applications in the Social Sciences, No. 92). Thousand Oaks, CA: Sage.

Bodner, T. E. (2016). Tumble graphs: Avoiding misleading end point extrapolation when graphing interactions from a moderated interaction analysis. *Journal of Educational and Behavioral Sciences*, *41*(6), 593–604. doi:10.3102/1076998616657080

Bollen, K. A. (1989). *Structural equations with latent variables*. New York: John Wiley.

Boneau, C. A. (1960). The effects of violations of assumptions underlying the *t* test. *Psychological Bulletin*, *57*(1), 49–64.

Boomsma, A. (1982). Robustness of LISREL against small sample sizes in factor analysis models. In K. G. Jöreskog & H. Wold (Eds.), *Systems under indirect observation: Causality, structure, prediction* (Part 1, pp. 149-173). Amsterdam, the Netherlands: North Holland.

Boomsma, A. (1985). Nonconvergence, improper solutions, and starting values in LISREL maximum likelihood estimation. *Psychometrika*, *50*, 229–242.

Borenstein, M., Hedges, L. V., Higgins, J.P.T., & Rothstein, H. R. (2009). *Introduction to meta-analysis*. Hoboken, NJ: John Wiley.

Boutron, I. (2017). CONSORT statement for randomized trials of nonpharmacologic treatments: A 2017 update and a CONSORT extension for nonpharmacologic trial abstracts. *Annals of Internal Medicine*, *167*(1), 40–61. doi:10.7326/M17-0046

Box, G.E.P., Jenkins, G. M., Reisel, G. C., & Ljung, G. M. (2015). *Time series analysis: Forecasting and control* (5th ed.). Hoboken, NJ: John Wiley.

Brackett, M. A. (2004). Conceptualizing and measuring the life space and its relation to openness to experience (Dissertation at University of New Hampshire, Durham, NH). *Dissertation Abstracts International*, *64*(7), 3569B.

Bray, J. H., & Maxwell, S. E. (1985). *Multivariate analysis of variance* (Quantitative Applications in the Social Sciences, No. 54). Beverly Hills, CA: Sage.

Byrne, B. M. (1998). *Structural equation modeling with LISREL, PRELIS, and SIMPLIS: Basic concepts, applications, and programming*. New York: Routledge Academic.

Byrne, B. M. (2011). *Structural equation modeling with Mplus: Basic concepts, applications, and programming*. New York: Routledge Academic.

Byrne, B. M. (2013). *Structural equation modeling with EQS: Basic concepts, applications, and programming* (2nd ed.). New York: Routledge Academic.

Byrne, B. M. (2016). *Structural equation modeling with AMOS: Basic concepts, applications, and programming* (3rd ed.). New York: Routledge Academic.

Campbell, D. T., & Stanley, J. (1963). *Experimental and quasi-experimental designs for research*. New York: Wadsworth.

Carmines, E. G., & Zeller, R. A. (1979). *Reliability and validity assessment* (Quantitative applications in the social sciences, No. 17). Beverly Hills, CA: Sage.

Carpenter, J., & Bithell, J. (2000). Bootstrap confidence intervals: When, which, what? A practical guide for medical statisticians. *Statistics in Medicine*, 19(9), 1141–1164.

Christie, R., & Geis, F. (1970). *Studies in Machiavellianism*. New York: Academic Press.

Cohen, J. (1988). *Statistical power analysis for the behavioral sciences* (2nd ed.). Hillsdale, NJ: Lawrence Erlbaum.

Cohen, J. (1990). Things I have learned (so far). *American Psychologist*, *45*, 1304–1312.

Cohen, J. (1992). A power primer. *Psychological Bulletin*, *112*(1), 155–159. doi:10.1037/0033-2909.112.1.155

Cohen, J. (1994). The earth is round ($p < .05$). *American Psychologist*, *49*, 997–1003.

Cohen, J., Cohen, P., West, S. G., & Aiken, L. S. (2013). *Applied multiple regression/correlation analysis for the behavioral sciences* (3rd ed.). Hillsdale, NJ: Lawrence Erlbaum.

Cook, T. D., & Campbell, D. T. (1979). *Quasi-experimentation: Design and analysis issues for field settings*. Boston: Houghton Mifflin.

Costa, P. T., & McCrae, R. R. (1995). Domains and facets: Hierarchical personality assessment using the Revised NEO Personality Inventory. *Journal of Personality Assessment*, *64*, 21–50.

Costa, P. T., & McCrae, R. R. (1997). Stability and change in personality assessment: The Revised NEO Personality Inventory in the year 2000. *Journal of Personality Assessment*, *68*, 86–94.

Crawford, J. T., Jussim, L., & Pilanski, J. M. (2014). How (not) to interpret and report main effects and interactions in multiple regression: Why Crawford and Pilanski did not actually replicate Lindner and Nosek (2009). *Political Psychology*, *35*(6), 857–862. doi:10.1111/pops.12050

Cumming, G. (2012). *Understanding the new statistics: Effect sizes, confidence intervals, and meta-analysis*. New York: Routledge.

Cumming, G. (2014). The new statistics: Why and how. *Psychological Science*, *25*(1), 7–29. doi:10.1177/0956797613504966

Cumming, G., & Calin-Jageman, R. (2016). *Introduction to the New Statistics: Estimation, Open Science, and beyond*. London: Routledge.

Cumming, G., & Finch, S. (2005). Inference by eye: Confidence intervals and how to read pictures of data. *American Psychologist*, *60*(2), 170–180. doi:10.137/0003-066X.60.2.170

Daniel, L. G. (1998). Statistical significance testing: A historical overview of misuse and misinterpretation with implications for the editorial policies of educational journals. *Research in the Schools*, 5(2), 23–32.

Danner, D. D., Snowdon, D. A., & Friesen, W. V. (2001). Positive emotions in early life and longevity: Findings from the nun study. *Journal of Personality and Social Psychology*, *80*(5), 804–813.

Donlon, T. F. (Ed.). (1984). *The College Board technical handbook for the Scholastic Aptitude Test and Achievement Tests*. New York: College Entrance Examination Board.

Edwards, J. R., & Lambert, L. S. (2007). Methods for integrating moderation and mediation: A general analytical framework using moderated path analysis. *Psychological Methods*, *12*, 1–22.

Embretson, S. E., & Reise, S. P. (2000). *Item response theory for psychologists*. Mahwah, NJ: Lawrence Erlbaum.

Epstein, S., & O'Brien, E. J. (1985). The person-situation debate in historical and current perspective. *Psychological Bulletin*, *98*(3), 513–537.

Everitt, B. S., Landau, S., Leese, M., & Stahl, D. (2011). *Cluster analysis* (5th ed.) Hoboken, NJ: John Wiley.

Fava, J. L., & Velicer, W. F. (1992). An empirical comparison of factor, image, component, and scale scores. *Multivariate Behavioral Research*, *27*, 301–322.

Ferguson, C. J. (2009). Is psychological research really as good as medical research? Effect size comparisons between psychology and medicine. *Review of General Psychology*, *13*(2), 130–136. doi:10.1037/a0015103

Field, A. (2018). *Discovering statistics using IBM SPSS statistics* (5th ed.). Thousand Oaks, CA: Sage.

Field, A. P., & Gillett, R. (2010). How to do a meta-analysis. *British Journal of Mathematical and Statistical Psychology*, *63*(3), 665–694. doi:10.1348/000711010X502733

Field, A. P., & Wilcox, R. R. (2017). Robust statistical methods: A primer for clinical psychology and experimental psychopathology researchers. *Behaviour Research and Therapy*, *98*, 19–38. doi:10.1016/j.brat.2017.05.013

Freedman, J. L. (1975). *Crowding and behavior*. New York: Viking.

Friedmann, E., Katcher, A. H., Lynch, J. J., & Thomas, S. A. (1980). Animal companions and one year survival of patients after discharge from a coronary care unit. *Public Health Reports*, 95, 307–312.

Fritz, C. O., Morris, P. E., & Richler, J. J. (2012). Effect size estimates: Current use, calculations, and interpretation. *Journal of Experimental Psychology: General*, *141*(1), 2–18. doi:10.1037/a0024338

Fritz, M. S., & MacKinnon, D. P. (2007). Required sample size to detect the mediated effect. *Psychological Science*, *18*, 233–239.

Garcia-Pérez, M.A. (2017). Thou shalt not bear false witness against null hypothesis significance testing. *Educational and Psychological Measurement*, 77(4), 631–662. doi:10.1177/0013164416668232

Gelman, A., & Loken, E. (2013). *The garden of forking paths: Why multiple comparisons can be a problem even when there is no "fishing expedition" or "p-hacking" and the research hypothesis was posited ahead of time*. Unpublished manuscript. Retrieved from http://www.stat.columbia.edu/~gelman/research/unpublished/p_hacking.pdf

Gergen, K. J. (1973). Social psychology as history. *Journal of Personality and Social Psychology*, *26*, 309–320.

Gergen, K. J., Hepburn, A., & Fisher, D. C. (1986). Hermeneutics of personality description. *Journal of Personality and Social Psychology*, *50*, 1261–1270.

Gilbert, D. T., King, G., Pettigrew, S., & Wilson, T. D. (2016). Comment on "Estimating the reproducibility of psychological science." *Science*, *351*(6277), 1037. doi:10.1126/science.aad7243

Goldberg, L. R. (1999). A broad-bandwidth, public domain, personality inventory measuring the lower-level facets of several five-factor models. In I. Mervielde, I. J. Deary, F. De Fruyt, & F. Ostendorf (Eds.), *Personality psychology in Europe* (Vol. 7, pp. 7–28). Tilburg, the Netherlands: Tilburg University Press.

Goodman, L. A. (1960). On the exact variance of products. *Journal of the American Statistical Association*, *55*, 708–713.

Goodwin, D. W., Schylsinger, F., Hermansen, L., Guze, S. B., & Winokur, G. (1973). Alcohol problems in adoptees raised apart from alcoholic biological parents. *Archives of General Psychiatry*, *28*, 238–243.

Gould, S. J. (1996). *The mismeasure of man*. New York: Norton.

Graf, R. G., & Alf, E. E. (1999). Correlations redux: Asymptotic confidence limits for partial and squared multiple correlations. *Applied Psychological Measurement*, *23*(2), 116–119.

Graham, J. W. (2009). Missing data analysis: Making it work in the real world. *Annual Review of Psychology*, *60*, 549–576. doi:10.1146/annurev.psych.58.110405.085530

Green, S. B. (1991). How many subjects does it take to do a regression analysis? *Multivariate Behavioral Research*, *26*, 449–510.

Greenland, S., Maclure, M., Schlesselman, J. J., Poole, C., & Morgenstern, H. (1991). Standardized regression coefficients: A further critique and review of some alternatives. *Epidemiology*, *2*, 387–392.

Greenland, S., Senn, S. J., Rothman, K. J., Carlin, J. B., Poole, C., Goodman, S. N. & Altman, D. G. (2016). Statistical tests, *p* values, confidence intervals, and power: A guide to misconceptions. *European Journal of Epidemiology*, *31*, 337–350. doi:10.1007/s10654-016-0149-3

Grimm, K. J., & Ram, N. (2016). *Growth modeling: Structural equation and multilevel modeling approaches*. Thousand Oaks, CA: Sage.

Halsey, L. G., Curran-Everett, D., Vowler, S. L., & Drummond, G. B. (2015). The fickle *p* value generates irreproducible results. *Nature Methods, 12*, 179–185. doi:10.1038/nmeth.3288

Hardy, M. A. (1993). *Regression with dummy variables* (Sage University Papers Series on Quantitative Applications in the Social Sciences, No. 07-093). Thousand Oaks, CA: Sage.

Harman, H. H. (1976). *Modern factor analysis* (2nd ed.). Chicago: University of Chicago Press.

Harris, R. J. (2001). *A primer of multivariate statistics* (3rd ed.). Mahwah, NJ: Lawrence Erlbaum.

Hayes, A. F. (2009). Beyond Baron and Kenny: Statistical mediation analysis in the new millennium. *Communication Monographs, 76*, 408–420.

Hayes, A. F. (2017). *Introduction to mediation, moderation, and conditional process analysis: A regression-based approach* (2nd ed.). New York: Guilford.

Hayes, A. F. (2019). *The PROCESS macro for SPSS and SAS*. Retrieved from http://processmacro.org/index.html

Heck, R. H., Thomas, S. L., & Tabata, L. N. (2013). *Multilevel and longitudinal modeling with IBM SPSS* (2nd ed.). New York: Routledge.

Hoekstra, R., Kiers, H.A.L., & Johnson, A. (2012, May 14). Are assumptions of well-known statistical tests checked, and why (not)? *Frontiers in Psychology, 3*, Article 137. doi:10.3389/fpsy.2012.00137

Horton, R. L. (1978). *The general linear model: Data analysis in the social and behavioral sciences*. New York: McGraw-Hill.

Hosmer, D. W., & Lemeshow, S. (2000). *Applied logistic regression* (2nd ed.). New York: John Wiley.

Huang, P. H. (2017). Asymptotics of AIC, BIC, and RMSEA for model selection in structural equation modeling. *Psychometrika, 88*(2), 407–426.

Huberty, C. J. (1994). *Applied discriminant analysis*. New York: John Wiley.

Huberty, C. J., & Morris, D. J. (1989). Multivariate analysis versus multiple univariate analyses. *Psychological Bulletin, 105*, 302–308.

Humphrey, S. E., & LeBreton, J. M. (2018). *The handbook of multilevel theory, measurement, and analysis*. Washington, DC: American Psychological Association.

Iacobucci, D. (2008). *Mediation analysis* (Sage University Papers Series on Quantitative Applications in the Social Sciences, No. 07-156). Thousand Oaks, CA: Sage.

Ioannidis, J.P.A. (2005). Contradicted and initially stronger effects in highly cited clinical research. *JAMA, 294*(2), 218–229. doi:10.1001/jama.294.2.218

Irwig, L., Irwig, J., Sweet, M., & Trevena, L. (2007). *Smart health choices*. London: Hammersmith Press.

Jaccard, J., & Turrisi, R. (2003). *Interaction effects in multiple regression* (2nd ed.). Thousand Oaks, CA: Sage.

Jaccard, J., Turrisi, R., & Wan, C. K. (1990). *Interaction effects in multiple regression* (Quantitative applications in the social sciences, Vol. 118). Thousand Oaks, CA: Sage.

Jenkins, C. D., Zyzanski, S. J., & Rosenman, R. H. (1979). *Jenkins Activity Survey manual*. New York: Psychological Corporation.

Johnsen, T. J., & Friborg, O. (2015). The effects of cognitive behavioral therapy as an anti-depressive treatment is falling: A meta-analysis. *Psychological Bulletin, 141*(4), 747–768. doi:10.1037/bu10000015

Johnson, D. R., & Young, R. (2011). Toward best practices in analyzing datasets with missing data: Comparisons and recommendations. *Journal of Marriage and Family, 73*(5), 926–945. doi:10.1111/j.1741-03737.2011.00861.x

Johnson, J. E., O'Leary, C. C., Striley, C. W., Abdallah, A. B., Bradford, S., & Cottler, L. B. (2011). Effects of major depression on crack use and arrests among women in drug court. *Addiction, 106*(7), 1279–1286. doi:10.1111/j.1360-0443.2011.03389.x

Jöreskog, K. G. (1969). A general approach to confirmatory maximum likelihood factor analysis. *Psychometrika, 34*, 183–202.

Judge, T., & Cable, D. (2004). The effect of physical height on workplace success and income: Preliminary test of a theoretical model. *Journal of Applied Psychology, 89*(3), 428–441.

Kahn, J. H. (2011). Multilevel modeling: Overview and applications to research in counseling psychology. *Journal of Counseling Psychology, 58*(2), 257–271.

Kaufmann, L., & Rousseeuw, P. J. (2009). *Finding groups in data: An introduction to cluster analysis*. Hoboken, NJ: John Wiley.

Kelley, T. L. (1927). *Interpretation of educational measurements*. Yonkers-on-Hudson, NY: World Book.

Kenny, D. A. (1979). *Correlation and causality*. New York: John Wiley.

Kenny, D. A., & Judd, C. M. (1986). Consequences of violating the independence assumption in analysis of variance. *Psychological Bulletin, 99*(3), 422–431.

Kenny, D. A., & Judd, C. M. (1996). A general method for the estimation of interdependence. *Psychological Bulletin, 119*, 138–148.

Kenny, D. A., & Milan, S. (2012). Identification: A nontechnical discussion of a technical issue. In R. Hoyle (Ed.), *Handbook of structural equation modeling* (pp. 159–177). New York: Guilford.

Kenrick, D. T., & Gutierres, S. E. (1980). Contrast effects and judgments of physical attractiveness: When beauty becomes a social problem. *Journal of Personality and Social Psychology, 38*(1), 131–140. doi:10.103u7/0022-3514.38.1.131

Keppel, G., & Zedeck, S. (1989). *Data analysis for research designs: Analysis of variance and multiple regression/correlation approaches*. New York: W. H. Freeman.

Kerr, N. L. (1998). HARKing: Hypothesizing after the results are known. *Personality and Social Psychology Review, 2*(3), 196–217. doi:10.1207/s15327957pspr0203_4

King, J. E. (2003). Running a best-subsets logistic regression: An alternative to stepwise methods. *Educational and Psychological Measurement, 63*, 393–403.

Kirk, R. (1996). Practical significance: A concept whose time has come. *Educational and Psychological Measurement, 56*, 746–759.

Kirk, R. E. (1995). *Experimental design: Procedures for the behavioral sciences* (3rd ed.). New York: Wadsworth.

Kline, R. B. (2013). *Beyond significance testing: Reforming data analysis in behavioral research* (2nd ed.). Washington, DC: American Psychological Association.

Kline, R. B. (2016). *Principles and practice of structural equation modeling* (4th ed.). New York: Guilford.

Knapp, T. R. (2017). Significance test, confidence interval, or neither? *Clinical Nursing Research, 26*(3), 259–265. doi:10.1177/1054773817708652

Knezevic, A. (2008). Overlapping confidence intervals and statistical significance. *StatNews* #73. Cornell Statistical Consulting Unit. Retrieved from http://www.cscu.cornell/news/statnews/Stnews73insert.pdf

Kristman, V. L., Manno, M., & Côté, P. (2005). Methods to account for attrition in longitudinal data: Do they work? *European Journal of Epidemiology, 20*(8), 657–662. doi:10.1007/s10654-005-7919-7

Krueger, J. (2001). Null hypothesis significance testing: On the survival of a flawed method. *American Psychologist, 56*, 16–26.

Kruschke, J. K., & Liddell, T. M. (2018). The Bayesian new statistics: Hypothesis testing, estimation, meta-analysis, and power analysis from a Bayesian perspective. *Psychonomic Bulletin and Review, 25*, 178–206. doi:10.3758/s13423-016-1221-4

Landis, J. R., & Koch, G. G. (1977). The measurement of observer agreement for categorical data. *Biometrics, 33*, 159–174.

Larsen, R. J., & Kasimatis, M. (1990). Individual differences in entrainment of mood to the weekly calendar. *Journal of Personality and Social Psychology, 58*(1), 164–171.

Lauter, J. (1978). Sample size requirements for the t^2 test of MANOVA (tables for one-way classification). *Biometrical Journal, 20*, 389–406.

Lee, T., Cai, L., & McCallum, R. C. (2012). Power analysis for tests of structural equation models. In R. Hoyle (Ed.), *Handbook of structural equation modeling* (pp. 181–194). New York: Guilford.

Lenhard, J. (2006). Models and statistical inference: The controversy between Fisher and Neyman-Pearson. *British Journal of Philosophy of Science, 57*(1), 69–91. doi:10.1093/bjps/axi152

Lester, B. M., Hoffman, J., & Brazelton, T. B. (1985). The rhythmic structure of mother-infant interaction in term and preterm infants. *Child Development, 56*, 15–27.

Lewis, G. F., Furman, S., McCool, M. F., & Porges, S. W. (2011). Statistical strategies used to quantify respiratory sinus arrhythmia: Are commonly used metrics equivalent? *Biological Psychology, 89*(2), 349–364.

Liberati, A., Altman, D. G., Tetzlaff, J., Mulrow, C., Gøtzsche, P. C., Ioannidis, J. P. A., . . . Moher, D. (2009). The PRISMA statement for reporting systematic reviews of meta-analyses of studies that evaluate health care interventions: Explanation and elaboration. *PLoS Medicine, 6*, e1000100. doi:10.1371/journal/pmed/1000100

Linden, W. (1987). On the impending death of the Type A construct: Or is there a phoenix rising from the ashes? *Canadian Journal of Behavioural Science, 19*, 177–190.

Little, R. J. (2006). Calibrated Bayes. *American Statistician, 60*(3), 213–223. doi:10.1198/000313006X117837

Little, R.J.A. (1988). A test of missing completely at random for multivariate data with missing values. *Journal of the American Statistical Association, 83*(404), 1198–1202.

Lohnes, P. R. (1966). *Measuring adolescent personality*. Pittsburgh, PA: University of Pittsburgh Press.

Lord, F. M. (1967). A paradox in the interpretation of group comparisons. *Psychological Bulletin, 72*, 304–305.

Lowry, R. (2019). *Bayes' theorem: Conditional probabilities*. Retrieved from http://vassarstats.net/bayes.html

Lyon, D., & Greenberg, J. (1991). Evidence of codependency in women with an alcoholic parent: Helping out Mr. Wrong. *Journal of Personality and Social Psychology, 61*, 435–439.

MacCallum, R. C., & Austin, J. T. (2000). Applications of structural equation modeling in psychological research. *Annual Review of Psychology, 51*, 201–226.

MacKinnon, D. P. (2008). *Introduction to statistical mediation analysis*. Hillsdale, NJ: Lawrence Erlbaum.

MacKinnon, D. P., Fairchild, A. J., & Fritz, M. (2007). Mediation analysis. *Annual Review of Psychology, 58*, 593–614.

MacKinnon, D. P., Krull, J. L., & Lockwood, C. M. (2000). Equivalence of the mediation, confounding and suppression effect. *Prevention Science, 1*(4), 173–181.

Malakoff, D. (1999). Bayes offers a "new" way to make sense of numbers. *Science, 286*(5444), 1460–1464.

Malenka, D. J., Baron, J. A., Johansen, S. Wahrenberger, J. W., & Ross, J. M. (1993). The framing effect of relative and absolute risk. *Journal of General Internal Medicine*, *8*(10), 543–548.

Manly, C., & Wells, R. S. (2015). Reporting the use of multiple imputation for missing data in higher education research. *Research in Higher Education*, *56*(4), 397–409. doi:10.1007/s11162-014-9344-9

Maronna, R. A., Martin, R. D., Yohai, V. J., & Salibián-Barrera, M. (2019). *Robust statistics: Theory and methods (with R)*. Hoboken, NJ: John Wiley.

Maslow, A. (1966). *The psychology of science*. New York: Harper & Row.

McDowall, D., McCleary, R., & Bartos, B. J. (2019). *Interrupted time series analysis*. Oxford, UK: Oxford University Press.

Meehl, P. E. (1978). Theoretical risks and tabular asterisks: Sir Karl, Sir Ronald, and the slow progress of soft psychology. *Journal of Consulting and Clinical Psychology*, *46*, 806–834.

Mills, J. L. (1993). Data torturing. *New England Journal of Medicine*, *329*, 1196–1199.

Mischel, W. (1968). *Personality and assessment*. New York: John Wiley.

Mittag, K. C., & Thompson, B. (2000). A national survey of AERA members' perceptions of statistical significance tests and other statistical issues. *Educational Researcher*, *29*(4), 14–20.

Mooney, K. M. (1990). Assertiveness, family history of hypertension, and other psychological and biophysical variables as predictors of cardiovascular reactivity to social stress. *Dissertation Abstracts International*, *51*(3-B), 1548–1549.

Morizot, J., Ainsworth, A. T., & Reise, S. P. (2007). Toward modern psychometrics: Application of item response theory models in personality research. In R. W. Robins, R. C. Fraley, & R. F. Krueger (Eds.), *Handbook of research methods in personality psychology* (pp. 407–421). New York: Guilford.

Morrison, D. E., & Henkel, R. E. (Eds.). (1970). *The significance test controversy*. Chicago: Aldine.

Muller, D., Judd, C. M., & Yzerbyt, V. Y. (2005). When moderation is mediated and mediation is moderated. *Journal of Personality and Social Psychology*, *89*, 852–863.

Muthén, B., Asparouhov, T., Hunter, A. M., & Leuchter, A. F. (2011). Growth modeling with nonignorable dropout: Alternative analysis of the STAR*D antidepressant trial. *Psychological Methods*, *16*(1), 17–33. doi:10.1037/a0022634

Nelson, L. D., Simmons, J. P., & Simonsohn, U. (2012). Let's publish fewer papers. *Psychological Inquiry*, *23*(3), 291–293. doi:10.1080/1047840X.2012.7052-45

Newsom, J. (2017). *Nested models, model identifications, and correlated errors*. Retrieved from http://web.pdx.edu/~newsomj/semclass/ho_nested.pdf

Nieminen, P., Lehtiniemi, H., Vähäkangas, K., Huusko, A., & Rautio, A. (2013). Standardised regression coefficient as an effect size index in summarizing findings in epidemiological studies. *Epidemiology, Biostatistics, and Public Health*, *10*(4), e8854-3. doi:10.2427/8854

Noordzij, M., van Diepen, M., Caskey, F. C., & Jager, K. J. (2017). Relative risk versus absolute risk: One cannot be interpreted without the other. *Nephrology Dialysis Transplantation*, *32*(2), ii13–ii18.

Nunnally, J. C. (1978). *Psychometric theory* (2nd ed.). New York: McGraw-Hill.

Nunnally, J. C., & Bernstein, I. (1994). *Psychometric theory* (3rd ed.). New York: McGraw-Hill.

O'Hara, R. B., & Kotze, D. J. (2010). Do not log-transform count data. *Methods in Ethology and Evolution*, *1*(2), 118–122. doi:10.1111/j.2041-210X.2010.00021.x

Olkin, I., & Finn, J. D. (1995). Correlations redux. *Psychological Bulletin*, *118*, 155–164.

Open Science Collaboration. (2015). Estimating the reproducibility of psychological science. *Science*, *349*, aac4716. doi:10.1126/science.aac4716

Palij, M. (2012, January). New statistical rituals for old. *PsycCRITIQUES* 57(24).

Palmer, M. (n.d.). Ordination methods for ecologists. Retrieved from http://ordination.okstate.edu/

Pampel, F. C. (2000). *Logistic regression: A primer* (Quantitative applications in the social sciences, No. 07-132). Thousand Oaks, CA: Sage.

Parent, M. C. (2013). Handling item-level missing data: Simpler is just as good. *Counseling Psychologist*, *41*(4), 568–600. doi:10.1177/0011000012445176

Paulhus, D. L., Robins, R. W., Trzesniewski, K. H., & Tracy, J. L. (2004). Two replicable suppressor situations in personality research. *Multivariate Behavioral Research*, *39*, 303–328.

Peduzzi, P., Concato, J., Kemper, E., Holford, T. R., & Feinstein, A. (1996). A simulation of the number of events per variable in logistic regression analysis. *Journal of Clinical Epidemiology*, *99*, 1373–1379.

Pek, J., & Flora, D. B. (2018). Reporting effect sizes in original psychological research: A discussion and tutorial. *Psychological Methods*, *23*(2), 208–225. doi:10.1037/met0000126

Peng, C. J., Lee, K. L., & Ingersoll, G. M. (2002). An introduction to logistic regression analysis and reporting. *Journal of Educational Research*, *96*(1), 3–14.

Peterson, E. A., Augenstein, J. S., Hazelton, C. L., Hetrick, D., Levene, R. M., & Tanis, D. C. (1984). Some cardiovascular effects of noise. *Journal of Auditory Research*, *24*, 35–62.

Peterson, R. A., & Brown, S. P. (2005). On the use of beta coefficients in meta-analysis. *Journal of Applied Psychology*, *90*(1), 175–181. doi:10.1037/0021-9010.90.1.175

Pigott, T. D. (2001). A review of methods for missing data. *Educational Research and Evaluation*, 7(4), 353–383. doi:1380-3611/01/0704-353

Porges, S. W. (2011). *The polyvagal theory: Neurophysiological foundations of emotions, attachment, communication, and self-regulation*. New York: W. W. Norton.

Preacher, K. J., & Coffman, D. L. (2006, May). *Computing power and minimum sample size for RMSEA* [Computer software]. Retrieved from http://www.quantpsy.org/rmsea/rmsea.htm

Preacher, K. J., & Hayes, A. F. (2008). Asymptotic and resampling strategies for assessing and comparing indirect effects in multiple mediator models. *Behavior Research Methods*, *40*, 879–891.

Preacher, K. J., Rucker, D. D., & Hayes, A. F. (2007). Addressing moderated mediation hypotheses: Theory, methods, and prescriptions. *Multivariate Behavioral Research*, *42*, 185–227.

Prentice, D. A., & Miller, D. T. (1992). When small effects are impressive. *Psychological Bulletin*, *112*(1), 160–164.

Preregistration of research plans. (n.d.) Retrieved from https://www.psychologicalscience.org/publications/psychological_science/preregistration

Radloff, L. S. (1977). The CES-D scale: A self-report depression scale for research in the general population. *Applied Psychological Measurement*, *1*, 385–401.

Recommended data repositories. (n.d.). Retrieved from https://www.nature.com/sdata/policies/repositories

Ried, K. (2008). Interpreting and understanding meta-analysis graphs. *Australian Family Physician*, *35*(8), 635–638.

Robinson, J. P., Shaver, P. R., & Wrightsman, L. S. (Eds.). (1991). *Measures of personality and social psychological attitudes*. San Diego, CA: Academic Press.

Robson, K., & Pevalin, D. (2015). *Multilevel modeling in plain language*. Thousand Oaks, CA: Sage.

Rosenthal, R. (1995). Writing meta-analytic reviews. *Psychological Bulletin*, *118*(2), 183–192. doi:10.1037/0033-2909.118.2.183

Rosenthal, R., & Rosnow, R. L. (1975). *The volunteer subject*. Oxford, UK: Wiley.

Rosenthal, R., & Rosnow, R. L. (1991). *Essentials of behavioral research: Methods and data analysis* (2nd ed.). New York: McGraw-Hill.

Rosnow, R. L., & Rosenthal, R. (1989). Statistical procedures and the justification of knowledge in psychological science. *American Psychologist*, *44*(10), 1276–1284.

Royal Statistical Society (n.d.) *Timeline of statistics*. Retrieved from http://www.statslife.org.uk/images/pdf/timeline-of-statistics.pdf

Rozeboom, W. W. (1960). The fallacy of the null-hypothesis significance test. *Psychological Bulletin*, *57*, 416–428.

Rubin, Z. (1970). Measurement of romantic love. *Journal of Personality and Social Psychology*, *16*, 265–273.

Rubin, Z. (1976). On studying love: Notes on the researcher-subject relationship. In M. P. Golden (Ed.), *The research experience* (pp. 508–513). Itasca, IL: Peacock.

Sarawitz, D. (2016). The pressure to publish pushes down quality. *Nature*, *553*(7602), 147. doi:10.1038/533147a

Savalei, V., & Dunn, E. (2015). Is the call to abandon *p*-values the red herring of the replicability crisis? *Frontiers in Psychology*, *6*, 245. doi:10.3389/fpsyg.2015.00245

Sawilowsky, S. S., & Blair, R. C. (1992). A more realistic look at the robustness and Type II error properties of the *t* test to departures from population normality. *Psychological Bulletin*, *111*(2), 352–360. doi:10.1037/0033-2909.111.2.352

Schafer, J. L., & Graham, J. W. (2002). Missing data: Our view of state of the art. *Psychological Methods*, 7(2), 147–177. doi:10.1037//1082-989X.7.2.147

Schimmack, U. (August 2, 2017). *What would Cohen say? A comment on* p < *.005*. Retrieved from https://replicationindex.wordpress.com/2017/08/02/what-would-cohen-say-a-comment-on-p-005/

Schlomer, G. L., Bauman, S., & Card, N. A. (2010). Best practices for missing data management in counseling psychology. *Journal of Counseling Psychology*, *57*(1), 1–10. doi:10.1037/a0018082

Schmidt, C. M. (2004). *David Hume: Reason in history*. Philadelphia: Pennsylvania State University Press.

Schreiber, J. B., Stage, F. K., King, J., Nora, A., & Barlow, E. A. (2006). Reporting structural equation modeling and confirmatory factor analysis results: A review. *Journal of Educational Research*, *99*(6), 323–338.

Schulman, K. A., Berlin, J. A., Harless, W., Kerner, J. F., Sistrunk, S., Gersh, B. J., . . . Escarce, J. J. (1999). The effect of race and sex on physicians' recommendations for cardiac catheterization. *New England Journal of Medicine*, *340*, 618–626.

Schwartz, A. J. (2003). A note on logistic regression and odds ratios. *Journal of American College Health*, *51*, 169–170.

Schwartz, L. M., Woloshin, S., & Welch, H. G. (1999). Misunderstandings about the effects of race and sex on physicians' referrals for cardiac catheterization. *New England Journal of Medicine*, *341*, 279–283.

Sharpe, D. (2013). Why the resistance to statistical innovations? Bridging the communication gap. *Psychological Methods*, *18*(4), 572–582. doi:10.1037/a0034177

Simmons, J. P., Nelson, L. D., & Simonsohn, U. (2011). False-positive psychology: Undisclosed flexibility in data collection and analysis allows presenting anything as significant. *Psychological Science, 22*, 1359–1366.

Singer, J. D., & Willett, J. B. (2003). *Applied longitudinal data analysis: Modeling change and event occurrence.* Oxford, UK: Oxford University Press.

Snijders, T., & Bosker, R. (2011). *Multilevel analysis: An introduction to basic and advanced multilevel modeling* (2nd ed.). Thousand Oaks, CA: Sage.

Sobel, M. E. (1982). Asymptotic confidence intervals for indirect effects in structural equation models. In S. Leinhardt (Ed.), *Sociological methodology* (pp. 290–312). Washington, DC: American Sociological Association.

Soper, D. (2019). *Calculator: R-square confidence interval.* Retrieved from https://www.danielsoper.com/statcalc/calculator.aspx?id=28

Soumerai, S. B., Ceccarelli, R., & Koppel, R. (2017). False dichotomies and health policy research: Randomized trials are not always the answer. *Journal of General Internal Medicine, 32*(2), 204–209.

Spring, B., Chiodo, J., & Bowen, D. J. (1987). Carbohydrates, tryptophan, and behavior: A methodological review. *Psychological Bulletin, 102*, 234–256.

Steering Committee of the Physicians' Health Study Research Group. (1989). Final report on the aspirin component of the ongoing Physicians' Health Study. *New England Journal of Medicine, 321*, 129–135.

Stevens, J. P. (2009). *Applied multivariate statistics for the social sciences* (5th ed.). New York: Routledge.

Stigler, S. M. (1990). *The history of statistics: The measurement of uncertainty before 1900.* Cambridge, MA: Belknap.

Stricker, L. J. (1997). *Using just noticeable differences to interpret test of spoken English scores.* Research Reports, Report 58, RR-97-4. Princeton, NJ: Educational Testing Services.

Sullivan, G. M., & Feinn, R. (2012). Using effect size—Or why the *p* value is not enough. *Journal of Graduate Medical Education, 4*(3), 279–282. doi:10.4300/JGME-D-12-00156.1

Tabachnick, B. G., & Fidell, L. S. (2018). *Using multivariate statistics* (7th ed.). Boston: Pearson.

Tankard, J. W. (1984). *The statistical pioneers.* Cambridge, MA: Schenkman.

Tatsuoka, M. M. (1988). *Multivariate analysis: Techniques for educational and psychological research* (2nd ed.). New York: Macmillan.

Taylor, A. B., MacKinnon, D. P., & Tein, J.-Y. (2008). Test of the three-path mediated effect. *Organizational Research Methods, 11*, 241–269.

Thompson, B. (1999). Statistical significance tests, effect size reporting, and the vain pursuit of pseudo-objectivity. *Theory and Psychology, 9*, 191–196.

Thompson, B. (2002a). "Statistical," "practical," and "clinical": How many kinds of significance do counselors need to consider? *Journal of Counseling and Development, 80*(1), 64–71. doi:10.1002/j.1556-6678.2002.tb00167.x

Thompson, B. (2002b). What future quantitative social science research could look like: Confidence intervals for effect sizes. *Educational Researcher, 31*(3), 25–32. doi:10.3102/0013189X031003025

Tkach, C., & Lyubomirsky, S. (2006). How do people pursue happiness? Relating personality, happiness-increasing strategies, and well-being. *Journal of Happiness Studies, 7*, 183–225.

Trafimow, D., & Marks, M. (2015). Editorial. *Basic and Applied Social Psychology, 37*, 1–2. doi:10.1080/01973533.2015.1012991

Trochim, W. M. (2006). *The research methods knowledge base* (2nd ed.). Retrieved from http://www.socialresearchmethods.net/kb/

Twisk, J., & de Vente, W. (2002). Attrition in longitudinal studies: How to deal with missing data. *Journal of Clinical Epidemiology, 55*(4), 329–337. doi:10.1016/S0895-4356(01)00476-0

UCLA Institute for Digital Research & Education. 2019. *Negative binomial regression | SPSS data analysis examples.* Retrieved from https://stats.idre.ucla.edu/spss/dae/negative-binomial-regression/

Ullman, J. B. (2007). Structural equation modeling. In B. G. Tabachnick & L. S. Fidell (Eds.), *Using multivariate statistics* (5th ed., pp. 676–780). Boston: Pearson.

University of Texas at Austin, Department of Statistics and Data Science. (n.d.) Handling non-normal data in AMOS. Retrieved from https://stat.utexas.edu/software-faqs/amos#handlingnonnormdata

Vigen, T. (n.d.) *Spurious correlations.* Retrieved from https://www.tylervigen.com/spurious-correlations

Vogt, W. P. (1999). *Dictionary of statistics and methodology: A nontechnical guide for the social sciences.* Thousand Oaks, CA: Sage.

Wagner, B., Riggs, P., & Mikulich-Gilbertson, S. (2015). The importance of distribution-choice in modeling substance use data: A comparison of negative binomial, beta binomial, and zero-inflated distributions. *American Journal of Drug and Alcohol Abuse, 41*(6), 489–497. doi:10.3109/00952990.2015.1056447

Wainer, H. (1976). Estimating coefficients in multivariate models: It don't make no nevermind. *Psychological Bulletin, 83*, 213–217.

Wainer, H. (2000). *Computerized adaptive testing: A primer* (2nd ed.). Mahwah, NJ: Lawrence Erlbaum.

Warner, R. M. (1979). Periodic rhythms in conversational speech. *Language and Speech, 22*, 381–396.

Warner, R. M. (1991). Incorporating time. In B. Montgomery & S. Duck (Eds.), *Studying interpersonal interaction.* New York: Guilford.

Warner, R. M. (1992). Sequential analysis of social interaction: Assessing internal versus social determinants of behavior. *Journal of Personality and Social Psychology, 63*, 51–60.

Warner, R. M. (1996). Coordinated cycles in behavior and physiology during face-to-face social interaction. In J. Watt & A. Van Lear (Eds.), *Cycles and dynamic patterns in communication processes*. Newbury Park, CA: Sage.

Warner, R. M. (1998). *Spectral analysis of time-series data*. New York: Guilford Press.

Warner, R. M. (2002a). Rhythms of dialogue in infancy: Comments on Jaffe, Beebe, Feldstein, Crown and Jasnow (2001). *Journal of Psycholinguistic Research*, *31*, 409–420.

Warner, R. M. (2002b). What microanalysis of behavior in social situations can tell us about relationships over the life span. In A. L. Vangelisti, H. T. Reis, & M. A. Fitzpatrick (Eds.), *Stability and change in relationships* (pp. 207–227). Cambridge, UK: Cambridge University Press.

Warner, R. M. (2020). *Applied Statistics I: Basic bivariate techniques*. Thousand Oaks, CA: Sage.

Warner, R. M., & Mooney, K. (1988). Individual differences in vocal activity rhythm: Fourier analysis of cyclicity in amount of talk. *Journal of Psycholinguistic Research*, *17*, 99–111.

Warner, R. M., & Rasco, D. (2014). Structural equation models for prediction of subjective well-being: Modeling negative affect as a separate outcome. *Journal of Happiness and Well-Being*, *2*(1), 161–176.

Warner, R. M., & Sugarman, D. B. (1986). Attributions of personality based on physical appearance, speech, and handwriting. *Journal of Personality and Social Psychology*, *50*, 792–799.

Warner, R. M., & Vroman, K. G. (2011). Happiness inducing behaviors in everyday life: An empirical assessment of "The How of Happiness." *Journal of Happiness Studies*, *12*, 1–10.

Warner, R. M., Frye, K., Morrell, J. S., & Carey, G. (2017). Fruit and vegetable intake predicts positive affect. *Journal of Happiness Studies*, *18*(3), 809–826. doi:10.1007/s10902-016-9749-6

Warner, R. M., Malloy, D., Schneider, K., Knoth, R., & Wilder, B. (1987). Rhythmic organization of social interaction and observer ratings of affect and involvement. *Journal of Nonverbal Behavior*, *11*, 57–74.

Warner, R. M., Waggener, T. B., & Kronauer, R. E. (1983). Synchronized cycles in ventilation and vocal activity during spontaneous conversational speech. *Journal of Applied Physiology: Respiration, Environmental and Exercise Physiology*, *54*, 1324–1334.

Watson, J. C., Lenz, A. S., Schmit, M. K., & Schmit, E. L. (2016). Calculating and reporting estimates of effect size in counseling outcome research. *Counseling Outcome Research and Evaluation*, *7*(2), 111–123. doi:10.1177/2150137816660584

Westen, D., & Rosenthal, R. (2003). Quantifying construct validity: Two simple measures. *Journal of Personality and Social Psychology*, *84*, 608–618.

Whisman, M. A., & McClelland, G. H. (2005). Designing, testing, and interpreting interactions and moderator effects in family research. *Journal of Family Psychology*, *19*, 111–120.

Wicherts, J. M., Veldkamp, C.L.S., Augusteijn, H.E.M., Bakker, M., van Aert, R.C.M., & van Assen, M.A.L.M. (2016). Degrees of freedom in planning, running, analyzing, and reporting psychological studies: A checklist to avoid *p*-hacking. *Frontiers in Psychology*, *7*, Article 1832. doi:10.3390/fpsyg.2016.01832

Wiggins, J. S. (1973). *Personality and prediction: Principles of personality assessment*. New York: Random House.

Wilkinson, L., & Dallal, G. E. (1981). Tests of significance in forward selection regression with an *F*-to-enter stopping rule. *Technometrics*, *23*, 377–380.

Wilkinson, L., & Task Force on Statistical Inference, APA Board of Scientific Affairs. (1999). Statistical methods in psychology journals: Guidelines and explanations. *American Psychologist*, *54*, 594–604.

Williamson, J. (2013). Why frequentists and Bayesians need each other. *Erkenn*, *78*(2), 293–318. doi:10.1007/s10670-011-9317-8

Wolf, E. J., Harrington, K. M., Clark, S. L., & Miller, M. W. (2013). Sample size requirements for structural equation models: An evaluation of power, bias, and solution propriety. *Educational and Psychological Measurement*, *73*(6), 913–934. doi:10.1177/0013164413495237

Woloshin, S., Schwartz, L. M., Casella, S. L., Kennedy, A. T., & Larson, R. J. (2009). Press releases by academic medical centers: Not so academic? *Annals of Internal Medicine*, *150*(9), 613–618.

Wright, R. E. (1995). Logistic regression. In L. G. Grimm & P. R. Yarnold (Eds.), *Reading and understanding multivariate statistics* (pp. 217–244). Washington, DC: American Psychological Association.

INDEX

ab, 296–297, 300
Additivity, 510
Adjusted degrees of freedom, 516
Adjusted effects in ANCOVA (α^*_i), 266, 278
Adjusted means in ANCOVA, 257, 262, 267, 278, 281
Adjusted R^2, 113
Akaike information criterion, 564
Alpha level (α)
  $\alpha = .005$, 10
  Bonferroni corrections, 328, 333
  experiment-wise (EW_a), 328
  per-comparison (PC_a), 328, 355, 386
A matrix, 410, 456
American Psychological Association (APA)
  online resources, 505–506
  Task Force on Statistical Inference, 4
Amos Graphics program, 305, 540–578
  analysis, 554–557
  analysis properties, 552–554
  bootstrapped confidence intervals, 557–561
  results, 554–557
  variable names and paths, 547–551
Analysis of covariance (ANCOVA), 212, 254
  adjusted means on outcome variable, 257, 262, 267, 278, 281
  advantages of controlling for covariate variables, 256
  ANOVA outcomes versus, 283
  assessing pretest/posttest differences, 264, 281–284
  assumptions, 257, 259–262, 270, 281, 357
  computing adjusted effects and means, 266, 278
  covariate selection, 264
  effect size, 267, 273, 279
  empirical example and sample data, 257–259
  error variance suppression, 257
  example results, 275–279
  formulas, 265–266
  F ratios, 256–257, 266
  interaction terms, 273–275
  MANCOVA, 394
  matrix algebra for computing sums of squares, 265
  multiple factors or covariates, 281
  nonequivalent comparison groups and, 254
  no treatment by covariate interaction assumption, 259, 265, 270, 358
  null hypotheses, 255, 266
  outcomes of controlling for covariate variables, 266–267
  preliminary data screening, 259–262, 269–270
  repeated-measures ANOVA versus, 529–530
  research situations, 254–257, 279–281
  sample Results section, 275–277
  SPSS procedures and output, 259, 269–279
  statistical power and, 268
  summary, 279–281
  treatment group means rank ordering, 266–267
  unadjusted *Y* means, 278–279
  variable order of entry, 264
  variance partitioning, 263–264
Analysis of variance (ANOVA), 329, 353, 362
  ANCOVA outcomes versus, 283
  assumptions for, 6
  comparison with discriminant analysis and multiple linear regression, 349
  discriminant analysis and, 315, 328–329, 332–333, 338, 345–349
  dummy variables. *See* Dummy variables, ANOVA application
  equivalence to regression analysis, 188, 190, 210–212
  factorial. *See* Factorial ANOVA
  F ratio in, 7
  interaction terms, ANOVA versus regression, 212
  MANOVA versus, 355–356, 362–364, 395
  mixed-model, 521–530
  null hypothesis and, 353
  post hoc ANOVAs for discriminant analysis, 328
  post hoc protected tests, 212
  preliminary data screening for ANCOVA, 260–262
  regression versus, relative advantages and disadvantages, 212
  repeated-measures. *See* Repeated-measures ANOVA
  research example, 194–197
  significance testing, MANOVA comparison, 362
  sums of squared deviations, 361–362

Type I error risk from running multiple ANOVAs, 355
unequal group sizes and, 194
univariate, 521
variance partitioning, 362
Analysis of variance (ANOVA), using dummy predictor variables
assumptions, 190
empirical example and sample data, 190
reporting results, 210–211
research example, 194–197
sample size and statistical power, 190
screening for violations of assumptions, 191–195
study design issues, 193–194
Aroian test, 300
Attenuation of correlation, 474–475
Authors, 23–24
Autoregressive model, 631
Available case analysis, 56

Backward method of entry, 142
Bar graphs, 12–13
Bayesian statistics, 9–10
Bayes' theorem, 639–640
b coefficient for logistic regression, 595, 603
Beck Depression Inventory (BDI), 465, 503
Bem Sex Role Inventory, 404–405, 567
Bentler comparative fit index, 565
Beta coefficients
equations for, 120–121
magnitude of, 117–118
multiple regression with multiple predictors, 117–118, 133, 176
raw-score data matrices used to calculate, 177–180
standard-score, 111
Between-*S* factors, 522–530
See also Independent-samples *t* test
Bias, 466
confirmation, 23
definition of, 31
faking, 361, 498
missing values as cause of, 32
publication, 24
reverse-worded questions and, 479–480
social desirability, 361, 478
sources of, 31–34
yea-saying, 479
"Big Five" model of personality, 403, 407, 445
Binary logistic regression, 190, 583
assumptions, 584
B coefficient, 595, 603
causal interpretations and, 604
chi-square test versus, 605–606
coding binary scores, 608–610
coefficient estimation, 592
correctly specified model, 584
discriminant analysis versus, 620
empirical example, 584–585
example using categorical and quantitative predictors, 611–619
expected cell frequencies, 607–608
generalization for multiple outcomes, 611
group classification accuracy, 595–596
interaction terms, 611
linear regression versus, 585–587
logit as outcome variable, 588, 590–591
maximum likelihood estimation, 592
more complex models, 610–611
multiple linear regression and, 583
multiple predictor variables, 611
null model, 591, 600
odds and odds ratios, 588–590, 620–621
outliers and, 607
overall goodness of fit, 592–594, 601, 613
preliminary data screening, 584, 607–608
quantitative predictor variables, 610
reporting results, 604, 619
research situations and questions, 583
sample Results section, 606–607
sample size and statistical power, 608
significance of individual predictors, 595, 602–603
slope coefficient interpretation, 595, 603
SPSS procedures and output, 596–607
summary, 620–622
terms involved in, 591–592
variables, 583
Binomial regression, 638–639
Bivariate correlation, 64
Bivariate linear regression
binary logistic regression versus, 584
dummy-coded dummy variable example, 197–199
effect-coded dummy variable example, 200–205
Bivariate outliers, 35, 38–41, 108
Bivariate Pearson correlation. *See* Pearson's *r*
Bivariate regression
coefficients for, 101
definition of, 99
two-dimensional graph for, 102
used to remove variance predictable by X_2 from both X_1 and Y, 74–77
Bonferroni procedure, 328–329, 333
Bootstrapped confidence interval, 300, 553, 557–561
Bootstrapping, 300, 553, 577–578
Boundary rules, for discriminant functions, 313
Box's *M* test, 322, 331, 335, 359–360
MANOVA and, 368
b slope coefficient, confidence interval for, 114

C (level of confidence), 1
Canonical correlation (r_c), 325, 327, 337, 365

Case identification numbers, 29
Categorical predictor, quantitative predictor and, 219–226
Categorical variables
  coding for binary logistic regression, 608–610
  interaction, factorial ANOVA, 217–219
  interobserver reliability assessment, 470–471
  multiple-group predictors in regression, 187–188
  as predictor, 219
  X_2, 66–70
Causal hypotheses, 85, 544
Causal inference, logistic regression and, 604
Causal models
  definition of, 82, 544
  limitations of, 290–291
  See also Path models
Causal path, 540
Causal-steps approach, 298–299
Centering, 233
Centroid, 341
CES-D scale, 465, 468, 476, 478–481, 505–506
  Cronbach's alpha for items, 490–494
  factor analysis of items, 488
  validity, 498
CFA. *See* Confirmatory factor analysis
CFI. *See* Comparative fit index
Change scores, 281
  potential problems with, 282–284
Chi-square (χ^2) tests
  effect size for, 14
  Hosmer and Lemeshow goodness-of-fit test, 613
  logistic regression and, 606
  logistic regression and, Wald statistic for, 595, 619
  –2 log likelihood statistic (–2LL), 593
  Wilks's Λ and, 315, 323, 363
Chi-square goodness-of-fit index, 565
Classical suppression, 91–94
Classification errors, 311, 320
Classification table, 618
Clinical significance, effect size used to evaluate, 18–19
Cluster analyses, 628–630
Cohen's *d*
  clinical significance of, 18
  definition of, 14
  verbal labeling of, 15
Cohen's kappa (κ), 468, 471
Communality, 410
Comparative fit index, 565
Comparison groups, nonequivalent, 254
Complete case analysis, 56
Concurrent validity, 499, 503
Confidence intervals
  bootstrapped, 300, 553, 557–561
  for *b* slope coefficient, 114, 126
  C for, 1
  effect sizes and, 16
  error bar graphs, 12–13
  graphing of, 12
  interpretation of, 11–12
  for partial *r*, 81
  p values versus, 13–14
  R^2, 183
  review of, 10–14
  setting up, 10–11
  structural equation modeling, 553
Confirmation bias in publication, 23
Confirmatory factor analysis
  characteristics of, 565–569
  comparison of, 575
  description of, 398, 540
  latent variables in, 457–458
  measurement invariance, 575–576
  models of, 461
  structural equation modeling using, 461
Conservative tests, 360, 391
Consolidated Standards of Reporting Trials. *See* CONSORT
CONSORT, 30, 59
Constraints on SEM model parameters, 567
Construct validity, 496–497
Content validity, 497–498
Contingency tables
  chi-square analysis of. *See* Chi-square (χ^2) tests
  discriminant analysis, 319
  group classification accuracy for logistic regression, 595–596
Contrast coefficients, 205–206
Control groups, nonequivalent, 255, 281
Control variables, 65, 124
  See also Covariates
Convenience samples, 6
Convergent validity, 498–499
Coordinated interpersonal timing, 637
Coordination, 637
Correctly specified model, 584
Correlation
  attenuation of, 474–475
  data screening for, 35
  factor loading as, 434
  partial. *See* Partial correlation
  R matrix of, 399
  semipartial. *See* Semipartial (or "part") correlation (sr and sr^2)
  spurious. *See* Spurious correlation
  zero-order, 70, 72–73, 82, 87, 90
Correlation coefficients, dichotomous variables and, 188
Correlation matrix (**R**)
  computation of, 411–412, 424–425
  diagonalization of, 455
  eigenvalues and, 455

eigenvectors and, 455
factor analysis and, 405
factor loading, 408–409
MANOVA and, 363
multiple-item scale internal consistency reliability assessment, 494
principal-component analysis and, 405
reproduced, 410
Covariates, 254, 256
definition of, 65
multiple, 281
noise or error variance suppressors, 257
no treatment by covariate interaction assumption, 259, 265, 270, 358
outcomes of controlling for, 266–267
selection of, 264
significance tests, 266
Cox and Snell's R^2, 594, 618
Criterion-oriented validity types, 498–499
Cronbach's alpha (α), 468, 478, 489–495
factors affecting size of, 490, 494
formula, 490
improving of, by dropping "poor" item, 494
KR-20 reliability coefficient, 490
multiple-item scale example, 490–494
proportion of variance interpretation, 493
split-half reliability and, 495
Crossover suppression, 94–95
Cross products, sum of, 206
Cross-tabulation, SPSS Crosstabs, 375
Cutoff value, 318, 338
Cycles, in time series, 634–637

Data-driven regression, 135, 140–142
Data files, 30–31
"Data fishing," 124, 453
Data reduction, factor analysis and, 401–404
Data screening
in all situations, 34
ANCOVA, 259–262, 269–270
ANOVA using dummy predictor variables, 191–193
binary logistic regression, 584
checklist for, 59–60
for correlation, 35
discriminant analysis, 320–322
file tracking, 30–31
group means comparison with, 34–35
histograms, 259–260
MANOVA, 358–360
mediation analysis, 293–294
missing values, coding for, 29–30
one categorical and one quantitative predictor, 220–221
for partial correlation between X_1 and Y, controlling for X_2, 72
for regression, 35
variable names, 29–31
Decision error, 3
Dendrogram, 628–629
Determinants, 324
describing variance, 363
of matrix, 174–177
Diagnostic and Statistical Manual of Mental Disorders-IV (DSM-IV), 498
Diagonal matrix, 405
Dichotomous variables, 188, 583
coding for binary logistic regression, 608–610
Cronbach's alpha assessment of response homogeneity, 490
Difference scores, 281, 529–530
Dimension reduction analysis, 317
Direct difference *t* test, 282
Direct effect, 297
Discriminant analysis, 309
ANOVA and, comparison between, 349
assessing contributions of individual predictors, 317–318
assumptions, 320–322, 357–358, 584
binary logistic regression versus, 620
Box's *M* test, 322, 331, 335
classification errors and, 311, 320
contingency table, 319
describing differences between groups, 311
dimension reduction, 317
discriminating variable selection, 322
effect size, 327
effect size, Wilks's Λ conversion, 324
eigenvalue/eigenvector problem, 349–351, 365
empirical example and sample data, 320
equations, 323–326, 350
example results, 329–345
factors affecting Wilks's Λ magnitude, 326–327
factor scores, 487–488
finding optimal weighted linear combination of scores, 309–312, 487
follow-up tests and, 318–319, 328–329, 345–348
F ratios, 316, 318, 331
group differences versus controlling for predictor correlations, 333
group membership outcomes, predicting, 310–311, 313–315, 327, 341–342
group numbers and sizes, selecting, 322
handling violations of assumptions, 321–322, 335
MANOVA and, 349, 354, 356–357, 365, 394–395
MANOVA and, post hoc testing, 372, 378–386, 390
meaningful interpretations or labels, 319
multiple-item scale development, 502
multiple linear regression versus, 309–311, 349
nonsignificant outcomes, interpreting, 328
null hypotheses, 316, 322
post hoc ANOVAs for, 328

preliminary data screening, 322
research questions for, 310–311, 316–320
reporting results, 329
research situations, 309–315
sample Results section, 343–345
sample size and statistical power, 321, 327, 336
separate-groups covariance matrices, 331
significance of discriminating variables, 325–326, 338
significance tests, 315, 317, 323–324, 328–329, 331, 333
SPSS procedures and results, 317, 329–345, 356–357
stepwise entry of predictor variables, 348
study design issues, 322–323
summary, 348–349
sum of cross-products matrix, 321–323, 350
unequal group sizes and, 620
unexplained variance estimate (Wilks's Λ), 315
unit-weighted composite variable, 320
univariate ANOVA and, 315, 328, 332–333, 338, 345–348
variance/covariance matrices, 321–323
Discriminant function coefficients, 313, 325
eigenvalues/eigenvectors and canonical correlation, 325, 337
interpreting, 325, 338
sample size and, 327
SPSS and structure coefficients, 326
standardized, 318, 325
Discriminant functions, 310, 312
boundary rules and territorial maps, 313
equation, 312
Fisher linear discriminant function, 329
MANOVA and, 365, 394–395
meaningful interpretations or labels, 317
number of, 311, 317, 337
one-way ANOVA on scores on, 345–348
optimally weighted composites of scores, 487–488
Discriminant validity, 499, 503
Disordinal interaction, 219
Distribution
normal. *See* Normal distribution assumptions
skewness of, 34
Distribution shape, 477
Dosage, 193
statistical power and, 608
Double negatives, 5
Doubly multivariate repeated measures, 518–522
Dummy-coded dummy variables, 197
multiple group example, 200
two-group example, 197–199
Dummy variables
dummy-coded, 197–200
multiple-group categorical variables versus, 187–188
practical applications, 211–212
study design issues, 193
summary, 211–212
trend analysis and, 207
when to use, 187–189
Dummy variables, ANOVA application
assumptions, 190
empirical example and sample data, 190
research example, 194–197
sample size and statistical power, 190
screening for violations of assumptions, 191–195
study design issues, 193
Dummy variables, coding methods, 197–207
dummy coding, multiple group example, 200
dummy coding, two-group example, 197–199
effect coding, multiple group example, 203–205
effect coding, two-group example, 200–203
orthogonal coding, 205–207
Dummy variables, regression applications
dummy coding, multiple group example, 200
dummy coding, two-group example, 197–199
effect coding, multiple group example, 203–205
effect coding, two-group example, 200–203
effect size, 210
models including quantitative predictors, 208–210
parameter estimates, 194
reporting results, 211
sample Results section, 211
sample size and statistical power, 210
significance tests, 194
slope coefficient interpretation, 189, 194, 198
study design issues, 193

e, 589
Educational Testing Service (ETS), 506
Effect
direct, 297
indirect, 298
mediated. *See* Mediated effect
total, 297
Effect-coded dummy variables, 200–201
multiple group example, 203–205
two-group example, 200–203
Effect sizes
ANCOVA, 267, 273, 279
in biomedical research, 18
chi-square tests, 14
clinical significance evaluated using, 18–19
confidence intervals for, 16
discriminant analysis, 327
discriminant analysis, Wilks's Λ conversion, 324
eta squared, 217
generalizations about, 14–17
interpretation of, 16
MANOVA, 367
mediation analysis, 301–302
multiple regression with multiple predictors, 122–123, 148–149
for partial *r*, 81–82

practical significance evaluated using, 18–19
reasons for not reporting, 17
regressions using dummy variables, 210
regression with two predictors, 122–123
sample size versus, 14, 16–17
SPSS and, 17
squared semipartial correlation, 194
theoretical significance evaluated using, 17–18
uses of, 19–20
verbal labeling of, 15
See also Cohen's *d*
Eigenproblem, 454
Eigenvalues (λ), 349–351
discriminant analysis and, 325, 337
MANOVA and, 365
Eigenvectors, 325, 349–351, 454–455
Empirical keying, 498
Entrainment, 637
Equivalent models, 564
Error
classification, 311, 320
measurement. *See* Measurement error
Type I. *See* Type I error
Error bar graphs, 12–13
Error variance, 93
Error variance suppressor, 257
Eta squared (η^2)
ANCOVA and, 267, 279
effect sizes, 217
partial η^2, 267, 279, 367
Wilks's Λ and, 324, 327, 333, 346, 363, 367
Exp(*B*), 595, 600, 603, 609, 619
Exp(*X*), 589
Expectancy effects, 361
Expected cell frequencies, 607–608
Experimental control, 65
Experimental design, ANCOVA and, 255, 264–265
Experimenter expectancy effects, 361
Experiment-wise alpha (EW_a), 328
Exploratory factor analysis, 404, 448, 457
Exponential functions, 589, 603
Extraction, 400
Extraneous variables, 476
Extreme outliers, 547
Extreme scores, 322, 607
Eysenck Personality Inventory, 496

Face validity, 497–498
Factor analysis
components or factors, number of, 445–446
computation of, 410–414
computation of loadings, 411–412
correlation matrix computation in, 411
as data reduction method, 401–404
definition of, 398, 410
dimensionality of a set of items assessed using, 480–482
empirical example, 404–405
exploratory, 404, 448, 457
factor loading computations, 408–410, 455
factor rotation, 434–437
factor scores versus unit-weighted composites, 449–451
importance of components or factors, 446
interpretation of, questions to address in, 445–447
issues in, summary of, 451–454
limiting the number of components or factors, 412–413
matrix algebra of, 454–457
multiple-item scales, 403, 478, 501–502
naming or labeling components or factors, 414, 446–447
orthogonal, 400
path model for, 400–401
questions to address in interpretation of, 445–447
results of, 453
retained components or factors, 447
rotation of factors, 413–414
screening for violations of assumptions, 405–407
study design issues, 407–408
two sets of multiple regressions, 438–440
Factorial ANOVA, 360
categorical predictors, interaction between, 217–219
unequal group sizes and (nonorthogonal factorial ANOVA), 360
Factorial MANOVA, 360, 375–389
Factor loadings. *See* Loadings
Factor rotation, 434–437
Factors, 398
Factor score, 487–488
definition of, 403
unit-weighted composites versus, 449–451
unit-weighted sums of items and, correlation between, 488–489
Factor score coefficients, 487–488, 565
definition of, 438
problems associated with, 449
unit-weighted composites versus, 488
Faking, 361, 498
False negatives, 320
False positives, 25, 320
File tracking, 30–31
First-order partial correlation, 73
Fisher classification coefficients, 340
Fisher linear discriminant function, 329
Five-point rating scales, 465
Follow-up (post hoc) tests
analyses using dummy variables, 210
discriminant analysis and, 318–319, 328
discriminant analysis and, ANOVA scores on discriminant functions, 345–348
MANOVA, 368–374, 378–390
MANOVA, comparison of univariate and multivariate analyses, 393–394

multivariate, inflated Type I error risk and, 372, 393
protected tests, ANOVA, 212
Forest plots, 21
Forward method of entry, 142
Forward statistical regression, multiple R^2 significance in, 180–183
F ratios, 324, 362
ANCOVA and, 256–257, 266
ANOVA and, 7
critical values, 362
discriminant analysis, 316, 318, 331
Wilks's Λ and, 315, 323–324, 364
Frequentist methods, 640
F test for null hypothesis (H_0): $R = 0$, 113
F-to-enter, 145
Fully identified model, 555

Gain scores (or change scores), 281
potential problems with, 282–284
General intelligence, 399
General linear model (GLM)
assumptions, 358
description of, 31, 188, 357–358
Goodman test, 300
Goodness-of-fit tests
chi-square, 565
Cox and Snell's R^2, 594, 618
Hosmer and Lemeshow, 613
logistic regression models, 592–594, 601, 613
log likelihood model, 601
structural equation modeling, 564–565
Wilks's Λ, 315
Grand mean, 202, 205
Graphing interactions
between quantitative variables "by hand," 244–249
for two quantitative predictors, 236–242
Greenhouse-Geisser test, 515
Group means, 34–35
Growth curve analysis, 282, 530

Hamilton Depression Rating Scale, 465
Hierarchical modeling, 640
Hierarchical regression
description of, 135, 140–142
SPSS menu selections, output, and results for, 156–162
variance partitioning in, 142–144
Histograms, 259–260
Homogeneity of regression assumption, 259
Homogeneity of variance assumption
ANCOVA and, 260
ANOVA using dummy predictor variables and, 190, 194–195
Homogeneity of variance/covariance matrices assumption, 322, 331, 336, 357, 359, 371, 584
Homoscedasticity of variance assumption, 587
Hosmer and Lemeshow goodness-of-fit test, 613
Hotelling's T^2, 366
Hotelling's trace, 364–366
Hypothesis, null. *See* Null hypothesis (H_0)

Inconsistent mediation, 299
Incremental sr^2, sr^2_{inc}, 141
Increment in R^2, R^2_{inc}, 141
Independence model, 552
Independent-samples t test
assumptions for, 6
general linear model, 357
sample size for, 16
Indicators, 566
Indirect effect, 298
Inflated risk of Type I error, 328, 333, 355, 393
Institutional incentives and norms, 25
Integrated model, 631
Intelligence quotient (IQ), 472
description of, 399
measurement reliability, 472
Interaction
analysis of, 243
definition of, 215
disordinal, 219
graphing of, 236–242
moderation, 215
more than three categories, 226–228
nonadditivity, 215
one categorical and one quantitative predictor, 219–226
sex-by-years, 228–230
summary of, 243
two categorical variables, factorial ANOVA, 217–219
Interaction effects, MANOVA and, 378, 385–392
Interaction terms
ANCOVA model, 273–275
ANOVA versus regression, 212
binary logistic regression and, 611
Intercept (b_0), 198, 202
Internal consistency reliability
assessment for multiple-item measures, 489–496
multiple-item scale assumptions, 477
Interobserver reliability (interrater reliability)
assessment for categorical variable scores, 470–471
description of, 468
Invasive measures, 466–467
Inverse matrix, 350
Item response theory, 503–504
Iterations, 412

Jenkins Activity Survey, 501
Joint significance test, 299
Journal editors, 24
Just-identified model, 563

Kuder-Richardson 20 (KR-20) reliability coefficient, 490

Lag function, 631
Latent variables
  in confirmatory factor analysis, 457–458
  description of, 398–399, 565
  extraversion of, measurement model for, 458
  in structural equation modeling, 457–461
Latin square design, 531–532
Levels of a factor, 217
Levene *F* test, 33
  MANOVA and, 368
Likert scales, 465
Linearity assumptions, 41–43, 60
Linear regression, SPSS procedure for, 114–117
Listwise deletion, 44, 56
Little's test of MCAR, 54–55
Loadings
  A matrix of, 410, 456
  computation of, 408–412, 455
  as correlation, 434
  rotated, 437, 442–444
  sum of squared, 410
Logarithm function, natural log, 589
Logistic regression, 292, 311, 583–624
Logit (L_i), 348, 588, 590–591
Log likelihood (LL), 592–594, 601
Log-linear analysis, 311
Log transformations, 36
Lord's paradox, 283, 530

M_1, 16
M_2, 16
Mahalanobis distance, 38–39
Main effects, 230–234
MANCOVA, 394
MAR missingness, 51–52
Matrix
  determinant of, 174–177
  raw-score data, 177–180
Matrix algebra
  addition, 170
  determinants, 323, 362
  eigenvalue/eigenvector problem, 325, 349–351
  factor analysis, 454–457
  inverse (division), 172–173, 350
  MANOVA null hypothesis, 353
  multiplication, 170–171
  regression coefficients for multiple regression with multiple predictors estimated using, 168–180
  subtraction, 170
  sums of squares computation for ANCOVA, 265
  transpose, 173–174
Maximum likelihood estimation (MLE), 592, 620
MCAR missingness, 51–52, 54
Measured variables, 539
Measurement error
  attenuation of correlation, 474–475
  description of, 464, 472
Measurement invariance, 575–576
Measurement issues, 464–466, 504–505
  cost, 466
  invasiveness, 466–467
  multiple outcome measures, 361
  reactivity, 467
  sensitivity, 465–466
  survey size and inflated risk of Type I error, 466
  triangulation, 361, 505
Measurement models, 539, 565
Measurement theory, 471–475
  modern, 503–504
  path models, 473–474
Mediated effect
  magnitude of, 298
  unit-free index of, 297
Mediated moderation, 303–305
Mediated path, 296–297, 303
Mediating variables
  multiple, 302–303
  research example, 290
Mediation
  definition, 289
  inconsistent, 299
  path models, 289–290, 563
  as reasonable hypothesis, 290
  summary of, 306
Mediation analysis
  assumptions, 293–294
  causal-steps approach, 298–299
  data screening, 293–294
  designing of, 292–294
  effect size information, 301–302
  joint significance test, 299
  logistic regression, 292
  mediated effect, 298
  mediated path, 296–297
  path coefficient estimation, 294–295
  questions in, 292
  sample size and statistical power, 302
  Sobel test, 299–301
  statistical significance testing, 298–301
  temporal precedence of variables in, 293
  variables in, 292–294
Mediation models
  Amos Graphics program, 305
  example Results section, 305–306
  multiple mediating variables, 302–303
  multiple-step mediated paths, 303
  structural equation modeling programs to test, 305
  X_1,Y is completely mediated by X_2, 89

Mediation structural model, 544–545
Meta-analysis
  forest plots, 21
  goals of, 20–21
  graphic summaries of, 21–22
  information needed for, 20
Minnesota Multiphasic Personality Inventory (MMPI), 498
Missing at random, 51
Missing completely at random, 51
Missingness
  assessing amount of, 45–46
  MAR, 51–52
  MCAR, 51–52, 54
  MNAR, 51–52
  Type A, 51–56
  Type B, 51–53
Missing not at random, 51
Missing values and data
  with Amos, 547
  bias caused by, 32
  coding for, 29–30
  data screening, 34
  describing amount of, 44–50
  for each case, 47–48
  for each variable, 47
  in entire data set, 46–47
  handling of, 56
  imputation of, 48
  multiple imputation, 51, 57–59
  outliers replaced with, 51
  patterns in, 51–53
  planned, 50
  problems created by, 44–45
  reasons for, 50–51
  refusal to participate as cause of, 50
  remedies for, 56–57
  SPSS, 45–46, 48–49, 482–485
Mixed-model ANOVA, 521–530
MNAR missingness, 51–52
Model
  equivalent, 564
  just-identified, 563
  nested, 564
  overidentified, 563
  path. *See* Path models
  recursive, 564
  terminology associated with, 562–564
  underidentified, 563
Model fit, 563
Model identification, 562–563
Moderated mediation, 303–305
Moderation
  definition of, 215
  path model about, 216
Moderator variable, 215
Modern measurement theory, 503–504
Modification indexes, 547
Moments, 563
Monotonic relationship, 187
Monte Carlo simulations, 33, 56
Moving average model, 632
Multicollinearity, 124–125, 151, 175
Multilevel modeling, 640–641
Multiple imputation, 51, 57–59
Multiple-item scales
  assessment of internal homogeneity, 489–496
  computing summated scales, 478–489
  computing summated scales, optimally weighted linear combinations, 487–488
  computing summated scales, unit-weighted sum of raw scores, 486
  computing summated scales, unit-weighted sum of *z* scores, 486–488
  discriminant analysis and development of, 502
  factor analysis and development of, 478, 501–502
  generating candidate items, 500–501
  initial scores assigned to individual responses, 478–479
  internal consistency assumptions, 477
  measurement reliability and, 476–478
  online resources, 506
  pilot testing, 501
  published tests, 501
  reverse-worded questions, 479–480
  scale development process, 501–503
  summated scale development, 502
  summing scores across items to compute total score, 482–485
Multiple linear regression
  ANOVA and, 188, 190, 210, 212, 348–349
  assessing pretest/posttest differences, 282
  assumptions, 587
  binary logistic regression and, 583
  binary output variables and, 585–587
  curvilinear relations and, 587
  discriminant analysis versus, 309–311, 348–349
  dummy predictor variables and, 188–189
  general linear model, 357
  interaction terms, ANOVA versus regression, 212
  optimal weighted linear combination of scores, 309–311, 487–488
  sums of cross-products, 206
Multiple linear regression using dummy predictor variables
  dummy coding example, 200
  effect coding example, 203–205
  effect size, 210
  models including quantitative predictors, 208–210
  parameter estimates, 194
  sample Results section, 211

sample size and statistical power, 210
significance tests, 194
slope coefficient interpretation, 189, 194
study design issues, 193
Multiple linear regression with two predictor variables, 211
Multiple logistic regression, 611–619
Multiple *R*
as effect size measure, 210
formula for, 112–113
reporting, 593
Multiple R^2
as effect size measure, 210
formulas for, 112–113
logistic regression goodness-of-fit and, 594
significance of, in forward statistical regression, 180–183
See also R^2
Multiple regression with multiple predictors
beta coefficients, 117–118, 133
effect size for, 122–123, 148–149
empirical example, 136
equations for ß, *b*, *pr*, and *sr*, 120–121
F and *R* changes as additional predictors, 149–150
individual predictors in, significance tests for, 145–148
methods of entry for predictor variables, 140–142
multivariate outliers in, 151–152
path diagrams, 118
predictor variables, 137–142
purpose of, 134
regression coefficients for, computed with *K* predictor variables, 138–140
regression coefficients for, matrix algebra used to estimate, 168–180
research questions, 133–135
screening for violations of assumptions, 136
SPSS examples, 152–197
statistical power, 150
study design issues, 136–138
summary of, 167
tracing rules for path models, 118–120
variance partitioning, 142–144
X predictor and *Y*, 150–151
See also Regression with two predictors
Multivariate analysis of covariance (MANCOVA), 394
Multivariate analysis of variance (MANOVA), 353–354, 361–362
assumptions and data screening, 358–360
comparison with univariate ANOVA, 355–356, 362–364, 395
discriminant analysis and, 349, 354, 356–357, 365, 394–395
discriminant analysis and, post hoc testing, 372, 378–386, 390
effect size, 367
eigenvalues and canonical correlation, 365
empirical example, 2 × 3 factorial design, 375–389
empirical example, 3 × 6 factorial design, 389–392
empirical example, sample Results section, 390–392
factorial designs, 360, 375–392
group characteristics, 360
handling violations of assumptions, 359, 365, 371
matrix algebra and, 353
multivariate test statistics, 364–366
nonsignificant outcome variables and, 367
null hypotheses, 353–354
outcome measure interactions, 354, 358, 378, 385–389
outcome variables, 520
outcome variable selection, 361, 393
post hoc tests, 368–374
post hoc tests, assessing interactions, 385–389
post hoc tests, assessing main effects, 378–385, 390
post hoc tests, comparison of univariate and multivariate analyses, 393–394
reasons for using multiple outcome measures, 355–356
for repeated measures, 516–518
research questions addressed by, 358, 394
robustness to outliers, 359
sample size and statistical power, 361, 367–368
SCP matrix, 359, 361–362
significance tests, 359, 362–364
SPSS GLM procedures, 358, 360
statistical power and, 367–368
study design issues, 360–361
sums of cross-products, 359, 361
Multivariate outliers, 35, 38–41, 108, 151–152

Nagelkerke's R^2, 594, 601
Natural log, 589
Nested models, 564, 574
New Statistics, 2, 9
NHST. *See* Null-hypothesis statistical testing
Noise suppressor, 257, 265
Nonadditivity, 215, 510
Noncausal path, 540
Nonequivalent control group, 254–255, 281
Nonlinear (curvilinear) relations
multiple linear regression and, 587
sigmoidal function, 587
Nonlinear data transformations, 36–37
Non-nested models, 574–575
Nonorthogonal factorial ANOVA, 360
Normal distribution assumptions
ANCOVA and, 259
ANOVA using dummy predictor variables and, 190
discriminant analysis, 321
MANOVA and, 359
Null hypothesis (H_0), 353
ANCOVA, 255, 266
ANOVA, 353
discriminant analysis, 316, 322

F test for, 113
individual predictors for logistic regression, 594
MANOVA, 353–354
t test for, 113–114
Null-hypothesis statistical testing
assumptions, violations of, 6
description of, 1
double negatives and, 5
logic of, 4–5
misuses of, 5–8
remedies for problems with, 9–10
Null model, binary logistic regression, 591, 600

Oblique rotation, 457
Odds, 311, 588–590
converting to probability estimates, 620
Odds ratios, 589–590, 620–621
natural log of (logit), 590–591
probability versus, 589
Omnibus test
discriminant analysis (Wilks's Λ), 316–320, 324
MANOVA (Wilks's Λ), 363, 520
null hypothesis for, 516
test statistic for, 144
One-way between-subjects (between-*S*) ANOVA. *See* Analysis of variance (ANOVA)
One-way repeated-measures ANOVA. *See* Repeated-measures ANOVA
Optimal weighted linear composite of scores
description of, 309–311, 325, 338, 348
summated scores for multiple-item scales, 487–488
Order effects, 530–536
Ordinary least squares (OLS), 290, 540, 594
Orthogonal coding for dummy variables, 205–207
Orthogonal factor analysis, 400, 565
Orthogonal rotation, 437
Outcome variables (*Y*)
intercorrelations between, MANOVA, 356, 359, 378, 385–389
logit for logistic regression, 588
multiple, MANOVA and, 353, 355–356
selection for MANOVA, 393
variance of, 104
Outliers
assessment and handling, 322, 607
binary logistic regression and, 607
bivariate, 35, 38–41, 108
extreme, 547
handling of, 40–41
identification of, 37–40
MANOVA and, 359
missing values used to replace, 51
multivariate, 35, 38–41, 108, 151–152
robustness versus, 359
univariate, 37–38, 41
Overidentified model, 563
Overlapping circles graph, 263

Paired-samples *t* test (correlated samples or direct difference), 516
assessing pretest/posttest differences, 281–282
description of, 529
test-retest reliability assessment, 469
Pairwise deletion, 44
Parallel-forms reliability, 496
Parametric statistics, 40
Partial correlation
between X_1 and Y, controlling for X_2, 71–72, 99
bivariate regressions used to remove variance predictable by X_2 from both X_1 and Y, 74–77
first-order, 73
notation for, 72–73
results section, 95–96
second-order, 73
summary of, 96–97
understanding of, 74–77
X_2 controlled, correlation between X_1 and Y becomes smaller, 90–91
$X_1 \times X_2$ interaction, 77–78
zero-order correlation versus, 90
See also Partial *r*
Partial eta-squared (η^2), 267, 279, 367
Partial *r*
computation of, from Pearson's *r*, 79–81
confidence intervals for, 81
effect size for, 81–82
equation for, 120–121
sample size for, 81–82
statistical power for, 81–82
statistical significance of, 81
See also Partial correlation
Partial slope coefficients
ANCOVA and, 265
dummy predictor variables and, 189, 194
Partition of variance
definition of, 103
in hierarchical regression, 142–144
with highly correlated predictors, 125
in standard regression, 142–144
in statistical regression, 142–144
in *Y* in regression with two predictors, 104–107
See also Variance partitioning
Path coefficients, 100, 294–295
Path diagrams, 117–118
Path models
among X_1, Y, and X_2, 83–86
causal hypotheses in, 85
causal path, 540
classic measurement theory, 473–474
complex, 542–544

definition of, 82, 540
factor analysis, 400–401
measured variables, 577
mediation, 289–290, 563
moderation, 216
multiple regression with *k* variables, 139
noncausal path, 540
notation, 289–290
tracing rules for, 118–120
unidirectional arrows in, 85
X_1 and X_2 are completely redundant predictors of *Y*, 87–88
X_1 and *Y* are not related whether you control for X_2 or not, 86
Pearson product–moment correlation. *See* Pearson's *r*
Pearson's *r*
general linear model, 357
magnitude of, 15
parallel-forms reliability, 496
partial *r* computation from, 79–81
regression coefficient and, 121–122
test-retest reliability assessment, 468–470, 473
zero-order, 117
Per-comparison alpha (PC_a), 328, 355
p-hacking, 23–25
Pillai's trace, 360, 364–366, 371, 518
Planned contrasts, orthogonally-coded dummy predictor variables, 205–206
Poisson regression, 61–62, 638–639
Polytomous logistic regression, 611
Population variance/covariance matrix (**S**), 321, 359, 362
Posterior probability, 640
Practical significance, 18–19
Prediction errors, 32
Predictive validity, 499, 503
Predictor variables (*X*), 350
bivariate correlations among, 100
coding binary scores, 608–610
complex logistic regressions, 610
correlation between, 216
effect size for, 122–123
individual predictor significance for logistic regression, 595, 602–603
k, 138–140
as latent variables, 401
method of entry for, 140–142
multicollinearity, 124–125, 333–334
multiple regression with multiple predictors, 137–142
number of and Type I error risk, 611
tolerance for, 151
z scores for, 244
Pretest/posttest design
description of, 264, 281–284
problems with gain or change scores, 282–284
Principal-axis factoring (PAF)
correlation matrix computation, 411–412, 424
description of, 398
limiting the number of components or factors, 412–413
nine-item two-retained model, 425–434
nine-item two-retained model, item communalities, 428–432
nine-item two-retained model, partial reproduction of correlations from loadings on only two factors, 433–434
nine-item two-retained model, variance reproduction, 432–433
principal-components analysis versus, 424–425
Varimax rotation, 440–444
Varimax rotation, Results section, 447–449
Principal-components (PC) analysis
computation of, 410–414
correlation matrix computation, 411–412, 424
description of, 398–399
limiting the number of components or factors, 412–413
one-component model, 420–424
one-component model, correlations not perfectly reproduced from loadings on only one component, 423–424
one-component model, item communalities, 420–423
one-component model, variance reproduction, 423
principal-axis factoring versus, 424–425
reproduction of correlations from loadings, 419–420
three-component model, 414–420
three-component model, reproduction of correlations from loadings, 419–420
three-component model, variance reproduced by each component, 419
Probability
description of, 9
"frequentist" understanding of, 9
log odds conversion, 620
odds ratios versus, 589
p values and, 4
Projective tests, 498
Psychological testing resources, 506
Publication bias, 24
Published tests, 501
p values
artifacts that affect, 7
confidence intervals versus, 13–14
data problems that affect, 7
decision error and, 3
misconceptions about, 3–4
misinterpretations of, 2–4
probability statements and, 4
SPSS output, 333, 611
Type I error and, 5

Quadratic regression, regression coefficients for, 43
Quantitative predictor(s)
categorical predictor and, 219–226
two, interaction between, 232–234
Quantitative variables
binary logistic regression and, 610–619
graphing interactions between, 244–249

regressions using dummy variables and, 208–210
test-retest reliability assessment, 468–470
Quasi-experimental design, 255

R^2
computation of, 122
confidence interval for, 183
Cox and Snell's, 594, 618
description of, 107
multiple. *See* Multiple R^2
Nagelkerke's, 594, 601
Random assignment to treatment groups
ANCOVA and nonequivalent group comparisons, 255
unlucky randomization, 255–256
Rank scores, multiple-group categorical variables and, 187–189
Raw-score coefficients, 112, 120
Reactivity of measurement, 467
Recursive models, 564
Regression
binary logistic. *See* Binary logistic regression
binomial, 638–639
data-driven, 135, 140–142
data screening for, 35
dummy variables. *See* Dummy variables, regression applications
effect size for, 148
hierarchical. *See* Hierarchical regression
interaction in. *See* Interaction
logistic, 311
normal equations for, 541
Poisson, 61–62
prediction equation for, 102
predictive relationships, 121–122
sequential, 135, 140–141, 146, 156–162
simultaneous, 135, 140–141
standard, 135, 140–141, 152–156
statistical. *See* Statistical regression
stepwise, 135, 141
test of significance for, 113
user-determined order of entry in, 135, 140, 142
zero-inflated binomial, 61–62
Regression analysis
hypothetical research example for, 101–102
multiple, 104
Regression coefficients
k predictor variables used to compute, 138–140
for multiple regression with multiple predictors, matrix algebra used to estimate, 168–180
Pearson's *r* and, 121–122
for quadratic regression, 43
Regression plane, 102–103
Regression slope coefficients, 120
Regression slope coefficients, raw score (*b*), 603
B coefficient for logistic regression, 595, 603
dummy predictor variables and, 189, 194, 198
as weights for optimally weighted composites of scores, 487
Regression with two predictors
assumptions for, 107–111
control variables, 124
effect size for, 122–123
formulas for, 111–114
multiple *R* and R^2, 112–113
partition of variance in *Y* in, 104–105
predictor variables, 124
range of scores, 125–126
raw-score coefficients, 112
results, 126–129
sample size for, 123–124
standard-score beta coefficients, 111
statistical power for, 123
study design issues, 124–126
summary of, 129–130
See also Multiple regression with multiple predictors
Reliability, 464
attenuation of correlation, 474–475
classic measurement theory, 473–475
Cohen's kappa (κ), 468, 471
consequences of unreliable measurements, 475
Cronbach's alpha (α), 489–495
definition and related terms, 467–468
improving, 475
interobserver or interrater, 468, 470–471
Pearson's *r* as coefficient of, 468–470, 473
SPSS procedures, 490–495
summary, 504–505
test-retest, 467–470
variance ratio interpretation, 473
Reliability assessment, multiple-item measures, 476–478, 502–503
assessment of internal homogeneity, 489–496
Cronbach's alpha, 489–495
internal consistency assumption, 477
parallel-forms reliability, 496
split-half reliability, 496
typical scale development study, 502–503
Repeated-measures ANOVA
ANCOVA versus, 529–530
assessing pretest/posttest differences, 282
assumptions for, 509–510
difference-scores approaches versus, 529–530
doubly multivariate, 518–522
MANOVA for, 516–518
order effects, 530–536
participant-by-time interaction, 510–513
participant-by-treatment interaction, 510–513
of simple main effects, 528
sphericity assumption, 513–514

sphericity assumption, testing, 516
sphericity assumption, violation, 515
SPSS GLM procedure, 513–516, 527
univariate, 513
Replication, 8–9
Reporting guidelines, 60
Reproduced correlation matrix (**R***), 410
Research design
ANCOVA, 264–265
discriminant analysis, 322–323
MANOVA, 360–361
pretest/posttest designs, 281–282
quasi-experimental design, 255
recommendations for, 22–23
using dummy predictor variables, 193
Research reports, 10
Residuals
normal distribution of, 32
plots of, 35
Results section
ANCOVA, 275–277
discriminant analysis, 343–345
dummy variables, regression applications, 211
multiple linear regression, using dummy predictor variables, 211
principal-axis factoring with Varimax rotation, 447–449
reporting guidelines, 60
two quantitative predictor interaction, 242–243
Reverse-worded questions
example of, 444
names for, 31
self-report measures with, 31, 479–480, 482
SPSS procedure for recoding scoring, 482
variable naming for, 479–480
Reviewers, 24
Robustness
ANOVA and, 190
group sizes and, 190, 360
multivariate test statistics (Pillai's trace), 360, 364–366, 371
outliers and, 359
Root mean square error of approximation, 547, 565
Rorschach test, 498
Rotated factor loadings, 437, 442–444
Roy's largest root (or greatest characteristic root), 365–366
r_{Y1}, 82, 91, 94
$r_{Y1.2}$, 82, 87, 91, 94

Sample mean (*M*), 592
Sample size and statistical power
ANOVA using dummy predictor variables and, 190
binary logistic regression and, 608
decision making about, 123
discriminant analysis and, 321, 327, 336
effect size versus, 14, 16–17
independent-samples *t* test, 16
MANOVA and, 361, 367–368
mediation analysis, 302
partial *r*, 81–82
regressions using dummy variables, 210
regression with two predictors, 123–124
SCP matrix and, 361
structural equation modeling, 546–547
See also Statistical power
Saturated model, 552
Saved factor scores, 485–487
Scatterplots, 67, 109–110
ANCOVA assumptions and, 260
bivariate, 41
one categorical and one quantitative predictor, 221–223
Scholastic Aptitude Test (SAT), 472
verbal SAT (VSAT), 311
SCP matrix, 169
discriminant analysis and, 321–323, 335, 350
MANOVA and, 359, 361–362
sample size and statistical power, 361
S matrix from, 169
Scree plot, 412
SE_{ab}, 299–300
SE_{est}, 113
Self-report measures
data quality issues, 361, 464
reverse-worded items, 31, 479–480
validity assessment, 497–499
Semipartial (or "part") correlation (sr and sr^2)
description of, 103–104
regressions using dummy variables, 194
squared, 105
Sensitivity of measurements, 465–466
Separate-groups covariance matrices, 331
Sequential regression
description of, 135, 140–142, 146
effect size for, 149
SPSS menu selections, output, and results for, 156–162
Sigma matrix (Σ), 321, 362
Sigmoidal function, 587, 590
Significance testing
ANCOVA, 266
discriminant analysis, 315, 317, 323–324, 328, 331, 333
for individual predictors in multiple regression, 145–148
MANOVA, 359, 362–364
misuse of, 2
multiple tests and inflated Type I error risk, 328, 333, 355, 393
for overall regression model, 144–145
post hoc ANOVAs for discriminant analysis, 328
regressions using dummy variables, 194
univariate ANOVA versus MANOVA, 362
Simple main effects, 230–234, 526

Simultaneous multiple regression, 130
Simultaneous regression, 135, 140–141
effect size for, 148
reporting results for, 147
Skewness
description of, 33–34
nonlinear data transformations for, 36–37
Sobel test, 298–301
Social desirability bias
Crown-Marlowe Social Desirability scale, 499
description of, 361, 478
Spearman-Brown prophecy formula, 490, 496
Sphericity assumption
testing of, 516
violation of, 515
Split-half reliability, 496
SPSS GLM procedures
MANOVA, 358–359, 364–366, 386–392
Type III sum of squares, 264, 271, 360
Type I sum of squares, 264, 360
SPSS procedures and applications
advantages of ANOVA versus regression, 212
ANCOVA, 259, 269–279
ANCOVA, interaction terms, 273–275
ANCOVA, variance partitioning, 264
ANOVA for ANCOVA data screening, 262
conversion of Wilks's Λ to χ^2, 316
correlation (Crosstabs), 375
discriminant analysis, 319, 329–345
discriminant analysis, classification methods, 331
discriminant analysis, MANOVA and, 358
discriminant analysis, significance tests, 317
discriminant function coefficients, 326
effect size, 17
hierarchical regression, 156–162
linear regression, 114–117
logistic regression, 596–607, 616–619
log likelihood statistic (–2LL), 593
Mahalanobis distance, 39
missing values and data, 45–46, 48–49, 482–485
multiple regression with multiple predictors, 152–197
partial correlations, 81
p value output, 333, 611
recoding scoring for reverse-worded questions, 482
reliability procedure, Cronbach's alpha, 490–495
sequential regression, 156–162
standard regression, 152–156
statistical regression, 163–167
structure coefficients, 326
syntax, 31, 522, 534–535
two quantitative predictor interaction, 234
Wilks's Λ and, 333
Spurious correlation
definition of, 70
time series, 636
X_1, Y, 88–89
Spuriousness, 66
Squared semipartial correlations, 105
SRMR. *See* Standardized root mean square residual
SSL. *See* Sum of squared loadings
Standardized achievement tests, 472
Standardized discrimination function coefficients, 318, 326, 488
Standardized root mean square residual, 565
Standard multiple regression, 130
Standard regression, 135, 140–141
path model for, 100
SPSS menu selections, output, and results for, 152–156
variance partitioning in, 142–144
Standard score, 399
See also *z* scores (standard scores)
Standard-score beta coefficients, 111
Standard-score regression slope, 129
Standard-score slope coefficient, 120
Statistical control
categorical X_2 variable, 66–70
definition of, 64
overview of, 64–66
Statistical power
ANCOVA and, 268
ANOVA using dummy predictor variables and, 190
binary logistic regression and, 608
description of, 123
discriminant analysis and, 321, 327, 336
dosage level and, 608
listwise deletion effects on, 56
MANOVA and, 367–368
multiple regression with multiple predictors, 150
partial *r*, 81–82
regressions using dummy variables, 210
See also Sample size and statistical power
Statistical regression, 135, 140–142, 146, 163–167
effect size for, 149
reporting results for, 147
SPSS menu selections, output, and results for, 163–167
variance partitioning in, 142–144
Statistical significance
definition of, 18
of interaction between one categorical and one quantitative predictor, 224–226
mediation analysis, 298–301
Statistical significance tests
derivations of, 33
selection of, 7
Statistics
cluster analyses, 628–630
goal of, 642
historical developments in, 625–627
New, 2, 9
replication issues in, 8–9

survival analysis, 627–628
time-series analyses, 630–638
Stepwise regression, 135, 141
Structural equation modeling (SEM)
Amos, 540–578
Amos, analysis, 554–557
Amos, analysis properties, 552–554
Amos, bootstrapped confidence intervals, 557–561
Amos, results, 554–557
Amos, variable names and paths, 547–551
assumptions for, 547
confidence intervals, 553
confirmatory factor analysis using, 461
constraints on SEM model parameters, 567
data screening and preparation for, 546–547
definition of, 539
description of, 398
goodness-of-fit indexes, 564–565
latent variables in, 457–461
measurement and structural components, 569–573
mediation models, 305, 544–545, 561–562
model terminology, 562–564
nested models, 574
non-nested models, 574–575
path models, 540–544
reporting of, 576–577
sample size for, 546–547
SPSS data file, 550–551
summary of, 577–578
unreliability of measurements and, 475
variables, 547
variance/covariance matrix, 539
Structural model, 539
Structure coefficients, 326
Sum of cross products
Box's *M* test and, 359–360
discriminant analysis, 264, 321–323, 335
MANOVA and, 359, 361
multiple regression, 206
Sum of squared loadings, 410
Sums of squared deviations from the mean (SS)
ANCOVA variance partitioning, 263–264
ANCOVA variance partitioning, matrix algebra, 265
ANOVA variance partitioning, comparison with MANOVA, 362
computation methods for MANOVA, 360
SCP matrix, 321
SPSS GLM procedures, 264, 271, 360
SPSS GLM procedures, Type I, 264, 360
SPSS GLM procedures, Type III, 264, 271, 360
Suppression
classical, 91–94
crossover, 94–95
definition of, 91
of error variance, 257
negative, 299
Suppressor variables, 91, 299
Survival analysis, 627–628
Synchrony, 637

Territorial maps, 313, 331, 341
Test-retest reliability
assessment for quantitative variable, 468–470
description of, 467–468
Thematic apperception test, 498
Time series
coordination or interdependence between, 636–638
cycles in, 634–637
interrupted, 632–634
Time-series analyses, 630–638
Tolerance, 151
Total effect, 297
Tracing rule, 118–120, 292
Transpose, matrix, 173–174
t ratios, 594–595
Treatment by covariate interaction, 271, 358
Trend analysis, orthogonally-coded dummy predictor variables and, 207
Triangulation of measurement, 361, 505
True score, 472–474
t test
independent-samples. *See* Independent-samples *t* test
for null hypothesis, 113–114
paired-samples. *See* Paired-samples *t* test
Tukey honestly significant difference (HSD), 369, 371–372, 379
Type A missingness, 51–56
Type A personality, 403
Type B missingness, 51–53
Type I error
Bonferroni corrections, 328–329, 333
follow-up tests and, 372, 393
multiple significance tests and inflated risk of, 328, 333, 355, 393
number of predictors and, 611
p values and, 5
statistically significant interaction as, 243
survey size and inflated risk of, 466
violations of homogeneity of variance/covariance matrices and, 359
Type I sum of squares, 264, 360
Type III sum of squares, 264, 271, 360

Underidentified model, 563–564
Underpowered studies, 23
Unit-weighted composite of scores
description of, 320, 485–487
factor scores versus, 449–451, 488

Unit-weighted sums of items, correlation with
factor score, 488–489
Univariate outliers, 37–38, 41
Univariate repeated measures, 513
Unlucky randomization, 255–256
Unweighted means, 202
User-determined order of entry in regression, 135, 140, 142
Utility, 19

Validity
concurrent, 499, 503
content, 497–498
convergent, 498–499
criterion-oriented types, 498–499
description of, 464–465, 496–499
difficulty in assessing, 472
discriminant, 499, 503
face, 497–498
multiple-item scale development, 502–503
poor measurement reliability and, 475
predictive, 499, 503
Variable names, 29–31
Variables
categorical. *See* Categorical variables
dummy. *See* Dummy variables
latent. *See* Latent variables
measured, 539
moderator, 215
naming of, 479–480
quantitative. *See* Quantitative variables
structural equation modeling, 547
Variance
matrix determinants and, 363
partition of. *See* Partition of variance
Variance/covariance matrix (**S**)
homogeneity across groups assumption, 322–323, 331, 335, 358–359, 371, 584
separate-groups covariance matrices, 331
structural equation modeling use of, 539
Variance partitioning
ANCOVA, 263–264
conservative approaches, 360, 391
one-way between-S ANOVA, 361
SPSS GLM SS Type I, 264, 360
SPSS GLM SS Type III, 264, 271, 360
See also Partition of variance
Varimax rotation
definition of, 413, 437
principal-axis factoring with, 440–444
rotated factor loadings, 442–444
Verbal SAT (VSAT), 311

Wald chi-square statistic, 595, 619
Weighted linear combination of scores, 309–311, 325, 487–488
Welch's *t*, 33
Wilcoxon rank sum test, 40
Wilks's Lambda (Λ), 316–320, 323–324, 518
as effect size, 518
factors affecting size of, 326–327, 367
F and, 364
η^2 (eta squared) and, 324, 327, 333, 346, 363, 367
MANOVA and, 363–367
SPSS procedures and results, 333–334
Within-*S* factors, 522–530
See also Repeated-measures ANOVA

$X_1 \times X_2$ interaction, 77–78

Yea-saying bias, 479

Zero-inflated binomial regression, 61–62
Zero-inflated count data, 638
Zero-order correlations, 70, 72–73, 82, 87, 90
z scores (standard scores), 439–440, 487–488
definition of, 399
discriminant analysis and optimal weighted linear combination of, 309–311, 325, 338, 487–488
factor score coefficients and, 487–488
unit-weighted composites, 486–488
Zung Self-Rating Depression Scale, 465

Suggested Verbal Labels for Cohen's *d* and Other Common Effect Sizes

Verbal Label Suggested by Cohen (1988)	Cohen's d	r, r_{pb},[a] b, Partial r, R, or β	r^2, R^2, or η^2
Large effect	0.8	.371	.138
(In-between area)	0.7	.330	.109
	0.6	.287	.083
Medium effect	0.5	.243	.059
(In-between area)	0.4	.196	.038
	0.3	.148	.022
Small effect	0.2	.100	.010
(In-between area)	0.1	.050	.002
No effect	0.0	.000	.000

Source: Adapted from Cohen (1988).

a. Point biserial r is denoted r_{pb}. For an independent-samples t test, r_{pb} is the Pearson's r between the dichotomous variable that represents group membership and the Y quantitative dependent variable.

Effect Size Interpretations

Research Question	Effect Sizes	Minimum Reportable Effect[a]	Moderate Effect	Large Effect
Difference between two group means	Cohen's d	.41	1.15	2.70
Strength of association: linear	r, r_{pb}, R, partial r, β, tau	.2	.5	.8
Squared linear association estimates	r^2, partial r^2, R^2, adjusted R^2, sr^2	.04	.25	.64
Squared association (not necessarily linear)	η^2 and partial η^2	.04	.25	.64
Risk estimates[b]	RR, OR	2.0	3.0	4.0

Source: Adapted from Fritz et al. (2012).

a. The minimum values suggested by Fritz et al. are much higher than the ones proposed by Cohen (1988).

b. Analyses such as logistic regression (in which the dependent variable is a group membership, such as alive vs. dead) provide information about relative or comparative risk, for example, how much more likely is a smoker to die than a nonsmoker? This may be in the form of relative risk (RR) and an odds ratio (OR). See Chapter 16.

Note: Verbal labels for effect sizes are arbitrary. The second table is more conservative; that is, larger numeric values are required before an effect size is labeled as "large."